Perspectives on Animal Behavior

Perspectives on Animal Behavior

Second Edition

Judith Goodenough

Biology Department / University of Massachusetts, Amherst

Betty McGuire

Department of Biological Sciences / Smith College, Northampton

Robert A. Wallace

John Wiley & Sons, Inc.

New York • Chichester • Weinheim • Brisbane • Singapore • Toronto

ACQUISITIONS EDITOR   Keri L. Witman
MARKETING MANAGER   Clay Stone
PRODUCTION EDITOR   Barbara Russiello
COVER AND TEXT DESIGNER   Karin Kincheloe
PHOTO EDITOR   Lisa Gee
ILLUSTRATION EDITOR   Sandra Rigby
ILLUSTRATION   Second edition updatd by Hadel Studio
COVER PHOTO   Stephen J. Krasemann/DRK Photo

This book was set in 10/12 Janson by Nesbitt Graphics, Inc.
and printed and bound by Courier Westford.
The cover was printed by Lehigh Press Lithographers, Inc.

This book is printed on acid free paper. ∞

Library of Congress Cataloging-in Publication Data
Goodenough, Judith.
        Perspectives on animal behavior / Judith Goodenough, Betty
McGuire, Robert A Wallace.—2nd ed.
                p. cm.
        Includes bibliographical references (p.).
        ISBN 0-471-29502-7 (cloth : alk. paper)
        1. Animal behavior. I. McGuire, Betty. II. Wallace, Robert A.
III. Title.

QL751.G59 2000
591.5—dc21   00-043552

Printed in the United States of America

10 9 8 7 6 5 4 3 2 1

•

In loving memory of my father, Raymond Levrat;
Better known as Papa, he was wise, witty and loving,
as well as an outstanding role model

To my husband, Stephen Goodenough;
my daughters, Aimee and Heather;
and my mother, Betty Levrat

•

Preface

The stated goals of the first edition—to make it current, balanced, and integrated—remain the goals of the second edition. Many questions about animal behavior have yet to be answered. This adds excitement to research. This also reminds us of the need to think critically about existing data and to test alternative hypotheses for explanations of behaviors. For these reasons, I have developed more topics with a hypothesis testing approach.

Although I strove to make this edition current, I was keenly aware that students of animal behavior will better understand the ideas and controversies of today if they are familiar with the background of the topic. Since animal behavior is often taught as an upper level course in both biology and psychology departments, I elected to keep the basic background material in genetics, neurophysiology, and endocrinology. My hope is that this will allow a student who is unfamiliar with the basics to move to the cutting edge of modern research in animal behavior without having to bone up through independent study. Furthermore, I have retained the classic studies but have substantially updated the content. For example, many of the molecular approaches to the study of behavior, such as the use of knock out mice and the molecular basis of the circadian clock, are now included. Molecular techniques have also proven to be useful for determining parentage, which has raised some interesting questions about the evolution of mating systems.

The title of this book, *Perspectives on Animal Behavior*, stems from my awareness that there are varied viewpoints within the field of animal behavior. Furthermore, whereas some researchers are primarily interested in the mechanisms underlying behavior, others are more concerned with the evolution of animal behavior. I have tried to maintain a balanced presentation that includes information on both the mechanisms and the evolution of behavior. To that end, I have included all the major approaches to the field. One issue that continues to stir up debate is the question of animal awareness. I have addressed this question in several chapters, showing how studies in learning, orientation mechanisms, and communication might help us grapple with the issue.

Although a specific discussion often focused on one approach to the study of behavior, I integrated the information as often as possible. For example, this edition has a new section on the hormonal basis of selected adaptive behaviors, including mate choice, alternative reproductive strategies, mating systems, and helping at the nest. As we will see, the endocrine system is just one proximate mechanism through which natural selection can act to increase an individual's fitness.

The organization of the second edition is similar to that of the first. The first chapter discusses some of the reasons for studying animal behavior. It then looks at cooperative breeding among dwarf mongooses, an example of behavior that illustrates how our fullest understanding of behavior results from the integration of information on both its mechanisms and it evolution. Finally, the chapter emphasizes the importance of testing alternative hypotheses to explain behavior. Part One introduces various approaches to the study of behavior. It begins with a discussion of the history of the study of animal behavior because that is the foundation on which current and future studies rest. It then considers specific approaches to study, including genetics, evolution and ecology, learning, neurobiology, endocrinology, and development. Part Two discusses some of the problems faced by the individual surviving in its environment and Part Three considers some aspects of social living.

There are also some important changes in the second edition, most of which were suggested by reviewers.

- Parts of Chapter 10 and Chapter 11 have been combined to form a single chapter on the Mechanisms of Orientation.

- Natal philopatry and dispersal, the ecology of territoriality (Chapter 16), and the costs and benefits of migration (Chapter 10) have been combined to form a new chapter entitled The Ecology and Evolution of Spatial Distribution.

- Chapter 16 (Sociality, Conflict, and Resolution) is a new combination of topics, which was largely sug-gested by reviewers. Sociality and territoriality are from the original Chapter 16. The discussion of dominance relationships is new. The discussion of game theory and single conflict encounters has been moved from Chapter 4.

- The chapters on Communication have been reorganized. Chapter 18 is a description of communication and its functions. Chapter 19 discusses the evolution of communication.

Judith Goodenough
Biology Department, University of Massachusetts
Amherst, Massachusetts

Betty McGuire
Department of Biological Sciences
Smith College
Northampton, Massachusetts

Robert A. Wallace

Acknowledgments

The creation of the second edition was a solo effort. However, it could not have been done without all that my co-authors brought to the first edition. Although Betty McGuire decided that her schedule would not allow her to actively participate in the revision, her chapters were so clearly written and well organized that updating them was not an onerous chore. She was always there to answer my many queries and to provide current references. I also owe a great deal to my late friend and colleague Bob Wallace. Not only did he improve my skill as a writer, but he also made the process *fun*. He encouraged me and made me laugh. His presence during this revision was sorely missed. I hope that his spirit is still expressed in the prose of the text.

My editor, David Harris, had confidence that I could revise the text alone. Together we developed a schedule that was realistic. I am grateful for his encouragement and support.

Many other dedicated people at John Wiley & Sons helped get this book to your hands. Barbara Russiello, the production supervisor, kept track of everything with careful diligence. She provided prompt answers to my questions, helped me set priorities, and kept me on schedule. The text was greatly improved through the efforts of the copyeditor Barbara Connor and was refined even more by the skills of the proofreader Gloria Hamilton. The appearance of this text helps make it a useful learning tool, and for this I thank the designer, Karin Kincheloe, the photo editor, Lisa Gee, and the illustration editor, Sandra Rigby. Each person is a professional, a pleasure to work with, and a team player.

I would like to thank many of my friends and colleagues at UMass for lending their expertise to various aspects of this book. Jack Palmer, my close friend and critic, read drafts of several chapters. His comments and the current references he provided greatly improved the manuscript. During our numerous conversations, he often offered viewpoints that differed from my own, thereby expanding my perspective on the subject. The skill of James Craig and the other members of the staff of the Biological Sciences library at UMass were invaluable in helping me find information, references, and addresses for permissions. The cheerfulness of Margaret Tillson helped me cope when things got tense, as they often did.

Although all the reviewers on this text have been excellent, one truly stands out—Daniel Cristol. During this revision, I would frequently e-mail him and ask for advice, references, or explanations of comments. He was always quick to respond and very helpful.

Most of all, I would like to thank my family who encouraged and supported me at every stage of this revision. My husband Steve bore the brunt of my obsession. After month's of my spending many more hours with the computer than with him, Steve merely quipped that he was getting rather fond of the back of my head and continued to do what was necessary to keep the household running smoothly. He's the funniest person I know, and his wit kept me sane. After 29 years of marriage, he still keeps me on my toes by introducing me as his *first* wife. My daughters, Aimee and Heather, still the joy of my life, remind me that the people you love must always come first. They'll find time to spend with me when they know I need a break. Their courage and confidence in reaching for their dreams, encourages me to keep stretching toward mine—the most immediate of which is the completion of this revision. I'd also like to thank Aimee and her friend Shamus Sullivan for helping me proof the index. My parents, Betty and Ray Levrat, have been my cheering squad since I was a young child. They instilled in me a love of learning. They were always willing to do whatever needed to be done to free time for my writing. My father died during the production of this edition but even during his illness he showed keen interest in this project.

Contents

Part 3 • Behavior of Groups— Social Behavior   363

Chapter 16 • Sociality, Conflict, and Resolution   365

Chapter 17 • Cooperation and Altruism   383

Chapter 18 • Maintaining Group Cohesion: Description and Functions of Communication and Contact   421

Chapter 19 • Maintaining Group Cohesion: The Evolution of Communication   447

1

Introduction

Four Questions About Animal Behavior
Animal Behavior as an Interdisciplinary Study
Hypothesis Testing

In one way or another, people have been studying animal behavior for thousands of years. This knowledge was important in earlier ages for some very practical reasons. The most skillful hunters and fishermen are usually those who can make predictions about the behavior of their prey (Figure 1.1). It is important to know that when salmon are spawning, they will not respond to a fisherman's bait; that many rodents escape toward the dark, whereas most birds escape toward the light; and that many kinds of animals will fight, some ferociously, if they are trapped.

The study of animal behavior may have occupied the fringes of human consciousness for centuries for just such a practical reason. Later, when animals were domesticated and put to work, it was necessary to learn new things about them. Horses could be trained for riding or for pulling wagons or tools. Dogs could be trained to track prey or to protect individual humans; cats could not.

In time, the study of animal behavior took on new dimensions. The goals, as well as the techniques, changed. Animals are no longer studied simply so that we can exploit them better, although this may still be one reason for our attention. Now, though, we have become aware that increased knowledge of the behavior of specific species in their natural habitats may help us save some endangered groups from extinction. In addition, information on their normal behavior may help us ensure their welfare, not just in the wild, but also in laboratories or zoos (Sutherland 1998). We may be interested in behavior as an example of a broader intellectual concern, such as evolutionary theory. Or, we may be interested in studying animal behavior because it may serve as a model to help us understand human behavior. And, sometimes we study animal behavior simply because our curiosity prompts us to ask questions.

FOUR QUESTIONS
ABOUT ANIMAL BEHAVIOR

As casual observations of animal behavior crystallized into a field of scientific study, Niko Tinbergen (1963) identified four types of questions that should be asked about behavior: What are the mechanisms that cause it? How does it develop? What is its survival value? How did it evolve? Tinbergen believed that the biological study of behavior, ethology, should "give equal attention to each of them and to their integration." It was

FIGURE 1.1 People have been studying animal behavior for centuries, sometimes for very practical reasons. Knowledge of the behavior of game species may make it easier to put food on the table.

some of us may be asking about immediate causes (the machinery underlying the response) and others may be asking about the evolutionary causes.

No one type of question is better than the others. Answers to all types are necessary as we weave the fabric of our understanding. These are not competing avenues of investigation. Rather, they are complementary. Each may feed back on the others, deepening our understanding and broadening our avenues of investigation (Armstrong 1991; Halpin 1991; Stamps 1991).

ANIMAL BEHAVIOR AS AN INTERDISCIPLINARY STUDY

Marion Stamp Dawkins (1989) has drawn an analogy between N. Tinbergen's (1963) four aspects of investigation of behavior and the four legs of an animal. An animal that lacks one of its legs can only hobble along. Similarly, progress in the study of animal behavior is hampered by a lack of information in any one of these areas of study. This is not to imply that each investigator must ask all types of questions. Often we find that individuals are more excited by one type of question than by others. However, each investigator will be more successful in finding the answer to the question of personal interest if he or she is armed with information and techniques from all four areas of study.

Consider, for example, the variability of reproductive suppression in dwarf mongooses (*Helogale parvula*), a phenomenon that shows the interplay of mechanisms that control behavior and its evolution (Creel and Waser 1997). The smallest African carnivore, an adult dwarf mongoose weighs about 11 or 12 ounces (Figure 1.2). Its size and color are similar to those of a red squirrel: It is about 43 cm (16 inches) from head to tail, but roughly half of that length is tail. Dwarf mongooses are found in woodland and wooded savanna in sub-Saharan Africa.

Unlike most other mongoose species, which are solitary creatures, the dwarf mongoose is social. It lives in packs that may contain as many as 40 individuals, but the average size is 10 to 12. Theirs is a matriarchal society; that is, each pack is founded and dominated by an old female. She has priority of access to food and initiates pack movements. Next in charge is her mate, an old male. The others in the pack have varying degrees of authority, each taking its place in a dominance hierarchy. In many dwarf mongoose packs only the top-ranking female and male breed.

Nonetheless, the subordinates help the dominant pair raise their offspring. They baby-sit, attack predators, drive away intruding mongoose groups, and warn others of a predator's approach. If a subordinate female does give birth, she will nurse the young of the top-ranking female along with her own, even though she

later suggested that consideration of the mechanisms of behavior should, at least in some cases, include both cognitive and emotional mental processes (Sherman 1988).

To better appreciate the types of questions we may ask about animal behavior, consider those that may be raised when we watch in amazement as moonlight seems to flicker because thousands of tiny songbirds are flying through it during migration. We may each become curious about migration, but depending on our personal interests, we may each ask different types of questions. How do they "know" it is time? How do they find their way? Such questions focus on the mechanisms that underlie the behavior. Must those making this journey for the first time learn the route from experienced travelers? Do they inherit a directional tendency from their parents? Questions such as these concern development. Why do they do it? What do they gain that outweighs the risks and demands of such a journey? These are questions about the survival value, or adaptiveness, of migration. Finally, how did it all begin? Were the advancing glaciers responsible? Were the migratory paths modified during the thousands of years each species has been migrating? These questions center on the evolution of the behavior. So we see that when we ask *why* an animal behaves in a certain way,

FIGURE 1.2 **The dwarf mongoose lives in groups in which the members cooperate in raising the young. The dominant male and female are often the only group members that breed. Reproduction by other group members is usually suppressed. However, other high-ranking individuals are sometimes allowed to breed. The variability of reproductive suppression raises many "why" questions. The answers illustrate the interaction of physiological, behavioral, and evolutionary mechanisms.**

has fewer young than the dominant female. The efforts of these helpers allow the breeders to raise more off-spring than they could without help.

Dwarf mongoose behavior raises a host of "why" questions, and the answers involve both the evolution and the mechanisms of the behavior. Why are dwarf mongooses social although most mongoose species are not? Why do subordinates help the breeders raise young? Why don't all adults breed? Why is the degree of reproductive suppression variable?

When attempting to answer questions about why sociality has evolved in a particular species, we usually look at the ecological factors that affect the costs and benefits of sociality. In this case, two important factors favor sociality. The first is diet. Dwarf mongooses feed primarily on insects—crickets, grasshoppers, termites, spiders, and scorpions—a food source that is plentiful and renews rapidly. Consequently, there is little cost in sharing a territory with conspecifics. And, by living in groups, dwarf mongooses are less vulnerable to preda-tors, to which they are especially susceptible because of their small size and daytime activity. More group mem-bers means increased vigilance. Furthermore, the costs of leaving the group are high—51% of males and 78% of females that disperse die in the process, primarily because of predation.

We assume that behavior evolves such that an indi-vidual maximizes the copies of its kinds of genes in the

population, either in offspring or in relatives. There are two possible ways in which a subordinate dwarf mon-goose can leave copies of its genes, and the best way depends on the circumstances. It can stay with the pack in which it was born and wait to attain dominance and breeding privileges, or it can leave in search of breeding opportunities elsewhere. If an adult of either sex suc-cessfully leaves the pack and joins another one, its chances of breeding are greater than if it stays with its natal pack. The most common path to breeding status for dispersers is to form a new pack. But, dispersers can also become breeders by joining a new pack and waiting for the breeders to die, by joining a new pack without a breeder of their own sex, or by driving out the adult breeders of the new pack (Rood 1990). Breeders will then pass their genes directly to their offspring. However, more than half of those who attempt disper-sal die in the process. On the other hand, if a subordi-nate stays with its pack, it may never have the chance to breed. Pack members are closely related, however. So, it could leave copies of its genes in relatives' bodies if its assistance in raising the young allowed the breeding pairs to raise more offspring than they otherwise could.

What should a subordinate dwarf mongoose do? That depends to a great extent on its chances of attain-ing breeding status in its natal pack and the closeness of its genetic relationship to other group members. Let's consider how mechanisms that control breeding oppor-tunities and evolution interact to see how the dispersal decision is made.

We would expect evolution to favor reproductive suppression of subordinates when the number of young that can be raised in a group is limited by the scarcity of resources. This seems to be the case for dwarf mon-gooses because in groups that were provided with extra food, more subordinate females became pregnant than in groups without supplemental food.

Reproductive suppression is not absolute. In one study of dwarf mongooses, 12% of the subordinate females became pregnant. Furthermore, DNA finger-printing to determine the parents of the young born in a group revealed that 15% of the offspring had subor-dinate mothers and 25% had subordinate fathers.

We may wonder, then, what determines whether a particular subordinate will be permitted to breed. In trying to understand how behavior evolves, behaviorists often create models that weigh the factors thought to favor one behavioral pattern over another. One such model for reproductive suppression predicts that the dominants will bias reproduction in their own favor, up to the point at which a subordinate could gain more fit-ness (leave more copies of its kinds of genes) by leaving the group in search of breeding opportunities than it could by staying and helping to raise relatives. When the subordinate could benefit more by leaving, the dominant may offer breeding opportunities to entice it to stay. In other words, the model predicts that the dominant would vary the degree of reproductive suppression to maximize its own fitness.

Reproductive suppression can occur either hormonally or through the aggressive actions of the dominants toward subordinates who attempt to breed, or both mechanisms may play a role. Reproduction is costly, and most reproductive attempts of subordinates fail either because the dominant individuals will kill the subordinates' offspring or because their young may starve—being allowed to eat only what is left over after the young of the dominants are fed. Thus, the evolution of neuroendocrine controls that prevent reproduction in the first place, before the costs were paid, would be favored.

Reproductive suppression in male and female dwarf mongooses involves different mechanisms. In females, it occurs through hormonal, as well as behavioral, controls. The ability to breed depends on the female's peak level of the hormone estrogen. Estrogen level is correlated with dominance, age, and ability to breed. High-ranking subordinates have estrogen levels similar to those of the breeder. Older females have higher estrogen and a better chance of breeding if they disperse. As a result, the top-ranking female must ease her behavioral suppression of the older females to keep them as helpers. Reproductive suppression of males is accomplished entirely by the aggression of the dominants.

There was very good agreement between the observed degree of reproductive suppression of females in a group and the degree of suppression that was predicted by the model referred to earlier. The dominant female enforced the degree of reproductive suppression of other females that allowed her to maximize her own reproductive success. However, optimal reproductive suppression was not observed among males. Subordinate males had better luck at breeding in large groups than in small ones, presumably because the dominants had more subordinates to control.

Why should optimal reproductive suppression evolve for females but not for males? The answer may lie in the certainty of parentage and the ultimate means of suppressing reproduction by subordinates—infanticide. The top female is likely to be able to identify her own offspring by odor cues learned at the time of birth. However, the young of a single litter can have different fathers, so a male cannot easily determine which of them are his. If resources become too scarce to support the young of a subordinate, a dominant can kill them. Since a female can recognize her own young, she can selectively kill those of subordinates. A dominant male practicing the same infanticidal policy would risk killing some of his own offspring. So, a top-ranking female will allow reproduction of subordinate females who are most likely to leave and breed elsewhere because she can veto that decision later if need be.

Thus, the variation in reproductive suppression in dwarf mongooses results from an interplay between the mechanisms that control it and evolution.

In the pages that follow, we will consider aspects of behavior related to all four of Tinbergen's questions. We will treat each aspect as interesting in its own right, and in some discussions we will consider how the various aspects may act synergistically to broaden our understanding of animal behavior. Before we begin, however, we should take a moment to consider how scientists answer questions about animal behavior.

HYPOTHESIS TESTING

The study of animal behavior usually begins with an observation that prompts a question, which is followed by forming hypotheses about a possible answer. It is necessary to be able to test each hypothesis. Generally, the hypothesis leads to a prediction, which will support the hypothesis if it holds true when tested. Depending on the hypothesis, the test may involve further observations, comparisons of behavior among species, or experimental manipulation.

Different hypotheses can sometimes lead to identical predictions, and then both hypotheses are supported or refuted, depending on the outcome of the test. In this event, it is necessary to make other predictions that will allow us to reject one of the hypotheses.

The studies of sentinel behavior of meerkats (*Suricata suricatta*) by Timothy Clutton-Brock and his colleagues (1999) provide an example of the process of testing alternative hypotheses. A meerkat is a cooperatively breeding mongoose related to the dwarf mongoose. While serving as a sentinel, a meerkat stands upright on its hind legs and watches for the approach of a predator (Figure 1.3). If the sentinel spots a predator, it warns other group members with repeated alarm calls. This is an important job because predation rates are high and a foraging meerkat can't see a predator coming. A meerkat must forage for five to eight hours a day, and meerkats forage as a group. Each digs in the ground to look for prey, primarily small arthropods and lizards. It digs so deeply that only its hind end sticks out of the hole. In this position, it can't detect a predator's approach, and so the sentinel's calls save the lives of many group members. After a while, the sentinel on duty will return to foraging or will rest, and another group member will take over the guard duty.

Why do group members take turns acting as sentinels? One hypothesis is that sentinel behavior has evolved because it saves the lives of family members, that is, by kin selection. Family members share copies of one's genes. Kin selection includes fitness gained by helping one's own offspring (direct fitness), as well as fitness gained by helping nondescendant kin, including

FIGURE 1.3   A meerkat serving as sentinel, watching for predators. Meerkats find their food by digging in the ground. With their head stuck in the ground, they can't see a predator's approach. Group members take turns standing guard. Rather than being an altruistic action, however, sentinel behavior seems to be selfish. A sentinel increases its personal safety. An individual is more likely to go on guard duty when it has had enough to eat and when no one else is serving as sentinel.

siblings, nieces, nephews, aunts, and uncles (indirect fitness). Thus, if group members are closely related, a sentinel ensures that copies of its genes reach future generations by saving many family members, even if it dies while alerting others (Hamilton 1964). (Kin selection is discussed more thoroughly in Chapter 17.)

The next step in hypothesis testing is to make predictions that will hold true only if the hypothesis that sentinel behavior evolved through kin selection is correct. The first prediction might be that group members have a close genetic relationship. If they do not, kin selection could not work. But, this prediction does hold true. One dominant, breeding female is the mother of 75% of all the litters born to the group, and one dominant male fathers 75% of all the pups born. Although

the group usually contains a few immigrants, most of the subordinate adults are siblings or half siblings. As a result, subordinate adults are likely to have 50% or 25% of the same kinds of genes in common.

Because most meerkat group members are family members, it is possible that kin selection has favored sentinel behavior. However, by itself, a close genetic relationship is not enough evidence to conclude that kin selection has played a role. We need further evidence, and therefore we must refine the prediction. We see, then, that a hypothesis can lead to tiers of increasingly refined predictions.

A more specific prediction based on the hypothesis that sentinel behavior evolved by kin selection is that individuals should stand guard more frequently when they are near family members. To test the prediction, the group was observed to determine which members stand guard and when. Immigrants unrelated to other group members and individuals with many relatives nearby were equally likely to act as sentinels. Thus, the test of this prediction does not support the kin selection hypothesis (Clutton-Brock et al. 1999).

A second hypothesis that is often suggested to explain cooperative behavior such as this is that it results from reciprocal altruism—individuals take turns standing guard. Reciprocal altruism can only work when group members can identify those who cheat by shirking guard duty and punish them (Trivers 1971). This hypothesis produces the prediction that there should be a regular rotation of guard duty within the group and that slackers should be chastised. But this is not observed. The group members do take turns on guard duty but not in any predetermined order. Furthermore, when some individuals reduce the time they spend guarding, other group members increase their own contributions. These observations are not consistent with the predictions based on the reciprocal altruism hypothesis.

A third hypothesis for the evolution of sentinel behavior is that it results from selfish antipredator behavior. The idea is that watching for predators increases a meerkat's personal safety, and warning others isn't costly. So, when a meerkat has had enough to eat, it should watch for predators. The sentinel on duty can then return to foraging (Bednekoff 1997). One prediction produced by this hypothesis is that serving as a sentinel is not a risky business. This does seem to be true. No sentinels were attacked or killed by predators during over 2000 hours of observation. Sentinels may actually be safer than others because they are the first to detect the predator. Furthermore, they usually stand guard within 5 meters of a burrow and are among the first animals below ground when a predator approaches.

If serving as a sentinel increases a meerkat's personal safety, we would predict that an individual would spend a proportion of its time guarding, whether it was solitary or part of a group. As predicted, solitary individuals do spend about the same amount of time on guard duty as members of large groups. Large groups suffer less predation because there is a guard on duty for a greater proportion of the foraging time than in small groups.

A third prediction based on this hypothesis is that the probability that an individual will go on guard should depend on its own energetic state. A meerkat should be more likely to act as sentinel when its belly is full. This prediction was tested by interrupting a sentinel's guard duty after 2 minutes and feeding it 25 grams of hard-boiled egg, which is the equivalent of their average morning weight gain. The length of time between guarding bouts is normally 38.9 minutes, but supplemental feeding reduced the interval to 8.77 minutes. After 30 days of supplemental feeding, an individual served as sentinel three times more often than those that had not received extra food.

Since all group members benefit from a single sentinel's warning calls, the benefit accrued by a particular individual when it stands guard will depend on whether other group members are guarding or foraging. Thus, a fourth prediction based on this hypothesis is that an individual should be more likely to go on guard duty if there is no other guard. This also turned out to be true. It seems then that the alternation of guarding among group members is the result of independent, selfish decisions that depend on the individual's nutritional status, as well as the actions of others.

When we find that the results of various tests are more consistent with one hypothesis than with others, we must still be cautious. New evidence may come to light that will disprove the hypothesis, or a new hypothesis may be proposed that is also consistent with the observations.

Furthermore, the hypothesis that proves to be the correct explanation for a particular behavior in one species may not be correct for another species. For example, sentinel behavior in meerkats and alarm calling by marmots and ground squirrels are similar behaviors that have been selected by different forces. We've seen that sentinel behavior in meerkats appears to have evolved for personal benefit. However, alarm calling in marmots seems to be a form of parental care that has been selected because it saves the lives of offspring. There is no evidence that the rate of marmot alarm calling increases in the presence of nondescendant kin if offspring are not also present. Thus, it seems that direct selection alone has played a role in the evolution of alarm calling in marmots (Blumstein et al. 1997). In Belding's ground squirrels (*Spermophilus beldingi*), however, both direct and indirect selection have played a part in the evolution of alarm calling. Females with relatives living nearby are more likely to alarm-call than are females without kin close by. In this species, even females without offspring will sound an alarm if nondescendant kin are in the vicinity (Sherman 1977).

In the pages that follow, we will see many examples of hypothesis testing. Keep these general procedures in mind as you read them. Instead of passively accepting the given explanation for a particular behavior, be critical of the evidence. Try to think of alternate hypotheses for the behaviors described. Make predictions and design your own tests of those predictions.

SUMMARY

Animal behavior is studied for many reasons, both practical and intellectual. A full understanding of behavior requires answers to four types of questions, those about (1) immediate mechanisms, (2) development, (3) survival value, and (4) evolution. Our progress in understanding the behavior of animals will be enhanced by considering all four types of questions.

The study of animal behavior usually begins with an observation that prompts a question. The next step is to think of tentative explanations, called hypotheses, to answer that question. Each hypothesis should produce testable predictions. The tests of those predictions support or refute the hypothesis.

PART ONE

Approaches to the Study of Animal Behavior

2

History of the Study of Animal Behavior

To understand a field of study today, we must know something about its past. In this chapter we consider the history of the study of animal behavior. Our focus is on the development of key concepts in the field. We begin with some of the earliest ideas about the minds and emotions of animals and describe the conceptual framework given to the study of behavior by Darwin's theory of evolution. We then describe the development during the early part of the twentieth century of ethol-

ogy and comparative psychology, two approaches to animal behavior. Despite initial hostility between proponents of the two approaches—hostility that eventually culminated in the nature/nurture controversy—interactions between the two camps modified in positive ways the types of questions that we ask about animal behavior today. Indeed, many of the contemporary approaches to the study of animal behavior that are discussed in subsequent chapters were inspired by the arguments of ethologists and comparative psychologists. We conclude the chapter with a brief examination of sociobiology, an approach to the study of behavior that achieved prominence in the mid-1970s. Today, sociobiology is often called behavioral ecology.

THE BEGINNINGS

INTELLECTUAL CONTINUITY IN THE ANIMAL WORLD

It is difficult, perhaps impossible, to pinpoint the precise beginnings of the study of animal behavior. Rather than attempting this feat, we will simply consider some of the highlights in the development of the discipline. One idea, that of intellectual continuity among animals, was important in shaping some of the earliest views of animal behavior. Although the idea of intellectual continuity was summarized in 1855 by Herbert Spencer in his book *Principles of Psychology*, its roots can be traced back to the ideas of the ancient Greek philosophers. It focused on a continuity in mental states between "lower" and "higher" animals and was based on a picture of evolution similar to Aristotle's *scala naturae*, the great chain of being. In the *scala naturae*, the evolution of species was viewed as linear and continuous. This classification system was hierarchical, with animals ranked according to their degree of relationship, the highest point of evolution being humans (Hodos and Campbell 1969). At the bottom of the scale are creatures such as sponges; then further up the scale are insects, fish, amphibians, reptiles, birds, nonhuman mammals, and finally humans. Because evolution was seen as a linear process, with each higher species evolving from a lower one until, finally, humans emerged, it was thought that the animal mind and the human mind were simply points on a continuum.

DARWIN'S EVOLUTIONARY FRAMEWORK

A few years after the publication of Spencer's book, Charles Darwin (Figure 2.1) published his thoughts on evolution by natural selection in *The Origin of Species* (1859). Although Darwin's focus in this book was not on animal behavior, his ideas provided a conceptual framework within which the field of animal behavior could develop. Discussed in more detail in Chapter 4, these ideas can be briefly summarized as follows:

1. Within a species, there is usually variation among individuals, and some of this variation is inherited.

2. Most of the offspring produced by animals do not survive to reproduce.

3. Some individuals survive longer and produce more offspring than others, as a consequence of their particular inherited characteristics.

4. Natural selection is the differential survival and reproduction of individuals that results from genetically-based variation in their behavior, morphology, physiology, and so on.

5. Evolutionary change occurs as the heritable traits of successful individuals (i.e., those that survive and reproduce) are spread throughout the population, whereas those traits of less successful individuals are lost.

In two later books, *The Descent of Man, and Selection in Relation to Sex* (1871) and *Expression of the Emotions in Man and Animals* (1873), Darwin applied his evolutionary theory to behavior. In these volumes he recorded his careful and thorough observations on animal behavior, but his records were anecdotal and often anthropomorphic. This was not, however, sloppy science. In the

FIGURE 2.1   Charles Darwin. His ideas on evolution by natural selection provided an evolutionary framework for the study of animal behavior.

tradition of his day, he believed that careful observations were useless unless they were connected to a general theory. Darwin's general theory was evolution by natural selection. Because humans evolved from other animals, he considered the minds of humans and animals to be similar in kind and to differ only in complexity. Because of this belief, he described the behavior of animals by using terms that denote human emotions and feelings: Ants despaired, and dogs expressed pleasure, shame, and love. Darwin's opinion was influential, and it is not surprising that others interested in animal behavior followed his lead. Both ethologists and comparative psychologists trace the beginnings of their respective fields to the ideas of Darwin.

For about a decade after the publication of *Expression of the Emotions in Man and Animals* (1873), descriptions of animal behavior usually took the form of stories about the accomplishments of individual animals that were believed to think and experience emotions as humans do. For example, on the basis of his subjective interpretation of what he observed, George J. Romanes, a protégé of Darwin, constructed a table of emotions that charted the evolution of the mind and listed the emotions in order of their historical or evolutionary appearance (Figure 2.2). In his books, *Animal Intelligence* (1882), *Mental Evolution in Animals* (1884), and *Mental Evolution in Man* (1889), Romanes examined the implications of Darwinian thinking about the continuity of species for the behavior of nonhuman and human animals.

In addition to Romanes, several other scientists made notable contributions to the study of animal

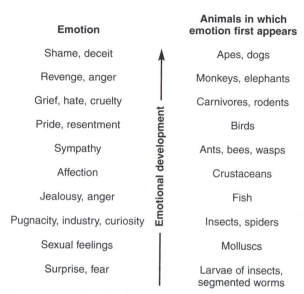

Emotion	Animals in which emotion first appears
Shame, deceit	Apes, dogs
Revenge, anger	Monkeys, elephants
Grief, hate, cruelty	Carnivores, rodents
Pride, resentment	Birds
Sympathy	Ants, bees, wasps
Affection	Crustaceans
Jealousy, anger	Fish
Pugnacity, industry, curiosity	Insects, spiders
Sexual feelings	Molluscs
Surprise, fear	Larvae of insects, segmented worms

FIGURE 2.2   The ideas of George J. Romanes on the evolutionary appearance of emotions in animals. (Modified from Romanes 1889.)

behavior at the turn of the century. Jacques Loeb (1918) believed that all patterns of behavior were simply "forced movements" or tropisms, physiochemical reactions toward or away from stimuli. Herbert Spencer Jennings, perhaps best known for his book *Behavior of the Lower Organisms* (1906), disagreed with Loeb's ideas and emphasized instead the variability and modifiability of behavior. Of course there were many other pioneers in the study of animal behavior, but we will move on to the turmoil that developed in the discipline during the twentieth century.

As interest in animal behavior grew, differences in opinion developed. These differences eventually led to the development of two major approaches to the study of animal behavior, ethology and comparative psychology, centered in Europe and the United States, respectively. The split that developed between the two approaches seemed at times quite severe. Indeed, one must wonder why two groups of scientists, each striving for a greater understanding of the marvels of animal behavior, could have stood worlds apart. The European ethologists and the American comparative psychologists were separated by more than the Atlantic Ocean. In fact, the basic questions they asked about animal behavior were different. (Recall from Chapter 1 that the four questions outlined by Niko Tinbergen in 1963 concerned the mechanism, function, development, and evolution of behavior.) Whereas the ethologists focused their attention on the evolution, function, and mechanism of behavior, the comparative psychologists concentrated on the mechanism and development of behavior. Because they asked different questions, the types of behavior they studied and even their experi-

mental organisms differed. Whereas ethologists, by and large, studied innate behavior in birds, fish, and insects, comparative psychologists, particularly those of the behaviorist school (see later discussion), emphasized learned behavior in mammals such as the Norway rat. Furthermore, to determine the normal function of a behavior, ethologists often attempted to observe the animal in its natural habitat or in environments designed to simulate that habitat. On the other hand, comparative psychologists believed that learning was best studied in the laboratory, where variables could be controlled. Finally, ethologists were interested in species' differences, whereas comparative psychologists searched for general "laws" of behavior.

Of course there were many exceptions to this characterization. Some ethologists made remarkable discoveries indoors in their homes and laboratories and explored the influence of experience on behavioral development (Barlow 1989), and some comparative psychologists studied a wide range of species and patterns of behavior and conducted their studies in the field (Dewsbury 1989). Indeed, within each field there were individuals interested in all four questions of animal behavior. Although traditionally ethology and comparative psychology have been portrayed as very different approaches, some recent accounts of the history of animal behavior have tended to downplay the differences between them (e.g., Dewsbury 1984).

CLASSICAL ETHOLOGY

THE APPROACH: EVOLUTIONARY, COMPARATIVE, DESCRIPTIVE, FIELD-ORIENTED

"Why is that animal doing that?" is perhaps the fundamental question of ethology, the approach to the study of behavior founded largely by Konrad Lorenz, Niko Tinbergen, and Karl von Frisch, European zoologists who shared the Nobel Prize in medicine and psychology in 1973 (Figure 2.3). Traditionally, ethology concentrated on the evolutionary basis of animal behavior. Because natural selection can act only on traits that are genetically determined, it seems a logical outcome of the ethologists' basic interest in the evolution of behavior to focus on those behavior patterns that are inherited. An emphasis on phylogeny (the evolutionary history of a species) is particularly characteristic of the work of Konrad Lorenz.

The studies of ethologists often involve comparisons among closely related species. In the words of Lorenz (1958), "Every time a biologist seeks to know why an organism looks and acts as it does, he must resort to the comparative method." The method Lorenz was referring to is that employed by comparative anatomists when they ask the same question about morphology. If comparative anatomists were to ask why a whale's flipper is structured as it is, they might compare the skeleton of the flipper with that of the forelimb of other vertebrates. They could then see that the typical vertebrate forelimb has been specialized for the aquatic life of this mammal. Likewise, if one were to wonder why a male fly of the species *Hilara sartor* spins an elaborate silken balloon to present to a female before mating (Figure 2.4), the significance of the behavior would become apparent after comparing it with the mating behavior of the other species of flies in the family Empididae. Let us consider the gift-giving behavior of *H. sartor* in more detail to illustrate the ethologist's comparative method.

Among the empidid flies there are species that display almost every imaginable evolutionary step in the progression toward the balloon display. By observing the manner in which the male empidid fly approaches the female for mating, one sees that at the heart of the problem is the fact that the male may be a meal, rather than a mate, for the predacious female. In one species, *Empis trigramma*, the male approaches the female while she is eating. Because her mouth is already full, his well-timed approach increases his chances of surviving the encounter. In another species, *E. poplitea*, the male captures a prey, perhaps a fly, and gives it to the female, providing her with a meal before attempting to copulate. Males of the species *H. quadrivittata* gift-wrap the meal in a silky cocoon before offering it to the female. In another species, *H. thoracica*, the cocoon, or case, is large and elaborate, but the food inside is small and of little value. In yet another species, *H. maura*, only some males place food inside cocoons; others enclose something meaningless, such as a daisy petal. Finally, the

FIGURE 2.3 Konrad Lorenz (above), Niko Tinbergen (top right), and Karl von Frisch (right), three ethologists who shared the Nobel Prize in 1973.

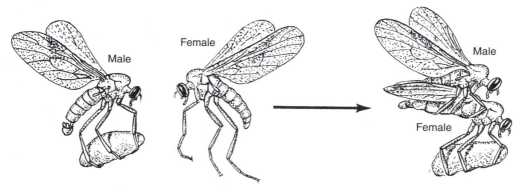

FIGURE 2.4   **Male flies of the species** *Hilara sartor* **present females with an empty silken balloon before mating. The evolution and function of this behavior can be understood by comparing the behavior of closely related species, that is, by using the comparative method characteristic of the ethological approach. (Drawn from descriptions in Kessel, 1955.)**

males of *H. sartor* present the females with an empty gauze case that turns off the predatory behavior of the female, thereby allowing them to mate (Kessel 1955). Without a comparison of the behavior with that of other species, an observer would be hard pressed to explain why males offer silken balloons to females.

In addition to utilizing the comparative method, ethologists often work in the field rather than in the laboratory. After all, they reason, it is in the natural setting that the normal context in which the behavior is displayed is apparent. From this, the function of the behavior may be deduced, and knowledge of the function may allow us to understand why the behavior has been shaped to its present form by natural selection. Tinbergen and his students conducted much of their research in the field. Lorenz and his followers, on the other hand, studied captive animals, but they often attempted to simulate in captivity some characteristics of the animal's natural habitat (Barlow 1991). As we will see in Chapter 18, von Frisch's carefully designed field experiments reveal that scout honeybees communicate the distance and direction of a rich food source to recruits by "dancing" within the hive.

CLASSICAL ETHOLOGICAL CONCEPTS

The Fixed Action Pattern

At the turn of the twentieth century, Charles Otis Whitman of the University of Chicago and Oskar Heinroth of the Berlin Aquarium were pioneering the field of ethology (Lorenz 1981). Both scientists were interested in the behavior of birds, and each independently came to the conclusion that the displays of different species are often exceptionally constant. In fact, they considered these patterns of movement to be just as reliable as morphological characters in defining a particular group.

Wallace Craig (1918), a student of Whitman, pursued the study of avian displays and examined the behavioral sequences of ring doves. He is perhaps best known for making the distinction between appetitive and consummatory components of behavior. He noted that in the beginning of a behavioral sequence, the acts are often variable and serve to bring the animal into the proper stimulus situation for a simpler, more stereotyped response. Impressed by the plasticity and apparent purposefulness of the initial motor patterns, Craig referred to these actions as appetitive behavior. As the name implies, appetitive behavior is thought of as a striving for the correct stimulus situation that will release the more rigid pattern of behavior that concludes the sequence, as if the animal had an appetite for that specific stimulus. If we continue with the epicurean designation, the stereotyped behavior that completes the sequence is called the consummatory act. Just as the consumption of a meal "satisfies" the appetite, the consummatory act brings about a sudden drop in the animal's motivation to perform those behaviors. Using the reproductive behavior of mammals as an example, the appetitive phase would be the search for a mate and the consummatory phase would be copulation. Craig published his ideas in 1918 in a paper entitled "Appetites and Aversions as Constituents of Instincts."

The stereotyped patterns of behavior that intrigued ethologists such as Whitman, Heinroth, and Craig were named fixed action patterns by Lorenz. By definition, a fixed action pattern (FAP) is a motor response that may be initiated by some environmental stimulus but can continue to completion without the influence of external stimuli. For example, Lorenz and Tinbergen (Lorenz and Tinbergen 1938) showed that a female greylag goose (*Anser anser*) will retrieve an egg that has rolled just outside her nest by reaching beyond it with her bill and rolling it toward her with the underside of

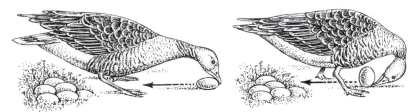

FIGURE 2.5   The egg retrieval response of the greylag goose. The chin-tucking movements used by the female as she rolls the egg back to the nest are highly stereotyped and are an example of a fixed action pattern. The side-to-side movements that correct for any deviations in the path of the egg are called the taxis component of the response. If the egg is removed, the female will continue to roll an imaginary egg back to the nest. One defining characteristic of a fixed action pattern is that it will continue to completion even in the absence of guiding stimuli. (Drawn from a photograph in Lorenz and Tinbergen 1938.)

the bill (Figure 2.5). If the egg is experimentally removed once the rolling behavior has begun, the goose will continue the retrieval response until the now imaginary egg is safely returned to the nest. We have emphasized the fact that once initiated, FAPs continue to completion. There is little consensus, however, on their defining attributes. Other characteristics that have been used to describe them include the following: (1) the sequence of component acts of an FAP is unalterable; (2) an FAP is not learned; (3) it may be triggered under inappropriate circumstances; and (4) it is performed by all appropriate members of a species (Dewsbury 1978).

Although it is true that individuals of the same species execute an FAP in much the same way, there is some variation because, in some cases, the FAP must be oriented according to the conditions under which it is performed. These orientation movements are called taxes. As the greylag goose retrieves an egg, for example, the egg might roll slightly to one side or the other. The female must adjust her retrieval motions according to which way the egg rolls. The chin-tucking movements constitute the FAP, and the orientation movements the taxis component.

The concept of a fixed action pattern was questioned in the years following Lorenz's first introduction of the term. George Barlow (1968) suggested that in reality, most patterns of behavior are not so stereotyped as suggested by the notion of the FAP, and furthermore, most cannot easily be separated into fixed and orientation components. He suggested calling the behavior a modal action pattern (MAP) instead of a fixed action pattern. In specific cases, however, the former term may be appropriate. Finley and colleagues (1983) examined the courtship displays of mallard ducks (*Anas platyrhyncos*) and concluded that the patterns of behavior were indeed as highly stereotyped as suggested by the notion of FAP. We will continue to use the traditional term

here, keeping in mind that there is some debate over the appropriateness of its use.

Sign Stimuli and Releasers as Triggers

A fixed action pattern is obviously produced in response to something in the environment. Let's consider the nature of the stimulus that might trigger the behavior. Ethologists called such a stimulus a sign stimulus. If the sign stimulus is emitted by a member of the same species, it is called a social releaser or, simply, a releaser. Releasers are important in communication among animals, as we will see in Chapter 18. Although releasers are technically a type of sign stimulus, the terms are often used interchangeably.

Sign stimuli may be only a small part of any environmental situation. For example, a male European robin (*Erithacus rubecula*) will attack another male robin that enters its territory. Experiments have shown, however, that a tuft of red feathers is attacked as vigorously as an intruding male (Lack 1943). The attack is not stimulated by the sight of another bird but only by the sight of red feathers. Of course, in the world of male robins, red feathers usually appear on the breast of a competitor.

Sign stimuli, simple cues that may be indicative of very complex situations, get through to the animal's nervous system, where they release patterns of behavior that may consist, in large part, of fixed action patterns. For example, the attack of the male European robin may be composed of FAPs that involve pecking, clutching, and wing fluttering. The end result is that when an intruding male robin appears, it is immediately identified and effectively attacked.

Types of Sign Stimuli   Any of the traits possessed by an animal or an object may serve as a sign stimulus. It may have a certain color, a special shape, or a pattern of movement. At the same time, it may produce a sound or have an odor. Perhaps it even touches the responding animal in a certain way. Any one of these aspects may be the one that releases the response—that is, that serves as the sign stimulus. How, then, are we to know which character serves as the sign stimulus? One way in which ethologists can find the sign stimuli in the barrage of information that reaches the responding animal is with models of the signaling animal with only one trait (or a few) presented at a time.

Many sign stimuli and releasers are visual. The male common flicker (*Colaptes auratus*), a type of woodpecker found throughout North America, has a black stripe, or moustache, extending from its bill along the sides of its head (Figure 2.6). It is the presence of this moustache on a male entering another's territory that is the specific releaser for aggressive behavior. If such a moustache is painted on a female of a mated pair, domestic harmony disappears to the point at which the male actually attacks his mate. When the moustache is removed, his aggression ceases and she is once more allowed to enter the territory (Noble 1936).

In some cases, color is a sign stimulus. During the early spring, a male three-spined stickleback, *Gasterosteus aculeatus*, concentrates his efforts on reproduction. After constructing a nest in shallow fresh water, he begins to defend his territory from any intruding males. At this time the underside of this fish turns a brilliant red. Niko Tinbergen and his coworkers at the University of Leiden, Netherlands, demonstrated—by constructing models of sticklebacks of varying degrees of likeness to the real male and painting them red, pale silver, or green—that it is the red tint of the undersurface of the trespasser that releases an aggressive response from the territory's defender (Pelkwijk and Tinbergen 1937). A very realistic replica in everything but red color was not attacked, but models barely resembling a fish, on which the underside was painted red, did provoke assault (Figure 2.7). Red models were always more effective than those of other colors. More recent studies, however, suggest substantial individual variation in the response of male sticklebacks to models with red undersides (e.g., Rowland and Sevenster 1985). This has led some researchers to suggest that the motivational state of the test fish may play an important role in determining its response (Baerends 1985). Thus, the issue of releasers may not be as simple as the early experiments of ethologists suggested.

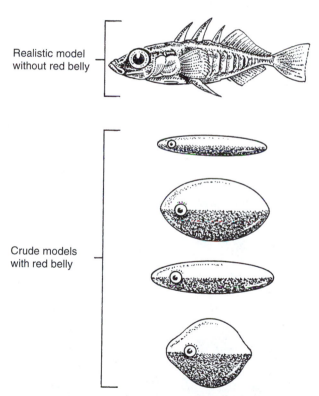

Realistic model without red belly

Crude models with red belly

FIGURE 2.7 Models of male three-spined sticklebacks presented to other males of the species to test their effectiveness as releasers. The crude models with red bellies were more effective in eliciting aggression from males than was the more realistic replica that lacked the characteristic red belly. Experiments by Tinbergen and colleagues, using models such as these, demonstrated that it is the red color on the underside of the male that releases aggression from other males. More recent experiments, however, suggest substantial individual variation in the responses of males to red bellies. (After N. Tinbergen, 1948.)

Female

Male

FIGURE 2.6 A male (bottom) and female flicker. It is the black moustache, present only on the male, that is the visual releaser for aggression among males of this species.

Supernormal Stimuli   When one constructs a model that isolates the effective characteristic of the stimulus, it is sometimes possible to exaggerate that characteristic and create an unrealistic model that elicits the behavior more strongly than the natural stimulus. This model serves as a supernormal stimulus. A well-known example comes from the incubation behavior of the oystercatcher (*Haematopus ostralegus*). Given a choice of one of two eggs of the appropriate shape, a female oystercatcher will choose the larger (N. Tinbergen 1951). Although she may topple off, she will persistently attempt to incubate an "egg" 20 times the normal size, even if there is an egg of a more appropriate size available. Similar behavior has been noted in other shore birds (Figure 2.8).

Sign Stimuli and Mimicry   Some animals exploit the reliability of the response to a sign stimulus for their own advantage. Brood parasites are birds that lay their eggs in the nest of other species, leaving the care of the offspring to the foster parents. In some such cases, the eggs of the brood parasite must have the appropriate sign stimuli to elicit incubation behavior by the female of the host species; otherwise, she will eject them from the nest. The eggs of several species of brood parasitic cuckoos mimic the eggs of their host species, being similar in size, color, and markings (e.g., Southern 1954).

Not only must the host accept the eggs of the brood parasite, it must also feed the nestlings. The releasers for feeding include markings on the inside of the mouth of the young that are apparent as they gape. The markings and gaping behavior of the brood parasites may mimic those of the host, ensuring that the offspring will be cared for. The similarity of these markings in one brood parasite, the long-tailed paradise widow bird (*Steganura paradisaea*), and its host, the finch *Pytilia melba*, are shown in Figure 2.9 (Nicolai 1964).

Chain of Reactions

So far we have considered only relatively simple behaviors, but a great deal of complexity can be added to the behavioral repertoire by building sequences of FAPs. The final product is an intricate pattern called a chain of reactions by ethologists. Here, each component FAP brings the animal into the situation that evokes the next FAP.

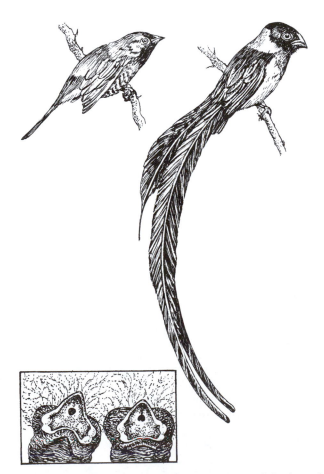

FIGURE 2.8   A gull attempting to incubate a super-egg instead of her own egg (foreground). A model, such as this monstrous egg, that emphasizes the aspect of the natural stimulus that triggers a behavior may be more effective than the natural stimulus in eliciting the response. Such a model is called a supernormal stimulus.

FIGURE 2.9   The finch (top left) is the host of the brood parasitic long-tailed paradise widow bird (top right). The gaping signal and markings on the inside of the mouth of the host species (bottom left) are mimicked by nestlings of the brood parasite (bottom right). These are the key stimuli that release the feeding behavior of the parents. (Modified from Wickler 1968.)

One of the earliest analyses of a chain of reactions was conducted by N. Tinbergen (1951) on the courtship ritual of the three-spined stickleback. This complex sequence of behaviors culminates in the synchronization of gamete release, an event of obvious adaptive value in an aquatic environment. Each female behavior is triggered by the preceding male behavior, which in turn was triggered by the preceding female behavior (Figure 2.10).

A male stickleback in reproductive condition may sometimes attack a female entering his territory. If the female does not flee and begins to display the appropriate head-up posture in which she hangs obliquely in the water, exposing her egg-swollen abdomen, the male will begin his courtship with a zigzag dance. He repeatedly alternates a quick movement toward her with a sideways turn away. This dance releases approach behavior of the female. Her movement induces the male to turn and swim rapidly toward the nest, an action that entices the female to follow. At the nest he lies on his side and makes a series of rapid thrusts with his snout into the entrance while raising his dorsal spines toward his mate. This behavior is the releaser for the female to enter the nest. The presence of the female in the nest is the releaser for the male to begin to rhythmically prod the base of her tail with his snout, causing the female to release her eggs. She then swims out of the nest, making room for the male to enter and fertilize the eggs. At the completion of this ritual, the male chases the female away and continues to defend his territory against other males until another female can be enticed into the courtship routine. The male mates with three to five females and then cares for the developing eggs by guarding them from predators and fanning water over them for aeration. We see, then, that this complex sequence is largely a chain of FAPs, each triggered by its own sign stimulus, or releaser.

The chain of reactions is not as rigid as the above description of courtship in the three-spined stickleback implies. There are actually many deviations in the precise order of the events in the ritual (Figure 2.11), and some actions must be repeated several times if one partner is less motivated than the other (Morris 1958). Such flexibility begins to make sense when the function for which the ritual evolved is considered. For the stickleback, courtship is important to time the release of the gametes, and thus males and females seem to adjust their activities so that they are physiologically ready for gamete release at the same time. Despite some flexibility, however, the component behaviors do not occur randomly. In the display, a particular behavior is more likely to be followed by certain actions than by others.

Action Specific Energy

Lorenz introduced the concept of action specific energy in 1937 and renamed it action specific potential

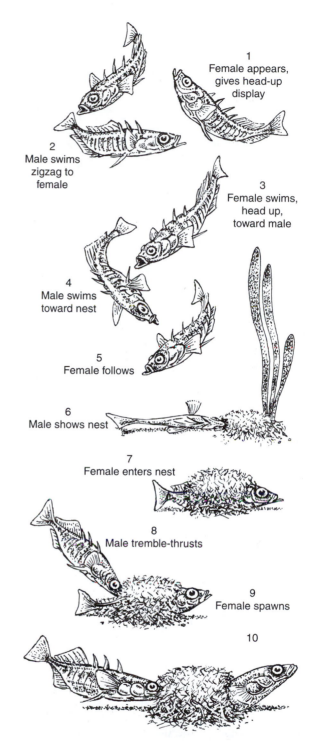

1 Female appears, gives head-up display

2 Male swims zigzag to female

3 Female swims, head up, toward male

4 Male swims toward nest

5 Female follows

6 Male shows nest

7 Female enters nest

8 Male tremble-thrusts

9 Female spawns

10

FIGURE 2.10 Courtship behavior in the three-spined stickleback. (From N. Tinbergen, 1989.)

in 1981. The concept has not gained wide acceptance because of the difficulties of explaining it in concrete neurological terms. Lorenz used the concept to explain his observation that the ease with which a stimulus released an action depended on how long it had been since the act was previously released. He postulated that action specific energy (energy for a specific action)

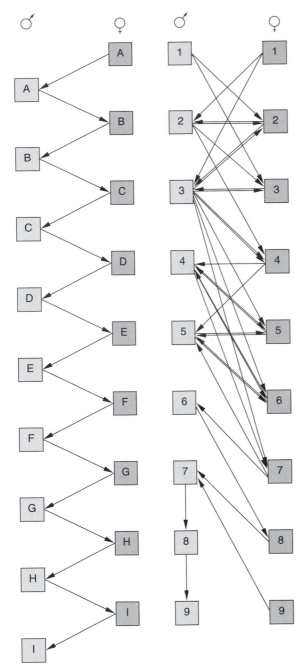

FIGURE 2.11 The ideal courtship sequence of the three-spined stickleback (left) and the observed sequence (right). The units of behavior in the ideal sequence are staggered because one action invariably elicits a particular response. In real life, the sequence of behaviors is not so consistent as it is in the ideal sequence. Although there is a general pattern to the observed courtship ritual, there is some variability in the specific reactions to the behavior of the partner. (From Morris 1958.)

is constantly being produced in the animal's nervous system. The energy is held in check by an inhibitory mechanism until the appropriate stimulus releases it. The released energy activates certain muscle systems, and a particular behavior pattern results. The concept of action specific energy is consistent with observations of two types of behavioral activity, vacuum activity and displacement activity.

Vacuum Activity   In vacuum activity, a fixed action pattern is displayed in the absence of a stimulus. Basically, ethologists hypothesized that if the proper stimulus is not found, action specific energy continues to build up until it can be contained no more, and then it goes off on its own, without being triggered by a stimulus.

One example of vacuum activity was described in the reproductive behavior of ring doves (*Streptopelia risoria*). In an experiment, the male dove, isolated from females of his own species, eventually courted a stuffed model of a female that he had previously ignored. As time went on, he became even less discriminating in his choice of a mate and would bow and coo to a rolled-up cloth. Finally, even the corner of his cage elicited courtship behavior (W. Craig 1918).

Displacement Activity   During times of conflict, animals are often observed performing behaviors that seem to be totally irrelevant to the situation. For example, a male stickleback within his own territorial boundaries will unfailingly attack any intruding male. Within his neighbor's territory, however, the same male will typically flee from the territory owner. So what happens if two neighboring males meet just at the borders of their territories? Strangely enough, they may both start digging a pit as if for a nest. Why would they do this—an act that appears to have nothing to do with the situation at hand? According to the ideas of Lorenz and Tinbergen, because each fish cannot express its urge to attack nor its desire to flee, it is driven by tension to find an outlet in another activity. In other words, within each of the male sticklebacks at the territorial boundary, there is a high level of energy that drives them to attack their rival and an equally high level of energy that drives them to flee. As the nervous energy for the two incompatible behaviors peaks within each male, it spills over into a third system, one that controls nesting behavior. Behavior patterns such as the nest-digging behavior in this example are called displacement activities because they are caused by the displacement of nervous energy from one system to another.

Models for the Organization of Instinctive Behavior

Models are hypothetical constructs that help us understand complex phenomena by framing their essential

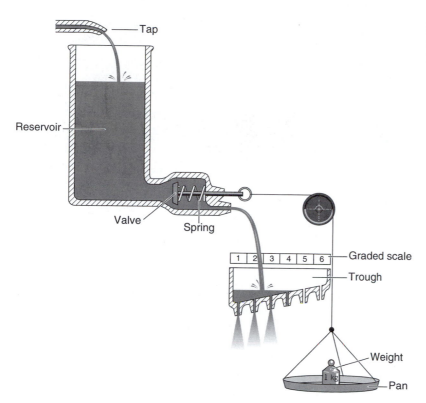

Tap

Reservoir

Valve

Spring

Graded scale

Trough

Weight

1 kg

Pan

relationships in simpler terms, with simpler structure. Good models force us to organize our observations and our thinking about the natural phenomena on which they are based. Ideally, they should lead to experiments.

Lorenz's Psycho-hydraulic Model   Lorenz (1950) organized his observations on animal behavior into the psycho-hydraulic model, also called the flush toilet model (Figure 2.12). Vacuum behavior and displacement activity are among the observations that he hoped to explain in this model. The continuous production and storage of action specific energy during the interim when an animal is not performing some FAP is represented by the damming of water within the reservoir as it pours in from the tap. The idea that action specific energy accumulates until the particular FAP is performed is depicted as the filling of the reservoir while a valve is held shut by the spring. The valve can be opened to permit the outpouring of water. In the real world, FAPs are generally released by the appropriate sign stimuli, but sometimes when the animal has not had an opportunity to perform the FAP in a long time, the behavior seems to appear spontaneously. The model is consistent with the observation of vacuum activity because it was designed with two ways in which the valve can be opened: (1) weights of various strengths, symbolizing stimuli of varying degrees of effectiveness, can be added to the pan connected to the valve, or (2) the pressure of the water in the reservoir, representing the action specific energy, may become

intense enough to compress the spring, thus forcing open the valve. The higher the water level, the smaller the weight required, and eventually hydraulic pressure alone pushes the valve open, which is analogous to a vacuum activity. This model emphasizes that the appearance of an FAP will depend on both the level of action specific energy and the strength of the releaser.

The motor activity itself is represented by the water spouting from the spigot. The graded trough was added to account for the observation that there are often a number of components to a motor pattern, some of which are only displayed at the highest levels of motivation. If the valve is only slightly open, a small trickle of water enters the trough and will drip through the first and lowest hole in the trough, symbolizing the motor activity with the lowest threshold. As the valve opens wider, an increasingly greater amount of water is discharged, allowing it to reach higher trough holes, which represent activities with higher thresholds for elicitation. If the valve is fully open and the reservoir is full, water gushes forth and pours from all openings in the trough. This is analogous to the full-blown response of the animal. In this way, the model accounts for observations of behaviors such as the aggressive display of the cichlid fish, *Astatotilipia strigigena*, which includes five successive behavior patterns of increasing threshold: display coloration, fin spreading, distending the covering of gills, beating the caudal fin sideways, and ramming the adversary. Ramming, the final and most difficult component of the total display to elicit, would be represented by the highest hole in the trough (Seitz 1940).

According to the hydraulic model, one might describe displacement activity as motor patterns that occur because the trough that represents one set of FAPs, such as those for aggression, is blocked while the valve is open. This would cause water to spill over into another trough, one symbolizing a different set of FAPs, such as grooming.

The model also accounts for the decrease in the probability that an FAP will be performed after it has been elicited many times in succession. In hydraulic terms, the reservoir is empty. As time passes and the reservoir fills, the likelihood of eliciting an FAP increases until the repeated performance of the behavior pattern drains the tank again. This is consistent with the observation of cyclical changes in responsiveness.

Lorenz's hydraulic model has been criticized, in part because it does not fit with what we know about the neurophysiology of the brain (Lehrman 1953). There is no physiological correlate of action specific energy, and the brain does not control behavior in the same way that a flush toilet operates. The hydraulic model has been useful in organizing observations on behavior into a conceptual framework, but now that the ability to determine the actual physiological mechanisms that control behavior is beginning to be technologically possible, less schematic models will be more useful. Lorenz (1950) himself recognized the limitations of his model. In his words, "This contraption is, of course, still a very crude simplification of the real processes it is symbolizing, but experience has taught us that even the crudest simplisms often prove a valuable stimulus to investigation."

Tinbergen's Hierarchical Model Tinbergen observed the same regularities of behavior that Lorenz did and was particularly struck by certain features that are not addressed in the hydraulic model. He developed a hierarchical model that emphasizes some different features of behavior and is easier to relate to the nervous structures that control behavior (N. Tinbergen 1951). The model makes use of the ideas of appetitive and consummatory components of behavior and includes the notion of the innate releasing mechanism (IRM), an idea developed jointly by Lorenz and Tinbergen but elaborated on largely by Tinbergen (Lorenz 1981). According to their scheme, when a sufficiently motivated animal encounters a certain key stimulus, the stimulus is identified by an innate releasing mechanism. The IRM then removes the inhibition from a particular brain center and allows it to direct the performance of a stereotyped behavior pattern, the fixed action pattern. As defined by Tinbergen (1951), an IRM is a "special neurosensory mechanism that releases the reaction and is responsible for its [the reac-

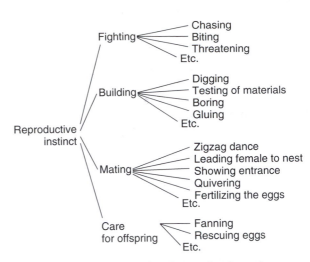

FIGURE 2.13 The hierarchical organization of an instinct as visualized by Tinbergen. (From N. Tinbergen 1942.)

tion's] selective susceptibility to a very special combination of sign stimuli." Thus, at least part of its job is to function as a stimulus filter, identifying the sign stimulus and ignoring the irrelevant information.

According to Tinbergen's (1951) model, each general instinct is hierarchically organized (Figure 2.13), and as such it is susceptible to certain priming, releasing, and directing influences. These stimuli may be of internal, as well as external, origin. In other words, the general instinct can be broken down into several levels of consummatory acts. Each level has its own appetitive behaviors and is made up of a number of specific FAPs, and each is considered to be controlled by a different neural center. When appetitive behavior brings the animal into the correct stimulus situation, a neural block is removed so that the "energy" of the general drive can flow to the next hierarchical level, allowing that consummatory act to be performed. The part of the nervous system that identifies the correct releasing stimulus and removes the neural block is the IRM. As the energy flows to lower levels, it progresses along a path determined by the particular stimuli actually encountered that can remove the blocks for specific behaviors. If the appropriate stimulus is not encountered, the energy activates the centers that control the appetitive behavior for that level of consummatory behavior, and the animal searches for the specific stimuli that will remove the block and allow the behaviors to be performed.

Although the behavior patterns involved are not FAPs, a simple analogy of this hierarchical model would be a man walking along a path until he faces a closed door. If he looks around (appetitive behavior), he will find the key (releaser) to open the door. When the key is turned, the lock mechanism (IRM) allows the door to swing open. The man walks through the portal (the first-level consummatory response). Once he is on the other side, he faces a choice of several doors. The specific key he finds will determine which door he proceeds through. Having passed through the door (the second-level behavior), he confronts another series of doors, and so on.

Although Tinbergen's hierarchical model has been criticized because it, like Lorenz's hydraulic model, incorporates the concept of "energy" that develops within the nervous system and drives certain behaviors, it is easier to make the transition from hypothetical construct to real neurophysiology with the hierarchical model. Nevertheless, there are problems associated with the notion of innate releasing mechanisms. Historically the IRM has been associated with central filtering in the minds of ethologists, and this has led to certain problems. It has given rise to the notion that there is a special device in the central nervous system of animals that is waiting to be switched on or off. Such a notion is undoubtedly simplistic. The concept of IRM is useful, however, if we keep in mind that we do not know how central filtering is accomplished, that it may (and probably does) involve a multitude of complex neuronal interactions at various levels (from peripheral to central), and that the systems may differ in various kinds of animals. The concept of IRM also has value if it stimulates research into the nature of central filters.

COMPARATIVE PSYCHOLOGY

THE APPROACH: PHYSIOLOGICAL, DEVELOPMENTAL, QUANTITATIVE, LABORATORY-ORIENTED

The comparative psychologists' emphasis on laboratory studies of observable, quantifiable patterns of behavior distinguished them from the European ethologists during the first half of the twentieth century. Recall that at this time, many ethologists preferred to study animal behavior under natural conditions. This meant that they went into the field and observed behavior. The problem was that in the field, the unexpected is expected; one cannot control all the variables. The comparative psychologists argued that good, experimental science cannot be done under such uncontrolled conditions. The ethologists were further criticized because, although they described changes in behavior, they often neglected to quantify their results and rarely analyzed the data with statistical procedures. Given the psychologists' penchant for laboratory studies that produce quantifiable results, it is not too surprising that much of their early research focused on learning and

the physiological basis of behavior. Again, however, we wish to emphasize that although learning and physiology were popular areas of study among comparative psychologists, the evolution and function of behavior were also examined by some of comparative psychology's practitioners. We will now consider some of the major conceptual developments in the field of comparative psychology.

EARLY CONCEPTS OF COMPARATIVE PSYCHOLOGY

Morgan's Canon

Recall from our previous discussion of the ideas and writings of Darwin and Romanes that the early descriptions of animal behavior were often subjective, anthropomorphic accounts. C. Lloyd Morgan helped stop the anecdotal tradition, thereby helping to send comparative psychology toward the objective science it is today. He argued that behavior must be explained in the simplest way that is consistent with the evidence and without the assumption that human emotions or mental abilities are involved. This idea was crystallized in Morgan's Canon (1894): "In no case may we interpret an action as the outcome of the exercise of a higher psychical faculty if it can be interpreted as the outcome of the exercise of one which stands lower in the psychological scale." In other words, when two explanations for a behavior appear equally valid, the simpler is preferred. People were urged to offer explanations of an animal's behavior without referring to the animal's presumed feelings or thought processes.

Learning and Reinforcement

We have already mentioned that many of the early comparative psychologists focused their research efforts on learning. The early days of these studies were exciting times, indeed, and some of the most important work was done by scientists in America. E. L. Thorndike (1898), for example, devised experimental techniques to study learning in the laboratory. He was a pioneer in research on what was called trial-and-error learning, now usually called operant conditioning. In operant conditioning the animal is required to perform a behavior to receive a reward. In one series of experiments, Thorndike invented boxes that presented different problems to animals. For instance, one problem box was a crate with a trapdoor at the top through which an experimenter might drop a cat to the inside of the box. A hungry cat was left in the box until it accidentally operated a mechanism, perhaps pulling a loop or pressing a lever, that opened an escape door on the side of the box, allowing it access to food that had been placed nearby. The length of time it took for each escape provides an objective, quantifiable measure of learning

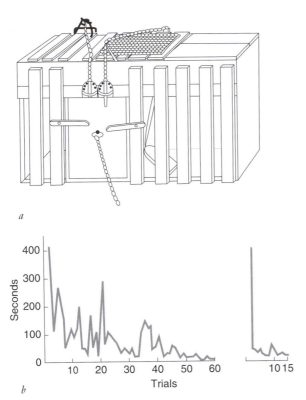

a

b

FIGURE 2.14   (*a*) A problem box. Thorndike invented many "problem boxes" to measure the learning ability of animals. An animal would be placed inside the box and would have to learn how to operate an escape mechanism. (*b*) The time required for escape on successive trials was a measure of how quickly the animals mastered the task. (From Thorndike 1911.)

FIGURE 2.15   Ivan Pavlov described a conditioned reflex in the dog.

progress (Figure 2.14). During repeated trials, the animal became more efficient and required less time to hit the escape latch. Thorndike's studies led him to develop the Law of Effect, a cornerstone of operant conditioning. His basic notion was that responses that are rewarded, that is, followed by a "satisfying" state of affairs, will tend to be repeated (this idea was also described by C. Lloyd Morgan and other investigators of animal behavior). Thorndike began publishing studies on animal intellect and behavior in the late 1800s, and in 1911 published a collection of his writings in a book entitled *Animal Intelligence: Experimental Studies* .

Just a few years after Thorndike introduced the idea of trial-and-error learning, Ivan Pavlov (Figure 2.15), a Russian physiologist, described the conditioned reflex. Pavlov noticed that a dog begins to salivate at the sight of food and he reasoned that the sight of food must have come to signal the presence of food. (In science, the key observations that trigger great ideas are often quite commonplace, as in this case. It is not what you observe; it is what you make of it.) In his well-known experiment, Pavlov rang a bell immediately before feeding a dog and found that, in time, the dog

came to salivate at the sound of the bell alone (Pavlov 1927).

Learning to salivate at the sound of a bell is now recognized as an example of a more general phenomenon called classical conditioning, or establishing a conditioned reflex. In general terms, a conditioned reflex begins with a regular and measurable unlearned response, called the unconditioned response (UR—here, salivation), that is normally elicited by a stimulus called the unconditioned stimulus (UCS—here, food). During conditioning, a neutral stimulus, one that does not ordinarily cause the unconditioned response, is presented just before the UCS is offered. This is the conditioned stimulus (CS—here, the bell). Once the association is made, the conditioned stimulus alone is sufficient to elicit the response. In this case, the sound of the bell is sufficient to cause the dog to salivate.

At first, comparative psychologists used operant and classical conditioning techniques to study the learning abilities of a wide variety of species. Thorndike, for example, examined learning in fish, chickens, cats, dogs, and monkeys and noted striking similarities in the learning processes of these animals. His results were, thus, consistent with the idea of intellectual continuity. Thorndike concluded that although animals might differ in what they learned or in how rapidly they learned it, the process of learning must be the same in all species. In his 1911 collection of papers he summarized his idea of intellectual continuity as follows (p. 294):

> [Intellect's] general law is that when in a certain situation an animal acts so that pleasure results, that act is selected from all those performed and associated with that situation, so that, when that situation recurs, the act will be more likely to follow than it was before. . . . The intellectual evolution of the race consists of an increase in the number, delicacy, complexity, permanence and speed of formation of such associations. In man his increase reaches such a point that an apparently new type of mind results, which conceals the real continuity of the process. . . . Amongst the minds of animals that of man leads, not as a demigod from another planet, but as a king from the same race.

Behaviorism

Another important event that steered comparative psychology toward objectivity and laboratory analysis was the birth of behaviorism, a school of psychology that restricts the study of behavior to events that can be seen—a description of the stimulus and the response it elicits. Behaviorists sought to eliminate subjectivity from their studies by concentrating their research efforts on identifying the stimuli that elicit responses and the rewards and punishments that maintain them. This was, indeed, a step toward better science. They designed experiments that would yield quantifiable data, invented apparatus to measure and record responses, and developed statistical techniques that could be used to analyze behavioral data. The assumptions that an animal's mental capacity could not be measured directly, but its ability to solve a problem could, again focused attention on learning ability as a popular research subject. A learned response could be described objectively, and experiments could be conducted under the controlled conditions of the laboratory.

B. F. Skinner, one of the most famous behaviorists, devised an apparatus that was similar to Thorndike's problem box but lacked the Houdini quality. Instead of learning to operate a contrivance that provides a means of escape, a hungry animal placed in a "Skinner box" must manipulate a mechanism that provides a small food reward (Figure 2.16). A rat may learn to press a lever, and a pigeon may learn to peck at a key. Patterns of behavior that are rewarded tend to be repeated, or to increase in frequency, and so learning can be measured as the number of responses over time. Skinner believed that the control of behavior was basically a matter of reinforcement.

Behaviorists began to see basic principles underlying learning that were common to all species. They expected to see similarities in the learning process because at that time, the minds of all species were con-

FIGURE 2.16   B. F. Skinner and his apparatus, the Skinner box. Animals placed in the box learned to operate a mechanism to obtain a food reward.

sidered similar in kind. Thus, according to the traditional view of learning held by the followers of behaviorism, the minds of humans and animals were similar in kind and differed only in complexity. In short, there were general laws of learning that transcended all species and problems. If this was true, then it was reasonable to study the laws of acquisition, extinction, delay of reinforcement, or any other aspect of the learning process in a simple and convenient animal, such as the domesticated form of the Norway rat (*Rattus norvegicus*), and the results could then be broadly applied to other species.

THE ROOTS OF PHYSIOLOGICAL PSYCHOLOGY

Although learning was a dominant focus of research during the middle of this century, it was not the only research interest of comparative psychologists. Another research topic was the physiological basis of behavior. Part of the psychological foundation of behavior is, of course, the nervous system. The comparative psychologists' interest in the neurological mechanisms of behavior can be traced back to Pierre Flourens (Jaynes 1969), a protégé of Baron Cuvier, a famous scientist of nineteenth-century France who stressed the importance of laboratory research. Flourens's reputation was earned for his studies of the relationship between behavior and brain structure. For example, he did experiments in which parts of the brain were removed, such as the cerebral hemispheres from a pigeon, to look for the effect on the animal's behavior.

Karl Lashley was one comparative psychologist who maintained an interest in physiology, as well as a comparative base of study, during the years when learning by the laboratory rat dominated the field. His attempts to localize learning in the cerebral cortex resulted in the rejection of some hypotheses that were widely accepted at the time. For example, based on Pavlov's ideas, it was assumed that learning depended on the growth or strengthening of neural connections between one part of the cerebral cortex and another. To test this idea, Lashley (1950) trained rats on a variety of mazes and discrimination tests and then tried to disrupt the memory by making a cut into the cerebral cortex in a different place in each animal. After destroying varying amounts of brain tissue, he would then retest the animals to see if their behavior changed. In general, he found that when it came to complex problem solving, the entire cerebral cortex was involved, and any particular area was just as important as any other. He also experimentally addressed questions such as whether a learned response depends on the stimulation of the same receptors that were activated during learning, and whether the learned response depended on a fixed pattern of muscle movements. Contrary to expectations, he found that they do not. But Lashley was not solely concerned with the neurological basis of learning in the rat. He also examined the role of the brain in emotion and in vision.

The comparative psychologist Frank Beach began his career by using brain surgery to determine the effects of lesions on the maternal behavior of the rat, but he later went on to study the effects of hormones on behavior. He analyzed the roles of nerves, hormones, and experience in the sexual behavior of fishes, amphibians, reptiles, birds, and mammals. We will discuss some of his work in the field of behavioral endocrinology in more detail in Chapter 7.

Recognizing that animal behavior is concerned with the activities of groups of animals, as well as of individuals, some comparative psychologists studied social behavior. Robert Yerkes, for example, established a research facility (later named the Yerkes Laboratory of Primate Biology) at Orange Park, Florida, to study a wide range of behavior in primates. Some researchers also began to see that although it is often easier to make measurements in the laboratory, it is not impossible to get good measurements in the field. C. R. Carpenter

studied a variety of primate species, each in its natural setting: howler monkeys in Panama, spider monkeys in Central America, and gibbons in Thailand, to name a few. T. C. Schneirla used both field observation and laboratory experimentation to investigate the social behavior in army ants. In doing so, he applied the rigorous methodology of laboratory researchers to his field studies. Such pioneering studies began to help weave the two independent sciences of ethology and comparative psychology together.

THE NATURE/NURTURE CONTROVERSY AND CURRENT VIEWS

THE NATURE/NURTURE CONTROVERSY

Differences in beliefs about the importance of inheritance or experience in behavior were one cause for heated debate between the European ethologists and the American comparative psychologists. The early ethologists, concerned with the evolution of behavior, concentrated on the adaptiveness of behavior. They believed that behavior is inherited in a form that is suited to the life circumstances of the organism because it is shaped by natural selection. Their experiments were designed and their results were interpreted within that theoretical framework. Because the ethologists perceived behavior as having evolved to suit the environment, it seemed best to study behavior in as natural a setting as possible. The American psychologists were more interested in how an individual's behavior develops than in explaining its adaptiveness. Furthermore, the majority of psychologists believed that given the uncontrolled conditions of the natural environment, the laboratory was the best place to assess the effects of various experiences on behavioral development.

The result of these differences was that a battle, often called the nature/nurture controversy, raged for about 20 years. (The nature/nurture controversy actually has a much longer history than the argument between ethologists and comparative psychologists, but we will concentrate on their debate here.) Behavior must be inherited, the antagonists said, or it must be learned. Behavior is shaped by natural selection, or it is shaped by the experience of the individual. The classic quote that represents the behaviorist's position comes from John B. Watson (1930, p. 104; Figure 2.17):

Give me a dozen healthy infants, well-formed, and my own specified world to bring them up in and I'll guarantee to take any one at random and train him to become any type of specialist I might select—doctor, lawyer, merchant, chief and yes, even beggar man and

FIGURE 2.17   John B. Watson, a prominent behaviorist.

thief, regardless of his talents, peculiarities, tendencies, abilities, vocations, and race of his ancestors.

The behaviorist position was clear—learning is the ultimate basis of behavior, so instinct and innate tendencies are irrelevant and unnecessary. It is interesting to note that although the extreme statements made by Watson in his later years came to symbolize the behaviorist's position and resulted in his becoming perhaps the most unpopular of all psychologists among the ethologists, his work as a young man—comparative studies of the behavior of two species of shorebirds on islands off the coast of Florida—was quite compatible with the ethologist's approach (Dewsbury 1989). It should also be noted that most of his extreme statements were made with specific reference to human behavior.

In contrast to the attitudes of behaviorists, the ethologists' position made learning subordinate to instinct. This feeling is reflected in the words of Niko Tinbergen (1952): ". . . learning and many other higher processes are secondary modifications of innate mechanisms, and . . . therefore a study of learning processes has to be preceded by a study of the innate foundations of behavior." Tinbergen saw learning as "superimposed on the innate patterns and their mechanisms."

One of the most striking events of this time period was the publication in 1953 of Daniel Lehrman's article "A Critique of Konrad Lorenz's Theory of Instinctive

Behavior." Lehrman harshly criticized many of the ideas that had formed the foundation of ethology. He argued against Lorenz's belief that all behavior could be divided into the two categories of innate and learned. According to Lehrman, the innate/learned dichotomy was totally inappropriate for an analysis of behavioral development, and he proposed instead an interactionist approach. All behavior, Lehrman argued, arises from interactions between a developing animal and its environment. Lehrman was not the first comparative psychologist to criticize the fundamental concepts of ethology. Indeed, it was his mentor, T. C. Schneirla, who initially led the attack. However, because Schneirla's style of writing was somewhat difficult and his critiques of ethological ideas were often published in journals that were not commonly read by the European ethologists (Hinde 1982), the 1953 article published by his pupil Lehrman in the *Quarterly Review of Biology* caused the greatest stir.

In the debate that followed, each camp, with its own dogma as a shield, attacked the other with the vehemence that springs from seeing issues as black or white. Even if the primary proponents of each position did not see nature and nurture as a dichotomy, others were perceiving their views in that light. In 1970, as tempers were cooling, Lehrman said, "I have heard it said so often that I believed that all species-specific behavior develops through individual learning that I almost came to believe that I *had* said it."

Persuaded by cogent arguments and experimental data, each side began to see some merit in the opposing view and reformulated key concepts. Although some differences in emphasis still exist and accusations of dichotomous thinking about behavioral development are still being made in certain research areas (e.g., Johnston 1988 and accompanying commentary), compromises have been reached. In the modern synthesis of classical ethology and comparative psychology, both environment and inheritance are seen as important in shaping behavior. But before discussing how the ideas about nature and nurture were modified, we will focus on the issues that fired the controversy.

Difficulties with the Concept of Innate

Problems with the Definition   At the heart of the nature/nurture controversy was the word *innate*. The first problem was in defining the term. An innate behavior came to be regarded as one that appears as the animal develops, apparently without having been learned. It is often stereotyped (always performed the same way by members of a particular species) and complete (performed in its entirety). The problem was that other factors could also account for such behaviors. Two specific objections were raised.

First, stereotyped behavior may result from similarities in environment. Daniel Lehrman (1953) pointed out that constancy in the form of a behavior does not mean that it is genetically programmed. Instead, a behavior may be performed in the same way by all the members of a species because they all developed within the same environment. He cited the example of the Ichneumon fly. This parasitic fly provides food for its offspring by laying its eggs on the larvae of a moth. In nature, flour moth larvae are the victims. For this reason, the parasite's preference could easily be interpreted as a response to stimuli that were determined by its genes. However, when the larvae were raised on the larvae of moths other than the flour moth, these individuals, when adult, preferred to lay their eggs on the larvae of the moth species that had nourished them during development. This was one of the first demonstrations that a species-specific stereotyped behavior—the choice of flour moth larvae for an oviposition site—was influenced by experience (Thorpe and Jones 1937).

Another characteristic of innate behavior traditionally included in the definition is that it appears in a complete form without practice the first time it is performed. The second objection concerns this characteristic. Numerous studies, including the one just described, have shown that learning or environmental effects can take place so early, before hatching or birth, that the completeness of a new behavior does not demonstrate that learning was unimportant. Prenatal learning has been demonstrated, for example, in humans (Kolata 1984) and rats (Smotherman 1982).

Modifying Innate Behavior by Experience   Jack Hailman's (1967b, 1969) investigation of the feeding behavior of gull chicks demonstrated that an innate

behavior can be modified by learning. Before Hailman's study, Tinbergen and Perdeck (1950) had described the behavior of herring gull chicks (*Larus argentatus argentatus*) and had described their food begging and feeding actions as innate, or genetically based. The chicks peck at their parent's bill, causing the parent to regurgitate food. The behavior was called innate because it was performed in the same stereotyped way by all herring gull chicks soon after hatching.

However, Hailman (1967b, 1969) showed that the innate feeding responses of gull chicks are modified by learning. Laughing gull chicks (*L. articilla*) that are at least one week old beg for food from their parent in much the same manner as do herring gull chicks. The pattern involves the three separate actions, the last of which is shown in Figure 2.18: (1) The chick opens and closes its bill, (2) the chick moves toward the parent, and (3) the chick rotates its head so that it can grasp the bill of the parent. But the behavior of newly hatched chicks is slightly different from that of one-week-old chicks; they do not rotate their heads and are therefore unable to grasp the parent's bill. Instead, they end up just pecking at the bill, if they are lucky enough to actually hit it. Hailman demonstrated that the chicks learn to rotate their heads and grasp the parent's bill. He raised some chicks in the dark so that their visual experience would be limited and force-fed them so that they would not have experience in begging from a parent. At an age when wild chicks showed the rotational component, the hand-reared chicks did not. Obviously, that component of the response does not develop without practice in begging.

In addition, the chick's experience during its first week of life refines its image of a parent, the stimulus for the begging response. Laughing gull chicks that had

FIGURE 2.18   An example of an innate behavior pattern that is modified by experience. Right after hatching, a laughing gull chick begs by striking at the parent's bill and, with luck, hitting it. By the time the chick is three days old, its begging is nearly perfected. The begging response involves the correct timing of opening and closing the mandibles around the parent's bill, the movement of the head toward the parent, and the rotation of the head so that the chick can grasp the parent's bill. This last component of the response requires the experience of begging from a parent. (Redrawn from Hailman 1969.)

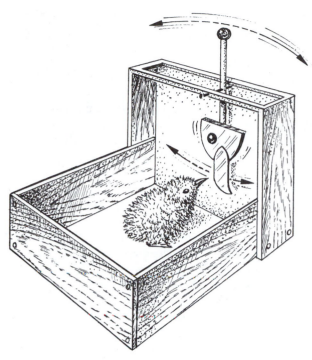

FIGURE 2.19 **The laboratory apparatus used to test the effectiveness of various models in eliciting pecking (the begging response) from gull chicks. The model was attached to a pivot arm and moved in time with a metronome in front of the chick. The number of pecks per unit of time, usually 30 seconds, was recorded. (Redrawn from a photograph in Hailman 1967b.)**

been kept in the dark for the first 24 hours after hatching were presented with a set of five models of gull heads, one at a time. The models were attached to a metronome and moved with the same rate and pattern in front of a gull chick (Figure 2.19). The number of pecks aimed at the model in a given amount of time, usually 30 seconds, was recorded. As you can see in Figure 2.20, newly hatched chicks pecked with almost the same frequency at anything that vaguely resembled a bill. However, as the chicks matured, the rate at which they pecked at a model increased with the realism of the model. At one week old, chicks pecked only at models

that closely resembled a parent. We can see, then, that the chicks responded to more lifelike models as they learned the characteristics of their parents.

Again, just because a behavior is innate does not mean that learning plays no role in its development. Conversely, we will see next that the involvement of learning in the shaping of a behavior is not an indication that the behavior has no genetic basis.

Problems with Early Learning Theory

If it was difficult for the early ethologists to accept that what they considered to be innate behavior could be modified by experience, it was equally difficult for the comparative psychologists (particularly the behaviorists) to accept that learning may be influenced by, or constrained by, genetic factors. Let's see how ideas progressed in understanding the development of behavior by looking at the position of the behaviorists and some of the problems they encountered.

Interference of Innate Behavior with a Conditioned Response According to learning theory, a response that produces a rewarding outcome becomes more frequent or more probable over time. This idea, which dates back to Thorndike's Law of Effect, is the basis of operant conditioning, the means of training described earlier, in which a behavior is encouraged through reward. The reward is said to reinforce the appropriate response.

The idea that reinforced responses become more frequent was questioned by Keller and Marion Breland (1961), students of B. F. Skinner who used operant conditioning techniques to train animals for commercial purposes. The Brelands discovered that instinctive behavior patterns would sometimes disrupt a trained animal's performance by appearing without reinforcement in place of desired actions that were rewarded. Because their animals were often trained with a food reward, the disruptive innate behavior patterns were those associated with eating or obtaining food. In one case, a raccoon was being trained to perform in a window display, where it was to deposit money in a "bank,"

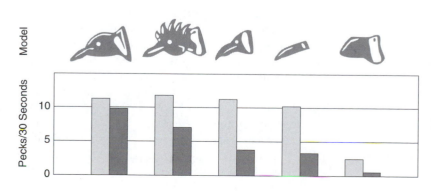

FIGURE 2.20 **Representative models of parent heads and the responses of laughing gull chicks to them. The light gray bars indicate the responses of newly hatched chicks and the black bars the responses of older chicks. The newly hatched chicks pecked at almost anything that resembles a bill, but older chicks preferred a more realistic model of a laughing gull head. (Modified from Hailman 1969.)**

FIGURE 2.21 A raccoon trained by the Brelands to deposit money in a bank often handled the coin as it would a prey item. In this instance, an innate behavior interfered with a conditioned response.

thereby encouraging observers to save their money (Figure 2.21). The raccoon would pick up a coin and bring it to the bank, but occasionally it appeared to be reluctant to make the deposit. Instead, it would dip a coin into the bank, retrieve it, and fondle it. The coin was handled in much the way a raccoon might prepare a crayfish by rubbing it in a stream before eating it. After playing with the coin for awhile, the raccoon would eventually deposit the money and receive a food reward. To make matters worse, when the raccoon was required to deposit two coins before being rewarded, it had even greater difficulty parting with its money. The masked miser would spend minutes rubbing the coins between its fingers and dipping them into the bank. Because this behavior was not likely to inspire spectators to invest, the raccoon soon lost its job. These behaviors—rubbing the coins with the hands and dipping them into the bank—are instinctive actions for the preparation of food. The raccoon had made the link between performance and food, but it was born with the link between cleaning and food. The researchers were often unable to disconnect the first from the second.

In this example, nonreinforced innate behaviors were interfering with those conditioned by operant techniques. The Brelands (1961) concluded that "the behavior of any species cannot be adequately understood, predicted or controlled without knowledge of its instinctive patterns, evolutionary history, and ecological niche."

Learning Predispositions   Another important principle in classical learning theory has been poetically called the premise of equivalence of associability: Through the process of reward and punishment, any stimulus can come to be associated with any response. In classical conditioning, this would mean that any conditioned stimulus could be associated with any unconditioned stimulus with equal ease. The premise is also relevant in operant conditioning; in that case it means that any reinforcer should be equally effective in increasing the frequency of any response.

Martin Seligman (1970) was among the first in the twentieth century to reevaluate this idea (ideas contrary to the premise of equivalence of associability can be found at least as far back as the writings of C. L. Morgan, 1894). Seligman argued that natural selection not only shapes the sensory and motor abilities of the species to suit its ecological niche but also influences the organism's "associative apparatus." As a result, individuals of different species may differ in their abilities to associate different events because evolution prepared them to make certain associations more easily than others. The number of trials or pairings of stimuli necessary before a response occurs routinely can serve as an operational measure of preparedness. A comparison of the abilities of different species to learn certain associations reveals a continuum of preparedness. At one end of the preparedness spectrum are associations that individuals of a particular species are prepared to learn (associations that they learn easily). At the other end of the continuum are those associations that an animal seems unable to learn. The animal is said to be contraprepared to learn that association. In between are the associations that the animal is prepared to learn, but which are not part of its normal repertoire.

There are some clear lines of evidence of a genetic predisposition to make certain associations. One of the more interesting experiments that demonstrate such preparedness was done by John Garcia and R. A. Koelling (1966) and involved "bright, noisy water" and rats. The experimenters had two groups of rats. The first group was trained to drink bright, noisy water created by circuitry that caused a flash of light and a click when a rat's tongue touched the drinking spout in its cage. The second group of rats drank water sweetened with saccharin. Both groups were then divided in half, and one-half was given an electric shock after drinking, whereas the other was made sick by X-ray irradiation. The animals were then offered the type of water they had been accustomed to drinking, either bright, noisy water or sweet water. The results are summarized in Table 2.1. Notice that the animals that had been given a shock (external pain) developed an aversion to the

TABLE 2.1 Associations Between Cue and Consequence in the Domestic Strain of the Norway Rat: Evidence for Selective Learning (Garcia and Koelling 1966)

	Consequences	
Cues	Nausea	Electric Shock
Taste (saccharin sweet)	Drinking inhibited	No effect
Sound and light (click and light flash)	No effect	Drinking inhibited

bright, noisy water (external cues) but not to the sweet water (internal cue). Conversely, a distinctive saccharin taste (internal cue) could be associated with nausea (internal discomfort) but not with the external pain of electric shock. Such results clearly support the concept that not every stimulus can be associated with every response.

Experiments with avoidance learning provide additional evidence of genetic restrictions on learning. In avoidance learning, an animal must learn to perform some action to prevent the occurrence of an unpleasant stimulus. These experiments showed that an animal may be able to learn one response to a stimulus but not another. For example, Robert Bolles (1969) found that rats would readily learn to run to avoid an electric shock but would not learn to rear up and stand on their hind legs to avoid the same stimulus.

So we can see that behind the concept of a preparedness continuum is the idea that each species evolved so that it is exquisitely designed for its specific niche, and this specialization process may result in predispositions to learn some responses and not others. This means that it should be possible to interpret species differences in learning ability as adaptations to particular habitats or lifestyles. For example, the fact that parent-offspring recognition is better developed in bank swallows (*Riparia riparia*) and cliff swallows (*Hirundo pyrrhonata*) than in northern rough-winged swallows (*Stelgidopteryx serripennis*) and barn swallows (*H. rustica*) makes sense when we consider that the first two species nest in densely crowded colonies and the latter two do not (Beecher 1990). In noncolonial species, the chances that parents will provide care to young other than their own are fairly slim (there are not several other swallow nests nearby), and thus the ability to recognize their own young is less critical. The often-cited example that parental recognition of young is well developed in herring gulls, a colonial species in which the young can mix at an early age, and poorly developed in kittiwakes, a cliff-nesting species in which the young cannot safely wander from their nests, is

incorrect. This generalization began with the writings of E. Cullen (1957b), but more recent studies have shown that gull parents fail to recognize their young during the first few weeks (e.g., Holley 1984), although chicks appear to recognize their parents.

CONTEMPORARY VIEWS ON NATURE AND NURTURE

It is important to understand that the terms *innate* and *learned* are not invalidated by examples of the modification of innate behavior by learning or by the constraints placed on learning by genetic factors. There are, indeed, species-specific behaviors that are expressed in a highly stereotyped manner in a complete form the first time the situation is appropriate. A male fruit fly that has been isolated from the larval stage to adulthood still exhibits species-specific courtship behavior. It is also true that some behaviors are learned. No one would argue that a surgeon was born with the motor patterns for performing a coronary bypass. The point is that it can be appropriate to describe a behavior as innate or learned if certain cautions are kept in mind. First, labeling a behavior as innate or learned should never be interpreted as an explanation of its development, the mechanisms that allow its expression, or the forces active in its selection. If a label is accepted as an explanation, the research required for a complete understanding of the behavior is stifled. Furthermore, innate and learned should not be considered as mutually exclusive descriptions; just as innate behaviors are not purely genetic, learned behaviors are not entirely environmental. Genes and environment interact in the development of every behavior.

We have learned that there is an intimate relationship between the genetic and learned contributions to behavior. One way to get at the question of the contribution of genes versus environment is to learn when genes and experience have their greatest relative effects. Although genes and experience work together throughout the organism's development to produce most behaviors, as the organism changes over time the nature of the interaction will vary. For example, as a result of its genetic makeup, an animal may be more or less responsive to certain environmental stimuli at different times of its life. Conversely, the presence or absence of certain stimuli at critical times during development may alter the expression of the genes.

As the previous discussion indicates, perhaps the most important result of the nature/nurture controversy was the increasing focus on behavioral development in both ethology and comparative psychology. Indeed, although the two approaches to the study of animal behavior moved through a period of extreme divergence and debate, the distinction between them has blurred in recent years.

SOCIOBIOLOGY/BEHAVIORAL ECOLOGY

The field of animal behavior has grown enormously since the time of reconciliation between the ethologists and comparative psychologists, and there have been several developments in the years that followed. In the 1960s and early 1970s, for example, field researchers such as John Crook (1964, 1970) and John Eisenberg and colleagues (1972) suggested that ecological context was sometimes a better correlate of social behavior than was phylogeny (remember that ethologists often focused on phylogenetic analyses of behavior). Another dramatic development was the birth of a new discipline that focused on the application of evolutionary theory to social behavior. This new discipline was called sociobiology, which is more commonly called behavioral ecology today. Many scientists do not distinguish between sociobiology and behavioral ecology. However, narrower definitions might confine sociobiology to the study of social interactions and behavioral ecology to the study of behavior involved in obtaining and using resources (Barlow 1989; J. R. Krebs 1985).

In addition to some debate over how narrow or broad the definition of sociobiology should be, there has also been some question concerning the uniqueness of its approach. Some scientists, for example, questioned whether sociobiology is really a new discipline or simply part of contemporary ethology (e.g., M. S. Dawkins 1989). In contrast, others believe that at least early on, ethology and sociobiology could be separated in several ways (Barlow 1989). For example, whereas classical ethologists tended to derive hypotheses from detailed observations (i.e., through induction), sociobiologists tended to be more deductive, typically deriving hypotheses from larger theoretical frameworks. Whereas classical ethologists were interested in species differences, sociobiologists began to investigate individual differences, examining the costs and benefits of a particular act. Although ethologists often did not focus their research on the mechanisms of behavior, when they did seek reductionist explanations they typically considered physiological factors; recall, for example, Tinbergen's (1951) hierarchical model for the organization of instinctive behavior. In contrast, sociobiologists sought reductionist explanations in the area of genetics. Having mentioned some of the questions concerning the precise relationship between sociobiology and other fields of animal behavior, let us consider the relatively recent "history" of sociobiology.

During the late 1960s and early 1970s, most scientists were quite comfortable with the idea that natural selection acted primarily on individuals. Despite the existence of widespread agreement, however, some nagging issues that seemed inconsistent with selection at the level of the individual remained (Hinde 1982). For example, how could one explain the evolution of sterile castes in species of ants, bees, and wasps? How could the evolution of nonreproducing individuals be consistent with Darwinian selection? Similarly, how was one to explain the evolution of certain patterns of behavior, called altruistic behavior, that seemed to benefit others but were costly (with respect to survival and reproduction) to the performer? Why, for example, do some animals give alarm calls when they spot a predator, when calling may actually increase their own chances of being detected? The answer to these questions came in 1964 when W. D. Hamilton published his seminal papers "The Genetical Evolution of Social Behaviour, I, II." Hamilton showed that evolutionary success (the contribution of genes to subsequent generations) should be measured not only by the number of surviving offspring produced by an individual but also by the effects of that individual's actions on nondescendant kin (e.g., siblings, nieces, and nephews). He coined the term *inclusive fitness* to describe an individual's collective genetic success, that is, a combination of direct fitness (own reproduction) and indirect fitness (effects on reproduction by nondescendant kin). When quantifying an individual's inclusive fitness, we count—to varying degrees, depending on how closely they are related—all the offspring, personal or of relatives, that are alive because of the actions of that individual. This concept of inclusive fitness paved the way toward an understanding of the evolution of sterile castes and altruistic acts (these topics are discussed in more detail in Chapter 17).

In the years following Hamilton's (1964) papers, many studies were conducted in which the idea of inclusive fitness was used to interpret social behavior. However, it was not until 1975, when E. O. Wilson published his landmark text, *Sociobiology*, that the true impact of sociobiological ideas was felt. The text, an engaging integration of ideas from fields such as ethology, ecology, and population biology, gained almost instant notoriety from both within and outside the scientific community (see later discussion). Wilson defined sociobiology as the "systematic study of the biological basis of all social behavior" and proposed that a knowledge of demography (e.g., information on population growth and age structure) and of the genetic structure of populations was essential in understanding the evolution of social behavior.

Although sociobiological ideas had been developing for several years before the publication of Wilson's book, the text crystallized many of the relevant issues and soon became the focal point for proponents and critics alike. Criticism arose from both the scientific and political arenas. First, in attempting to establish sociobiology, Wilson attacked fields such as ethology

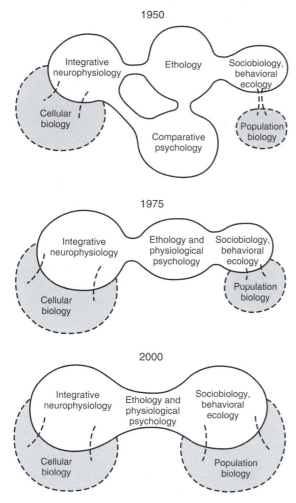

1950

1975

2000

FIGURE 2.22   E. O. Wilson's dire predictions for the fields of ethology and comparative psychology. (From E. O. Wilson 1975.)

and comparative psychology and made the bold prediction that in due time sociobiology would engulf these disciplines (Figure 2.22). He specifically predicted that ethology and comparative psychology would be "cannibalized by neurophysiology and sensory physiology from one end and sociobiology and behavioral ecology from the other" (E. O. Wilson 1975). Another area of great concern, this time from the political arena, was the extension of sociobiological thinking, in the absence of sound evidence, to human social behavior (Cooper 1985). Opponents of sociobiology claimed that Wilson advocated biological determinism, the idea that the present conditions of human societies are simply the result of the biology of the human species and therefore cannot be altered. Although only the final chapter of his text was devoted specifically to humans, heated debate over the social and political implications of sociobiological theories ensued (e.g., see the collection of papers in Caplan 1978).

During the 1970s and the early 1980s, research on sociobiological topics in animal behavior flourished. George Barlow (1989) suggested, "The study of animal behavior had indeed begun to stagnate by 1975, and the advent of sociobiology was just the kick in the pants the field needed to get moving again." A revitalization of the field of animal behavior occurred because sociobiology provided a framework that could be used to test hypotheses about the adaptiveness or survival value of behavior. But, the "kick in the pants" was so strong that for a time almost *all* research in animal behavior was done under the banner of sociobiology (Bateson and Klopfer 1989). Of all possible questions about animal behavior, one—its function, or survival value—had come to dominate the field.

By the end of the 1980s, however, many researchers began to notice the imbalance in the study of animal behavior. It became apparent that our understanding of animal behavior will be fuller if both its immediate causes and its evolutionary causes are considered. As Marian Stamp Dawkins (1989, p. 53) has said,

Genes operate through making bodies do things. These bodies have to develop and they need machinery (sense organs, decision centers, and means of executing action) to be able to pass their genes on to the next generation. To understand this process fully, we need a science that is not only aware of the evolutionary ebb and flow of genotypes over evolutionary time, but can look at the bridge between generations, at the bodies that grow and move and court and find food and pass their genetic cargo on through time with the frailest and most marvelous of flesh-and-blood machinery.

Today, the study of animal behavior is alive and well. Interactions among ethologists, comparative psychologists, and sociobiologists have stimulated research and advanced the study of animal behavior. As a result, the field has grown and changed so much since its beginning that the foundation laid by the pioneers is not always immediately apparent. In the words of Konrad Lorenz (1981), " . . . the development of a science resembles that of a coral colony. The more it thrives and the faster it grows, the quicker its first beginnings . . . the vestiges of the founders and the contributions of the early discoverers . . . become overgrown and obscured by their own progeny."

It has always been assumed that natural selection shaped behavior, and so we should expect that behavior observed in the field should increase the animal's chance of survival. Thus, it is not surprising that today many studies of animal behavior are long-term field studies. However, unlike many of the pioneering field studies, a hypothesis to be tested is clearly stated at the outset and guides data collection. Also, the data collected are often the outcome or consequence of a behavior, such as owning a nest site or the number of females inseminated (Barlow 1989). As we will discuss in later chapters of this book, particularly in Chapters 4 and 12, today's field studies of animal behavior usually focus on the costs and benefits of a particular behavior, with the common currency being reproductive success. Natural selection is often assumed to have shaped not just an efficient but also an optimal form of behavior. For example, we would expect a starling to select the type of prey that will maximize the amount of food (energy) that can be delivered to its brood. Also, because an animal's environment includes competitors, an individual's best choice of action often depends on what other members of the population are doing. In such situations, it is often helpful to consider whether an individual's choice of action is an evolutionarily stable strategy (ESS). An ESS is a strategy that cannot be invaded by the spread of any rare alternative strategy when it is adopted by most members of the population. The concept of ESS has been applied in studies of mating systems, communication, conflict, and cooperation (Krebs and Davies 1997).

We are also making great strides in understanding the mechanisms of behavior, largely because of tools and techniques that were not available even a few years ago. New recording techniques have made it possible to map the nervous systems of several invertebrates. In some animals, specific neurons have been linked to specific behaviors. For instance, in the grasshopper (*Omocestrus viridulus*) three different hind leg movements (FAPs) are involved in producing the courtship sound signals. By using microinjection techniques and intracellular recording, it has been shown that a specific type of brain nerve cell is responsible for each of these three FAPs. During courtship, these nerve cell types are activated in a specific sequence (Hedwig and Heinrich 1997). We will consider many other examples of how physiology, neurobiology, and molecular biology have enhanced the study of behavior in the chapters that follow.

Today there is a sense of rejuvenation in the study of animal behavior, largely because many disciplines are now contributing to its study (van Staaden 1998). New techniques and interactions among disciplines allow us to ask and answer many questions about behavior that could not be addressed previously. The early ethologists wondered how an animal's behavior is organized so that it behaves appropriately in each situation. Those problems have yet to be answered, not because they are unimportant, but because answers require more knowledge of neuroscience. As we advance on that front, we may soon be able to answer some of the questions raised by early ethologists. If we learn from the history of animal behavior, we will see that whether our primary interest is the mechanism or the function of behavior, our future efforts will be most fruitful if we keep a clear focus on *behavior* as the driving interest of research.

SUMMARY

Perhaps the most important concept in the study of animal behavior was Darwin's idea of evolution through natural selection, which provided the evolutionary framework necessary for the development of animal behavior.

In the early 1900s, two approaches to the study of animal behavior were ethology, centered in Europe, and comparative psychology, centered in America. Ethologists focused primarily on the function and evolution of behavior. Because the context in which a behavior is displayed is sometimes a clue to its function, ethologists often studied behavior under field conditions. It followed from their interest in evolution that ethologists used a comparative approach and studied primarily innate behaviors.

Early ethologists such as Charles Whitman, Oskar Heinroth, and Wallace Craig were interested in stereotyped patterns of behavior, considering them to be just as reliable as morphological characters in defining a particular group. These stereotyped behaviors were called fixed action patterns (FAPs) by Konrad Lorenz. An FAP is triggered by a very specific stimulus. That portion of the total stimulus that releases the FAP is called the sign stimulus or releaser. Because most behaviors are not so stereotyped as implied by the notion of FAP, they have more recently been described as modal action patterns (MAPs).

Action specific energy is a hypothetical construct proposed by Lorenz to describe the specific motivation or drive for a particular action. A buildup of action specific energy can be used to explain vacuum activity (the appearance of an FAP in the absence of a releaser) and displacement activity (behavior that is seemingly inappropriate in a particular context).

In contrast to the early ethologists, comparative psychologists emphasized laboratory studies of observable, quantifiable patterns of behavior. In general they asked questions that concerned the development or causation of behavior. Learning and the physiological bases of behavior were the focus of much of their research.

Many exciting advances were made in the study of learning. Thorndike developed the techniques for studying trial-and-error learning, and Pavlov provided the methodology for classical conditioning.

Behaviorism is a school of psychology that proposes limiting the study of behavior to actions that can be observed. B. F. Skinner, a prominent behaviorist, found that patterns of behavior that are rewarded tend to be repeated or to increase in frequency, and he concluded that the control of behavior was largely a matter of reinforcement.

The physiological basis of behavior is another traditional subject investigated by comparative psychologists. Karl Lashley made many contributions to the study of the neurobiology of behavior, and Frank Beach pioneered the study of its endocrine basis. Despite their emphasis on learning and physiology in the laboratory, some comparative psychologists, such as C. R. Carpenter and T. C. Schneirla, studied the social behavior of animals in the field.

The ethologists' emphasis on instinctive behavior and the comparative psychologists' focus on learning soon became cause for heated debate between the two groups over the importance of inheritance versus experience in behavior. Central to the nature/nurture controversy was the concept of the innate. Part of the problem was defining the term. Ethologists had applied the word to behaviors that were stereotyped and appeared in a functional form the first time the appropriate stimulus was encountered. It was argued, however, that because members of a given species develop within essentially the same environment, the similarity in the form of behavior could be a product of the uniformity of experience. Furthermore, even though an animal shows the appropriate response on the first suitable occasion, it does not mean that learning is unimportant.

It cannot be concluded that a behavior with a genetic basis is immune to environmental influences. For example, although the begging responses of young gulls have a genetic basis and are present in a functional form soon after hatching, they are modified by the experience of begging from a parent.

Just as it was difficult for the early ethologists to accept that what they considered to be innate behavior could be modified by experience, it was also difficult for comparative psychologists to accept that learning could be constrained by genetic factors. Initially, it was believed that any rewarded response would persist. However, Keller and Marion Breland noticed that some of the animals they trained by using operant conditioning methods interrupted their performances with innate behaviors used for obtaining or eating food.

A second principle of classical learning theory is the premise of equivalence of associability: Any stimulus can be associated with any response through the process of reward and punishment. However, it soon became apparent that members of different species may vary in their abilities to associate different events because, through evolution, different species have adapted to different ecological niches. A given species may be prepared to make some associations, unprepared but capable of making other associations, and unable to associate some stimuli.

In spite of these difficulties, it has proven useful to describe certain behaviors as innate or as learned. There are differences in the degree of flexibility of behaviors—some are stereotyped, and others are obviously acquired or easily modified. Nature and nurture are both important to the development of every behavior.

In the 1960s, a new discipline emerged in the study of animal behavior. Called sociobiology (sometimes called behavioral ecology), it focused on the application of evolutionary theory to social behavior. One of its central concepts, that of inclusive fitness, was articulated in 1964 by W. D. Hamilton. According to Hamilton, individuals behave in such a manner as to maximize their inclusive fitness (i.e., their own survival and reproduction plus that of their relatives), rather than acting simply to maximize their own fitness. Suddenly, certain issues that seemed inconsistent with selection at the level of the individual, such as the evolution of sterile castes in insects and altruistic behavior (behavior that benefits others at the expense of the performer), were explainable.

Approximately ten years after Hamilton's paper, E. O. Wilson crystallized sociobiological ideas in his landmark text, *Sociobiology*. The book became a focal point for both critics and proponents of the new discipline. Whereas some critics from within the scientific community were irritated by Wilson's prediction that sociobiology would swallow up ethology and comparative psychology, critics from both within and outside this community attacked the extension of sociobiological ideas to human behavior.

Sociobiology and an interest in the survival value of behavior dominated the study of animal behavior for approximately a decade, but it soon became apparent that knowledge of both mechanism and function is necessary for a complete understanding of behavior. Today the function of behavior is often considered in long-term field studies that are investigating the costs and benefits of actions in terms of reproductive success. New techniques in neurobiology and molecular biology have advanced our understanding of the mechanisms of behavior.

3

Genetic Analysis of Behavior

Oh what a tangled web we'd weave! And yet, a spider with its minuscule brain fashions a lacy masterpiece (Figure 3.1). Furthermore, it must create this orb without parental guidance on its first attempt because, right after hatching, it floats away from its parent on a gos-samer thread. How can it create such a complex and delicate thing on its first try? It is obvious that its behavior is strongly influenced by genes.

THE RELATIONSHIP BETWEEN GENES AND BEHAVIOR

Does that mean that the spider inherited a gene containing a crash course in web weaving? The answer is no, and the reason is that genes do not contain instructions for performing specific behaviors. Nor are genes blueprints for behavior or even for the circuitry of the nervous system.

So, then, if a gene is not an instruction manual or a blueprint, what does it mean to say that a behavior has a genetic basis? It simply means that the probability that a behavior will be performed is due to the presence of a particular form of a gene. Put another way, if the animal has a certain form of a gene it will be able to perform a behavior, but if it lacks that form it may be unable to perform the behavior, the expression of the behavior may be altered, or the frequency of the behavior may change. Within a population, then, we will find that variation in one or more genes often underlies some of the variation in behavior.

FIGURE 3.1   A spider's web. The ability to construct a web is a behavior that is strongly influenced by genes.

BASICS OF GENE ACTION

What do genes really do, then? How do they work? As you may already know, genes simply direct the synthesis of proteins. Each protein is specified by a different gene, or if the protein consists of more than one chain of amino acids, a gene specifies one of those chains. The protein may be structural and be used as a building block of the organism, or it may be regulatory. The protein produced by a regulatory gene may modify the activity of other genes, or it may be an enzyme that facilitates chemical reactions. In other words, genes code for specific proteins, and the proteins affect the composition and organization of the animal in ways that influence how it behaves. An animal's sensory receptors detect stimuli and send information to the nervous system, where it is interpreted and analyzed. The nervous system may then initiate a response by effectors (muscles and glands) that results in behavior. The structure and function of all these cells—receptors, nerves, muscles, and glands—require information from the genes. Alterations in genes change the proteins they code for, with the result that anatomy or physiology may be altered in a manner that changes behavior. We will look at a few examples of the links between genes and behavior later in this chapter.

To understand the relationship between genes and

proteins, it is helpful to know a little biochemistry. Genes are made up of DNA (deoxyribonucleic acid). The structure of DNA is somewhat like a long ladder, twisted about itself like a spiral staircase. The DNA ladder is composed of two long strings of smaller molecules called nucleotides. Each nucleotide chain makes up one side of the ladder and half of each rung. A nucleotide consists of a phosphate, a nitrogenous base, and a sugar called deoxyribose. There are only four different nitrogenous bases used in DNA: adenine (A), thymine (T), cytosine (C), and guanine (G). Although DNA has only four different nucleotides, a DNA molecule is very long and has thousands of nucleotides. When forming a rung of the ladder, adenine must pair with thymine and cytosine must pair with guanine (Figure 3.2). The specificity of these base pairs is important, not only for the accurate production of new DNA molecules, but also for converting the information in the gene into a protein.

The instructions for each protein are written as the sequence of bases in the DNA molecule. The first step is to transcribe the information in DNA into messenger RNA (ribonucleic acid) (mRNA). The DNA unzips for part of its length so that an mRNA molecule can be formed. The mRNA has a structure similar to DNA except for three differences: It is single-stranded, its sugar is ribose, and the base uracil substitutes for thymine. Since adenine must pair with uracil and cytosine must pair with guanine, the sequence of bases on RNA is specified by the sequence of bases on DNA. Now you see why the specificity of base pairing is important in getting the information in a gene translated into a protein. Because the DNA molecule is so long, there is a multitude of ways in which the four bases can be ordered. Therefore, with only four bases, DNA can direct the synthesis of a myriad of different proteins.

The next step is to translate the order of bases on the mRNA molecule into a protein. Proteins are long chains of amino acids. Each different protein has a unique order of amino acids. A group of three bases on the mRNA molecule is translated, three bases at a time, into a protein. The order of bases on DNA specifies the sequence of bases in mRNA, which can be translated into only one array of amino acids. Thus, the information in the gene is its sequence of bases that codes for a specific protein. The "chain of command" can be summarized as follows:

Sequence of bases in DNA → Sequence of bases in mRNA → Sequence of amino acids in a protein

EXAMPLES OF LINKS BETWEEN GENES AND BEHAVIORS

A fundamental question in behavior genetics is how genes interact with one another and the environment to produce an organism's behavior. Certain genes code for proteins that bring about a relatively constant and per-

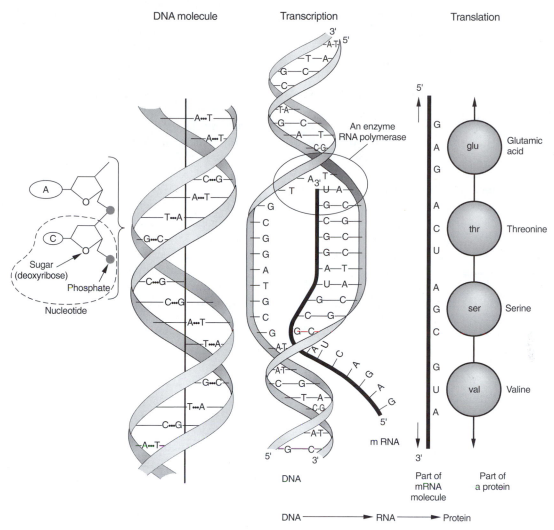

FIGURE 3.2   A diagrammatic representation of the biochemistry of gene expression. A gene is a region of a DNA molecule that has the information needed to make a specific protein. A DNA molecule is made up of nucleotides. A nucleotide, shown on the extreme left, is a nitrogenous base, a sugar (deoxyribose), and a phosphate. In the ladderlike DNA molecule, the two uprights are composed of alternating sugar and phosphate groups and the rungs are paired nitrogenous bases. The pairing of bases is specific—adenine with thymine and cytosine with guanine. During transcription, the synthesis of mRNA, the sequence of bases of the DNA molecule is converted to the complementary sequence of bases of the mRNA molecule. Each unit of three bases on the mRNA molecule signifies a particular amino acid. The message of messenger RNA is, therefore, its sequence of bases that determines the order and kinds of amino acids in the protein product.

manent change in a structure or function. Consider, for example, the *Shaker* gene in the fruit fly, *Drosophila melanogaster*. A particular mutation—that is, a specific change in the sequence of bases in the DNA—in the *Shaker* gene results in flies that shake violently under anesthesia. It turns out that the mutation changes a protein used in the formation of channels in the membranes of nerve cells that allow potassium ions to pass through. Potassium channels are critical to normal nerve cell function, and so the mutation causes a nerve cell defect and abnormal behavior (Kaplan and Trout 1969).

It should be kept in mind that the possession of "a gene for some behavior" in the sense in which we have just described the relationship is a necessary but not always sufficient condition for the expression of that behavior. An animal may have all the genes that are needed to perform a behavior and still not perform it because an essential gene is not active. Although all the cells of an animal's body have the same genes, some of them are turned off in the normal process of things and do not produce a protein. If these are then turned on, however, their proteins are produced, and in some cases they will modify a structure or function in a way that will alter behavior. Thus, an organism's behavior may *change* as specific genes are turned on or off.

So, then, what factors might regulate gene activity? Important among the answers to this question are steroid hormones. One example of hormonal regulation of gene activity may be familiar to you—puberty. Puberty is the time when, if you are male, your voice deepened and you had to start shaving or when, if you are female, your breasts developed and you started menstruating. These physical changes were probably accompanied by behavioral changes, such as beginning to associate with members of the opposite sex. Studies show that human males have the most interest in sex between the ages of 15 and 25 years, when the level of their steroid sex hormones, the androgens, is peaking (Davidson, Camargo, and Smith 1979). The physiological and behavioral changes experienced at puberty are triggered by the steroid sex hormones that your body began to produce. Only certain body parts change at puberty because a steroid hormone affects the protein production of only the cells that have receptors that can bind to it. In other words, the steroid sex hormones that bring on the anatomical, physiological, and behavioral changes that signify sexual maturity in vertebrates work by stimulating the activity of specific genes, but only in those cells that have receptors that can bind to the hormones.

How can changes in gene activity influence something as complicated as sexual behavior? Although we do not know all the answers, some specific links between gene activity and sexual behavior are known in nonhuman mammals. Receptors for steroid sex hormones are most concentrated in brain regions known to be essential to the control of sexual behavior, including the preoptic area, hypothalamus, and amygdala. Both testosterone and estrogen, two important sex steroids, are known to turn on specific genes in certain brain regions (Sagrillo et al. 1996). We know that gene activity is essential to the performance of certain sexual behaviors because those actions are blocked if chemical inhibitors of RNA or protein synthesis are administered to female rats (Meisel and Pfaff 1984). Changes in gene activity in nerve cells in these brain areas could, in turn, alter the level of certain neurotransmitters, chemicals used for communication among neurons. For instance, changes in gene activity could cause the level of the neurotransmitter dopamine to rise, and elevated dopamine concentration is known to stimulate sexual behavior in a male rat (Bitran et al. 1988). In addition, the activity of certain genes stimulates the formation of new connections between nerve cells. The new connections could then affect sexual behavior by increasing appropriate sensory input or by linking to memories of previous sexual encounters (R. J. Nelson 1995).

Nerve activity can also turn on specific genes, lead-

ing to changes in behavior. We see this, for example, when a male zebra finch (*Taeniopygia guttata*) or canary (*Serinus canaria*) is first exposed to the song of its own species. Young males learn to sing their species song by imitating the songs of adult males. It is important that they learn the correct song because, like other songbirds, these birds sing to defend territories and to attract mates.

But what changes in the nervous system underlie a male songbird's learning his species song? Claudio Mello, David Clayton, and their colleagues have uncovered some answers to this question. They suspected that changes, if they occurred, would be found in the bird's forebrain because this region is important in auditory processing. They further guessed that if gene activity changed, it would most likely be the activity of one of the so-called immediate-early genes, which code for proteins that regulate the activity of other genes. These other genes then produce proteins that are thought to be important in the growth of nerve cells and to affect nerve cell activity. In this way, immediate-early genes are involved in the formation of long-term memories. *Zenk* is one of these genes. If *zenk* were turned on by exposure to the song, the levels of zenk mRNA would be expected to rise. So, Mello and his colleagues (1992) played a 45-minute tape recording to young male zebra finches or canaries. The recording was either of the song of its own species, one of another species, or simply bursts of sound. They then measured the level of zenk mRNA in the birds' forebrains. It turns out that *zenk* activity is increased greatly when males hear the song of their own species and much less so in response to the song of another species. Exposure to simple bursts of sound did not increase *zenk* activity above that found in birds that did not experience any auditory stimulation (Figure 3.3). The results of this experiment are consistent with the idea that *zenk* is activated when a male songbird learns the song of its species.

One function of bird song is to defend territories, and it would be a waste of energy to continuously defend borders that are not contested. It should not be surprising, then, that male songbirds can discriminate among the songs of various singers. Such discrimination requires a male to learn the characteristics of songs of specific males, its neighbors for instance. Mello and his colleagues (1995) hypothesized that *zenk* activity underlies the *formation* of long-term memories. Therefore, they predicted that *zenk* activity would increase in response to the songs of unfamiliar males but would not increase in response to the songs of familiar males. To test this hypothesis, they measured the levels of zenk mRNA in auditory forebrain regions of the brains of male zebra finches following repeated exposure to the song of the same zebra finch. As expected, zenk mRNA increased during the first 30

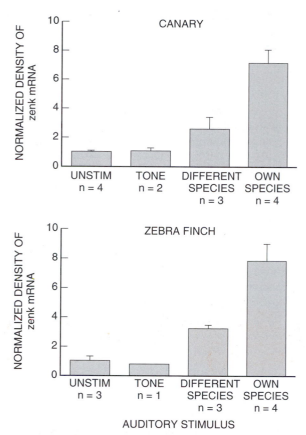

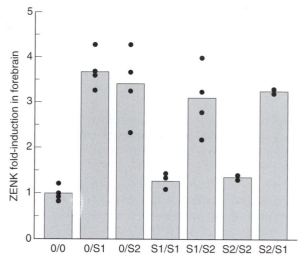

FIGURE 3.3　*Zenk* gene activity in the forebrains of male songbirds following exposure to songs of their own species, songs of another species, bursts of tones that are not song, or no auditory stimulation. The density of zenk mRNA, which reflects gene activity, increases following exposure to birdsong. The increase following exposure to the song of the listener's own species is significantly greater than that following the song of another species. Zenk activity is not a simple response to noise because it is not greater following non-song tones than when the bird is not stimulated by any sound. (Data from Mello et al. 1992.)

FIGURE 3.4　*Zenk* activity in auditory regions of the forebrain of a male zebra finch drops following repeated exposure to the song of one male, but it can be reactivated by a new song. The birds were exposed to repeated song stimulus for 2.5 hours, which was immediately followed by a second "test" stimulus lasting 30 minutes. The levels of zenk mRNA in the birds' forebrains were then measured. Each dot represents the amount of zenk mRNA in a single bird. The bars represent the mean zenk mRNA level for the experimental group. These results are consistent with the hypothesis that *zenk* activity underlies the formation of long-term memories. We would not expect *zenk* to be activated after the memories have been formed. The 0 represents silence. S1 and S2 indicate specific songs of two different individuals from another aviary. (Data from Mello et al. 1995.)

minutes. However, in spite of continued stimulation by the same song, zenk mRNA fell to baseline levels. Furthermore, when the same male's song was played after a full day of silence, there was no increase in *zenk* activity. Nonetheless, *zenk* activity did increase following exposure to the species song of another, unfamiliar zebra finch individual (Figure 3.4).

Similar changes in *zenk* activity have been shown in freely ranging song sparrows. When the species' song was played through a loudspeaker within a male's territory, he approached the speaker, searched for the intruder, and began actively singing in defense of his territory. The activity of *zenk* in several brain auditory structures was higher in the males whose territories had been challenged in this way than in unstimulated controls. Thus, natural behaviors in the field affect gene activity in much the same way they do in the laboratory (Jarvis et al. 1997).

We see, then, that the relationship between genes and behavior need not be static. Instead, the environment can continuously alter gene activity and, therefore, modify an organism's behavior.

GOALS OF BEHAVIOR GENETICS

When we observe an animal in nature, its actions are generally a result of many genes interacting with one another and the environment. Also, in most cases, an animal's behavior has been modified by its previous experiences.

We might wonder, then, how we can ever understand the relationship between genes and behavior. One goal of behavior genetics is to determine the strength of the genetic influence on a particular behavior. This goal is usually met through selection experiments or crossing experiments. (When studying human behavior, studying the behavior of twins is a way to see the influence of genes on certain behaviors.) The results of such experiments can shed light on the heritability of the behavior. In simple terms, heritability is a statistical measure that suggests how strongly a behavior is influenced by genes. Variation in behavior is caused by differences in both genes and environment. Since heritability is a statistical measure, we must measure the differences in the behavior of a sufficiently large number of individuals. Then we must determine how much of the observed variation is due to genetic differences among the individuals and how much is caused by differences in their environments. The heritability of a particular trait in a specific population is the ratio of the variation caused by genetic differences to the total amount of variability in the trait in that population. Heritability can vary from 0 to 1. A value of 0.5 indicates that 50% of the variability in the population studied is due to genetic differences. After a genetic influence on the behavior has been demonstrated, it is often of interest to identify the genes involved in producing the behavior in question. This usually involves mapping the genes, that is, identifying their locations along a specific chromosome. A second goal may be to determine the mechanisms by which the genes influence the behavior.

EXPERIMENTAL METHODS THAT DEMONSTRATE THE INFLUENCE OF GENES ON BEHAVIOR

As we have seen, the observed variation in behavior among individuals results from differences in genes and in environments. Therefore, when we are interested in exploring the influence of genes on behavior, it is important to rule out environmental effects by raising the animals in the same environment. If environmental conditions are identical for all animals, then observed behavioral differences are due to genetic differences.

INBREEDING

Inbred lines, strains created by mating close family members with one another, are useful in behavioral genetics because they provide a way to hold constant the genetic input. By using inbred strains, therefore, it is possible to separate the effects of genes from those of the environment. To show the effects of genes, the behavior of members of two inbred strains is compared in the same environment. In this case, any observed difference in behavior must be caused by a difference in the genes of the strain. However, the influence of the environment on behavior can also be shown by using inbred strains. If members of the same inbred strain, individuals who are almost genetically identical, behave differently when they are raised under different conditions, then the variation must be caused by environmental effects.

Before considering how inbreeding reduces genetic diversity, we should think about why a population of individuals is dissimilar. Usually a genetically controlled characteristic or trait can be expressed in more than one way. For example, coat color in mice might be black or brown. In other words, this gene for coat color can be expressed in two different ways; one produces intense black pigment granules, and the other produces chocolate brown pigment granules. These alternative forms of a gene are called alleles. Sometimes, as in the eye color of fruit flies, there are many possible alleles. The wild type, or most common eye color of fruit flies, is red, but other alleles of this gene can result in white or vermilion eyes. Furthermore, most of the animal species used in genetic studies are diploid, meaning that an individual possesses two alleles of each gene, one from each parent. If the two alleles of a gene are identical, the individual is said to be homozygous for the trait. However, an individual may inherit different alleles for a gene from its mother and father. Such an individual is heterozygous for the gene. The genetic diversity among unrelated individuals results from the particular alleles that they possess and from heterozygosity.

Inbreeding reduces genetic diversity because it creates a population that is homozygous for almost all of its genes. How does it do this? In the laboratory, an inbred strain is usually created by mating brothers with sisters for many generations. If the original parents of an inbred line lacked certain alleles found in the general population, those alleles would also be missing in their offspring. So when close family members mate, the alleles present in the parents will be retained and will increase in frequency relative to the general population; those not present in the founders will be lost. In addition, each parent passes on only one of its two alleles for each gene to the offspring. Therefore, if a parent is heterozygous for a trait, one of its alleles will be missing in the offspring unless the other parent supplies it. With each successive generation of inbreeding, genetic variability is lost. Each generation of brother–sister matings reduces heterozygosity by 25%. After 20 generations of sibling matings, all individuals are expected to have identical alleles for 98% of their genes. As a result, no matter which member of each chromosome pair is transmitted from parent to offspring, the same alleles will be passed on. Inbred lines, therefore, are virtually genetically identical (Plomin, DeFries, and McClearn 1980).

Comparisons of Inbred Strains to Show the Role of Genes

As previously mentioned, inbred lines can be used to demonstrate that differences in genetic makeup may result in differences in behavior. According to the scientific method, when you want to explore the effect of one variable, you hold the others constant. Inbred lines provide a way to do this. Inbreeding minimizes genetic diversity within strains but maintains diversity between strains and may even enhance it. If individuals of different inbred strains are raised in the same environment, any difference in their behavior is due to a difference in their genes. So the comparison of inbred strains raised under the same conditions is a way to demonstrate that a particular behavior has a genetic basis.

Comparisons of the behavior of different inbred strains have been a popular way to show that genes play a role in behavior in a wide variety of animals. The work of Ádám Miklósi and his colleagues (Miklósi, Csányi, and Gerlai 1997) on the antipredator behavior of paradise fish (*Macropodus opercularis*) larvae is an example. Paradise fish live in densely vegetated, shallow marshes and rice fields in Southeast Asia, along with several predator fish species. Avoiding predation is crucial to survival, and so one might suspect it has a genetic basis. Having raised strains of paradise fish that had been inbred for over 30 generations, Miklósi could investigate this possibility. He crafted model predators from plastic centrifuge tubes. Since eyes are known to be an important cue in predator recognition in many species, some models were created with black eyespots. The larvae of two inbred strains (S and P) of paradise fish were placed individually into an experimental tank, and their responses to model predators were observed for three minutes. There are two common antipredator responses: fleeing, in which the larva suddenly darts by slapping with its caudal fin, and backing, in which the larva swims backward with its body curved. Larvae of strain P were significantly more likely to show antipredator responses, both fleeing and backing, than were the larvae of strain S (Figure 3.5).

Comparisons of Inbred Strains to Show the Role of Environment

Because individuals of inbred strains are identical in 98% of their genes, such strains, as we mentioned, can provide a way to hold the genetic input constant while varying the environment. Thus, if differences in behavior are found, they must be due to the environment.

Some of the environmental effects on behavior take place very early in life. Eliminating the effects of early learning is tricky, but it can be done. The simplest way is by a reciprocal cross, one in which males of inbred strain A are mated with females of inbred strain B and males of strain B are mated with females of strain A. Since the individuals of each strain are homozygous for almost every gene, the hybrid offspring of both of these crosses have the same genotype even though their mothers are from different strains. So if the behavior of the hybrid offspring of reciprocal crosses differs, it

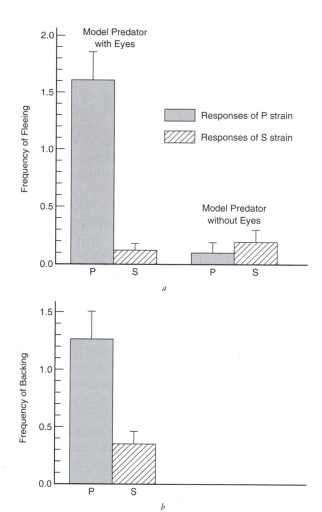

FIGURE 3.5 The antipredator responses of larvae of two inbred strains of paradise fish when presented with model predators. Fleeing consists of darting by slapping the caudal fin. Backing involves swimming backward with a curved body in a direction perpendicular to the body axes of the larva. Larvae of strain P showed a significantly higher frequency of both fleeing and backing than did larvae of strain S, indicating genetic differences in antipredator behavior between the strains. (Data modified from Miklósi, Csányi, and Gerlai 1997.)

must be an effect of the parental environment. If the young are raised solely by their mothers, as they often are in the laboratory, the difference must be an effect of the maternal environment.

Theoretically, it is even possible to determine whether the maternal environment had its greatest influence on the offspring before or after birth. Cross-fostering, transferring the offspring shortly after birth to a mother of a different strain, is a technique for detecting maternal influences that occur after birth. If offspring that were transferred to a foster mother immediately after birth behave more like individuals of the foster mother's strain than like those of their own strain, postnatal maternal influences are implicated.

A cross-fostering experiment involving two different species of voles helped in separating the influences of genes and parental environment in the expression of a particular behavior. Prairie voles, *Microtus ochrogaster*, show more parental care than do meadow voles, *M. pennsylvanicus*. For example, prairie vole females spend more time in the nest with their young, contacting their offspring by huddling over them and nursing them. Male prairie voles also show more parental care than do male meadow voles. In contrast to meadow vole males, who nest separately and rarely enter the female's nest, prairie vole males share a nest with the female and frequently groom and huddle over the young.

To determine whether the species difference in parental care was due to genes or early experience, Betty McGuire (1988) fostered meadow vole pups to prairie vole parents. As a control, she fostered meadow vole pups to other meadow vole parents. When the foster pups became adults and had their own families, she measured the amount of parental care they gave to their second litters. The meadow vole pups that were raised by prairie voles gave more care to their own offspring than did those pups that had been fostered to other meadow vole parents. Cross-fostered females, for example, spent more time huddling over and nursing their young. The early experience of male voles influenced their parental behavior in much the same way as it affected the females' behavior. Male meadow vole pups that were fostered to parents of their own species behaved as meadow vole males usually do, in that they rarely entered the nest with the young. However, four of the eight meadow vole males raised by prairie vole parents nested with their mates and spent time in contact with the young. This cross-fostering experiment shows that the experience a vole has with its own mother can influence the way it treats its own offspring. However, not all behaviors were modified by early experience. Nonsocial behaviors such as food caching and tunnel building and overall activity level were unaffected by the species of the foster parents. Thus, using the technique of cross-fostering, it is possible to determine whether a particular behavior is influenced by the parental environment.

ARTIFICIAL SELECTION

Artificial selection is another means of demonstrating that a behavior has a genetic basis. It differs from natural selection in that an experimenter, not "nature," decides which individuals will breed and leave offspring. The rationale for artificial selection is that if the frequency of a trait in a population can be altered by choosing the appropriate breeders, it must have a genetic basis. Usually the first step is to test individuals of a genetically variable population for a particular behavior trait. Those individuals who show the desired attribute are mated with one another, and those who lack the trait are prevented from breeding. If the character has a genetic basis, the alleles responsible for it will increase in frequency in the population because only those possessing them are producing offspring. As a result, the behavior becomes more common or exaggerated with each successive generation. When it proves possible to select for a trait, it is concluded that genetic inheritance is important to the expression of that trait. Furthermore, the environment is held constant, so it is reasonable to conclude that any change in the frequency of a behavior is due to a change in the frequency of alleles.

Humans often use artificial selection to create breeds of animals with traits that they consider useful. For example, even if you are not a dog lover, you have probably noticed that different breeds of dogs have different personalities. All dogs belong to the same species, but various behavior traits and hunting skills have been selected for in different breeds (strains). Terriers, for example, are fighters. Aggressiveness was selected for so they could be used to attack small game. The beagle, on the other hand, was developed as a scent hound. Since beagles usually work in packs to sniff out game, they must be more tolerant with companions. Therefore, their aggressiveness was reduced by selective breeding. Shetland sheep dogs were bred for their ability to learn to herd sheep. Spaniels, used for hunting birds, are "people dogs." They are more affectionate than aggressive. These behavioral differences among breeds of dogs were brought about by people who placed a premium on certain traits and arranged matings between individual dogs that showed the desired behavior. In other words, the frequency of particular behavior patterns present in all breeds has been modified through artificial selection; new behaviors were not created (Scott and Fuller 1965).

Artificial selection for nesting behavior in house mice (*Mus domesticus*) has not only demonstrated a genetic basis for the behavior but has also shed some light on how natural selection might work on the trait in wild populations. House mice usually live in fields, where they build nests of grasses and other soft plant material. In the laboratory, both male and female house mice will use cotton as nesting material, which makes it easy to quantify the size of the nest constructed. Carol Lynch (1980) noticed that some mice built larger nests than others did (Figure 3.6). These differences could be due to genetic or environmental factors or both. Lynch suspected that there was some genetic basis in nesting behavior. To separate the genetic and environmental influences, she began to selectively mate mice, based on the size of the nest they built, and she raised all mice under the same environmental conditions. She began with a population of house mice that gathered between 13 and 18 grams of cotton over a four-day period to

FIGURE 3.6   Individual differences in the size of nests built by house mice in the laboratory. The genetic basis of these differences in nest-building behavior has been demonstrated by artificial selection experiments. Whereas the mouse on the left was selected to build a small nest, the one on the right was selected to build a large nest.

build their nests. Then, she selectively mated mice to create lines of high and low nest-building behavior. She mated males and females that built large nests to create high lines and males and females that built small nests to create low lines. In addition, she created control lines by randomly mating males and females from each generation. After 15 generations of artificial selection, the mice of the high nest-building line used an average of 40 grams of cotton for their nests. In contrast, the mice in the low nest-building line used an average of only 5 grams of cotton in nest construction. The mice of the

control line built nests about the same size as those built by the mice in the initial population—15 grams of cotton (Figure 3.7). Selection for high and low nesting behavior was continued for over 40 generations, at which time mice from the high nest-building line collected more than 40 times the amount of cotton than did the mice of the low line.

The results of this experiment confirm that there is, indeed, a genetic basis to nest building in house mice. Thus, natural selection could have a similar influence on nest building if nest size influences fitness (the number of offspring successfully raised). House mice do, in fact, build larger nests in the north than in the south (Lynch 1992). This suggests that large nests may be a factor that helps mice in cold environments raise more offspring. In the laboratory, Lynch bred groups of mice from both lines at 22°C or at 4°C and counted the number of offspring that survived to 40 days of age. Pup survival in both lines was reduced at the lower temperature. Nonetheless, mice from the lines that built larger nests raised more pups that lived to be 40 days old at both environmental temperatures. Thus, nest building is an important component of fitness, and its genetic basis allows it to be shaped by natural selection (Bult and Lynch 1997).

It is interesting to note, however, that selection can lead to the same observed behavior—constructing large or small nests—by favoring different sets of genes. Notice in Figure 3.7 that two high lines and two low strains were created through artificial selection. Crosses between mice of different high strains (or between mice of different low strains) revealed that there were still genetic differences between strains that expressed the behavior in similar ways. This suggests that natural selection can follow different paths in different popula-

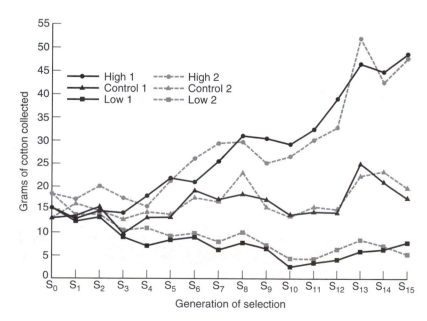

FIGURE 3.7   Artificial selection for large or small nest size in house mice. The nests of mice in the original population consisted of between 13 and 18 grams of cotton. Individuals who built the largest nests were bred with others who built large nests to create a high nest-building line. Individuals who built the smallest nests were bred with others who built small nests to create a low nest-building line. A control line was developed by randomly mating individuals of each generation. After 15 generations, the nests of the high line were an average of 8 times larger than those of the low line—40 grams and 5 grams, respectively. The nests of mice in the control line were roughly the same size as those built by mice in the initial population (15 grams). (Data from Lynch 1980.)

tions to produce the same adaptive behavior (Bult and Lynch 1996).

In some cases, selection affects behavior by affecting genes that code for proteins that affect the structure or function of the nervous system. For example, it is possible to selectively breed mice to be very active and to explore an open test arena or to be less active. Such selection results in specific differences in brain structure between the high and low activity lines: Specific regions of the hippocampus are more developed in the mice that are selected for high activity (Hausheer-Zarmakupi et al. 1996).

HYBRIDIZATION

Hybridization is a third way to demonstrate a genetic influence on behavior. The usual procedure is to mate two individuals that display a particular type of behavior in distinct but different ways. Once interbreeding has occurred, the behavior in the hybrid offspring is then observed. Depending on the number of genes influencing the behavior, the hybrids may perform the action in the same manner as one of the parents or as a combination of the parental types.

Rover and Sitter Fruit Fly Larvae

Most of us would not find foraging fruit fly larvae of great interest, unless it was taking place in a fruit bowl on the kitchen table. But, it is fascinating to many behavioral geneticists. The interest began when Marla Sokolowski noticed two forms of feeding behavior in natural populations of larval fruit flies (*D. melanogaster*) in the vicinity of Toronto, Canada. Whereas the so-called "rover" larvae move around continually on their food, "sitter" larvae travel only short distances. In fact, when these larvae were brought into the laboratory and allowed to feed on yeast paste in a petri dish, the distances traveled by rovers were nearly four times longer than those of sitters. Sokolowski immediately suspected that a behavior with two distinct forms would be genetically controlled.

To investigate the genetic basis of these foraging strategies, Sokolowski performed a series of cross-breeding experiments using adult rovers and adult sitters. The results are consistent with the idea that the trait is controlled by a single gene and that *rover* is dominant to *sitter*, as shown in Figure 3.8. Sokolowski began by crossing adult rovers with other rovers and adult sitters with other sitters, creating parental strains of sitters and rovers. The responses of male larvae resulting from these crosses are shown in Figure 3.8*a*. Notice that the frequency distribution of path lengths of rovers is easily distinguished from that of rovers. Next, Sokolowski crossed adults of sitter larvae with adults of rover larvae. The path lengths of the F_1 offspring of these crosses are shown in Figure 3.8*b*. Nearly all the larvae were rovers. When the F_1 adults were crossed with one another, the resulting F_2 larvae consisted of both rovers and sitters in a ratio of three rovers to one sitter (Figure 3.8*c*). Thus, most of the variation in foraging strategies could be explained by variation in a single gene, dubbed *foraging* (*for*) (de Belle and Sokolowski 1987).

Several aspects of the *for* gene make it attractive to study. First, behavioral variants of the gene occur in natural populations; It isn't necessary to use mutants created in a laboratory. Second, the gene is subject to natural selection, and the form of the trait favored depends on environmental conditions. Rovers, who will search farther for food, are favored in crowded conditions. On the other hand, sitters are favored when the population density is low because they conserve energy by limiting movement (Barinaga 1994). Sitters don't move slowly because they are energy-deficient or sick. When they aren't feeding, there are no significant differences in locomotion between sitters and rovers (Sokolowski and Hansell 1992). Also, the developmental time and growth rates of sitters and rovers are essentially identical (Graf and Sokolowski 1989). Third, the expression of the trait illustrates that genes and environment interact to produce the observed behavior. For example, if a rover is deprived of food before it has an opportunity to forage, the chances are good that it will behave as a sitter and conserve energy (Barinaga 1994). Fourth, we are beginning to see glimpses of the mechanisms by which *for* alters the feeding strategy. *For* is located on chromosome 2 (de Belle, Hilliker, and Sokolowski 1989), and its location seems to match that of a gene for an enzyme (cyclic GMP*-dependent kinase) that is known to be important in signaling pathways within the cell (reviewed in Barinaga 1994). Moreover, it is speculated that the change in this enzyme affects behavior by altering some aspect of information transfer between sensory input and motor output that influences the way in which the larvae perceive and respond to the quality of food in the area where they are foraging (Pereira et al. 1995).

LOCATING THE EFFECTS OF GENES THAT INFLUENCE BEHAVIOR

The breeding experiments we have discussed so far were the primary tools used by classical behavior geneticists, but they are actually not very powerful tools. Inbreeding, selection, and hybridization may indicate that the behavior in question has a genetic basis and perhaps that it is controlled by few or many

---

*Guanosine monophosphate

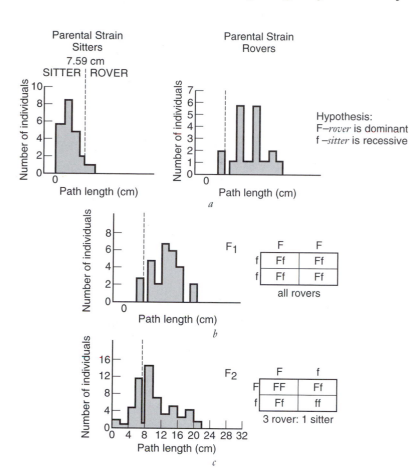

FIGURE 3.8   The results of mating experiments on fruit flies showing "rover" or "sitter" foraging strategies as larvae. Rover larvae forage over significantly longer distances than do sitter larvae. (*a*) The parental strains were created by crossing rovers with rovers or sitters with sitters. (*b*) The frequency distributions show foraging path distances of male (F_1) larvae of crosses between adults of the parental rover strain and adults of the parental sitter strain. Almost all the resulting larvae were rovers. (*c*) The frequency distributions show foraging path distances of male (F_2) larvae resulting from crosses between F_1 flies. The results are consistent with the hypothesis that the trait is controlled by a single gene, called *foraging (for)*, and that the *rover* allele is dominant to the *sitter* allele. Punnett squares of the crosses are shown on the right. (Data modified from de Belle and Sokolowski, 1987.)

genes, but it tells us nothing about where or how the genes are exerting their effects. More modern techniques such as inducing single-gene mutations, mutation and mosaic analysis, and the use of recombinant DNA have filled in some of the gaps in our understanding of the relationship between genes and behavior.

SINGLE-GENE MUTATIONS

One way to find the links between behavior and genes is to induce a mutation in a single gene, screen the population for individuals who behave abnormally, and then search for the anatomical or physiological differences between the mutant and the normal organisms. A mutation, a change in a gene's instructions for producing a protein, can be induced by agents that change the DNA bases. When organisms are exposed to mutagenic agents, some of them become behaviorally aberrant and can be separated from the population of normal individuals on this basis. Appropriate genetic crosses can then determine whether the behavioral change is caused by an alteration in a single gene. Even a small change may result in a difference in a specific aspect of an anatomical structure or a physiological process that mediates a behavior. Identifying the anatomical or physiological differences between mutant

and normal individuals brings us closer to understanding how genes can influence behavior.

Studies on learning in the fruit fly, *D. melanogaster*, have helped fill in a few of the missing links between genes and behavior. You have probably never met a fruit fly with remarkable intelligence, but some mutant strains are so poor at olfactory learning that they have earned the epithet *dunce*. Before considering the deficiency caused by the *dunce* mutation, olfactory learning in normal fruit flies and the way of demonstrating it should be described. Normal fruit flies can associate an odor with an unpleasant event, such as an electric shock, and learn to avoid that odor if it is encountered again. This should not be surprising because odors are important in the daily life of fruit flies for locating both food and appropriate mates. William Quinn and his colleagues (1974) demonstrated this avoidance conditioning by shocking a group of flies for 15 seconds in the presence of one odor but not in the presence of a second odor. Then they presented the odors, one at a time, to the flies without shocking them to see how many would avoid each odor. Most normal flies avoid only the odor connected with the electric shock, an association that lasts for three to six hours (Dudai et al. 1976).

Two mutants, *dunce*[1] and *dunce*[2], were isolated by

exposing a population of *D. melanogaster* to a chemical known to cause mutations and then screening the flies on the basis of their learning ability. The *dunce* mutations are alleles, or alternate forms, of a single gene on the X chromosome. In contrast to normal flies, *dunces* fail to learn to avoid odors associated with shock when they are taught with Quinn's experimental design (Dudai et al. 1976). Even larval *dunces* are deficient in olfactory learning (Aceves-Pina and Quinn 1979).

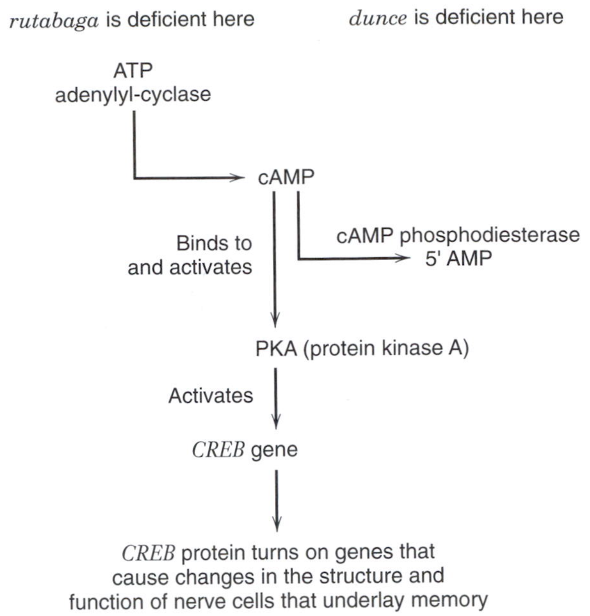

Why don't *dunce* fruit flies learn as well as normal flies? The first guess, that the sensory system was defective so that the *dunce* flies could not detect either the odors or the shock, was incorrect. Experiments revealed that the mutants are able to detect both (Dudai et al. 1976). In spite of this, mutant flies are unable to remember the association between the shock and an odor.

Apparently the *dunce* mutants have a problem with the early stages of memory formation. If they are shocked in the presence of an odor without subsequent exposure to a second odor, they do associate the odor with the aversive stimulus and avoid it if tested immediately after training, but the association fades quickly. The association between the shock and an odor is short-lived, and the experience with a control odor during the interval between the training and the test of learning seems to eliminate associations that may have formed. For these two reasons, it is concluded that an early stage of memory formation is defective in the *dunce* mutants (Dudai 1979).

What does the *dunce* gene do? It codes for a form of an enzyme called cyclic AMP (adenosine monophosphate) phosphodiesterase. This enzyme is important because it breaks down cyclic AMP (cAMP), which is a mediator of many processes in different types of cells. Thus, we see that the *dunce* mutation causes a reduced level of cAMP phosphodiesterase and, therefore, an increased level of cAMP. In addition, the mutation impairs an early stage of memory formation in olfactory learning. This should lead you to suspect that the enzyme or cAMP might play a role in this type of memory formation, an idea that has been tested by inhibiting cAMP phosphodiesterase in normal flies and testing the olfactory learning abilities. When treated in this way, normal flies learn no better than *dunces* (Byers, Davis, and Kiger 1981). Thus, the behavior of *dunce* flies suggests that cAMP has a role in learning and memory.

Another single-gene mutation in fruit flies, *rutabaga*, also causes poor learning and memory. This mutation has filled in some of the details of the connection between cAMP and memory formation. The level of cAMP within a cell actually depends on two enzymes. As we've seen, cAMP phosphodiesterase breaks down cAMP, lowering its concentration. In contrast, another enzyme, called adenylyl cyclase, raises

FIGURE 3.9   A summary of the molecular events that accompany memory formation in fruit flies. Single-gene mutations that result in poor olfactory learning were helpful in uncovering many of the details of memory formation. The *dunce* mutant has a defective form of cAMP phosphodiesterase, an enzyme that breaks down cAMP. *Rutabaga* has a defective form of adenylyl cyclase, an enzyme that forms cAMP from ATP. Thus both mutations affect the levels of cAMP within nerve cells. The cAMP binds to another enzyme, PKA (protein kinase A), and activates it. Active PKA then turns on the *CREB* gene, whose protein regulates the activity of other genes so that new connections can be made among nerve cells. These connections are responsible, in part, for long-term memory.

cAMP levels by causing the formation of cAMP from ATP (adenosine triphosphate) (Figure 3.9). *Rutabaga* mutants have a defective form of adenylyl cyclase (Levin et al. 1992). As a result, *rutabaga*'s adenylyl cyclase is not activated by the stimulus involved in learning in the way it would be in a normal fly. Therefore, the learning stimulus doesn't cause cAMP levels to rise. Thus *dunce* and *rutabaga* have opposite effects on the level of cAMP within a cell.

As is often the case in science, the more we learn, the more questions we have. In this case, the *dunce* and *rutabaga* mutants prompt us to wonder how cAMP is involved in memory formation. A pathway involving cAMP that is thought to be important for memory formation is summarized in Figure 3.9. Notice that cAMP binds to another enzyme, protein kinase A (PKA), and activates it. All protein kinases work by adding a phosphate group to another molecule, activating or inactivating the other molecule. In this case, PKA activates the *CREB* gene, which codes for the protein CREB (*cAMP* *re*sponse *b*inding protein) that, in turn, activates other genes. These other genes then control the growth of connections between brain cells, changes in the nervous system that are responsible for memory (Davis et al. 1995).

It is interesting to note that cAMP and the *CREB* gene play a key part in memory formation in other organisms. Marc Klein and Eric Kandel (1978) have shown that in the sea hare *Aplysia californica*, changes in the levels of cyclic AMP in certain nerve cells play a role in sensitization of the gill-withdrawal reflex. Sensitization, an increase in neuronal responsiveness to a stimulus because of previous exposure to another intense stimulus, might be considered a simple form of memory. Also, preventing CREB proteins from activating other genes blocked long-term memory formation in *Aplysia* (Dash, Hochner, and Kandel 1990). In mice, the story is similar—long-term memory formation depends on normal *CREB* gene activity (Bourtchuladze et al. 1994).

MUTATION AND MOSAIC ANALYSIS

In the study of *dunce* fruit flies, luck was on the side of the investigators. When a gene for the mutant behavior was mapped, it turned out to be in the same chromosome region as a gene whose ultimate product was known. This helped to pinpoint cyclic AMP phosphodiesterase as a link between the mutant behavior and the gene. However, we are not always that fortunate, so other techniques must be used to determine how or where particular genes are acting.

Why is it so difficult to determine where a gene is acting? First, examining a population of mutants to look for the defective structure or physiological problem is like looking for the proverbial needle in a haystack. Where do you begin? It would be helpful to know which part of the haystack to search first. Furthermore, knowing where to look for a problem is crucial when the effect of the mutation cannot be seen directly, as would be the case if the gene's activity resulted in abnormal neural connections; an inability to absorb an important precursor; or an inability to produce a neurotransmitter, enzyme, or hormone. Knowing where to look for a defect is also helpful when there are several structures that logic would tell us could be affected by a mutation and cause a behavioral anomaly. For example, there is a mutant fruit fly called *wings up* that cannot fly because its wings are always held straight up. Think for a moment about where the fly's problem might be. Obviously the wings are the immediate problem, but what causes the wings to be held in an abnormal position? Is the defect in the wing structure itself, in the wing muscles, in the nerves that control the wing muscles, or in the messages sent from the brain? Not only is it difficult to determine which of

several possible sites is the source of the problem, it is also hard to find the site of gene action when it is not in the structure that is outwardly affected. In the *wings up* fruit fly, for example, the wings are up because the indirect flight muscles that attach to them are either abnormal or missing (Hotta and Benzer, 1972). Another example, one in which a gene has its primary effect on a structure distant from the site of the problem, is seen in some humans. An inability of the cells of the gut to absorb enough vitamin A may cause the retina of the eye to degenerate, resulting in a sensory deficiency with obvious consequences for behavior. In this case, the gene's action is in the small intestines even though the defect is seen in the eye. In both of these examples, it would be easy to misidentify the site of gene action as the structure with the obvious defect.

It is possible to locate the primary site of gene action by using mosaic analysis, which is a comparative study of mosaics, individuals composed of some sections of normal wild-type tissue and some pieces of mutant tissue. Although mosaic analysis does not always reveal what the gene is doing—that is, the nature of the problem—it does locate where the gene is working. The procedure is to compare the distribution of normal and mutant tissue in all the mosaics displaying the mutant form of a behavior to identify the region of mutant tissue common to them all. Those regions are then implicated as the sites of action of that gene.

You have probably used the same reasoning used in mosaic analysis in solving other problems. Suppose, for example, you sit down to read in your favorite seat but, much to your frustration, the lamp next to the chair will not go on. How do you locate the source of the problem? First, you might suspect that there was no power in the outlet you were using. So you plug the lamp into an outlet that you know is functional. The lamp still does not work. Second, you check the bulb and find that it was not screwed in tightly, but screwing in the bulb does not solve the problem. In this analogy, the electrical connection between the bulb and the socket is considered defective, or "mutant." However, correcting that problem or "mutation" did not make the light go on, so that defect is not the primary source of the problem. Third, you replace the bulb with a new one. No luck. Finally, you try replacing the plug. The light goes on! Even if you could not see anything wrong with the plug, you know it was the cause of the problem because when it was replaced by a good plug, the light worked. What you did was to determine which of several components of the system was the source of the problem by replacing possible defective parts with good ones until you found the one that caused the problem.

When you want to determine which part of an animal showing mutant behavior is the source of the problem, you look at the distribution of normal and mutant tissue in mosaics and determine which structures must be mutant for the aberrant behavior to be seen. For example, suppose you compare all the mosaic fruit flies that show the mutant behavior and find that in some, but not all, the brain is made up of mutant tissue. In some mosaics, "nature" replaces a mutant part with a normal part. Since the animal still behaves abnormally, even when the brain is made up of normal tissue, clearly the brain cannot be the source of the problem. However, if the thoracic ganglion is made up of mutant tissue in every individual showing the aberrant form of the action, and if the presence of normal tissue in the thoracic ganglion is associated with the normal form of the action, then the conclusion is that the thoracic ganglion is the structure altered by the gene. When the thoracic ganglion is defective, the behavior is abnormal. Conversely, when nature puts in a normal, rather than mutant, thoracic ganglion, the animal behaves normally. This is the equivalent to replacing the plug in the lamp analogy. Furthermore, as in the analogy, the site of the problem can be determined even if the specific anatomical or physiological defect in that structure cannot be detected without further testing. Mosaic analysis may not identify all the structures necessary for the expression of a particular behavior, but it does point out those that are always altered in individuals showing the mutant, rather than the wild-type, behavior. This technique has been useful in locating the sites of gene action in both fruit flies and mice.

Drosophila

When tracing the path between a gene and behavior by mosaic analysis, the first step must be to obtain mosaic animals. A common way to acquire a mosaic fruit fly (*Drosophila*) is by producing a gynandromorph, a fly that has some sections of male tissue and some pieces of female tissue. A *Drosophila* cell is female if it contains two X chromosomes. Normal male cells contain one X and one Y chromosome, but cells containing only a single X chromosome are also male. A gynandromorph results from an XX (female) zygote when one of the X chromosomes is lost from one of the two nuclei created by the first cell division. The chances that the X chromosome will be lost are increased if it has an abnormal ring shape. When an X chromosome is lost, one of the resulting nuclei and all its descendants have two X chromosomes and express female characteristics, but the other daughter nucleus and all its descendants have only one X chromosome and express male characteristics. The fly that develops from this embryo is, therefore, half male and half female.

To produce a gynandromorph for behavioral analysis, a female who contains a ring X chromosome and who is also homozygous for a dominant allele producing the normal wild-type behavior is mated with a male who has a recessive mutation for that behavioral trait on his X chromosome. For example, a female with nor-

mal wings and flight might be mated with a male bearing the *wings up* mutation. Remember, male fruit flies have only one X chromosome. As a result, a male with a recessive mutation can be identified, because he displays the mutant form of the behavior. In this example, he would be identified by the position of his wings. It is necessary to have the female bear a dominant allele and the male a recessive allele so that female (normal) tissue can be distinguished from male (mutant) tissue in the mosaic. Notice in Figure 3.10 that when this male's X-containing sperm joins with an ovum containing a ring X, a female zygote results. If the ring X chromosome is lost during the first cell division, half the cells—those containing only the X chromosome that came from the father's gamete—are male. The remaining cells, those containing an X chromosome from the father in addition to the ring X chromosome from the mother, are female. The mutation can be expressed only in male cells because they lack an X chromosome with the dominant wild-type allele. None of the female cells of the gynandromorph expresses the mutation. In other words, the gynandromorph is a special type of mosaic, a fly with patches of both male (mutant) and female (normal) tissue.

Although all gynandromorphs are produced in the same way, by the loss of an X chromosome during one

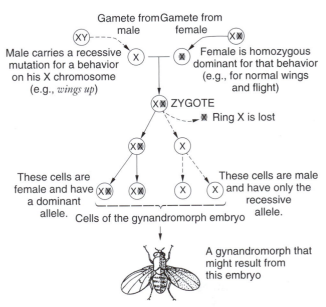

FIGURE 3.10   The development of a fruit fly gynandromorph because of the loss of a ring X chromosome during the first cell division of a female zygote. When the ring X chromosome is lost, one of the daughter nuclei then has only one X chromosome and is, therefore, male. Male cells are shaded. The other daughter nucleus has two X chromosomes and is female. The unshaded cells are female. These cells develop into a gynandromorph, an individual composed of half male cells and half female cells.

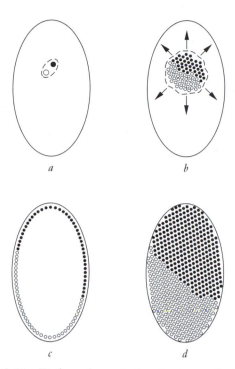

FIGURE 3.11   **Early embryonic development of a fruit fly gynandromorph. (*a*) The products of the first nuclear division of the zygote. The male nucleus is indicated as a black circle, and the female nucleus is indicated as a white circle. (*b*) A view of the surface of the early embryo. The nuclei continue to divide and migrate to the surface of the egg. During the divisions and the migration, nuclei tend to stay near their neighbors. As a result, all the female nuclei are on one half of the embryo and all the male nuclei are on the other half. (*c*) A cross section of a slightly older embryo. The nuclei are enclosed in cells at this point. The embryo is still divided so that one half has male cells and the other half has female cells. (*d*) A later stage of development. An irregular imaginary line divides the embryo into a male half and a female half.**

Furthermore, as in all fruit fly embryos, the destiny of each cell in this single layer is determined by its position. As a result, the orientation at the first cell division determines which structures will consist of male cells and which of female cells. A fate map indicating the structures that will eventually be formed is shown in Figure 3.12. If we mentally draw a line to separate the fate map in half, the orientation of this line will determine which structures will be composed of male cells and which of female cells. In a real embryo, the orientation of the initial cell division varies, so there are innumerable ways in which the adult fly may be divided into male and female chunks (Figure 3.13).

The sex of the cells making up each structure is important because it will determine whether the mutation under investigation can be expressed in those cells. If the cells of the anatomical structure in which the gene is normally expressed are female, the mutation will be hidden by the wild-type allele on the second X chromosome and the tissue will be normal. For example, the *wings up* mutation mentioned earlier will not be expressed in female cells, so when the wing muscles consist of female cells the gynandromorph has normal use of its wings. However, if the anatomical structure affected by the mutant gene falls in a male region of the adult, the mutation can be expressed and the tissue will be mutant. In the *wings up* example, the mutant allele can be expressed in the flight muscles of gynandromorphs that hold their wings up because those muscle cells are male. The next step is to select the gynandromorphs that show the abnormal behavior and compare the distribution of male and female tissue in each animal. If the insect shows the mutant behavior, each anatomical structure altered by the gene must be made up of male cells. The structure that is male in every gynandromorph displaying the mutant behavior is considered to be the site of action for the gene in question.

To make this comparison, there must be a way to identify male or mutant cells. These cells can be distin-

of the divisions of the embryo, they are not all identical. The male (mutant) and female (normal) tissue may be divided in a variety of ways because of the way insects develop. In Figure 3.11 you can see that daughter nuclei from each mitotic division tend to remain near one another. After several divisions, the nuclei migrate to the surface, forming a single layer of cells around the yolk. In a gynandromorph, approximately half the cells of this layer are female and the remainder are male. Because the nuclei stay near their neighbors even as they migrate, all the male cells are on one side of the embryo and the female cells on the other. However, the orientation of the dividing line between male and female cells varies because the first cell division can be in any direction on the surface of the egg. As a result, there are many ways in which the embryo may be divided into male (mutant) and female (wild-type) parts.

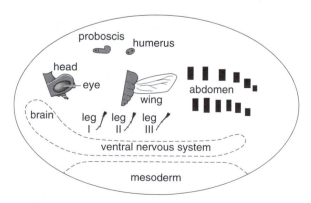

FIGURE 3.12   **A diagrammatic representation of a fate map of a fruit fly embryo, indicating the structures that will eventually be formed from cells in various locations in the embryo.**

Initial nuclear division →

Resulting gynandromorph →

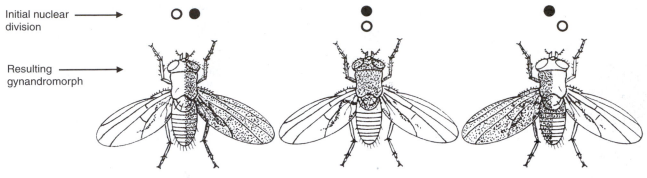

FIGURE 3.13   Diagrams of the gynandromorphs that might develop if the first division of the zygote nucleus were oriented as shown above the individual.

guished if the male who fathered the gynandromorphs had another recessive allele on his X chromosome, one that produces a trait we can see in each male cell. Genetic variants of enzymes are particularly useful tissue markers. Although these variants catalyze the same reaction within the cell, they have different physical and chemical properties. Because each variant has different biochemical properties, it is sometimes possible to find a stain that reacts with only one form of the enzyme. Then the cells producing that genetic variant can be distinguished from cells producing other forms. Variants of enzymes such as acid phosphatase (Kankel and Hall 1976) and succinate dehydrogenase (Lawrence 1981) are examples of tissue markers useful for labeling internal tissues. When histological markers such as these are used, the fly is sectioned and stained to highlight the form of the enzyme present in each cell. If the female cells have one form of the enzyme and the male cells another, the sex of the cells (and, therefore, the presence of the mutant tissue) can be determined at a glance. An example of a section of the brain of a gynandromorph that showed courtship behavior typical of males is shown in Figure 3.14. The stained cells are female, and the unstained cells are male.

Before looking at examples of the use of mosaic analysis, it may be useful to summarize the steps in the procedure. A female whose ring-shaped X chromosome bears a dominant allele that produces a normal behavior is mated with a homozygous recessive male who behaves abnormally. If the ring X chromosome is lost during one of the first cell divisions of a female embryo, a gynandromorph is produced. Because the mutation can be expressed only in male cells, the gynandromorph is also a mosaic of mutant and normal tissue. The mutant cells are also labeled with a tissue marker so that they can be distinguished from normal ones. Then all

FIGURE 3.14   A section through the brain of a gynandromorph. Variants of an enzyme were used as tissue markers. The female cells were set up genetically to produce a form of the enzyme that leads to a staining reaction. The male cells are unstained because they carry a mutant form of the enzyme gene.

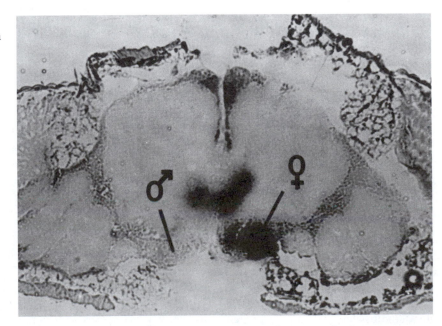

the mosaic flies showing the mutant behavior are compared. The region that is male in every gynandromorph that behaves abnormally is where the gene is acting.

Mosaic analysis has been used to locate the primary site of action of genes that influence a variety of behaviors. This information is often useful in directing future investigations to determine the exact nature of the anatomical or physiological cause for the mutation's effect on behavior. For example, mosaic analysis has been used to pinpoint particular regions of the nervous system that are key to specific components of courtship and mating in the fruit fly.

Courtship and mating are activities that fruit flies do well, even though the choreography is fairly complex. The dance begins with orientation, during which the male faces the female and taps her on the abdomen with his foreleg. If she wanders away, he follows her. Next, he begins to "sing" a courtship song by fluttering a single outstretched wing. If the female does not show interest, he will repeat these actions. When the female seems receptive, he extends his proboscis (a tubular structure, bearing the mouthparts) and licks the female's genitalia. Next, he will try to copulate with her. If the attempt fails, he will wait a few moments before starting the ritual from the beginning (Hall 1994). It sounds simple, but remember, the minuscule nervous system of a fruit fly is orchestrating the whole thing.

Mosaic analysis has been helpful in identifying brain regions important in aspects of courtship. Male behavior could be expressed only in brain regions that consisted of male cells. Jeffrey Hall examined many mosaic flies and identified brain regions important in the beginning of courtship (orienting, tapping, following, and extending a wing). These stages of courtship required male cells, on one side of the brain or the other, in a small region near the top rear of the fly's brain. This brain region is important in integrating signals from the various sensory systems (Hall 1977, 1979). Tapping through wing extension will occur only if the dorsal posterior brain consists of male tissue. The fly will attempt to copulate only if its thoracic nervous system is male. And even if a fly has female genitalia, if its brain has the appropriate male regions, it will court any fly whose abdomen is genetically female. Furthermore, the antennae of a fly can be male or female, but the brain must be male if courtship is to occur. We can conclude, then, that the antennae of both males and females detect olfactory courtship cues, but genetically male brain cells process these cues differently from genetically female brain cells (Siegel et al. 1984).

More recently, mosaic fruit flies have been produced in another manner. By inserting a transposon (a segment of DNA that causes change in nearby genes) into specific regions of the nervous system of male flies, it is possible to convert those regions into female tissue.

When certain regions of the nervous system that are involved in processing olfactory information were feminized, otherwise male fruit flies would court males or females with equal vigor. Thus these regions of the nervous system (portions of the antennal lobes or of the mushroom bodies) are important in discriminating males from females, probably on the basis of odors (Ferveur et al. 1995).

Mice

Although mosaic mice are used in the same way as mosaic fruit flies to determine where genes act to control behavior, they are generated through different procedures. A mosaic mouse, like a mosaic fruit fly, is made up of two genetically different types of cells, cells that were originally from two different embryos. A mouse begins its life as a single cell, called a zygote, but that cell soon begins to divide rapidly and forms a cluster of cells called a morula. A morula can be removed from the mother's oviduct and separated into individual cells by treatment with the appropriate enzyme. To produce a mosaic mouse, a morula is removed from each of two females with different genetic makeups. There must be two differences in the genotypes of the donor females. First, they must differ in the alleles for the behavior under investigation. The genetic information in one of the morulae would cause a normal behavior pattern, and the genes in the other would cause an aberrant action. Second, the morulae would have to differ in some trait that could be used as a marker to identify which morula gave rise to each cell in the resulting mouse. The morula removed from each female is dissociated, and the cells from both are mixed together. While this mixture is incubated for several hours, the cells reassociate to form a new morula composed of cells from both of the original morulae. Amazingly, this new embryo continues its development. When it reaches a slightly more advanced stage of development, the blastocyst stage, the embryo is implanted in the uterus of another female, where it will continue to develop into a genetically mixed-up mouse.

Mosaic mice differ from mosaic fruit flies in two ways. The first difference is in the percentage of normal and mutant cells in the mosaic. About half of the cells in a mosaic fruit fly are normal and the remainder are mutant. No such predictions can be made about the proportion of cells in a mosaic mouse that are derived from a particular morula. Although different morulae sometimes make equal contributions to a mosaic mouse, it is also possible that the resulting mouse is not a mosaic at all. Its cells may be all one genotype. Any mixture is possible. One reason for the difficulty in predicting the percentage of cells derived from each morula is that some of the cells of the mosaic morula will develop into extraembryonic structures and others will give rise to the fetus. As a result, there is an even greater potential for variety in mosaic mice than there was in mosaic flies. A second difference in mosaics of these two species is in the way cells of different genetic composition are divided. Remember that mosaic fruit flies were made up of chunks of normal or mutant tissue. Usually a whole structure such as a limb or wing is composed of one genotype. In contrast, the cells from different morulae are more thoroughly intermingled in mosaic mice (Stent 1981).

Mosaic analysis provides a way to locate the primary site of action of some of the genes that result in neurological defects in the cerebellum of mice. Nerve cells interact with one another more than cells of most other systems do, so it is particularly difficult to determine which cells in the nervous system are being affected by a particular gene. When a nerve cell degenerates or behaves in an abnormal manner, the problem could originate within that cell or within another cell that then fails to interact with it properly. Mosaic mice have been used to distinguish between these possibilities.

One recessive mutation that results in neurological difficulty for mice is called *pcd* because it causes *P*urkinje *c*ell *d*egeneration. Purkinje cells are large neurons that provide the only output from the cerebellum, the part of the brain that coordinates bodily movement. Animals that are homozygous for the *pcd* alleles seem normal at birth because they have Purkinje cells. However, these neurons gradually die, and as they degenerate the mice become uncoordinated. Richard Mullen (1977b) wondered whether the *pcd* mutation affects the Purkinje cells directly or whether the Purkinje cells are responding to an abnormality in some other cell. To answer this question, Mullen studied mosaic mice in which some cells were homozygous normal and other cells were homozygous for the *pcd* mutation. An enzyme tissue marker was used so that the normal cells could be distinguished from the *pcd* cells. Mullen reasoned that if the *pcd* mutation had its primary effect on the Purkinje cells, any Purkinje cells bearing the mutation would degenerate and all those with the normal dominant allele would survive. However, if the *pcd* mutation directly affected some other cell type, making it incapable of interacting with Purkinje cells in the proper manner, then the population of surviving Purkinje cells would have mixed genotypes. When the brains of mosaic *pcd*/normal mice were sectioned and stained, Mullen found that all the surviving Purkinje cells were normal, but those with the *pcd* allele had degenerated. Therefore, the Purkinje cells were directly affected by the *pcd* gene. We now know that *pcd* is located on chromosome 13, which will help us determine the biochemical mechanism by which the gene causes the problem (Campbell and Hess 1996). In any event, with the *pcd* gene, it seems that justice prevails. The fate of the Purkinje cells depends on their own genotype.

However, the *reeler* mutation shows that life, even for a neuron, is not always fair; the fate of the Purkinje cells of *reeler* mice is determined by the genotype of other neurons. By the time *reeler* mice are two weeks old, they can be identified because they walk as if they were drunk, similar to the way normal mice move while they are recovering from anesthesia (Sidman, Green, and Appel 1965). A *reeler* mouse has trouble walking because during its development, the Purkinje cells fail to migrate to the proper positions in the cerebellum. In contrast to the direct effect of the *pcd* mutation, mosaic analysis suggests that the *reeler* allele affects the Purkinje cells indirectly. Some of the Purkinje cells in the *reeler*/normal mosaic mice do migrate properly, but others do not. Of those that move to the correct locations, some have the *reeler* allele and others have the normal allele. In addition, many of the Purkinje cells that were in the proper position had the *reeler* genotype (Mullen 1977a). The best explanation for this is that the *reeler* mutation has its effect on some other type of cell.

Indeed, it is true—the neurons in brains of reeler mice are scrambled because of a defect in another type of cerebellar cell (the Cajal-Retzius cells). These cells normally produce an extracellular protein called Reelin, which acts as a signpost that guides Purkinje cells to their proper locations within the brain. Cajal-Retzius cells with the *reeler* mutation do not produce this protein, and therefore the Purkinje cells never migrate to their proper locations (D'Arcangelo et al. 1995). Furthermore, there is another protein involved in the migration of the Purkinje cells. This protein is the product of a gene called *mouse disabled-1* (*mdab1*) that was identified in *scrambler* mice, which behave like *reeler* mice and have similarly scrambled brains. It is thought that the mDab protein may act in migrating Purkinje cells to relay the message from Reelin (Howell et al. 1997; Sheldon et al. 1997).

GENETIC ENGINEERING

The relationships between genes and behavior are now being considered by using techniques of genetic engineering. Of primary importance is the production of recombinant DNA, that is, DNA from two different organisms. Specifically, it is possible to insert certain genes into organisms to determine their effect on behavior. In addition, recombinant research has revealed some of the ways in which genes control behavior.

The *Drosophila per* Gene

In nature, *Drosophila* eclosion, the emergence of adult flies from their pupal cases, occurs around dawn each day. Any pupae that fail to develop enough to eclose during the dawn hours must wait until the following morning. Eclosion is described as rhythmic because flies in a population tend to emerge in a repeating pattern—many emerging at dawn and then very few until the following dawn. It is interesting that the pupae do not require environmental time cues to remain rhythmic. When pupae are maintained in a laboratory without normal time cues such as light or temperature cycles, they still emerge at approximately 24-hour intervals. Since the ability to measure the passage of time is not dependent on these obvious environmental cues, we attribute it to a biological clock (see Chapter 9). The biological clock is still running in adult fruit flies; their locomotor activity is cyclic, with a 24-hour period. The *per* gene affects the biological clock, governing both eclosion and activity.

The genetic roots of the biological clock of *Drosophila* have been studied by Ronald Konopka and Seymour Benzer (1971). They induced mutations and screened the population for individuals whose biological clocks had been altered. As you can see in Figure 3.15, one mutation created flies whose biological clock runs fast, resulting in eclosion and activity rhythms with shorter period lengths. This mutation, *per^s*, results

in rhythms with a 19-hour period. Another mutation, *per^l*, has the opposite effect—a rhythm with a long, 28-hour period. The third mutation, *per^o* creates arrhythmic flies; they emerge from their pupal cases and are active at random times throughout the day. These three mutations seem to be in the same functional gene, the *per* gene, which is located on the X chromosome. To determine where this gene has its primary effect, mosaic flies were created. The head region in flies with aberrant rhythms was always composed of mutant cells, indicating that the *per* gene has its primary effect in the brain.

Now let us see how recombinant DNA techniques have been used to study behavior. Recombinant DNA has made it possible to insert the *per⁺* allele into a *per^o* fly. A normal *per⁺* fly has 24-hour rhythms, but a *per^o* fly is arrhythmic. In these experiments, a fragment of DNA containing the *per* gene along with the dominant allele for rosy eye color was inserted into a plasmid (a small circular piece of DNA). These recombinant plasmids were then injected into the pole cell region of early embryos of *per^o* flies that also had the recessive alleles for eye color. Some of the cells in the pole region develop into the cells that go on to form gametes when the embryo becomes an adult. The flies that developed from embryos treated in this way were mated with arrhythmic (*per^o*) flies that were homozygous recessive for the rosy eye color gene. In the resulting offspring, rosy-eyed individuals could have gotten their eye color only from the DNA in the recombinant plasmid since both of their parents were homozygous recessive for that eye color gene. Therefore, the presence of rosy eyes indicates not only that the plasmid is present in those individuals but also that the inserted genes could be expressed. The eye color serves as a marker that can be used to identify flies that have incorporated the plasmid. But our interest is really in the *per⁺* gene. Flies that incorporated the plasmid were not only rosy-eyed but also rhythmic. Their eclosion and activity rhythms resembled those of *per⁺* flies (Bargiello, Jackson, and Young 1984). Using recombinant DNA is, as you can see, a rather precise way to demonstrate that a particular gene is necessary for the expression of a complex behavior pattern, in this case biological timing.

Is the *per* gene needed for the daily functioning of the clock or does it influence the clock during development, perhaps by creating a brain-damaged fly? This question has also been answered by using recombinant DNA techniques. Recall from the discussion at the beginning of this chapter that genes are not always active; they can be turned on or off. One way in which this may occur is through the activity of another gene, which is called a promoter gene. Through recombinant techniques, a segment of DNA was created that contained the *per⁺* gene, as well as a promoter gene to regulate its activity. When induced to do so by a temperature shock of 27°C, the promoter turned on the *per⁺*

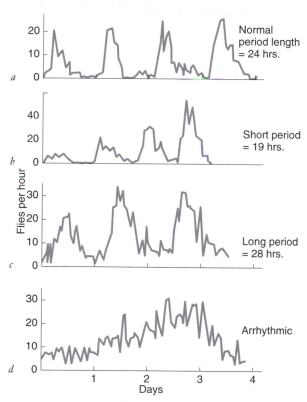

FIGURE 3.15   The eclosion rhythms of normal and three mutant strains of fruit flies studied in constant conditions. The top curve (*a*) shows the number of flies emerging each hour in a population of normal fruit flies. Peaks of eclosion occur every 24 hours. The lower curves depict the number of flies that emerge each hour in a population of (*b*) short-period (19-hour) flies, (*c*) long-period (28-hour) flies, and (*d*) arrhythmic flies. (Data from Konopka and Benzer 1971.)

gene. This segment of DNA was injected into a *per⁰* embryo, where, as in the previous example, it was incorporated into almost all the adult cells. When these adult flies were kept at 18°C, they were arrhythmic. However, when the temperature was raised to 27°C, the promoter turned on the *per⁺* gene, so the flies became rhythmic. The animals quickly became arrhythmic again when the temperature was lowered. This indicates that the product of the *per* gene is not very stable. Rhythmicity can be turned on or off by changing the temperature. The conclusion is that the product of the *per* gene must be present for the clock to function (Ewer, Rosbash, and Hall 1988).

Another question that has been addressed by recombinant DNA techniques concerns the number of clocks present in a multicellular organism. As previously mentioned, mosaic analysis revealed that the *per* gene affects a certain cluster of cells in the brain. Certainly, the brain is the site of one clock. Are there more? Research using recombinant techniques suggests that there are many clocks. A clock might be located wherever the *per* gene is active. Since there is no stain for the *per* gene product, the *per* gene was fused with a gene that would produce a product that could be stained, the β-galactosidase gene taken from the bacterium *E. coli*. The fused genes were inserted into a *per⁰* embryo and became incorporated into adult cells. The activity of fused genes is linked, so the product of the bacterial gene, galactose, could be stained green to reveal the tissues that were expressing the *per* gene. In this way, it was shown that the *per* gene becomes active shortly before eclosion. In the adult, the *per* gene is expressed in many tissues, including the antennae, eyes, optic lobe, central brain, thoracic ganglia, Malpighian tubules, and ovarian follicle cells (Liu et al. 1988). Another study used antibodies specific for the *per* gene product to determine where the gene is expressed. Similar results were obtained (Siwicki et al. 1988). Because the *per* gene is expressed in so many tissues, it seems likely that there are multiple independent clocks. (The role of the *per* gene in biological timing is discussed further in Chapter 9.)

Inactivating and Adding Genes— Knockout and Knock-in Mice

It is now possible to explore the mechanisms that underlie behavior by deleting or adding extra copies of specific genes. Mice in which a specific gene has been inactivated, or knocked out, are aptly named "knockout mice." Knockout mice completely lack the product of the disabled gene. Thus, by selectively disrupting the function of a single gene, this technique should allow one to determine the gene's function.

The production of a knockout mouse is tricky. The gene of interest must first be identified, isolated, and disabled by mutating it. The gene is then delivered to a target cell from an early mouse embryo (Figure 3.16). These target cells, called embryonic stem cells, will take up foreign DNA, such as the inactivated mutant gene. The mutant gene finds its complement on the stem cell chromosome and switches places with it. In this way, the mutant stem cell incorporates a copy of the inactivated gene. Furthermore, embryonic stem cells are still capable of differentiating into any tissue type. Thus, they can be injected into another, otherwise normal early embryo and continue developing as part of it. The resulting embryo is then implanted into a surrogate mother.

The newborn mice will have some cells that contain the inactivated gene and some cells that contain the functional gene. This type of animal, one that contains a mixture of cells with different types of genes, is called a chimera. If the mutant stem cells differentiate into cells that will become eggs or sperm, some of the gametes will contain the mutant inactivated gene. When the chimeric mice are bred with wild-type mice, some of the offspring will be heterozygous for the gene of interest. The heterozygous mice are bred with one another, and according to the laws of genetics one-fourth of the offspring will be homozygous for the inactivated gene; these are the knockout mice.

The use of knockout mice in behavioral research has some drawbacks. It is not uncommon for all of the knockouts to die as early embryos or soon after birth because the product of the inactivated gene is essential. Even when inactivating a gene isn't lethal, it can cause gross anatomical and physiological abnormalities, making it difficult to determine the direct cause of the resulting changes in behavior (R. J. Nelson 1997). Genes can interact in complex ways, making it difficult to pinpoint the effects of one of them. Thus, another problem is that the mutant organism might have physiological or developmental processes that have been altered and compensate for the missing gene. And, knocking out the same gene in different species may have different effects on behavior (Crawley 1996).

For technical reasons, up to now most knockout mice have been created by using embryonic stem cells from one strain of mice (the 129/SV strain), but the chimeric mice were mated to mice of a different strain (the C57BL/6 strain). During the genetic recombination that occurs as gametes are formed (discussed in Chapter 4), the genes from chromosomes of two different strains may be mixed. The genes that contain alleles from the 129/SV strain and those that contain alleles from the C57BL/6 strain may differ among littermates. Put simply, the genetic backgrounds of the knockout mice littermates will differ from one another and from the wild-type control mice. Thus, observed differences in behavior between the knockout mice and mice with a functional gene may actually be caused by differences in other genes (Gerlai 1996). As newer techniques are developed that use the same strain of mice for stem cells

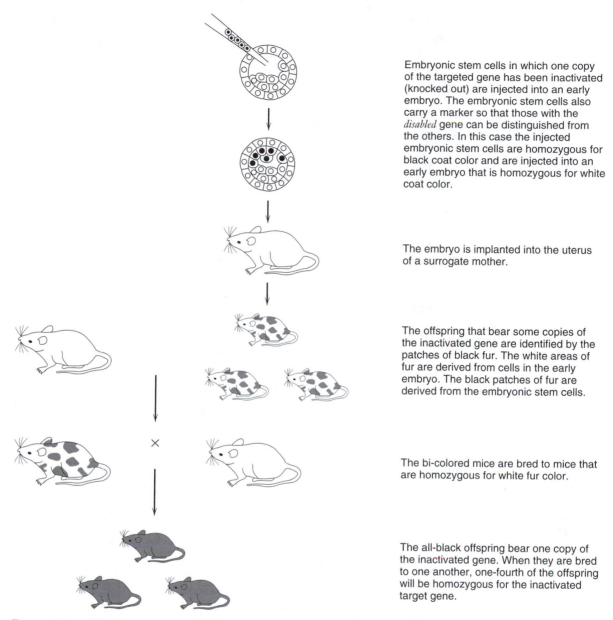

Embryonic stem cells in which one copy of the targeted gene has been inactivated (knocked out) are injected into an early embryo. The embryonic stem cells also carry a marker so that those with the *disabled* gene can be distinguished from the others. In this case the injected embryonic stem cells are homozygous for black coat color and are injected into an early embryo that is homozygous for white coat color.

The embryo is implanted into the uterus of a surrogate mother.

The offspring that bear some copies of the inactivated gene are identified by the patches of black fur. The white areas of fur are derived from cells in the early embryo. The black patches of fur are derived from the embryonic stem cells.

The bi-colored mice are bred to mice that are homozygous for white fur color.

The all-black offspring bear one copy of the inactivated gene. When they are bred to one another, one-fourth of the offspring will be homozygous for the inactivated target gene.

FIGURE 3.16   The creation of a knockout mouse—a mouse in which a specific gene has been inactivated.

and as donors, differences in the genetic background of mice will be less of a problem (R. J. Nelson 1997)

Currently, there are knockout mice for many genes that influence behavior (e. g. reviewed in Nelson and Young 1998). We will consider how knockout mice have been used to explore the role of hormones in regulating behavior in Chapter 7. Later in Chapter 3, we will see how knockout mice have been useful in understanding the way that genes influence nurturing behavior in mice. So, here we will look at the way knockout mice have been used to shed light on the mechanisms of learning and on the role of the hippocampus in mammalian learning and memory.

Although we are still far from completely under-

standing how we, or even mice, learn and remember, recent studies using knockout mice and knock-in mice to which extra copies of genes have been added have made significant advances. Many studies have demonstrated that a brain region called the hippocampus is important in the formation of long-lasting memories. According to one hypothesis a molecular mechanism by which memories are acquired and stored involves a protein that spans the membrane of nerve cells and serves as a receptor for glutamate, a chemical signal used by certain nerve cells to communicate with one another. Like most receptors for such neural chemical signals, this one, called NMDA (N-methyl-D-aspartate), opens molecular gates to tiny channels through the nerve cell membrane, allowing ions to pass through. Specifically, when the neurotransmitter glutamate binds to NMDA, calcium channels are opened. However, NMDA differs from most other receptors in an important way. It requires *two* signals before it opens the calcium channel. One signal is glutamate, which is released by a neighboring cell. The other signal is electrical, generated within the cell. When the two stimuli occur at about the same time, the calcium channels are opened and calcium ions flow into the cell. As a result, it is easier to stimulate the cell the next time the stimuli occur. Because it requires two signals, NMDA is a molecular mechanism for associating stimuli, which is the basis for a memory.

The hypothesis that the NMDA receptor is a keystone for learning and memory was tested using genetically engineered mice. Susumu Tonegawa created knockout mice in which the gene that codes for NMDA was inactivated in the nerve cells of the hippocampus. These knockout mice had difficulty passing a standardized test of spatial learning called the hidden-platform water test. In this test, the mice are placed in a pool of an opaque liquid. They must swim until they locate a platform on which they can rest, which is hidden beneath the liquid's surface. Not only were the knockout mice slower than normal mice to learn the location of the platform in the initial trials, they also had difficulty remembering the location of the platform in subsequent trials (Silva et al. 1992; Tonegawa 1994, 1995).

More recently, Joseph Tsien and his colleagues (Tang et al. 1999) provided additional support for the hypothesis by creating mice that overexpress the *NR2B* gene that codes for the NMDA receptor. The ability of these mice to learn and remember was then tested in a variety of ways and their performance compared to their normal, wild-type littermates. In the hidden-platform test of spatial memory described earlier, mice with the enhanced *NR2B* genes escaped to the hidden platform more quickly than normal mice (Figure 3.17). After three training sessions, the mice were placed in the pool without a platform. The genetically-enhanced mice spent more time searching in the quadrant of the pool where the platform was previously located than did their normal littermates. In similar tests after six training sessions, the difference in performance between genetically-enhanced and normal mice was smaller. The enhanced mice were just learning more quickly than normal mice were.

The ability of the mice to remember a familiar object was also tested. The mice were placed in an area and allowed to explore two objects for five minutes, becoming familiar with them. Several days later, the mice were returned to the test area, but one of the objects had been replaced by a new one. Normal mice spend about as much time exploring each object. Apparently, normal mice forget their previous experi-

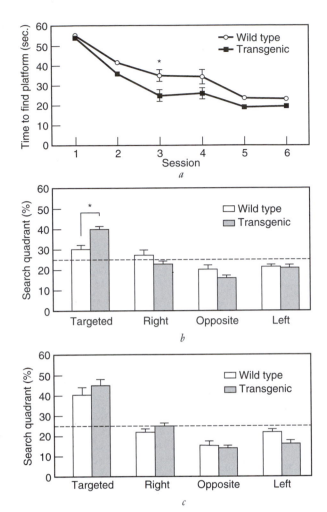

FIGURE 3.17   **Transgenic mice that overexpress the gene for the NMDA receptor on nerve cells in the hippocampus of the brain perform better than their normal, wild-type littermates in the hidden-platform water test. The mice are released into a pool of an opaque liquid and must swim until they find a platform hidden under the surface. (a) The genetically-enhanced mice are able to escape to the hidden platform more quickly than normal mice. However, after six training sessions, the normal mice perform almost as well as the enhanced mice. The mice were then tested in a pool without a platform. (b) After three training sessions with a hidden platform, the genetically-enhanced mice spent more time swimming in the quadrant of the pool where the platform had been hidden in previous tests than did normal mice. (c) The differences in performance between enhanced and normal mice were smaller after six training sessions. (Data from Tang et al. 1999.)**

ence with one of the objects. In contrast, the mice with the enhanced *NR2B* genes spend most of their time exploring the novel object. They don't waste time exploring the familiar object.

The emotional memory of the genetically-enhanced mice was also compared with that of normal mice. The mice were placed in a chamber where they received electrical shocks to their feet. After a few trials, the mice associated the chamber with shock and showed signs of fear when placed in it. When they were returned to the chamber after one hour, one day, or ten days, the genetically-enhanced mice showed more pronounced fear responses than did normal mice. The mice with enhanced *NR2B* genes also remembered an association between an audible tone and an electric shock better than normal mice. We see, then, how the use of knockout and knock-in mice have helped us understand one of the mechanisms underlying learning and memory.

MECHANISMS OF CONTROL

Keeping what we have learned about the relationship between genes and behavior in mind, we can now reconsider the mechanisms through which genes orchestrate specific behaviors.

APLYSIA EGG LAYING

The reproductive behavior of the sea hare *Aplysia* involves a number of fixed action patterns, such as the coordinated actions involved in the laying of eggs (Figure 3.18). When the snail is about to produce an egg string, it stops normal activities such as moving and eating. There are also several physiological changes at this time: The heart and respiratory rates increase, and the muscles of the reproductive ducts contract as the animal expels a string of eggs. The record-holding sea hare laid a string 17,520 cm long at a rate of 41,000 eggs per minute! When the egg string first appears, the snail grasps it in a fold of its upper lip and, waving its head in a stereotyped pattern, helps to pull the string out, winding it into an irregular mass. During this process, the egg string is coated with sticky mucus. Finally, the tangled egg string is attached to a rock or other firm support with a characteristic wave of the head (Kandel 1976).

The neuroendocrine link between the genes and egg-laying behavior is known. Two small proteins, peptide A and peptide B, are produced by the atrial gland in the reproductive system. When these peptides are injected into an animal, they stimulate two clusters of neurons on top of the abdominal ganglion, called bag cells, to release other peptides (Figure 3.19). One of the bag cell products is egg-laying hormone (ELH), which acts as an excitatory neurotransmitter and increases the firing rate of neurons in the abdominal ganglion. In addition, ELH acts as a hormone that causes the smooth muscle of the reproductive ducts to contract and expel the egg string. It may also cause the changes in heart and respiratory rates. Simultaneously, the bag

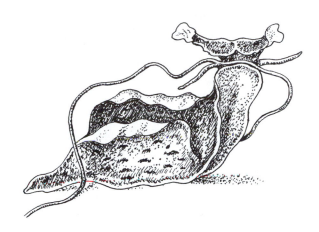
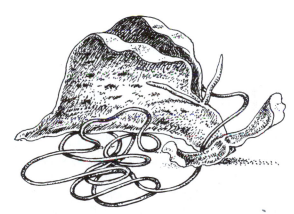

FIGURE 3.18 The egg-laying behavior of the sea hare *Aplysia*. (*a*) The snail grasps the end of the egg string in a fold of its upper lip. (*b*) Waving its head, it helps pull the egg string out. (*c*) The tangled mass of egg string is attached to a solid surface. (Scheller and Axel 1984.)

cells release alpha bag cell factor, an inhibitory transmitter that decreases the rate of specific neurons (L2, L3, L4, and L6), and beta bag cell factor, an excitatory transmitter that increases the firing rate of certain other neurons (L1 and R1). The stimulation or inhibition of particular neurons presumably coordinates the egg-laying behavior pattern (Scheller and Axel 1984).

How do genes fit into this scenario? The *ELH* gene is translated into a large protein that is broken down into the shorter functional proteins, ELH, alpha bag

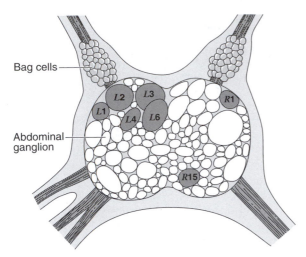

FIGURE 3.19 The abdominal ganglion and the bag cells of *Aplysia*. The *ELH* (egg-laying hormone) gene is expressed in the bag cells. Bag cells produce several substances, including ELH, alpha bag cell factor, and beta bag cell factor, that coordinate the actions involved in egg laying. Some of the large neurons in the abdominal ganglion that are affected by bag cell products are labeled. (Redrawn from Scheller and Axel 1984.)

factor, and beta bag factor. Thus a single gene is activated, and because its products act as neurotransmitters and hormones, that gene can orchestrate a complex behavioral sequence. Furthermore, if the gene is activated, all the components of the behavior pattern appear, as is typical of a fixed action pattern.

It was the sequence of bases on the *ELH* gene that revealed the mechanism of this gene's control of egg-laying behavior, but sequencing the *ELH* gene involved many steps. First, the *ELH* gene had to be identified and cloned. To accomplish this an *Aplysia* DNA library was created. A DNA library is made up of clones of different recombinant plasmids, containing fragments of all of the DNA of an organism. Second, the plasmids that bear the *ELH* gene had to be identified. This was done by making cDNA from mRNA in the bag cells and from mRNA in nonneural cells and then labeling each with a radioactive marker so that it could be identified. The cDNA from both sources was hybridized with the *Aplysia* DNA in each clone of recombinant plasmids. The key to hybridization is the specificity of base pairing. When a DNA molecule is heated, the molecule unzips down the middle by breaking the bonds that hold the two strands together. This separates the base pairs and leaves two single strands of DNA. As the DNA cools, each strand usually finds another strand with the complementary sequence of bases, and these match up to re-form double-stranded DNA. Because the bag cells produce ELH, they will have many mRNA molecules coding for ELH. Therefore, the cDNA from bag cells will have the gene for ELH. Nonneural cells do not produce ELH, and

consequently the cDNA made from their mRNAs will lack the *ELH* gene. As a result, the recombinant plasmids that hybridized with cDNA from bag cells, but not with cDNA from other cells, were identified as those containing the *ELH* gene. Third, the exact piece of DNA comprising the *ELH* gene was identified by hybridizing the clones known to contain the gene with mRNA from the bag cells. The predominant mRNA in the bag cells would code for ELH. The *ELH* gene could be spotted on an electron micrograph as a region of hybridization. Once the gene was isolated, it was sequenced.

The sequence of bases in the *ELH* gene revealed that it codes for a large polyprotein precursor that is chopped into several functional proteins. The sequence of bases on the *ELH* gene codes for 271 amino acids, but ELH itself has only 36 amino acids. The sequence of bases that would be translated into ELH could be seen in the precursor, and it was bordered by signals that indicated that ELH would be clipped out of the larger protein (Figure 3.20). What came as a surprise was that the alpha bag factor and beta bag factor were also flanked by cleavage signals. In fact, the gene codes for 11 different proteins, each bordered by cleavage signals, and 4 of these are known to be released by the bag cells. In other words, a single gene codes for a large protein that is broken down into at least three smaller active proteins with different roles in producing a behavior pattern. The role of the genes in generating this fixed action pattern is to produce peptides that act locally as neurotransmitters or distantly as hormones. These peptides stimulate or inhibit specific structures to coordinate a stereotyped behavior pattern (Scheller and Axel 1984).

NURTURING IN MICE

Genes are also involved in behaviors less stereotyped than egg laying in *Aplysia*. For example, consider the role of the *fosB* gene in nurturing in mice. After giving birth, a female mouse normally nurtures her young, which involves creating a nest, cleaning the pups, bringing the pups back to the nest if they become scattered, and huddling over the young to keep them warm and to nurse them. These activities are crucial to the survival of the pups.

Jennifer Brown and her colleagues (1996) demonstrated the importance of the *fosB* gene in nurturing by creating mice with a knockout mutation in that gene. Brown noticed that the pups of mutant mice with an inactive *fosB* died when they were one or two days old. Furthermore, the pups of mutant mice were scattered throughout the cage and neglected (Figure 3.21). The mutant mothers investigated pups but seemed unable to process the cues that bring about nurturing in normal mothers.

Clearly, *fosB* is important in nurturing behavior, but

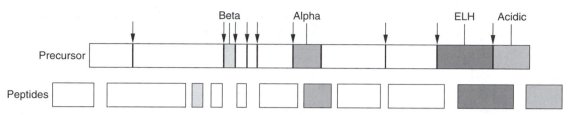

FIGURE 3.20 The products of the *ELH* gene. This gene codes for a large polyprotein precursor that is later cleaved into several smaller functional peptides. (Redrawn from Scheller and Axel 1984.)

what is the mechanism of its action? The development of normal nurturing has two components, both involving the same brain regions. One component depends on sensory input from touching and hearing newborn pups, and especially from smelling them. We know that in mice who nurture normally, *fosB* becomes active following experience with pups. The other component of nurturing depends on hormones. The hormone levels of *fosB* mutant mice are normal. Therefore, it is thought that the mutant *fosB* interferes with the first component of the nurturing response, the one built through experience with the pups.

We are certain of only a few of the details in the link between nurturing and *fosB*. *FosB* is an immediate-early gene. You may recall from our previous discussion of *zenk* and song learning that immediate-early genes produce proteins that activate other genes, which are then directly involved in changing behavior. The pre-optic area of the hypothalamus (POA) is a brain region critical to the nurturing response, and it is here that input from several senses is integrated. *FosB* is activated in the POA neurons following experience with pups. In turn, the protein coded for by *fosB* activates a program

of gene expression that leads to structural changes in the neuron circuits responsible for nurturing. In mutant mice, however, *fosB* cannot be activated. Therefore, normal nurturing does not develop.

It is possible that one of the genes activated by *fosB* codes for oxytocin receptors. If so, a chain of events that strengthens the neural circuit involved in nurturing might be this: Sensory input from experience with pups raises oxytocin levels; oxytocin activates *fosB*; *fosB* then turns on other genes and leads to the production of more oxytocin receptors; the increase in oxytocin receptors would then make the neurons more sensitive to the hormone, strengthening the neural pathway for nurturing (Brown et al. 1996).

RETROSPECTIVE

We have seen that each gene codes for a specific protein. Because proteins play so many functional and structural roles in the body, genes can influence behavior in a myriad of ways. The gene may alter a structural protein, as we saw in *Shaker* fruit flies, which have

FIGURE 3.21 Mice with an inactive *fosB* gene fail to nurture pups appropriately. Normal mice nurture their pups after birth—they clean, retrieve, and huddle over pups (*left*). Mice with an inactive *fosB* gene do not show appropriate nurturing behavior and their pups are scattered (*right*). The pups of mutant mothers usually die when they are one or two days old (Modified from Brown et al, 1996).

altered channels through nerve cell membranes, and *reeler* mice, which have altered extracellular proteins that fail to guide nerve cells to their correct positions. Alternatively, the gene may code for an enzyme, as we saw in *dunce* and *rutabaga* fruit flies. The altered enzymes influenced levels of cAMP and, therefore, learning. Furthermore, the immediate-early genes, *zenk* and *fosB*, illustrated that a gene may even code for a protein that regulates the activity of other genes. We see, then, that we can look for behavioral effects of genes virtually anywhere proteins are found (which is everywhere).

In some cases, a single gene influences several different behaviors, a phenomenon called pleiotropy. For instance, the *dunce* mutation in fruit flies not only alters olfactory learning but also impairs courtship. A female fruit fly who has already mated produces a pheromone (a chemical that sends a message to other members of the species) that inhibits courtship by normal males. This makes sense in the evolutionary scheme of things. Why should a male waste time and energy in courting an already inseminated female when she can block most attempts at copulation? Nonetheless, *dunce* and *rutabaga* males never learn to cease courting a mated female, in spite of her persistent signals (Hall, 1994). Furthermore, the *per* gene, which alters the period length of a fruit fly's circadian rhythms, also changes the rhythm of the male's courtship song (Kyriacou and Hall 1980).

Complex behaviors result from the combined activities of many genes. Consider, for instance, the genes that have effects on aggressive behavior in mice. The list of genes with demonstrated effects include the genes for (1) monoamine oxidase on the X chromosome, (2) a serotonin receptor on chromosome 9, (3) the estrogen receptor on chromosome 10, (4) α-calcium calmodulin kinase II on chromosome 18, (5) the androgen on the X chromosome, and (6) a t-complex gene on chromosome 17 (reviewed in Maxson 1996).

Finally, different gene combinations can result in the same expression of a particular behavior. We saw this in the results of the selection experiments that led to the strains of mice that built large nests and small nests (Bult and Lynch 1996). As a result, populations that are adapted to specific environmental conditions will still retain genetic variability. This genetic variability will be important if environmental conditions change, as we will see in Chapter 4.

SUMMARY

Genes control behavior by directing the synthesis of proteins. The instructions for making a protein, which is a long chain of amino acids, are in the order of bases in the DNA. Many of those proteins influence

anatomy or physiology—the foundation for behavior. Some genes cause permanent changes in behavior. For example, the *Shaker* gene in fruit flies changes a protein in the ion channels through nerve cell membranes. The change affects nerve activity and causes flies to shake under anesthesia.

If an animal has "a gene for some behavior," it means that the gene makes a difference in the performance of a behavior. However, even if an individual has a gene for a behavior, the action may not be observed. One reason is that not all genes are expressed all the time. They are turned on and off by several regulatory mechanisms. An important class of regulatory substances is the steroid hormones. In addition, nerve activity in response to stimuli can affect gene activity. For example, hearing the species song turns on the *zenk* gene in zebra finches. The resulting protein then regulates the activity of other genes, which are important in the growth of nerve cells that is essential for the formation of long-term memories of the species song.

The goals of behavior genetics are to demonstrate that there is a genetic basis to a behavior and then to determine the mechanisms by which the gene affects behavior. The classical methods of experimentally demonstrating the genetic basis of behavior include inbreeding, artificial selection, and hybridization. Inbreeding is the mating of close relatives. When mating is arranged in this way, the siblings of successive generations become increasingly similar genetically until 98% of their genes are identical. The genetic similarity of individuals in inbred strains is the key to their usefulness in behavioral genetics. When inbred strains that have been raised in similar environments are compared, any difference in behavior must be due to differences in their genes. Therefore strain comparisons are a means of demonstrating that genes can make a difference in the form of a behavior. For example, two inbred strains of paradise fish larvae respond differently to a predator model. This difference in behavior is due to genetic differences in the inbred strains.

Artificial selection is a different breeding regimen. Here individuals showing a desired behavior are bred with one another. If the frequency of the trait in the population increases when the appropriate breeders are mated, then the behavior must have a genetic basis. Artificial selection is the way breeds of dogs are created. Artificial selection for nesting behavior in house mice has demonstrated a genetic basis to nest building. These experiments have also revealed that natural selection can select for different combinations of alleles in different populations to produce the same observed adaptive behavior. It is important to keep in mind that natural selection alters behavior by acting on the structure or function of the nervous system.

Hybridization is another breeding system used to demonstrate that a given behavior has a genetic basis. Individuals that display the behavior in distinct but different ways are mated with one another, and the behavior of the hybrid offspring is observed. One such series of experiments demonstrated that foraging strategies of fruit fly larvae have a genetic basis. These strategies, called rover and sitter, occur in natural populations. Furthermore, the strategy favored by natural selection varies with environmental conditions.

Although these breeding programs demonstrate that genes influence the expression of many behaviors, they do not show where or how the genes act. Techniques that have been useful in pointing out what genes do to alter behavior include inducing mutations, mosaic analysis, and producing recombinant DNA.

A mutation is a change in the DNA. Mutations often result in the production of different proteins. Some mutations cause the individuals to show abnormal behavior patterns. When these mutants are separated from the general population, they can be studied to learn how anatomy or physiology has been altered to modify the behavior. The poor memory of the *dunce Drosophila* mutant is thought to be related to its lack of an enzyme, cyclic AMP phosphodiesterase. The learning problems of the *rutabaga* mutant are caused by a defective form of another enzyme, adenylyl cyclase. These mutants have helped us discover some of the links between genes, cAMP, and memory.

Mosaic analysis is another way to pinpoint the primary site of a gene's action. A mosaic is an individual made up of some mutant and some normal cells. When many mosaic animals are compared, the structure that is always composed of mutant tissue in the individuals that show the mutant form of a behavior is thought to be the primary site of action for that gene. The production of mosaics has been useful in studying certain behaviors in both fruit flies and mice. Mosaic analysis of mutant male fruit flies who had difficulties in courtship identified some specific regions of the nervous system involved in each component of courtship behavior. The primary site of action of genes that cause neurological disorders in the cerebellum of mice has also been studied in this way. The *pcd* mutation affects cerebellar Purkinje cells directly, but the *reeler* mutation seems to have its primary effect on cells that direct the migration of Purkinje cells during development of the brain.

Producing and cloning recombinant DNA is another way in which the relationship between genes and behavior has been clarified. For example, the *per* gene was shown to be an essential factor in the expression of circadian rhythmicity in fruit flies. When the *per+* allele was isolated and inserted into arrhythmic flies, it restored the rhythms in eclosion and activity.

The mechanisms through which genes influence behavior are varied. One example is provided by egg laying in *Aplysia*. The gene coding for egg-laying hormone (ELH) in *Aplysia* has also been cloned and identified. When it was sequenced, some interesting facts about genetic control of the fixed action pattern of egg laying were learned. The *ELH* gene codes for a polyprotein precursor that could be broken down into 11 separate proteins. Four of these are known to be released by bag cell neurons, cells that initiate egg laying. The roles of three of them, ELH, alpha bag cell factor, and beta bag cell factor, are known. They all act as neurotransmitters that either excite or inhibit specific cells in the abdominal ganglion. In addition, ELH serves as a hormone that causes the ducts of the reproductive system to contract and expel the eggs.

The *fosB* gene in mice affects nurturing. It is thought that *fosB* in the preoptic area of the hypothalamus is normally activated by experience with newborn pups. The protein coded for by *fosB* then activates other genes that are responsible for strengthening the neural circuits necessary for nurturing behavior. If *fosB* is mutant, proper nurturing does not develop.

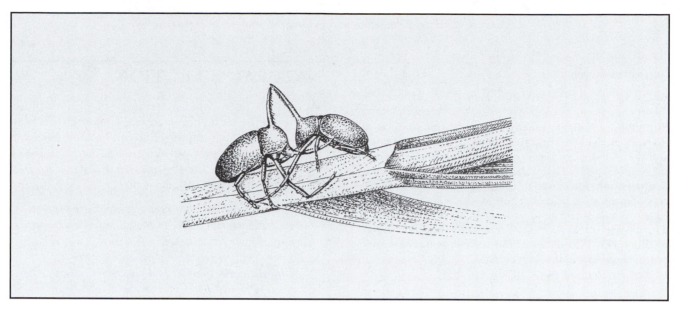

4

Natural Selection and Ecological Analysis of Behavior

Sex is the key to success.

Or, more precisely, sex is the key to success if sex results in reproduction* and if success refers to evolutionary success. Evolutionary success, in the sense we will consider it, involves an individual's having a proportionally greater genetic representation in the next generation's gene pool (the collection of all alleles of all genes of all individuals of a population). Evolution is a change in the composition of the gene pool—that is, a change in the frequency of certain alleles relative to others. We can see that the best reproducers contribute more to the gene pool and are, therefore, winners in the game of evolution (Figure 4.1).

Evolution measures the value of every trait in terms of its ability to enable its bearer to produce offspring, offspring that will, in turn, produce more offspring. But that does not mean that the anatomy and physiology of reproductive systems or sexual behaviors are the only traits that matter. Almost every anatomical, physiological, or behavioral trait can have an effect on reproductive ability. For example, before an individual can leave descendants, it must survive to reproductive age. Therefore, traits that are important in finding food and appropriate habitat or evading predators will influence reproductive success. In some species, males must be capable of winning the competition for a territory where mating can occur or where the young can be raised. Males may also have to compete for access to mates. However, even if an individual has traits that will enable it to build a better nest, find more food than others, or live to be 1000 years old, if it does not leave offspring the alleles for those traits are not passed on; an individual may be the most magnificent specimen of its species, but without offspring it is generally an evolutionary failure.

---

*Reproduction could, of course, be asexual.

Genes of generation A

Genes of generation B

FIGURE 4.1   A diagrammatic representation of hypothetical changes in allele frequencies. Ten individuals and their alleles are represented by each of the boxes in the top rectangle. Given the natural variation that exists in any population, some of the individuals will be better reproducers than others. Their alleles will occupy a larger chunk of the gene pool of the next generation. The alleles of less successful reproducers will dwindle or disappear. The alleles of the second generation are represented in the lower rectangle.

Although the phrase "survival of the fittest" is often cited as a description of natural selection, it is often misunderstood. One reason is that the emphasis is too often on the "survival" part. Of course, survival is important, but for evolution, unless survival results in the opportunity to reproduce, it does not alter fitness. After all, what counts is ensuring that one's alleles remain in the gene pool. The individuals who leave more alleles than others do earn a larger chunk of the future gene pool.

Even the term *fitness* is often misinterpreted. Who is the fittest? Is it the biggest, strongest, most aggressive individual? Perhaps it is, but only if those traits enhance reproductive success. Although fitness is often thought of as physical fitness, in a discussion of evolution, fitness is the reproductive success of a gene or an individual relative to that of the other members of the species. The fittest individual, then, is the one that leaves the greatest number of surviving offspring, or in other words, individual fitness is a measure of the relative success an individual has in getting its alleles into the next generation.

This concept of fitness was expanded by William D. Hamilton (1964). He said that because of common ancestry, close relatives share copies of alleles; therefore, by helping relatives to survive and reproduce, an individual increases its kind of alleles in the gene pool. In recognition of this, Hamilton broadened the concept of fitness to inclusive fitness—a measure of the proportion of an individual's kinds of alleles in the future gene pool. It does not stipulate whether the alleles arrive there through an individual's own offspring or those of a relative. Thus, inclusive fitness counts all offspring, an individual's own and those of relatives that exist as a result of that individual's effort. Of course, close relatives have more alleles in common than do distant relatives. Therefore, the offspring of relatives are devalued in their contribution to inclusive fitness according to the distance of the relationship. At this point in our discussion, we will simply note the nature of inclusive fitness. It will be discussed more fully in Chapter 17.

NATURAL SELECTION

We can ask, then, what determines which individuals are more fit, and the broad answer is natural selection. Darwin's idea of natural selection can be summarized in the following words from *On the Origin of Species by Natural Selection or The Preservation of Favored Races in the Struggle for Life* (1859, chap. 1):

> As many more individuals of a species are born than can possibly survive and, as there is a frequently recurring struggle for existence, it follows that any being, if it vary however slightly in any manner profitable to itself, under the complex and sometimes varying conditions of life, will have a better chance of surviving, and thus be naturally selected. From the strong principle of inheritance, any selected variety will tend to propagate its new and modified form.

From this quotation, we can make several generalizations based on more recent observations:

1. Populations have a great reproductive potential. If every individual produced as many offspring as it possibly could, population growth would be explosive (exponential).

2. Populations tend to outgrow the resources available to them.

3. Individuals in the population tend to compete for the limited resources.

4. Variation exists among individuals of a population. Some of the variable traits have been inherited from parents and can be passed on to offspring. Among the inherited traits are a few that improve an individual's chances of surviving and reproducing.

5. Individuals possessing these beneficial inherited traits have a competitive edge so they are likely to survive and leave more offspring than other individuals. Since the offspring are likely to inherit the beneficial traits, the frequency of these traits increases in the population relative to the traits borne by less successful reproducers.

Natural selection, then, is differential reproduction. That is, nature "selects" those traits that enhance reproductive success. Such traits are selected because the individuals possessing them leave more offspring, including many that will have inherited the traits that contributed to their parents' reproductive success. For example, if males that are victorious in head-to-head conflicts with other males leave more offspring than do the losers, as is the case among bighorn sheep (Figure 4.2, Geist 1971), alleles favoring the behavioral and anatomical traits that lead to fighting success will increase in the gene pool while the alternative alleles of the losers decrease. If every individual made an equal

FIGURE 4.2 Two bighorn rams, clashing heads. The winner will have priority in mating and be more likely to leave offspring.

contribution to the next generation, there would be no evolutionary change.

It should be made clear that although natural selection alters the frequencies of certain alleles in the gene pool, it does not operate directly on genotype (the genetic makeup) of individuals. It can operate only on phenotype, the expression of genotype. Thus, it has no direct effect on alleles for tallness, but it can operate on height—tall or short individuals being selected for, depending on the demands of the environment. It therefore cannot operate on recessive alleles in heterozygous individuals. Again, natural selection operates only on expressed characters.

It should be made equally clear, however, that evolution can occur only when natural selection operates on traits that are genetically determined. It is only when offspring inherit the alleles responsible for the selected trait that the gene pool is affected. A trait that is acquired during an animal's lifetime cannot be passed on to its offspring.

We know that in some cases variation may be disjunct, with variants falling into discrete classes, such as when a trait is due to the effect of a single gene. In other cases, variation is continuous. The continuum may exist as a gradient, changing gradually from one extreme to the other. It may also exist along a "normal," or bell-shaped, distribution, with most individuals falling about midway between the extremes, as in Figure 4.3. The reduced number of individuals at either extreme of a normal distribution is usually a result of stronger selection operation against those that are most different from the optimum. Under stable conditions, the optimum can be expected to remain unchanged as stabilizing selection operates (Figure 4.3*a*). However, as we know,

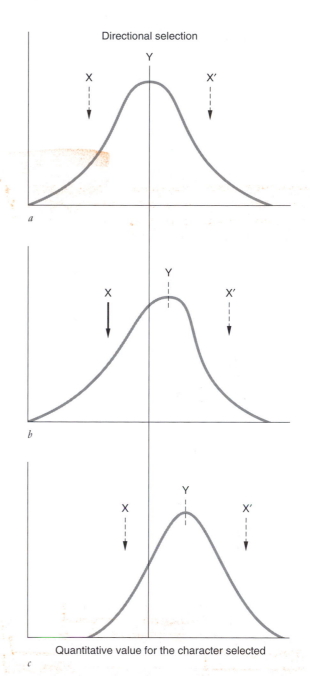

FIGURE 4.3 Directional selection. As long as the selection pressures (X and X') remain stable, the value for any trait in the population will tend to remain stable. When the pressures of X and X' change, the optimum Y may shift, causing the population to fall in line with the new optimum.

the optimum can shift as the environment changes. In this case, those at one extremity might come to be favored, and the curve that represents the population might be shifted in that direction. The phenomenon is called directional selection (Figure 4.3*b* and *c*).

Although we now know more about the mechanism of heredity and the source of variation than did Darwin, his ideas still provide the framework for our understanding of natural selection. Any explanation of evolution by natural selection must include at least these two ideas: (1) Genetically determined variation exists within a population, and (2) some variants leave more descendants than others.

GENETIC VARIATION

EXISTENCE OF VARIATION

All men are *not* created equal. Nor are all women. Nor dogs, cats, slugs, or fruit flies. Not even the offspring of the same parents are identical, as is easily demonstrated by a glance at your little brother or a look at your cat's kittens (Figure 4.4). In some cases, the differences are apparent; in other cases, less so. Nonetheless we are aware that variation generally exists within any population. In particular, we are interested in genetic variation, the ways in which the alleles differ from one individual to the next. So, the question may arise, what is the source of such variation?

SOURCES OF VARIATION

Where does genetic variability come from? There are two main sources: mutation and recombination. Mutation is a change in a gene or a chromosome that can be inherited. One type of mutation, a gene mutation, occurs when a mistake is made while the chromosome is being copied in preparation for cell division. The mistake may change the chromosome's instructions and result in an altered protein. Another type of mutation, a chromosomal mutation, occurs when there is a break in a chromosome or an incorrect number of chromosomes. Either type of mutation increases genetic variation.

Another source of genetic diversity is the recombination of alleles that occurs during meiosis, the type of cell division that results in the formation of gametes. During meiosis, the chromosomes the individual inherited from its mother and those inherited from its father are shuffled. Only one member of each chromosome pair, either the one from the individual's mother or the

one from its father, goes into a gamete, and the choice is random. As a result, each sibling of the same parents will have a different mixture of chromosomes. Besides mixing up the assortment of entire chromosomes, meiosis scrambles alleles between homologous* chromosomes. Toward the beginning of meiosis (during the stage called prophase I) the homologous chromosomes lie close to one another and pieces of chromosomes containing alleles for the same genes are exchanged. As a result of this exchange, which is called crossing over, a chromosome inherited from one's mother may have some alleles replaced by alleles from the chromosome inherited from one's father.

The resulting variability provides raw material on which natural selection can work. In more specific terms, an offspring may have a novel combination of maternal and paternal chromosomes and traits that allows it to survive and reproduce better than other members of the species.

FACTORS THAT PRESERVE GENETIC VARIATION

Since natural selection favors those individuals with certain traits, why hasn't it eliminated from the population those individuals that bear other traits? Why do populations remain variable?

Frequency-Dependent Selection

One way for variation to be maintained is by frequency-dependent selection, in which an allele has a greater selective advantage when it is rare than when it is common. As a result, the frequency of the allele fluctuates—increasing until the allele is common and then decreasing. Thus, the fitness of an individual changes with the prevalence of its genotype. There are many examples of frequency-dependent selection (Ayala and Campbell 1974). We will consider two types—frequency-dependent predation and frequency-dependent reproduction.

Frequency-Dependent Predation   It is easy to see how frequency-dependent predation could maintain variation in a population of prey. Although predators usually have a varied diet, they often attack one prey type in excess of its abundance in nature. For example, when the members of a prey species differ in some characteristic, color for instance, a predator often concen-

---

*Homologous chromosomes are those that carry genes for the same traits. One was originally obtained from the mother and the other from the father.

FIGURE 4.4 **Variation in offspring of the same two parents. During the meiotic cell division that forms gametes, the alleles of each parent are shuffled and recombined. Sexual reproduction would scramble any "perfect" combination of genes produced by natural selection.**

trates on the most common form. Individuals with this characteristic are attacked until their numbers decline in the population, carrying with them the alleles that coded for that trait. Then the predator often switches to the next most common form of prey, decreasing its numbers while the original variant restores its numbers. There is evidence from many species that when there is variation in a characteristic of prey, predators often choose the most common type, especially if the prey density is low (reviewed in Allen 1988). Recently, Alan Bond and Alan Kamil (1998) found experimental support for the hypothesis that frequency-dependent predation can maintain genetic diversity in a population. In these ingenious experiments, blue jays (*Cyanocitta cristata*) "preyed on" virtual moths. Photographs of the dark form of a moth (*Catocala relicta*) commonly eaten by blue jays were scanned to create digital images that were then modified to create four different forms of moths. A fifth form was generated in a similar manner, but with a different moth species (*C. retecta*). The virtual moths were presented on a computer screen against one of five different backgrounds that altered the difficulty of detecting the moths. The blue jays preyed on the moths by pecking at the screen. The "dead" moth was removed from the display, and the jay received a food reward. The relative numbers of moth forms that escaped detection determined the subsequent abundance of each prey type. The blue jays preyed on the most common form of moth and switched to alternative forms when that form became less common. This was true as long as the virtual moths were not too cryptic. Thus, frequency-dependent predation can maintain prey polymorphism. In nature, the maintenance of prey

polymorphism would also maintain the genetic variation underlying it.

Frequency-Dependent Reproduction   In this type of mating, sometimes called the rare-male effect, a male with a rare phenotype enjoys more than the expected share of matings. This type of frequency-dependent mating would maintain variation because as alleles become uncommon, their bearer's mating success is enhanced, thereby increasing the frequency of his alleles. The rare-male effect has been demonstrated in guppies (*Poecilia reticulata*), a fish species in which male coloration is extremely variable, even within a single population. Because female guppies may not be very discriminating the first time they mate, the females in this study were first mated to males with distinct markings (M1 males). In the subsequent choice tests, female guppies chose among males of two types, which differed in their color pattern. It is important that the alternative male types in these choice tests were presented in equal numbers.* However, before the choice tests, females were familiarized with males of one color type; each female could view the males through a glass partition in her aquarium. Females in the treatment group were exposed to males of one distinct color pattern (M7 males). Females in control group A were exposed to M1 males similar to those to which they had been previously mated, and females in control group B

---

*This experimental design avoids some of the earlier criticisms of tests of the rare-male effect (Partridge 1988; Partridge and Hill 1984).

were exposed to GP males, which have highly variable markings. Each female was then placed in a separate aquarium with a GP male and an M7 male. Females who had been previously exposed to the M7 males (treatment females) were significantly less likely to mate with M7 males than were females of either control group. In other words, the females chose males with novel color patterns—rare males—over males with a color pattern with which they were familiar (Hughes et al. 1999).

Negative-Assortative Mating

Negative-assortative mating preserves genetic variation in a population. This, in essence, means "opposites attract." The term describes the situation in which individuals tend to choose mates with a different phenotype than their own. Obviously, if the phenotypic difference has a genetic basis, genetic variability will be enhanced. Admittedly, such assortment is not common in nature.

Negative-assortative mating maintains the tan-striped and white-striped morphs of white-throated sparrows (*Zonotrichia albicollis*) in approximately equal numbers in a population. In this case, the tan-striped males are preferred by both female morphs. However, the white-striped females outcompete the tan-striped females for access to the tan-striped males. Tan-striped females then pair with the leftover white-striped males. As a result, 93% to 98% of the population mates with an individual of the opposite morph (Houtman and Falls 1994).

In some species, negative-assortative mating is a mechanism that prevents inbreeding. For example, in some strains of mice, individuals can determine whether others share certain alleles by the smell of urine. They then choose mates that are different from themselves (reviewed in Penn and Potts 1999). Patrick Bateson (1983) has suggested that some species are able to recognize kin, individuals sharing many common alleles, and then choose mates that differ from kin. Bateson's ideas are discussed further in Chapter 8.

Environmental Variation

Genetic diversity may also be maintained because some alleles are favored in one environment whereas others are more successful in other localities. Susan Riechert (1986a) has found, for instance, that there are genetically based differences in territorial behavior among populations of the funnel-web-building spider (*Agelenopsis aperta*) living in different habitats. These spiders compete for web sites and defend an area around the web as a territory (Figure 4.5). The territory must be large enough to provide sufficient food for survival and reproduction. This species occupies a wide variety of habitats in its range from northern Wyoming to southern Mexico. Spiders living in the relatively lush vegeta-

FIGURE 4.5   A funnel-web spider at the entrance to the funnel that extends from its web. There are differences in the expression of territorial behavior in populations of this spider that live under different ecological conditions.

tion along the rivers and lakes in Arizona (riparian populations) allow neighbors to build webs closer to their own than will spiders in the desert grassland populations of New Mexico (Figure 4.6). Furthermore, the intensity of territorial disputes between desert grassland spiders is greater than those between riparian spiders. Threat displays of grassland spiders are more likely to escalate into battles, and the fighting more often results in physical injury (Riechert 1979, 1981, 1982).

A series of interesting experiments has demonstrated the genetic basis of territorial behavior in these spiders (Riechert 1986a). Spiders were collected from a desert grassland environment in New Mexico and from a riparian environment in Arizona. Purebred lines were established by allowing individuals from a particular habitat to mate with one another. After the spiderlings emerged from the eggs, each was raised separately on a mixed diet of all they could eat. When they were mature, 16 females from each population line were placed into experimental enclosures where they could build webs. Just as is found in field populations, the average distance between laboratory-raised females from riparian populations was less than that between females from desert grassland populations. We see, then, that territory size is an inherited characteristic and not determined by recent feeding history or learned from previous territorial disputes. Riechert (1986a) also examined the degree of genetic diversity between the two populations of spiders. To do so she used electrophoresis, a technique that reveals the number of alleles that exist for a given gene. This measure, in turn, can be used to estimate the

FIGURE 4.6 Variation in ecological conditions in different regions of the range of the funnel-web spider. The desert grassland of New Mexico (top) has few suitable web sites and low prey abundance. In contrast, most of the woodlands near rivers in Arizona (bottom) have many suitable web sites and prey is plentiful.

degree of genetic variability among individuals in a population. In electrophoresis, molecules are placed in an electric field. Various components of a mixture separate because molecules of different sizes, shapes, and charges move at different rates in the electric field. In this case, the molecules were proteins, enzymes that are known to exist in several different forms. The number of different forms (allozymes) of particular enzymes within a population can be used as a measure of genetic diversity because the specific structure of an enzyme is genetically determined. Riechert found that members of the two populations were genetically quite different from each other.

Those genetic differences and the differences in territorial behavior presumably arose as natural selection favored traits in the populations that suited the local environments. In relatively lush riparian areas found along rivers and lakes, for instance, prey is more abundant and there are many more suitable web sites than in the desert grassland habitat. As a result, a spider living along a river can capture enough prey in a smaller area. In contrast, the desert grassland environment is severe. Prey is scarce and the scorching sun makes it too hot to forage during much of the day. Thus, larger territories are well suited to these stringent environmental conditions. It is interesting to note that desert grassland spiders defend large territories even when food is abundant. Territory size has been set genetically so that an individual's territory is large enough to provide sufficient energy during times of low prey abundance, presumably because prey may be plentiful today but scant tomorrow (Riechert 1986b). Why, we might wonder, are territorial disputes among grassland spiders so intense? The answer undoubtedly lies in the scarcity of web sites in the desert grassland habitat. Here, as in most places, the law of supply and demand increases the value of a web site. In this species, genetic diversity is increased as a response to the environment throughout its range.

NATURAL SELECTION AND ADAPTATION

Variation implies inequality. As we have seen, in a variable population, certain individuals are likely to be more reproductively successful than others. Ecological conditions will determine which traits are favored during evolution. Therefore, as a result of natural selection, we would expect the genetically determined traits of individuals to become gradually fitted to the environment. The traits that allowed individuals to survive and reproduce better than their competitors are called adaptations.

In the previous section, we saw that the expression of territoriality in the funnel-web spider is well suited to the existing ecological conditions. We assumed that natural selection was responsible for the appropriateness of the behavior and, therefore, that territoriality is an adaptation. However, these were assumptions. For any characteristic to be an adaptation, individuals bearing the trait must leave more offspring than those lacking it. Although there are no data to prove that owners of large territories in desert grassland population leave more offspring than individuals with small territories, there are data to support the suggestion that territory quality may influence reproductive success in another locality—a recent lava bed in central New Mexico. Here, spiders with high-quality web sites, those with the best ambient temperatures and prey abundance, have 13 times the reproductive potential of their neighbors in poor-quality areas (Riechert and Tracy 1975). Thus, it seems that our assumption of the adaptiveness of territoriality is correct.

When we ask questions about the adaptiveness of behavior, we are asking about its value for enhancing reproductive success. The aim in answering such a question is to understand why those animals that behave in a certain way survive and reproduce better than those who behave *in some other way*. In our consideration of territoriality in funnel-web spiders, the alternative forms of behavior were easy to identify. The fitness of spiders that build webs in areas that are hot and have few prey could be compared to the fitness of others that build webs in areas with more favorable thermal conditions and features that attract prey. But the alternatives are not always this easy to identify because the losers of the competition may be long gone. Nonetheless, if we are to demonstrate adaptiveness, we must always identify the alternatives that natural selection had to choose from (M. S. Dawkins 1986).

Questions about the evolution and adaptive significance of behavior have been central to ethology since its beginning. Recall that two of Tinbergen's (1963) four questions were what is its function (survival value) and how did it evolve? These questions still form the roots and part of the supporting trunk of ethology. However, many who focus on questions about the function and evolution of behavior now prefer to think of their studies as branches of the ethological tree, branches with labels such as "behavioral ecology" or "sociobiology." Many argue that any such branch ultimately deals with adaptation.

We must keep in mind that the notion that any specific trait is an adaptation is an assumption that must be tested. However, Stephen J. Gould and Richard Lewontin (1979) claim that the adaptiveness of a trait is too often accepted without testing. Indeed, they have ridiculed the adaptationist approach, which they claim breaks an individual into separate traits and assumes that each of those traits is adaptive. They named this assumption a Panglossian paradigm, after Dr. Pangloss in Voltaire's *Candide* who asserted, "Things cannot be other than they are. . . . Everything is made for the best purpose. Our noses were made to carry spectacles, so we have spectacles. Legs were clearly intended for breeches, and we wear them."

Although Gould and Lewontin suggest that Panglossian philosophy is inherent in the thinking of all adaptationists, Ernst Mayr (1983) argues that it is not. He asserts that adaptationists claim only that natural selection produces the best genotype possible given the many constraints placed on it, not that all traits are perfect. There are several reasons why a trait may not be adaptive. First, natural selection must act on the total individual, which is usually a mixed bag of traits—some good, some bad. Second, natural selection can act only on the available alternatives. Those alternatives will depend on the constraints imposed by the individual's evolutionary history and its present conditions—ecological, anatomical, and physiological. The "perfect" mutation or allele combination may not have arisen. Finally, natural selection works in a given environment, and conditions may vary from place to place or change over time.

Some evolutionary theorists restrict the term *adaptation* to those traits that have a genetic basis (G. C. Williams 1966). We will follow their lead and exclude learned behaviors from this chapter. However, we will return to the role of learning in producing appropriate responses in Chapter 5.

THE MAINTENANCE OF NONADAPTIVE TRAITS

We cannot expect that all observed characters, even when the data are in, will be found to be adaptive or at least neutral. There are undoubtedly nonadaptive characters in any population.

This question then arises: If certain traits are nonadaptive, how are they maintained in a population? One way in which nonadaptive traits can be maintained is through pleiotropy, multiple effects of a single gene. For example, a certain gene in mice that produces short ears also produces fewer ribs; and a single gene in *Drosophila* produces small wings, lowered fecundity, shorter life span, and other changes. If the *net* pleiotropic effect is beneficial, nonadaptive traits can be maintained in the population.

Nonadaptive characters may also continue in a population through linkage. In this case, a nonadaptive gene may be located very close to a highly adaptive one on the chromosome. So, a nonadaptive allele may be carried along on the coattails of the adaptive one.

Another way in which nonadaptive traits may be maintained in a population is by gene flow. In this case, animals with an assortment of alleles adjusted to different environmental conditions may migrate into an area from a nearby region. This is why a population of funnel-web-building spiders in a particular riparian woodland environment in Arizona shows highly aggressive territorial behavior typical of desert populations of these spiders. You may recall that spiders from riparian woodlands are usually nonaggressive in disputes because there are plenty of good web sites. In all areas except this riparian woodland, spiders experience limited prey and web sites, and in these arid desert regions, spiders are extremely aggressive in territorial disputes.

Drift fence collections indicate that spiders continually move into the riparian woodland from the surrounding, more stringent environments. Thus the population in this riparian woodland is actually a mixture of spiders—those originally from the woodlands and immigrants from the surrounding desert areas. The influx of genes from surrounding areas is responsible for the nonadaptive aggressive responses of spiders in the riparian woodlands (Riechert 1993a, 1993b).

In other cases, nonadaptive traits are not being maintained. They are being selected out, and we simply observe them on their way to oblivion. Thus, when we look at any population, there are those on a winning track and those being nudged out.

We must remember that because of the variation produced by mutation and meiotic recombination, winners and losers in the evolutionary game are constantly being generated. We should also keep in mind that variation produced by mutation is likely to be deleterious. After all, a random change in any finely tuned machine, whether it is a living body or an automobile engine, is unlikely to be an improvement. Furthermore, even if natural selection were somehow able to produce an individual with a perfect combination of alleles, that combination would be scrambled during the production of gametes and diluted in fertilization when the alleles are mixed with those of the other parent (G. C. Williams 1975).

Nonadaptive traits may survive in spite of natural selection because of evolutionary time lag, the time between selection and its effects. Today's genetic combinations were selected to suit yesterday's conditions. But today will almost certainly differ from yesterday or tomorrow. Paradoxically, natural selection may actually be the reason that yesterday's adaptation is antiquated. Adaptations selected in one species will change the conditions for others (Van Valen 1973). Maintaining the same advantage requires continuous change. The cat quick enough to snatch a bird successfully today might not be speedy enough to catch the next generation of birds because only the swiftest of yesterday's birds remained to leave offspring. Thus, natural selection always involves a lag as adaptation tracks conditions. It is like riding on the down escalator and trying to hug someone on the up escalator. To maintain the embrace, you would both have to change steps constantly.

THE SEARCH FOR ADAPTIVENESS

Where do such problems leave the adaptationist approach? Right back where it started—with the idea that the hypothesized adaptiveness of any trait must be tested. If the hypothesis is incorrect, a new one can be generated and tested as science proceeds (Brown 1982; Mayr 1983).

The hypothesis that traits are adaptive has, in some instances, stimulated interesting research that might have been neglected if one readily assumed the nonadaptiveness of traits. Consider, for instance, Niko Tinbergen's observation of a seemingly nonadaptive behavior of black-headed gulls (*Larus ridibundus*). Tinbergen observed that the gull parent does not immediately remove the broken eggshells from the nest. He and his colleagues had already demonstrated that the presence of eggshells in the nest attracted predators such as herring gulls and carrion crows. But here was the black-headed gull, sitting for hours among the conspicuous shell fragments. Tinbergen first thought that this delay must be dangerous and only explainable as a pleiotropic and nonadaptive effect associated with the removal behavior. He warned, however, that such assumptions are, in essence, a refusal to investigate.

Tinbergen then observed the black-headed gull colonies more carefully, looking for any adaptiveness of the delay. He noted that chicks were commonly eaten by neighboring gulls and that cannibalistic neighbors took many more chicks when they were newly hatched and still wet than when they had dried and become fluffy. Whereas a gull could swallow a wet chick within a few seconds, it took about ten minutes to down a dry chick. One might imagine the difference as similar to trying to swallow a peeled grape as opposed to a cotton ball. Robber gulls were observed to snatch the wet young within a fraction of a second if the parent was distracted by a predator. In fact, one chick was gulped down while its parent was carrying off some eggshells. So, Tinbergen surmised that although removing the shells reduced predation by other species of birds, delaying removal until the chicks were dry decreased the likelihood of the chicks' being cannibalized by neighboring gulls while their parents were away on their chores (N. Tinbergen et al. 1962).

In this case, Tinbergen's hypothesis was tested by simple observation, but the test for adaptiveness of other behaviors has involved two intriguing avenues of research—the comparative approach and the experimental approach. Each approach to testing hypotheses about adaptation involves ascertaining the success of different forms of the same trait. The observational approach compares the observed forms to other imagined forms; the comparative approach compares the behavior of the same or related species in different environments; and the experimental approach compares different forms of the behavior, at least one of which has been created by the investigator.

THE COMPARATIVE APPROACH

The comparative approach to the investigation of adaptiveness involves comparing individuals of the same or related species that inhabit different kinds of environments. The rationale for this approach is that related species will have inherited some common genes because they have a common ancestor. But if they live in different ecological conditions, they experience different selection pressures. Differences in their behavior, therefore, may be due to natural selection, which is molding their traits to suit their ecological conditions. For example, if one species suffers great risk of predation and a closely related species has few predators, we might interpret behaviors shown only by the first species as antipredator adaptations. However, we must not leap to unwarranted conclusions. There may be other explanations for the behavioral difference that were not considered. Conclusions about the function of a behavior gain validity as more species with similar differences in selection pressure are compared and found to have similar behavioral differences.

The Comparative Approach in Action

Esther Cullen's (1957a, b) comparisons of several gull species that descended from a common ground-dwelling ancestor but now live in very different ecological situations provide an excellent example of the comparative approach. She compared the behavior of typical ground-nesting gulls, the herring gull (*L. argentatus*) and the black-headed gull (*L. ridibundus*), with that of a cliff-dwelling species, the kittiwake (*Rissa tridactyla*). Most gulls, including the herring gull, nest in low-lying areas or on offshore islands (Figure 4.7) and are subject to more predation than their cliff-dwelling cousins. Many aspects of herring gull behavior have been interpreted as adaptations to reduce predation. They hide their nests, which are composed of bits of twigs and grass piled together among clumps of shore grasses. The eggs and young are also cryptic—the

young have certain behavioral characteristics that maximize the effect of their camouflage. At their parents' warning call, they may quickly run under the nearest cover and thus may escape any predator that has been attracted to the nest, or they "freeze" and rely on their markings to render them inconspicuous.

The kittiwake nests on cliff ledges (Figure 4.8) and displays a wide range of behavioral traits not found in other gulls. Cullen concluded that many of these patterns have developed as adaptations to a lifestyle in which the bird truly is "on edge." For example, the kittiwakes fight by grabbing each other's bills and twisting, thus causing the opponent to fall off the ledge, whereas ground-nesting gulls usually try to get above an opponent and peck downward or pull at its body (Figure 4.9). Probably because of their fighting patterns, ground nesters threaten by stretching into an "upright threat" posture. The kittiwake does not. Furthermore, female kittiwakes sit down to copulate, thus reducing the risk of falling off their narrow ledges during the process. Also, the kittiwake builds much tighter and deeper nests than do ground nesters because it would not be able to retrieve eggs that rolled out, as could its flatland fellows.

Later, when kittiwakes are nesting, conspicuous eggshells and droppings are left around the nest, a habit that in other gulls would attract predators. Young kittiwakes, of course, must remain in the nest until they can fly; therefore, all feeding is done at the nest. The young take the food directly from the parent, leaving no partly digested matter lying around the nest as a potential source of disease. Ground nesters, on the other hand, feed their free-roaming young anywhere in the nesting area, and they sometimes regurgitate food onto the ground near their young. Also, the young of ground

FIGURE 4.7   The nesting grounds of the herring gull. The nests of herring gulls are subject to predation because of their accessibility.

FIGURE 4.8 Nesting area of the kittiwake. The simple shift in nesting sites from the ground to cliffside has rendered the kittiwake nests inaccessible to predators and has triggered a whole gamut of behavioral changes.

nesters practice flying by jumping high into the wind. It is a risky business living on a cliff ledge, so the young kittiwakes never jump very high and always face the cliff when they do jump.

The aim of Cullen's studies, and of all comparative studies, is to determine the function of a behavior by identifying the selective pressures that led to it. However, by itself the method merely identifies correlations between ecological variables and behavior. The investigator must then explain the correlation. Critics of the adaptationist approach have suggested that these explanations have no more scientific validity than Rudyard Kipling's *Just So Stories*, which imaginatively explain how a camel got its hump or an elephant got its trunk (Gould and Lewontin 1979).

The comparative approach becomes more convincing when it is used to predict behavior in a previously unstudied group (Hailman 1967a). For example, the nest site of the Galápagos swallow-tailed gull, *L. (Creagus) furcatus*, is intermediate between the kittiwake and ground-nesting gulls. Its behavior is also intermediate between the two types, supporting the conclusion that the differences in behavior between the herring gull and kittiwake are due to selective pressures in their different environments.

If we accept that behavior in related species can diverge because of differences in selective pressure, then the converse should also be true—unrelated species that experience the same selective pressure should display similar behavior. We might expect, for instance, that other species that experience high predation would show antipredator behaviors similar to those of the herring gull and black-headed gull. Mobbing, a communal attack on an intruder, is shown by both species of ground-dwelling gulls, which experience high predation, but not by the kittiwake, which suffers little predation. This comparison suggests that the function of mobbing is to drive away potential predators. The bank swallow (*Riparia riparia*) is unrelated to the gulls but experiences high predation, as do the herring gull and black-headed gull. Bank swallows also mob potential predators and successfully drive them away (Hoogland and Sherman 1976).

FIGURE 4.9 Aggression in herring gulls. The form of fighting is more violent than that of kittiwakes.

Following the same reasoning, we might also expect that species that experience little predation might lack certain antipredator behaviors. The herring gull and black-headed gull remove broken eggshells from the nest after the chicks have hatched, but the kittiwake does not. It was hypothesized, based on this comparison, that the function of eggshell removal was to make the nest less noticeable to predators. Gannets (*Sulsa bassana*) are less closely related to kittiwakes than are the herring gull or black-headed gull. However, like the kittiwake they nest in places that are inaccessible to predators, and also like the kittiwake they leave broken eggshells in the nest (J. B. Nelson 1967).

Limitations of the Comparative Approach

Although the comparative method can be helpful in the study of adaptation, it must be applied carefully (Clutton-Brock and Harvey 1979, 1984; Gittleman 1989). We should, for example, always consider, test, and rule out alternative hypotheses. Sometimes this can be done by listing the competing hypotheses and developing predictions for each. The confirmation of the predictions should lend more weight to some hypotheses than to others.

We should also keep in mind that correlations between traits and ecological variables identified by the comparative approach do not demonstrate causality. Consider the difficulties in determining whether the diet was a cause or an effect of sociality in John Crook's (1964) comparative study of over 90 species of weaver birds (Ploceinae). He noticed that species living in the forest generally eat insects and forage alone. In contrast, species inhabiting the savannah eat seeds and feed in flocks (Figure 4.10). Crook identified correlations between the degree of sociality and two factors—diet and predation. But which is cause and which effect?

Seeds often have a patchy distribution, and groups of foragers are more likely to find a patch that can feed them all. One might infer, therefore, that diet is the cause of flocking in weaver birds. However, living in groups is also a good defense against predators, and seeds may be the only food source that could supply enough food for an entire flock. So, is sociality a cause or an effect of diet? The correlation does not answer the question.

In addition, we must be watchful for confounding variables and control for them. A confounding variable is a variable other than the factor of principal concern that may be a cause of the observed correlation. Common confounding variables are age and body size. The failure to correct for these variables can lead to incorrect conclusions. For example, antler size is correlated with reproductive success among red deer (*Cervus elaphus*) stags. Antler size is also correlated with age and body size. When one controls for age by comparing reproductive success among stags of the same age, no association is found between antler length and reproductive success (Clutton-Brock and Harvey 1979). So, antler size is not important to reproductive success, as it first appeared.

THE EXPERIMENTAL APPROACH

The comparative approach may suggest hypotheses to explain the function of particular traits that may be tested by using the experimental approach, in which conditions can be manipulated by the researcher.

Esther Cullen's (1957a, b) comparative study identified differences in the behavior of kittiwakes and ground-nesting gulls that were correlated with the degree of risk predation. She found that kittiwakes, which have low predation rates, leave eggshell pieces in the nest, but ground-nesting gulls, which have high

FIGURE 4.10   Weaver birds. John Crook compared the diet and degree of sociality of weaver birds that inhabit different environments. He observed that solitary species, such as *Ploceus nelicouri* (shown on the left; nest not in proportion) are usually insectivores that defend large territories in the forest. In contrast, social species, such as *P. phillipinus* (shown on the right) eat seeds and live in the open savannah.

FIGURE 4.11 A bittern removing pieces of eggshell from its nest. This behavior is typical of many bird species that rely heavily on nest concealment to reduce predation of the young. The white inner surface of the eggshell may make the nest more noticeable to a predator.

predation rates, generally remove broken eggshells. This correlation prompted Tinbergen to hypothesize that the survival value of eggshell removal was reduced predation on the young (Figure 4.11). However, several other hypotheses are possible: The sharp edges on shells might injure chicks or interfere with brooding; an empty shell might slip over an unhatched egg, encasing the chick in an impenetrable double layer of shell; or the egg remains might serve as a breeding ground for parasites, bacteria, or mold.

Tinbergen and his colleagues (1962) began experiments designed to test the hypothesis that eggshell removal reduced predation on chicks. The study began with observation. It was noted that the eggs, chicks, and nest were camouflaged and might be difficult for a predator to spot. However, the bright white inner surface of a piece of eggshell might catch a predator's eye and reveal the nest site. So they began by painting some black-headed gull eggs white to test the idea that white eggs might be more vulnerable to predators than the naturally camouflaged eggs. Of 68 naturally colored gull eggs, only 13 were taken by predators. However, 43 of the 69 white eggs were taken. The difference in predation rates lent credence to the idea that the white inner surface of eggshell pieces might endanger nearby eggs or chicks. Since all black-headed gulls remove eggshells, Tinbergen could not compare survival rates

in natural nests with and without eggshell pieces. Instead, he created artificial variation to observe the effects of natural selection. He made his own gull nests with eggs and placed white pieces of shell at various distances from some of the nests. The broken eggshell bits did attract predators. Furthermore, the risk of predation decreased with increasing distance of eggshell pieces from the nest.

OPTIMALITY

So far our discussion of adaptation has been one-sided. By focusing on the survival value of adaptation, we have considered only the benefits gained through certain behaviors. But most actions also have costs. For example, Tinbergen (1965) pointed out that in the black-headed gull, colonial living is advantageous as a means of predator defense. The drawback is that some birds lose their young to cannibalistic neighbors. Likewise, when a mother robin "decides" whether to search for one more worm to feed her young, the energy gained must be weighed against the risk of becoming cat food (G. C. Williams 1966).

Optimality theory recognizes both costs and benefits and suggests that traits have evolved to be optimal—the best they can be under existing conditions. Natural selection might be seen to act as a person might when faced with an important decision in life. When choosing the best course of action, it is not uncommon for someone to create lists of the pros and cons of each alternative. The best choice is the alternative in which the advantages outweigh the disadvantages by the greatest amount. The lists are helpful because they identify factors that should be considered in the decision. But the decision may still be difficult because it may require integration of concerns along different dimensions. For instance, if you were deciding whether to get your own apartment or live with your parents, you might have to weigh factors that represent two dimensions—money (the difference in rent) against freedom (gained by being on your own).

Optimality theory views natural selection as weighing the pros and cons, or to use the proper jargon, the costs and benefits, of each available alternative. First, natural selection translates all costs and benefits into common units—fitness. Then it chooses the behavioral alternative that maximizes the difference between the costs and benefits (Figure 4.12). In economic terms, the alternative that maximizes the difference between costs and benefits is the one that yields the greatest profit. In terms of evolution, this is the choice that maximizes fitness, and therefore it is assumed that this is the alternative that would continue into the next generation.

Measuring costs and benefits in terms of fitness is difficult at best. It might be easy to count surviving off-

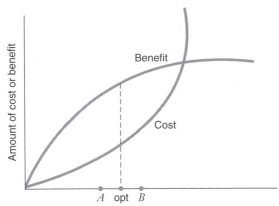

FIGURE 4.12   An optimality model for a hypothetical behavior. According to optimality theory, natural selection measures the costs and benefits of each behavioral alternative in units of fitness. It then chooses the alternative that maximizes the difference between the cost and benefit and, thus, maximizes fitness.

spring or those that are lost to predators. However, it is impossible to determine the number of offspring that never existed but could have *if* the parent had behaved in some other way. Could a mother robin have laid more eggs if she had been the earlier bird? Would she have saved any offspring from predation if she had stayed at the nest? The list of "what ifs" presents a problem for long-term optimality studies.

It has been argued, however, that animals might be expected to behave optimally, not just over the long haul, but also in daily activities. Consequently, optimality theory has used cost/benefit analyses to predict the form of many different short-term behaviors, such as degree of sociality, mating strategies, aggressiveness, and foraging. (Optimality theory has been extensively applied to foraging, and we will reconsider the concept in Chapter 12.) In each case, it is assumed that an animal's behavior maximizes the difference between costs and benefits. However, in studies of short-term optimality, profit is rarely considered in terms of fitness. Instead, costs and benefits are measured in terms such as time or energy, factors that are thought to be related to fitness but are more easily measured.

It is reasonable to assume that short-term optimality might influence long-term optimality. An animal that forages efficiently, for instance, might be expected to leave more offspring than one that does not. But maybe it wouldn't. Conceivably, it could become so intent on foraging to maximize energy gain that it would give up too quickly in an aggressive encounter to win a mate or it might fail to notice a stalking predator.

Whereas short-term optimality studies focus on a single activity as if it were the only determinant of evo-lutionary success, long-term optimality considers the larger picture. When we fail to get matches between predictions and observations in short-term optimality studies, it may be because natural selection must balance many different factors (M. S. Dawkins 1986). Deviations from predictions are often surprisingly useful because they encourage us to probe for explanations. For example, it was the deviations from predictions based on Copernicus's model of the solar system that prompted the astronomer Percival Lowell to consider how gravity from another planet might affect the system, and he postulated the existence of a ninth planet 14 years before there was photographic proof of Pluto's existence. Likewise, altruistic behavior seems to be a deviation from the predictions of natural selection, and this deviation led to the development of the concept of inclusive fitness. Therefore, optimality remains a useful construct (Barash 1982).

EVOLUTIONARILY STABLE STRATEGIES

We have been discussing optimality as if an individual's reproductive success depends only on its own behavior. In the examples mentioned this was generally true. Thus these are frequency-independent optimality models.

There are situations, however, in which the individual is not the sole master of its fate. Instead, the reproductive success gained by behaving in a certain way depends on what the other members of the population are doing. In this situation, we would say that the optimal behavior is frequency-dependent because the fitness gains it yields to one individual depends on how many other members of the population are behaving in the same way (Maynard Smith 1976, 1982).

Rock–paper–scissors is a familiar game in which you win or lose depending on what your opponents do. Recall the rules: Rock breaks scissors; scissors cut paper; paper covers rock. Any one of the moves could win or lose depending on the actions of the other players. Play scissors and you win *if* your opponent plays paper. But if you play scissors too often, your opponent will catch on and defeat you by playing rock. The only way to avoid certain loss and have a chance of breaking even is to play all three strategies—rock, paper, and scissors—in random order with equal frequency (Maynard Smith 1976).

Just as rock, paper, and scissors may be considered alternate strategies in a game, an animal's behavior may be described as a strategy. However, here strategy does not imply that the animal consciously plans the best course of action for maximizing reproductive success. "Strategy" is simply a description of what the animal does, and "winning" means that the individual's fitness increases more than its competitor's does.

For some behaviors, alternate strategies may exist within a species. Consider the following examples: A male cricket may call to attract a mate and risk being parasitized, or he may intercept a female on her way to a caller. A female digger wasp may build her own nest or take over one that is already constructed. A male may fight over a female or display and not fight.

Which is the best strategy? In the previous examples and in many others, the best course of action depends on what others are doing. The optimal strategy for an individual when the payoffs depend on what others are doing is called an evolutionarily stable strategy (ESS). By definition, an ESS is a strategy that cannot be bettered and, therefore, cannot be replaced by any other strategy when most of the members of the population have adopted it. It cannot be bettered because when it is adopted by most of the members of the population, it results in maximum reproductive success for the individuals employing it. As a result, an ESS is unbeatable and uncheatable.

ESS theory is able to cope with situations in which the payoffs of an action depend on the actions of others, but it does not require that the individuals employ different strategies. An ESS may be pure, that is, consist of a single strategy, or it may be mixed, that is, consist of several strategies in a stable equilibrium. We have discussed an example of a pure strategy among black-headed gulls—the removal of eggshells from the nest. It is an ESS because there is no alternative behavior that will yield greater reproductive success.

A mixed strategy is actually a set of strategies that exists in an evolutionarily stable ratio. Consider, for instance, a hypothetical group of fish-catching birds. There are two strategies for getting dinner—catch your own fish or steal one from another bird. Natural selection is assumed to favor the strategy that maximizes benefit. Since the thief minimizes cost and gets full benefit, thievery is favored initially. However, as the proportion of bandits in the population increases, so does the likelihood of encountering either another robber or a bird that had already had its fish stolen. Then honesty becomes the best policy. When hard-working birds become common, thievery once again becomes profitable (R. Dawkins 1980).

As the relative frequencies of alternative strategies fluctuate, they reach some ratio at which both strategies result in equal reproductive success. That particular mix of strategies will be an ESS. Should either strategy become more common, the fitness of those adopting it will drop, bringing it back to the frequency at which the individual's reproductive success equals that of those adopting the other strategy.

Mixed ESSs can arise in two ways. First, different genotypes could be responsible for producing each strategy. In this case, each member of the population would always adopt the same strategy and the relative proportions of pure strategists would remain stable. Second, each individual could vary its strategy and play each one with a certain frequency. Using the rock–paper–scissors example again to illustrate this difference, we see that stability would result if one-third of the population played rock, one-third played scissors, and one-third played paper. Alternatively, each member of the population could play rock, paper, and scissors with equal frequency.

EXAMPLES OF MIXED ESSs

Digger Wasps' Nesting Strategies

The nesting behavior of female digger wasps (*Sphex ichneumoneus*) is a clear illustration of a mixed ESS in which two strategies coexist (Brockmann, Grafen, and Dawkins 1979). A female lays her eggs in underground nests that consist of a burrow with one or more side tunnels, each ending in a brood chamber (Figure 4.13). She lays a single egg in a brood chamber, after provisioning it with one to six katydids, a process that can take as long as ten days.

To dig or not to dig, that is the question (Figure 4.14). On some occasions a female will dig her own nest and on others she will enter an existing burrow. These alternative strategies have been called "digging" and "entering." Digging has an obvious cost in time and energy since it takes a female an average of 100 minutes to dig a burrow. Furthermore, the investment is not risk-free. There is no guarantee that she will not be joined by another female, and if she is, she may lose her investment. In addition, there are temporary catastrophes, for instance, an invasion by ants or a centipede, that may force a female to abandon her burrow. Once the intruders have gone, the abandoned nest is quite suitable for nesting again. A female that finds an abandoned nest reaps the benefits without incurring the costs. So, it might seem that entering an existing burrow would be the favored strategy. Indeed, it is—*if* the burrow is actually abandoned. Unfortunately, there is no way to determine whether the nest is abandoned or whether the resident is just out hunting. A female who is provisioning a nest is gone most of the time; it may be hours or days before she learns that another female is occupying the nest. Eventually the two females will meet. When they do, they fight and the loser must leave behind all the katydids she placed in the nest. The winner takes all. She lays her eggs on the jointly provided supply of katydids.

Whether it is best to dig or to enter depends on what the other members of the population are doing. Entering saves time and energy. However, this strategy does best when it is rare. As entering becomes more common, there are fewer diggers. As a result, the chances of entering an occupied nest increase, and

FIGURE 4.13   A female digger wasp at the entrance to a burrow.

eventually digging becomes a better strategy. The mixture of strategies should reach an equilibrium with an evolutionarily stable ratio of strategies. For the digger wasps studied in New Hampshire, the ratio was 41 enter decisions to 59 dig decisions (Brockmann et al. 1979; Figure 4.15).

If that mix of strategies is to be evolutionarily stable, the reproductive success of females who enter must equal that of females who dig. If they are not, females choosing the more successful strategy would become more prevalent in the population. To determine the reproductive success of these strategies, Jane

Brockmann and her colleagues (1979) spent over 1500 hours observing 410 burrows. The reproductive success of dig and enter decisions was calculated. In a New Hampshire population, there was no significant difference in the reproductive success of individuals who adopted the digging versus entering strategies. Whereas the reproductive success of individuals who decided to dig a nest was an average of 0.96 eggs laid per 100 hours, the reproductive success of those who decided to enter an existing nest was 0.84 eggs per 100 hours. Therefore, the nesting strategies of female digger wasps are truly a mixed ESS.

Male Side-Blotched Lizard Reproductive Strategies

The changes in morph fitness among alternative male reproductive strategies of the side-blotched lizard (*Uta stansburiana*) provides an example in which the strategies of a mixed ESS cycle from one generation to the next. This small iguanid lizard lives in the inner coast range of California. Each of the three genetically determined color variations of males displays a different reproductive strategy. Notice the similarity between changes in morph frequency and the rock–paper–scissors game. Males with orange throats are very aggressive and defend large territories, within which live several females. But, a population of orange-throated males can be invaded by males with yellow stripes on their throats. Males with yellow stripes don't defend territories. Instead, these "sneaker" males, steal copulations with females on orange-throated male territories. They get away with this because orange-throated males cannot defend all their females. So, the number of males with yellow-striped throats increases. However, a population

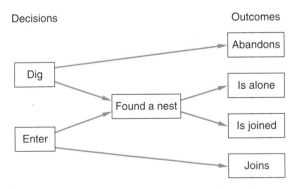

FIGURE 4.14   A female digger wasp's alternative nesting strategies and their outcomes. The two available strategies are to "dig" and to "enter." There are three possible outcomes to a decision to dig: the female may remain alone and retain exclusive use of the burrow, she may be joined by another female, or she may have to abandon the nest because of a temporary catastrophe. If a female decides to enter an existing burrow, she may be alone or she may be joining another female. (From Brockmann et al. 1979.)

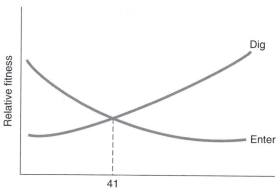

FIGURE 4.15 **The digger wasp's strategies to "dig" and to "enter" are a mixed ESS. The success of either strategy depends on that adopted by other members of the population. For a New Hampshire population of digger wasps, the strategies have equal fitness when they exist in the ratio of 41 "enter": 59 "dig."**

of males with yellow-striped throats can be invaded by males with blue throats. Blue-throated males defend territories that are big enough to hold a single female, and they successfully defend her against sneaky yellow-striped males. When the yellow-striped sneaker males are rare, it once again pays to defend large territories with several females: The reproductive success of orange-throated males exceeds that of blue-throated males. Thus, orange-throated males can successfully invade a population of blue-throated males, completing the dynamic cycle. In this mixed ESS, then, yellow beats orange, blue beats yellow, and orange beats blue. The predominant color morph in a population was observed to fluctuate in the manner predicted by these frequency-dependent changes in fitness: Blue was predominant in 1991, orange in 1992, yellow in 1993-1994, and blue again in 1995 (Sinervo and Lively 1996).

SUMMARY

Evolution is a change in the composition of the gene pool. Alleles that enhance the reproductive success of individuals become more common because the individuals that leave the most offspring make a larger contribution to the gene pool. Individual fitness is a measure of the relative reproductive success of an individual. Inclusive fitness includes all offspring that exist because of the effort of a particular individual. Inclusive fitness, therefore, counts not only an individual's own offspring but also relatives (devalued according to the degree of relationship) that exist because of the individual's efforts.

Natural selection is differential reproduction. Individuals with genetic traits that enhance survival

and reproduction leave more offspring than poorer competitors.

Some of the variation in populations is genetic. The sources of new variation are mutation and meiosis. Among the mechanisms that may preserve variation in a population are frequency-dependent selection, negative-assortative mating, and environmental variation. In frequency-dependent selection, the rare allele always has an advantage. Therefore, the allele increases in frequency until it is common, and then it decreases in frequency until it becomes rare enough to regain its advantage. These fluctuations maintain alternate alleles in the population. In frequency-dependent predation, for example, the most common form of prey is eaten in excess. This decreases the frequency of common alleles. In frequency-dependent reproduction, the rare male has a mating advantage. This increases the frequency of rare alleles. In negative-assortative mating, dissimilar individuals mate. If their differences are genetic, variation is preserved. Genetic diversity may also be maintained by variation in ecological conditions. If a species' range includes regions with different ecological conditions, some alleles may be favored in one environment whereas others are more successful in other regions.

Traits that allow individuals to survive and reproduce better than their competitors are called adaptations. Adaptive traits are appropriate for the prevailing environmental conditions.

Not all existing traits are adaptive. Some nonadaptive traits are maintained through pleiotropy, multiple effects of a single gene. The nonadaptive effects of the allele may be maintained in the population if the net effect of all its effects is beneficial. Other nonadaptive traits may be maintained because the genes responsible for them are located very close to a highly adaptive gene on the chromosome.

There are other reasons for the presence of nonadaptive traits. Natural selection may not have had time to eliminate them. Furthermore, natural selection is constantly being undermined by mutation and meiosis. Also, traits that are nonadaptive in existing conditions may have been adaptive in the previous environmental conditions.

Adaptation should always be treated as a hypothesis that must be tested. Methods of testing the hypothesis include comparisons and experimentation. In the comparative approach, traits of related species in different ecological conditions are compared. Differences between such species are generally interpreted as the result of different selective pressures. Thus, the comparative approach identifies correlations between traits and ecological variables. The correlations must be explained and the hypotheses for the explanations tested by additional observations and/or experimentation.

Optimality theory considers both the benefits and the costs of actions. Natural selection is viewed as measuring costs and benefits in units of fitness and choosing the alternative that maximizes inclusive fitness. Fitness may be difficult for ecologists to measure. Therefore, in studies of optimality some factor assumed to be related to fitness, such as time or energy, is often measured.

An evolutionary stable strategy (ESS) is one that cannot be bettered and, therefore, cannot be replaced by any other strategy once most of the members of the population have adopted it. A pure ESS is a single strategy adopted by all members of the population. A mixed ESS is a stable combination of several strategies.

5

Learning

Live and learn—a truism, it would seem, but the truth is, some do and some don't. Some species, tapeworms for instance, learn little in life. Others, say chimpanzees, learn a great deal throughout their lives, and if they are healthy, they are probably learning right up to the moment of their deaths. Without the ability to learn, some animals indeed would lose the struggle for survival in our changing world.

LEARNING AND ADAPTATION

When we observe an animal in nature, we generally find that its behavior enhances its chances of surviving and reproducing. In the previous chapter, we focused on natural selection as a mechanism for producing behavior suited to environmental conditions. Some evolutionary theorists apply the concept of adaptation only to those circumstances in which there is evidence that natural selection shaped the behavior for the stated function (G. C. Williams 1966). However, others define adaptation more broadly as all differences in traits that increase inclusive fitness. Adaptation may thus include not only traits with known genetic causes but also the inherited potential for learning and even the learned behaviors themselves (Clutton-Brock and Harvey 1979). Learning is a process in which the animal benefits from experience so that its behavior is better suited to environmental conditions (Rescorla 1988a). We will view learning here as adaptive, and we will consider some of the different ways in which it can shape behaviors to enhance an individual's ability to survive and reproduce.

Since learning ability is a product of natural selection, we expect to see differences in the mechanisms and processes of learning among species (Kamil and Mauldin 1988; Kamil and Yoerg 1982). We expect these differences because the environment and evolutionary history of a species will influence the degree to which a particular type of learning will alter fitness (the relative number of offspring left by an individual). In Chapter 2, we considered some biological constraints on learning and some ways in which individuals of some species may be biologically prepared to learn certain things and not others.

Here we consider an example of species-specific differences in spatial learning and memory among Clark's nutcrackers (*Nucifraga columbiana*), pinyon jays (*Gymnorhinus cyanocephalus*), and scrub jays (*Aphelocoma coerulescens*), illustrating how ecology and evolution may affect a species' learning skills (Figure 5.1). These birds are among the species that store seeds and later, when food is more difficult to find, recover them. So that their seeds will not be stolen by other animals, they hide them in small holes they dig in the ground and then cover over. The seeds are cached in this way in the autumn and used, as needed, through the winter and spring. Clark's nutcrackers may even use the cached seeds to feed to their fledglings in the summer (Vander Wall and Hutchins 1983). This means that the birds must be able to return to cache sites months after the seeds were buried. Recovery of the seeds is quite an impressive feat for all three species, but particularly for Clark's nutcrackers, which may have 9000 caches, covering many square kilometers (Balda 1980; Vander Wall and Balda 1977).

How do they find seeds buried so long ago? As surprising as it may seem to humans, many of whom have trouble remembering where they put their car keys, evidence is accumulating that the birds find the cached seeds simply because they remember exactly where they hid them (Balda 1980; Vander Wall 1982).

If natural selection shapes the learning ability of species, we might predict that those species that depend more heavily on cached food for survival would be better at recovering caches than other species that are less reliant on it. Clark's nutcrackers, pinyon jays, and scrub jays are related species, members of the same family, that differ in their dependence on cached food. The nutcrackers live at high altitudes, where there is little else but their stored seeds to eat during the winter and spring. Their winter diet consists almost entirely of cached pine seeds. Nutcrackers prepare for winter by storing as many as 33,000 seeds. Pinyon jays live at slightly lower altitudes, where food is a little easier to find during the winter. Nonetheless, 70% to 90% of the pinyon jays' winter diet consists of some of the 20,000 seeds they cached in preparation for winter. In contrast, winter stresses are not this severe for scrub jays. Furthermore, they live at much lower altitudes, where food is somewhat easier to find in the winter. Scrub jays store only about 6000 seeds a year, and these account for less than 60% of the winter diet.

A study that compared cache recovery by Clark's nutcrackers, pinyon jays, and scrub jays found species differences in this ability to be correlated with their relative dependence on stored seeds (Balda and Kamil 1989). In these experiments, the birds were permitted to store seeds in sand-filled holes in the floor of an indoor aviary. The floor had 90 holes, any of which could be filled with sand or blocked with a wooden plug. This arrangement allowed the experimenters to vary the position and number of the holes available for caching. Each bird's ability to recover caches was tested in two conditions. In one, only 15 holes were available for caching. In the other, all 90 holes were available. After a bird had placed 8 caches, it was removed from the room. One week later, when it was returned to the aviary, all 90 holes were filled with sand. The bird then had to remember where it had buried the seeds. The accuracy of recovery was measured as the proportion of holes probed that contained seeds. All three species performed better than expected by chance alone. However, the pinyon jays and nutcrackers, the species that depend most heavily on finding their stored seeds to survive the winter, did significantly better than the scrub jays in both experimental conditions (Figure 5.2). The pinyon jays actually performed better than the nutcrackers. They were only slightly better when there were 15 holes available for caching but much better when all 90 holes were available. It was suggested that pinyon jays tend to place their caches closer together than do nutcrackers, and so the jays had a smaller area to search. The clumping of the caches was unexpected, but the researchers surmised that it might be an evolutionary result of differences in social structure. Pinyon jays are the only species of those tested that remain in flocks throughout the year, so caches are formed and recovered in the presence of others, which might prevent them from spreading their caches over a larger area. This idea could be tested by comparing related species that remain in flocks to see whether they, too, place their caches in clumps. In any case, the experiments suggested an influence of evolution and ecology on learning ability.

The differences in spatial memory among Clark's nutcrackers, pinyon jays, and scrub jays are not limited to the recovery of cached seeds. Thus, spatial memory is an adaptive specialization that is not dedicated exclusively to food caching. Deborah Olson (1991) investigated the differences in spatial memory among Clark's nutcrackers, scrub jays, and pigeons (*Columba livia*), using a test of spatial memory that does not involve foraging, called a delayed operant nonmatching-to-sample procedure. In this experiment, the bird was first trained to peck at the illuminated key that was always in a particular location in an array of keys. Then two keys within the array were illuminated, and the bird was rewarded for pecking at a key located in a position

a

b

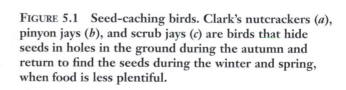

c

FIGURE 5.1   Seed-caching birds. Clark's nutcrackers (*a*), pinyon jays (*b*), and scrub jays (*c*) are birds that hide seeds in holes in the ground during the autumn and return to find the seeds during the winter and spring, when food is less plentiful.

other than the one that was previously rewarded. A delay between the training sessions and the testing sessions was gradually introduced and increased to determine how long the birds could remember the location of the key that was rewarded during training. The nutcrackers performed consistently better than the other species, and they retained information about spatial locations longer than the scrub jays or pigeons.

These results are, indeed, consistent with the hypothesis that better spatial memory has been selected for in species, such as Clark's nutcracker, that are highly dependent on remembering the location of seed caches. But other explanations for the results are plausible.

One might wonder, for example, whether nutcrackers and pinyon jays perform better during experiments than scrub jays simply because they adapt to life in captivity better or are more "intelligent." If this were true, then nutcrackers and pinyon jays would be expected to perform better than noncaching species on any learning task, even those that do not involve spatial memory. This possibility was explored in experiments that compared the performance of nutcrackers, pinyon

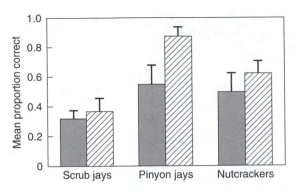

FIGURE 5.2 **Histograms that show the accuracy with which scrub jays, pinyon jays, and Clark's nutcrackers find their caches. Each bird first hid seeds in sand-filled holes in an indoor aviary. The aviary floor had 90 holes, each of which could be filled with sand or plugged. In one experimental condition, indicated with the solid bars, 15 holes were available for caching. In the other, indicated by striped bars, all 90 holes were available. After caching, the birds were removed from the room for a week. When they were returned, all 90 holes had been filled with sand. To recover the seeds the birds would probe the sand with their beaks. Accuracy was measured as the proportion of holes probed that contained seeds. Clark's nutcrackers and pinyon jays, the species most dependent on cached seeds for winter survival, were significantly better than scrub jays at recovering their caches. (From Balda and Kamil 1989.)**

jays, scrub jays, and Mexican jays (*A. ultramarina*). The Mexican jay is another corvid species that is known to cache food, especially in the fall. It lives at higher elevations than do scrub jays, suggesting that it may be more dependent on stored food. The procedure was nearly identical to the spatial nonmatching-to-sample procedure previously described. In experiments in which the birds had to use a spatial cue—the location of a stimulus circle—to receive a reward, the species that depend more heavily on seed caches performed better than the others. The nutcrackers learned the task more quickly than did the other species tested (Figure 5.3a). However, when a nonspatial cue, the color of the stimulus circle, was used in otherwise identical experiments, a different pattern of species differences in learning resulted. In these experiments, the birds were first trained to peck a stimulus circle of a particular color on a TV screen. After a period of time during which no circles were present, two circles were presented in random locations on the screen. The birds were rewarded for pecking a circle of a different color than the original one. This time the differences in learning were not correlated with dependence on stored food. The fastest learners of the nonspatial task were the cache-dependent pinyon jays and the less dependent Mexican jays, followed by the highly cache-dependent nutcrackers and then the least dependent scrub jays (Figure 5.3b). It

seems, then, that it is spatial memory, not memory in general, that is more highly developed in the species that are most dependent on stored food (Olson et al. 1995).

This correlation between spatial memory and food caching is consistent with our hypothesis that evolution may shape learning ability, but it is also possible that the observed differences in spatial memory are simply chance differences among species. However, food storing has also evolved among certain species of the Paridae (titmice and chickadees), a family of birds that is phylogenetically distinct from the Corvidae (nutcrackers and jays). Parids store seeds or insects in hundreds of widely scattered sites for shorter time periods than do the corvids, for hours to weeks as opposed to months. There are some differences in the spatial memory of food-storing and non-food-storing species of parids (Giraldeau 1997; Krebs et al. 1996; Shettleworth 1995). Whereas food-storing parid species seem to remember the sites where they previously found food, nonstorers remember the sites they previously visited. One interpretation of this difference is that food storers are more resistant to interference in memory. In other words, in nonstorers, but not in storers, the memory of visits to unrewarded sites interferes with the memory of the visit to the correct site (Clayton and Krebs 1994b). The spatial memory of the champion food-storing corvid, Clark's nutcracker, is also highly resistant to interference. When a nutcracker makes two sets of caches, neither set interferes with the bird's ability to recover the other set (Bednekoff, Kamil, and Balda 1997).

Other studies have also revealed that spatial memory cues are more highly developed in food-storing species than in nonstoring species. Unrelated species of food storers (or nonstorers) behave more similarly than they do to related species of nonstorers (or storers). Food-storing species of parids rely more heavily on spatial cues than do nonstoring species (Brodbeck 1994). The same reliance on spatial cues may also occur among corvids. The food-storing European jay (*Garrulus glandarius*) prefers spatial cues over cues directly related to the goal, but the nonstoring Eurasian jackdaw (*Corvus monedula*) does not (Clayton and Krebs 1994a).

However, a more recent test, which used landmarks as spatial cues, revealed no species differences among the food-storing corvids—Clark's nutcracker, pinyon jay, and western scrub jay. It may be that no species differences were found because the task in these experiments was so simple. The birds needed to remember only one location that remained the same throughout the tests (Gould-Beierle and Kamil 1998).

It is interesting to note that evolution and ecology seem to have produced learning differences between two mammals, golden lion tamarins (*Leontopithecus ros-*

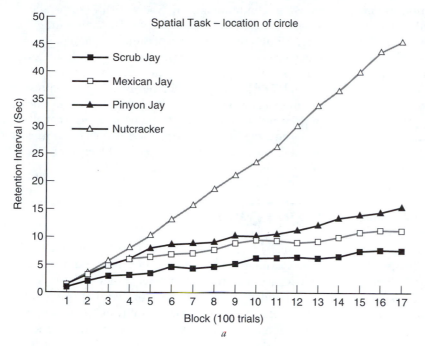

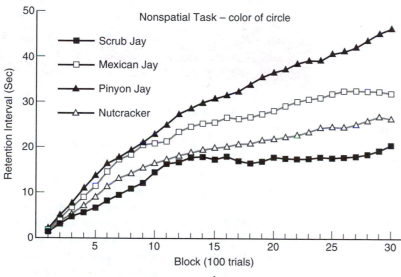

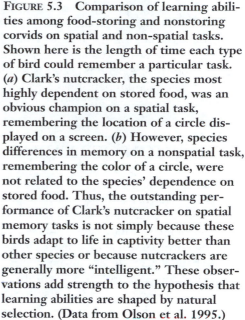

FIGURE 5.3 Comparison of learning abilities among food-storing and nonstoring corvids on spatial and non-spatial tasks. Shown here is the length of time each type of bird could remember a particular task. (*a*) Clark's nutcracker, the species most highly dependent on stored food, was an obvious champion on a spatial task, remembering the location of a circle displayed on a screen. (*b*) However, species differences in memory on a nonspatial task, remembering the color of a circle, were not related to the species' dependence on stored food. Thus, the outstanding performance of Clark's nutcracker on spatial memory tasks is not simply because these birds adapt to life in captivity better than other species or because nutcrackers are generally more "intelligent." These observations add strength to the hypothesis that learning abilities are shaped by natural selection. (Data from Olson et al. 1995.)

alia) and Wied's marmosets (*Callithrix kuhli*). Both of these species eat primarily fruit and insects, and to a lesser extent gums, seeds, and flowers. Marmosets rely more heavily on tree gums than do tamarins. Tree gums are replenished within minutes or hours. As a result, Wied's marmosets generally forage within a small core area that contains one to a few gum trees, which can be visited for food several times a day. In contrast, golden lion tamarins feed more heavily on insects, fruits, flowers, and seeds. These foods occur in patches spread over a wide area. Thus, tamarins have much larger home ranges than do marmosets. Furthermore, tamarin food sources are ephemeral. Once used, they take time to be renewed.

Differences in learning between Wied's marmosets and golden lion tamarins seem to suit the differences in their feeding ecologies. Whereas marmosets use spatial memory to remember the location of closely spaced gum trees for relatively short periods of time, tamarins must remember spatial cues for days to exploit widely separated, slowly renewing food sources. As might be predicted, then, marmosets remember spatial or visual (color) cues associated with food better than tamarins can over short time periods, such as 5 minutes. However, after 24 or 48 hours, only the tamarins performed above chance (Platt et al. 1996).

Adding strength to the hypothesis that natural selection can mold learning abilities is the observation that the hippocampal complex, a brain region important in spatial memory, is larger in food-storing species than in nonstoring species. We see this relationship not just among the parids (Krebs et al. 1989; Sherry et al. 1989) but also among the corvids (Basil et al. 1996). Indeed, comparative studies have shown differences between food storers and nonstorers in the size of the hippocampus, the number of neurons within it, and its neurochemistry.

DEFINITION OF LEARNING

We can usually identify learning when we see it. However, it can be difficult to define precisely. In the interest of simplicity, we begin with a broad definition and then refine it. First, then, let us say that learning is a process through which experience changes an individual's behavior. Clearly not all behavioral changes are due to learning. An athlete may run the last mile of a marathon at a slower speed than the first, but this is due to muscular fatigue, not learning. A person entering a dark movie theater from the sunlit street may at first trip over debris on the floor. However, within a few minutes the clutter is avoided. This change results from receptor adaptation. The receptors in the eyes had become less sensitive in the bright light, and it took a few minutes in dim light for their sensitivity to be restored so that the clutter could be seen. Maturation of the nervous system may also alter behavior. Our definition of learning must, therefore, specify that the behavioral changes it causes "cannot be understood in terms of maturational growth processes in the nervous system, fatigue or sensory adaptation" (Hinde 1970). Instead, learned changes in behavior are due to experience (Thorpe 1963). The experience may involve a certain stimulus situation (Hilgard 1956) or practice (Kimble 1961). Furthermore, although some types of learning may occur with a single trial, most learning takes place gradually over several trials.

The behavioral change that results from learning is not always expressed immediately. For example, a person may memorize facts for a test and not demonstrate that learning has occurred until the day of the exam. The possibility of a time lag between the process of learning and the resulting change in performance requires another modification of our definition of learning. The change in behavior that results from learning is more accurately a change in the probability that a certain behavior will occur.

For convenience, learning is often grouped into different categories, but keep in mind that the relationship among them may be more complex than first appears. For example, they may overlap and the distinctions between them may not be clear-cut. Those who study learning do not all agree on the nature of the categories or how many there should be. Nonetheless, categorizing types of learning is somewhat useful, if only

to assist communication. So, we will forge ahead, ignoring the arguments, and consider some of the distinguishing characteristics of some of the commonly recognized categories of learning. Our list of categories will not be complete. We will postpone discussions of examples of learning considered by some to represent other categories until later chapters. For example, imprinting, an interesting type of learning that illustrates how the lines of distinction between categories of learning are blurred, is discussed in the chapter on the development of behavior (Chapter 8).

HABITUATION

We usually imagine the change in behavior that results from learning as the addition of a new response to an individual's repertoire. However, in habituation, often considered the simplest form of learning, the animal learns *not* to show a characteristic response to a particular stimulus because, during repeated encounters, the stimulus was shown to be harmless. A bird must learn not to take flight at the sight of windblown leaves. Habituation has been defined more precisely as a "relatively permanent waning of a response as a result of repeated stimulation which is not followed by any kind of reinforcement. It is specific to the stimulus" (Hinde 1970). "Permanent" here means that the effects of habituation are rather long lasting, once they are acquired.

For an example that illustrates the essential characteristics of habituation we turn to the clamworm, *Nereis pelagica*. This marine polychaete lives in underwater burrows it constructs out of mud. *Nereis* partially emerges from the tube while feeding. However, certain sudden stimuli, such as a shadow that could herald the approach of a predator, cause it to withdraw quickly for protection. If the stimulus is repeated and there are no adverse consequences, the withdrawal response gradually wanes. It is obviously adaptive to withdraw when a predator is near, but a clamworm cannot capture its own prey while it is within its burrow. So, it is also adaptive to learn not to withdraw in the absence of danger. (The neural basis of habituation is discussed in Chapter 6, pp. 108–111.)

The reduction of responsiveness of *Nereis* to recurring stimuli was easily demonstrated in the laboratory. A group of clamworms was maintained in a shallow pan of water and provided with glass tubes as burrows. A shadow was used to trigger the withdrawal response. The first time the shadow appeared, there was a high probability that each worm would withdraw. The second time the shadow was presented, slightly fewer worms responded. The third presentation of the shadow elicited even fewer withdrawals. As shown in Figure 5.4, subsequent presentations resulted in

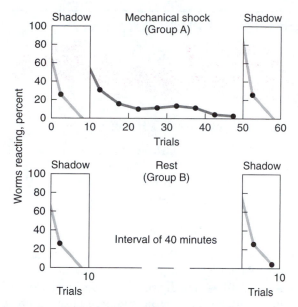

FIGURE 5.4 Habituation of the withdrawal response to a shadow by the marine worm, *Nereis*. In habituation, the simplest form of learning, the animal learns not to respond to frequently encountered stimuli that are not associated with reward or punishment. Notice that both groups of worms habituated to the shadow during trials 1–10. A mechanical shock was then administered to group A at 1-minute intervals. The worms initially responded to the new stimulus, showing that the loss of the response to the shadow was not a result of fatigue. Group B rested for 40 minutes while group A received the mechanical shocks. Both groups were equally responsive to the shadow during the last 10 stimulus presentations. One can conclude that the animals learned not to respond to a specific stimulus, the shadow in this case. (Modified from R. B. Clark 1960.)

responses from a smaller and smaller percentage of the group. The animals began to learn that the stimulus was not associated with a threat (R. B. Clark 1960). The time period for recovery from habituation took several hours, showing that the loss of responsiveness was not due to sensory adaptation, which would have occurred in seconds. A group of worms that was habituated to a recurring shadow remained fully responsive to a mechanical shock. Thus the decline in responsiveness was not due to motor fatigue.

One finds habituation everywhere, occurring in simple unicellular protozoans as well as in complex humans (Wyers, Peeke, and Herz 1973). Consider just a few examples. The cnidarian *Hydra*, living in a shallow dish of water, will quickly retract its tentacles and shorten its body if the glass is tapped. But after a few such taps, its reactions slow and may stop (Rushforth, 1965). If touched on the side of its body, the sea slug *Aplysia* immediately withdraws its gills. However, it stops withdrawal if it is touched repeatedly (Carew and

Kandel 1973; Carew, Pinsker, and Kandel 1972). The reversal response of *Caenorhabditis elegans*, a nematode with only 302 neurons, will also habituate (Beck and Rankin 1997). Habituation has also been demonstrated in honeybees (Bicker and Hähnlein 1994) and rats (Bonardi, Guthrie, and Hall 1991). Birds will at first be startled by a scarecrow but will come to ignore it or sit complacently on its arms, preening their feathers after a large meal of the farmer's grain.

THE ADAPTIVE VALUE OF HABITUATION

The benefit of habituation is that it eliminates responses to frequently occurring stimuli that have no bearing on the animal's welfare without diminishing reactions to significant stimuli. Obviously it is important for *Nereis* to withdraw to the safety of its burrow when a shadow is that of a predator. However, if the shadow is encountered often without a predator's attack, it is probably caused by something harmless, perhaps a patch of algae that is repeatedly blocking the sun as it undulates with the waves. In this case, responding to the shadow each time it appears would leave the worm little time for other essential activities such as feeding or reproducing. These nonessential responses would not only compete with vital activities, they would also be a waste of energy. Habituation is one of the mechanisms that focuses attention and energy on the important aspects of the environment (Leibrecht and Askew 1980).

As an example of how habituation is adaptive, consider the role it plays in shaping the natural pattern of escape responses of birds in the field. The young of birds such as turkeys, chickens, and pheasants need not be taught escape behaviors, such as crouching, at the sight of objects moving overhead. The chicks show protective responses to a great variety of objects, only a few of which are dangerous. On the other hand, the adults respond only to the image of a predator such as a hawk, flying overhead. The narrowing of the responsiveness of adults is due to habituation. This occurs because innocuous stimuli are so much more common in an animal's environment than harmful stimuli. As the young mature, they are often exposed to harmless stimuli such as falling acorns, leaves, or nonpredatory birds, and they gradually learn not to cringe at the sight of them. However, because predators are rare, the young do not habituate to them and continue to respond to their appearance.

Although this explanation is intuitively logical, Schleidt (1961a, b) verified that habituation could shape behavior in this way. He showed that models of various shapes—a gooselike silhouette, a circle, and a square—were all effective in eliciting alarm calls from young turkey chicks during the initial two days of testing.

When any one of these models was presented frequently, the number of alarm calls it elicited decreased. The chicks remained responsive to stimuli they encountered only occasionally.

The adaptiveness of habituation becomes clear when we consider the effect of environment on the duration of habituation of the escape responses in two closely related species of crabs. *Chasmagnathus* is a semiterrestrial crab that lives on the mudflats or amid the cord grass along the coast of South Brazil, Uruguay, and Argentina. *Pachygrapsus*, on the other hand, inhabits the rocky intertidal zone. Crabs of both species begin to run when a shadow passes overhead. The escape response habituates if a shadow is encountered frequently. Habituation lasts much longer in *Chasmagnathus* than in *Pachygrapsus*. This makes sense because the windblown grass of *Chasmagnathus'* habitat creates many shadows, but on the bare rocks of *Pachygrapsus'* habitat a shadow is more likely to signal the presence of a predator (Maldonado, Romano, and Tomsie 1997).

Let us next consider some of the ways in which habituation is adaptive in the three-spined stickleback (*Gasterosteus aculeatus*). For instance, when male stickleback fish first establish their territories, a great deal of aggressive display occurs at the boundaries. However, over time the neighbors habituate to one another and the frequency of aggressive display decreases (Van den Assem and van der Molen 1969). This is adaptive because aggressive behavior uses a great deal of energy and exposes the animal to potential harm. Because habituation is specific to the stimulus, in this case a particular neighbor, the fish remains ready to confront and fight unfamiliar males. Besides defending a territory, an evolutionarily successful male stickleback must court and mate with a female. When a male sees a gravid female, he often courts her, beginning with a zigzag dance. If the female is ready to spawn, she may then follow him into the nest, spawn, and leave, allowing the male to fertilize the eggs. However, not all gravid females are ready to spawn. When a male encounters an unresponsive female, he soon stops courting and directs his attention to another female. Experiments in which model female sticklebacks were presented to males have shown that habituation is important in reducing courtship to unresponsive females. In this case, habituation conserves energy and allows the male to direct his attention to another gravid female, which may be ready to spawn (Peeke and Figler 1997). Habituation also plays a role in reducing a stickleback's efforts to capture suitable prey when the prey is unobtainable. Although a stickleback presented with a tube filled with live brine shrimp will initially bite at the tube in an attempt to capture them, the response wanes within minutes (Peeke 1995). We see, then, that habituation is an adaptive learning process in the stickleback.

ASSOCIATIVE LEARNING

Associative learning, sometimes called conditioning, is a type of learning in which an association is made between a stimulus and a response.

CLASSICAL CONDITIONING

Even if it were true that "you can't teach an old dog new tricks," it should be possible to teach it to perform an old trick under a new set of circumstances. We see such a relationship in classical conditioning.

Characteristics of Classical Conditioning

The principle of classical conditioning was first stated by Ivan Pavlov (1927), a Russian physiologist whose main interest was actually the digestive activity of dogs. Pavlov's work began with the common observation that a dog salivates at the anticipation of food. Pavlov reasoned that it was unlikely that salivating before actually receiving the food was an inborn response. He thought, reasonably enough, that the animal had learned to associate the sight or smell of food with the food itself and that the salivation was due to such learning. Thus, in anticipation of food, the dog would begin to salivate to a new stimulus, such as a sound, that signaled the arrival of food: A connection had formed between an old response and a new stimulus.

Pavlov then began to experimentally investigate the association. To quantify the response, a small opening, or fistula, was made in the dog's cheek in such a manner that the saliva would drain into a funnel outside the dog's body, where it was collected and could be measured. The hungry dog was then harnessed into position on a stand, where it could be subjected to various stimuli. One stimulus tested was powdered food, and as expected, the dog salivated when the food was blown into its mouth. Then, immediately before the food was offered, Pavlov presented the dog with another stimulus, one that did not initially cause salivation, such as the sound of a bell. At intervals over several days, the dog was exposed to this pair of stimuli, first the sound and then immediately the food. Then Pavlov checked to see whether the dog salivated at the sound of the bell alone. It did. Although the tone originally elicited no salivation, after 30 paired stimulations, the dog's saliva indeed would flow when the sound was presented without the food. As the number of trials increased, there was an increase in the amount of saliva secreted and a decrease in the length of time it took the dog to respond (Pavlov, 1927).

In more general terms, what happens? To begin, an animal has a particular inborn response to a certain stimulus. This is called the unconditioned stimulus (US) because the animal did not have to learn the response to it. A second stimulus, one that does not initially elicit the response, is repeatedly presented immediately before the US. After several pairings, the second stimulus is able to elicit the response. (The conditioned response actually differs slightly from the original response, but this point need not concern us.) This new stimulus is now called the conditioned stimulus (CS) because the animal's response has become conditional upon its presentation. The whole phenomenon is called a conditioned reflex. Conditioned is an unfortunate translation from the Russian. A better translation would be conditional, implying that the behavior is conditional upon something occurring.

The Conditioned Stimulus as a Signal

The order of the presentation of the US and CS is important. The CS must precede the US. The CS serves as a signal that the US will appear; a cue is of little value if it occurs after the fact. Also, the two stimuli must occur fairly close together if an association between them is to be made.

Furthermore, useful signals are specific; they predict that a particular event or stimulus will follow. A signal is useless if it merely indicates that any one of a dozen events may follow. Therefore, it should not be surprising that for classical conditioning to occur, the CS must precede the US more often than it does other stimuli (Rescorla 1988a, b).

Extinction

If the CS (the tone in the experiment described) is presented frequently without being followed by the US (the food, in this case), the association between the stimuli is gradually lost. The loss of the classically conditioned response is called extinction.

The Adaptive Value of Classical Conditioning

Ivan Pavlov (1927) viewed conditioned responses as generally beneficial. Although the following quotation refers to a motor response, it could just as easily pertain to an internal response:

> It is pretty evident that under natural conditions the normal animal must respond not only to stimuli which themselves bring immediate benefit or harm, but also to other events which in themselves only signal the approach of these stimuli; though it is not the sight or sound of the beast of prey which is in itself harmful— but its teeth and claws.

Karen Hollis (1984) has experimentally tested the hypothesis that the adaptive function of classical conditioning is to prepare animals for important events. Her studies have centered on territorial and reproductive behaviors in blue gouramis (*Trichogaster trichopterus*),

fish that inhabit shallow pools and streams in Africa and Southeast Asia.

A male blue gourami begins its territorial defense against intruders with a frontal display in which it approaches the intruding fish rapidly, with all fins erect. If the intruder does not respond with a submissive posture or retreat, the contest escalates into a battle in which serious injury is possible. The males bite each other and flip their tails to beat water against the opponent's sensitive lateral line organ. Damaging aggressive contests are most likely to evolve when the value of the resource is great. In this case, it is essential for males to be successful in territorial defense because females rarely mate with a male without a territory. It is reasoned, therefore, that any mechanism that might increase the likelihood of successful defense would be favored by natural selection. A conditioned response to signals that indicate the approach of a rival might prepare a male for battle and give him a competitive edge. In nature, as the rival approached, he would inadvertently send visual, chemical, or mechanical signals that territorial invasion was imminent.

Hollis has shown that male blue gouramis that have been classically conditioned to a signal that predicts an encounter with a rival are more successful in aggressive contests. Pair members, chosen to be of similar body size and level of aggression, were housed on opposite sides of a divided aquarium. One member of each pair was trained so that a 10-second light (the CS) preceded a 15-second viewing of a rival (the US). The CS and US were also shown to the other member of the pair, but their presentations were not paired. During the test trials, the light signal was given before the barrier that separated the members of the pair was removed, allowing them to interact. The conditioned males were superior in territorial defense. They approached the territorial border with their fins already erect in the frontal display. During the territorial fights, they delivered significantly more tailbeats and bites than did their competitors (Hollis 1984). The conditioned males may have been the winners because the light (CS) increased the level of androgens, male sex hormones known to heighten aggressiveness in many species (Hollis 1990). A conditioned male is a more aggressive male, and he does indeed have a better chance of winning the battle and defending his territory, thereby increasing his chances of mating.

The conditioned male reaps a long-term benefit in addition to his immediate competitive edge: Winners in one battle are more likely to remain victors in later battles. In subsequent battles, a previous winner will fight longer and is more likely to dominate by fighting in an opponent's territory. Thus, winners are likely to remain winners and conditioned males are more likely to win the first battle. In contrast, fish that lose their first battle are likely to lose later battles as well. In one experiment, all fish that lost the first battle also lost the second one (Hollis et al. 1995).

Male blue gouramis that successfully defend a territory are more likely to attract females, but excessive aggressiveness could actually harm mating success. A territorial male is likely to attack any visitor to his territory, even females. Hollis and her colleagues (1997) have shown that a male blue gourami that is conditioned with a light signal of the availability of a gravid female enjoys greater reproductive success than an unconditioned male. As can be seen in Figure 5.5, conditioned males were less aggressive to females, as measured by the number of times the male bit the female. Conditioned males also spent more time building a nest. The shift in behavior from aggressive to reproductive activities paid off in reproductive success. Conditioned males spawned more quickly and produced more fry than did unconditioned males. In nature, it is not likely that a flash of light will reliably precede the appearance of a female. But there are other cues, the shape of a gravid belly, for instance, that might be a reliable cue of a female's approach.

OPERANT CONDITIONING

When a behavior has favorable consequences, the probability that the act will be repeated is increased. This relationship may result because the animal learns to perform the behavior in order to be rewarded. This type of learning has been named operant conditioning to emphasize that the animal operates on the environment to produce consequences. Again, just as in classical conditioning, the timing of events is critical. In this case though, the behavior must be spontaneously emitted—not elicited by a stimulus, as it is in classical conditioning—and the favorable result, or reinforcement, must follow it closely. In a sense, a cause-and-effect relationship develops between the performance of the act and the delivery of the reinforcer.

As long ago as 1855, before the term *operant conditioning* was coined, Herbert Spencer described an animal learning a response through operant conditioning. His account is interesting because it illustrates how operant conditioning may perfect motor skills in nature:

> *Suppose, now, that in putting out its head to seize prey scarcely within reach, a creature has repeatedly failed. Suppose that along with the group of motor actions approximately adapted to seize prey at this distance—a slight forward movement of the body is caused on some occasion. Success will occur instead of failure—On recurrence of the circumstances, these muscular movements that were followed by success are likely to be repeated: what was at first an accidental combination of motions will now be a combination having considerable probability.*

FIGURE 5.5 Blue gourami males that received classically conditioned pairing of a light signal and a receptive female enjoyed greater reproductive success than did unconditioned males. Conditioned males (*a*) bit approaching females less frequently, (*b*) spent more time building nests, (*c*) were quicker to spawn, and (*d*) produced more fry than did unconditioned males. (Data from Hollis et al. 1997.)

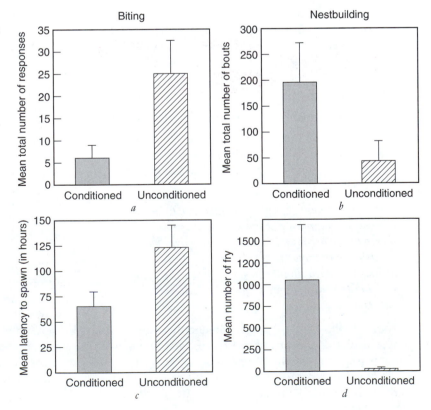

B. F. Skinner later devised an apparatus that is still used to study operant conditioning in the laboratory. Typically, a hungry animal is placed into a Skinner box, and it must learn to manipulate a mechanism that yields food. For example, a hungry rat placed in a Skinner box will move about randomly, investigating each nook and cranny. Eventually, it will put its weight on a lever provided in the box (Figure 5.6). When the lever is pressed, a bit of food drops into a tray. The rat will usually press the lever again within a few minutes. In other words, the rat first presses the lever as a random act, and when it is rewarded, the probability of the rat repeating the act increases. The apparatus has been modified for use with pigeons, which learn to peck a key for a food reward.

A stimulus that alters the probability that a behavior will be repeated is called a reinforcer. In the experiments described so far, positive reinforcers were used. Positive reinforcers are those that increase the probability of a behavior being repeated. Examples include food offered to a hungry rat or drink to a thirsty one. Negative reinforcers are those that increase the probability of a response when they are removed. If an unpleasant or painful stimulus stops when an animal

FIGURE 5.6 A rat in a Skinner box. The hungry animal explores the box and eventually presses the bar. This automatically results in the delivery of a small food pellet which the rat quickly consumes. The food reward increases the probability that the rat will press the bar again.

performs a certain act, it is likely to repeat that action. For example, a rat will learn to push a panel to stop an electric shock (Mowrer 1940) or push a bar to turn off a bright electric light for 60 seconds (Keller 1941). We generally think of reinforcers as rewards. Skinner, however, preferred the term *reinforcer* because *reward* implies sensations that might be intuitively inferred but are not measurable. Reinforcement, then, is best defined operationally; it alters the probability of a response.

Shaping

Animals can be taught novel and sometimes complex acts in order to receive a reward. The process by which this occurs, called shaping, has been likened to the way in which a sculptor molds a lump of clay (Skinner 1953). When creating a sculpture, the artist changes the formless mass into a final masterpiece through a series of minute changes in the former condition. At first, any gross approximation of the desired act is rewarded, but later reinforcements require closer and closer approximations to the desired goal.

For example, if you were to train a sea lion to jump from the water through a hoop, you would first reward it for approaching the hoop. When it learned to approach, it would be rewarded only when it swam through the hoop. You would raise the hoop on successive trials until it was clear of the water, and the sea lion would have to jump to receive the reward.

Extinction

When reinforcement is withheld, the response rate will gradually decline, just as the strength of the conditioned reflex decreases when the CS is presented many times without the US. The process is called extinction.

Reinforcement Schedules

In real life, reward seldom follows every performance of an act. Instead the reward is usually intermittent. For example, a honeybee can obtain nectar only during the restricted interval each day in which a particular flower is open. Therefore, it will not always be rewarded when visiting a particular flower. The frequency with which rewards are offered is called the reinforcement schedule. Partial reinforcement schedules vary either the ratio of nonreinforced to reinforced response or the time period between successive reinforcements. Alternatively, rewards may be doled out aperiodically (Ferster and Skinner 1957).

Careful studies have revealed the effect of the schedule of reinforcement on the response strength. Each reinforcement schedule has predictable effects on the rate of response and on how long the animal will continue responding when it is no longer rewarded. We will highlight just a few examples. A continuous reinforcement schedule, in which each occurrence of the behavior is rewarded, is best during the initial training to establish and shape a response. A fixed ratio schedule, one in which the animal must respond a set number of times before reinforcement is given, usually results in very high response rates because the individual, in essence, determines how quickly it will be rewarded. The faster it responds, the sooner it completes the number of responses required to receive the reward. A fixed ratio reinforcement schedule is similar to piecework in factories, in which the employee gets paid when a certain number of items are completed. The very high production rate generated by piecework is a reason that employers like the system and labor unions generally oppose it. In a variable ratio schedule, the number of responses required for reinforcement varies randomly. This also generates very high response rates because the individual is rewarded for fast responses. In addition, the response tends to persist even if the reward is withheld for a while because the variability prevents any obvious response patterns from becoming established. An individual who is gambling with a slot machine experiences a variable ratio reinforcement schedule. The slot machine averages a certain level of payoff, but the reward is unsystematic and variable. This type of reinforcement schedule has been blamed for the persistence of gambling in those who have become addicted to games of chance.

ANIMAL COGNITION AND LEARNING STUDIES

Beginning in the early 1980s, some scientists began to wonder whether animals have mental experiences—thoughts and feelings, for instance (Bateson and Klopfer 1991; Griffin 1981, 1982, 1984, 1991; Hoage and Goldman 1986; Mellgren 1983; Ristau 1991). But how could we ever *know* whether other animals think or whether they are aware? Some scientists believe that learning studies may shed light on the issue of animal cognition, so we will consider several types of learning and the studies that are interpreted by some researchers as indications of the mental activities of animals.

LATENT LEARNING

We know that rewards increase the probability that an act will be repeated. However, according to many learning theorists, there should be no change in the frequency of an unrewarded behavior. There are situations, though, in which animals seem to learn without any obvious reward. For instance, an animal can learn important characteristics of the environment during unrewarded explorations. Even though the knowledge

is not put to immediate use (i.e., it is latent), it may later prove to be lifesaving.

The value of latent learning seems intuitively obvious, and laboratory studies have shown that familiarity with the terrain is an asset for survival (Metzgar 1967). Pairs of white-footed mice (*Peromyscus leucopus*) were released into a room with a screech owl (*Otus asio*). One of the duo had the opportunity to explore the room for a few days before the test. The other mouse had no experience in the room. On 13 of the 17 trials, the owl caught a mouse. Eleven of these were from the group that was unfamiliar with the room, clearly showing that knowledge of the environment increased the ability to evade the predator.

Latent spatial learning may also be important for certain fish. For instance, blackeye gobies (*Coryphopterus nicholsi*) are members of the subtidal community along the West Coast of the United States. When threatened, these fish seek shelter in burrows. Blackeye gobies who are familiar with the location of burrows retreat to safety significantly faster than those without such knowledge (Markel 1994).

We are left wondering not just how latent learning takes place without reinforcement but also what exactly is learned. The animals in these examples didn't learn an appropriate *behavior*. But, they did learn something that allowed them to respond appropriately in a new situation. One explanation is that the animals had formed a mental representation of the spatial relationship of objects in their environment, called a cognitive map. Is this evidence for mental activity in animals? We will return to this question in Chapter 10 and examine the evidence that animals form cognitive maps.

INSIGHT LEARNING

In cartoons a sudden solution to a problem is suggested by a light bulb going on over someone's head. People often express the experience as "Aha, I get it." You may have puzzled over a problem for several days or weeks, and suddenly the answer flashes into your mind. Some learning theorists call this insight learning. It is characterized by its suddenness; it seems to occur too quickly to be the result of a trial-and-error process (Thorpe 1963).

A famous example of insight learning comes from the behavior of Wolfgang Köhler's (1927) chimpanzees, particularly one called Sultan. In one experiment, Sultan first learned to use a stick as a tool to extend his reach and rake in a banana on the ground outside his cage. Having mastered this trick, he was given two sticks that when put together end to end were just long enough to reach the fruit. Sultan tried unsuccessfully to reach the reward with each of the sticks. He even managed to prod one stick with the tip of the other until it touched the banana, but since the sticks were not joined, he could not retrieve the fruit. For over an hour, Sultan persistently tried, and failed, to get the banana. Finally, he seemed to give up and began to play with the sticks. Later an intellectual flash apparently struck Sultan. As the chimp was playing with the sticks (Figure 5.7), he happened to hold one in each hand so the ends were pointed toward one another. At this point he realized that the end of one stick could be fitted into the other, thus lengthening the tool. Immediately, he ran to the bars of his cage and began to rake in the banana. As he was drawing the banana toward him, the two sticks separated. That Sultan quickly recovered the sticks and rejoined them was evidence to Köhler that the chimp understood that fitting two bamboo poles together was an effective way to increase his reach far enough to obtain the fruit. Köhler believed that the chimp displayed insight behavior in that he was able to apply the information gained from the experience of playing with the sticks to solve a problem, getting the bananas.

Köhler's explanation of chimps' problem-solving abilities was that they saw new relationships among events, relationships that were not specifically learned in the past, and they were able to consider the problem as a whole, not just the stimulus–response association between certain elements of the problem. It has been suggested, for instance, that chimps form a mental representation of the problem and then mentally apply trial-and-error patterns to it. An animal could be "thinking" through the possible responses and evaluating the possibility of success of each trial based on its past experience. The solution might seem sudden to

observers because they do not have access to the animal's mental processes.

Other researchers explain sudden problem solving, such as that shown by Köhler's chimps, as the result of associations among previously learned components. It has been argued, for instance, that chimps that moved boxes and then climbed on them to reach a banana had previously acquired two separate behaviors—moving boxes toward targets and climbing on an object to reach another object. This idea was tested on pigeons. Pigeons do not usually do either of these actions, but they can learn to do them. When a similar insight-learning situation was set up for pigeons, it was shown that *only* pigeons that had learned both of these two actions were able to solve the problem of reaching the banana. In one experiment, pigeons were reinforced for pushing a box toward a green spot on the floor but not for pushing a box if there was no green spot on the floor. In a separate experiment the pigeons learned to sit on a box in order to peck at a banana that was suspended overhead. Pigeons trained to do both of these activities were then placed in a room that lacked a green spot on the floor but with both a box and a banana suspended from the ceiling. The behavior of these birds was remarkably similar to that of Köhler's chimps. Although they stretched and turned beneath the banana at first, they suddenly pushed the box under the banana, climbed on it, and pecked the banana. However, when pigeons that were trained to peck a banana but not to climb on a box to do so were placed in the same situation, they repeatedly stretched toward the banana but never reached it. Thus, learning to climb was an important component of the behavior. Other pigeons were trained to push a box in a certain direction for a reward, but they were never trained to climb and peck at a banana. When these birds were then placed in the same situation, they pushed the box around the room aimlessly, never climbing atop it to peck the banana. Therefore, learning to push a box in a certain direction for a reward was also an important component of the response. It was concluded, then, that seemingly insightful behavior might be built from specific stimulus–response relationships (Epstein et al. 1984). So we see that although no one disagrees about the importance of prior experience in insight learning, there is controversy about its role.

Keeping in mind the difficulty of clearly demonstrating insight, let's consider another example of apparently insightful behavior in a bird, this time in the common raven (*Corvus corax*). Bernd Heinrich (1995) presented common ravens with food suspended from string, which was a problem they had never encountered before. To reach the suspended food, a bird had to pull up a loop of string, step on the loop to hold it in place, and then reach down and pull up another loop. The bird had to repeat this cycle six to eight times to

FIGURE 5.7 Sultan playing with sticks. After playing with sticks, Sultan gained insight into how to obtain the banana placed beyond his reach. The sticks could be fitted together end to end to increase his effective reach.

obtain the meat. Since ravens are able to grasp objects with either their bill or their feet, they are physically capable of solving the problem. At least ten species of birds can be taught by operant conditioning to pull up food dangling on a string if the distance between the food and the perch is gradually lengthened. However, a few of the ravens in this study solved the problem without any indication of going through a learning process. In fact, one of the wild birds went through the entire sequence of 30 steps and obtained the food the first time it approached the string, even though no other bird in the group had previously shown the behavior. In nature, ravens don't normally pull on one object to obtain another one. When they pull on food, entrails, for example, they eat the food while pulling. Thus, it is unlikely that this complicated behavior was learned, was genetically programmed, or occurred by random chance. Is it an example of insight?

As we see, the investigation of insight learning has proven very difficult. For one thing, it occurs very quickly and it is impossible to predict when it will happen. Thus, the experimenter can easily miss the moment of enlightenment. In addition, small changes in the arrangement of the problem situation can alter the outcome greatly. But an even greater difficulty is in the interpretation of the observation. All the observer sees is the solution to the problem. It is difficult to know *how* the animal achieved that solution. So, the interpretation of insightful behavior can lead to considerable controversy.

Why is insight learning controversial? As we have seen, some workers believe that insight learning shows that the animal is *thinking*, and an animal that thinks about objects or events can be said to experience a simple level of consciousness (Griffin 1991). An animal that thinks must also form mental representations of objects or events. Therefore, insight has been used as evidence of animal awareness or cognition. But not everyone agrees that animals, or even that some animals, might be aware. Some might be willing to accept the idea of awareness in a chimp but not in a pigeon. However, in the experiments described earlier, if the chimp solved the problem by thinking, weren't the birds also thinking?

RULE AND CONCEPT LEARNING

Sometimes an animal solves a problem in a flash as a result of prior experience with similar tasks. The animal seems to have learned the *rule* or *principle* of that type of problem. This "learning how to learn" has been called a learning set. For example, humans establish learning sets in solving math problems when they have solved other problems of a similar type.

For an illustration of the formation of a learning set, we turn to the experiments of Harry Harlow (1949). Harlow repeatedly presented a monkey with two objects, each covering a small food well. However, the well under only one of these objects contained food. The positions of the objects were alternated randomly, but the monkey quickly learned which object covered the food. Then two new objects were introduced, and once again the animal had to learn which object hid the food. As new pairs of objects were introduced, the monkey was provided with new but similar problems. With each new challenge, it took fewer presentations before the animal reliably chose the correct object. After several hundred object-discrimination problems of this sort, the monkey was able to choose the rewarded object by the second try about 97% of the time. The monkey had adopted the strategy of win–stay, lose–shift: If the object is rewarded, stick with it; but if it is not, choose the other one and stick with *it*. The monkey had formed a learning set.

A variation on this learning-set experiment is a repeated reversal problem. In this case, if the objects to be discriminated were a square and a circle, the square might be rewarded until the animal learned that association, and then the circle would yield the reward. Once the animal learned to choose the circle, the square would be designated correct. The strategy for this type of problem would be win–shift, lose–stay.

The ability to form a learning set would be adaptive. In real life, an animal encounters a multitude of problems that it must learn to solve. Many of them are variations on a theme, just as are the trials in the learning-set experiments just described. A learning set would reduce tremendously the amount of time wasted in solving each similar problem separately.

Some researchers suggest that the ability to form a learning set might be evidence of awareness. According to Harlow (1949), learning-set formation "transforms the organization from a creature that adapts to a changing environment by trial-and-error to one that adapts by seeming hypothesis and insight." Learning sets, he said, "are the mechanisms which, in part, transform the organism from a conditioned response robot to a reasonable rational creature."

Individuals of a few species have been able to learn certain concepts (abstract ideas). Surely this must show that animals form mental representations of objects and events. Pigeons (*Columba livia*) are able to form natural concepts such as "tree" or "water" or "human." They can recognize water, for instance, in various forms—a droplet, a river, a lake (Herrnstein, Loveland, and Cable 1976; Mallot and Siddall 1972; Siegel and Honig 1970). Perhaps more surprising is their ability to understand the ideas of sameness and difference, an ability previously thought to characterize humans and maybe their close primate relatives (Premack 1978). Pigeons can indicate whether an array of 16 icons are all identical or whether some of them are different (Wasserman 1995; Wasserman, Hugart, and Kirkpatrick-Steger 1995).

Pigeons can even indicate the degree of difference, that is, the extent of variability among icons in an array (Young and Wasserman 1997). When humans look at a photograph, they understand that the photo is a representation of the real object. Pigeons may have a similar understanding, because they can also discriminate between a real object and a photograph of that object, but they are still able to combine the object and its photograph into the same category (Watanabe 1997).

A nonhuman animal who has vocally demonstrated the ability to form concepts is Alex, an African grey parrot (*Psittacus erithacus*) (Figure 5.8). We all know that parrots can be trained to talk, but most of us would guess that they are mimicking their trainers. This is certainly not true of Alex. Irene Pepperberg (2000) has detailed more than 20 years of research on Alex in *The Alex Studies*. He has learned labels (names) for over 35 different objects. By combining labels, he can identify, request, refuse, or comment on more than 100 different objects. Furthermore, he has used language to show that he understands certain concepts. One such concept is quantitative. He can tell you how many items are in a group for collections of up to 6 items. The objects don't have to be placed in any particular pattern or be familiar to Alex (Pepperberg 1987b). Even more incredible is Alex's ability to accurately count specific items in what is called a confounded number set, which consists of four groups of items that vary in two characteristics. In other words, the set could consist of two types of objects, say cars and keys, that appear in two colors, red or blue, for instance. Alex can tell you the number of items of a specific type and color, such as the number of blue keys. He responds correctly to these types of questions 83.3% of the time (Pepperberg 1994).

Alex also understands certain abstract concepts—the concepts of same and different. He demonstrated this ability in experiments in which he was shown two objects at a time. The objects would differ in one of three qualities: color, shape, or material. He might be shown a yellow, rawhide pentagon and a gray, wooden pentagon or a green, wooden triangle and a blue, wooden triangle. Then Alex would be asked, "What's same?" or "What's different?" A correct answer to the first question is to name the category of the similar characteristic. When he saw the first of the previous examples, Alex would have to answer "shape," not "pentagon." In the second example, a correct response to "What's different?" would be "color," not "green." When shown objects he had seen before, Alex correctly identified the characteristic that was the same or different 76.6% of the time. He was also shown pairs of objects that he had never seen before, and 85% of the time he correctly identified the characteristic that was the same or different (Pepperberg 1987a). The experiments with Alex suggest that nonprimates can learn concepts and that we should continue our efforts to understand how other species might handle ideas.

SOCIAL LEARNING

Some organisms are able to learn from others. The possibility for learning in this way is much greater in social species because they spend more time close to others. Social learning by observation and imitation is assumed to be the basis of many human behaviors (Bandura 1962; Meltzoff 1988). You can probably think of many examples of this from your own family life. For instance, a child may mimic a parent's technique for

FIGURE 5.8 Alex, an African grey parrot who has learned several concepts. Alex knows the concept of same/different, an idea once thought to characterize only humans and their closest primate relatives.

caring for a younger sibling by handling a doll in the same way.

The adaptive value of social learning is fairly clear. It saves some of the time and energy that might be wasted as an individual learned the business of survival by trial and error. We see this advantage in many situations. Consider, for example, learning what to eat or how to obtain food. Individuals of some species, Norway rats (*Rattus norvegicus*), for instance, learn what and where to eat from others of their kind (Galef 1990). Indeed, when a rat observes another eating an unfamiliar food, the observer is more likely to eat the unfamiliar food than if it observed another rat eating a familiar food. This is beneficial because an unfamiliar food that is eaten by another rat is likely to be nutritious, and so this is a safe way to add breadth to the diet (Galef 1993). As a result, groups of rats will learn to select a nutritionally balanced diet more quickly than do rats that are housed alone (Galef and Wright 1995). Other species may learn where to look for food by observing others. For instance, shoaling, swimming in large groups, is one way that guppies (*Poecilia reticulata*) quickly learn a safe route to a food source (Laland and Williams 1997). Individuals of other species, particularly those whose next meal depends on finding and obtaining food more quickly than competitors, can benefit by watching how successful competitors of the same or different species look for and handle food (Lefebvre et al. 1997). It should not be surprising, then, that the foraging ecology of a species may affect social learning. We see this in different populations of Zenaida doves (*Zenaida aurita*) in Barbados, which live only a few hundred meters apart. It turns out that whereas group-foraging doves learn more quickly from other doves, territorial doves learn more readily from Carib grackles (*Quiscalus lugubris*), the species they most often feed with in mixed groups (Carlier and Lefebvre 1997).

Other animals may learn to avoid dangerous situations by watching their fellows. For example, rhesus monkeys (*Macaca mulatta*) can learn to fear and avoid snakes by watching other monkeys show their fear (Mineka and Cook 1988). Similarly, fathead minnows (*Pimephales promelas*) who have never had experience with northern pike (*Esox lucius*), their natural predator, learn to show fear responses to the odor of northern pike when they are paired with other fathead minnows who have had experience with northern pike, but not when paired with inexperienced minnows. After learning to recognize the pike as a predator, the minnows have a better chance of surviving future encounters, and they are able to transmit the information to naive fathead minnows (Mathis, Chivers, and Smith 1996).

We see, then, that although each member of a population may have the capacity to learn appropriate responses for themselves, it is often more efficient and perhaps less dangerous to learn about the world from others (Galef 1976).

In other instances, interaction with adults is *critical* to learning appropriate behaviors by the young. An example is juvenile Bewick's wrens (*Thryomanes bewickii*), which perfect their crude rendition of the song used for territorial defense by countersinging with a neighboring adult (Kroodsma 1974).

Many observations of birds and mammals in nature seem to involve social learning. Some of the most commonly cited examples involve traditions, learned behaviors that appear in only one of several social groups of a single species in their natural habitat. For instance, a larcenous tradition was begun in England around 1921 when certain birds learned to break into milk bottles to steal the cream, which, in the days before homogenization, floated to the top. This nefarious technique spread throughout the continent as other birds acquired the habit (Fisher and Hinde 1949) (Figure 5.9). There are many examples of habits spreading through groups of primates. Jane van Lawick-Goodall (1968) observed young chimpanzees in nature learning to use sticks and stems to gather termites from their holes after watching their mothers or other adults.

A frequently cited example of social learning is the food-washing habit that spread within a particular group of snow monkeys. As the story goes, a young female snow monkey of Japan, named Imo, developed new techniques for the treatment of sweet potatoes and wheat, food provided by the researchers who study the social behavior of the snow monkeys. First Imo discovered that washing the sweet potato in the sea not only cleaned it but also enhanced the flavor by lightly salting it (Figure 5.10). One of Imo's playmates observed her and followed suit. Then Imo's mother caught on. And so the tradition spread, usually from youngsters to mothers and siblings. When the youngsters became mothers, their offspring imitated the behavior as if food had always been cleaned in this way. Several years later Imo started a new custom. The researchers spread wheat on the sand, from which the snow monkeys had to painstakingly pick each grain. One day Imo tossed a handful of sand and wheat into the sea. The sand sank but the wheat floated so that it could be scooped up from the surface. This trick was also picked up by most monkeys in the troop during the next few years (Kawai 1965; Kawamura 1959; Lefebvre 1995).

Observations such as these are often interpreted as examples of social learning, but before we leap to unwarranted conclusions we should consider what is meant by social learning more carefully. It is generally agreed that an individual can learn through its own experience, that is, through individual learning, or by observing or interacting with another animal or its products, that is, through social learning (Heyes 1994). These types of learning can interact to produce and

FIGURE 5.9   The tradition among birds of opening milk bottles to steal sips of cream spread rapidly from one area in England. This trick may have been spread by social learning.

maintain adaptive behaviors. In other words, a behavior that is initially learned socially can be maintained because the individual is rewarded for performing it (Galef 1995, 1996).

Interaction between individual and social learning has led to disagreements over the mechanism of the development and maintenance of the food-washing behavior in the troop of snow monkeys. The caretakers of the monkeys may have inadvertently rewarded food washing. The monkeys' only source of sweet potatoes is the caretaker. Since the food washing interests researchers and amuses tourists, the caretakers give more sweet potatoes to those members of the troop that were known to wash them than to those that did not. Thus, differential reinforcement may maintain the

FIGURE 5.10   The tradition of washing sweet potatoes in the sea was begun by a young Japanese snow monkey, and it spread rapidly to other members of the troop.

behavior. The habit may have spread because the monkeys near those who washed their sweet potatoes (who, by the way, were likely to be relatives) were also close to the caretaker and the source of the reinforcement (S. Green 1975). Although the habit clearly spread throughout the population, we do not know whether the snow monkeys learned to wash food by imitating others or because they were trained to do so by the caretaker.

Some behaviors are learned socially, but not because the individual observed another performing the activity and then imitated it. In local enhancement, for example, an individual learns a behavior because its attention is directed to a particular place or object by the activity of others (Galef 1988; Giraldeau 1997; Marler 1996). We see this when rats can learn dietary preferences from other rats. They learn what to eat, not by watching others, but by smelling their breath. In one experiment, a "demonstrator" rat ate food flavored with cocoa or cinnamon. The demonstrator was then anesthetized and placed two inches away from the wire cage of an awake "observer" rat. Although the demonstrator slept through the demonstration, the observer later showed a preference for the food the demonstrator had eaten (Galef 1990).

Local enhancement can also explain sophisticated problem solving by organ grinder monkeys (*Cebus apella*). When a monkey was permitted to watch another one use a stick to retrieve food from inside a glass tube, the observer was then able to solve the problem quite quickly. But, the observer did not use the motor pattern that the demonstrator had used to solve the problem, so imitation cannot be the explanation for this example of social learning. Instead, it seems that the observer monkey approached the apparatus because of the presence of the other monkey. Then the observer ran through its existing motor patterns until it solved the problem (Visalberghi and DeLillo 1995).

There are surprisingly few examples of animals that learn to do something just by seeing it done. In one such example, budgerigars learned a specific technique for removing the cover from a food dish. In this experiment, one budgie would watch a demonstrator remove the cover in one of three ways: using its feet, pecking with its bill, or pulling with its bill. When presented with a similar food dish, the observer used the same technique it had just witnessed (Dawson and Foss 1965; Galef, Manzig, and Field 1986).

Another example of social learning, this time by rats, illustrates that even within a single species social learning can take place through different mechanisms. We have already seen that rats learn to eat certain foods by smelling the food on the breath of other rats through local enhancement. More recent studies have shown that rats are also able to learn by imitation. In these studies, a demonstrator rat was trained to push a joystick hanging from the ceiling either to the right or to the left for a food reward. After observing 50 responses by the demonstrator, an observer rat was allowed to push the joystick. Whereas the rats that observed left-pushing demonstrators made 86% of their responses to the left, rats that observed right-pushing demonstrators made 71% of their responses to the right (Heyes and Dawson 1990). Further studies showed that observers learned to push the joystick in a particular direction relative to their own bodies, even if that meant moving the joystick toward a different section of the cage than was demonstrated (Heyes, Dawson, and Nokes 1992). However, when rats observed the joystick moving by itself to the left or right without another rat demonstrating, the observers didn't show a left or right response preference (cited in Heyes 1993).

The experimental design in this study of rats moving joysticks is a good one for demonstrating imitation for several reasons. It involves two responses of equal difficulty, novelty, and probability of being shown spontaneously. In tests, the observers are expected to use the variant that their demonstrators used. More recently, this design has been used to show imitation in pigeons (Zentall, Sutton, and Shelburne 1996), Japanese quail (*Coturnix japonica*) (Akins and Zentall 1996), chimpanzees (*Pan troglodytes*) (Whiten et al. 1996), and perhaps in marmosets (*Callithrix jacchus*) (Bugnyar and Huber 1997).

MEMORY OF PAST EXPERIENCES

Do you remember your first kiss or learning to drive? Do you remember what happened, where, and when? These features are key to the memory of past experience, called episodic memory, which has been consid-

ered to be uniquely human. However, recent experiments of food-caching scrub jays suggest that these birds can remember the what, where, and when of past events. Scrub jays prefer to eat wax worms (wax-moth larvae) more than peanuts. However, wax worms are perishable, and peanuts are not. The scrub jays learned to avoid recovering wax worms when 124 hours had elapsed since the worms were stored because they had become unpalatable. The jays were trained to cache peanuts on one side of a sand-filled storage tray and wax worms on the other side. Then the jays were allowed two caching sessions 120 hours apart. During

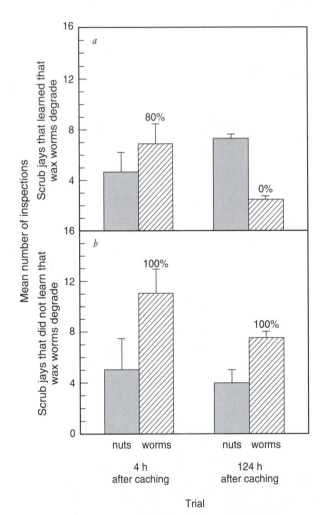

FIGURE 5.11   The cache recovery pattern of scrub jays is consistent with the hypothesis that they recall the what, where, and when of past experiences. (*a*) Scrub jays that previously learned that their favorite food, wax worms, will degrade and become unpalatable after 124 hours will search preferentially for wax worms if only 4 hours have passed since caching, but they spend more time searching for peanuts if 124 hours have elapsed. (*b*) Scrub jays that did not learn that wax worms degrade with time continue to search preferentially for wax worms, even after 124 hours. (Data from Clayton and Dickinson 1998.)

each session they could cache only one food item, either peanuts or wax worms. One group of jays stored peanuts during the first session, and the other group stored wax worms first. Before the recovery phase, when the jays searched for cached food, all food items were removed and fresh sand was placed in the tray. Then, 4 hours after the second caching session, the jays were allowed to search either side of the caching tray for food. The jays that had cached the wax worms during the second caching session searched preferentially for wax worms. After this time interval, the wax worms would still be fresh. However, the jays that had cached wax worms during the first caching session (124 hours before the recovery session) inspected the peanut side of the tray more frequently than the wax worm side (Figure 5.11a).

The scrub jays in a similar series of tests did not learn that wax worms degrade over time because in the initial training trials, fresh wax worms were placed in the tray before recovery. These scrub jays had two caching sessions, as in the previous tests. During the recovery phase, they searched for their favored food, wax worms, even if 124 hours had elapsed since caching (Figure 5.11b). This cache recovery pattern suggests that the scrub jays remembered which food item was cached in which part of the tray and when. Thus, the recovery pattern seems to fulfill the what, where, and when criteria of episodic memory (Clayton and Dickinson 1998).

Keep in mind the question of animal cognition as you become more familiar with other aspects of animal behavior. We will return to this issue again, especially in the discussion of animal communication.

SUMMARY

Learning is adaptive and is shaped by natural selection. Corvid species that are the most dependent on food caches for survival have the best spatial memory.

Learning is a change in behavior as a result of experience.

Habituation.   The animal learns *not* to respond to a specific stimulus because it has been encountered frequently without important consequences. The loss of responsiveness can be distinguished from sensory adaptation and muscular fatigue. Habituation is adaptive because it conserves energy and leaves more time for other important activities.

Associative learning occurs when an animal learns to associate a stimulus with a response.

(a) Classical Conditioning.   The animal learns to give a response normally elicited by one stimulus (the unconditioned stimulus) to a new stimulus (the conditioned stimulus) because the two are repeatedly paired. The conditioned stimulus (CS) must precede the unconditioned stimulus (US). If the CS is presented many times without the US, the response to the new stimulus will be gradually lost. This is called extinction.

(b) Operant Conditioning.   The frequency of some action is increased because it is reinforced. Novel behaviors can be introduced into the repertoire through shaping. During shaping, the reward is made contingent upon closer and closer approximation to the desired action. Sometimes not every response is reinforced. The frequency with which the reward is doled out is called the reinforcement schedule.

Many scientists have begun to wonder whether animals have mental experiences. Some studies of learning can be interpreted as supporting the idea that animals form mental representations of external objects or events and that they can form concepts.

Latent Learning.   Latent learning occurs without any obvious reinforcement and is not obvious until sometime later in life. The information gained through exploration is an example.

Insight Learning.   This type of learning occurs rapidly and without any obvious trial-and-error responses. The animal seems to draw on information gained in previous similar situations to arrive at a solution to the problem.

Rule and Concept Learning.   During the formation of a learning set, the animal is learning how to learn. The subject learns to solve problems more quickly because it has an idea of the principle of the problem as a result of experience with other similar tasks.

Social Learning.   The animal learns from others. Social learning may occur by watching the behavior of another, but it may also occur by simpler means. Some customs spread rapidly throughout populations of animals by social learning, but some traditions may arise through individual learning.

Memory of Past Experiences.   Scrub jays seem to demonstrate a type of memory previously thought to be uniquely human. They remember past experiences. Their cache recovery patterns suggest that they remember what type of food is cached, where it is, and how much time has passed since it was cached.

6

Physiological Analysis—Nerve Cells and Behavior

Concepts from Cellular Neurobiology

Types of Neurons and Their Jobs

The Message of a Neuron

Ions, Membrane Permeability, and Behavior

Behavioral Change and Synaptic Transmission

The Structure of the Synapse

Aplysia Learning and Synaptic Change

Neuromodulators and Leech Feeding Behavior

Specializations for Perception of Biologically Relevant Stimuli—Sensory Processing

Stimulus Filtering in the Little Skate

Selective Processing of Visual Information in the Common Toad

Processing of Sensory Information for Sound Localization

Predators and Prey: The Neuroethology of Life-and-Death Struggles

Responding—Motor Systems

Neural Control in Motor Systems

Locust Flight

When you stop to think about it, it is amazing that animals exist at all. The survival of each depends on its ability to solve a myriad of problems. To name just a few of the challenges they face, they must find food, avoid predators, find a mate, and orient in and move through the environment. And, their behavior must be modified by experience in adaptive ways. Although bats and moths, barn owls and toads, fish and locusts face similar challenges, they have found varied solutions because their worlds can be quite different from one another. Through natural selection, nervous systems have been shaped by the demands and constraints imposed by the habitat in which they live and their role in it (Wagner and Luksch 1998).

Neurons and neural circuits underlie all behavior, but adaptive behavior depends on interactions among the components of the nervous system, the body, and the environment. Sensory receptors must detect critical stimuli, and sensory input must be filtered to extract the most biologically relevant information. A toad must recognize that a worm is food but a twig is not, for instance. In the real world, critical stimuli are often masked in a "noisy" environment. A little skate, for example, must filter out the electric fields generated by its own breathing movements and yet remain sensitive to the weak electric fields produced by the movements of its prey. Sensory input must then be processed to produce adaptive responses. A moth's simple nervous system must process information from a hunting bat's echolocation calls to effectively avoid predation. The nervous system of a barn owl processes auditory information so that the sounds of a scurrying mouse can be precisely located and a direct strike executed in complete darkness. The movements involved in an animal's response are also sculpted by interactions among nerve cells. One of the most interesting questions in the field of motor control is how rhythmic motor patterns, such as locust flight, are generated by a group of neurons called a central pattern generator.

These are just a few of the issues we will address in this chapter. Although we will focus on mechanism here, keep in mind that mechanism cannot be considered apart from evolution. Nervous systems are the product of evolution, and they in turn affect the direction of evolution. Furthermore, an understanding of the mechanisms underlying behavior can help us answer some of the questions about the evolution of behavior that will be the focus of later chapters in this book. For example, what cues are used to assess a rival or a mate and how are kin recognized?

Before we can understand the selective behavioral responses of animals, we must first understand how the nervous system is put together. Our examples should demonstrate that nervous systems have evolved so that animals can quickly (1) detect pertinent events in their environment, (2) choose appropriate responses to such events, and (3) coordinate the parts of their bodies necessary to execute the responses.

CONCEPTS FROM CELLULAR NEUROBIOLOGY

TYPES OF NEURONS AND THEIR JOBS

Just as the house cat raises its paw to strike, the cockroach (*Periplaneta americana*) dashes across the floor and disappears into a tiny crevice. If we were to film this sequence and then replay it in slow motion, we would see that the cat's paw was still several centimeters away from the cockroach when the intended victim turned its body away from the cat and ran.

How did the cockroach detect the predator in time to take evasive action? Kenneth Roeder, and later Jeffrey Camhi and his colleagues, studied the escape response of cockroaches and discovered that these remarkable, albeit unlovable, house guests respond to gusts of air that are created by even the slightest movements of their enemies (Camhi 1984, 1988; Camhi, Tom, and Volman 1978; Roeder 1967). Cockroaches, it turns out, have numerous hairlike receptors that are sensitive to wind, and these receptors are located on two posterior appendages called cerci (singular, cercus; Figure 6.1). When these wind-sensitive receptors are stimulated, they alert the nervous system of the cockroach, and within a matter of milliseconds the cockroach turns away from the direction of the wind and starts to run.

The escape of the cockroach was orchestrated by

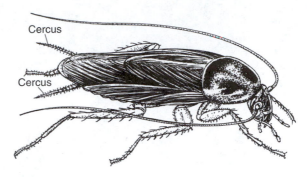

FIGURE 6.1 A cockroach, showing the two cerci, each with approximately 200 wind-sensitive hairs.

the interactions among nerve cells, which are also called neurons. They can be classified on the basis of their function into three groups. Neurons that carry signals from a receptor organ at the periphery toward the central nervous system (in vertebrates, the brain and spinal cord, and in invertebrates, the brain and nerve cord) are called sensory or afferent neurons. Those that carry signals away from the central nervous system to muscles and glands are called efferent or motor neurons. Interneurons, found within the central nervous system, connect neurons to each other. Interactions among interneurons process sensory input and determine the motor response.

Look at these three types of neurons in the cockroach as we consider the role each plays in the escape response (Figure 6.2). At the base of each of the wind-sensitive hairs on the cercus is a single sensory neuron that relays pertinent information from the external environment into the central nervous system. In the central nervous system, the sensory neuron makes contact with an interneuron; in this case, the interneuron is described as a giant because of its exceptionally large diameter. This giant interneuron ascends the nerve cord to the head. Before reaching the head, however, the giant interneuron makes contact with an interneuron in the thoracic area, which in turn connects with motor neurons that relay messages to the hind leg muscles (Schaefer, Kondagunta, and Ritzman 1994). (An advantage of studying the neural basis of behavior in an invertebrate animal such as the cockroach is that it is sometimes possible to identify the individual neurons involved in a specific behavior, particularly a pattern of behavior associated with escape. Because escape requires fast action, the neurons involved in escape responses are often large in diameter to permit the rapid conduction of messages. The end result is that these large neurons are somewhat easier to identify than their smaller counterparts.)

How does the cockroach determine the direction of a wind gust, so that it can run away from cats rather than straight into them? Most of the segments on each

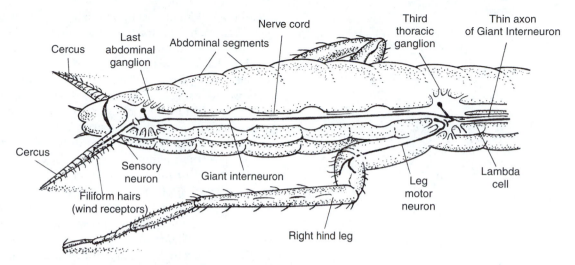

FIGURE 6.2 Some of the cells in the underlying neural circuitry of the cockroach escape response. Note that the sensory neuron, also called the wind-receptor neuron, that is leaving the cercus makes contact with a giant interneuron in the central nervous system, which in turn makes contact with another interneuron that synapses with a motor neuron in the leg. (Modified from Camhi 1980 with new information from Ritzmann and Pollack 1986)

cercus have a row of sensory hairs that can be deflected slightly by wind or touch. Each wind-sensitive hair responds differently to a gust of wind from a particular direction. Thus, the pattern of output from the sensory hairs encodes information about the direction of the wind. This information is sent to the giant interneurons. Seven giant interneurons run along each side of the cockroach's ventral nerve cord. The pattern of output from the sensory hairs caused by a wind gust will stimulate the giant interneurons on one side of the cockroach more than those on the other side. A running cockroach will turn away from the side on which the giant interneurons are most highly stimulated, and so it can scurry away from a predator or a well-aimed shoe (Liebenthal, Uhlmann, and Camhi 1994).

Although there is no such thing as a typical neuron, it is possible to identify characteristics common to some neurons. We will use a motor neuron (Figure 6.3), in this case from a mammal, as our example. The nucleus of a motor neuron is contained in the cell body (soma), from which small-diameter processes (neurites) typically extend. In the traditional view, information enters a neuron via a collection of branching neurites and then travels down a single, long neurite to be passed on to other neurons. The neurites that receive the information are called dendrites; the single, long, cablelike neurite that transmits the information to other neurons is called an axon. In vertebrates, some axons have a fatty wrapping called the myelin sheath. In our example of the motor neuron, the axon ends on a muscle or a gland (an effector), which brings about the animal's behavioral response. Although the terms dendrite and axon are well established in the literature, it is now recog-

nized that the flow of information through a neuron is often not so neatly divided into separate receiving and transmitting processes. We will continue to use the terms, keeping in mind that in many cases the specific direction of the informational flow has not actually been demonstrated.

The myelin sheath is formed by the plasma membranes of glial cells, which are supporting cells in the nervous system that become wrapped around the axon many times. Since a single glial cell encloses only a small region, about 1 mm, of an axon, the myelin sheath is not continuous. The regions along an axon between adjacent glial cells are exposed to the extracellular environment. This arrangement is important to the speed at which the nerve cell conducts messages. The message "jumps" successively from one exposed region to the next, increasing the rate of transmission as much as 100 times. For this reason, axons that conduct signals over long distances are usually myelinated.

THE MESSAGE OF A NEURON

Let us now delve more deeply into the details of the how and why of ion movements that are responsible for a neuron's message, called an action potential. An action potential is an electrochemical signal caused by electrically charged atoms, called ions, moving across the membrane. Ions can cross the membrane of a nerve cell by means of either the sodium-potassium pump or ion channels. The pump uses cellular energy to move three sodium ions (Na^+) outward while transporting two potassium ions (K^+) inward. An ion channel, on the other hand, is a small pore that extends through the

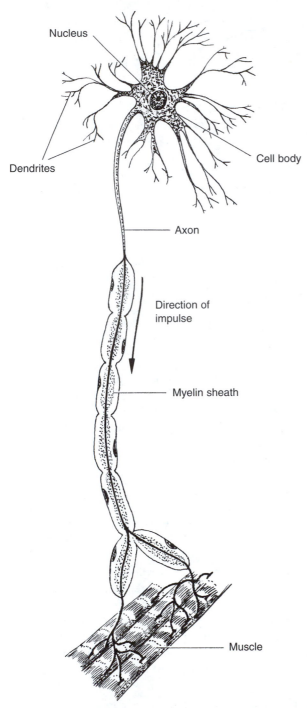

FIGURE 6.3 A motor neuron. The soma or cell body maintains the cell. The dendrites are extensions specialized for receiving input from other cells. The axon is specialized to conduct the message away from the cell and to release a chemical that will communicate with another cell. In vertebrates, some axons are covered by a fatty myelin sheath.

membrane of a nerve cell. There are different types of channels, each type forming a specific passageway for only one or a few kinds of ions. Whereas some are passive ion channels that are always open, others are active ion channels (also called gated channels) that open in response to a specific triggering signal. Triggering signals may include the presence of chemicals (neurotransmitters) in the space between the membranes of neurons, changes in the charge difference across the membrane (the membrane potential), changes in the concentration of intracellular calcium ions (Ca^{++}), or any combination of these factors (see later discussion). Once an ion channel is open, the ion it admits may move across the membrane in response to either a concentration gradient (ions tend to move from an area where they are highly concentrated toward an area of lesser concentration) or an electrical gradient (because ions are charged atoms, they tend to move away from an area with a similar charge and toward an area of the opposite charge).

The Resting Potential

In a resting nerve cell, one that is not relaying a message, the area just inside the membrane is about 60 millivolts (mV) more negative than the fluid immediately outside the membrane. This charge difference across the membrane is called the resting potential of the neuron. A membrane in this resting state is described as polarized (Figure 6.4).

The resting potential results from the unequal distribution of certain ions across the membrane. The concentration of Na^+ is much greater outside the neuron than within. The concentration of K^+ shows just the opposite pattern, being greater inside the cell than outside. Certain large, negatively charged proteins are held within the neuron either because the membrane is impermeable to them or because they are bound to intracellular structures. These proteins are primarily responsible for the negative charge within the neuron.

Most of the active ion channels in the membrane of a resting neuron are closed, but passive channels are, of course, open. Because most of the passive channels are specific for K^+, the membrane is much more permeable to K^+ than it is to other ions. Drawn by the negative charge within, positively charged K^+ will enter the neuron and accumulate there. At some point, when there is roughly 20 to 30 times more K^+ inside than outside, the concentration gradient counteracts the electrical gradient. At some point, the two forces—an electrical gradient that draws K^+ inward and a concentration gradient that pushes K^+ outward—are equally balanced and there is no further net movement of K^+. At this point, the cell has reached its resting potential.

Why is Na^+ more concentrated outside the neuron? Although Na^+, like K^+, is attracted by the negative

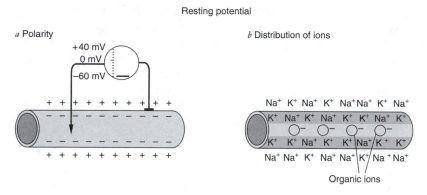

FIGURE 6.4 **The resting potential. (*a*) In the resting state, the inside of an axon is more negative than the outside. (*b*) This charge difference is caused by the unequal distribution of ions inside and outside the cell. There are more sodium ions outside and more potassium ions inside. In addition, there are large, negatively charged proteins (organic ions) held inside the cell, giving the interior an overall negative charge.**

charge inside the neuron, the membrane is relatively impermeable to it, and so only a few Na^+ can leak through. Furthermore, the sodium-potassium pump actively removes Na^+ from within the cell, transporting it outward against electrical and concentration gradients.

The Action Potential

The nerve impulse (action potential) is an electrical event that lasts for about 1 millisecond. The action potential consists of a wave of depolarization followed by repolarization that spreads along the axon. The depolarization, or loss of the negative charge within, is caused by the inward movement of Na^+. However, the repolarization, or restoration of the negative charge within the neuron, is caused by K^+ leaving the cell (Figure 6.5).

Let's see how depolarization and repolarization occur. The membrane becomes slightly depolarized when some of the active Na^+ channels open and Na^+ enters the cell, drawn by both electrical and concentration gradients. The positive charge on Na^+ slightly offsets the negative charge inside the cell, and the membrane becomes slightly depolarized. If the depolarization is great enough, that is, if threshold is reached, voltage-sensitive sodium channels open and Na^+ ions rush to the interior of the cell. At roughly the peak of the depolar-

ization, about 0.5 milliseconds after the voltage-sensitive sodium gates open, they close and cannot reopen again for a few milliseconds. Almost simultaneously, voltage-sensitive potassium channels open, greatly increasing the membrane's permeability to K^+. Potassium ions then leave the cell, driven by the temporary positive charge within and by the concentration gradient. The exodus of K^+ restores the negative charge to the inner boundary of the membrane. In fact, enough potassium ions may leave to temporarily make the cell's interior even more negatively charged than usual, a condition called hyperpolarization. Notice that although the original resting potential is eventually restored, the distribution of ions is different. This situation is corrected by the sodium-potassium pump.

This depolarization and repolarization of the neuronal membrane spreads rapidly along the axon, generated at each spot in the same manner in which it was started. The local depolarization at one point of the membrane opens the voltage-sensitive sodium channels in the neighboring region of membrane, thereby triggering its depolarization. The net result is that a wave of excitation travels down the axon as each membrane section passes on its excitation, in the form of increased sodium permeability, to its neighbor.

Immediately after an action potential, the neuron cannot be stimulated to fire again for 0.5–2 millisec-

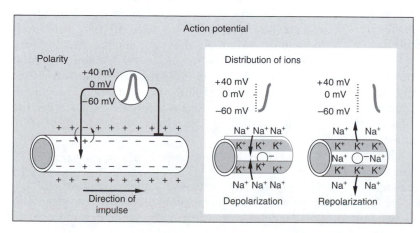

FIGURE 6.5 **The action potential of a neuron. Depolarization is caused by sodium ions entering the cell and repolarization is caused by potassium ions exiting the cell.**

onds because the sodium channels cannot be reopened right away. At the start of this absolute refractory period, no amount of stimulation can generate an action potential. During the latter part of this interval, the relative refractory period, stimulation must be greater than the usual threshold value to generate an action potential. Although the refractory period is brief, it is biologically significant because it determines the maximum rate of firing.

The action potential is a neuron's long-distance signal. Once it is triggered, it travels along the axon with no loss in magnitude. Therefore, we describe the action potential as an all-or-none phenomenon.

If an action potential is always the same, how can differences in the intensity of stimuli be sensed? The intensity is encoded in the firing rate of the neuron and by the number of neurons responding. For example, if we place our hand on a hot stove rather than on a warm one, the firing rates of neurons in our hand that respond to heat or pain may be increased. Also, touching a very hot stove will activate more neurons than will touching a warm stovetop because the thresholds of neurons that register heat vary—more neurons reach their thresholds and fire at higher temperatures.

IONS, MEMBRANE PERMEABILITY, AND BEHAVIOR

Although ions and their movements through the channels of nerve cell membranes may seem, at best, to be only remotely related to an animal's behavior, we will see that this is not the case. Let us consider how ions and changes in membrane permeability relate directly to what we see an animal doing. Here we describe an example of how changes in the membrane permeability caused by mutation in the *Shaker* gene of the fruit fly (*Drosophila melanogaster*) produce atypical behavior.

When fruit flies are anesthetized with ether, one occasionally sees a mutant fly that shakes its legs, wings, or abdomen. Among the mutations that result in shaking under ether anesthesia are *Shaker*, *Hyperkinetic*, and *Ether-a-go-go*. All this shaking is apparently a result of neurons with mutations that make them exceptionally excitable. The most is known about what makes the *Shaker* mutants so jittery, so we will concentrate on them.

It was first shown that the *Shaker* larvae were jittery because an excessive amount of neurotransmitter, a chemical released by a neuron that allows communication with another neuron or a muscle cell (discussed shortly), was released at the junction between a motor neuron and a muscle cell, causing extreme muscle contractions. Furthermore, the muscle contractions were uncoordinated because the release of the transmitter from different neurons was asynchronous (Jan, Jan, and Dennis 1977). A neuron's message is an electrochemical signal consisting of a wave of depolarization caused by sodium ions (Na^+) entering the nerve cell followed by a wave of repolarization caused by potassium ions (K^+) leaving the nerve cell. By recording from the neurons of adult *Shaker* flies, it was shown that the mutant neurons do not repolarize as quickly as normal neurons (Tanouye, Ferrus, and Fujita 1981). As a result, an excess of calcium ions enters the neuron and causes the release of neurotransmitter to continue longer than is typical.

Why doesn't the cell repolarize normally? Apparently, the *Shaker* gene codes for a protein that forms part of the potassium channels involved in repolarization; a mutation at the *Shaker* locus results in certain potassium channels not being formed, and this disrupts the process of repolarization (Kaplan and Trout 1969; Molina, Castellano, and López-Barneo 1997). That the behavioral defect, that is, shaking, results from the absence of K^+ channels was elegantly shown by experiments in which a functional *Shaker* gene was inserted into mutant flies. This experimental procedure resulted in a normal flow of potassium across the membranes of nerve cells and an end to the jittery behavior caused by the mutations (Zagotta et al. 1989).

BEHAVIORAL CHANGE AND SYNAPTIC TRANSMISSION

Synapses are important structures because they are decision and integration points within the nervous system. The molecular events that occur at synapses determine whether the message of one neuron will generate an action potential in the next neuron. Typically, a neuron receives input through thousands of synapses. That input is integrated in ways that make complex behaviors possible. Changes in the functioning of synapses or in the number of synaptic connections often explain why behavior can change as a result of experience or maturation. As we will see shortly, the gill-withdrawal reflex of the sea hare *Aplysia* can be modified through various forms of learning, each involving changes in the ways in which neurons communicate with one another at synapses. We will also learn how the changes in leech locomotor behavior that accompany feeding are brought about by the modification of synaptic function caused by chemicals called neuromodulators. But before exploring the cellular mechanisms for behavioral change, we should become more familiar with the structure of synapses.

THE STRUCTURE OF THE SYNAPSE

The gap between neurons is called a synapse, and at a specific synapse information is typically transferred in one direction, from the presynaptic neuron to the post-

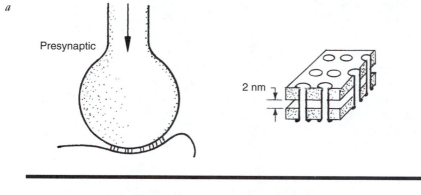

a

Presynaptic

2 nm

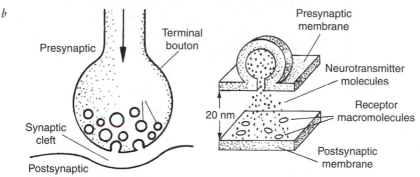

b

Presynaptic

Terminal
bouton

Presynaptic
membrane

Neurotransmitter
molecules

Receptor
macromolecules

20 nm

Synaptic
cleft

Postsynaptic

Postsynaptic
membrane

FIGURE 6.6 **The gap between neurons is called a synapse. There are two types of synapses, electrical (*a*) and chemical (*b*). In an electrical synapse, the distance between neurons is very small, and ions can flow directly from one neuron to the next. In a chemical synapse, however, the distance between neurons is greater, and one neuron affects the activity of the other by releasing chemicals into the gap between them. The original distribution of ions is later restored by the sodium-potassium pump.**

synaptic neuron. (The descriptor *presynaptic* or *postsynaptic* refers to the direction of information flow at a specific synapse.) There are two major categories of synapses, electrical and chemical (Figure 6.6).

Electrical Synapses

We will discuss electrical synapses first because they are much less common and the behavioral examples that follow pertain to chemical synapses. Electrical synapses are known for their speed of transmission. Whereas a signal may cross an electrical synapse in about 0.1 milliseconds, durations on the order of 0.5 or 1 millisecond are typical at chemical synapses. Not surprisingly, then, electrical synapses are often part of the neural circuitry that underlies patterns of behavior when sheer speed is essential, such as the escape responses exhibited by animals confronted with a predator.

In an electrical synapse, the gap between the neurons is small, only about 2 nanometers (1 nm = 10^{-9} m), and is bridged by tiny connecting tubes that allow ions to flow directly from one neuron to the other. When an action potential arrives at the axon terminal of the presynaptic neuron, Na^+ enters this terminal, causing a potential difference between the inside of this cell and the postsynaptic cell. As a result of the difference, positively charged ions, mostly K^+, move from the presynaptic cell through the tiny tubular connections into the postsynaptic neuron. These newly arriving ions may sufficiently depolarize the postsynaptic cell to induce an action potential.

Chemical Synapses

Chemical synapses are characterized by a larger space between the membranes of the two neurons (typically 20–30 nm) than is found in electrical synapses. Rather than information being transmitted from one neuron to the next via direct electrical connections, it is transmitted across a chemical synapse in the form of a chemical called a neurotransmitter. There are several steps in this process, and these steps account for the slower speed of transmission at a chemical synapse than at an electrical one. First, the action potential travels down the axon to small swellings (terminal boutons) at the end of the axon. Second, at the terminal boutons, the action potential causes the neuron to release a neurotransmitter from small storage sacs called synaptic vesicles. Third, the release of neurotransmitter occurs because the action potential opens voltage-sensitive calcium ion channels. Fourth, the calcium ions that flood to the inside initiate events that cause the synaptic vesicles to move toward the membrane of the terminal bouton. Fifth, once there, the vesicles fuse with the membrane and dump their contents into the gap between the cells (the synaptic cleft). Finally, the neurotransmitter then diffuses the short distance across the cleft and binds to receptors on either another neuron or a muscle cell.

When the neurotransmitter binds to a receptor on the postsynaptic cell, it either directly opens ion gates or indirectly affects ion gates through biochemical mechanisms. In either case, the neurotransmitter will increase the permeability of the membrane of the post-synaptic cell to specific ions. This results in either excitation or inhibition of the postsynaptic neuron, depending on the particular ions involved. In the case of excitation, the neurotransmitter causes the opening of channels that allow both Na^+ and K^+ to pass through. Although some K^+ ions move out, they are outnumbered by the Na^+ ions moving in. This causes a slight temporary depolarization of the membrane of the postsynaptic cell. This depolarization is called an excitatory postsynaptic potential, or EPSP (Figure 6.7). If this depolarization reaches a certain point, the threshold, an action potential is generated in the postsynaptic cell.

On the other hand, a neurotransmitter may act in an inhibitory fashion, making it less likely that an action potential will be generated in the postsynaptic neuron. When the neurotransmitter binds to the receptors in an inhibitory synapse, either K^+ channels or K^+/Cl^- channels open. The exit of positively charged potassium ions or the influx of negatively charged chloride ions makes the inside of the postsynaptic membrane even more negative than usual. In other words, an inhibitory neurotransmitter momentarily hyperpolarizes the membrane. The hyperpolarization is called an inhibitory postsynaptic potential, or IPSP (Figure 6.8).

As long as the neurotransmitter remains in the synapse, it will continue to excite or inhibit the postsynaptic cell. There are, however, a variety of ways in which the effect of a neurotransmitter can be halted. In some cases, the neurotransmitter is broken down by an enzyme and its component parts are absorbed for resynthesis. In other instances, however, the molecules of neurotransmitter are released intact after acting on

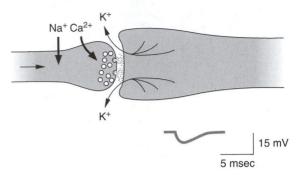

FIGURE 6.8   The events at an inhibitory synapse. The neurotransmitter binds to a receptor and causes the opening of channels that allow potassium ions to leave the postsynaptic cell, and thus it becomes hyperpolarized. This hyperpolarization is called an IPSP.

the postsynaptic cell, absorbed by the presynaptic cell, and repackaged for subsequent release.

Integration

Most neurons receive input from many other neurons. In fact, a given neuron may communicate with hundreds or thousands of other cells. The slight depolarization (EPSPs) and hyperpolarization (IPSPs) that result from input from all the synapses are summed on the postsynaptic membrane, either because one neuron sends a repeated signal or because many neurons send messages to one postsynaptic cell. In other words, the EPSPs and IPSPs combine with one another as they arrive at the soma. If these interacting changes in membrane potential combine to produce a large enough depolarization, voltage-sensitive Na^+ gates are opened and an action potential is triggered.

APLYSIA LEARNING AND SYNAPTIC CHANGE

Now that we are familiar with the molecular events that occur at synapses, we will apply the information to help us understand how behavior can change with experience, as occurs when the sea hare *Aplysia* learns. While *Aplysia* moves across the ocean bottom eating seaweed, its siphon is extended and its gills, the respiratory organs, are spread out on the dorsal side. The gills are partly covered by a protective sheet called the mantle shelf, which terminates in the siphon, a fleshy spout through which *Aplysia* can squirt out excess seawater and wastes. When the siphon is touched, the gills, the siphon, and the mantle shelf withdraw into the mantle cavity. This defensive response, called the gill-withdrawal reflex, protects the gills from predators (Figure 6.9).

The gill-withdrawal response can be modified by experience, that is, through learning. One form of

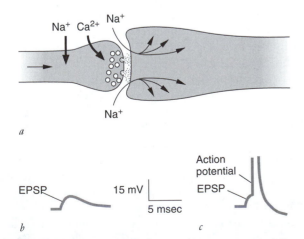

FIGURE 6.7   The events at an excitatory synapse. The neurotransmitter binds to a receptor and causes the opening of channels that allow sodium ions to enter the cell (*a*), thereby slightly depolarizing the postsynaptic cell. This slight depolarization is called an EPSP (*b*). If the depolarization reaches the threshold, an action potential is generated (*c*).

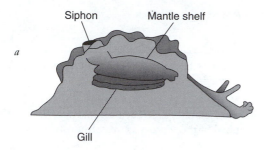

FIGURE 6.9 **The gill-withdrawal reflex in the sea hare,** *Aplysia.* **The gills, mantle shelf, and siphon are drawn here as if the animal were transparent. Normally, the gills are spread out and are only partially protected by the mantle shelf. The siphon, through which water is drawn in over the gills and excess water is expelled, is extended so just the tip is visible when the animal is seen from the side (*a*). If the siphon is touched, the gills, mantle shelf, and siphon are withdrawn into the mantle cavity (*b*). The gill-withdrawal reflex can be modified by learning.**

learning, called habituation, occurs when an animal learns not to respond to a repeated stimulus that proves to be harmless. In other words, *Aplysia* learns to ignore an irrelevant stimulus. Habituation is adaptive because it saves energy. We can demonstrate habituation by disturbing an *Aplysia*'s siphon many times, by touch or by a brief jet of seawater, for instance. After 15 such stimuli administered 10 minutes apart, the reflexive responses are only half of their initial value. In sensitization, a second form of learning in *Aplysia*, the withdrawal reflex becomes stronger when a stimulus that elicits withdrawal is preceded by a strong, noxious stimulus, such as an electric shock, administered almost anywhere on *Aplysia*'s surface. Depending on the number and strength of the noxious stimuli, sensitization can last for seconds to days (Carew, Pinsker, and Kandel 1972; Kandel 1979b; Pinsker et al. 1970).

Through years of intensive research, Eric Kandel and other investigators have determined the neural circuits and many of the molecular mechanisms that underlie these forms of learning in *Aplysia* (Kandel 1976, 1979a). *Aplysia* is an ideal organism for a neurobiologist because it has just 20,000 neurons and the neurons are large—10 to 50 times larger than neurons in a mammalian brain. As a result, neurobiologists know many of these nerve cells by name. A diagram of the neural circuitry for the gill-withdrawal reflex is shown

in Figure 6.10. Although there are actually 24 sensory neurons serving the siphon skin that terminate on 6 motor neurons to the muscle for gill retraction, this simplified diagram shows only one of each type of neuron. You can see that the sensory neuron from the siphon skin synapses directly on a motor neuron for gill withdrawal.

Beginning in the late 1960s, neurobiologists began to work out the details of the neural mechanism of habituation. (The adaptive value of habituation is discussed in Chapter 5.) One guess was that the sensory neuron was giving a weaker response to repeated stimuli. This idea was proven wrong by inserting a microelectrode into a sensory neuron to measure its electrical responses to stimulation. After repeated stimulation, the sensory neuron still responds normally. But, it fails to excite the motor neuron as it initially did. Could habituation be due to a decrease in the motor neuron's responsiveness? No. If a motor neuron is repeatedly stimulated directly with an electrical current, it remains fully responsive. Another guess was muscle fatigue, but this, too, was ruled out. Even after habituation, direct stimulation of the motor neuron causes full contraction of the gill muscle (Kupfermann et al. 1970).

It turns out that habituation occurs because the sensory neuron releases less neurotransmitter as a result of repeated stimulation (Figure 6.11). This, in turn, results in fewer action potentials in the motor neuron for gill withdrawal. Recall that many molecules of neurotransmitter are released from a synaptic vesicle. This multimolecular packet is called a quantum. The change in the EPSP caused by the release of neurotransmitter from one synaptic vesicle is called a quantum of response. During habituation, the EPSP decreases with repeated stimulation of the sensory neuron, and it does so in integral multiples of a quantum of response

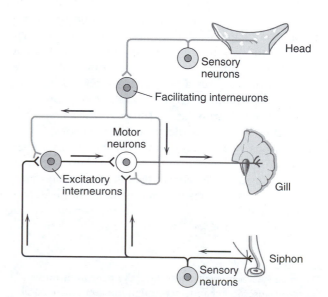

FIGURE 6.10 **Neural circuitry for habituation and sensitization of the gill-withdrawal reflex in *Aplysia*.**

FIGURE 6.11 Changes in synaptic functioning that accompany habituation and sensitization.

Habituation	Sensitization
Repeated stimulation of the sensory neuron from siphon skin	Strong, noxious stimulation to skin stimulates a sensory neuron from skin
↓	↓
Less effective calcium channels in axon terminal of sensory neuron	Stimulates a facilitating interneuron that synapses on the axon terminal of the sensory neuron from siphon skin
↓	↓
Decreased Ca^{++} inflow	Release of serotonin from facilitating neuron
↓	↓
Decreased neurotransmitter released by sensory neuron	Increased cAMP levels within axon terminal of sensory neuron from siphon skin
↓	↓
Decreased firing rate of gill motor neuron	cAMP causes closing of a certain type of K^+ channel and K^+ increases in cell, prolonging depolarization of sensory neuron
	↓
	More Ca^{++} than usual entering the axon terminal
	↓
	Increased release of neurotransmitter by sensory neuron
	↓
	Increased firing rate of gill motor neuron

(Castellucci and Kandel 1974). In other words, during habituation, successive action potentials cause increasingly fewer synaptic vesicles to fuse with the membrane and release their contents into the synaptic cleft. The reason is that calcium channels become less effective because of repeated stimulation, and so they allow less Ca^{++} into the axon terminal (Byrne 1987; Hochner et al. 1986). Calcium ions are needed for the synaptic vesicles to fuse with the presynaptic membrane.

Sensitization also involves changes in the functioning of synapses, but in this case the amount of neurotransmitter released by the sensory neuron from the siphon onto the gill motor neuron is *increased*, thereby increasing the motor neuron's rate of firing. Sensitization requires a facilitating interneuron. The process begins when strong stimulation to the body surface of *Aplysia* stimulates a sensory neuron, which in turn stimulates a facilitating interneuron. These interneurons release serotonin onto the axon terminal of the sensory neuron from the siphon skin. Serotonin increases intracellular concentration of a second messenger, in this case cyclic adenosine monophosphate (cAMP), which causes the closing of a certain type of K^+ channel of the cell membrane.* This keeps K^+

inside the cell and keeps the sensory neuron depolarized longer than normal, allowing additional Ca^{++} to enter the cell. The elevated Ca^{++} levels cause more neurotransmitter to be released (Kandel and Schwartz 1982).

Although short-term memories involve strengthening synaptic connections, long-term memories involve the growth of new synapses between cells. Growing new synaptic connections requires protein synthesis, and this means that the critical genes must be turned on. In Chapter 3 we discussed how experience can turn on specific genes that lead to song learning in zebra finches, as well as the molecular events that turn on the genes that cause structural and functional changes in the neurons responsible for olfactory learning in *Drosophila*. We noted at the time that similar events accompany long-term learning in *Aplysia*. You can look back to the previous discussions for details, but the gist of the story is this: (1) an experience causes a nerve cell to fire, (2) Ca^{++} enters the axon terminal of that nerve, (3) the elevated Ca^{++} leads to the formation of cAMP, which (4) activates certain enzymes (protein kinases) that (5) activate a protein called CREB. (6) The activated CREB then leads to protein synthesis by immediate-early genes (IEGs). (7) The proteins produced by IEGs turn on late effector genes (LEGs). (7) The protein products of LEGs are necessary for the growth of new synaptic connections. These products might include cell adhesion molecules (CAMs) and growth factors. For instance, CAMs are known to play

---

*At some synapses, such as this one, the neurotransmitter opens or closes ion channels through indirect biochemical mechanisms. These indirect mechanisms are similar to those used by neuromodulators, as discussed shortly.

an important role in forming connections between neurons during development. Changes in *CAM* gene expression also accompany learning in *Aplysia*. Although the role of the resulting CAM proteins isn't certain, the proteins do help new synaptic connections to form by controlling cell adhesion, cytoskeletal structure, and signaling within the cell (Fields and Itoh 1996). Recent experiments have shown that growth factor TGF-β, another chemical known to be important in the early development of the nervous system, is also necessary for long-term memory formation in *Aplysia* (Zhang et al. 1997).

NEUROMODULATORS AND LEECH FEEDING BEHAVIOR

Learning isn't the only change in behavior that can be brought about by modifications in the way synapses function. For example, the behavior of the European leech (*Hirudo medicinalis*) depends on its nutritional status. The leech feeds on the blood of other animals, mammals in particular. Before describing some of the neurophysiological aspects of feeding, let us first review the patterns of behavior involved (Figure 6.12). A hungry leech typically rests quietly at the water's edge, with both its anterior and posterior suckers attached to a surface. If aroused by shadows or ripples in the water, the leech releases its anterior sucker, then the posterior sucker, and swims in an undulating fashion in the direction of the stimulus. Once the leech reaches the target, it explores the surface of the animal, moving in an inchworm-like manner. When a warm region is encountered, the leech bites with its three jaws, which are lined with approximately 70 pairs of sharp teeth. If blood flows from the wound, ingestion begins. Rhythmic contractions of the muscular pharynx pump blood into the crop for the next half hour or so, and then the leech, enormously distended, detaches from its host. (Leeches often ingest from seven to nine times their initial weight and then stop feeding when they sense the extreme distention of their body wall.) Following one of these massive blood meals, the leech crawls into a rock crevice to begin digestion, a process that may take up to one year to complete. Leeches in this satiated state behave very differently from their hungry counterparts, tending to crawl rather than to swim, and when on a warm surface to lift their heads rapidly and repeatedly rather than to bite. Sometime in the weeks following a blood meal, the leech will reproduce.

Many aspects of the leech's feeding behavior are influenced by the neuromodulator serotonin (Lent, Dickinson, and Marshall 1989). Neuromodulators cause voltage changes that occur over a slower time course than those caused by "traditional" neurotransmitters—seconds, minutes, hours, and perhaps even days. The fast changes are brought about by traditional neurotransmitters, those chemicals, previously dis-

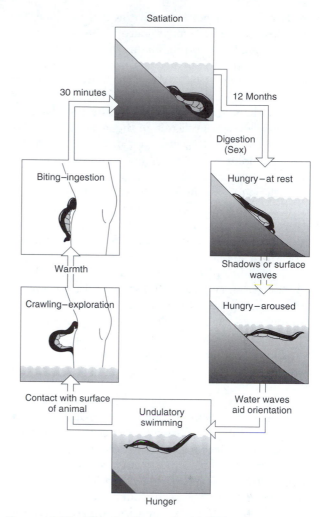

FIGURE 6.12   The feeding cycle of the leech. (Modified from Lent, Dickinson, and Marshall 1989.)

cussed, that open ion gates, causing EPSPs or IPSPs. In contrast, neuromodulators alter neuronal activity slowly, by biochemical means.

The effects of neuromodulators appear to be mediated by substances within the postsynaptic neuron called second messengers (Gillette et al. 1989). These second messengers (e.g., calcium and the cyclic nucleotides cAMP and cGMP*) couple the membrane receptors of the postsynaptic cell with the movements of ions through one or more enzymatic steps. Neuromodulators may, for example, upon reaching the receptor on the postsynaptic neuron, trigger the formation of the second messenger cAMP within the neuron, which in turn activates an enzyme that changes the shape of proteins in certain ion channels. Once the ion channels have been altered in this manner, the permeability of the membrane to specific ions is also changed, thereby affecting the activity of the neuron. It is the relatively

---

*cyclic guanosine monophosphate

slow pace of the enzymatic activities that produces the typically prolonged effects of neuromodulators.

Functionally, neuromodulators appear to be intermediate to classic neurotransmitters and hormones (Lent and Watson 1989). Whereas neurotransmitters are released at specific synaptic clefts and hormones are broadcast throughout the body via the bloodstream, neuromodulators are released in the general vicinity of their target tissue. It is, however, difficult to establish the precise point at which a neurotransmitter becomes a neuromodulator, and a neuromodulator a hormone. In fact, the same chemical may have different functions in different places. Some chemicals, dopamine, for instance, act as neurotransmitters at some synapses and as neuromodulators at others. Likewise, a chemical may act locally in the nervous system as a neurotransmitter, whereas in other places in the body it is released into the bloodstream and acts at a distant site as a hormone does (see Chapter 7 for a discussion of hormones). Despite some fuzziness in definition, there is no question that neuromodulators, through their actions on neurons, glands, and muscles, can produce profound effects on behavior. Consider, for example, the case of neuromodulation of feeding behavior in the leech.

Using a variety of techniques, Charles Lent and his colleagues (1989) demonstrated that serotonin modulates the physiology and patterns of behavior associated with feeding in leeches. Here we discuss a subset of their experiments—the experiments are meant not only to demonstrate how a particular neuromodulator influences behavior but also to introduce some techniques that are commonly used when investigating nervous systems.

The first step in determining whether or not serotonin was involved in the feeding behavior of leeches entailed an examination of its effects on intact animals. Bathing leeches in serotonin (a method that does not cause injury to the leech and is therefore preferable to injecting serotonin) produced profound effects on feeding behavior. A serotonin bath reduced the time taken by leeches to initiate swimming toward potential prey, increased the frequency of biting by 40% and pharyngeal contractions by 25%, and increased the volume of blood ingested. Exposure to serotonin also increased both heart rate and mucus secretion, the first response being necessary to cope with the increased metabolic demands of swimming, and the second to aid in adhering to mammalian blood "donors." In addition, bathing in serotonin caused satiated leeches to bite (recall that satiated leeches typically do not bite but instead lift their heads rapidly and repeatedly when on a warm surface). Clearly, then, serotonin influences the feeding behavior and physiology of *H. medicinalis*.

In a second series of experiments, microelectrodes were inserted into identified neurons—neurons known to contain serotonin—to artificially cause these neurons to fire. A microelectrode consists of a fine wire inside a glass capillary tube that ends in a fine point so that it can pierce a neuron. The capillary tube is filled with a conducting solution that creates an electrical connection between the electrode and the neuron. Microelectrodes can be used to record electrical activity from neurons or to stimulate neurons. Frequently, researchers pair recording and stimulating electrodes; for example, action potentials can be stimulated and then recorded at a specific neuron (Figure 6.13). Alternatively, by stimulating one neuron while recording from a second, the functional relationship between the two neurons can be identified. In this particular experiment, electrical current was passed through a microelectrode, causing a specific serotonin neuron to become depolarized (i.e., to fire), and then the researchers examined whether such firing evoked the physiological responses associated with feeding. The physiological responses were monitored by recording with a second microelectrode the membrane responses of the effector cells (salivary and muscle cells). Stimulation of the serotonin-containing neurons induced salivation, peristaltic (wavelike) movements of the pharynx, and twitches in jaw muscles. Thus, impulse activity of identified serotonin-containing neurons was sufficient for the major physiological components of feeding. It is important to note that because the membrane responses of the salivary and muscle cells occurred five to ten seconds after the serotonin neurons were stimulated and then continued beyond the period of stimulation, Lent and his coworkers (1989) inferred from this delay that serotonin was released some distance from the effector cells and not at discrete synapses with these cells. In other words, serotonin was acting as a neuromodulator and not as a neurotransmitter (remember that neurotransmitters are released at a specific synapse and produce their effects in a matter of milliseconds).

In a third group of experiments, the researchers showed that serotonin neurons are absolutely essential for feeding behavior and its physiological components.

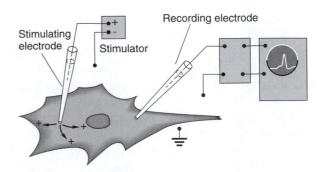

FIGURE 6.13   Microelectrodes can be used to stimulate and record from neurons and thereby assess neuronal function.

In this case, they examined the effects of selectively removing the serotonin neurons from the underlying neural circuits. Hungry leeches were treated with 5,7-DHT, a neurotoxin that specifically destroys serotonin within neurons without destroying the neurons themselves (unlike many neurotoxins that actually destroy the nerve cells, a technique called lesion or ablation). When tested on a warm surface, these hungry, toxin-treated leeches did not bite and instead lifted their heads and behaved as though they had just finished a blood meal. The leeches also failed to show any of the physiological responses associated with feeding. Removal of serotonin from leeches thus eliminated feeding behavior and its physiological components. However, when serotonin was subsequently administered to these animals, biting, salivation, and pharyngeal peristalsis returned.

A final experiment revealed the interaction between feeding behavior and the neuromodulator serotonin. In this particular experiment, Lent and his coworkers (1989) demonstrated that not only is the leech's feeding behavior driven by serotonin but also that the expression of feeding behavior (i.e., ingestion) decreases the amount of serotonin in the central nervous system. Hungry leeches were set up in pairs, and one member of each pair was fed to satiation. The amount of serotonin in the central nervous system of each member of the pair was then measured. Leeches that had been allowed to feed had less serotonin in their central nervous system than did hungry leeches. From these results we can see that serotonin drives feeding, and feeding in turn reduces the amount of serotonin in the central nervous system.

These experiments established the pivotal role of the neuromodulator serotonin in activating leech feeding. Not all of serotonin's effects on behavior, however, are activational. Serotonin inhibits the leech's sexual behavior, a response that makes sense given that reproductive activity would most likely reduce the feeding efficiency of these remarkable gluttons (Leake 1986). The modulatory effects of serotonin on feeding are summarized in Figure 6.14.

SPECIALIZATIONS FOR PERCEPTION OF BIOLOGICALLY RELEVANT STIMULI—SENSORY PROCESSING

In spite of the wide variety of stimuli bombarding an animal from its environment, it is able to detect only a limited range, and of those that it detects, it may ignore all but a few key stimuli.

In some cases, the receptors themselves are "tuned" so that they detect only biologically relevant stimuli.

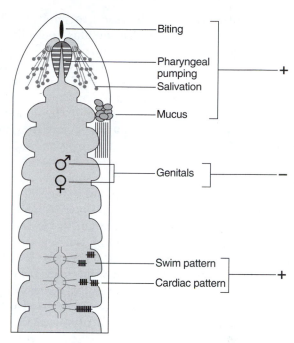

FIGURE 6.14   A summary of the effects of the neuromodulator serotonin on the behavior and physiology of the leech. (Modified from Lent, Dickinson, and Marshall 1989.)

For example, the auditory receptors of the caterpillar *Barathra* respond best to 150 Hz sound, the major frequency of the wingbeat of the parasitic wasp, *Dolichovespula*, which lays its eggs in the body of this caterpillar (Tautz 1979). Frequencies outside the 150 Hz range are most likely associated with events of less significance to the caterpillar than the arrival of the wasp. The wasp's buzzing wings can be detected within the distance of a meter. Low-intensity sounds of this frequency, which indicate that the wasp is still at a distance, cause the caterpillar to freeze in place, making it less noticeable to the wasp. However, as the wasp approaches, the buzz of its wings gets louder. High-intensity sounds of this frequency cause the caterpillar to drop from the branch on which it had been resting (Tautz and Markl 1978).

Another example of an animal's sensory receptors being most sensitive to the most biologically relevant stimuli is the relationship between the sensitivity of photoreceptors and the dominant wavelengths of light in the habitat of certain teleost fish. The wavelengths of light that actually reach the eye of a fish will depend on many factors. The color of the water is one such factor that varies among habitats. When the sensitivity of the eyes of fish from different habitats is compared, we see that the sensitivity to different wavelengths (colors) of light has been adjusted by natural selection so that peak sensitivity occurs at the most common wavelength of light in a given habitat (Bowmaker et al. 1994; Lythgoe et al. 1994).

Besides detecting stimuli critical to survival, animals must be able to pick out that stimulus from a background of "noise" in the same sensory modality. For example, recall that the cockroach runs when exposed to a gust of wind from a predator. Yet, it ignores nonthreatening or irrelevant sources of wind, such as the wind that it creates itself while walking. How is such selectivity possible? In the laboratory, cockroaches run when exposed to wind with a peak velocity of only 12 mm per second, the approximate velocity of wind created by the lunge of a predator such as a toad. Cockroaches, however, do not respond to the 100 mm-per-second wind that they create by normal walking. How is it that cockroaches manage not to respond to the relatively strong wind created by walking and yet run when exposed to much softer wind signals? It turns out that it is not the velocity of the wind that is the critical factor but the acceleration (rate of change of wind speed), and a strike by a predator delivers wind with greater acceleration than the wind produced by the stepping legs of a cockroach. In fact, when cockroaches were tested with wind puffs that had the same peak velocity but differed in acceleration, they ran more frequently when exposed to winds of higher acceleration (Plummer and Camhi 1981). Winds with low acceleration typically produced no response. Thus, cockroaches appear to pay particular attention to the acceleration of the wind stimulus, and this allows them to ignore irrelevant wind signals and to focus on important information in their environment.

We can see, then, that the job of an animal's sensory system is not to transmit all available information but rather to be selective and provide only information that is vital to the animal's lifestyle or, more to the point, information that influences its reproductive success. How does an animal's nervous system enable it to focus on important events and to ignore irrelevant ones? We will consider the selective filtering action of nervous systems in the little skate and in the common toad.

STIMULUS FILTERING IN THE LITTLE SKATE

So far in our coverage of sensory systems, we have focused on two systems that are quite familiar to us—hearing and vision. Some animals, however, gather and process information from the environment by using sensory systems that, from our perspective, can only be described as remarkable. The little skate (*Raja erinacea*) is one such animal, having receptors and brain structures that are specialized for detecting and analyzing weak electric fields. These fields are so weak that we, as humans, are totally unable to detect them. Little skates, however, derive a wealth of information from these fields; particularly critical is information relating to the whereabouts of their next meal. This specialized sensory ability of skates is called electroreception (Bullock and Szabo 1986).

The ability to detect weak electrical fields is not unique to little skates, or even to other elasmobranch fishes (sharks, skates, and rays). Indeed, electroreception seems to have evolved several times in a number of families of fish and amphibians (although its occurrence is still fairly rare), and it has been reported in at least one species of mammal, the duck-billed platypus (Scheich et al. 1986). What do these animals have in common? Electroreception, as you might guess from our list of animals, is associated with aquatic habitats because water, unlike air, serves as a good conductor of electricity.

There are two general categories of electroreception, active and passive. Some animals, such as the electric fish of Africa (Mormyriformes) and South America (Gymnotiformes), use specialized electric organs to generate an electrical field around their body—this is active electroreception (discussed in Chapter 10). Electric fish have two different types of receptor organs, called ampullary and tuberous receptors. Little skates, however, do not generate their own electrical fields by means of special electric organs. Instead, skates respond to incidental electrical fields in their environment such as those fields produced by prey or the earth's magnetic field—this is passive electroreception. Animals with this system have only one type of receptor, the ampullary organ. In the case of elasmobranchs, ampullary organs are called ampullae of Lorenzini, after their seventeenth-century discoverer.

The ampullae of Lorenzini of the little skate are distributed in clusters over the head and expanded pectoral fins (Figure 6.15). Eighty percent of the receptors are located on the ventral body surface and are especially dense in the area of the snout, just in front of the mouth. Skates feed on bottom-dwelling invertebrates, and thus it makes sense that receptor organs are concentrated in areas where they permit accurate detection and localization of prey. Now let's consider the structure of ampullary organs in order to understand how they function.

Ampullary organs consist of a canal that opens by way of a pore on the skin surface. (Note that within a cluster of ampullary organs, the canals radiate in all directions; see Figure 6.15.) Each canal runs some distance beneath the skin and then terminates in a bulb-shaped structure, called the ampulla, which contains numerous pouches, each with hundreds of receptor cells. The canal is filled with a jellylike substance that is a good conductor of electricity; the walls of the canal are nonconductive. The jelly thus serves as a connection between the receptor cells at the base of the canal and the skin surface. The receptors can detect a difference in electrical potential between the water at the

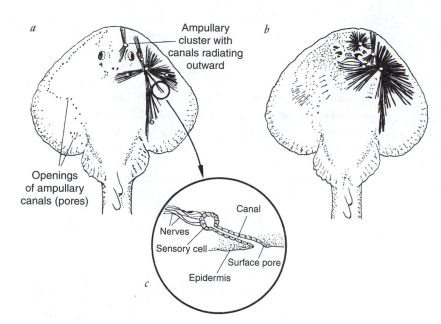

FIGURE 6.15   Distribution of canals and ampullary clusters in the thornback ray, a relative of the little skate. (*a*) Dorsal surface. (*b*) Ventral surface. (*c*) A single ampullary organ. Elasmobranch fish such as skates and rays use electroreception to locate prey—note the concentration of receptors on the ventral surface, especially around the mouth. (*a* and *b*: Redrawn from Montgomery 1984.)

skin pore and the interior of the body, at the ampullae. This property of ampullary organs permits the detection of electrical fields that are basically changes in electrical potential in space. Signals from the receptor cells are passed on to the anterior lateral line nerve, which projects to the brain, specifically to an area of the medulla called the dorsal octavolateralis nucleus (DON). From there, messages are sent via ascending efferent neurons to other areas of the brain.

One problem faced by little skates, as well as by other animals with similar electrosensory systems, is how to deal with self-generated "noise." The skate's normal respiratory movements do, in fact, generate weak electrical fields. These self-generated fields stimulate the skate's own electroreceptors, and in so doing they could interfere with the detection of extrinsic electrical signals, such as those produced by prey.

How is the nervous system of the little skate designed so that it filters out irrelevant information pertaining to the animal's own respiratory movements? Apparently, within the central nervous system of the skate, there is a neural mechanism that, in effect, cancels out signals generated by respiratory movements from opposite sides of the body (New and Bodznick 1990). How does the cancellation occur? First, all receptors, regardless of their location on the body surface, respond in a similar fashion to the electrical fields generated by respiratory movements; thus, the signals coming from receptors on each side of the body are equal. Second, as a result of the synaptic connections within the central nervous system, the signals from one side of the body are excitatory, whereas those from the other side of the body are inhibitory. As a result of this arrangement, the equal and opposite signals essentially cancel each other out. It is only when an electrical field

stimulates a localized area of electroreceptors on the skate's body, such as when a small crustacean edges along the ocean bottom, that the skate pays attention to the signal. The cancellation occurs in the DON because interference from respiratory movements is substantial in the anterior lateral line nerve fibers that lead from the electroreceptors into the DON and is negligible in the fibers that leave the DON. By suppressing the electrosensory signals caused by respiratory movements, the neural mechanism in the DON allows the skate to focus on biologically relevant electrical stimuli, such as those associated with invertebrate prey.

SELECTIVE PROCESSING OF VISUAL INFORMATION IN THE COMMON TOAD

The nervous systems of amphibians, like that of the little skate, must filter out irrelevant information from the environment. In addition to this general filtering action, many nervous systems have cells that are specialized to detect specific, relevant features of certain stimuli. Frogs and toads, for example, recognize prey (e.g., earthworms, slugs, and beetles) and predators (e.g., large wading birds such as herons in the case of frogs, and hedgehogs in the case of toads) by sight, using what may seem to us to be rather simple criteria to make this important distinction (see the following discussion). Note that this distinction is made by specialized cells in the brain.

Over the past 20 or so years, Jörg-Peter Ewert and his coworkers have examined how common toads (*Bufo bufo*) use visual information to recognize prey. Their approach combines behavioral observations with investigations of the underlying neural circuitry, and the results provide a remarkable example of how an animal's sensory system is made up of specialized features that promote the detection of critical stimuli (Ewert 1987; Ewert et al. 1990; Wachowitz and Ewert 1996).

When a toad is sitting in a garden and a slug moves into its visual field, the toad turns toward the prey, fixes it in its sight, leans forward, flips out its sticky tongue, captures the prey, and draws it back into its mouth—where it is swallowed with an eye-closing gulp. All this, of course, is followed by a ceremonial wipe of the mouth (Figure 6.16a). Part of this behavioral sequence—the initial orientation toward prey—was used by Ewert (1987) to determine exactly what stimulus parameters toads use to recognize prey. To get at this problem, he placed a hungry toad in a glass container and moved cardboard models that emphasized various stimulus characteristics horizontally around the toad (Figure 6.16b). If, on the one hand, the toad interpreted the model as prey, it would orient toward the model, the frequency of turns increasing as the resemblance between the model and prey increased. If, on the other hand, the toad interpreted the model as predator, it might freeze or crouch. As it turns out, toads distinguish between predator and prey simply on the relationship between two characteristics—the length of the object in relation to its direction of movement. Whereas a small horizontal stripe (called the "worm configuration" by Ewert) is interpreted as prey, a vertical stripe (the "antiworm configuration") is interpreted as predator, or at least as something inedible. To understand how such a distinction is made by toads, we need to consider how visual information is processed, first in the retina and then in the brain.

Processing of Visual Information in the Retina

Considerable processing of information occurs in the retina of the common toad. From our description of the visual system of vertebrates, you will recall that light energy hits the retina, which contains a layer of photoreceptors. Receptor potentials are then relayed to bipolar cells, which in turn send their messages to ganglion cells. The axons of the ganglion cells collectively form the optic nerve, which carries impulses to the brain (in the case of the toad, to the optic tectum and thalamus). How does the retina process visual information before passing it to the brain?

The ganglion cells in the toad's retina participate in one level of stimulus filtering. Such filtering can be examined by recording with microelectrodes the responses of individual ganglion cells to various stimuli moved in front of a stationary toad. Each ganglion cell receives input from several bipolar cells, and thus each ganglion cell receives information from a large number of receptors. The region of the retina from which

FIGURE 6.16   (a) The behavioral sequence of prey catching in the common toad. (b) Analysis of prey recognition in the toad. Here, the toad is confined to a glass container and a prey model is moved horizontally around it. If the toad interprets the model as prey, it orients to the model. (a: Modified from Ewert 1983, based on descriptions in Ewert 1967. b: Modified from Ewert 1969.)

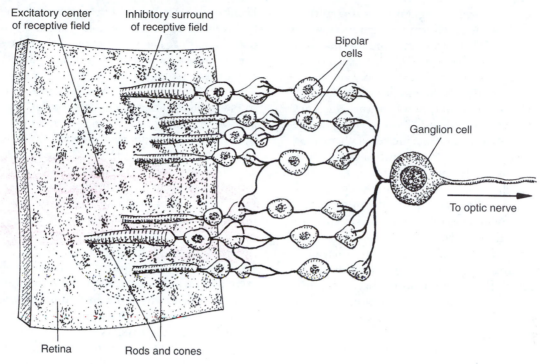

FIGURE 6.17   **The receptive field of a ganglion cell.**

receptors send messages to bipolar cells and then on to a specific ganglion cell is called the receptive field of that ganglion cell (Figure 6.17). Receptive fields in the retina of the toad are roughly circular in shape and have an outer inhibitory area that surrounds a central excitatory region. If a small object, let's say a slug, moves in front of a toad, the small image created on the retina will cover groups of receptors, some of which may make up the excitatory region of a ganglion cell's receptive field. In this instance, then, the ganglion cells may fire and send impulses to the brain. If, however, the toad encounters a large stationary object, perhaps a garden wall, the image on the retina may cover both the inhibitory and excitatory regions of many ganglion cells, and the ganglion cells probably will not fire. (When an inhibitory influence of a stimulus offsets an excitatory stimulus in a nearby region, the phenomenon is called lateral inhibition.) Because of their special properties, then, ganglion cells detect movement and pass this information on to the brain, where it will be used to determine whether such movement is associated with prey or predator.

Not all ganglion cells of the common toad transmit exactly the same type of information. Indeed, one of the discoveries from the recording experiments is that in *B. bufo*, there are several different classes of ganglion cells. These classes differ in the size of the centers of their receptive fields and in their responses to movement and contrast. Nevertheless, the important point here is that the ganglion cells filter the information provided to them by the bipolar cells, selecting only the most important aspects (e.g., information that pertains

to movement, contrast, and changes in illumination) and passing this information on to the brain for further processing. It is in the brain, and not in the retina, that the actual distinction between prey and predator is made.

Processing of Visual Information in the Brain

From the ganglion cells the toad's brain receives information on the basic parameters of the visual stimulus (e.g., dimensions, and direction and rate of movement). In the brain, specifically in the optic tectum, there is a class of neurons—called Class T5(2) neurons—that responds only to certain combinations of features, those combinations that normally would identify a prey item. How were these feature-detecting neurons identified?

In a manner similar to that described for ganglion cells, microelectrodes can be used to monitor the responses of individual neurons in the brain of a stationary toad presented with various moving stimuli. Such recording experiments have shown that neurons in the optic tectum and thalamus can also be categorized into different classes, classes that are based, in part, on characteristics of their receptive fields. Because neurons in the brain receive information from a cluster of ganglion cells, each neuron in the brain also has a receptive field—that is, an area of the retina monitored by all of the ganglion cells connecting to that T5(2) neuron. Among the neurons that have been studied so far, the responses of the T5(2) neurons of the optic tectum show the best correlation with stimuli associated with prey recognition. Specifically, they show a greater response when the worm configuration is presented to the toad than when the antiworm configuration is presented. Thus, the T5(2) neurons respond specifically to combinations of visual characteristics that usually allow a toad to correctly distinguish between prey and predator or, more generally, between prey and inedible objects. We should point out, however, that the T5(2) neurons do not precisely identify worms, slugs, or beetles; rather they respond to a specific set of characteristics that normally serves to distinguish such delicacies from other objects in the environment.

Of course, the neural circuitry for prey-catching behavior in the toad does not end in the optic tectum. As is true for the neural circuitry that underlies patterns of behavior in all animals, the sensory system must interact with the motor system to generate behavior. In toads, how is information about prey processed after the optic tectum? Where do neurons that carry information on the visual features of prey eventually interact with neurons involved in motor output? Although the answers to these questions are not entirely clear at present, it is known that the axons of some of the neurons in the optic tectum project to the medulla, an area of the brain near the spinal cord. By recording from neurons in the medulla, Ewert and his colleagues have found some neurons that are responsive to visual stimuli associated with prey (Ewert et al. 1990; Schwippert, Beneke, and Ewert 1990). Thus, this region of the brain may be an important premotor- or motor-processing station for information pertaining to prey. We see, then, that the neural base for prey recognition in the common toad is actually a multilevel operation with processing of information occurring in the retina, optic tectum, and medulla.

PROCESSING OF SENSORY INFORMATION FOR SOUND LOCALIZATION

It is often important for an organism to locate the source of sound. For example, the sound may be produced by a potential mate. A male mosquito finds a female by the sound of her beating wings. It would do a male little good to know that a female was present and be unable to locate her. Likewise, many predators determine their prey's position by localizing sounds generated by the prey. Locating the source of a sound has obvious importance to prey animals as well—the crunching sound of brush under a leopard's foot has fixed its position for many a wary baboon.

What properties of sound enable its source to be located? Actually, part of the answer is remarkably simple: Sounds can be localized by how loud they are, that is, by their intensity. A simple rule might be that sound seems louder when the receptor is closer to the sound.

If only one ear is involved in locating the sound source, however, the rule may not hold (Camhi 1984). Let's say that the left ear hears a soft sound; was the sound soft because it was produced by a weak source on the left side or by a strong source on the right side? To eliminate such confusion, both ears must be used in the sound localization process—this is called binaural comparison. Binaural comparison of sound intensity is used by some animals to locate the source of sound.

Timing is also important in locating the source of a sound. There are two differences in timing that could be of potential use, and both rely on binaural comparison. The first occurs at the onset of sound—sound begins and ends sooner in the ear that is closest to the source. The second difference in timing occurs during an ongoing sound. During a continuing sound, there are differences in the phase (the point in the wave of compression or rarefaction) of the sound wave reaching each ear. The extent of the phase difference will depend both on the wavelength of the sound and on the distance between the ears. When the wavelength of the sound is twice the width of the head, the peak of a sound wave arrives at one ear and the trough arrives at the other. Thus, this ratio between the wavelength of the sound and the width of the head is most useful for localizing the sound. In contrast, when the wavelength of the sound equals the head width, the phase of the sound wave is the same in each ear (Figure 6.18).

PREDATORS AND PREY: THE NEUROETHOLOGY OF LIFE-AND-DEATH STRUGGLES

Now let's consider how nervous systems gather and process information about the source of sounds to produce adaptive behaviors—escape behavior by prey and prey localization by a predator. We will first consider

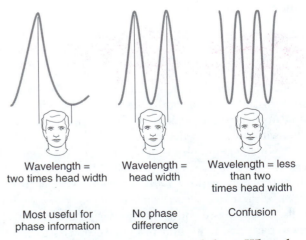

Wavelength = two times head width

Wavelength = head width

Wavelength = less than two times head width

Most useful for phase information

No phase difference

Confusion

FIGURE 6.18   Binaural comparison of phase. When the sound is prolonged, differences in the phase of the sound wave at each ear may indicate the direction of the source. The usefulness of this cue depends on the wavelength and the distance between the ears.

how sound information is processed by the relatively simple nervous system of a night-flying (noctuid) moth, allowing it to escape from an echolocating bat. Then we will consider how a barn owl obtains and processes sound information to locate its prey.

Escape Responses of Noctuid Moths

Noctuid moths are a favorite prey of certain bats. Indeed, moths typically make up more than half of a bat's diet. The bats capture their prey on the wing, locating these flying insects by echolocation, that is, by emitting high-frequency sounds that bounce back to the bat from any structure in the environment. Bat echolocation is discussed in Chapter 10, so here we will focus on how moths escape predation.

Kenneth Roeder (1967) has provided a fascinating account of how the relatively simple auditory apparatus of the moth is used to detect an approaching bat and how the moth then takes evasive action. When the bat's ultrasonic echolocation pulses are soft, indicating that the bat is still at a distance, the moth turns and flies directly away. However, loud ultrasonic pulses mean that the bat is very close, and emergency actions are needed—erratic, unpredictable looping and wingfolding to produce a free fall. Moths that hear a bat's approach and take evasive action are about 40% less likely to be eaten.

Roeder found that these moths have two ears, one on either side of the thorax (the insect's midsection), and each ear has only two auditory receptor cells (Roeder and Treat 1957). The receptors are tuned to the frequencies of the echolocation calls of species of bats living in their vicinity, which is generally between 20 and 50 kHz. One, called the A1 cell, is about 10 times more sensitive than the other cell. The A1 cell begins to respond when the sound is soft, indicating that the bat is still at a distance. The sensitivity of this cell is important because it will determine how much time the moth will have to take appropriate evasive action. The other, the A2 cell, responds only to loud sounds (Pérez and Coro 1984; Roeder and Payne 1966), as would come from a nearby bat. Based on these differences in threshold, Roeder suggested that the A1 cell functions as an "early warning" cell, and the A2 cell as an "emergency" neuron.

The very sensitive A1 neuron responds to bat sounds long before the bat can detect the moth (Roeder and Payne 1966). North American moths can detect a hunting big brown bat (*Eptesicus fuscus*) from a distance of nearly 100 feet, whereas the bat must be within about 15 feet to detect a moth-size target (Fenton 1992). The A1 cell, then, warns the moth that there is a hunting bat in the vicinity in much the same way that your car's radar detector alerts you of a police radar trap.

The moth can determine not only the distance of the bat by the intensity of its sound but also whether it is coming nearer. One clue is that the sound of an approaching bat grows louder. However, a more important clue is provided by the type of echolocation sounds the bat produces because these change during the hunt (discussed in Chapter 10). While the bat is searching for prey, its pulses are relatively long (about 10 ms) and are repeated slowly (about 10 per second). When prey has been detected, the bat switches to the approach phase of the hunt. The sound pulses get shorter (about 5 ms) and are repeated more rapidly (about 20 per second). In the final approach, which begins when the bat is within a meter of its prey, the bat begins a feeding buzz, consisting of short pulses (0.5 to 2 ms) repeated rapidly (100 to 200 per second) (Boyan and Miller 1991).

How does the moth's two-receptor nervous system analyze the available information and direct effective evasive maneuvers? The A1 cell is first to respond, and its input reveals the direction and distance of the bat (Figure 6.19). If the bat, for example, is on the left side, the left A1 cell is exposed to louder sounds because the A1 cell on the right is somewhat shielded by the moth's body. Therefore, the left receptor fires sooner and more frequently upon receiving each sound of the bat. When the bat is directly behind or in front of the moth, both neurons will fire simultaneously. A slight turn of the moth's body will then result in differences in the right and left receptors, which will reveal whether the bat is approaching from the front or rear. What about its altitude? If the bat is above the moth, the bat's sounds are louder during the upward beat of the moth's wings when the moth's ears are uncovered than when the moth's wings are down, covering the ears and muffling the bat's cries. However, if the bat is beneath the moth, the bat's echolocation cries will reach the moth's ears unimpeded regardless of the position of the moth's wings. Therefore, the moth's wingbeats will have no effect on the pattern of neural firing. The moth, then, is able to decode the incoming data, so it detects both the presence and precise location of the bat.

How is this information processed to produce an appropriate escape pattern? If the bat is passing some distance away, only the A1 cell is sensitive enough to fire. Its firing rate will increase as the bat gets closer and its cries become louder. As long as only the A1 cell is firing, the distance between predator and prey is too great for the moth to be detected by the bat. Therefore, the most adaptive response of the moth would be to turn and fly directly away, thus decreasing the likelihood of detection by increasing the intervening distance and by exposing less surface area to the bat. This escape pattern results when the moth turns its body until the A1 firing from each ear is equalized. When the bat changes direction, so does the moth (Roeder and Treat 1961).

Bats fly faster than moths, though, and it would obviously be helpful for a moth to be able to determine whether the bat was gaining on it. This information

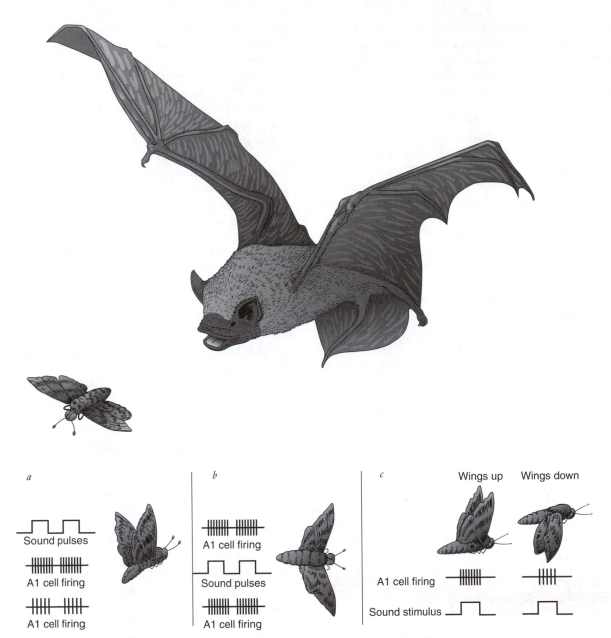

FIGURE 6.19   The relationship of sound pulses from a hunting bat and auditory neural firing in the hunted moth. (*a*) When a hunting bat, emitting its high-pitched sounds, approaches a noctuid moth from the side, the receptors on that side fire slightly sooner and more rapidly than those on the shielded side. (*b*) When the bat is behind the moth, the moth's receptors on both sides fire with a similar rapid pattern. (*c*) When the bat is above the moth, the moth's auditory receptors fire when its wings are up but not when its wings cover the receptors on the down stroke. (Redrawn from Alcock 1975.)

comes from two sources. The first is the way in which input from the A1 cell is processed by interneurons. The response of the A1 cell can follow the bat's call rates at all phases of the hunt. Since the call rate changes as the bat gets closer to its prey, the output of the A1 cell provides information about the distance of the bat. The A1 cell sends this information directly to two interneurons, called 501 and 504. These interneu- rons respond differently to the same input from A1. The differences in interneuron responses somehow encode information about the distance of the bat (Boyan and Fullard 1986; Boyan and Miller 1991).

The second source of information about the dis- tance of the bat is from the A2 cell. If the moth is about to be caught, the sounds of the onrushing bat will be very loud. At this point the A2 fiber begins to fire, and more drastic measures are in order. The messages from the A2 cell apparently shut off the thoracic ganglion that had been coordinating the antidetection behavior. The moth's wings begin to beat in either peculiar, irregular patterns or not at all. The insect itself probably has no way of knowing where it is going as it begins a series of loops, rolls, and dives. But it is also very difficult for the bat to pilot a course to intercept the moth. If the moth crashes into the ground, so much the better. It is safe here because the earth will mask its echoes.

We are beginning to learn more details about how the processing of information in the moth's ears leads to appropriate evasive action (Boyan and Fullard 1988; Fullard, Forrest, and Surlykke 1998; Tougaard 1996; Waters 1996). But many questions remain. How does a moth distinguish a bat's calls from the cacophony of sounds, mostly male insects calling for mates, that fill the night? The chorus of calling male insects often has some sound components in the ultrasonic range. Since moths can hear only in the ultrasonic range, they can't use other frequencies in the insects' calls to distinguish them as nonbat. How, then, do moths make this distinction? Also, how is input from the moth's two ears integrated to locate the bat exactly in three-dimensional space? This is still an active, exciting area of research.

Prey Localization by Barn Owls

Silently and suddenly, a barn owl (*Tyto alba*) sweeps from the sky to strike its prey with astonishing accuracy (Figure 6.20). How does it find its prey? Although in nature the barn owl's keen night vision is important in locating prey, the sounds of a scurrying mouse are sufficient for the owl to strike with deadly precision. Laboratory tests have revealed that birds such as the barn owl are able to locate the source of sounds within 1° or 2° in both the horizontal and vertical planes (1° is approximately the width of your little finger held at arm's length). Because of its astounding ability to detect and locate the source of sound, this nocturnal predator can pinpoint its prey by the rustlings the prey makes, and it can precisely determine not only the prey's location along the ground but also its own angle of elevation above the prey.

How do we know that the hunting owl uses the prey's sound? For one thing, we know that barn owls can catch a mouse in a completely darkened room (Payne 1962). In experiments, a barn owl was able to capture a skittering leaf pulled along the floor by a string in a dark room (indicating that sight and smell are not involved), and if unable to see, it will leap into the middle of an expensive loudspeaker from which mouse sounds emanate.

To locate its prey by using sound cues, the barn owl must place the source of the sound on a horizontal plane from left to right (i.e., its azimuth), as well as on a vertical plane (i.e., its elevation). We now know that a barn owl uses different cues for locating sound cues in horizontal and vertical planes.

The owl uses time differences in the arrival of sound in each ear to place it on a horizontal plane and differences in intensity between the two ears to determine the elevation of the sound source (Konishi

FIGURE 6.20   A hunting barn owl. A barn owl can locate its prey by using sound cues alone.

1993a, b). Masakazu (Mark) Konishi learned this by placing small earphones in a barn owl's ear through which sound could be played. Because an owl's eyes are fixed in their sockets, it turns its entire head to face the direction from which it perceives the sound source. When the sound in one ear preceded that in the other, the owl turned its head in the direction of the leading ear. The longer the time difference, the further the owl turned its head.

The intensity differences in the two ears vary with the elevation of the sound source largely because of the arrangement of the ear canals and facial feathers. The two ear canals that channel the sound toward the inner ears are, oddly enough, situated asymmetrically with the right one higher than the left. As a result of this difference in ear placement, each ear responds differently to a sound at a given elevation. This helps the owl determine its own elevation above the sound source, information critical to an aerial predator. Also, the face of the barn owl is composed of rows of densely packed feathers, called the facial ruff, that act as a focusing apparatus for sound (Figure 6.21). Troughs in the facial ruff, like a hand cupped behind the ear, both amplify the sound and make the ear more sensitive to sound from certain directions.

The facial ruff assists the owl in localizing sounds by creating differences in intensity of the sound in both ears. Loudness is a cue to localizing the sound in both the horizontal and the vertical dimensions. Sound is generally louder in the ear closer to the source. Because of the structure of the facial ruff, the left ear collects low-frequency sounds primarily from the left side and the right ear collects low-frequency sounds from the right side. A comparison of the intensity of low-frequency sounds in each ear helps the owl determine from which side of the head the sound originates. However, the facial ruff channels high-frequency sound to each ear differently, depending on the elevation of the sound source. As a result, the right ear is more sensitive to high-frequency sounds that originate above the head and the left ear is more sensitive to high-frequency sounds from below the head. The owl compares the loudness of high-frequency sounds in each ear to determine its position above or below the sound source. As a sound source moves upward from below the bird to a position above the owl's head, the high-frequency sounds would first be loudest in the left ear and then gradually become louder in the right ear (Knudsen 1981).

Information on the timing and loudness of sounds in each ear is then sent to the central nervous system of the owl over the auditory nerve in a pattern of nerve impulses. The information is first sent to the cochlear nuclei. Each side of the brain (cerebral hemisphere) has two cochlear nuclei, the magnocellular nucleus and the

FIGURE 6.21   The barn owl, a night hunter, has a facial disc of densely packed feathers that may gather sounds and aid in detecting their source.

angular nucleus. Every axon in the auditory nerve sends a branch to both of these nuclei. Whereas the branch of the auditory nerve that goes to the magnocellular nuclei conveys timing information, the branch to the angular nuclei conveys intensity information. Thus, the timing data that place the sound on a horizontal plane are processed separately from the intensity data that place the sound on a vertical plane. These different features of sound are processed in parallel along nearly independent pathways to higher processing stations, where a map of auditory space is eventually formed (Konishi, 1993a).

The map of auditory space is formed in the mesencephalicus lateralis dorsalis (MLD). Within the MLD are certain neurons that respond selectively to specific degrees of binaural differences in sound (Figure 6.22). For example, one MLD neuron may respond maximally to differences that correspond to a sound originating 30° to the right of the owl. The sound would arrive just a certain amount sooner and be a certain degree louder in the right ear than in the left. Those exact differences in timing and loudness stimulate that

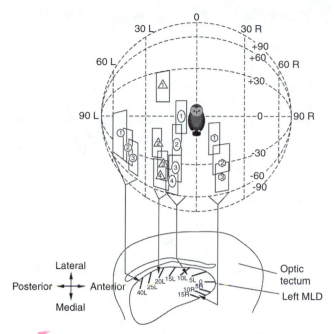

FIGURE 6.22 **Auditory neurons in the midbrain area (MLD) of a barn owl. The top figure shows a hemisphere of space in front of the owl's head. Neurons in the MLD respond to sounds that originate at different points. The numbered rectangles indicate 14 areas to which specific MLD neurons are tuned. The lower figure indicates the manner in which the auditory space of the owl is represented in the MLD. A horizontal section of the MLD is shown with bars, indicating the position of specific neurons. The point in space to which that neuron responds is indicated. Notice that the neurons in the MLD are spatially organized. (From Knudsen and Konishi 1978.)**

particular MLD cell. The degree of binaural difference varies with the location of the sound, and the binaural difference that stimulates cells of the MLD varies from neuron to neuron (Knudsen 1982).

RESPONDING—MOTOR SYSTEMS

Animals receive information about their environment via their sensory systems and then respond by means of their motor systems. Two principal components of motor systems are motor organs (typically muscles) and the neural circuits that control them. Like sensory receptors, muscles are often described as biological transducers. There are, however, some important distinctions between the two. Whereas sensory receptors transduce environmental energy such as light into the electrochemical signals of the nervous system, muscles convert the signals of the nervous system into the movements of the body. Furthermore, whereas sensory

receptors are concerned with input, muscles are involved in output.

NEURAL CONTROL IN MOTOR SYSTEMS

A particular movement, or behavior, is produced by muscles, whose activity is controlled by motor neurons. Recall that motor neurons, in turn, usually receive their information from interneurons. Thus, each movement is ultimately controlled by the activity of a neural circuit. It turns out that there are three major ways in which neural circuits control and coordinate movement: (1) the sensory reflex, (2) the central pattern generator, and (3) motor command.

In sensory reflexes, sensory neurons initiate activity in motor neurons, sometimes through direct synaptic connections but more typically through connections with a small number of interneurons. We have already considered the gill-withdrawal reflex in *Aplysia*, so here we will consider the ways in which a central pattern generator and motor command control the locust's flight.

It is generally accepted that rhythmic activities, such as walking, swimming, or flying, are driven by a group of neurons called a central pattern generator, in which motor neurons are activated by rhythms generated within the central nervous system itself. More specifically, within the central nervous system there is a neuron or network of neurons that is capable of generating patterned activity in motor neurons, even when all sensory input has been removed from the system. The idea of central control of motor patterns is obviously quite different from that of the peripheral control exemplified by the role of sensory neurons in the behaviors we have considered so far.

When one or more interneurons descending from the brain initiate activity in motor neurons, the neural mechanism is called motor command. These interneurons, sometimes called command neurons, act in a switchlike manner, determining, on the basis of incoming sensory information, whether or not a particular pattern of behavior will be initiated. One job of command neurons appears to be the turning on and off of central pattern generators. Command neurons are involved in patterns of behavior such as the escape response of crayfish and the startle response of teleost fish.

We will describe these sources of neural control in more detail in the context of the flying behavior of locusts. The precise role and relative importance of sensory feedback, central pattern generation, and motor command have been hotly debated for motor patterns in general and for locust flight in particular (e.g., Pearson, Reye, and Robertson 1983; Stevenson and Kutsch 1987). A more detailed review of the neuromuscular basis for locust flight can be found in Young (1989).

LOCUST FLIGHT

Locusts, those species of "short-horned" grasshoppers found in the family Acrididae, exhibit legendary mass migrations. In fact, accounts of locust plagues date back to the Book of Exodus, written about 1500 B.C. (C. B. Williams 1965). Swarms of locusts have also been recorded in recent times (Figure 6.23). Although representatives of the family are found throughout the world, the migratory species are found in the tropics or subtropics, typically in the drier regions of these areas. As evidence of the amazing flight behavior of locusts, consider this account written by the entomologist C. B. Williams (1965, pp. 77–78) while working in East Africa:

> A most spectacular flight occurred on 29th January, 1929, at Amani, in the Usambra Hills in north-eastern Tanganyika. . . . We received a telephone warning shortly after breakfast that an immense swarm of locusts was passing in our direction over an estate about six miles to the north. . . . An hour or so later the first outfliers began to appear—gigantic grasshoppers about six inches across the wings, and of a deep purple-brown. Minute by minute the numbers increased, like a brown mist over the tops of the trees. When they settled they changed the colour of the forest; by the weight of their numbers they broke branches of trees up to three inches in diameter; the noise of their slipping up and down on the corrugated iron roofs of the houses made conversation difficult. . . . The swarm was over a mile wide, over a hundred feet deep, and passed for nine hours at a speed of about six miles per hour.

The species of locusts described in Williams's account was the desert locust, *Schistocerca gregaria* (Figure 6.24). Although this species has been used in some laboratory studies of flight behavior, the migratory locust (*Locusta migratoria*) is more commonly studied. Let us now look at the neural control of flight behavior in these remarkable insects.

The Motor Pattern

Locusts have two pairs of wings, the forewings, located on the second thoracic segment, and the hind wings, located on the third thoracic segment. The wings of free-flying locusts move up and down, in a rhythmic manner, about 20 times per second. Because locusts maintained in the laboratory and tethered to a holder exhibit close to normal flight, it is possible to obtain a detailed analysis of the motor pattern (Figure 6.25). During flight, the two pairs of wings do not move in a precisely synchronous manner, but rather the hind wings lead the forewings. The entire cycle of movement lasts about 50 milliseconds.

Two sets of muscles act on each wing. One set, the elevators, raises the wing; the other set, the depressors, lowers it. Because these two sets of muscles have opposite effects on the wing, they are called antagonists (muscles with the same effects on a specific structure are called synergists).

How do signals from the locust's nervous system initiate activity in the flight muscles? A motor neuron and the muscle fibers (muscle cells) that it innervates

FIGURE 6.23   A swarm of desert locusts in Ethiopia.

FIGURE 6.24 A desert locust.

make up a motor unit; the junctions between motor neurons and muscle fibers are called neuromuscular synapses. When an action potential travels down the axon of a motor neuron, it moves across the neuromuscular synapse to the fibers in the motor unit. The resulting postsynaptic depolarization produces a muscle action potential that spreads across the surfaces of the muscle fibers and is then conducted to the insides of the fibers, where it produces twitch contractions.

Patterns of activity in the flight muscles of a tethered locust can be recorded by inserting tiny wire electrodes into the muscles (see Figure 6.25a). The signals (a recording of muscle action potentials) from the elevator and depressor muscles of the flying insect are then displayed on an oscilloscope; this type of recording is called a myogram. Myograms have shown that the depressors are activated when the wings are up, and the elevators are activated when the wings are down (Figure 6.25c). The relative timing of the activation of these muscles is critical, and we will now examine how such timing may be controlled by the central nervous system.

The Role of the Central Pattern Generator

The flying behavior of locusts was one of the first motor systems for which a central pattern generator was shown to be an important source of neural control (D. M. Wilson 1961). The role of the central pattern generator was highlighted by experiments in which the other two major sources of neural control, namely, motor command and sensory feedback, were eliminated. We will consider the results of each of these experimental procedures in turn.

In the laboratory, a stream of air directed at the head of a tethered locust can initiate flight. Wind-sensitive hairs on the head are stimulated, and the excitation is passed along brain interneurons that descend to ganglia in the thoracic area (a ganglion is an area of the central nervous system where large numbers of interneurons occur in closely packed aggregations). These interneurons that descend from the brain constitute the control system of motor command. Decapitation eliminates this system of neural control and results in an animal that can still fly when tethered. (Obviously, in the decapitated locust, flight cannot be initiated by blowing air on its head; flight can, however, be started by either gently pinching the abdomen of the animal or by removing the platform on which it is resting.) The important point here is that the basic system for controlling flight can operate independently of motor command (D. M. Wilson, 1961).

Locusts have mechanoreceptors on their wings that send information on wing position, via sensory neurons, to the central nervous system. This control system can be eliminated by sectioning the sensory nerve (e.g., D. M. Wilson 1961) or, in a technique developed more recently, by injecting the chemical phentolamine, which blocks the activation of mechanoreceptors without affecting the central nervous system (Ramirez and Pearson 1990). The removal of sensory input by either technique is called deafferentation. (Recall that sensory neurons are also called afferent neurons.) Although some changes in the flight pattern are observed after the surgery (e.g., the frequency of wing beats drops from 20 to about 10 per second), the basic pattern of wing movements is still present. The removal of sensory feedback from the wings can be taken one step further than simply sectioning the sensory nerves. The wings and thoracic muscles can be completely removed. Even after such extensive dissection, a pattern of neural activity similar to that seen during flight occurs in the motor nerves that emerge from the thoracic ganglia. Taken together, these results suggest that the basic flight pattern of the locust is not dependent on sensory feedback but is generated in the central nervous system. In other words, a neuronal circuit in the central nervous system generates the alternating bursts of activity in motor neurons that innervate the elevator and depressor muscles.

At the level of the neuron, what is a central pattern generator? The central pattern generators for rhythmic patterns of behavior such as flying, walking, and breathing are called oscillators, and there are two basic types, cellular and network. A cellular oscillator is a neuron that generates rhythmic activity on its own. In contrast, a network oscillator is a collection of neurons (none of which alone qualifies as a cellular oscillator) whose interaction produces temporally patterned output. Sometimes central pattern generators combine the characteristics of cellular and network oscillators, and these are called hybrid or mixed oscillators. At present, the central pattern generator of locust flight appears to be a network oscillator.

How are neurons that are part of a network responsible for generating rhythmic output, such as the move-

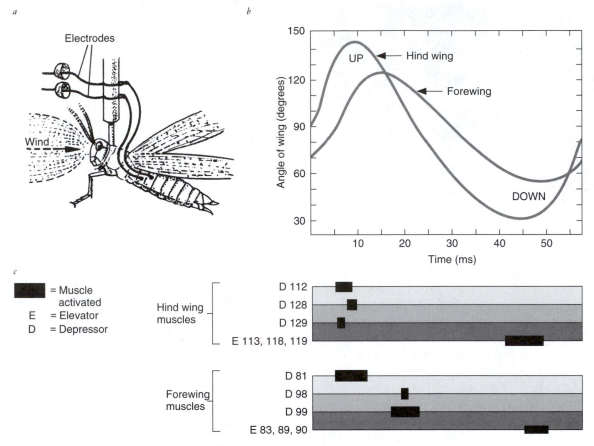

FIGURE 6.25   Analysis of the flight pattern of locusts. (*a*) If wind is directed at the head, a tethered locust can be induced to fly. The activity of flight muscles of the second and third thoracic segments can then be measured by inserting small wire electrodes in them. (*b*) The pattern of wing movements during flight in a tethered locust. (*c*) The pattern of activity in individual flight muscles as recorded with microelectrodes. (*a*: Modified from Horsmann, Heinzel, and Wendler 1983. *b* and *c*: Modified from Young 1989. Original data from Wilson 1961 and Wilson and Weis-Fogh 1962.)

ments involved in locust flight, identified? The following three criteria must be met for a neuron to be considered part of a particular oscillator: (1) The neuron must have synaptic connections, either directly or indirectly, with the appropriate motor neurons—in the case of the locust, those motor neurons involved in flight; (2) the neuron must be active at the appropriate frequency—here, the flight frequency; and, in a relatively simple network such as that involved in locust flight, (3) the neuron must be capable of resetting the flight rhythm when it is stimulated with current from a microelectrode. The motor neurons involved in flight do not meet these criteria. Some interneurons, however, have been discovered that do meet all three criteria, and they are considered to form part of the network of neurons that makes up the central pattern generator (Figure 6.26; Robertson and Pearson 1983, 1985). There remains, however, some debate about whether there are separate oscillators to drive the movements of

the forewings and hind wings, or whether the generator of the flight pattern, although distributed among several segmental ganglia, functions as a single unit (Robertson and Pearson 1984; Stevenson and Kutsch 1987; D. M. Wilson 1961)

The Role of Sensory Feedback

Locusts have sensory receptors on their wings that monitor wing movement. At the base of each forewing and hind wing, for example, there is a stretch receptor. These receptors consist of a single sensory neuron that is attached to a strand of connective tissue that runs across the hinge of the wing. When the wings are raised, the receptors are stretched and stimulated. The axon of each sensory neuron sends branches to both the second and third thoracic ganglia and synapses with interneurons and motor neurons involved in flight.

Do stretch receptors meet the three criteria used to

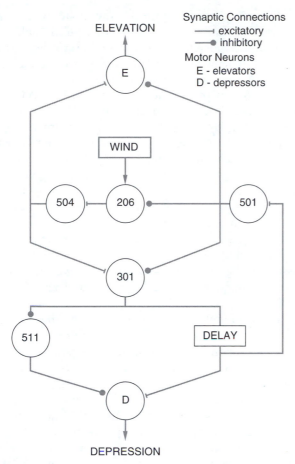

FIGURE 6.26 **A schematic circuit of the known elements of the central pattern generator for flight in the locust. The rhythmic activity produced by this network results from the interactions among component interneurons (identified interneurons have been assigned numbers). (From Robertson and Pearson 1985.)**

identify components of a rhythm-generating system? We have already mentioned that the sensory neuron of each stretch receptor synapses with the interneurons and motor neurons involved in flight—thus, stretch receptors meet the first criterion. With respect to the second criterion, stretch receptors are, in fact, active at the flight frequency. This can be shown by tethering a locust and preparing it in such a manner that both the wing position and the activity of a flight muscle are monitored. At the same time, the activity of a stretch receptor is examined by placing a fine wire around the sensory nerve that houses its axon. Such monitoring has revealed that stretch receptors produce rhythmic bursts of impulses that coincide with activation of the depressor muscles (Figure 6.27). In addition, the responses of stretch receptors are closely tied to wing movements. Specifically, the burst of impulses of the stretch receptor occurs near the peak of wing elevation. Thus, data pertaining to both wing position and wing muscle activ-

ity confirm that stretch receptors are active at the flight frequency (Mohl 1985). The final criterion is also met—stretch receptors are capable of resetting the flight rhythm. This has been shown in experiments with deafferented locusts (animals in which all sensory nerves from the wings have been cut). Such animals, as previously mentioned, typically show a drop in their flight frequency from about 20 to 10 beats per second. Stimulation of the axon of a stretch receptor, however, can reset the normal flight rhythm (Pearson, Reye, and Robertson 1983; Reye and Pearson 1988).

Having met all three criteria, stretch receptors thus appear to play an integral role in the pattern of motor output associated with flight in locusts. Because of their link with the outside world (e.g., sudden updrafts or downdrafts of air will cause changes in wing position that stimulate the receptors), stretch receptors impart some flexibility to the flight control system, allowing the animal to adjust its flight to prevailing environmental conditions on a cycle-by-cycle basis.

The Role of Motor Command

On the head of a locust, back between the compound eyes, are several fine hairs. As described earlier, these hairs are sensitive to wind, and when stimulated they initiate and maintain flight. Within each hair is a sensory neuron whose axon projects to the brain, where it synapses with interneurons whose axons extend down to the thoracic ganglia. These descending interneurons make up the system of neural control called motor command. Responding not only to wind stimuli but also to visual stimuli, these descending interneurons are described as multimodal because they are sensitive to stimuli from more than one sensory modality. It is thought that these descending interneurons are involved in modulating wing beats during corrective steering maneuvers following air turbulence or errors in flight performance (Reichert, Rowell, and Griss 1985; C. H. F. Rowell 1989). (During steering, locusts actually appear to use information from several sources—their simple and compound eyes, antennae, cerci, and hairs on the head.)

FIGURE 6.27 **The activity of a forewing stretch receptor compared with forewing movements and the myogram of a hind wing depressor muscle. The forewing depressor muscle fires a few milliseconds after the hind wing depressor and is therefore in phase with the receptor burst. (From Mohl 1985.)**

This descending pathway plays an important role in the flight system of locusts, being responsible for turning on and off the central pattern generator and for modulating flight. The axons of the interneurons synapse with flight motor neurons and interneurons in the thoracic ganglia, and those interneurons that have been studied are rhythmically active during flight. In fact, because the locust's own wing movements produce air currents that stimulate its wind-sensitive hairs, the activity of the interneurons coincides with the flight frequency. Finally, activity of the interneurons can reset the flight rhythm. This has been demonstrated by either manipulating air currents directed at the head (Horsmann, Heinzel, and Wendler 1983) or by electrically stimulating one of the interneurons (Bacon and Mohl 1983). Thus, in their modulatory role, the descending interneurons, like the sensory neurons of the stretch receptors, ensure that the flight system is able to respond to environmental perturbations.

We see, then, that for the flight system of the locust, all three sources of neural control interact to affect the pattern of motor output. Although the basic rhythm of flight is generated by a central pattern generator in the thoracic ganglia, information gathered by sense organs located on the wings enables the locust to adjust its flight pattern on a cycle-by-cycle basis. Sensory information thus imparts some flexibility to the system of motor control, allowing the animal to respond to air turbulence and other environmental uncertainties. The system of motor command plays a particularly important role in that the interneurons that descend from the brain initiate (and modulate) activity in the central pattern generator. These interneurons appear to play a role in correctional steering, when the locust compensates for deviations in the flight course. The combination of central and peripheral control observed in locust flight probably characterizes most patterns of rhythmic behavior in animals (Delcomyn 1980).

Neuromodulation of Central Pattern Generators

At one time, it was thought that there was a different central pattern generator for each distinct form of rhythmic behavior, such as walking, running, and trotting. In other words, the view was that there was a neural circuit dedicated to the control of each distinct behavior. But that view has changed. We now know that a single neuron or a group of neurons can be part of more than one central pattern generator and that two or more neural networks can work together, forming a completely new central pattern generator. Thus, neural circuits can be functionally rearranged. An important way of doing so is through neuromodulators (Selverston 1995).

Consider the behavior of the male blue crab (*Callinectes sapidus*) as an example of neuromodulation of rhythmic movements of the swimming legs. Just before a female blue crab matures, she releases a pheromone, a chemical used to communicate with other members of the species, in her urine. When a male blue crab senses the pheromone, he begins his courtship display. He spreads his claws apart in front, extends his walking legs, and raises his swimming legs in the rear. The swimming legs are then waved from side to side above the carapace. Besides courtship, two other distinct stereotyped behaviors, sideway swimming and backward swimming, involve the rhythmic movement of swimming legs. In each behavior, the swimming legs are waved in a slightly different way and the crab assumes a different posture. However, because these three behaviors are so similar, it is likely that they share common neural elements (Wood and Derby 1995).

Neuromodulators, combined with the proper olfactory stimulation, affect whether the crab will perform the courtship display instead of the two swimming behaviors. When the neuromodulators proctolin, dopamine, octopamine, serotonin, and norepinephrine are separately injected into blue crabs, each drug produces a unique posture or combination of limb movements. The postures and limb movements are the same as those observed in freely moving, untreated crabs. Injection of dopamine produces the posture of the courtship display, and injection of proctolin produces the rhythmic leg movements characteristic of courtship (Wood, Gleeson, and Derby 1995).

Electrical stimulation of specific neurons under different conditions has identified interneurons in the esophageal connectives that trigger rhythmic waving of the swimming legs. Some of these interneurons trigger rhythmic leg waving when sex pheromone is applied to the antennule of the crab. Under these conditions the leg waving is not distinctly characteristic of any of the three rhythmic behavior patterns. However, when proctolin is applied while these interneurons are being stimulated, the motor output changes to the rhythmic waving of the courtship display (Wood 1995). The natural source of the proctolin that initiates courtship leg movements is thought to be a cluster of nerve cells in the subesophageal ganglia (Wood, Nishikawa, and Derby 1996).

SUMMARY

The basic unit of the nervous system is the neuron, a cell that usually has three parts: a cell body, or soma, which maintains the cell; dendrites, which receive information and conduct it toward the soma; and an axon, which conducts the nerve impulse away from the soma. There are three functional classifications of neurons. A sensory neuron is specialized to detect stimuli or to receive information from sensory receptors and conduct it to the central nervous system. Interneurons, located in the central nervous system, link one neuron to another. A motor neuron carries the information from the central nervous system to a muscle or gland.

When a neuron is resting, that is, not conducting an impulse, it is more negative inside than outside. This resting potential is approximately –60 millivolts and results from an unequal distribution of ions. Many large, negatively charged proteins are held in the cell and give the interior the overall negative charge. The concentration of potassium ions (K$^+$) is roughly 30 times greater inside the neuron's membrane than outside. In contrast, sodium ions (Na$^+$) are more concentrated outside the membrane.

The message of a neuron is called an action potential or nerve impulse. The action potential is a wave of depolarization followed rapidly by repolarization that travels to the end of the axon with no loss in strength. Should a stimulus open ion channels that will allow sodium ions to cross the membrane, the ions will be drawn inside, attracted by the negative charge within and driven inward by their concentration gradient. The positively charged sodium ions depolarize the membrane. At the peak of depolarization, the sodium channels close and potassium channels open. Now potassium ions exit the cell, leaving that region of the membrane more negative on the inside once again (i.e., it is repolarized). The depolarization in one spot of the membrane triggers depolarization in the adjacent region so that the impulse spreads along the axon.

Information travels from one neuron to the next by crossing the gap, or synapse, between them. There are two types of synapses, electrical and chemical. Electrical synapses involve such tight connections between the neurons that the impulse spreads directly from cell to cell. Electrical synapses are characterized by a rapid speed of transmission and, as a result, are often part of the neural circuitry of escape responses. In contrast to the direct transfer of information that occurs at an electrical synapse, several steps are involved in transferring information across a chemical synapse: (1) The action potential travels down the axon to small swellings called terminal boutons. (2) Here, the action potential causes a neurotransmitter to be released from storage sacs (synaptic vesicles) into the gap between cells (synaptic cleft). (3) The neurotransmitter diffuses across the gap and binds to a special receptor. (4) When this occurs, ion channels open, changing the charge difference across the membrane. (5) If the synapse is excitatory, the postsynaptic cell is slightly depolarized. However, if the synapse is inhibitory, the postsynaptic cell is slightly hyperpolarized, and no new action potential will be generated.

When the siphon of *Aplysia* is touched, the animal withdraws its gills, mantle shelf, and siphon into the mantle cavity. This is called the gill-withdrawal reflex. This reflex shows habituation, that is, a decrease in responsiveness because of repeated stimulation. It also shows sensitization, in which a strong stimulus anywhere on *Aplysia's* surface will cause an exaggerated gill-withdrawal reflex from a light touch on the siphon.

Short-term memory involves changes in the strength of synaptic connections. In *Aplysia*, habituation occurs because the sensory neuron releases less neurotransmitter when it is repeatedly stimulated. As a result, the gill motor neuron is less likely to fire. Sensitization occurs when a strong stimulus causes a sensory neuron to activate a facilitating interneuron to release serotonin on its synaptic ending near the axon terminal of the sensory neuron from the siphon skin. Serotonin then causes biochemical changes that cause the sensory neuron to release more neurotransmitter than usual. This increases the likelihood that the gill motor neuron will fire.

Long-term memory involves the growth of new synaptic connections. This growth requires critical genes to be turned on.

The membrane potential of the postsynaptic cell can also be changed by neuromodulators. These substances work more slowly than neurotransmitters and bring about changes by biochemical means. The effects of neuromodulators appear to be mediated by substances, called second messengers, within the postsynaptic neuron. These second messengers (e.g., cAMP and cGMP) activate enzymes that alter the ion channels of the nerve cell and thereby affect neural activity. Neuromodulators can have profound effects on behavior, as demonstrated by the effects of the neuromodulator serotonin on feeding in the leech.

Animals receive information from the environment at their sense organs. Sense organs, such as eyes and ears, are selective, containing specific receptor cells that respond to a particular form of environmental energy.

Despite the diverse types of stimuli that bombard an animal from its environment, it is able to detect only a limited range, and of those, it may ignore all but a few key stimuli. Sensory filtering may occur in the receptor itself (e.g., the sensory hairs of the caterpillar *Barathra* are particularly sensitive to sounds in the frequency produced by its predator, the parasitic wasp *Dolichovespula*) or at any step in the processing of information in the nervous system. Little skates have a neural mechanism in their central nervous system that filters out the electrical fields generated by their own respiratory movements, allowing them to better detect fields generated by prey. In addition to filtering out irrelevant information from the environment, many animals have cells in their nervous system that are specialized to detect specific features of critical stimuli. In the brain of the common toad, some neurons respond only to certain combinations of features, those that normally identify prey.

Animals respond with their motor systems to information picked up by their sensory systems. Motor systems consist of motor organs (muscles) and the neural circuits that control them. There are three major sources of neural control for motor patterns: (1) the sensory reflex, (2) the central pattern generator, and (3) the motor command. In sensory reflexes, sensory neurons initiate activity in motor neurons, sometimes through direct synaptic connections, but more typically through connections with one or more interneurons. In the case of central pattern generators, motor neurons are activated by rhythms generated by a neuron or network of neurons in the central nervous system. Central pattern generators can activate motor neurons in the absence of all sensory feedback. Finally, the motor command consists of interneurons that descend from the brain and that initiate activity in motor neurons. Such command neurons turn central pattern generators on and off. A close examination of the neural mechanisms for locusts' flight suggests that a combination of all three sources of control operates to produce rhythmic patterns of behavior.

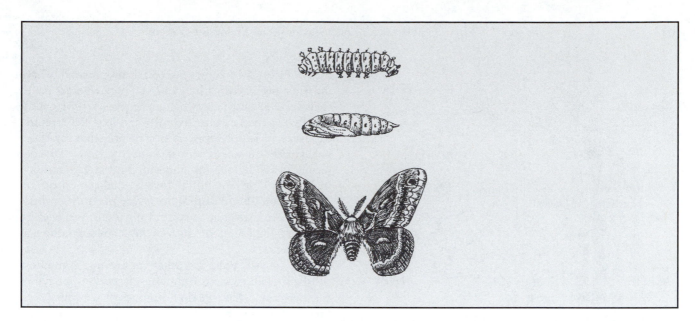

7

Physiological Analysis of Behavior— The Endocrine System

Like peas in a pod, house mouse fetuses (*Mus musculus*) line the uterine horns of their mother (Figure 7.1). Each fetus has its own personal placenta (vascular connection to the mother) and floats within a fluid-filled compartment called the amniotic sac. Even before birth, the endocrine glands of these tiny individuals are producing hormones, chemical substances that may permanently alter not only their own but also their neighbor's morphology, physiology, and behavior. In a fascinating series of experiments, Frederick vom Saal (1981) has shown that development of mouse fetuses can be modified by exposure to hormones secreted by contiguous littermates. As shown in Figure 7.1, there are three intrauterine positions that fetuses can occupy relative to siblings of the opposite sex. Females can be positioned between two male fetuses (2M females), next to one male (1M females), or not next to a male (0M females). But what does intrauterine position have to do with adult patterns of behavior? Vom Saal and Bronson (1980a) have shown that by day 17 of gestation, levels

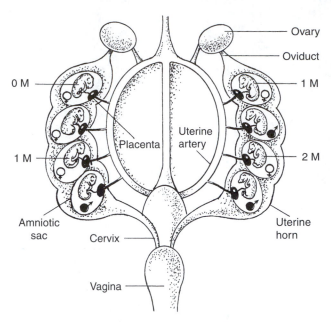

FIGURE 7.1   Mouse fetuses line the uterine horns of a pregnant female. Because the fetuses are in such close quarters, hormones from one fetus can influence behavioral development of contiguous fetuses. Female fetuses can occupy the following three positions relative to male fetuses: 2M, between two males; 1M, next to one male; 0M, not next to a male. The 2M females differ substantially from 0M females in their adult behavior and physiology as a result of proximity to male fetuses in the intrauterine environment. (Modified from McLaren and Michie 1960.)

of testosterone (a steroid hormone secreted by the testes) are three times higher in the blood of male fetuses than in the blood of female fetuses. Even more intriguing is the finding that on this same day, 2M female fetuses (i.e., those females nestled between two male littermates) have significantly higher concentrations of testosterone in their blood and amniotic fluid than do female fetuses not next to males (i.e., 0M females). Apparently, hormones pass through either the amniotic fluid or uterine blood vessels to contiguous littermates. As a result of prenatal exposure to testosterone, adult 2M females display a host of traits that distinguish them from 0M females: 2M females (1) are less attractive to males, (2) are more aggressive to female intruders, (3) mark a novel environment at a higher rate, and (4) have longer and more irregular estrous cycles (vom Saal and Bronson 1978, 1980a, 1980b). These differences in physiology and adult behavior exist despite the fact that after birth, testosterone levels do not differ between the two groups of female mice. Thus, behavioral differences in adulthood result from differential exposure to hormones in the intrauterine environment.

Position in the uterus also alters the adult behavior of male mice. Males that develop in utero between two

male fetuses (2M males) tend to exhibit parental behavior when presented with a newborn pup, whereas males that develop between two female fetuses (0M males) usually kill the newborn (vom Saal 1983). Differential exposure to testosterone is not responsible for these behavioral differences because concentrations of testosterone in the blood and amniotic fluid of 2M and 0M male fetuses do not differ. Instead, behavioral differences in adulthood appear to result from differential exposure to estrogen, a steroid hormone secreted by the ovary and found in relatively high concentrations in female fetuses.

We can ask, then, just what are hormones and how do they produce such dramatic effects on behavior? Furthermore, what role do hormones play in the development and display of behavioral differences between the sexes? Do hormonal effects on behavior vary as a function of species, season, experience, or social interactions? Can behavior, in turn, influence the presence of hormones? It is to these issues that we now turn.

THE ENDOCRINE SYSTEM

DEFINITION OF THE ENDOCRINE GLAND AND HORMONE

We begin with a definition. Hormones are substances secreted in one part of the body that cause changes in other parts of the body. Hormones are secreted either by endocrine glands or by neurons. Unlike exocrine glands (e.g., sweat or salivary glands), which have specialized ducts for secretion of products, endocrine glands lack ducts and secrete their products into the spaces between cells, from which the hormones diffuse into the bloodstream. Once in the blood, hormones travel along the vast network of vessels to virtually every part of the body. Hormones secreted by nerve cells are called neurohormones or neurosecretions. These are produced in the cell body of the nerve cell, travel along the axon, and are released at the axon tip. Then they diffuse into the blood and travel through the circulatory system to their target cells. Most of the hormones of invertebrates are secreted by neurons rather than endocrine glands. Functioning as chemical messengers in an elaborate system of internal communication, hormones and neurohormones exert their effects at the cellular level by altering metabolic activity or by inducing growth and differentiation. Changes at the cellular level can ultimately influence behavior.

HORMONAL VERSUS NEURAL COMMUNICATION

The endocrine system of an animal is closely associated with its nervous system. As mentioned, some hormones, in fact, are made by nerve cells. In addition, neurons and hormones often work together to control a single process. For example, in some interactions neurons respond to hormones, whereas in others endocrine glands receive information and directions from the brain. Nervous and endocrine systems are so closely associated that they are often discussed as a single system, the neuroendocrine system. Despite this close association, neural and hormonal modes of information transfer have different purposes within the body, and each system is essential in its own right.

In comparing communication through nervous or endocrine pathways, we should first briefly review how neurons transfer information. In the nervous system, information is transmitted along distinct pathways (chains of neurons) at speeds of up to 100 meters per second. After the impulse arrives at its destination in the body, neural information is transmitted via a series of electrical events that culminates in the release of neurotransmitter near its target, such as an effector tissue—a muscle, for example. Neurotransmitters are rapidly destroyed after they are secreted. As a result, information delivered by the nervous system usually produces a response that is rapid in onset, short in duration, and highly localized (Bentley 1982).

In contrast to neural communication, hormonal transfer of information tends to occur in a more leisurely, persistent manner. Typically, hormones are secreted slowly and remain in the bloodstream for some time. Rather than traveling to a precise location, these chemical messengers contact virtually all cells in the body, although only some cells are able to respond to the particular hormonal stimulus. The cells that respond, called target cells, have receptor molecules that recognize and bind to specific hormones. This binding activates the receptor and initiates the cell's response to the hormone, perhaps by turning on certain genes or by altering the cell's secretory activity or the properties of its plasma membrane. The precise nature of the response depends on the type of target tissue because different types of cells are specialized to perform specific functions in the body. Only cells with receptors for a particular hormone can respond to it. The concentration of receptors for a particular hormone determines the cell's sensitivity to it. In short, transfer of information by the endocrine system occurs more slowly than that of the nervous system, and it usually produces effects that are more general and long-lasting.

TYPES OF HORMONES AND THEIR MODES OF ACTION

Animals produce two major classes of hormones, peptides and steroids. Although differing in structure and mode of action, both types cause changes within the cell that eventually influence behavior.

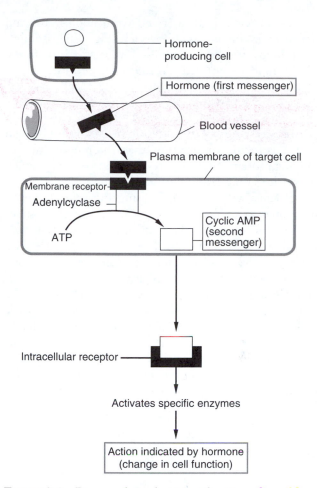

FIGURE 7.2   Proposed mechanism of action of peptide hormones.

Peptide Hormones and Amino Acid Derivatives

Peptide hormones are amino acid chains, ranging in size from 3 to 300 amino acids (Roberts 1984). These hormones, along with amino acid derivatives, are water-soluble and usually affect cells by binding to receptor molecules on the cell surface (Figure 7.2). Through a complex sequence of molecular interactions, often including the use of a secondary messenger (such as cyclic adenosine monophosphate, or cAMP), peptide hormones create short-term changes in cell membrane properties and long-term changes in protein function, often by activating enzymes. Examples of peptide hormones are luteinizing hormone (LH) and follicle-stimulating hormone (FSH) produced by the anterior pituitary.

Steroid Hormones

Steroids are a group of closely related hormones secreted primarily by the gonads and adrenal glands in vertebrates. The four major classes of steroids include progestins, androgens, estrogens, and corticosteroids. The first three classes are secreted by the gonads and are often referred to as the sex steroids. All steroid hormones are chemically derived from cholesterol and hence are highly fat-soluble. As a result of their solubility in lipids, steroid hormones move easily through the lipid boundaries of cells and into the cell interior, or cytoplasm (Figure 7.3). Once inside a cell, steroids combine with receptor molecules; move to the nucleus of the cell, where they attach to DNA; and affect subsequent gene expression and protein synthesis. Protein synthesis requires more time than the modification of existing proteins, and so steroid hormones are generally slower in action than are protein hormones.

Because we will be discussing the role of steroid hormones in reproductive behavior, two points about the sex steroids should be emphasized here. First, hormonal output is not rigidly determined by sex. Females generally have small amounts of "male hormones" such as testosterone, and males typically have low levels of "female hormones." However, we often think of the testes as sources only of androgens and overlook their ability to secrete estrogens. (Obviously, then, we should use the terms "male" and "female" hormones with caution.) A second point to emphasize is that although different hormones produce different effects, the sex steroids are chemically very similar (Figure 7.4). Some

hormones lie along the pathway of synthesis of other hormones. Testosterone, for example, is an intermediate step in the synthesis of estradiol. As a result of the common structure of these two steroid hormones, some behavior patterns may be activated by injections of either testosterone or estradiol. In other words, a certain degree of substitutability is associated with steroid hormones.

HOW HORMONES INFLUENCE BEHAVIOR

There are several pathways through which hormones can influence behavior. Generally speaking, hormones modify behavior by affecting one or more of the following: (1) sensory or perceptual mechanisms, (2) development or activity of the central nervous system, and (3) effector mechanisms important in the execution of behavior.

EFFECTS ON SENSATION AND PERCEPTION

Hormones influence the ability to detect certain stimuli, as well as preferences for particular stimuli. In some species, mate choice is at least partially based on hormone-mediated differences in the ability to detect stimuli. In domestic pigs (*Sus scrofa*), for instance, females are able to detect lower quantities of the boar pheromone, 16-androsterone, than are males (Dorries, Adkins-Regan, and Halpern 1995). Females are attracted to this chemical and assume a sexually receptive posture in response to it. Males are not attracted by the boar pheromone. However, if a male is castrated to remove the source of testosterone before the age of five months and is given the female hormone estradiol as an adult, he shows the usual female responses to a boar (Adkins-Regan, Signoret, and Orgeur 1989).

Hormones can modify preference, rather than the ability to detect certain stimuli. The hormone thyroxin is secreted by the thyroid gland and plays a critical role in the behavior of three-spined sticklebacks (*Gasterosteus aculeatus*). These fish normally spend the autumn and winter in the sea and in the spring migrate into rivers to breed. Successful migration and breeding require, among other things, a switch in preference from salt water to fresh water. This change in salinity preference appears to result from increases in the circulating levels of the hormone thyroxin (Baggerman 1962).

Hormones are also responsible for an adaptive change in odor preference among female meadow voles, *Microtus pennsylvanicus*. During the winter, female meadow voles nest communally with other females and, at this time, prefer female odors to male odors. However, in the spring and summer, when they defend territories against other females and mate with males,

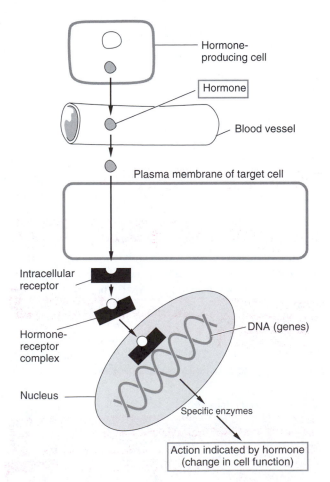

FIGURE 7.3   Proposed mechanism of action of steroid hormones.

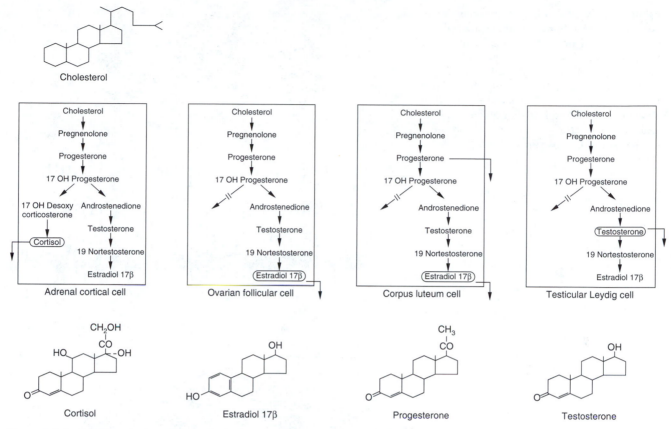

FIGURE 7.4   Biochemical pathways by which steroid hormones are synthesized. Note that many of the steroid hormones are chemically very similar. (Bentley 1982; modified from Daly and Wilson 1983; Tepperman 1980.)

they prefer the scent of a male to that of a female. Thus, changing odor preferences helps female voles choose their company so that they can successfully raise as many offspring as possible. The reversal in odor preference is caused by changes in the amount of estrogen the female produces. Estrogen levels are affected by the seasonal changes in the length of daylight (Ferkin and Zucker 1991).

EFFECTS ON DEVELOPMENT AND ACTIVITY OF THE CENTRAL NERVOUS SYSTEM

Circulating hormones can influence behavior by altering the morphology, physiological activity, or neurotransmitter function of the central nervous system. In fact, hormones have been found to affect a variety of characteristics of different regions of the brain, including (1) the volume of brain tissue, (2) the number of cells in brain tissue, (3) the size of cell bodies of neurons, (4) the extent of dendritic branching of neurons, and (5) the percentage of neurons sensitive to particular hormones.

An example in which hormones influence the morphology of the nervous system, in this case the brain, involves the development of singing behavior in birds. In the zebra finch (*Taeniopygia guttata*), sex differences in the brain nuclei that control song are established around the time of hatching. Early exposure to hormones regulates the size of song nuclei (Gurney and Konishi 1980) and the number of neurons contained within these areas (DeVoogd 1990; Nordeen, Nordeen, and Arnold 1986). Thus in the zebra finch, sex differences in adult singing behavior (males sing and females do not) are linked to morphological differences that are established in the brains of males and females as a result of the hormonal milieu soon after hatching. (See Chapter 8 for a detailed discussion of bird song.)

As previously mentioned, some hormones affect behavior by altering the activity or neurotransmitter function of the central nervous system. Certain hormones secreted by the adrenal cortex, for example, are known to affect behavior by influencing brain excitability and neurotransmitter metabolism (e.g., Joels and De Kloet 1989; Rees and Gray 1984).

EFFECTS ON EFFECTOR MECHANISMS

Hormones can influence behavior by affecting motor neurons and muscles. Consider, for example, two cases

FIGURE 7.5 (*a*) Male clawed frog clasping a sexually receptive female that responded to his mating call (shown below). The male's call consists of slow and fast trills. (*b*) Male clasping a sexually unreceptive female that responds to his clasp by emitting the ticking call (shown below). Differences in the calling behavior of male and female clawed frogs result from the effects of hormones on the muscles of the larynx. (From Kelley and Gorlick 1990.)

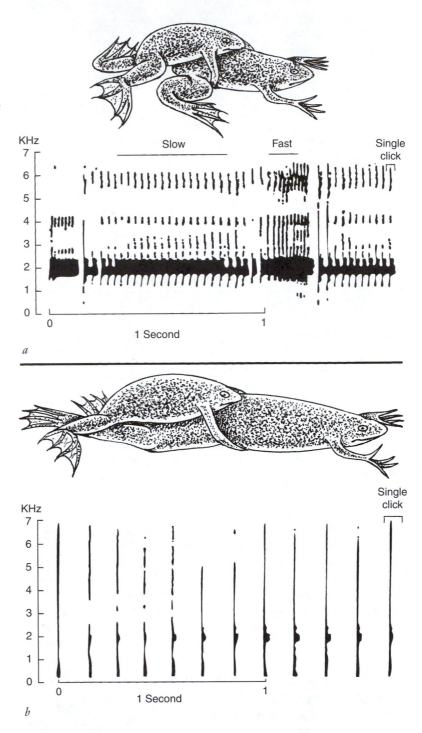

of sexually dimorphic patterns of behavior—copulatory movements in rats and calling behavior in frogs—that illustrate hormonal influences on effectors.

The levator ani/bulbocavernosus muscles control copulatory reflexes in male rats (Hart 1980). Although these muscles are present in both sexes at birth, they are completely absent in adult females. Breedlove and Arnold (1983) have shown that the levator ani/bulbocavernosus muscles shrink and fold inward in females unless supplied with androgen. The lack of androgen in females during the perinatal period (the time surround-

ing birth) also results in the death of the motor neurons that supply these muscles. Thus, sex differences in the copulatory movements of adult rats result, in part, from early hormonal influences on the growth and mainte-nance of the specific muscles and motor neurons involved in mating.

A second example of sex differences in behavior resulting from hormonal influences on effector mecha-nisms involves the calling behavior of the clawed frog, *Xenopus laevis* (Kelley 1988; Kelley and Gorlick 1990). Clawed frogs occur in sub-Saharan Africa, where they inhabit shallow, and often murky, bodies of water, ranging from ponds and lakes to sewers. Males of this species attract and excite females by emitting metallic-sounding calls in which they alternate fast and slow trills (Wetzel and Kelley 1983). These calls allow females to find males in their typically soupy locations. Sexually receptive females do not vocalize, and they permit males to clasp them around the waist, in a position called amplexus, for several hours while their eggs are released and fertilized. Although receptive females are silent, females that are not sexually ready produce a ticking call of slow, monotonous clicks. Females that are declining or terminating clasping attempts by males also typically produce the ticking call (Weintraub, Bockman, and Kelley 1985). The calls of male and female clawed frogs are summarized in Figure 7.5.

How does it come about that male clawed frogs produce mating calls of rapid trills, whereas females are capable of producing only the slow ticking call? Darcy Kelley and coworkers demonstrated that characteristics of the muscles and neuromuscular junctions of the larynx are responsible for sex differences in the rate at which calls are produced (in males the muscles of the larynx contract and relax 71 times per second and only 6 times per second in females). Adult males have 8 times as many muscle fibers in their larynx as do females. Also, male muscle cells are of the fast-twitch, fatigue-resistant type, whereas most muscle cells in the larynx of females are slow-twitch and fatigue-prone.

What causes these changes in laryngeal muscles? Sassoon and Kelley (1986) demonstrated that at the time of metamorphosis (the change from tadpole to adult frog), the number of muscle fibers in the larynx of males and females is identical to the number in the larynx of adult females. Apparently, as males mature and their levels of androgens rise, new fibers are added. In addition to increasing the number of muscle fibers, androgens also influence the type of fiber, promoting expression of the fast-twitch cells. In short, sex differences in the calling behavior of *X. laevis* can be traced, in part, to hormone-induced changes in the muscles of the larynx.

PULLING IT ALL TOGETHER— HORMONAL CONTROL OF METAMORPHOSIS AND ECDYSIS IN INSECTS

We have covered a lot of ground concerning the ways in which hormones influence behavior. So, before going on to explore other important concepts in the hormonal regulation of behavior, we will take time to see how the information we have learned so far fits together. The example we will use is the hormonal control of development and behavior during insect metamorphosis.

Because there are substantial interspecific differences in hormonal control of development among insects (a class with more than 750,000 described species), we have to generalize a bit here. Growth in insects typically involves a series of steps in which the rigid, nonexpansible exoskeleton is periodically discarded and replaced with a shiny, new, and (most important) larger outfit. In addition, almost all insects undergo a process of metamorphosis, proceeding through a series of juvenile stages—each requiring the formation of a new exoskeleton and ending with a molt—before adulthood is reached. Some insects, such as grasshoppers and bugs, undergo a metamorphosis in which adult organs are gradually formed (a process called incomplete metamorphosis); other insects, such as flies, beetles, moths, and butterflies, undergo a sudden and dramatic transformation between larval and adult stages. During this latter process, called complete metamorphosis, the old body of the larva is systematically destroyed during pupation and new adult organs develop from undifferentiated groups of cells. Figure 7.6 contrasts incomplete and complete metamorphosis.

In most species of insects, hormones and neurosecretions that affect growth, development, and behavior are secreted at five locations: (1) neurosecretory cells, located in the brain; (2) paired glands, corpora cardiaca, located just behind the brain; (3) paired glands, corpora allata, located alongside the esophagus; (4) a single prothoracic gland, located at the rear of the head; and (5)

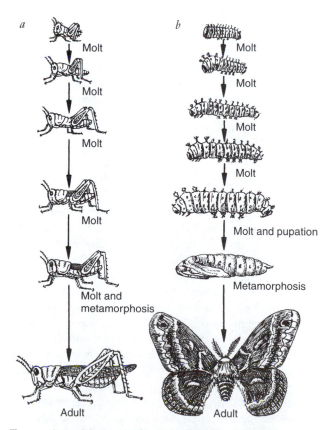

FIGURE 7.6 Metamorphosis in insects may be incomplete (*a*) or complete (*b*). (Modified from Gilbert 1988.)

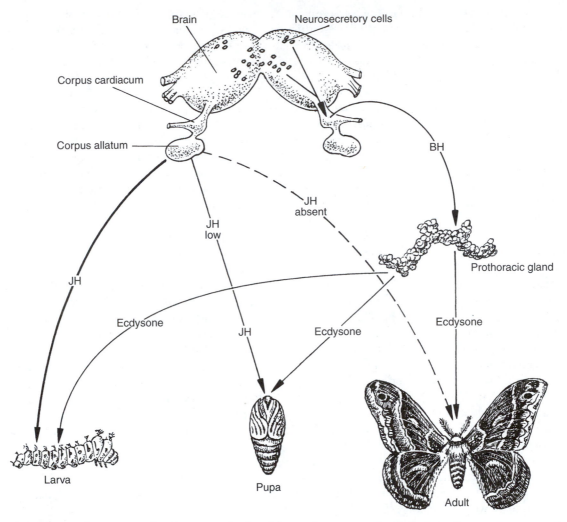

FIGURE 7.7   The neuroendocrine control of molting and metamorphosis in a moth. Neurosecretory cells in the brain secrete brain hormone (BH), which stimulates the prothoracic gland to secrete ecdysone. In the immature insect, the corpora allata secrete juvenile hormone (JH), which suppresses metamorphosis at each molt. Metamorphosis to the adult form occurs when ecdysone acts in the absence of JH. (Modified from Villee, Solomon, and Davis 1985.)

male and female gonads, located toward the rear of the body.

Molting and metamorphosis are controlled by the interaction of two hormones (Figure 7.7), one favoring the growth and differentiation of adult structures (ecdysone, secreted by the prothoracic gland) and the other favoring the retention of juvenile characteristics (JH, or juvenile hormone, secreted by the corpora allata). In the case of molting, some environmental factor (e.g., a temperature change) typically activates neurosecretory cells in the brain to secrete brain hormone (BH) into the corpora cardiaca, where it is stored. Once released from the corpora cardiaca, BH stimulates secretion of ecdysone from the prothoracic gland, and growth and molting ensue. In immature insects, the endocrine glands, called the corpora allata, secrete JH. At each larval molt, JH suppresses metamorphosis and

ensures that the insect retains its immature characteristics. However, as the concentration of JH decreases over time, metamorphosis occurs and the insect is transformed from a larva into a pupa, the next developmental stage. Finally, once secretion of JH has ended altogether, the pupa molts and becomes an adult. Thus, metamorphosis to the adult form occurs when molting hormone acts in the absence of juvenile hormone.

Each stage of metamorphosis ends with the shedding of the old cuticle, allowing the insect to emerge into a more mature stage of life. This process, called molting or ecdysis, is generally similar among all insects. We will describe molting as it occurs when the fifth and final larval stage emerges from the old cuticle in the tobacco hornworm, *Manduca sexta* (Figure 7.8). Before the larva can crawl out of the old cuticle, it must loosen this casing. To accomplish this, the larva repeat-

a

b

c

FIGURE 7.8   Metamorphosis in the tobacco hornworm, *Manduca sexta*: (*a*) **larva or caterpillar,** (*b*) **pupa, and** (*c*) **adult.**

edly contracts the muscles along the entire length of its abdomen in a particular sequence. First the muscles on the dorsal surface of the larva are contracted. Then the muscles on each side of the body are alternately contracted. This pre-ecdysis behavior pattern begins about an hour and a half before the actual shedding process.

The motor pattern involved in ecdysis is distinctly different. The larva moves the cuticle posteriorly by lifting and contracting each body segment from the back toward the front. The linings of the tracheae, the air tubules necessary for oxygen delivery, must be carefully withdrawn from the old cuticle. Next, the anterior legs are pulled out of the cuticle. These movements are accompanied by changes in blood distribution and air swallowing, which together help expand the anterior end of the larva and cause a split to form in the head region of the old cuticle. The larva can then emerge from the cuticle. Everything must go perfectly on the first attempt. Once ecdysis begins, irreversible changes in the new cuticle take place that can trap the larva in its old skin or cause crippling deformities if something stops the shedding.

Hormones coordinate the pre-ecdysis and ecdysis behavior patterns with one another and trigger them at the appropriate time relative to the developmental changes of metamorphosis. A successful molt depends on peptide hormones from three sets of cells: eclosion hormone (EH) from the ventromedial (VM) neurons in the brain, eclosion-triggering hormone (ETH) from Inka cells located on the trunks of the tracheal air tube, and crustacean cardioactive peptide (CCAP) from clusters of neurosecretory cells found in each body segment.

How do these hormones choreograph the behavior patterns of a successful molt? Before the molt, the levels of ecdysone begin to fall, and this probably serves as a starting signal by triggering the release of low levels of ETH from the Inka cells (Figure 7.9). As the level of ETH slowly rises, pre-ecdysis begins. By stimulating the abdominal ganglion, ETH initiates abdominal contractions. One of the most critical steps of ecdysis is the withdrawal of the tracheal linings from the old cuticle. If they are torn or left behind, the tracheae in that region will be blocked and oxygen delivery impaired. The Inka cells are located on the tracheae and can serve as peripheral checkpoints to ensure that the necessary developmental changes have occurred before the insect commits itself to molting.

When the level of ETH climbs slightly higher, it begins to excite the VM neurons in the brain, and these release EH. These two sets of neurosecretory cells (the Inka cells and the VM neurons) excite one another, creating a positive feedback loop—ETH stimulates the release of EH, which in turn stimulates the release of ETH. This mutual excitation eventually causes a massive surge of EH and ETH. Indeed, the Inka cells and VM neurons become completely drained of their hormones. This EH

FIGURE 7.9   Hormonal control of pre-ecdysis and ecdysis (molting). Molting is controlled by hormones: eclosion hormone (EH) from the brain, eclosion-triggering hormone (ETH) from the Inka cells on the tracheae (air tubules), and crustacean cardioactive peptide (CCAP) from segmental neurosecretory cells. ETH is produced as the level of ecdysone drops. It initiates pre-ecdysis behavior, which loosens the cuticle, and it stimulates the release of EH. Then EH stimulates the release of more ETH, forming a positive feedback loop. The resulting surge in EH and ETH stimulates the release of CCAP, which acts on the central nervous system to trigger and maintain ecdysis, shedding the cuticle. CCAP acts peripherally to cause circulatory changes and air swallowing.

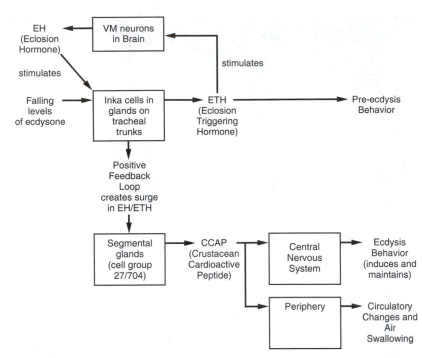

surge causes the release of CCAP, which acts in the central nervous system to initiate and maintain the motor patterns of ecdysis. In the periphery, CCAP is involved in the changes in blood distribution and air swallowing that help rupture the old cuticle (Ewer, Gammie, and Truman 1997; Gammie and Truman 1997).

Metamorphosis entails a remarkable change, not only in the animal's morphology, but also in its behavioral repertoire. During this dramatic transformation, the nervous system of the hornworm must sequentially control three very different stages: the larva or caterpillar, the pupa, and the adult moth. The animal is transformed from a crawling, eating machine to a flying, reproductive machine. Although some patterns of behavior are exhibited in all three stages (e.g., behavior associated with shedding of the cuticle), many behaviors are restricted to a single stage (e.g., crawling in the larva and flight in the adult). The neural circuitry controlling stage-specific patterns of behavior is assembled and dismantled during development.

In the tobacco hornworm, much of the remodeling of the nervous system is controlled hormonally. Figuring prominently in this remodeling process are the ecdysteroids, various forms of ecdysone, which are released by the prothoracic glands. Responses of the nervous system to ecdysteroid hormones include the production and programmed death of neurons and alterations in the structure of neurons, specifically the growth and regression of dendrites (Weeks, Jacobs, and Miles 1989; Weeks and Levine 1990). A number of questions are associated with such changes in behavior. Is the production of neurons involved in the appearance of new behaviors? Is neuronal death responsible for the

disappearance of outmoded patterns of behavior? And, how does the growth or regression of dendrites relate to the behavioral change? We can approach these questions by examining how the hormonally mediated modification of the nervous system is responsible for the loss of specific behaviors in the tobacco hornworm.

Caterpillars of the tobacco hornworm have abdominal prolegs (Figure 7.10a) that act in simple withdrawal reflexes, as well as in more complex behavior such as crawling, shedding the cuticle, and helping the animal grasp the substrate. Although these behaviors are important to the caterpillar, they are not to the pupa; and the proleg behaviors gradually disappear during the larval–pupal transformation. The question is, then, what causes their disappearance?

Most proleg movements are accomplished by retractor muscles (Figure 7.10b) that are innervated by motor neurons with densely branching arbors, or dendrites. During the larval–pupal transformation, substantial regression of the dendrites of the motor neurons occurs (Figure 7.10c). Many of the motor neurons die, and the proleg muscles degenerate and become nonfunctional.

We now know that the demise of the proleg neuromuscular system, and hence proleg behaviors, is caused by a peak in ecdysteroid hormones just before the transition to the pupal stage (Figure 7.10d). At this time, high levels of ecdysteroids trigger regression of the dendrites of the motor neurons that innervate the proleg muscles. As a result, the motor neurons are removed from behavioral circuits and proleg behaviors are lost in the pupa (Weeks, Jacobs, and Miles 1989).

To gain a better appreciation for how ecdysone

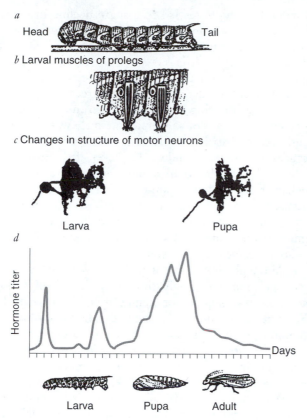

a Head   Tail

b Larval muscles of prolegs

c Changes in structure of motor neurons

Larva   Pupa

d

Hormone titer

Days

Larva   Pupa   Adult

FIGURE 7.10 **During metamorphosis in the tobacco hornworm, modifications of the nervous system that produce or eliminate patterns of behavior are controlled hormonally. (*a*) The abdominal prolegs of the caterpillar are involved in withdrawal reflexes, crawling, and grasping the substrate. These behaviors are important to the larva but disappear from the animal's repertoire once it reaches the pupal stages. (*b*) Cutaway view showing the retractor muscles of the prolegs. (*c*) The dendrites of the motor neurons that innervate the retractor muscles regress during the larval–pupal transformation. (*d*) The peak in ecdysteroid hormones just before transformation to the pupa triggers dendritic regression. (Modified from Weeks et al. 1989; data from Bollenbacher et al. 1981; Weeks and Truman 1984.)**

brings about such changes, we will look more closely at how it acts. Ecdysone is a steroid hormone. It enters the nucleus, binds with a receptor, and turns on a hierarchy of genes. There are several forms of ecdysone receptors, and their relative levels in specific tissues will determine how each tissue responds to ecdysone. The ratio of the various receptor types changes throughout development and is responsible for changes in the response of tissues during development (Riddiford and Truman 1993).

The pattern of receptor types serves as an important switch during development. For instance, some of the remodeling of the nervous system occurs during metamorphosis because neurons with high levels of one

receptor type will mature and form new synapses, whereas neurons with another receptor type will lose synapses and their processes will wither (Truman 1996). The pattern of receptor types also influences development by triggering programmed cell death in specific cells. High levels of one particular form of ecdysone receptor will turn on the genes that lead to programmed cell death. This is how larval tissues are destroyed at the appropriate time in development. Indeed, the fate of certain neurons depends on their predominate form of ecdysone receptor (Robinow et al. 1993). Likewise, particular muscle fibers within a muscle may be destroyed and others built up according to the receptor types they carry (Hegstrom, Riddiford, and Truman 1998). As larval muscles degenerate, the motor neurons that serve them regress. Then, when the new muscles form, the motor neurons regrow and new functional synapses form (Consoulas and Levine 1998). These changes in the nervous and muscular systems underlie the behavioral changes that accompany metamorphosis.

METHODS OF STUDYING HORMONE–BEHAVIOR RELATIONSHIPS

Several techniques are available for the study of hormonal influences on behavior. Here we examine two experimental approaches to questions about behavioral endocrinology. The first type of approach might be called interventional because the experimenter manipulates the hormones of the animal. This often involves the removal of a gland, followed by hormone replacement therapy. In the second approach, researchers look for changes in behavior that parallel fluctuations in hormone levels. These are called correlational studies.

INTERVENTIONAL STUDIES

Fairly conclusive evidence of the function of a hormone can be gained by removing its source, that is, the endocrine gland, and recording the subsequent effects. The hormone is then replaced by implanting a new gland or by administering the hormone. If the effects of gland removal are reversed by replacing the hormone, we conclude that the hormone was responsible for the changes.

David Crews (1974, 1979) used gland removal (castration) and hormone (androgen) replacement therapy in the study of hormonal control of sexual and aggressive behavior in lizards. Among his favorite subjects is *Anolis carolinensis*, the green anole. This small iguanid lizard inhabits the southeastern United States and displays a social system in which males fiercely defend their territories against male intruders. The territory of a sin-

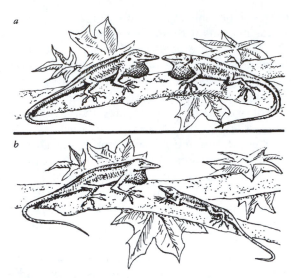

FIGURE 7.11   Displays of the male green anole. (*a*) Aggressive posturing between two males. (*b*) Courtship display directed by a male to a female (smaller individual). (Redrawn from Crews 1979.)

gle male often encompasses the home ranges of two or three females. As you might expect from these living arrangements, male anoles have an interesting repertoire of aggressive and sexual behaviors (Figure 7.11).

Both agonistic and sexual displays of male *A. carolinensis* share a species-typical bobbing movement, which is made even more dramatic by extension of the red throat fan, or dewlap. When confronted by a male intruder, a resident male anole immediately begins to display—usually by compressing his body and adjusting his posture in such a way as to present the intruder with a lateral view of his impressive physique. As if this were not enough, the resident then lowers his hyoid apparatus (a structure in the back of the throat that is responsible for movements of the tongue) and exhibits a highly stereotyped bobbing pattern. The display ends at this point if the intruder rapidly nods his head, thereby acknowledging his subservient position. However, if the intruder fails to display the submissive posture, the display of the resident male escalates to ever-increasing frequencies. In the heat of confrontation, the two combatants acquire a crest along the back and neck and a black spot behind each eye. A wrestling match ensues as the resident and intruder circle, with locked jaws, in an attempt to dislodge the other from the prized perch. For male anoles, courtship behavior is very similar to aggressive behavior. Typically, however, the body is not laterally compressed, and the bobbing dewlap display is less stereotyped in courtship; in effect, each male has his own version of how best to attract females.

Once Crews had documented the display repertoire of the feisty *A. carolinensis*, he set out to examine hormonal control of male aggressive and sexual behavior through castration and androgen replacement therapy.

Removal of the testes led to a sharp decline in sexual behavior, but administration of testosterone implants reinstated this behavior to precastration levels (Figure 7.12) (Crews 1974; Crews et al. 1978). Thus, Crews and his coworkers concluded that testosterone regulates courtship and copulation in the male green anole.

The relationship between testosterone and aggressive behavior was not so simple. If a male was castrated and returned to his home cage, he continued to be aggressive toward intruders. However, if the male was castrated and placed in a new cage, his aggressive behavior declined in a manner similar to that noted for sexual behavior. Thus, unlike sexual behavior, aggressive behavior appears to be only partially dependent on gonadal hormones and subject to influence by social factors such as residence status.

Although simple in concept, interventional studies have become quite sophisticated as a result of major advances in techniques for manipulating hormone levels. For example, cannulation techniques now allow administration of minute amounts of hormone to specific brain regions. In some species, direct introduction of gonadal hormones into appropriate areas of the brain induces sexual behavior in gonadectomized (i.e., castrated or ovariectomized) adults. Other advances utilize techniques whereby hormones are labeled with radioactivity and their paths traced through the body. The discovery of antihormones, drugs that can temporarily and reversibly suppress the actions of specific hormones, has also aided investigation of hormonal influences on behavior.

Another example of a technique used to alter the hormone levels of an animal is cross transfusion (Terkel 1972). This technique allows blood to be cross-transfused between two freely moving animals (Figure 7.13). In a now famous study, Terkel and Rosenblatt (1972) cross-transfused blood between a mother rat and a virgin female. They found that by linking blood supplies, maternal behavior could be induced in the virgin rat. These data were instrumental in demonstrating the hormonal basis of maternal behavior.

Genetic "knockout" mice offer new opportunities to manipulate hormone levels to study the relationship between hormones and behavior. A knockout mouse is one in which a specific gene is targeted and inactivated to eliminate the gene product. In this case, the gene product may be a hormone or a hormone receptor. The use of knockout mice to study the role of hormones in behavior is not without problems, but it shows great promise. Soon we may even be able to turn specific genes on or off in precise regions of the body (R. J. Nelson 1997). This will make it possible to manipulate hormone levels to distinguish the effects of the hormone during development from those that trigger behavior.

Knockout mice have already proven useful in exploring the ways in which oxytocin influences behavior. Oxytocin is a hormone produced by the posterior

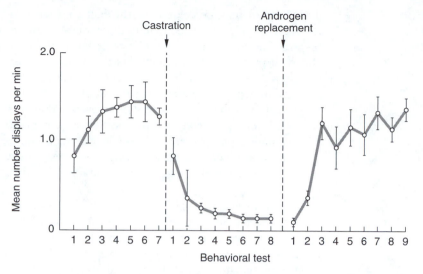

FIGURE 7.12   **Effect of castration and testosterone replacement therapy on the courtship behavior of the male green anole. (Modified from Crews 1979.)**

pituitary gland in the brain that has been thought to be important in mating and parental care (as discussed shortly). However, studies of mice in which the gene for oxytocin has been made nonfunctional shed some doubt on its importance in maternal care. When the gene for oxytocin is disrupted, the mice can mate, give birth, care for their young, and produce milk. However, when the pups nurse, milk is not released from the teats unless the mother is injected with oxytocin. It isn't clear why the maternal behavior of these knockout mice is so normal.

It may be that another hormone, vasopressin, binds to the oxytocin receptors and causes the same behaviors as oxytocin would. To test this idea, researchers plan to create knockout mice that lack the receptor for oxytocin. If these mice also show normal maternal behavior, it would be convincing evidence that oxytocin is not important in maternal care (Young et al. 1998).

CORRELATIONAL STUDIES

Hormonal influences on behavior can also be approached by correlational studies. In using this approach, researchers look for changes in behavior that parallel fluctuations in hormone levels. Correlational studies are useful, but they are not as conclusive as experimental work because there is no evidence of causation. Consider, for example, a correlational study that revealed the relationship between the level of testosterone and aggressive behavior in a songbird.

John Wingfield has examined the behavioral endocrinology of birds under natural conditions (for a review, see Wingfield and Moore 1987). In one study of song sparrows, *Melospiza melodia* (Figure 7.14), Wingfield (1984a) captured males in mist nets or traps baited with seed, collected a small blood sample from the wing vein, and marked each individual with a unique combination of leg bands. Birds were released at the site of capture and seemed relatively unperturbed by the sampling procedure; in fact, some individuals sang within 15 to 30 minutes of release. A given male was sampled from 5 to 10 times during a single breeding season, and each sample was analyzed for testosterone. Wingfield also observed their behavior during this period.

As shown in Figure 7.15, he found a close correlation between peak levels of male territorial and aggressive behavior and maximum levels of testosterone. A male song sparrow defends his territory most intensely

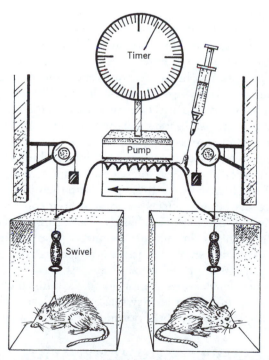

FIGURE 7.13   **Cross-transfusion apparatus used by Terkel and Rosenblatt (1972) to demonstrate the hormonal basis of material behavior in rodents. (From Terkel 1972.)**

FIGURE 7.14   Male song sparrow.

during its initial establishment and when his mate is laying the first clutch. At the time of egg laying, females are sexually receptive and males aggressively guard them from other would-be suitors. Testosterone reaches peak levels during the initial period of territory establishment and during the laying of eggs for the first brood. It is interesting that testosterone does not peak during the period when the female is sexually receptive and laying the second clutch. Aggressiveness is correlated with testosterone levels. The enthusiasm with which a male guards his mate during the second laying period is less than that exhibited during the first period.

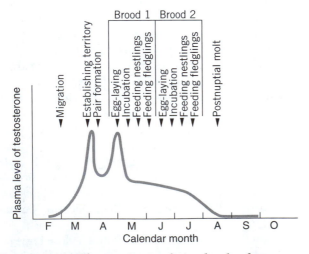

FIGURE 7.15   Changes in circulating levels of testosterone in free-living male song sparrows as a function of the stage of the breeding cycle. (From Wingfield 1984a.)

This pattern may be related to the fact that while a female is laying the second clutch, a male is often responsible for feeding fledglings from their first clutch. Wingfield (1984a) speculates that high levels of testosterone and the resulting heightened levels of male aggression would interfere with parental behavior. Field studies of the song sparrow demonstrate that circulating levels of testosterone wax and wane in parallel with changing patterns of male territorial aggression. This correlational evidence strongly suggests that testosterone regulates aggressive behavior in this species.

In recent years, it has become possible to monitor hormone levels through the analysis of urine and feces rather than blood. These less invasive procedures are often used in field studies and when repeated sampling is necessary (Whitten, Brockman, and Stavisky 1998).

ORGANIZATIONAL AND ACTIVATIONAL EFFECTS OF HORMONES—SEXUAL BEHAVIOR: A CASE IN POINT

DEFINING THE DICHOTOMY

The modes by which steroid hormones influence behavior may be classified as organizational and activational (Phoenix et al. 1959). In organizational effects, steroids organize neural pathways responsible for certain patterns of behavior. Organizational effects occur early in life, usually just before or after birth, and tend to be permanent. This permanence implies structural changes in the brain or other long-term cellular changes, such as in the responsiveness of neurons to steroid hormones (Arnold and Breedlove 1985). Steroid hormones may also affect behavior by activating neural systems responsible for mediating specific patterns of behavior. In contrast to organizational effects, activational effects usually occur in adulthood and tend to be transient, lasting only as long as the hormone is present at relatively high levels. In keeping with their impermanence, activational effects are thought to involve subtle changes in previously established connections (such as slight changes in neurotransmitter production or release along established pathways) rather than gross reorganization of neural pathways. At this point, we will consider the organizational and activational effects of steroid hormones as they relate to the development and display of sexual behavior in the Norway rat.

AN EXAMPLE—SEXUAL BEHAVIOR OF THE NORWAY RAT

Historically, the sexual behavior of rodents has been the most popular research topic in behavioral endocrinology, and the trend continues today (Svare 1988). Because of the large amount of information available, we have chosen the Norway rat to consider the effects of sex steroids on sexual behavior. We will first consider sex differences in copulatory behavior, then the organizational effects produced by testosterone secretion in young animals, and finally the activational effects produced by secretion of sex steroids in adults.

Not surprisingly, adult male and female rats differ in their sexual behavior (Figure 7.16). Whereas mounting, intromission, and ejaculation typify mating in males, behavioral patterns associated with solicitation and acceptance characterize the sexual behavior of females. The lordosis posture, for example, is a copulatory position that female rats assume when grasped on the flanks by an interested male. The intensity of the lordosis response varies across the ovulatory cycle, being most pronounced when mature eggs are ready to be fertilized. The sexual behavior of female rats also includes a variety of solicitation behaviors, such as ear wiggling and a hopping and darting gait, that typically precede display of the lordosis posture (Beach 1976). Although mounting is almost always associated with males and lordosis with females, these behavior patterns occasionally occur in the other sex. Every once in a while, females will mount other females and, similarly, males will occasionally accept mounts from their cagemates. However, by and large, males display mounting and females assume the lordosis posture. These differences in patterns of adult copulatory behavior are due to differences in the brains of male and female rats, differences that are induced by the irreversible actions of androgens in late fetal and early neonatal life. Let's now consider the organizational effects of gonadal steroids on sexual behavior.

It is clear that testosterone in the bloodstream of neonatal rats produces organizational effects. During perinatal life, male and female rats have the potential to develop neural control mechanisms for both masculine and feminine sexual behavior. Certain neurons in the brains of both males and females have the capacity to bind sex hormones. During a brief period, starting about two weeks after conception and extending until approximately four to five days after birth, however, testosterone secreted by the testes of developing males is bound to receptors in the target neurons. Once there, testosterone initiates the production of enzymes that will switch development onto the "male track." The neonatal testosterone causes males to (1) develop the capacity to express masculine sexual behavior and (2) lose the capacity to express feminine copulatory behavior.

Experiments involving castration and hormone replacement techniques have demonstrated the organizational effects of early secretion of testosterone. Removal of the testes in a rat soon after birth results in an adult with a reduced capacity to display masculine patterns of sexual behavior and an enhanced capacity to display feminine patterns. These males are capable of high levels of female solicitation and lordosis as adults. However, if removal of the testes is followed by an experimental injection of testosterone before five days of age and the proper male hormones are administered in adulthood, the rat will display normal male sexual behavior. Normal female fetuses produce low levels of testosterone, so the male developmental pattern is not initiated. A single injection of testosterone into a female rat soon after birth, however, produces irreversible effects on her adult sexual behavior. The testosterone-treated female shows fewer feminine and more masculine patterns of copulatory behavior than does a normal female. Thus, the development of a "male" brain requires the presence of testosterone around the time of birth. In the absence of testosterone, a "female" brain develops. The effects of perinatal testosterone secretion on adult sexual behavior are organizational in that they occur early in life and involve permanent structural changes in the brain. Figure 7.17 summarizes sexual differentiation in the brain and behavior of the neonatal rat.

Before moving to the activational effects of sex steroids, we should mention that masculinization of the brain may be somewhat more complex than just

FIGURE 7.16 Male and female Norway rats differ in their sexual behavior. Whereas mounting is characteristic of males, the acceptance posture, called lordosis, is characteristic of females. These sex differences in adult behavior are established through the action of steroid hormones around the time of birth.

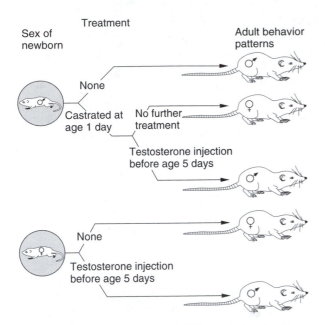

FIGURE 7.17   Pattern of sexual differentiation in the brain and behavior of the Norway rat.

described. In laboratory rats, testosterone appears to be only an intermediate chemical in the process, and it is estradiol, a hormone usually associated with females, that actually directs development along the masculine track. Testosterone enters neurons in specific regions of the brain and is converted intracellularly to estradiol, which in turn causes masculinization. Look again at Figure 7.4 to see that steroid hormones are chemically very similar and that testosterone lies along the pathway of synthesis of estradiol.

The main question that arises from estradiol's role in the masculinization process is this: Why doesn't estradiol in young female rats have the same effect? To begin with, the levels of estradiol in young females are very low. In addition, during this critical period of brain development, an estrogen-binding protein, called alpha-fetoprotein, is produced in the livers of the fetuses. This protein, found in the cerebrospinal fluid of newborn males and females, persists in ever-decreasing amounts during the first three weeks of life. During this time, alpha-fetoprotein prevents estradiol from reaching target neurons in the brain. In female rats, then, alpha-fetoprotein binds any circulating estradiol and thereby prevents it from initiating the male pattern (McEwen 1976). Alpha-fetoprotein does not, however, bind testosterone. Thus, in male rats, testosterone produced by the testes can reach the brain, be converted to estradiol, and result in sexual differentiation. The extent to which this information can be generalized to other mammalian species remains to be established.

In adulthood, steroid hormones produce activational effects on sexual behavior in male and female rats. Female rats with high blood levels of estrogen and prog-

esterone display feminine sexual behaviors in the presence of a sexually active male, but these patterns rarely occur when levels of these ovarian hormones are low. In fact, an adult female whose ovaries have been removed will not copulate unless she receives injections of estrogen and progesterone. Similarly, removal of the testes in an adult male eventually eliminates copulatory behavior, unless he is given injections of testosterone. In these cases, the effects of steroid hormones on sexual behavior are described as activational because estrogen and progesterone in females and testosterone in males presumably exert their effects by activating existing neural pathways. High levels of the gonadal steroids activate specific patterns of sexual behavior. Thus, in contrast to permanent changes in sexual behavior caused by administration of testosterone during the neonatal period, only a transient activational effect on copulatory behavior is produced by sex steroids in adulthood.

One final point will help to distinguish organizational and activational effects of steroid hormones on sexual behavior. Males and females that have had their reproductive organs removed in adulthood generally cannot be induced to behave like members of the opposite sex. For example, a female rat whose ovaries have been removed in adulthood cannot, through injections of testosterone, be induced to show mounting behavior. By adulthood, the nervous systems of adult males and females have already differentiated (i.e., the organizational effects of early steroid secretion have long since occurred), and the mature brains are not capable of responding to hormonal signals of the opposite sex.

QUESTIONING THE DICHOTOMY

In 1985 Arthur Arnold and S. Marc Breedlove questioned the traditional division of steroid influences on behavior into organizational and activational effects. They reviewed experimental findings from the previous decade and concluded that the organizational–activational distinction was too restrictive. How would one classify, for example, effects produced by steroid hormones that were both organizational and activational in nature, such as the production of permanent effects in adulthood? Also, although acknowledging the wealth of behavioral evidence supporting the organizational–activational dichotomy, their attempts to uncover biochemical, anatomical, or physiological evidence of two fundamentally different ways in which steroid hormones act on the nervous system were unsuccessful. In their opinion, failure to find specific cellular processes uniquely associated with each type of effect further blurs the organizational–activational distinction. Although it is important to keep such concerns in mind when discussing steroid influences, we believe that the traditional distinction of organizational and activational effects is still useful in categorizing hormonal effects on behavior.

THE DYNAMIC RELATIONSHIP BETWEEN BEHAVIOR AND HORMONES

The interaction between behavior and hormones is a dynamic one. Whereas hormones can activate specific forms of behavior, social or behavioral stimuli can, in turn, induce rapid changes in the levels of those hormones. For example, sexual stimuli have been shown to trigger rapid increases in plasma androgen levels in many male vertebrates. The marine toad (*Bufo marinus*), a native of Central and South America, is an explosive breeder and the first amphibian species in which it was shown that sexual behavior could affect hormonal state.

Orchinik, Licht, and Crews (1988) studied two populations of marine toads in Hawaii, where the species breeds year-round, with bursts of mating activity following heavy rainfall. During these breeding explosions, males typically compete to clasp the limited number of females, and mating involves prolonged amplexus. When male toads were allowed to clasp stimulus females for zero, one, two, or three hours, concentrations of androgens (testosterone and a form of testosterone called 5-alpha-dihydrotestosterone, or 5-DHT) increased with the number of hours spent in amplexus (Figure 7.18). In addition, in field-sampled males, androgen concentrations were higher in amplexing males than in unpaired, "bachelor" males. The apparent rise in androgens during amplexus suggests

that mating behavior induced the hormonal response rather than vice versa.

Social factors also interact with hormone levels and mating in squirrel monkeys (*Siamiri sciureus*). During the nonbreeding season, the social organization of this species consists of a cohesive nucleus of females and a group of peripheral males. As the breeding season approaches, both males and females undergo "fatting," a phenomenon that is characterized by increases in body weight of up to 35%! At the same time, the sex-segregated pattern of living begins to change, and males and females engage in more social activities.

Several studies of squirrel monkeys have revealed interesting interactions between social and hormonal factors and seasonal patterns of reproduction. For example, in group-housed animals, manipulation of the female's physiological state during the nonbreeding season through the administration of hormones is sufficient to initiate behavioral and physiological changes in both males and females that normally occur only during the breeding season (Jarosz, Kuehl, and Dukelow 1977).

Another experiment demonstrated that changes in social environment can produce different effects on the behavior and physiology of males and females. As mentioned, during the breeding season squirrel monkeys typically live in social groups that contain individuals of both sexes and all ages. Not surprisingly, then, in the laboratory this species reproduces more successfully when housed in large mixed-sex groups rather than in male–female pairs. Mendoza and Mason (1989) attempted to induce breeding activity in pair-housed monkeys by forming new heterosexual pairs just before the start of the breeding season. The change in social partners induced breeding readiness, as measured by social behavior and gonadal hormones, in males but not in females. Although the social stimulation provided by a single novel female was sufficient to produce increases in the levels of plasma testosterone in males, the levels of estrogen and progesterone were not altered in females by presentation of a new male. However, the male response to a new female was transitory, lasting only a few days. This transient response by males is in sharp contrast to the changes in sexual behavior induced by group formation, which we will now consider.

The type of social group appears to be an important influence on hormone–behavior interactions in squirrel monkeys. Mendoza et al. (1979) examined the physiological response of adult male squirrel monkeys that were first housed individually, then in groups of three males, and finally in heterosexual groups formed by the addition of five females to each group of males. When males were moved from individual housing into all-male groups, a linear dominance hierarchy was established within each group: The alpha male was the most dominant individual, the beta male was next in

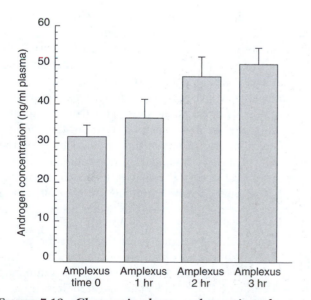

FIGURE 7.18 **Changes in plasma androgen in male marine toads as a function of the time spent in amplexus. The rise in androgens during amplexus suggests that mating behavior induced the hormonal response rather than vice versa. (From Orchinik, Licht, and Crews 1988.)**

line; and the gamma male was the most subordinate group member.

The relative dominance status that emerged following the formation of male groups was critical in determining the physiological response of each male to group formation (Figure 7.19). The alpha male in each group had the highest level of plasma testosterone and the gamma level male the lowest. However, testosterone levels prior to group formation (i.e., when the males were individually housed) could not be used to predict the subsequent dominance status of the males.

In short, the social relationships among males seemed to influence each individual's level of testosterone, rather than the individual's physiological state determining the nature of social relationships. The impact of dominance status on testosterone levels was further accentuated when females were added to each group. Whereas dominant males showed dramatic increases in testosterone, beta and gamma males exhibited declines (Figure 7.19). The physiological responses of males to group formation lasted several weeks.

HORMONES AND ADAPTIVE BEHAVIOR

Hormones provide a mechanism through which an animal can adjust its behavior and physiology so that it is appropriate for the environmental and social situation. So far, we have considered the ways in which hormones influence behavior. We have seen that hormones act on some combination of the sensory system, the central nervous system, and the effector system (e.g., muscles) to alter the probability of particular behaviors. We have also seen that hormones can affect an animal's anatomy and physiology during development in ways that can be long-lasting and that hormones can have immediate effects by triggering certain behaviors. It is important, however, that behavior be tuned to the physical and social environment of the animal. Next we will consider some ways in which hormones, behavior, and environment interact to generate adaptive behavior.

ENVIRONMENTAL INTERACTIONS

Adjusting to Seasonal Changes— Photoperiodism

It is important that certain behaviors occur at the right time of year. In Chapter 9 we will discuss the control of migration and hibernation by an annual biological clock. This clock is especially important in timing behavior in animals that do not have access to environmental, seasonal cues. For example, annual clocks time the start of migration in birds that winter near the equator, and they regulate hibernation of animals that winter in burrows or other constantly dark places.

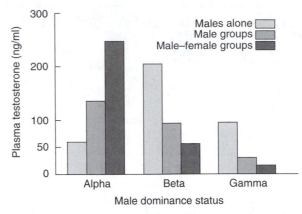

FIGURE 7.19   Mean levels of plasma testosterone in male squirrel monkeys during each of three phases of social group formation. Males are grouped according to their rank in the dominance hierarchy, alpha being highest and gamma being lowest. Although relative dominance status was an important factor in determining a male's physiological response to changes in the social environment, testosterone levels during the phase of individual housing could not be used to predict his subsequent dominance status. Thus, social interactions among the males in each group influenced individual levels of testosterone, rather than levels of testosterone dictating social behavior. The influence of dominance status on levels of testosterone was especially evident when five females were added to each group: Testosterone increased in dominant males but decreased in subordinate males. (From Mendoza et al. 1979.)

Most animals, however, have an obvious cue to signal the changing seasons—day length. Days are long in the summer and short during the winter (Figure 7.20). Many animals, especially those in the temperate zone, use the changing day length to time behavior to occur at the most appropriate time of the year. Reproduction, for instance, is often timed so that births occur during the spring, the optimal time for survival of the young.

The number of hours of light in a 24-hour period is called the photoperiod, and photoperiodic responses gear the annual pattern of behavior to the changing day length. Curiously, however, animals measure the length of the night, or dark part of the cycle, instead of the light period.

Melatonin, a hormone primarily from the pineal gland in vertebrates, is often the link between the season and reproduction in animals living in the temperate zone. In all vertebrates and some invertebrates, there is a daily rhythm in melatonin production, the peak occurring at night. Melatonin synthesis begins shortly after nightfall and continues until dawn. As a result, the daily duration of melatonin synthesis is brief during the long days of summer and long during the long nights of winter (Figure 7.21). Animals use the nightly duration

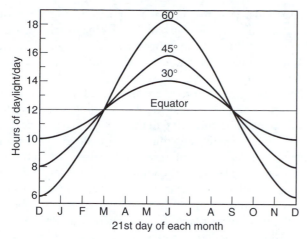

FIGURE 7.20 **The changing day length at selected latitudes throughout the year. Most places on earth have longer days during the summer than in the winter. These changes in day length provide cues to the time of year.**

of melatonin synthesis as an indication of the season of the year (Reiter 1993).

Adjusting to the Harshness of the Environment—Associated and Dissociated Reproductive Patterns

The habitats of different species differ in the number of mating opportunities they provide. As a result, the association between gonadal hormones and sexual behavior varies among species in ways that allow animals to produce the greatest number of surviving offspring.

David Crews (1984, 1987) compared patterns of reproduction in a wide variety of vertebrates and found numerous exceptions to the "rule" of hormone-dependence of mating behavior that we observed in the

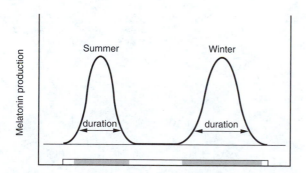

FIGURE 7.21 **In animals, the duration of elevated melatonin production indicates the time of year. Melatonin production begins shortly after dark and continues until the light of dawn. Nights are short during the long days of summer, and so melatonin production is brief. However, nights are long during the short days of winter. Melatonin production is, therefore, prolonged during the winter. (Modified from Reiter 1993.)**

Norway rat. In his survey, Crews considered relationships among the following three components of the reproductive process: (1) production of gametes, (2) secretion of sex steroids by the gonads, and (3) timing of mating behavior. Amid the diversity of reproductive tactics of vertebrates, the following three general patterns of reproduction emerged: associated, dissociated, and constant (Figure 7.22).

Some animals, such as the Norway rat, exhibit a close temporal association between gonadal activity and mating; specifically, gonadal growth and an increase in circulating levels of sex steroids activate mating behavior. This pattern of gonadal activity in relation to mating has been termed an associated reproductive pattern (Figure 7.22*a*) and has been found in numerous laboratory and domesticated species.

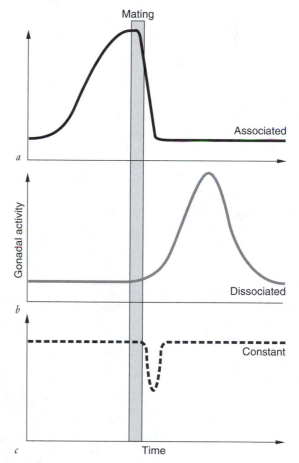

FIGURE 7.22 **Vertebrates exhibit three general patterns of reproduction. In species exhibiting the associated reproductive pattern (*a*), mating occurs at the time of maximum gonadal activity as measured by the maturation of gametes and peak levels of sex steroids. In species exhibiting the dissociated pattern (*b*), mating occurs at a time of minimal gonadal activity. In species exhibiting the constant reproductive pattern (*c*), gonadal activity is maintained at or near maximum levels at all times. (From Crews 1987.)**

Other species, however, exhibit a dramatically different pattern of reproduction in which mating behavior is completely uncoupled from gamete maturation and secretion of sex steroids. In species that exhibit the dissociated reproductive pattern (Figure 7.22*b*), gonadal activity occurs only after all breeding activity for the current season has ceased, and gametes are thus produced and stored for the next breeding season. Gonadal hormones may not play any role in the activation of sexual behavior in species that display the dissociated pattern.

Typically, species with a dissociated reproductive pattern inhabit harsh environments in which there is a predictable, but narrow, window of opportunity to breed, and a specific physical or behavioral cue triggers mating behavior. Consider, for example, the red-sided garter snake (*Thamnophis sirtalis parietalis*), a species that ranges farther north than any other reptile in the Western Hemisphere. The courtship behavior of adult male garter snakes is activated by an increase in ambient temperature following winter dormancy, rather than by a surge in testicular hormones (reviewed by Crews 1983).

Garter snakes in western Canada emerge in early spring from subterranean limestone caverns, where they have hibernated in aggregations of up to 10,000 individuals. Males emerge first, en masse, and congregate at the den opening. Females emerge singly or in small groups over the next one to three weeks and mate with males that are hanging out at the entrance. Because of this timing difference in the emergence of males and females, males greatly outnumber females at the den opening, sometimes 500 to 1 (Figure 7.23). In view of these odds, it is not unusual for a writhing mass of snakes, called a "mating ball," to form, in which over 100 males attempt to mate with a single female. Against all odds, females usually mate with only one male and immediately disperse to summer feeding grounds, where they give birth to live young in August. Males, on the other hand, remain at the den opening and move to feeding grounds only after all the females have emerged.

Testicular activity is minimal in male garter snakes during the period of emerging and mating. In fact, it is five to ten weeks later, after the males have left the den site and will no longer court females, that the testes grow and androgen levels increase. Sperm produced during this time is stored for use during the next spring.

Male red-sided garter snakes use environmental cues instead of circulating levels of sex hormones to determine the appropriate season for mating. Numerous experiments, utilizing castration and replacement techniques and destruction of either the temperature-sensing areas of the brain or the pineal gland, have revealed that rather than relying on surges of sex steroids, the neural mechanisms that activate sexual behavior in male garter snakes are triggered by a shift in temperature (Crews, Hingorani, and Nelson 1988; Krohmer and Crews 1987). This is not to say that courtship behavior is completely independent of hormonal control or that sex steroids do not play an organizational role in the development of sexual behavior, but rather that mating does not occur at the time of maximum gonadal activity.

Like the male, the female red-sided garter snake mates when her gonads are small, gametes immature, and circulating levels of sex steroids low. In the case of

FIGURE 7.23 Male red-sided garter snakes wait at den openings for emerging females. Dense mating aggregations form as females emerge singly or in small groups. The activation of sexual behavior in this species is independent of sex steroids.

the female, however, changes in sexual attractivity and receptivity are mediated by physiological changes that occur as a consequence of mating.

Thus, although both male and female red-sided garter snakes display a dissociated pattern of reproduction, they differ in the type of stimulus that triggers breeding behavior. Whereas a change in ambient temperature triggers courtship behavior in males, stimuli associated with mating appear to activate physiological changes in females.

The third type of reproductive tactic, described by Crews (1987) as the constant reproductive pattern (Figure 7.22c), is characteristic of species that inhabit harsh environments, such as certain deserts, where suitable breeding conditions occur suddenly and unpredictably. In the case of desert-dwelling animals, reproduction is often initiated by rainfall. While waiting for suitable circumstances in which to breed, these species maintain large gonads, mature gametes, and high circulating levels of sex steroids for prolonged periods of time.

The zebra finch, *Taeniopygia guttata castanotis*, lives in the deserts of Australia, where rainfall occurs rarely and unpredictably. Through droughts that can last for years, males and females maintain their reproductive systems in a constant state of readiness. No matter how long the wait, each sex is poised, prepared for the opportunity to breed (Serventy 1971).

Courtship among males and females in desert populations begins shortly after the rain starts to fall, copulation occurs within hours, and nest building can begin as early as the next day. To maintain this accelerated pace, both males and females carry material to the nest. It is interesting that in more humid areas of their range, where the reproductive needs are not so immediate, males and females exhibit the division of labor characteristic of finches—that is, the male alone carries grass to the nest, and the female waits at the nest for each delivery and arranges the new material as it arrives (Immelmann 1963).

SOCIAL INTERACTIONS

The social or territorial system of a species also affects hormone-behavior relationships. For example, testosterone affects the survival rate differently in song sparrows, which defend territories, and cowbirds (*Molothrus ater*), which establish dominance hierarchies.

Dufty (1989) reported that adult male cowbirds that were subcutaneously implanted with testosterone for extended periods of time showed lower survival than control males, which were either implanted with empty capsules or not implanted at all. Whereas only 1 of 16 (6.3%) testosterone-implanted males returned to the breeding grounds the following year, 33 of 81

(40.7%) males without implants and 3 of 6 (50.0%) males with empty implants returned. Testosterone-implanted males also suffered more injuries than did controls. In contrast to Dufty's findings for cowbirds, Wingfield (1984b) reported no difference in the survival of testosterone-implanted and control song sparrow males.

According to Dufty (1989), differences in the system of mating and territoriality help explain the differential effects of testosterone on survival. Male song sparrows are territorial, and territory boundaries may provide a "psychological" barrier to intruding males, something similar to the home court advantage in basketball. In addition, testosterone-implanted males of this species do not differ from control males in the frequency of intrusion into neighboring territories (Wingfield 1984a). In short, the system of territoriality in song sparrows may shield males from the high risks associated with aggressive interactions with other males. Cowbird males, on the other hand, do not defend territories but establish dominance hierarchies instead. The increased aggression induced by testosterone in the absence of the buffering effects of territory boundaries could result in escalation of aggressive interactions and increased risk of injury. In addition, given the existence of dominance hierarchies in male cowbirds, a male whose behavior is inconsistent with his position within the hierarchy may be a favorite target for aggression from other males. Species differences in territoriality may thus explain the differential effects of testosterone on survival of song sparrows and cowbirds.

THE HORMONAL BASIS OF SELECTED ADAPTIVE BEHAVIORS

In many chapters of this text we will discuss the evolution of behavior, often in terms of the costs and benefits of certain actions. We will also describe circumstances required for a particular behavior to evolve and consider how it contributes to fitness. In other words, we will consider the ultimate or evolutionary causes of behavior. But, natural selection affects behavior by selecting genes that affect the nervous and endocrine systems. Here we will consider one of the proximate causes of selected behaviors—the hormonal basis.

MATE CHOICE

Mate choice, which is obviously an important determinant of reproductive success, is affected by hormones in many ways. In some species, hormones play a role in sexual imprinting, which helps an animal choose a mate of the correct species. (Sexual imprinting will be discussed in Chapter 8.) We have seen that testosterone serves as a switch during a rat's development that organizes the brain to be masculine. In the absence of testosterone, the neonatal brain becomes female. Similar organizational effects are seen in many other species. In addition, sex hormones can have activational effects on sexual preference that help an animal to choose a mate of the opposite sex (Adkins-Regan 1998).

Hormones also influence how attractive an individual is to members of the opposite sex. Estradiol makes females more attractive to males. Among primates, for instance, males spend more time close to estrous females than to anestrous females. Although ovariectomized females are rarely attractive to males, estradiol injections make them more attractive. A female monkey whose ovaries have been removed is rarely mounted unless given an injection of estradiol. After estradiol treatment, she is mounted at a much higher frequency (reviewed in R. J. Nelson 1995). In a similar manner, higher androgen levels can make males more attractive to females. For example, the higher the androgen level of a male meadow vole, the more attractive his odor is to a female (Ferkin et al. 1994).

According to sexual selection theory (discussed in Chapter 14), a female is expected to choose a mate quite carefully. When compared with a sperm, each egg represents a much greater proportion of a female's lifetime production of gametes. She therefore has more at stake when she mates. Consequently, we would expect a female to be fussy about the male she chooses; we would expect her to assess a male's quality based on cues that he could provide "good genes" for her offspring.

Because hormone levels are related to traits that allow a female to assess the quality of her suitor, they are one of the mechanisms of sexual selection. Throughout this chapter we have seen examples of species in which androgens boost a male's competitive and aggressive abilities, helping him to defend a territory or rise in dominance status. Indeed, it has been suggested that in male white-crowned sparrows, testosterone functions primarily to arouse aggressiveness, not to elevate sex drive (Wingfield et al. 1987). Territory holders and dominant males are more likely to mate than their less aggressive rivals. Androgens are also responsible for the development and maintenance of many of the elaborate ornaments, including antlers, combs, spurs, and bright coloration, that attract females and enhance reproductive success.

Although testosterone may boost a male's mating success, it also has costs—it can interfere with paternal behavior and increase susceptibility to disease and parasites. Testosterone's interference with paternal behavior is a result of heightened aggressiveness. Studies have shown that in some species of birds, testosterone implants cause males to neglect their nestlings while they continue to sing and defend territories (Ketterson

and Nolan 1992; Silverin 1980). If the neglected young die, it cancels the benefit of increased mating success. Testosterone weakens the immune system indirectly and directly. It works indirectly by driving the stressful activities of the mating season—courtship displays, competition, and the development of elaborate ornaments. Stress is known to impair the immune response that occurs when the body is invaded by parasites or disease-causing organisms. Testosterone also directly suppresses the immune response (Zuk 1994).

ALTERNATIVE REPRODUCTIVE STRATEGIES

After nightfall, from late spring through summer, along the western coast of North America, one can hear the male plainfin midshipman fish (*Porichthys notatus*) "singing" to attract females to his nest. The song, which is more of a droning hum than a melody, is produced when the male contracts a pair of sonic muscles against the swimbladder. The nests are built under rocky shelters in the intertidal zone. Although a female mates with only one male each breeding season, a male may mate with five or six females. He then guards 1000 to 1200 eggs.

There are actually two types of male plainfin midshipman fish, and they display different reproductive strategies. The males we've described, those that build nests and sing to court females, are type I males. In contrast, type II males don't build nests or hum to attract females. Instead, they either sneak into the type I's nest and spawn or lie outside the entrance and deposit sperm there while fanning water toward the nest's opening. The sperm of the type II male will then compete with that of the type I male, who did all the work (Bass, Bodnar, and McKibben 1997).

There are other differences between type I and type II males. Type I males are larger and they take longer to reach sexual maturity. Although type II males become sexually mature sooner, they can't attract mates. The sonic muscles of type I males used to produce the droning hum that is so attractive to females are much more developed than those of a type II male. The brain of a type I male is also specialized to allow him to hum his courtship song. The size of the motor neurons to the sonic muscles and the brain center controlling those muscles are also larger in a type I male. Surprisingly, though, the ratio of testis to body weight of a type II male is nine times that of a type I. Because of his bulging gonads, a type II male resembles a gravid female (Figure 7.24). The coloration of a type II male is also more similar to that of a female than a type I male.

Hormonal differences are responsible for the physical and behavioral differences among type I males, type II males, and females. Although there are differences in the levels of testosterone and estrogen among these

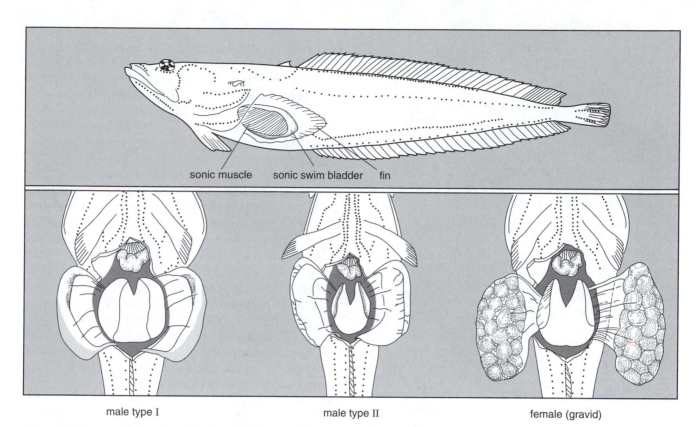

male type I   male type II   female (gravid)

FIGURE 7.24   A type I male plainfin midshipman fish contracts his sonic muscles against the swim bladder to create the droning hum of his courtship song. There are two types of males of this species, each with a distinct mating strategy. Type I males build nests, sing to attract females, and guard nests. A type II male waits for a type I male to attract a female and then sneaks into the nest and spawns or releases sperm just outside the nest and fans water into the nest opening. The sonic muscles of a type I male are much more developed than those of a type II male or a female. The ratio of sonic muscle to body weight is six times greater in a type I male than in a type II. Type I males are larger, but the ratio of gonads to body weight is nine times smaller than that of a type II male. The large testes make a type II male look like a gravid female (From Bass 1996.)

morphs, it is thought that the most important difference is in the level of a particular form of testosterone, 11-ketotestosterone. Type I males have five times more 11-ketotestosterone than testosterone, but 11-ketotestosterone is undetectable in either type II males or females.

There is a tradeoff of costs and benefits for these two mating strategies. Type II males develop sooner than type I males, and therefore have a greater chance of surviving to reproduce. Their costs include greater gonad development and an expanded hormonal system. Type I males take longer to mature than type II males. The delay allows them to grow larger by the time of maturity, which is advantageous in combat and nest competition. It also provides more time for their sonic muscles to develop, which allows them to hum courtship songs to attract females and grunt to intimidate rival males. However, delayed maturity brings with it a costly investment in physical and behavioral development, as well as the increased possibility that the male will not survive to reproduce.

It is interesting to note that some preliminary studies suggest that the number of type II males produced varies with population density. When population density is low, more type I males are produced; when density is high, the percentage of type II males increases (Bass 1996). This makes adaptive sense since competition increases with population density, and type II males avoid competition. The physiological mechanism that makes a male's developmental fate sensitive to population density is not yet known. However, it has recently been suggested that a physiological mechanism that might favor the development of different reproductive strategies under different environmental or social conditions might involve stress hormones (Moore, Hews, and Knapp 1998).

MATING SYSTEMS

Prairie voles (*Microtus ochrogaster*), common residents in midwestern grasslands, are monogamous (Figure 7.25). They form strong pair bonds, usually for life. In addition, after mating they spend most of their time with each another, preferring each other's company to that of other voles. In fact, when one of the pair dies, the survivor rarely pairs again, even though mates are available. If a stranger threatens the nest, the male becomes fiercely protective. Males also share in caring for the young.

Oxytocin and vasopressin, short chains of amino acids, are the hormonal basis for monogamy. These chemicals function in two ways—as traditional hormones and through local effects on nearby nerve cells. When they act as traditional hormones, they are produced by neurosecretory cells of the hypothalamus and released from the pituitary gland in the brain. Vasopressin regulates body water content by stimulating the absorption of water by the kidneys, decreasing the volume of urine produced. Oxytocin is involved with many reproductive functions, including stimulation of uterine contractions during the birth process and the release of milk from the mammary glands. Oxytocin and vasopressin are produced by certain nerve cells in the hypothalamus and other regions of the brain. They can also have local effects in the brain, where these neurohormones bind to receptors on the surface of nearby neurons and alter neural activity. In this manner, oxytocin is thought to stimulate maternal interest in newborn pups in rats and mice.

FIGURE 7.25 The prairie vole is a monogamous species. Following mating, a long-lasting pair-bond forms. If one member of the pair dies, the survivor rarely takes a new mate. The male guards his partner, protects the nest from any male intruders, and helps care for the young. In females, the formation of a pair-bond and the development of maternal care are caused by oxytocin, which is released during mating. Instead of oxytocin, vasopressin causes a male to bond with his partner, protect her and the nest, and care for the young.

Pair-bonding in prairie voles begins when a female sniffs the genital area of an appropriate male. In this way, she detects a male pheromone, and it triggers hormonal changes that make her sexually receptive. Within 48 hours of sniffing a male, a female comes into estrus and is sexually receptive. The pair begins to mate, and they continue to engage in bouts of mating, sometimes more than 50, for about 30 to 40 hours (Carter and Getz 1993).

The intense mating during this "honeymoon" triggers hormonal changes in the brain that are responsible for monogamous behavior. Specifically, it stimulates the release of oxytocin or vasopressin in key areas of the brain. Oxytocin causes females (but not males) to form a long-lasting pair-bond with their partners (Insel and Hulihan 1995; Williams et al. 1994). In contrast, vasopressin is responsible for the changes in male behavior. After mating, he prefers to spend time close to his mate, is aggressive toward strangers, and will care for the young (Young, Wang, and Insel 1998).

Vasopressin is the basis of social attachment in male prairie voles. Even without mating, a boost of brain vasopressin will cause a male to form a pair-bond. This was demonstrated by pumping small amounts of vasopressin, oxytocin, or a hormone-free fluid directly into the brains of male prairie voles for 24 hours. Each male was housed with a female vole whose ovaries had been removed. Although an ovariectomized female will not mate, the couple did cuddle and groom each other during the infusion. The male was then given a choice of partners. He was placed in a large cage with several chambers. The familiar female was tethered in one chamber and a female stranger was tethered in another. The male freely explored both females. Preference was measured as the amount of time spent with each female. Males who had been infused with oxytocin or the hormone-free fluid spent roughly the same amount of time with each female. However, the males who had been infused with vasopressin spent about 75% of the time with the familiar partner. It has been suggested that vasopressin leads to pair-bond formation by helping the male remember the time spent with his partner.

The vasopressin released during prolonged mating also transforms a nonaggressive male into one that will attack an unfamiliar male, a stranger that might threaten the nest. James Winslow and his colleagues (1993) gave male prairie voles an injection of a vasopressin antagonist (a chemical that binds to the receptor of the hormone and prevents the hormone from acting) just before the male vole was allowed to mate for 24 hours. The pair still mated intensely, but afterward the male did not attack a strange male vole placed in his cage. A male will attack if he is injected with an oxytocin antagonist or with a hormone-free fluid, showing that the effect is specific to vasopressin.

The vasopressin released at mating also makes the male prairie vole a good father. Like the mother, he spends a lot of time grooming and cuddling the pups. If newborn pups are placed in a cage with a virgin male, he shows little or no parental care. However, if vasopressin is first injected into the lateral septum of that male's brain, he immediately approaches and crouches over the pups (Wang, Ferris, and DeVries 1994).

In contrast to the monogamous behavior of prairie voles, their cousins the montane voles (*M. montanus*) are promiscuous. They don't form pair-bonds after mating. Indeed, the pair parts soon after copulating. For the male montane vole, parenting ends with conception.

What makes prairie vole males behave so differently from montane vole males? The answer probably lies in differences in the distribution and density of receptors for oxytocin and vasopressin. It is also interesting that brain maps of receptors for these hormones are similar in the prairie vole and another monogamous species, the pine vole (*M. pinetorium*). Likewise, the receptor distributions are similar in the montane vole and the polygamous meadow vole (*M. pennsylvanicus*). Because of these differences in the pattern of their receptors, oxytocin and vasopressin activate different brain circuits in monogamous and polygamous voles, causing the species to respond differently to these brain hormones. The differences in the brain maps of receptor sites arise because of differences in the regulation of the genes for the receptors during development (Young, Wang, and Insel 1998). The species differences in vasopressin receptor distribution appear quite early in development, and so they could cause differences in how the central nervous system is organized in prairie and montane voles (Wang et al. 1997).

HELPING IN FLORIDA SCRUB JAYS

Helpers are nonbreeding animals that assist the breeding pair in rearing their young. They have assorted parental duties, including providing food and protecting the young. Helping has been reported in over 200 species of birds and more than 100 species of mammals.

One of the favorite species to study is the Florida scrub jay (*Aphelocoma coerulescens*), a bird that lives in south central Florida, usually in the dry oak scrub. Although a group can range from two to six members, it usually has three. As we will discuss in Chapter 17, the helpers are offspring of the breeding pair from a previous year who remain on the territory and help raise their siblings. Because the siblings share some of the helpers' genes, the helpers manage to get some of their genes into the population even though the lack of a territory prevents them from breeding. In this way, the helpers make the best of a bad situation. Thus, we can see the evolutionary causes of helping.

Ronald Mumme and Stephen Schoech (Schoech 1998) have been investigating the physiological basis of helping in a population of Florida scrub jays at the

Archbold Biological Station. Apparently, there are no physiological reasons that cause helpers to delay reproduction. Helpers do produce the hormones important to reproduction. In both males and females, breeders and nonbreeders have the same levels of luteinizing hormone (LH), the hormone from the anterior pituitary gland in the brain that stimulates the growth and development of the ovaries and testes at the start of each breeding season. Although male breeders have slightly higher levels of testosterone than do male helpers, the seasonal pattern of testosterone production is essentially the same. In females, breeders and helpers have the same level of estradiol, their primary sex steroid. However, the seasonal pattern of estradiol production is different in female breeders and female helpers. These observations reveal that the testes and ovaries of helpers are functional, at least for hormone production.

Helpers don't delay reproduction because they are unable to gather enough food to become breeders. Helpers are slightly smaller than breeders, probably because they are younger. When the size difference is taken into account, the weights of male helpers and breeders are equivalent. Helper females are apparently able to forage as well as female breeders during the winter months because they weigh about the same at the start of the breeding season. If the helpers delay breeding because they cannot gather enough food, we would expect that supplying supplemental food to the population would allow more helpers than usual to switch to being breeders. But, when this was tried, the additional food did not increase the number of helpers that became breeders.

Although it is possible that the presence of the dominant breeding pair could stimulate the helper's adrenal gland to produce the stress hormone corticosteroid, which is known to suppress the production of the hormones needed for reproduction, this doesn't seem to occur. Helpers and breeders have equivalent amounts of corticosterone throughout the breeding season.

Lack of opportunity, not physiology, may cause helpers to delay reproduction, but why do they perform parental behaviors such as feeding the nestlings? Parental behaviors may be caused by the pituitary hormone prolactin in both helpers and breeders. Prolactin production is stimulated by cues from the nest, eggs, and nestlings. For this reason, prolactin levels increase throughout the breeding season. Birds that spend the most time caring for the eggs and young produce the most prolactin. In general, females produce more prolactin than males, and breeders of either sex produce more than helpers. Prolactin levels are lower in helpers because breeders won't allow them near the nest until the young have hatched. But, both breeders and helpers feed the young. There is a direct relationship between a helper's level of prolactin and the feeding score (a

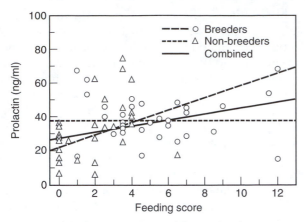

FIGURE 7.26 **There is a direct relationship between how much a bird feeds nestlings and its prolactin levels. (Data from Schoech, Mumme, and Wingfield 1996.)**

measure of how much a bird fed the nestlings). Notice that some helpers didn't actually help; they have feeding scores of zero. The prolactin levels of the helpers that *did* help are much higher than those of birds that did not help. It is interesting that there is no correlation between the prolactin levels of breeders and their feeding score. Prolactin is known to affect many aspects of parental behavior besides feeding the young. A relationship between prolactin and parental behavior among breeders may not be seen in these data because parental behaviors other than feeding were not measured (Figure 7.26) (Schoech, Mumme, and Wingfield 1996).

SUMMARY

Animals have two systems of internal communication, the nervous system and the endocrine system. These two systems are so closely associated that they are often referred to as a single system—the neuro-endocrine system. Despite this close association, neural and hormonal modes of information transfer differ in many respects. Typically, transfer of information by the endocrine system occurs more slowly than that by the nervous system, and the effects that are produced are more general and long-lasting. Whereas neural information is transmitted via a series of electrical events, communication by the endocrine system occurs through hormones, chemical substances that are secreted by either endocrine glands or neurons.

Both hormones and neurohormones produce changes at the cellular level that ultimately influence behavior. There are two types of hormones. Peptide hormones are water-soluble amino acid chains that bind to receptors at the cell surface, which activates a cascade of chemical reactions within the cell. In contrast, steroid hormones are derived from cholesterol, and because they are fat-soluble, they move through the cell boundaries and bind to receptors inside the cell. The hormone-receptor complex then enters the nucleus and turns on certain genes.

The mechanisms by which hormones influence behavior include alterations in (1) sensation or perception, (2) development and activity of the central nervous system, or (3) effector mechanisms.

During development, the tobacco hornworm goes through several larval stages, a pupal stage, and finally becomes a winged adult. The physical and behavioral changes that accompany metamorphosis are controlled by hormones.

Changes in the hormonal state can also trigger behavioral change, as when the tobacco hornworm loses behaviors associated with the prolegs as a result of a peak in the level of ecdysteroid hormones. This peak in ecdysteroids causes regression in the underlying neural circuitry and hence loss of proleg behaviors.

Traditionally, the effects of steroid hormones on behavior have been divided into organizational and activational effects. Organizational effects occur early in life and tend to be permanent. In contrast, activational effects occur in adulthood and tend to be transient, lasting only as long as the hormone is present in relatively high concentrations. In this case, steroids serve only to activate existing neural pathways responsible for a specific behavior rather than to organize neural pathways. The traditional distinction between organizational and activational effects has been questioned in recent years because of the lack of biochemical, anatomical, and physiological evidence for two fundamentally different ways in which steroids produce their effects.

Hormonal effects on behavior can be studied by the removal of the gland and hormone replacement or by correlational studies. In the latter, researchers look for changes in behavior that parallel fluctuations in hormone levels.

There are dynamic interactions between hormones and behavior. In some instances, hormones initiate changes in behavior; in others, behavior causes changes in the levels of circulating hormones.

Hormones produce adaptive behaviors through interactions that make it appropriate for environmental and social conditions. Hormones are a physiological mechanism through which natural selection affects behavior. Hormones can affect mate choice, reproductive strategies, mating systems, and parental care.

8

The Development
of Behavior

A mallard duckling (*Anas platyrhynchos*) hatches. After spending approximately four weeks in the egg, the youngster spends only a single day free from the confines of its eggshell, yet still beneath its mother, before confronting the world beyond the nest. Should a predator, sometime during the first day, wander into the area around the nest, the duckling, in concert with its eight or so siblings, responds rapidly to the alarm call of its mother by "freezing"—crouching low and ceasing all movement and vocalization (D. B. Miller 1980). If we assume a happy ending to this encounter (i.e., the nest and its contents go unnoticed), the very next morning the duckling responds promptly to yet another of its mother's calls, this time the assembly call, by following her from the nest to a nearby pond or lake (Miller and Gottlieb 1978). Here, the duckling will string along behind its mother and siblings for some time to come. As the weeks pass and the young bird continues to associate with family members, it learns the characteristics of an appropriate mate (Schutz 1965). This information, although obtained early in life, will not come in handy until the first breeding season. Indeed, the duckling learns, soon after hatching, most of the things it needs to survive and reproduce. We see, then, that the changes that occur during development may contribute to fitness immediately (as in the duckling's freeze response to its mother's alarm call) and in adulthood (as in mate preference).

Several questions arise from this brief description of the early behavioral development of mallard ducks. How does the genetic makeup of the duckling interact with its internal and external environment to produce such behavior? How are the nervous and endocrine systems involved in behavioral development? What role do visual, auditory, or social stimuli play in the development of freezing, following, and sexual behaviors? Are experiences prior to hatching important to the development of posthatching behavior? What happens when a behavior, such as the following response, ceases to be a part of the individual's behavioral repertoire? If we look beyond the single duckling in our example and consider the species as a whole, does behavioral development always proceed in a predictable and reliable fashion? As we see, many important questions are associated with the development of behavior in animals.

CAUSES
OF BEHAVIORAL CHANGE
DURING DEVELOPMENT

Patterns of behavior come and go throughout development. A behavior may appear in an animal's repertoire, only to disappear or change shortly thereafter. Here we consider four of the major causes of behavioral change during development. Keep in mind, however, that these causes are not mutually exclusive.

DEVELOPMENT
OF THE NERVOUS SYSTEM

Behavior sometimes changes in response to the development of the nervous system. Often, correlations between the state of the nervous system and the behavior of a developing animal are most obvious during embryonic life. Consider, for example, the neural and behavioral development of embryonic Atlantic salmon, *Salmo salar* (Abu-Gideiri 1966; Huntingford 1986). Stages in the development of this fish are depicted in Figure 8.1.

The first movements of the embryo are seen in the feeble twitches of the heart, soon followed by movements in the dorsal musculature. It is interesting that the earliest twitches of the heart and dorsal musculature occur before the nervous system has formed. These movements are thus myogenic, or muscular in origin. The impulse begins in the muscle itself. Approximately halfway through embryonic life, the major motor systems appear in the spinal cord. A short time later, the motor neurons make contact with anterior muscles, giving the embryo the ability to flex its body. Soon, with the development of neural connections at different points and on both sides of the body, the embryo dis-

FIGURE 8.1 Behavioral development in the Atlantic salmon. Patterns of behavior emerge in parallel with the development of neural structures necessary for their performance. (Modified from Huntingford 1986; drawn from the data of Abu-Gideiri 1966; Dill 1977.)

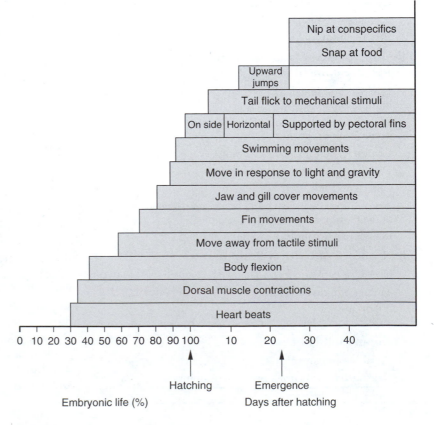

plays the first undulating movements associated with swimming. Development of the sensory system of the trunk and its connection to the skin occurs a short time later; after this, the embryo can move in response to tactile stimulation. Finally, the neural circuits that underlie both fin and jaw movements are complete, allowing independent and coordinated movement of these structures. Neural and behavioral development continues (in fact, the young salmon has not yet even reached hatching). But we can see from these examples that changes in the nervous system underlie the appearance of new patterns of behavior.

An associated question is this: What happens to neural circuits during development when an old behavior disappears from an animal's repertoire? In some cases, the neural circuits are dismantled or permanently altered once the behavior ceases to occur (see the example of the tobacco hornworm, described in Chapter 7). This alteration, however, does not occur in the patterns of behavior associated with hatching in chickens (*Gallus domesticus*).

Under normal circumstances, hatching behavior occurs only once in the life of a chicken, typically during a 45- to 90-minute period at the end of incubation. During hatching, the chick escapes from the confines of its shell through a highly stereotyped series of movements that involve rotation of the upper body and thrusts of the head and legs. Since patterns of behavior associated with hatching disappear from the repertoire of chickens, Anne Bekoff and Julie Kauer (1984) became interested in the fate of the neural circuitry underlying, in particular, the leg movements of hatching. Their method was an interesting one. Rather than asking the age-old question "Which came first, the chicken or the egg?" they asked, "What happens when you put a chicken back into an egg?" Bekoff and Kauer placed posthatching chicks up to 61 days of age (chicks at this age have molted their fuzzy down, are fully feathered, and basically resemble small chickens) in artificial glass eggs and recorded their behavior and muscle movements. Each chick was gently folded into the hatching position and placed into a ventilated glass egg of the appropriate size. Within two minutes of being placed in the artificial eggs, chicks of all ages began to produce a behavior that qualitatively and quantitatively resembled that of normal hatching. Rather than being dismantled or permanently altered after hatching, the neural circuitry for the leg movements of hatching clearly remains functional in older chickens. The question of why the circuitry remains functional is intriguing.

CHANGES IN HORMONAL STATE

Honeybees (*Apis mellifera*), generally considered to occupy one of the summits of insect social evolution, live in colonies made up of a single queen and up to 80,000 workers. Although the hive has no ruler who assesses what needs to be done and issues the commands, its activities are highly coordinated. The chores of a worker usually change dramatically with age, a phenomenon known as age polyethism. A typical honeybee worker that has emerged from the pupa as an adult, winged insect will live for approximately six weeks and during this time moves through an orderly progression of tasks: Young adults (4 to 12 days old) tend the brood and the queen, middle-aged bees (13 to 20 days old) perform a variety of tasks associated with nest maintenance and food storage, and the oldest bees (over 20 days of age) concentrate on foraging outside the nest (Winston 1987). In general, young workers stay inside the hive and tend to nursing and housekeeping, and older workers often leave the hive to forage. Keep in mind, however, that the actual changes in the duties of a worker usually involve only changes in the frequencies of particular behaviors with age, rather than the addition or loss of specific behaviors. Also, as we will see shortly, the sequence is flexible, allowing duties to be reassigned if the hive's needs change (Robinson 1998).

Age polyethism in worker honeybees is regulated by juvenile hormone (Figure 8.2); levels of JH increase with age and mediate changes in work habits. These levels vary with age because of changes in the rate of its synthesis by the corpora allata, a pair of glands near the brain. Not only are the naturally occurring levels of JH correlated with a worker's change in duties, but also experimental manipulations of JH level have resulted in the expected changes in behavior. Treating newly emerged worker bees with a chemical that mimics the effects of JH speeds up behavioral maturation (Robinson 1987; Robinson et al. 1987). On the other hand, removal of the corpora allata delays but does not prevent the development of a worker into a forager. But, that delay can be eliminated by juvenile hormone treatments (Sullivan et al. 1996; Sullivan et al. 2000).

How does JH cause an adult worker's behavior to change so dramatically? At least part of the answer is that JH affects brain structure. The most striking change in behavior during the usual progression of duties occurs when a worker becomes a forager. Unlike younger hive bees, foragers take long-distance flights, learn rapidly and form long-term memories, use the sun as a compass for orientation,* and communicate the distance to and direction of a food source to other foragers (Fahrbach 1997). As a honeybee matures, there is a 20% increase in the size of the mushroom bodies within the brain. The mushroom bodies are important in learning, memory, and the processing of sensory information. The increase in volume is due to increases

---

*Sun-compass orientation is discussed further in Chapter 10.

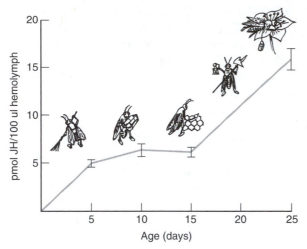

FIGURE 8.2   In honeybees, rising titers of juvenile hormone cause adult workers to switch tasks. Young workers concentrate on cleaning nest cells, middle-aged workers perform a variety of chores, and old workers focus on foraging outside the nest. (Data from Robinson et al. 1987.)

in the number of branches on the neurons, which increase the number of synapses and should boost the brain's ability to process information (Withers, Fahrbach, and Robinson 1993). Thus, these changes in the mushroom bodies may prepare a worker to learn to navigate during foraging flights and to find its way back to the hive.

Juvenile hormone, not experience, is responsible for the growth of the mushroom bodies. This was demonstrated by treating one-day-old workers with a chemical that mimics the effects of JH to cause them to begin foraging sooner than usual. Normally, young workers take orientation flights to learn the lay of the land before they begin to forage. But some of the treated bees were prevented from leaving the hive for orientation flights by a large tag attached to their backs. They were permitted to interact with other bees, however. Although the bees were different ages, the mushroom bodies of all treated bees grew to about the size of those of untreated workers that were beginning to forage. Furthermore, the growth of the mushroom bodies was equivalent in bees with and without foraging experience (Withers, Fahrbach, and Robinson 1995).

Rather than being a rigid system of control, hormonal regulation of the age-based division of labor in honeybees is a flexible process that permits ongoing adjustments in the proportions of workers involved in any one task. The rise in JH may be modulated by environmental conditions such as food availability or age structure of the colony. Workers from starved colonies begin foraging a few days earlier than workers from well-fed colonies (Schulz, Huang, and Robinson 1998). Furthermore, the rate of behavioral development is adjusted to the proportion of older foraging bees in the

colony. If foragers are removed, younger workers begin foraging sooner. On the other hand, if the number of foragers is increased, young workers remain in the hive longer than usual, caring for the young and maintaining the hive (Huang and Robinson 1996). Thus, worker bees adjust their rate of behavioral development so that there are enough workers to perform the tasks most in demand.

Social interactions among colony members modify behavioral development by altering the levels of JH. It has long been known that the queen bee's mandibular glands contain a pheromone (a chemical used in communication) that prevents workers from rearing new queens and the development of ovaries in worker bees. We now know that the queen's mandibular gland pheromone also influences the age-related division of labor and that it does so by lowering JH levels (Pankiw et al. 1998). However, social interactions among the workers themselves are more important in adjusting the rate of behavioral development than are queen–worker interactions. Foragers inhibit the behavioral development of younger workers, and this inhibition requires direct physical contact between older and younger workers, suggesting that a nonvolatile chemical may be involved (Huang 1992). Inhibitory chemicals are exchanged during physical contact among workers, and they adjust the rate of behavioral development by inhibiting JH synthesis. The total inhibition of JH and behavioral development requires direct physical contact between older and younger workers. As the proportion of older foraging workers in a colony increases, so does the frequency of social interactions between older and younger workers. These will slow the rate at which young workers become foragers. The mandibular glands of workers contain chemicals similar to those in the queen's mandibular gland pheromone and could be the source of the inhibitory signals (Huang, Plettner, and Robinson 1998).

CHANGES IN NONNEURAL MORPHOLOGY

Sometimes behavioral change is driven by morphological changes that are not neural. In this case, the development of a particular behavior may parallel the development of specific structures necessary for its performance. Consider the changes in feeding behavior that occur in the paddlefish *Polyodon spathula*. This fish, perhaps best known for its large, paddle-shaped snout, which in some individuals can be almost half the length of the body, occurs in the Mississippi and Ohio river drainages of North America. Although larval paddlefish feed by chasing and selectively plucking individual zooplankton from the water, adult paddlefish are indiscriminate filter feeders, dropping their lower jaw and consuming all material strained from the water as they plow through their environment (Figure 8.3).

FIGURE 8.3   A paddlefish exhibiting the adult feeding pattern, filter feeding. Rather than indiscriminate filter feeding, larval paddlefish selectively pick zooplankton out of the water column. The eventual development of the adult mode of feeding parallels the development of gill rakers in the young fish.

Changes in the feeding behavior of paddlefish parallel the development of gill rakers (Rosen and Hales 1981). These bony structures, comblike in appearance, project from the gill arches into the oral cavity and strain food particles from the water. Gill rakers first appear as small protuberances along the midsection of gill arches when the fish are approximately 100 millimeters long (about 4 inches). At this point in development, the fish still seek out and capture individual prey. By the time individuals are about 125 millimeters in length, the "buds" of gill rakers line all the gill arches and the projections themselves have increased in length. The young paddlers, however, are still not capable of efficient filter feeding. Paddlefish have well-developed gill rakers once they are approximately 300 millimeters in length, and it is at this stage that feeding behavior takes on the fully adult pattern of indiscriminate filter feeding. Here, then, we have an example of how changes in behavior are coordinated with the development of specific anatomical structures.

PLAY BEHAVIOR

Play is thought to be important in the normal development of behavior in many mammals, as well as in some birds and even a few reptiles. We know play when we do it and when we see animals doing it (Figure 8.4). During a visit to the zoo, we even recognize play in species we

FIGURE 8.4   Lion cubs playing. Play is thought to be vital to the development of behavior in mammals.

have never seen before. The attribute that pervades all play, and by which we most commonly identify it, is our subjective judgment of a lack of purposefulness.

Play *is* fun, and there is now some neurochemical evidence that supports this intuition. For example, when a rat anticipates an opportunity to play, its brain releases dopamine, a chemical associated with pleasure and reward. It is thought that the dopamine invigorates the animal and makes it more likely that the interactions that follow will be playful. Individual juvenile rats were placed in one of two plexiglass containers that were connected by a bridge. During the first five minutes of their stay in the apparatus, the rats typically explored the containers, crossing the bridge a number of times. The rats in the experimental group were then allowed to play with another rat in the apparatus for five minutes. Those in the control group were left in the apparatus without an opportunity to play. The control rats did get an equivalent amount of playtime, but at another time and place. After eight days, the experimental rats crossed the tunnel significantly more often than the control rats. This suggests that stimuli associated with the opportunity to play (coming from the apparatus) began to elicit an anticipatory response in the experimental rats. This expectation would then increase the likelihood of a playful interaction. The anticipatory excitement is due to the release of dopamine because it is prevented if the experimental rats are injected with a drug that blocks dopamine action (Figure 8.5). Other experiments have revealed that endorphin levels increase while rats play. Endorphins are pleasure-inducing chemicals sometimes called endogenous opiates. It has been suggested that the endorphins enhance the pleasure associated with playing (Siviy 1998).

Although play is easy to spot, it is difficult to define, mainly because there is no specific behavior pattern or series of activities that exclusively characterize it. Play borrows pieces of other behavior patterns, usually incomplete sequences and often in an exaggerated form. It consists of elements drawn from other, functionally different behavior patterns juxtaposed in new sequences. Polecats and badgers rapidly alternate prey-catching movements and aggressive behavior. A playful mongoose mixes components from hunting and sexual behavior (Rensch and Ducker 1959). In their normal context, behaviors usually involve several actions that follow one another in a predictable order. During play, however, there are often incomplete sequences of behavior. An adult cat would normally stalk a mouse and then strike, but a frolicking kitten may simply pounce on a leaf. An infant rhesus monkey may mount a companion without intromission. The play fights of polecats do not include the two extreme forms of attack ("sustained neck biting" and "sideways attack") or the two extreme fear responses ("defensive threat" or "screaming") (Poole 1966). When a dog is aggressive, it bares its teeth and growls. Its hair stands on end, adding to its ferocious appearance. However, in a play fight, the snarl is not accompanied by the raising of hair.

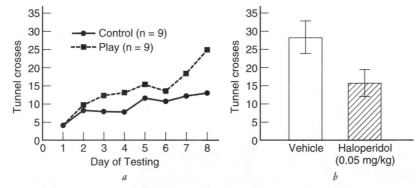

FIGURE 8.5   Rats about to be given an opportunity to play show an anticipatory response that depends on the release of the pleasure-mediating chemical dopamine in the brain. (*a*) The mean number of tunnel crossings between two plexiglass containers for experimental rats that were about to have a "play date" and control rats that did not have an immediate opportunity to play were significantly different after eight days of training. (*b*) The mean number of tunnel crossings by rats that were anticipating an opportunity to play when injected with a chemical that blocks the action of dopamine (haloperidol) and when injected with only the vehicle used to administer the dopamine blocker. There was a significant reduction in the number of tunnel crossings when the action of dopamine was prevented, showing that dopamine is a factor in the anticipatory excitement before play. (Data from Siviy 1998.)

Play is also difficult to define because there are several types. First, there is social play, which includes play fighting or play chasing, as well as sexual play. We have all been amused by the friendly tussles of kittens and puppies as they chase, wrestle, and pounce on one another. Sexual play includes the playful mounting of gazelles or precocious courtship in emydid turtles. A second form is locomotor play—that is, exercise. Foals kick up their heals and gallop. Young primates, including human children, may swing and roll and slide. Infant mountain gorillas are particularly fond of sliding. They begin on their mother's body and then graduate to dirt banks and tree trunks. Polar bear cubs climb ashore only to leap back into the water. Object play is the third form. In this case, objects are manipulated. When first presented with a novel object, a young animal typically explores it by touching, sniffing, or viewing it from different angles. After the initial sensory investigation, the object may become a toy (Fagan 1981). Sometimes a young animal will flit among the types of play in rapid succession, and the predominant form of play may change as the animal matures (Burghardt 1998).

But why do animals play? In other words, what function does this frolicking serve? What role does play serve in the development of behavior? The hypotheses for the long-term significance of play might be grouped into three categories (Thompson 1996):

1. Physical: training for strength, endurance, and muscular coordination, particularly the skills relating to intraspecific fighting and prey capture.

2. Social: practice of social skills such as grooming and sexual behavior; establishment and maintenance of social bonds. Play also helps develop an animal's ability to read and send signals to communicate with other members of its species.

3. Cognitive: learning specific skills or improving overall perceptual abilities.

As we explore these hypotheses further, we will see that some examples of play may fit into more than one category. One of the physical benefits of play may be that it helps form connections between neurons in the brain, especially in the cerebellum, a brain region important in motor coordination and memory of motor patterns. There is a limited time period, a sensitive period, during which synapses are being formed between neurons in the cerebellum. During this time, experience affects the number and pattern of the synapses. In house mice, Norway rats, and domestic cats, locomotor play coincides with the period when the cerebellar architecture is being shaped. In other words, locomotor play begins just when experience can modify the connections within the cerebellum and ends when the cerebellar structure is set (Byers and Walker 1995).

Another physical benefit of play is that it affords animals the opportunity to practice skills that will be essential to later survival (Caro 1988). Hunting games, for instance, may help perfect the movements of catching prey such as stalking, throwing, and shaking. Some examples of these actions are familiar. Kittens stalk leaves and pounce on balls of yarn. Puppies chase sticks and often "shake the life" from them as they would a prey animal. During object play, fledgling American kestrels (*Falco sparvericus*) prefer objects that resemble their natural prey (Negro et al. 1996). Play may have some immediate benefits to predatory skills for cheetah (*Acinonyx jubatus*) cubs. Cubs that playfully crouched and stalked littermates also crouched and stalked prey more often than less playful companions. And, when the mother released live prey for her cubs to catch, those that showed the highest rates of object play and contact social play were more likely to be successful (Caro 1995).

The play fighting of young animals may be practice for the battles of adults that establish dominance hierarchies and defend territories. In the fury of a play fight, there is no serious biting and no threat behavior. Larger, older, and more dominant animals seem to handicap themselves in tussles with weaker playmates. Strength and skill are often matched to those of the partner. Some animals seem to practice territory defense as well. Irenaus Eibl-Eibesfeldt (1975) had tame polecats that would defend wastebaskets and hide under a blanket only to leap out at playmates that passed. Young deer and goats vie for possession of an area in a game reminiscent of King of the Mountain.

On the other hand, it has also been suggested that rather than being training for serious adult fighting, play fighting develops beneficial cognitive and social skills. During play fights among squirrel monkeys (*Saimiri sciureus*), the young males may reverse dominant and submissive roles. Without experience in a dominant role, young monkeys may grow up to be overly submissive, and without experience in the submissive role, they might grow up to be bullies. The motor patterns associated with each of these roles may be easier to learn than the cognitive aspects. Play fighting may also help a juvenile learn to read the intentions of others. Is the opponent bluffing? How motivated is this opponent? These social and cognitive skills may, in fact, prove to be more important than physical skills (Biben 1998).

Play may be involved in cognitive behavior in other ways, too. Animals learn a great deal while manipulating objects as toys. You may recall from Chapter 5 that Köhler's (1927) chimp Sultan did not figure out how to retrieve a banana that was out of reach until it had played with the sticks. It was during play that the chimps at the Delta Primate Center learned to use a stick for pole vaulting, a skill that was later used to reach leaves on tall trees and to escape from their confines. The information gained from manipulating "toys" may provide the experience that will be used for insight learning.

We see, then, that factors such as play or changes in the animal's nervous system or hormonal state can cause behavioral change during development. Our next topic, the development of singing behavior in birds, illustrates the combined influences of neural, hormonal, and social control.

THE ROLE OF GENES AND ENVIRONMENT IN THE DEVELOPMENT OF BIRD SONG

All of us have probably, at one time or another, been struck by the beautiful and often complex songs produced by the birds around us. Consider, for example, the rich and melodious soliloquies of the canary (*Serinus canarius*) or the odd mix of rattles, whistles, and clicks woven throughout the songs of the starling (*Sturnus vulgaris*), a talented bird, indeed, but one that is largely unappreciated. This question arises, then: How does bird song develop?

The development of singing behavior in birds illustrates the complex interaction between genetic and environmental factors in the development of behavior. Song development has been examined from the standpoint of evolution and ecology, as well as from the more mechanistic approach of assessing neural, hormonal, and social influences on behavior. Together, these approaches have revealed the continuous interplay between the developing animal and its internal and external environment. As we will see, the factors that influence the development of song range from interactions between cells to those between individuals. We begin by describing the development of singing behavior in the zebra finch (*Taeniopygia guttata*), an extremely social species, native to Australia.

GENETIC, HORMONAL, AND NEURAL CONTROL OF SONG IN THE ZEBRA FINCH

As with most songbirds, only male zebra finches sing, and the song is thought to stimulate reproductive behavior in females (Slater, Eales, and Clayton 1988). Although males of many species of birds also use song to advertise the boundaries of their territories, zebra finches breed in colonies and males seem to tolerate other birds even short distances from their nests. Sex differences in the singing behavior of zebra finches are a reflection of sex differences in neuroanatomy, and researchers have now identified the relatively large portion of the brain that controls song (Figure 8.6*a*). The song pathway consists of nine areas of the brain; most are located in the forebrain and one innervates the

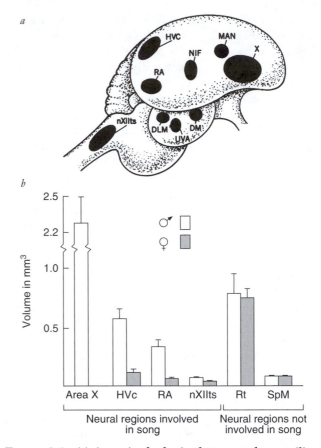

FIGURE 8.6 (*a*) **Areas in the brain that control song. (*b*) Sexual dimorphism in the brains of zebra finches. Note that four of the brain regions involved in song (Area X, HVc, RA, and nXII) are substantially larger in males than in females. No such difference is found in two regions (Rt and SpM) that are not involved in song. Sex differences in singing behavior (males sing and females do not) are thus reflected in brain structure. (*a*: Kroodsma and Konishi 1991. *b*: Nottebohn and Arnold 1976.)**

vocal organ called the syrinx. Male zebra finches have markedly larger song control areas than do females (Figure 8.6*b*) (Nottebohm and Arnold 1976). Sex differences in neuroanatomy are also evident at the cellular level; male song regions have more and larger neurons than do the same regions in females (Gurney 1981).

How, then, do such differences arise? Not surprisingly, the chromosomal difference between the sexes forms the basis for the dramatic differences in male and female brain structure. In birds, males have two X chromosomes, and females have one X and one Y chromosome. Absence of the Y chromosome results in the production of estrogen by the embryonic male gonads, and this hormone appears to be the organizing substance that stimulates growth and differentiation of the regions of the brain involved in song (Gurney and Konishi 1980). A remarkable group of experiments illustrates the powerful effects of estrogen on the neural system that controls singing behavior. If pellets containing estrogen are implanted in newly hatched male zebra finches, they have no effect (i.e., in adulthood the brains and songs of estrogen-treated males are similar to those of untreated males). In contrast, treatment of nestling females with estrogen results in enlarged song areas in the brain. These areas are larger than those of untreated females, though they are not as large as corresponding areas in the normal male brain (Figure 8.7). Estrogen, acting either directly or indirectly, appears to masculinize the brain.

Other experiments have revealed that the effects of hormones on the brain and singing behavior of zebra finches do not end around the time of hatching. Although the typical male has a masculinized brain, elevated levels of testosterone in the bloodstream are necessary to stimulate singing in adulthood. Similarly, those females that receive estrogen implants soon after hatching, and thus develop enlarged song areas, do not sing in adulthood unless testosterone is administered (Gurney and Konishi 1980). Females that do not receive estrogen implants at hatching, but do receive testosterone in adulthood, do not sing. Apparently, early exposure to estrogen establishes a sensitivity to testosterone in the brains of experimental females (and normal males), and exposure to testosterone in adulthood stimulates song. Whereas estrogen early in life organizes the development of a male song system, a high circulating level of testosterone in adulthood appears necessary to activate singing behavior (see Chapter 7 for a further discussion of the organizational and activational effects of steroid hormones).

The brains of birds are not always sensitive to the organizing effects of estrogen. For example, administering estrogen to adult females has no effect on the size of song areas in the brain (Gurney and Konishi 1980). Thus, there appears to be a critical period, around the time of hatching, in which the brain is particularly sensitive to hormonal influences. In this window of time, estrogen exerts its powerful effects on the developing nervous system. After that, the neural pathway that controls song cannot be switched to the male track. (We will discuss the concept of critical periods in behavioral development in more detail later in this chapter.)

ROLE OF LEARNING IN SONG DEVELOPMENT IN THE WHITE-CROWNED SPARROW

We have seen that genetic information, the hormonal milieu, and development of the nervous system all combine to produce sexual dimorphism in singing behavior. We can now ask what role learning plays in the development of song. Whereas zebra finches have frequently been used to study the physiological bases for song development, white-crowned sparrows (*Zonotrichia leucophrys*) have been a favorite subject for those curious about the role of learning.

The importance of learning to song development in adult male white-crowned sparrows was demonstrated in isolation experiments conducted by Peter Marler and his colleagues (Marler 1970; Marler and Tamura 1964). Under natural conditions, a young male sparrow hears the songs of his father and other adult males around him. During the first few months of life, the young male produces only subsong, a highly variable, rambling series of sounds with none of the syllables typical of the full song of adult males. As time passes, however, the song produced by the young male becomes more and more refined and soon contains elements recognizable as white-crowned sparrow syllables; this vocalization is called plastic song. Finally, the song of the male crystallizes, and he produces the full song characteristic of adult male white-crowned sparrows.

Young male sparrows reared in isolation develop abnormal song. However, if isolated males hear tapes of white-crown song during the period from 10 to 50 days of age, they will develop their normal species song. Young birds that hear white-crown song either before 10 days of age or after 50 days do not copy it. Thus, under the conditions of social isolation and "tape tutoring," the critical period for song learning occurs between 10 and 50 days posthatching. There is some flexibility in the timing and duration of this critical period (see later discussion). The important point, though, is that for song to develop normally, male sparrows must be exposed early in life to the songs of adult males of their species.

In addition to hearing adult conspecific song, young male white-crowned sparrows must also hear themselves

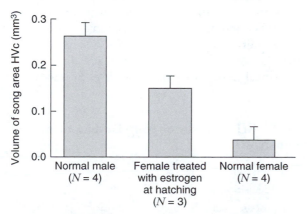

FIGURE 8.7 **Treatment with estrogen at hatching results in enlarged song areas in the brains of female zebra finches. Here, the volume of the song area HVc is substantially larger in estrogen-treated females than in normal females, but it is still somewhat smaller than that of normal males. Estrogen is thus involved in the masculinization of the brains of songbirds. (Modified from Gurney and Konishi 1980.)**

sing if they are to produce normal song. A series of classic experiments revealed the critical role of auditory feedback in song development (Konishi 1965). Birds that are exposed to the songs of adult males early in life and then deafened before the onset of subsong produce a rambling and variable song. Apparently, to develop full song, a bird must be able to hear his own voice and compare his vocal output to songs that he memorized some months previously. Further studies revealed that isolated males with intact hearing produce songs that are more normal in structure than those produced by birds deafened prior to subsong (Figure 8.8). Thus, if a bird can hear himself sing, he can produce a song with some of the normal species-specific qualities. Finally, although auditory feedback is important for song development, it is not essential for the maintenance of song. Birds deafened after their songs have crystallized continue to produce relatively normal songs for some time thereafter.

The results of these and other laboratory studies suggest that song development in sparrows consists of two phases—a sensory phase, during which songs are learned and stored, and a sensorimotor stage, when singing actually begins. In the sensory phase, sounds that are heard during the first few weeks of life are stored in memory for weeks or months without rehearsal. During this time, young males produce only subsong, which does not involve retrieval or rehearsal of previously learned material. In fact, young males probably use subsong in learning how to sing.

In most sparrows, the sensorimotor phase begins at approximately five months of age. At this time, birds retrieve a learned song from memory and rehearse it, constantly matching their sounds to the sounds they memorized. When males are about seven to nine months of age, song patterns crystallize, and they begin to produce full song. The final adult song will remain virtually unchanged for the rest of the bird's life. The phases of song development are outlined in Table 8.1.

Peter Marler and Douglas Nelson have suggested that during the sensitive phase, male sparrows learn more songs than they will need as adults. For instance, when hand-reared males of certain subspecies of white-crowned sparrows begin to practice singing, they may sing up to eight different songs. This interval, when many songs are rehearsed, is called the plastic song period. As the male practices his songs, he gradually discards all but one song. The song selection is made by listening to the songs of neighboring males. If a male hears one of the song types in his overproduced repertoire from a neighbor or through a researcher's loudspeaker, he retains that song as his "crystallized" song and discards the others (D. A. Nelson 1994; Nelson, Marler, and Morton 1996).

Earlier, we saw how sex differences in singing behavior are reflected in the neuroanatomy of songbirds. In most species it is only the male that sings, and song control areas in the brain are typically much larger in males than in females. One might wonder, then, do these song control areas also change during the phases of song development in males? The anatomy of the neural system that controls song indeed undergoes dramatic change over the course of song development (see review by Nordeen and Nordeen 1990). In fact, neurons in two song control areas (HVc and Area X) reproduce rapidly in male zebra finches during the period in which songs are being learned, and their proliferation may actually help define the critical period for vocal learning. Specifically, the

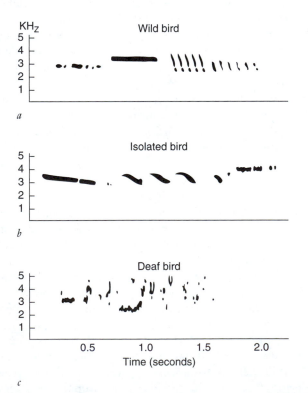

FIGURE 8.8 **Sonograms of songs that were produced by white-crowned sparrows (*a*) reared under natural conditions, (*b*) reared in social isolation, and (*c*) deafened at an early age. Note the great similarity between the first two. (Modified from Konishi 1965.)**

TABLE 8.1 **Phases of Song Development in Sparrows**

1. Sensory phase
   a. Acquisition—song learned
   b. Storage—song retained in memory
2. Sensorimotor phase
   a. Retrieval and production—motor rehearsal of learned song
   b. Motor stabilization—song crystallized into final adult song

Source: Marler (1987).

increase in the number of neurons in HVc and Area X occurs during the peak period of the sensory phase, when young males listen to and memorize songs that they hear around them. The number of neurons in these two areas does not change during the sensorimotor phase, when males rehearse material acquired earlier and eventually crystallize their song.

Now we will move from the level of the neuron to the level of behavior in our consideration of critical periods in song learning.

CRITICAL PERIODS IN SONG LEARNING

We know from experiments on male white-crowned sparrows raised in isolation and tutored with tape recordings of white-crown songs that the critical period for song learning extends from 10 to 50 days of age. How does this compare with what is known about the time course for learning in other songbirds or in white-crowned sparrows reared under different conditions?

The length and timing of the critical period for song learning vary greatly across species. Canaries, for example, have been called lifelong learners as a result of their ability to continually revise their songs throughout adulthood. White-crowned sparrows and zebra finches, however, have a restricted period of learning that occurs during the first weeks or months of life; these species are referred to as age-limited learners.

The critical periods for song learning vary not only from one species to another but within a species as well, from one individual to the next. Individual differences in the timing of song learning are commonly reported in both field and laboratory studies. In the laboratory, the length of the critical period can be manipulated by depriving male birds of optimal social and auditory stimuli, such as a singing adult male of their own species. Young male zebra finches copy sounds from adult males ("tutors") with whom they are housed when they are between 35 and 65 days of age. Males that are not permitted to hear any songs during this period, however, retain the capacity to learn songs well into adulthood (Eales 1985). Also, males may reproduce songs that they hear before 35 days of age if the tutor they are exposed to after this time provides inadequate stimulation, for example, is of a different species or is not visible to the young male (Slater, Eales, and Clayton 1988).

Critical periods are more flexible than was once thought and may be affected by social interactions. For example, several studies have shown that live, socially interactive tutors are more effective models than are tape recordings of songs for laboratory birds that are learning to sing. For example, in white-crowned sparrows, the critical period for song learning in males that are exposed to live tutors of their own species is

extended beyond the usual 50 days found with taped songs (Baptista and Petrinovich 1984). Furthermore, a song learned by males younger than 50 days can be modified after that age when a live tutor is used (Petrinovich and Baptista 1987). Thus, social variables are very important in song learning. However, an alternative explanation of these experiments is that the live tutor served not only as a model during the critical period but also played a role in selecting the song to be retained as the single adult song from a repertoire of songs learned during the critical period (D. A. Nelson 1998).

Aspects of the young birds' physical environment can also influence the timing of critical periods. In the marsh wren, *Cistothorus palustris*, the time course for vocal learning is influenced by the photoperiod (Kroodsma and Pickert 1980). All these examples show that both among and within species, there is variation in the timing and duration of the critical period for song learning.

OWN-SPECIES BIAS IN SONG LEARNING

When given a choice, most young male birds learn their songs from members of their own species. For example, male white-crowned sparrows copied only the songs of their own species when tutored with tapes of the song sparrow and white-crowned sparrow (Marler 1970). This preference is called own-species bias, and it also influences the speed and accuracy with which young males learn. Typically, birds learn the songs of their own species more rapidly and accurately than those of a different species, although this tendency varies across species and with the method of song presentation. When a young male is exposed to tape recordings of conspecifics, live tutors of another species can sometimes override the bias for learning conspecific songs. When young white-crowned sparrows are presented with visible, singing song sparrow tutors and can hear the songs of white-crowned sparrows in the background only, they learn the song of the song sparrow (Baptista and Petrinovich 1986). That young male sparrows learn the song of the adult that they are allowed to interact with, even if this adult belongs to another species, emphasizes the importance of social interactions in the song-learning process.

CHOICE OF TUTOR

In nature, male songbirds learn their songs from older conspecifics. One wonders, then, does a son keep it in the family and learn his songs from his father, or does he learn from neighboring males? The identity of a song tutor is related to the timing of critical periods. If the critical period for song learning occurs soon after hatching, then fathers are probably the song tutors. Alternatively, if the critical period occurs (or extends for) several months after hatching and includes the time during which young males are establishing territories of their own, then neighboring males may serve as tutors. The choice of tutors by zebra finches provides an example.

In the wild, zebra finches breed colonially, and because breeding is triggered by rainfall, pairs within a colony often breed synchronously (Slater, Eales, and Clayton 1988). The young fledge at approximately three weeks of age, gain independence from their parents at about five weeks, and then join flocks of other juveniles and nonbreeding adults. Thus, young males experience a complex social environment in which they are exposed to many conspecifics both before and after fledging.

We can ask, then, from which older male or males within this vast social network do young zebra finches learn their songs? Jorg Böhner (1983) examined this question by giving young males the choice of using either their father or another male as a song model. Two pairs of zebra finches were placed in the same cage but were separated into different compartments by means of a wire lattice. The adult males of these pairs had very different songs, although both songs were well within the bounds of typical zebra finch song. Young males hatched in these cages were raised by their parents and then at fledging were housed in a separate cage adjacent to that of their parents and neighbors. From this vantage point, the young males could see and hear their father and the neighboring male. When the young males were 100 days of age, their songs were recorded and compared with the song of their father and that of their neighbor. The results are summarized in Table 8.2.

When given the choice of using their father or a neighboring male as a song tutor, male zebra finches selected only one of the two available males, and the majority chose their fathers. Of the 11 males raised by Böhner, 9 of the males produced songs in which the majority of elements came from their fathers' songs, 1 copied all its song elements from the neighbor, and 1 produced a song that contained mostly new elements. Note that none of the males produced a "mixed song," that is, a song containing elements from both the father's and neighbor's songs.

In a later study, Böhner (1990) demonstrated that zebra finch males could learn all details of their species-specific song within the first 35 days of life, when still dependent on their parents. Young males exposed to their father for only 35 days posthatching developed copies of their father's song that were as complete as those of males that had contact with their father for 100 days. A father was thus a likely candidate for the role of tutor to the zebra finch.

Nicola S. Clayton (1987) found that if a young male has become independent and his father is no longer present, he will choose a tutor with a song similar to his father's. She raised young males with their parents, and once the males had fledged she housed each male with two adult zebra finch tutors. One tutor sang a song like that of the father, and the other sang a song that was quite different. The young males chose as their model for song learning the song most similar to that of their father, and again they selected only one tutor to learn from.

Many factors influence a young male zebra finch's choice of a song tutor, each biasing the choice in favor of the father. One such factor is proximity. Young males prefer to learn from the male they lived with before the

TABLE 8.2 Composition of the Songs of Male Zebra Finches
Given the Choice of Their Father or Male Neighbor as Song Tutor

Male	Percentage of Elements Copied from the Father's Song	Percentage of Elements Copied from the Neighbor's Song	Percentage of New Elements
1	100	——	——
2	100	——	——
3	67	——	33
4	100	——	——
5	75	——	25
6	83	——	17
7	90	——	10
8	—	100	——
9	100	——	——
10	17	——	83
11	64	——	36

Source: Data from Böhner (1983).

sensitive period. They also prefer the male that is paired with the female that raised them. And, they prefer a paired male over a single male as a song tutor (Mann and Slater 1994). In nature, these preferences will generally cause young males to choose their father as a song tutor.

However, social interactions, especially parental care, also affect the choice of tutor. A comparison of two experiments shows that the choice of song tutor depends on the extent of parental care. In both experiments, the birds were housed in an aviary, an environment that more closely simulated the colonial social conditions of zebra finches in the wild. In the first, Heather Williams (1990) raised young males in an aviary that contained ten breeding pairs of zebra finches and two nonbreeding males. Under these conditions, sons did not preferentially copy the songs of their fathers, and in fact most young males copied syllables from at least two adult males in the aviary population (Figure 8.9). The males chosen as tutors tended to be those that had shown the most parental care toward the young male before it left the nest. Perhaps because of the high density of birds in the aviary, in this study the juveniles tended to form a communal nest and parents fed the young birds indiscriminately. Thus, a father did not have closer contact with his sons than did other males. However, in a more recent experiment, the density of birds in the aviary was approximately half of that in Williams' study. In this aviary, there was almost no mixing of broods. Parents cared for their own young in their own nest. As a result, the adult male showing the greatest parental care was the father. Although it was not quite statistically significant, there was a definite preference for fathers as song tutors (Mann and Slater 1995).

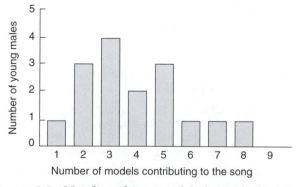

FIGURE 8.9   Number of song models from which male zebra finches raised in an aviary with several breeding pairs copied syllables for their songs. Note that in this complex social environment, males typically copied syllables from more than one model. (Modified from H. Williams 1990.)

NONAUDITORY EXPERIENCE AND SONG DEVELOPMENT IN THE COWBIRD

Young males of other species may learn to sing not only by listening to the songs of other males but also through direct social interaction. Brown-headed cowbirds (*Molothrus ater*) are brood parasites. Females of this species lay their eggs in the nests of other bird species, thereby relinquishing all parental duties to foster parents. If young male cowbirds relied on learning characteristics of their songs by listening to male birds in the vicinity of the nest during their first few weeks of life, they would most likely sing the song of their foster parents. But field and laboratory observations indicate that this is not the case at all; when male cowbirds become sexually mature, they sing cowbird song. Unlike the white-crowned sparrows previously described, male cowbirds that have never heard other cowbirds sing produce the correct cowbird song. Does this mean that experience is unimportant in the development of cowbird song? No—learning, as it turns out, does play an integral role. In a fascinating series of experiments, Meredith West and Andrew King have shown that the songs of male cowbirds are shaped, in part, by male-male dominance interactions and the response of females to their songs. Let us look at some of their experiments.

Interest in the role of social interactions in the development of song in cowbirds grew from the startling finding that although the songs of male cowbirds reared in isolation from male conspecifics were nearly identical to those of normally reared males, they were substantially more appealing to female cowbirds (Figure 8.10), as measured by their ability to release the female copulatory posture. In fact, the songs of males raised in either auditory and visual isolation from male conspecifics or visual isolation alone were twice as effective in releasing the copulatory posture in female cowbirds as those of males raised either in groups of both males and females or in visual contact with such groups (Figure 8.11) (West, King, and Eastzer 1981). Why are the songs of isolates more potent than those of normal males? The answer came from observations of what occurred when group-reared males and isolation-reared males were individually introduced into an aviary with an established colony of cowbirds. Group-reared males kept a low profile in their new environment; they did not sing in the presence of resident males and were not attacked. In contrast, the isolates repeatedly sang their high-potency song in the presence of resident males and females, and the resident males responded by attacking and sometimes killing the energetic songsters. Apparently, males living in mixed-sex groups, such as the flocks that occur in nature, learn

FIGURE 8.10 Female cowbirds prefer the songs of males reared in isolation to those of males reared socially.

to suppress their song and thereby avoid attacks from dominant males.

What role might female cowbirds play in shaping the songs of males? To examine this question, King and West (1983) used two different subspecies of cowbirds, one found in the area of North Carolina (*M. ater ater*) and the other around Texas (*M. ater obscurus*). We will refer to these cowbirds as subspecies A and subspecies O, respectively. Male cowbirds of subspecies A were hand-reared in acoustic isolation until they were 50 days old. At this time they were housed for an entire year with (1) individuals of another species (canaries or starlings), (2) adult female cowbirds of subspecies A, or (3) adult female cowbirds of subspecies O. At the end of this period, the researchers examined the songs of the males. The three groups of males of subspecies A had developed remarkably different songs (Figure 8.12).

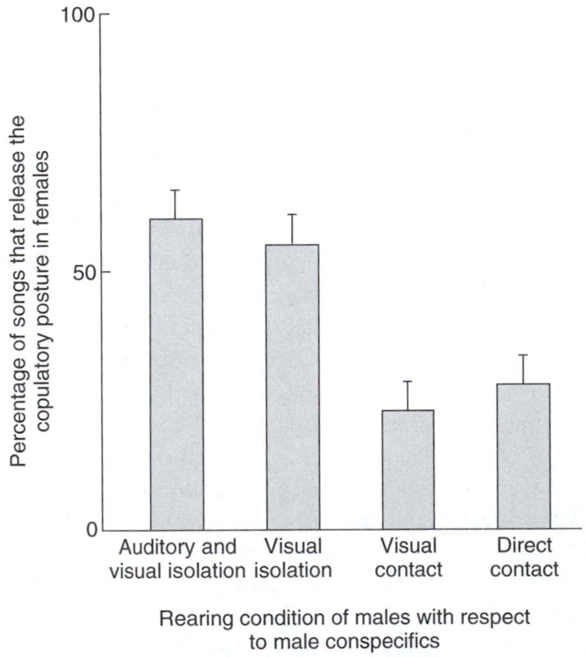

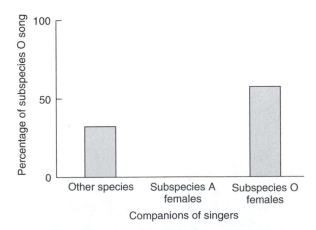

FIGURE 8.11 **The effects of interactions with dominant males on the development of song in cowbirds. Males that do not have visual or direct contact with dominant males produce more potent songs than do males that have visual or direct contact. In cowbirds, then, young males learn to modify their songs to avoid attack; this is an example of nonauditory stimuli influencing song learning. (Drawn from the data of West, King, and Eastzer 1981.)**

FIGURE 8.12 **Female cowbirds influence the development of song in males. When males of subspecies A are housed with either subspecies A or subspecies O females, they mold their song to the preferences of their female companion. Note that subspecies A males housed with females of their own subspecies do not sing any subspecies O song, but those housed with subspecies O females sing subspecies O songs predominantly. (Drawn from the data of King and West 1983.)**

Males that were housed with canaries or starlings had diverse repertoires, singing a mix of songs from the two cowbird subspecies, along with imitations of the songs of other species in their cage. Whereas males housed with cowbird females of subspecies A sang all A songs, those housed with cowbird females of subspecies O sang predominantly O songs. Male cowbirds thus develop song repertoires that are biased toward the preferences of their female companions.

Because female cowbirds do not sing, their influence on male song development cannot be through song. Instead, they may encourage the male's singing by a simple display called the wing stroke, in which one or both wings are moved rapidly away from the body (West and King 1988). These wing strokes are not doled out indiscriminately: Approximately 94% of the males' songs elicit no visible change in the behavior of female cowbirds, and on average, a single wing stroke occurs for every 100 songs. It seems, then, that male cowbirds adjust their songs to the whims (or is it wings?) of their audience, molding their song structure to avoid conflict with dominant males and to stimulate available females. Just how they achieve this delicate balance in natural flocks is an interesting question.

FUNCTIONAL SIGNIFICANCE OF SONG LEARNING

Much of our discussion of the development of singing behavior in birds has been about the role of learning. Before leaving the topic of bird song, however, we might ask, why must birds learn to sing? Wouldn't it be simpler, and indeed safer in terms of evolution, for male zebra finches, white-crowned sparrows, and cowbirds to produce a song that could not be changed by experience? Apparently not, but the precise reasons are still being hotly debated.

Three hypotheses have been proposed for the function of song learning. First, it may have evolved as a means of transferring complex patterns of behavior from one generation to the next (Andrew 1986). This hypothesis is based on the commonly held, but unverified, belief that more information can be stored in the memory than in the genes. Learning, it is believed, allows for all the exquisite details associated with species-specific song.

A second hypothesis states that learning allows individuals to conform more precisely to the particular social setting in which they find themselves. As a specific example of the advantage of learning to match the songs of neighbors, Petrinovich (1988) suggested that conforming songs might elicit fewer aggressive attacks than would completely novel songs.

A third hypothesis is that song learning may function as a population recognition mechanism (Marler and Tamura 1962) and therefore must be strongly influ-

enced by learning. This hypothesis was developed to explain the existence of regional song dialects, acoustic themes that are shared by males in the same geographic area and believed to be important in mate choice. According to this view, birds of both sexes could learn the song of their father and then, in adulthood, females would choose as mates males that sing that song (Baker and Cunningham 1985). In this way, learned regional dialects could function in the maintenance of local populations that are well adapted to their particular breeding habitat. This notion has been debated, however. For example, white-crowned sparrows in the field do not show a simple relationship between the type of song and mate choice (Petrinovich 1988). Obviously, a greater emphasis on the study of song development in free-living birds is essential to our understanding of its functional significance.

THE CONCEPT OF CRITICAL PERIODS

Earlier in this chapter we described a critical period around the time of hatching when the hormone estrogen exerts a powerful effect on the developing nervous system of zebra finches. We also described the critical period associated with song learning by white-crowned sparrows, a time when young males are particularly sensitive to the songs of older conspecifics around them. We should add now that critical periods are important in many other kinds of learning. Because such windows of opportunity for learning are so obviously important in the development of behavior, let's look at the phenomenon more closely.

HOW CRITICAL PERIOD WAS DEFINED

Konrad Lorenz (1935) borrowed the term *critical period* from embryology, where it was used to describe times in early development that were characterized by rapid changes in organization. During these brief, well-defined periods, an experimental interruption of the normal sequence of events produced profound and irreversible effects on the developing embryo. Thus, as first used by Lorenz, a critical period was a phase of susceptibility to environmental stimuli that was brief, well defined, and within which exposure to certain stimuli produced irreversible effects on subsequent behavior.

Recently, descriptions such as sensitive period, sensitive phase, susceptible period, and optimal phase have been used in place of critical period, but we will continue to use the traditional term here. We should be aware, however, that the newer terms indeed reflect certain modifications in the definition of this period in light of more recent research. In fact, many of Lorenz's (1935) basic precepts have now been modified; some of these more contemporary ideas should be familiar from our discussion of critical periods in learning bird songs. Specifically, we now know that critical periods (1) are fairly extended; (2) are not sharply defined but gradual in their onset and termination; (3) differ in duration among species, individuals, and functional systems; and (4) depend on the nature and intensity of environmental stimuli both before and during the critical period. Moreover, most phenomena based on critical periods are not irreversible. Instead, patterns of behavior developed during critical periods can usually be altered or suppressed under certain conditions, especially those associated with high levels of stress. Deprivation (e.g., rearing an animal in isolation or in darkness) can reverse or destroy a pattern of behavior established during a critical period. It is important, however, not to overemphasize the reversibility of patterns of behavoir established during critical periods. Conditions such as rearing in complete isolation or total darkness are unlikely to be encountered by most animals outside the laboratory environment. Furthermore, even in the laboratory, behaviors established in critical periods are usually more resistant to change than those learned at other times (Immelmann and Suomi 1981).

TIMING OF CRITICAL PERIODS

In most animals, as with zebra finches and white-crowned sparrows, critical periods occur early in life. Why is this so? We usually assume that this is the time when animals have the greatest opportunity to gain knowledge from parents and close relatives, knowledge that is particularly important in species recognition. Later, they might not interact with them so intimately, and in some cases they will, in fact, be exposed to intense stimuli from other species (Immelmann and Suomi 1981). For example, in some species of birds, the young remain in the nest for only a few weeks after hatching and then leave to join mixed-species flocks. It would not be surprising if these young learn to recognize conspecifics during a brief critical period before leaving the nest. Otherwise, choosing an appropriate mate could later be a confusing exercise indeed since birds that waited too long to learn the defining qualities of their species might very well learn the plumage and song characteristics of another species.

Some animals have little or no contact with their parents or other close relatives after birth or hatching. One might wonder, then, would critical periods occur early in development in these species? Consider the case of Pacific salmon in the genus *Oncorhynchus*. Adult salmon spawn in fresh water, usually streams, and depending on the species and population, they may or may not die after spawning. When the juveniles emerge from their gravel nests, they may reside in their home streams for awhile, but in many species the young fish migrate thousands of kilometers to oceanic feeding grounds. After a time at sea, virtually all surviving fish return to their home stream to spawn. It is a remarkable feat of navigation, and when they reach the freshwater inlets, they unerringly swim up the appropriate tributaries, making all the correct decisions at every fork until they reach the very stream where they were born—and they seem to do it by smell. Apparently, before their migration to the sea and during a critical period, juvenile salmon learn the odor of their home stream. Upon returning to this vicinity, they are stimulated to swim upstream by the familiar odor (Scholz et al. 1976). Why is it important that salmon learn the precise location of their natal stream? The answer is that each population is finely adapted to its home water, so much so that salmon experimentally introduced into other streams show a higher mortality rate than locally adapted individuals (Quinn and Dittman 1990). The period of early learning thus ensures that it is the odor of the fish's own birthplace that is remembered (for a more detailed discussion of salmon homing see Chapter 10). We see, then, that during critical periods, animals may learn the appropriate cues, not only of conspecifics, but also of the physical environment itself.

Onset of Sensitivity

It is clear that animals have a heightened sensitivity to certain environmental stimuli such as the song of a neighboring male or the smell of a home stream. One might ask, then, what causes this increased sensitivity to certain cues? It seems that the onset of sensitivity may be due both to endogenous (internal) and exogenous (external) factors (Bateson 1979). Increases in sensitivity generally begin as soon as the relevant motor and sensory capacities of the young animal are developed. Changes in the internal state, such as fluctuations in hormone levels, may also influence sensitivity. Then, endogenous factors interact with environmental variables to produce the start of the critical period. For example, although the visual component of filial imprinting in birds (the response of newly hatched young to follow their mother; see later discussion) begins once hatchlings are able to perceive and process optical stimuli, experience with light appears to interact with the internal conditions to initiate this particular critical period (e.g., Bateson 1976). Here, then, we have endogenous factors (ability of the nervous system to deal with optical stimuli) interacting with environmental factors (exposure to light) to produce the start of the critical period.

Decline in Sensitivity

Two hypotheses, the internal clock and the competitive exclusion, have been proposed to account for the termination of critical periods. The internal clock hypothesis assumes that the decline in sensitivity is under endogenous control. According to this idea, the termination of critical periods is determined by the ticking of some internal clock and has little, if anything, to do with external factors such as experience. In other words, some physiological factor, intrinsic to the animal, ends the period of receptivity to external stimulation.

In the competitive exclusion hypothesis, filial imprinting is viewed as a self-terminating process in which external factors play a critical role. According to this idea, a given external experience and its impact on neural growth prevent subsequent experiences from modifying the behavior (Bateson 1979, 1982). Suppose that a behavior, such as the response of ducklings to follow their mother, is based on the growth of neural connections into an area of finite size. When the growth of these connections has proceeded to some point, the first experience (e.g., a view of the mother) is better able to direct the behavior than is a later form of input (e.g., a view of other ducks on the pond). The competitive exclusion hypothesis states that the first experience effectively excludes subsequent experiences from affecting the behavior, and in so doing, it terminates the critical period.

IMPORTANCE OF CRITICAL PERIODS IN BEHAVIORAL DEVELOPMENT AND SOME EXAMPLES

Now that we have discussed the definition, adaptiveness, and control of critical periods, we should tell you that some people believe that they do not exist. In fact, there has been a vigorous argument over the importance of early experience in shaping subsequent behavior (reviewed by Klopfer 1988). Although strong proponents of critical periods believe that they characterize all developmental processes and that exposure to environmental stimuli during the sensitive phase produces irreversible effects on behavior (e.g., Scott 1962), critics downplay the influence of the early environment on subsequent behavior, claiming that virtually any effect on early experience can later be modified (e.g., Clarke and Clarke 1976). In this latter view, events that occur early in an organism's life are not of pivotal importance; instead, it is the accumulation of experiences, particularly those that occur time and again, that influences behavioral development.

As is often the case, the truth seems to encompass both views. In some instances early experience produces virtually unalterable effects on behavior, whereas in others environmental effects on behavior can be modified at a later time. Although critical periods do not characterize all developmental processes, they are important in many behavioral phenomena. Now let's consider selected examples of behavioral development that depend, to varying degrees, on specific experiences during a window of time.

Filial Imprinting in Birds

Anyone who has ever watched chicks in a farmyard or ducklings and goslings on a pond knows that the young generally follow their mother wherever she goes (Figure 8.13). How does such following behavior develop? Konrad Lorenz (1935), working with newly hatched goslings, was the first to systematically study this behavior. In one experiment, he divided a clutch of eggs laid by a greylag goose (*Anser anser*) into two groups. One group was hatched by the mother, and as expected, these goslings trailed behind her. The second group was hatched in an incubator. The first moving object these goslings encountered was Lorenz, and they responded to him as they normally would to their mother. Lorenz marked the goslings so that he could

FIGURE 8.13   Young Canada geese, following their mother.

FIGURE 8.14   Goslings following their "mother," Konrad Lorenz. Lorenz was one of the first scientists to study imprinting experimentally.

determine in which group they belonged and placed them all under a box. When the box was lifted, liberating all the goslings simultaneously, they streamed toward their respective "parents," normally reared goslings toward their mother and incubator-reared ones toward Lorenz. The goslings had developed a preference for characteristics associated with their "mother" and expressed this preference through their following behavior. The attachment was unfailing, and from that point on Lorenz had goslings following in his footsteps (Figure 8.14).

Because social attachment evidenced by following seemed to be immediate and irreversible, Lorenz named the process *Pragung*, which means "stamping." The English translation is "imprinting." Used in this context, the term suggests that during the first encounter with a moving object, its image was somehow permanently stamped on the nervous system of the young animal.

Today the process by which young birds develop a preference for following their mother is called filial imprinting. It is the archetypal example of the role of early experience in behavioral development. The biological function of filial imprinting is probably to allow young birds to recognize close relatives and thereby distinguish their parents from other adults that might attack them (Bateson 1990). We will focus our discussion on the development of the following response in mallard ducklings.

Mallard ducklings, like most young in the orders Anseriformes (ducks, geese, and swans) and Galliformes (chickens, turkeys, and quail) are precocial, quite capable of moving about and feeding on their own just a short time after hatching. Filial imprinting is usually studied in species with precocial young. The following response is nonexistent—or much less evident—in species, such as the songbirds discussed earlier, whose young are altricial, virtually helpless and therefore dependent on their parents for the first few weeks after hatching. We begin with a brief description of reproduction and early development in mallard ducklings.

Upon finding a suitable nesting site, typically a shallow crevice in the ground, the mallard hen begins to lay her eggs, at the rate of one egg per day (Miller and Gottlieb 1978). The average clutch size is 8 to 10, and after the last egg is laid the hen begins incubation, a process that lasts approximately 26 days. Encouraged by the warmth of the mother's body, the embryo inside each egg begins to develop. Two to three days before hatching, each embryo moves its head into the air space within its egg and begins to vocalize; these vocalizations are called contentment calls. About 24 hours later, the embryos pip the outer shell and then take another day to break through the rest of the shell and hatch. Most of the ducklings in a clutch will hatch within an interval of 10 hours. The hen broods her young for a day and then leaves the nest and emits calls to encourage the ducklings to follow. Although she calls during incubation and brooding, the frequency increases dramatically at the time of the nest exodus. Prompted by their mother's assembly calls, the young leave the nest and follow her to a nearby pond or lake, where they will paddle behind her.

Observations such as these stimulate many questions. What characteristics of the mother form the basis for the ducklings' attachment? Do the ducklings imprint on the mother's call, physical appearance, or some combination of the two? What role might siblings play in the development of the following response? Finally, is there a critical period during which exposure to certain cues must occur for the normal development of filial behavior?

Many of the answers to these questions have come

from the laboratory of Gilbert Gottlieb. Over the past 30 or so years, Gottlieb and his coworkers have examined the development of the following response in Peking ducks, a domestic form of the mallard (despite their domestication, Peking ducks are quite similar to their wild counterparts in their behavior).

In one experiment with ducklings that had never had contact with the mother, Gottlieb (1978) examined (1) the relative importance of the hen's auditory and visual cues in the development of the following response and (2) whether the parental call of a duckling's own species would be more effective than that of other species in inducing and maintaining attachment behavior. As part of the study, 224 eggs were hatched in an incubator; thus, the ducklings never came in contact with their mothers. The ducklings were divided into four groups and tested for their following response. In all four groups, the ducklings were tested with a stuffed replica of a Peking hen as it moved about a circular runway. Individuals in one of the groups, however, were tested with a silent hen, whereas individuals in the other three groups were tested with hens that emitted assembly calls through a speaker concealed on their undersides. Of the ducklings that heard assembly calls, one group was exposed to mallard calls (Peking calls), one group to wood duck calls, and one group to domestic chicken calls. Each duckling was given a 20-minute test to determine if it would follow the stuffed model around the circular arena.

The results are presented in Figure 8.15. As you can see, the auditory stimulus of the maternal call is important in filial imprinting: All conditions with calls were much more effective than no call at all. Furthermore, even though the ducklings had never before heard the maternal call of their species, they responded selectively to it. The maternal call of the mallard was far more effective than those of the wood duck or chicken in inducing following by the ducklings.

In a second experiment, incubator-reared ducklings were given a choice of following either a stuffed Peking hen that was emitting mallard calls or a stuffed Peking hen that was emitting chicken calls. When placed in this simultaneous choice situation approximately one day after hatching, the majority of ducklings (76%) followed the model that was emitting the mallard call.

Taken together, these experiments demonstrate that auditory stimuli from the mother play an important role in controlling the behavior of newly hatched ducklings. Furthermore, ducklings respond selectively to the maternal assembly call of their own species without any previous exposure to it (remember that all ducklings were reared in incubators and therefore had no contact with hens).

These findings do not, however, suggest that experience is unimportant in the development of a preference for the assembly call. Experience, as it turns out, is

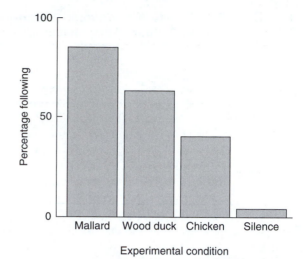

FIGURE 8.15 **A stuffed Peking hen emitting mallard calls was more effective in eliciting the following response in incubator-reared Peking ducklings than a hen emitting either wood duck or chicken calls (Peking ducks are a domestic form of the mallard duck). All hens with calls were more effective than a silent hen. Thus, auditory cues from the mother are important in controlling the early behavior of ducklings, and ducklings respond selectively to the call of their own species without previous exposure to it. (Drawn from the data of Gottlieb 1965.)**

critical, but it occurs prenatally (before hatching). For Peking ducklings to exhibit a preference for the mallard hen's call, they must hear their own contentment calls or those of their siblings before hatching (Gottlieb 1978). If ducklings are reared in isolation (and therefore are not exposed to the calls of their siblings) and made mute just before they begin to vocalize within the egg (and therefore are not exposed to their own calls), they no longer display their highly selective response to the maternal call of their species. When these ducklings without normal embryonic auditory experience are placed in a test apparatus equidistant between two speakers, one speaker emitting a mallard's maternal call and the other a chicken's maternal call, they choose the latter almost as often as they choose the former 48 hours after hatching. In contrast, ducklings with normal embryonic auditory experience always choose the mallard's call over the chicken's (Table 8.3). Thus, auditory experience before hatching is important to the development of the following response because it fine-tunes early call preferences in young mallards.

Is there a critical period during which exposure to contentment calls must occur for ducklings to exhibit a preference for the call of the mallard hen? If ducklings heard contentment calls after hatching, would they display the normal preference for mallard calls? Gottlieb (1985) set out to answer these questions. He began by raising ducklings in isolation and making them mute

TABLE 8.3 The Effects of Embryonic Auditory Experience on Call Preferences of Peking Ducklings 48 Hours After Hatching

Group	N	Mallard Call	Chicken Call	Both
		Preference		
Vocal-communal	24	24	0	0
Devocal-isolated				
First experiment	22	12	9	1
Replication	21	14	6	1
Total	43	26	15	2

Ducklings raised in the vocal-communal group could hear themselves and the calls of siblings prior to hatching; devocal-isolated ducklings had no such auditory experience. *N* = number of ducklings that responded to calls.

Source: Data from Gottlieb (1978).

before they began to vocalize within the egg (again, these manipulations ensure that the ducklings are not exposed to their own calls or the calls of their siblings). One group was exposed to contentment calls during the embryonic period (approximately 24 hours prior to hatching), and another group was exposed to contentment calls during the postnatal period (approximately 24 hours after hatching). Ducklings in each group were then tested for their preference for the mallard hen's call; this time, however, they were given the choice of approaching a speaker that emitted normal calls of a mallard hen or a speaker that emitted artificially slowed calls of a mallard hen. As shown in Table 8.4, ducklings must be exposed to contentment calls before hatching if they are to show a preference for the normal call; exposure after hatching is ineffective in producing the preference for the normal call of their species. Here, then, we have an example of an embryonic critical period.

What role, if any, might visual stimuli play in the development of the following response? Several experiments have shown that two conditions must be simultaneously met if visual imprinting on a mallard hen is to occur in the ducklings. The ducklings must be reared with other ducklings and allowed to actively follow a mallard hen, or a model, if they are to prefer the appearance of a hen of their own species to that of a hen of another species at later testing (e.g., Dyer, Lickliter, and Gottlieb 1989; Lickliter and Gottlieb 1988). Ducklings that are reared in isolation and given passive exposure to a stuffed mallard hen (i.e., housed with a stationary model) do not develop a preference for the mallard hen. These results suggest that under natural conditions, ducklings learn the visual characteristics of their mother after leaving the nest and following her around.

We see, then, that both auditory and visual stimuli are important in filial imprinting. Auditory stimulation from the mother appears to be largely responsible for prompting the ducklings to leave the nest and for influencing their earliest following behavior. However, the hen's appearance becomes important after the nest exodus.

The importance in filial imprinting of auditory cues, visual cues, and active following was well summarized by Lorenz (1952, pp. 42–43). In an experiment with Peking ducklings, he found himself in a rather embarrassing position for one destined to become a Nobel Laureate. In his words,

TABLE 8.4 Effects on Preferences of Devocal-Isolated Ducklings of Embryonic Versus Postnatal Exposure to Contentment Calls

Time of Exposure to Contentment Calls	N	Normal Mallard	Slowed Mallard	Both
		Preference		
Embryonic	32	21	8	3
Postnatal	37	14	20	3

N = number of ducklings that responded to calls.

Source: Data from Gottlieb (1985).

The freshly hatched ducklings have an inborn reaction to the call-note, but not to the optical picture of the mother. Anything that emits the right quack note will be considered as mother, whether it is a fat white Peking duck or a still fatter man. However, the substituted object must not exceed a certain height. At the beginning of these experiments, I had sat myself down in the grass amongst the ducklings and, in order to make them follow me, had dragged myself, sitting, away from them. So it came about, on a certain Whit-Sunday, that, in company with my ducklings, I was wandering about, squatting and quacking, in a May-green meadow at the upper part of our garden. I was congratulating myself on the obedience and exactitude with which my ducklings came waddling after me, when I suddenly looked up and saw the garden fence framed by a row of dead-white faces: a group of tourists was standing at the fence and staring horrified in my direction. Forgivable! For all they could see was a big man with a beard dragging himself, crouching, round the meadow, in figures of eight, glancing constantly over his shoulder and quacking—but the ducklings, the all-revealing and all-explaining ducklings were hidden in the tall spring grass from the view of the astonished crowd!

To summarize, the experiments of Gilbert Gottlieb and his coworkers have demonstrated that filial imprinting in Peking ducklings results from a complex interaction of auditory, visual, and social stimuli provided by the hen and siblings. Several important generalizations about early behavioral development arise from their work. First, we can no longer think of experience as occurring only after birth or hatching; embryonic experience can also influence behavior. In the case of Peking ducklings, listening to the contentment calls of siblings before hatching is critical to their preference for the maternal call of their own species after hatching. Second, the experiments with ducklings demonstrate that a variety of stimuli may be involved in the development of a single pattern of behavior and that the relative importance of different stimuli may change as the young animal matures. Although the following behavior of ducklings soon after hatching is influenced largely by auditory cues from their mother, only a few days later visual stimuli (in combination with auditory stimuli) become important in the following response. The relative priorities of different cues match the timing for development of the auditory and visual systems; in ducklings, as in all birds, the auditory system develops before of the visual system (Gottlieb 1968). These results emphasize the close interaction between physical maturation and experience in early behavioral development. Finally, the study of filial imprinting in Peking ducklings illustrates that we must be open-minded when trying to sort out just which experiences affect a

given behavior. Who would have thought that listening to siblings before hatching or interaction with siblings after hatching would be critical in the development of the ducklings' attachment to their mother? We now know that such nonobvious experiential factors are indeed essential to the development of following behavior in this species.

Sexual Imprinting in Birds

We have seen that early experience influences a duckling's attachment to its mother, an attachment that the young bird demonstrates by trailing behind her in the days following the exodus from the nest. Early experience also has important consequences for the development of mate preferences in birds. In many species, experience with parents and siblings early in life influences sexual preferences in adulthood. The learning process in this case is called sexual imprinting. Typically, sexual preferences develop after filial preferences, although the critical periods may overlap to some degree (Bateson 1979). Whereas filial imprinting is indicated by the following response of young birds, sexual imprinting is typically shown in the preferences of sexually mature birds for individuals of the opposite sex. It is important to note that unlike filial imprinting, sexual imprinting occurs in both altricial and precocial birds.

One dramatic demonstration of the importance of early experience to subsequent mate preference came from cross-fostering experiments with finches. Klaus Immelmann (1969) placed eggs of zebra finches in clutches belonging to Bengalese finches (*Lonchura striata*). The Bengalese foster parents raised the entire brood until the young could feed themselves. From then on, young zebra finch males were reared in isolation until sexually mature. When they were later given a choice between a zebra finch female and a Bengalese finch female, they courted Bengalese females almost exclusively.

A second study demonstrated that brief contact with foster parents early in life can exert a more powerful, longer-lasting influence on mate preference than long-term social contact in adulthood. Cross-fostered zebra finch males were again separated from their foster Bengalese parents, but this time they were provided with a conspecific female and nesting supplies. Most of these males eventually mated with conspecific females and successfully produced young. When they were tested several months or years later, however, the males still displayed a preference for Bengalese females (Immelmann 1972).

More recent experiments have shown that sexual imprinting is a two-stage process. During the acquisition phase, which occurs during the critical period that begins when a zebra finch first opens its eyes and ends at about 30 to 40 days of age, a young male forms a social bond to its parents. Because of this bond, the male prefers to socialize with members of its parents' species, and this social preference guides his first courtship attempts. The consolidation phase occurs when the male actually courts a female for the first time. During courtship, the social preference for the parental species becomes linked to sexual behavior and is stabilized (Bischof 1994).

The acquisition and consolidation phases are different processes, and each can be modified by different factors. The strength of the social preference for the parental species that develops during the acquisition phase is influenced by the amount of food that the young male is given by his parents. However, consolidation, the linking of the social preference for the parental species to sexual behavior, is most affected by the degree to which the male is aroused at the time of courtship (Oetting, Pröve, and Bischof 1995).

At first glance, it might seem curious that an individual has to learn to identify an appropriate mate. Wouldn't it be a safer evolutionary strategy to have a mating preference that cannot be modified by early social experience? Apparently not. For now, though, we can only hypothesize about the importance of early learning in choosing mates.

One idea, put forth by Patrick Bateson (1983), provides an interesting explanation for the functional significance of sexual imprinting. Bateson suggested that animals learn to identify and selectively respond to kin. Armed with information on what their relatives look like, individuals then choose mates similar but not identical to their family members. Given that both extreme inbreeding and outbreeding may have costs (see Chapter 14), sexual imprinting provides information that allows animals to strike a balance between the two. Bateson used evidence from studies of quail to support his argument.

Japanese quail (*Coturnix coturnix japonica*) prefer to mate with individuals that are similar to yet different from members of their family (Bateson 1982). In one study, chicks were reared with siblings for the first 30 days after hatching and then socially isolated until they became sexually mature. At 60 days of age, males and females were tested for mate preference in an apparatus that permitted viewing of several other Japanese quail (Figure 8.16a). The birds that were viewed were of the opposite sex of the test animals and belonged to one of the following five groups: (1) familiar sibling, (2) novel (unfamiliar) sibling, (3) novel first cousin, (4) novel third cousin, or (5) novel unrelated individual. Similarity in plumage between Japanese quail is considered to be proportional to genetic relatedness, and thus test animals could presumably judge genetic distance on the basis of plumage characteristics. As we see in

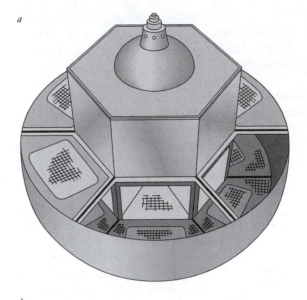

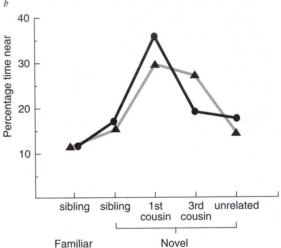

FIGURE 8.16   Japanese quail prefer to spend time near individuals that are similar to yet slightly different from members of their family. (*a*) Apparatus used by Patrick Bateson to test preferences of adult quail. (*b*) Both male (shown as triangles) and female quail (shown as circles) prefer first cousins, possibly striking an optimal balance between inbreeding and outbreeding. (*a*: From Bateson 1982. *b*: Modified from Bateson 1982.)

Figure 8.16*b*, both males and females preferred to spend time near first cousins. The sexual preferences displayed by quail reflect a choice slightly displaced from the familiar characteristics of relatives: Siblings are too familiar, novel unrelated animals are too different, but first cousins are the perfect mix of familiarity and novelty.

Bateson (1982) suggested that such fine-tuned preferences for first cousins result from a combination of two learning processes, imprinting and habituation. Whereas imprinting would tend to restrict preferences to the familiar, habituation would tend to reduce responsiveness to familiar animals. When one form of learning is superimposed on the other, the result is a preference slightly displaced from the familiar. (Habituation is discussed further in Chapters 5 and 6.)

The results from studies of Japanese quail certainly seem to support Bateson's contention that through sexual imprinting, some young animals learn the characteristics of their close relatives and then, in adulthood, choose a mate similar but not identical to their family members. A note of caution, however, is in order. As Bateson (1983) himself points out, we must be careful in generalizing results from laboratory studies to animals in their natural environment. Clearly, laboratory conditions of rearing and testing animals are vastly different from the natural conditions under which animals live and choose mates. We should also be aware that our choice of measures, such as the number of approaches an animal makes toward another, may not reflect mate preference or actual mating in nature. In the best of all possible worlds, we would compare results from the controlled environment of the laboratory with field observations on the impact of early experience on subsequent mate choice. In the field, however, it is often difficult, if not impossible, to know the precise genetic relationships of animals and to track animals from birth to adulthood, chronicling their early experiences and subsequent mating behavior. Although many factors preclude, or at least make very difficult, the study of sexual imprinting and later mate choice in animals in their natural environment, we should make every attempt to design laboratory experiments and to evaluate our results from such experiments in the context of what is known about a species' environment and social system.

Imprinting-like Processes in Mammals

Most of the research on imprinting has been done with birds, but the phenomenon is not restricted to birds. Just as ducklings will string along behind their mother, a litter of young shrews of the genus *Crocidura* will line up behind their mother, each youngster grasping the fur of the shrew in front, as the mother leads the caravan from place to place. As is typical of imprinting in most mammals, olfactory cues, rather than visual or auditory information, provide the relevant stimulus. From 8 to 14 days of age, young shrews imprint on the odor of their mother while nursing. If a shrew is nursed by a foster mother during this time period, it imprints on the odor of the substitute mother. When returned to its biological mother on day 15, the young shrew will not tag along behind her or any of its siblings that remained with her. The youngster will, however, follow a cloth impregnated with the odor of its foster mother (Zippelius 1972).

It is likely that odor preferences change with age, and mammals may not always find the odor of their mother and siblings quite so attractive. Christopher Janus (1988) examined the development of olfactory preferences in spiny mice, *Acomys cahirinus* (Figure 8.17a). The genus *Acomys* is unique among murid rodents (the family Muridae includes rodents such as New and Old World rats and mice, hamsters, gerbils, and lemmings) in that newborns are precocial—they are born relatively well developed. In contrast to altricial rodents, which are born naked with an incompletely developed olfactory system and closed eyes and ears, newborn spiny mice can hear and smell at birth. Shortly after that, their eyes open and they are capable of coordinated movement. Hair appears shortly thereafter. As a result of their rapid rate of physical development and tendency to be active at night, when visual cues are limited, the early development of attachment to specific olfactory cues might be of great importance to spiny mice. Janus tested spiny mice from 1 to 35 days of age in a multichoice preference apparatus: Mice had the option of spending time near dishes that were empty or dishes that contained shavings that were clean, soiled by family members, or soiled by unrelated conspecifics. An interesting pattern in the development of olfactory preferences emerged (Figure 8.17b). Youngsters from 1 to 10 days of age showed a strong preference for the familiar home cage odor. This preference declined with age, and by day 25 the pups no longer found the familiar odor of parents and siblings attractive. At about this time, the pups began to show an increasing preference for odors from unfamiliar conspecifics.

Janus speculated that decreased attachment to familiar odors and increased attachment to unfamiliar odors facilitate dispersal from the home area and emigration to other social groups. In addition, increased preference for odors different from those of parents and siblings might help spiny mice avoid breeding with close relatives. We have already mentioned the relationship between early exposure to family cues and subsequent mate preference in the section on sexual imprinting in birds. Now we see that experience with parents and siblings can also influence adult sexual preferences in mammals.

Another imprinting-like process that occurs in mammals involves maternal attachment. In some species of ungulates, a lasting bond between the mother and young is established rapidly after birth and results in the mother directing her care exclusively toward her own offspring. The rejection of the young of other mothers is a no-nonsense affair and can often be quite violent.

Consider the development of maternal attachment in the domestic goat (*Capra hircus*), a species known for its intolerance of alien young. Young goats (kids) are precocial, capable of wandering away from their mother soon after birth. Thus, it is important that a bond between mother and offspring be established

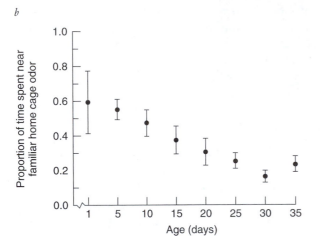

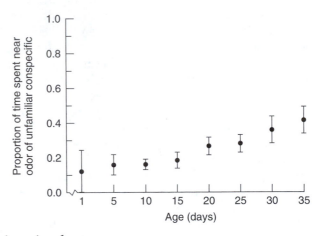

a

b

FIGURE 8.17   (*a*) A spiny mouse. (*b*) Olfactory preferences of spiny mice change over time. Preference for odors from family members declines with age, but preference for odors from unfamiliar conspecifics increases. (Modified from Janus 1988.)

early. Furthermore, kids will initially approach any mother to nurse, even though there is no reciprocity among mothers for care of their offspring. It also pays, then, for a mother to be able to tell her own young from others (Gubernick 1980). Although the formation of mother-young relationships in goats depends on mutual bonding between the mother and her offspring, we will focus our discussion on the development of the mother's attachment to her kids.

Domestic goat mothers, also called nannies, become strongly attached to their offspring and butt away young that are not theirs (Figure 8.18). Peter Klopfer was among the first people to investigate the formation of the maternal bond in this species. He showed that there is a short period of time shortly after parturition that is critical to the development of the mother-young relationship. As it turns out, however, the relationship is not built as much on social attachment as it is on the rejection of young that have been

labeled by other mothers as belonging to them. Let's see how this story unfolded.

In one experiment, Klopfer (1971) examined the influence of early contact with a goat's offspring on subsequent acceptance or rejection of her own and alien young after a period of separation. There were two treatment groups, each containing 15 mother goats. In the first group, kids were removed from mothers immediately after birth and were wrapped in toweling to keep them warm during the separation. Mothers in the second group were permitted five minutes of contact with their firstborn kids before the young, and all subsequent littermates, were removed. Kids were presented to females after a separation of between one and three hours and were introduced one at a time for a ten-minute test period. Females were scored according to whether they accepted or rejected the kid. Acceptance was characterized by the mother's licking the youngster and allowing it to nurse; rejection typi-

FIGURE 8.18   A mother goat rejecting an unfamiliar kid. Mother goats become attached to their own young shortly after birth and butt away alien young.

cally involved withdrawing from the kid and butting it when it persisted in trying to nurse. Fourteen of the 15 mothers who were allowed contact with their firstborn for five minutes after birth accepted and nursed all their own offspring, while vigorously rejecting all alien young. Of the 15 mothers that had been immediately separated from their kids at birth, only 2 allowed their kids to nurse and all rejected alien young. It seems, then, that a short period of contact with their young just after birth is essential for the development of maternal attachment in goats. We see in this example an exception to the rule that critical periods occur early in life. Here, it is the adult mother goat that imprints on her young.

Klopfer's early research indicated that maternal attachment in goats is specific, rapidly formed, and fairly stable. Maternal attachment to young seemed to occur during a critical period just after birth in which only five minutes of contact with at least one kid was necessary for the development of the maternal bond.

More recent findings, however, suggest that a longer period of initial mother–young contact is necessary for maternal imprinting in goats. In these studies, mothers given five minutes of contact with their own kids immediately after birth often failed to later discriminate between their own and alien young after a one-hour separation; 62% of the goats accepted both their own and alien young, and only 14% accepted their own and rejected alien kids (Gubernick 1980; Gubernick, Jones, and Klopfer 1979).

What could account for the mothers' acceptance of alien kids in these studies? The important factor turned out to be whether or not the alien kids had been kept with their mothers during the experiment. Alien kids that had never had contact with their own mother were accepted; alien young that had been kept with their mother for at least eight hours were routinely rejected by other mothers. These results suggested that mothers in contact with their kids label them in some way, perhaps by permeating them with their own scent, and that labeled kids are subsequently rejected by other mothers.

Additional studies revealed that mother goats do indeed label their youngsters, directly through licking and indirectly through their milk (Gubernick 1981). The alien kids used in Klopfer's (1971) study had all been kept with their own mothers, and thus maternal labeling could explain why they were subsequently rejected by mothers given only five minutes of contact with their newborns. The more recent findings suggest that if Klopfer had used kids that had not had contact with their mothers after birth (and thus had not been labeled), the mothers would probably have accepted the alien young.

In summary, then, mother goats must associate with their kids for a certain interval of time after birth for reliable recognition and discrimination to subsequently occur. During this critical period, the mother appears to label her kids and concurrently to learn the label. Although labels may initially be olfactory, we can expect that, in time, they may change to include other features of the young such as vocalizations and appearance. We might also expect different labels to be used for recognition at close quarters as opposed to at a distance. Finally, in line with current views on critical periods in behavioral development, the sensitive period for maternal attachment in goats does not appear to be as short and well defined as would have been previously believed. Indeed, mothers will accept alien kids that have not had contact with their own mother for at least one hour after birth.

The physiological basis of the critical period for maternal attachment in ungulates is the hormone oxytocin. The dilation of the mother's cervix during the final stages of the birth process causes a surge of oxytocin to be released from the hypothalamus of the brain. After five minutes, the oxytocin levels drop by nearly half their peak values. The timing of oxytocin release corresponds to the changes in the behavior of the mother. As the head of the offspring emerges, stretching the cervix, the mother becomes nurturing and caring. If the mother has contact with her infant for a few minutes immediately afterward, she will accept the young. If blood from a goat in labor is transfused into a virgin goat, the virgin will begin to show interest in kids when about one-quarter of her blood has been exchanged. Apparently, the oxytocin alters neurotransmitter release and electrical activity in the olfactory lobe of the brain. These changes in the olfactory lobe are thought to alter the sensory threshold to odor at the time of birth, allowing the maternal attachment to her offspring (Klopfer 1996).

Development of Social Behavior in Ants

Until now we have been considering the effects of experience during specific windows of time on the development of social behavior in birds and mammals. We can now ask, what evidence have we of critical periods in behavioral development in other classes of animals? Our next example comes from the insects, a class that contains roughly three-quarters of the approximately one million known species of animals on earth. The insect that will concern us here is the ant.

The life of an ant can often be quite complicated. In many species, individuals live in complex societies in which labor is divided among colony members, and their roles can change dramatically and repeatedly over the course of a few weeks. Given their intricate social system and frequent changes in job status, it is not surprising that ants have become popular subjects for researchers interested in behavioral development. Of particular interest is the neotropical ant *Ectatomma tuberculatum*, a species in which the individual's job changes with age.

Annette Champalbert and Jean-Paul Lachaud (1990) examined the role of early social experience in the development of behavior of *E. tuberculatum* workers. The two researchers were interested in what effects a 10-day period of social isolation would have on behavioral development in workers of different ages. They began their task on a coffee plantation in southern Mexico, where they collected four ant colonies from the bases of different coffee trees, each colony containing from 200 to 300 individuals. Each colony was placed into an artificial nest made of plaster that had several interconnected chambers and a foraging area. A glass pane permitted the ants to be observed. A few hours after their emergence from cocoons, 15 workers from each of the four colonies were individually labeled with a small numbered tag glued to their

thorax and then reintroduced into their respective colonies. These were the control workers. Other ants, the experimental workers, were labeled in the same manner and then isolated in a glass tube equipped with food and water. In one colony the isolation period began at emergence; in the second, third, and fourth colonies, the period of isolation began when workers were 2, 4, or 8 days old, respectively. Like the controls, there were 15 workers in each of the four isolation groups. Champalbert and Lachaud recorded the behavior of control and experimental ants for 45 days after emergence.

The question was, what effects, if any, would social isolation have on the behavioral development of workers in the experimental groups? Also, would the effects of isolation depend on when they were isolated? During the first week after emergence, the control workers spent most of their time feeding or being groomed by other colony members. By the second week, however, work began, and the ants started to specialize in nursing activities, first caring for larvae, then cocoons, and finally eggs. Sometime during the third week, the workers changed their job status again and began to explore the nest and engage in domestic tasks. Finally, about a month after emergence, the control workers focused on activities related to their new careers as either guards at the nest or foragers outside the nest. Thus, over the course of only a few weeks, *E. tuberculatum* workers typically underwent several changes in behavioral specialization, and in a very specific sequence.

So, how did the behavior of ants in the experimental groups compare with that of controls? What were the effects of isolation immediately at emergence, or 2, 4, or 8 days later? It is interesting that the workers isolated at 2 days after emergence showed the most abnormal behavioral development. They differed from controls in the order of appearance of their activities and in the level of performance of their tasks, being particularly lax in brood care (Figure 8.19). In contrast, the behavioral development of workers isolated either at emergence or 4 or 8 days later was more similar to that of the controls. In the case of workers isolated at emergence, behavioral development was simply delayed. The reintroduction of these workers into society seemed to correspond to a second emergence; although 11 days old at the time of reintroduction, these workers behaved like newly emerged ants. Soon after reintroduction into the colony, however, the various specializations in activities appeared progressively in the same sequence as that of the control workers. Only minor abnormalities were noted in the workers isolated at 4 days, and the behavioral development of those isolated at 8 days was almost identical to that of the controls. In summary, the extent of behavioral abnormalities observed in *E. tuberculatum* workers as a result of a 10-day period of social isolation

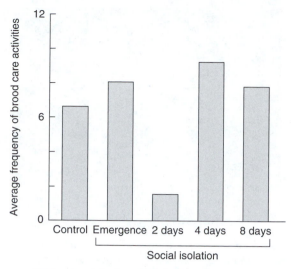

FIGURE 8.19 A critical period for the development of brood care behavior in worker ants. Here, a ten-day period of social isolation beginning two days after emergence produces more severe disruptions in the performance of brood care activities by *Ectatomma tuberculatum* workers than do periods of isolation beginning either at emergence or four or eight days later. (Drawn from the data of Champalbert and Lachaud 1990.)

depended on the age of the workers at the time of isolation, with the most serious abnormalities occurring in ants isolated 2 days after emergence.

Champalbert and Lachaud (1990) concluded from their observations of *E. tuberculatum* worker ants that there is a critical period during the first four days after emergence in which exposure to social stimuli affects the establishment of behavior, especially that related to brood care. One might ask, then, why would workers isolated at emergence (and therefore removed from social stimuli during the four-day critical period) display fairly normal behavioral development? The authors suggest that when workers are isolated immediately after emergence, their development is simply put on hold. The reintroduction of these workers into the colony after their period of social isolation mimics an emergence. During the four days following reintroduction, these workers receive the social stimuli important in the development of brood care activities. In other words, it is not the physiologically defined first four days of life that is important for the development of brood care behavior but rather the first four days of social contact with colony members. These four days may follow either natural emergence or the artificial second emergence produced by the reintroduction of workers into the colony after isolation. Here, again, we see an example of flexibility associated with critical periods in behavioral development; in ants isolated at emergence, the lack of exposure to appropriate social stimuli delays closure of the critical period until a few

days after reintroduction to the colony. In view of the poor performance of workers isolated at two days, it also appears that the entire sensitive period must occur in an uninterrupted fashion to ensure normal behavioral development. Because these workers received some portion of the necessary social stimulation during their first two days following emergence, their developmental system could not be put on hold until reintroduction into the colony.

DEVELOPMENTAL HOMEOSTASIS

Throughout this chapter we have seen that experience certainly does play an important role in behavioral development. Having said this, we might also note that the manipulations used to demonstrate the sensitivity of the developing animal to external influence are often quite severe, typically involving rearing in isolation or rearing by another species, two conditions that are not likely to be encountered by most animals under natural conditions. Indeed, despite the fact that in their natural environments, individuals within a species often develop under a diverse array of physical and social conditions, the vast majority of adults display species-typical, normal behavior. Under field conditions, after all, most white-crowned sparrows sing white-crown songs, most mallard ducklings follow their mother, and most worker ants perform their brood care responsibilities in a normal fashion. Why do such complex patterns of behavior develop so reliably when there seems to be so much room for error?

The developmental processes appear to be capable of buffering themselves against potentially harmful influences to produce functional adults. This buffering capacity is called developmental homeostasis. A buffer, of course, is something that dampens drastic changes. We can look around us and see that despite diverse experiences, behavioral development in most individuals proceeds in a very predictable and reliable manner. In effect, the developmental process demonstrates a certain stability and resilience in the face of a host of constantly changing variables, including, as we have seen, many of those we introduce experimentally. You may have already been struck by the virtually normal development of worker ants that were isolated from colony members for a ten-day period either at emergence or four or eight days later. Despite being removed from the hustle and bustle of normal colony life and placed in separate small glass tubes, individuals in these groups, for the most part, exhibited normal behavioral development. As we saw, only ants isolated two days after emergence displayed highly abnormal behavior, suggesting that the first four days of social contact with colony members are critical to normal development. Isn't it remarkable that in ants, highly social animals that undergo complex changes in behavioral specializations, only four days of contact with colony members are sufficient to produce normal behavioral development? As we see in our next example, social development in rhesus monkeys, often only a bare minimum of experience is necessary to promote virtually normal behavioral development, even under the harshest of circumstances.

SOCIAL DEVELOPMENT IN RHESUS MONKEYS

The ability of animals to exhibit normal development when provided with very little critical experience can be seen in the results of research with rhesus monkeys (*Macaca mulatta*). Experimenters devised highly abnormal environments in which to rear infant monkeys and then examined the effects of such environments on the subsequent social development of the subjects. In most cases, these highly contrived laboratory environments produced severe effects on social behavior. The environments were virtually devoid of the features necessary for the development of normal social behavior. The ability of rhesus monkeys to compensate for unfavorable rearing conditions was then shown by experiments that revealed just how little critical experience is needed to produce normal behavior.

Psychologists Margaret and Harry Harlow pioneered rearing-condition research with rhesus monkeys by examining the effects on social development of rearing infant monkeys in either total or partial isolation or with a surrogate mother. In the case of total isolation, the infants were raised alone in laboratory cages. Deprived of physical, visual, and auditory contact with other monkeys for the early part of their lives (typically for either 3, 6, or 12 months), these youngsters saw only the hand of the experimenter at feeding. Rhesus monkeys reared under these conditions exhibited stereotyped rocking movements and often clutched and pinched or bit themselves when huddled in a corner of the cage. When they were a few years old (still immature) and were placed with other infants, the experimental monkeys would sometimes freeze in apparent terror and at other times explode in bouts of extreme aggression. Once mature, they exhibited inappropriate sexual and parental behavior (Harlow, Harlow, and Suomi 1971). Females reared without mothers or peers were usually indifferent to their infants (Figure 8.20) or downright abusive, although they sometimes improved when caring for their second offspring.

In the partial isolation condition, infants were raised alone in wire cages, where they could see and hear other monkeys but had no physical contact with them (Figure 8.21). Despite the provision of some social stimulation, partial isolates developed a behavioral syndrome similar to that of total isolates. These findings under-

FIGURE 8.20   A rhesus monkey female that was reared in isolation during infancy is ignoring her infant.

score the importance of physical contact in the development of normal social behavior in rhesus monkeys.

In the third condition, infant rhesus monkeys were reared with inanimate surrogate mothers, typically a wire cylinder, often covered in terry cloth, and surmounted by a wooden head (Figure 8.22). These surrogates were designed by Harry Harlow to provide comfort for the infants from bodily contact, gaining for him the presti-

FIGURE 8.21   Rhesus monkey infants reared in partial isolation can see and hear other monkeys but receive no physical contact from conspecifics. These so-called partial isolates show all the severe social deficits associated with total social isolation, thus emphasizing the importance of physical contact to normal social development in this species.

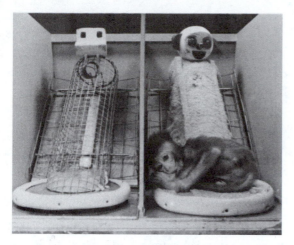

FIGURE 8.22 **A rhesus infant with an inanimate surrogate mother.**

gious title of "father of the cloth mother." Although infants became quite attached to their surrogate mother and exhibited less self-directed activities (e.g., clutching and biting themselves) than did total isolates, they still failed to show appropriate social behavior.

We must ask, then, what is so special about being raised by a rhesus monkey mother? To isolate the important factors in the bonding of infant to mother, surrogate mothers were modified in various ways to provide milk, warmth, and rocking movements. In one set of studies, infants preferred a nonlactating cloth mother to a lactating wire surrogate, suggesting that contact is more important in formation of the infant-to-mother bond than is the satisfaction of nutritional needs. This finding caused Harlow, Harlow, and Suomi (1971) to proclaim, "There is more than merely milk to human kindness." The fact remained, however, that rearing by an inanimate surrogate mother, no matter what its qualities, produced severely disturbed monkeys.

These experiments demonstrated that monkeys reared in highly abnormal physical and social environments display highly abnormal patterns of social behavior. Rearing by inanimate surrogate mothers or in total or partial isolation provided none of the critical social experience necessary for normal social development. The important question then became, what level of social contact would be necessary to produce normal social development? What would be the bare minimum of social experience that would lead to normal monkeys—monkeys that played with other monkeys rather than huddling and rocking in a corner, that displaced normal social and sexual interactions rather than extreme aggression or lack of interest, and that cared for their own infants rather than ignoring or abusing them? (It is important to keep in mind here that even the social settings in the laboratory that produce "normal" monkeys—i.e., settings in which infants are reared with their mother and also exposed to peers—are prob-

ably less complex than that which most infants would experience in the field. "Normal" behavior is thus in reference to laboratory, and not field, standards.) Amazingly enough, infants that were housed alone with a surrogate mother and given only 20 minutes of daily contact with peers exhibited the full range of social behavior, with no developmental delay (Harlow and Harlow 1962).

Although rhesus infants in nature are almost constantly exposed to conspecifics, even a few minutes of daily contact with peers in the laboratory are sufficient to produce normal social behavior. Obviously, a little social experience goes a long way. The ability of the infants to exhibit normal social development when provided with only a few minutes of daily contact with peers again illustrates the resilience of the developmental process.

Another interesting finding that arose from this research is that rearing with peers produces more normal social development than does rearing with mothers alone (Harlow and Harlow 1962). This finding must have come as a shock to those in the scientific community who, along with the famous psychoanalyst Sigmund Freud, believed that the relationship between an infant and its mother is the most important component of the early environment. It is not that mothers are unimportant to infant monkeys, but they are not, in the case of rhesus monkeys, as essential as peers to normal social development. Indeed, experience that would seem to be critical to the development of a particular behavior sometimes turns out to be less important or not important at all. We emphasize this point in the next section as we discuss the development of swimming behavior in an amphibian.

NEUROBEHAVIORAL DEVELOPMENT IN AMPHIBIANS

Behavioral Development in the Absence of Embryonic Experience

In a manner not unlike that of the embryonic salmon described earlier in this chapter, embryos of the South African clawed frog, *Xenopus laevis*, show certain behaviors prior to hatching. The onset of movement in the embryos begins about 25 hours after fertilization, and alternating flexure of the body and swimming behavior begin 10 and 20 hours later, respectively. A tadpole hatches approximately 50 hours after fertilization. Are embryonic behaviors necessary for the performance of normal swimming behavior by tadpoles? Lanny Haverkamp and Ronald Oppenheim (1986) sought to answer this question by immobilizing *Xenopus* embryos before the onset of movement and then observing the behavior of tadpoles once the period of immobility had ended. Immobilization was accomplished by immersing embryos in drug solutions (lidocaine or chloretone) or by injections of alpha-bungarotoxin before the onset of embryonic movement; all three treatments suppressed activity in the nervous system and thus all behavior in the embryo.

How did the complete absence of embryonic neural activity and behavior affect patterns of swimming after hatching? It is striking that once the suppressing drugs were removed, the swimming behavior of experimental tadpoles was virtually equivalent to that of normally reared tadpoles. In addition, the anatomical and neurophysiological substrates of swimming in the experimental tadpoles were indistinguishable from those of normal tadpoles (Haverkamp 1986). Thus, although embryonic experience may sometimes be critical to the development of species-typical postnatal behavior (e.g., mallard ducklings must hear the contentment calls of their siblings before hatching in order to follow the maternal assembly call after hatching), we cannot always assume that this is the case. In *X. laevis*, embryonic experience, in particular the patterns of behavior that lead up to swimming, are not necessary for the normal performance of swimming behavior by tadpoles. So, if at first glance we assumed that embryonic swimming movements were necessary precursors to the development of swimming behavior in tadpoles, we would certainly have been wrong.

Resilience in the Developing Nervous System

We began this chapter with examples of how behavioral change accompanies the development of the nervous system. Are there examples of developmental homeostasis at the level of the nervous system that might help to explain the tendency for behavior to develop normally? We will consider the same basic paradigm as that just described for *Xenopus* embryos, this time, however, focusing on the effects on the structure and function of the nervous system caused by eliminating all neural activity.

The ability of neurons to develop normally in the absence of neural activity was elegantly demonstrated by the transplantation experiments of William Harris (1980). Harris took advantage of the fact that tetrodotoxin (TTX), a powerful neurotoxin that blocks all propagation of neural impulses, is a normal constituent of the body fluids of the California newt, *Taricha torosa*. The newt's tissues, as you would expect, are insensitive to the effects of TTX. This is not the case, however, for the neural tissues of another amphibian, the Mexican axolotl, *Ambystoma mexicanum*. Upon exposure to TTX, all neural activity ceases in the axolotl. Harris transplanted an embryonic eye of the Mexican axolotl into the head region of newt embryos. The transplanted eye was a third eye to the newt, rather than a replacement eye. In the newt, all neural activity associated with the axolotl eye was inhibited because of the presence of the TTX. Once the newts had matured (Figure 8.23), Harris examined the neural projections

FIGURE 8.23   A newt with a third eye from an axolotl. The transplantation experiments of William Harris show that neural connections from the transplanted eye reach their appropriate targets in the newt's brain despite the complete inhibition of neural activity caused by tetrodotoxin in the tissues of the newt. (From Harris 1980.)

associated with the transplanted eye. Despite the complete inhibition of neural activity, projections from the transplanted eye found their way to their proper targets in the newt's brain and made functionally effective connections there.

The normal development of optic nerve fibers is even more surprising when we consider that the neural tissue was transplanted from a different genus and species and that the transplanted tissue was in direct competition with the normal host's eye at the target sites in the brain. The ability of the neural connections of the axolotl eye to develop normally in the face of all of these environmental perturbations illustrates developmental homeostasis. This is not to say that development of the nervous system can always proceed in a normal fashion in the absence of neural activity. Indeed, there are some developmental processes in which neural activity plays an important role. We simply use this example as a means of demonstrating the resilience associated with the developmental process, a resilience reflected in the structure and function of the system, as well as in the behavior of developing animals.

SUMMARY

Patterns of behavior appear, disappear, and alter in form as animals develop. There are several causes of behavioral change, and one of the most significant is the development of the nervous system. During embryonic life, patterns of behavior often emerge in parallel with development of the sensory and neural structures necessary for their performance. Sometimes neuronal growth, death, or alterations in structure can be linked to developmental changes in a particular behavior.

Changes in hormonal state can also trigger behavioral change, as in age polyethism in honeybees. As an adult honeybee matures, it assumes new tasks in a generally predictable order. Young workers assume chores within the hive, including caring for the brood and maintaining the hive. Older workers forage outside the hive. These changes are brought about by increasing juvenile hormone (JH) levels, which cause changes in the mushroom bodies of the brain that prepare the workers for foraging. JH levels can be modified by social interactions within the hive that spread chemicals that inhibit JH secretion. In this way, the rate of behavioral development can be adjusted to meet the changing needs of the colony.

Behavioral change may also come about through the development of specific morphological structures. In developing paddlefish, the change in feeding behavior from chasing and picking select items out of the water column to indiscriminate filter feeding parallels the development of gill rakers. Neural, hormonal, morphological, and experiential causes of behavioral change often interact during the continuous interplay between the developing animal and its internal and external environment.

Experience also affects the development of behavior. In addition to learning, play is an experience that is vital to the behavioral development in many species. It is expressed in a variety of ways: social play, including mock fighting and chasing and sexual play; locomotor play (exercise); and object manipulation. Hypotheses for the function of play include that it improves physical condition, is important in developing social skills and bonds, and helps develop cognitive skills.

Genetic factors also enter into the complex interaction between the developing animal and its environment. The interaction between genetic and environmental factors is perhaps best illustrated by the development of sexually dimorphic patterns of behavior, such as singing in birds. In general, only male birds sing, a fact reflected in their neuroanatomy. The developmental basis for sex differences in the brains and vocal behavior of songbirds is the chromosomal difference between the sexes. This genetic difference dictates patterns of secretion of steroid hormones that, in males, lead to the growth and differentiation of the regions of the brain involved in song. In addition to neural and hormonal influences, experience plays an important role in the development of singing behavior. If they are to produce normal songs in adulthood, young males must listen to the songs of adult conspecifics and must be able to hear themselves sing.

In male songbirds, the pulse of steroid hormones that affects the developing nervous system and the experience of hearing conspecific songs are not always capable of influencing the development of singing behavior. Indeed, to be effective, the change in hormone level and the experience of listening to conspecifics must occur during somewhat restricted periods of time, when the young bird is particularly sensitive to such influences. These periods of enhanced sensitivity to environmental stimuli are called critical periods, and they characterize a diverse array of behavioral phenomena. In some cases, critical periods occur before birth or hatching, and experience during the prenatal period influences the development of species-typical postnatal behavior. An embryonic critical period has been demonstrated for the development of filial imprinting, the response of ducklings to follow their mother. While still in the egg, mallard ducklings must hear the contentment calls of their siblings to respond to the assembly call of their mother after hatching. Although usually occurring early in life, critical periods occasionally occur in adulthood, as when mother goats learn the characteristics of their offspring just after birth, a learning process that results in the acceptance of their own young and the rejection of alien kids. There are tremendous differences among species in the timing of critical periods for the development of specific behaviors, and even within a species there is flexibility in the time course of critical periods. Social interactions, in particular, can influence the timing and duration of critical periods.

We have said that experience, especially that occurring during a critical period in the young animal's life, can profoundly influence behavioral development. Under normal circumstances, however, animals encounter a wide variety of physical and social conditions, and most develop normally. The ability of development to proceed in a normal fashion in the face of environmental perturbation is called developmental homeostasis. The buffering capacity of the developmental process can sometimes be seen in the surprisingly normal development of animals reared under the most deprived laboratory conditions. Even more striking, however, is how little critical experience is sometimes sufficient to induce normal behavioral development. In their natural environment, rhesus monkeys are usually exposed to numerous conspecifics, and yet in the laboratory only 20 minutes of daily contact with peers is sufficient to promote normal social development, at least by laboratory standards. The resilience associated with the developmental process can also be seen at the level of the nervous system. Here, in the face of apparently severe environmental change, such as that imposed by the complete inhibition of neural activity and by transplantation of neural tissue from one species to another, neural development proceeds in a predictable and reliable manner. The ability of the developing animal's behavior and nervous system to withstand the vagaries of environmental change results, in the vast majority of cases, in the production of fully functional adults.

PART TWO

Behavior of Individual Surviving in Environment

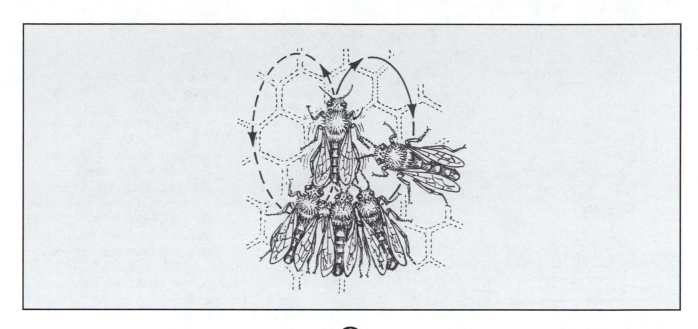

9

Biological Clocks

Imagine for a moment that you have a pet hamster, a friendly fellow who quietly shares your bedroom. However, late one Thursday night you are studying for a test but finding it hard to concentrate because a hamster is most active during night hours, and his running wheel is squeaking. So, you place the hamster and his cage in your closet and continue to study. The next morning, you take the test and then leave for the weekend. On Monday night you are back, and you notice a squeaking in your closet. You have forgotten about the hamster. He has had plenty of food and water, but he has been in the dark for three days. As you retrieve him, you notice that he begins to run on his wheel at about the same time as he normally did.

How could he know what time it was? We attribute the ability to measure time without any obvious environmental cues to an internal, living clock. When any hamster is sequestered in the constant darkness and temperature of the laboratory so that each turn of its running wheel can be recorded automatically for months or even years, a record similar to the one shown in Figure 9.1 usually results. Notice that the hamster woke up almost exactly 12 minutes later each day during the entire study. Its bouts of activity alternate with rest with such regularity that it is often described as an activity rhythm. The ability to measure time is common not just in hamsters but also in most animals. In fact, biological clocks have been found in every eukary-

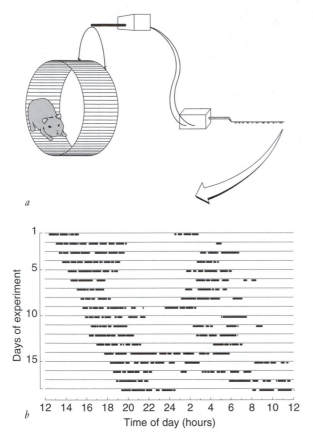

FIGURE 9.1   (*a*) **A hamster in a running wheel equipped to record each turn of the wheel shows periodic bouts of activity that alternate with rest.** (*b*) **In constant dim light, the cycle length of this activity rhythm is slightly longer than 24 hours. For each rotation of the running wheel, a vertical line is automatically made on a chart at the time of day when the activity occurred. In this record the bouts of activity were so intense that the vertical lines appear to have fused, forming dark horizontal bands. Notice that although the animal had no light or temperature cycle as a cue to the time, it would awaken about 12 minutes later each day.**

otic organism tested (Dunlap 1999), as well as in cyanobacteria (Golden, Johnson, and Kondo 1998).

The rhythmical nature of life may come as a surprise to some, although it should not in the light of evolutionary principles. Life evolved under cyclical conditions, and the differences in phases of the cycles are often so pronounced that they place a high adaptive value on being able to accommodate as specifically as possible to each phase.

Every living thing is subjected to the regularly varying environmental conditions on earth orchestrated by the relative movements of the heavenly bodies: the earth, the moon, and the sun. As the earth spins on its axis relative to the sun, life is exposed to rhythmic variations in light intensity, temperature, relative humidity, barometric pressure, geomagnetism, cosmic radiation,

and the electrostatic field. The earth also rotates relative to the moon once every lunar day (24.8 hours). The moon's gravitational pull draws the water on the earth's surface toward it, causing it to "pile up" and thus resulting in high tide. These tidal cycles cause dramatic changes in the environment of intertidal organisms—flooding followed by desiccation when exposed to air. The relative positions of the earth, moon, and sun result in the fortnightly alternation between spring and neap tides, as will be explained shortly. The moon revolves about the earth once every lunar month (29.5 days), generating changes in the intensity of nocturnal illumination and causing fluctuations in the earth's magnetic field. Finally, the earth, tilted on its axis, circles the sun, causing the progression of the seasons, with its sometimes dramatic alterations in photoperiod and temperature.

Although the environmental modifications may be extreme, they are generally predictable. Often it is advantageous to gear an activity to occur at a specific time relative to some rhythmic aspect of the environment. So, biological clocks are generally thought to have evolved as adaptations to these environmental cycles (Daan and Aschoff 1982; Enright 1970).

In this chapter we will see exemplified the various approaches to the study of behavior discussed in previous chapters. Rhythms are so pervasive that they have piqued the interest of those concerned with in the adaptive value of behavior and its evolution; others are interested in its genetic roots, and still others have focused on the hormonal or the neural basis of biological timing. As we will see, their interactions have been fruitful. First we will describe some rhythmic behaviors and the properties of the clock that drives them.

RHYTHMIC BEHAVIOR

Rhythmicity in behavior and physiology is so common that it must be considered by anyone studying animal behavior. An animal is not perpetually the same. Rather, its behavior may fluctuate so that it is appropriate to the time of day or the state of the tides or the phase of the moon or the season of the year. A description of a behavior at one point in a cycle may be totally inaccurate at another time. We will begin by describing a variety of biological rhythms.

DAILY RHYTHMS

Most animals restrict their activity to a specific portion of the day. The hamster, for example, is busiest at night, as are cockroaches, bats, mice, and rats. Other species, songbirds and humans, for instance, are active during the day.

Among the best "timers" are bees. Their time sense was experimentally demonstrated during the early part

a

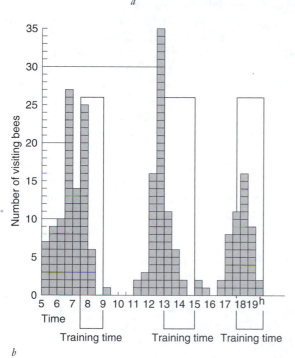

b

FIGURE 9.2 The time sense in bees. (*a*) Bees were marked for individual recognition and trained to come to a feeding dish only at the specific times at which food was made available. (*b*) After six days of training, the feeding dishes were left empty and the number of bees arriving throughout the day was recorded. The bees arrived at the feeding station only when food had been previously present. (Modified from Beling 1929.)

of this century by marking them for the purpose of individual recognition and offering them sugar water at a feeding station during a restricted time each day, between 10 A.M. and noon. After six to eight days of this training, most of the bees frequented the feeding station only during the learned hours. The real test, however, was on subsequent days, when no food was present at the feeding station. As seen in Figure 9.2, the greatest number of bees returned to the empty feeding station only at the time at which food had been previously available (Beling 1929). In subsequent tests, it was found that the bees' time sense is astonishingly accurate. Bees can be trained to go to nine different feeding stations at nine different times of the day. They are able to distinguish points in time separated by as few as 20

minutes (Koltermann 1971). The adaptiveness of such abilities for bees is clear. Flowers have a rhythm in nectar secretion, producing more at some times of the day than at others. The biological clock allows bees to time their visits to flowers so that they arrive when the flower is secreting nectar. This means that the bees can gather the maximum amount of food with the minimum effort.

Clocks are important in mating as well. For example, the fruit fly *Dacus tryoni* mates only during the evening twilight. Such a rhythm ensures reproductive synchrony between members of the same species, thereby increasing the chances of finding an appropriate mate. Other species of genus *Dacus* mate at different times of the day (Tychsen and Fletcher 1971). Thus, timing may also avoid matings with other selected species that are mating at the same season, and time, energy, and gametes are not wasted on doomed reproductive efforts. Even in moth species in which the female entices a male by emitting a potent sex attractant, the biological clock plays an important role (Figure 9.3). Female moths of the genus *Hyalophora* produce a sex attractant pheromone that is attractive to both *H. cecropia* and *H. promethea* males. Costly reproductive mistakes are avoided because these species are active at different times of the day (Wilson and Bossert 1963).

Certain patterns of learned behavior may be under temporal control. When an electrode is placed into certain brain regions (areas of the hypothalamus or certain midbrain nuclei) of a rat, the animal will quickly learn to press a lever to get electrical stimulation (Figure 9.4*a*). An observer is tempted to conclude that the effect of the stimulation is the rat's equivalent of some form of ecstasy since animals with electrodes in these brain areas would rather stimulate themselves than eat, drink, or even copulate. Some will press the lever as often as 5000 times an hour. When the electrodes are first implanted, a rat will typically begin a 2-day marathon of bar pressing at a high rate, resting infrequently for intervals of only a few minutes. After this time, quiet periods—during which the rat may still respond but with lower frequency—alternate with periods of rapid pressing, and soon a 24-hour cycle appears (Figure 9.4*b*) (Terman and Terman 1970). The animals have apparently learned to press the bar to receive a reward, but the reward has more value at some times of the day than at others. Likewise, the operant behavior

of a chimpanzee that is responding for a food reinforcement has been shown to be rhythmic (Ternes, Farner, and Deavors 1967). It has been suggested that since identical stimuli may vary in their effectiveness as reinforcers with the hour of the day, the ease of learning should also fluctuate daily.

LUNAR DAY RHYTHMS

Although the 24-hour daily rhythm in light and dark is probably the most familiar environmental cycle, there are many others of importance. For example, the interaction of the gravitational fields of the moon and the sun create other pronounced environmental changes—those associated with the tides.

As the moon passes over the surface of the earth, its gravitational field draws up a bulge in the ocean waters. One bulge occurs beneath the moon and another on the opposite side of the earth. These bulges sweep across the seas as the earth rotates beneath the moon, thus causing high tides when they reach the shoreline. Since there are two "heaps" of water, there are usually two high tides each lunar day, one every 12.4 hours. The tides may cause some rather dramatic changes in the environment, particularly for organisms living on the seashore.

The activity of the fiddler crab, *Uca pugnax*, a resident of the intertidal zone, is synchronized with the tidal changes. Fiddler crabs can be seen scurrying along the marsh during low tide in search of food and mates. Before the sea floods the area, the crabs return to their burrows to wait out the inundation. When a fiddler crab is removed from the beach and sequestered in the laboratory, away from tidal changes, its behavior remains rhythmic. Periods of activity alternate with quiescence every 12.4 hours, the usual interval between high tides (Figure 9.5) (Palmer 1995).

Living higher in the intertidal zone than the fiddler crab is a small isopod (a crustacean), *Excirolana chiltoni*. The open beach habitat of this organism is periodically exposed to the swirling surf. *Excirolana* is buried in the sand during low tide, but as the pounding waves reach its abode it emerges and begins swimming and feeding in the breaking waves. However, it is not simply that inundation presents an ultimatum of sink or swim. When the animals are brought into the laboratory and are maintained in a sandy-bottomed beaker of seawater, they begin to tread water only at the times of high tide on their local beaches (Figure 9.6). One isopod displayed a tidal rhythm for 65 days in constant conditions (Enright 1972).

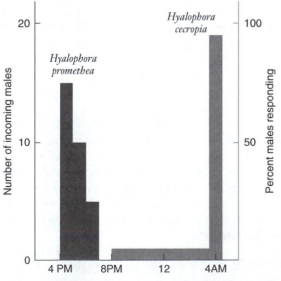

FIGURE 9.3   The flight times of moths of the species *Hyalophora cecropia* and *H. promethea*. The difference in the time of activity of these species helps prevent the males from mating with a female of the wrong species. (Modified from Wilson and Bossert 1963.)

SEMILUNAR RHYTHMS

The height of the tides is also influenced by the gravitational field of the sun. In fact, the highest tides are

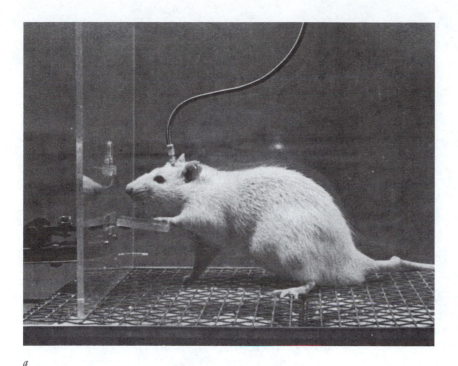

a

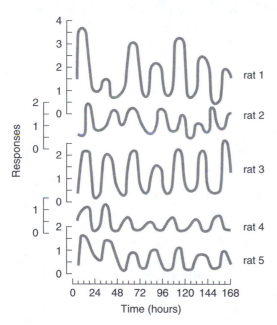

b

FIGURE 9.4   (*a*) A rat with an electrode implanted in the "pleasure center" of the brain. When the rat presses the lever, the pleasure center is stimulated and a record is made of the action. (*b*) The cyclic self-stimulation of the pleasure center by rats. Notice that periods of rapid stimulation alternate with rest in cycles of approximately 24 hours even when the rat has no obvious time cues. (From Terman and Terman 1970.)

caused when the gravitational fields of the moon and the sun are operating together. At new and full moons, the earth, the moon, and the sun are in line, causing the gravitational fields of the sun and the moon to augment each other (Figure 9.7). Thus the earth experiences the highest high tides and lowest low tides at new and full moons. These periods of greatest tidal exchange are referred to as the spring tides. At the quarters of the moon, the gravitational fields of the moon and the sun are at right angles to each other. Since their pulls are now antagonistic, the tidal exchange is smaller than at

other times of the month. These periods of lowest high tides and highest low tides are called the neap tides. Some organisms possess a biological clock that allows them to predict the times of spring tides or neap tides and gear their activities to these regular changes.

The grunion, *Leuresthes tenuis*, a small, silvery fish living in the waters off the coasts of California, precisely times its reproductive activities to occur at fortnightly intervals synchronized with the spring tides. This fish is the only one in the sea to spawn on land. Beginning in late February and continuing until early September, for

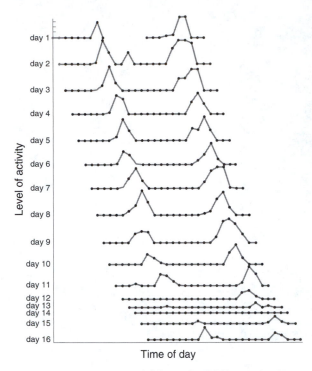

FIGURE 9.5 An activity rhythm of a fiddler crab (*Uca pugnax*). Although the crab was maintained in constant darkness and temperature (20°C), the animal was active at approximately the times of low tide at its home beach. (From Palmer 1990.)

three or four nights, right after the greatest spring tides, the adult grunion gather in the waters just offshore to wait until the tide has reached its peak and is just beginning to recede. Then they ride the waves ashore for a night of mating. At the height of the mating frenzy, the beach is carpeted with their quivering bodies. A female digs, tail first, into the sand so that only her head, from the gills up, is exposed. Several males may wrap around her and release sperm-containing milt that seeps through the sand and fertilizes the eggs discharged by the female (Figure 9.8). After mating, the fish catch a wave and disappear into the swirling sea, leaving the eggs buried in the sand to develop into the next generation.

The timing of this event is exquisite. Since the spring tides have the highest high tides of the fortnightly cycle, by waiting to mate until just after the spring tide, the grunion guarantee that unless there is a storm, their eggs will have about ten days to develop undisturbed in the sand before the pounding surf reaches them and triggers the hatching. If the fish mated before the peak of the spring tides, the next night's high tide would wash the eggs away. The same fate would meet the eggs if the grunion did not wait until the turn of the tide on their mating night (Ricciutti 1978).

Another example of a fortnightly rhythm is seen in the tiny chironomid midge, *Clunio marinus*. In *Clunio* it

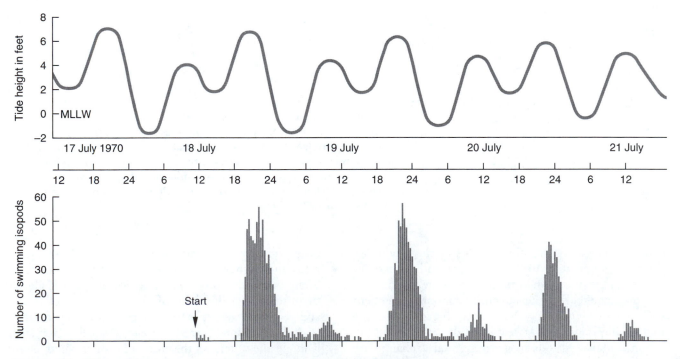

FIGURE 9.6 The activity pattern of isopods (*Excirolana chiltoni*) maintained in the laboratory. The pattern of activity mimics that of the height of the tide in the area of collection. (Modified from Klapow 1972.)

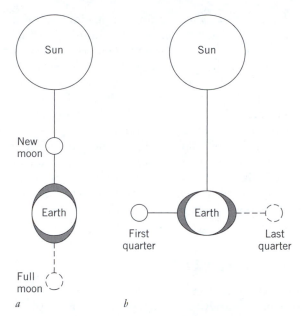

first to break free of their puparia, the cases in which they developed. Each one locates a female and assists in her emergence. They have little time to waste, for their habitat will soon be submerged again, so copulation follows quickly. Then the winged male carries his mate to where she will lay her eggs. All these activities must be precisely timed so that they occur during the short, two-hour period during which the habitat is exposed.

Dietrich Neumann (1976) has found that if a population is brought into the laboratory and maintained in a light-dark cycle in which 12 hours of light alternate with 12 hours of darkness, emergence from the puparium is random. If, however, one simulates the light of full moon by leaving a dim (0.4 lux) light on for 4 consecutive nights, the emergence of adults from the puparia becomes synchronized. Now, just as in nature, emergence occurs at approximately fortnightly intervals for about 2 months.

FIGURE 9.7 **The effect of the relative positions of the earth, moon, and sun on the amplitude of tidal exchange. (*a*) At the times of new and full moons, the gravitational fields of the moon and the sun assist each other, causing the spring tides. (*b*) During the first and last quarters of the moon, the gravitational fields of the moon and the sun are perpendicular to each other. This results in the smallest tidal exchange, the neap tides.**

MONTHLY RHYTHMS

The interval from full moon to full moon, a synodic lunar month (29.5 days), corresponds to the length of time it takes the moon to revolve once around the earth. Some organisms have a clock that allows them to program their activities to occur at specific times during this cycle.

An example of a clearly adaptive monthly rhythm is the timing of the reproductive activities of the marine polychaete *Eunice viridis*, the Samoan Palolo worm. Unlike the organisms previously mentioned, in which rhythms in reproduction are precisely synchronized with propitious environmental conditions, the Palolo worm restricts its procreation to a specific time so that

is the end of development, the emergence of adults from their pupal cases, that is programmed to coincide with tidal changes. These insects live at the lowest extreme of the intertidal zone so that they are exposed to the air for only a few hours once every two weeks, during each spring low tide. When the tide recedes, the males are

FIGURE 9.8 *Mating grunion. The female buries her posterior end in the sand and releases her eggs while a male wraps himself around her and releases his sperm.*

the population will release gametes simultaneously, increasing the probability of fertilization in the water.

The Palolo worm lives in crevices of the coral reefs off Samoa and Fiji. In preparation for reproduction, it elongates by budding new segments that become packed with gametes. This section of the worm, the epitoke, may be a foot long. Around sunrise on a day at or near the last quarter of the moon during October and November, the epitokes synchronously tear free from the rest of the body and rise to the surface, churning the water. Then they explode in unison, liberating gametes into the sea, where fertilization takes place. The timing of this self-destructive orgy is so predictable that the epicurean natives of Samoa and Fiji calculate the day of the event in advance and arrive at the coral reefs at dawn to await the frenetic swarming so that they can scoop up the epitokes, which they eat either raw or roasted (Figure 9.9) (Burrows 1945).

The ant lion (*Myrmeleon obscurus*) shows another sort of monthly rhythm. A lazy hunter, it builds a steep-sided conical pit in the sand and then lies in ambush at

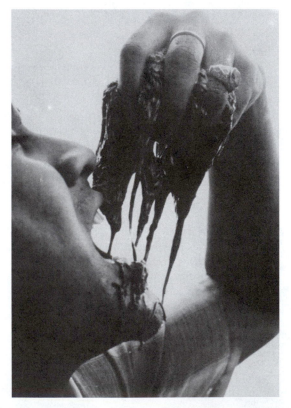

FIGURE 9.9 A feast of Palolo worms. The parts of the Palolo worms (*Eunice viridis*) specialized for reproduction (epitokes) break free from the rest of the worm and swarm synchronously to the surface of the sea, releasing their gametes. The monthly rhythm in swarming increases the probability of fertilization and allows the local human population to predict the event in advance so that they can scoop up handfuls of epitokes to eat.

the base, with all but its immense mandibles covered with sand, waiting for some small arthropod, such as an ant, to slide into the pit toward its outstretched jaws. The ant lion then sucks out the prey's body fluids (Figure 9.10). Perhaps the most interesting observation on the ant lion's behavior is that it is different at full moon, similar to the pattern of blood-sucking vampires in fictional (one hopes) literature. It constructs larger pits at the time of full moon than at new moon. Careful daily measurements of the size of the pits of ant lions cloistered in constant conditions in the laboratory have revealed that this is a clock-controlled rhythm and not a simple response to some aspect of the environment such as the amount of moonlight (Youthed and Moran 1969).

ANNUAL RHYTHMS

The seasonal changes in the environment can be quite dramatic, especially in the temperate zone. As the days shorten and the temperature drops, plants and animals prepare themselves for severe and frigid weather. Some species avoid the cold and limited food of winter by migrating. An annual biological clock is important in timing migration. We see this, for example, in the activity of garden warblers, *Sylvia borin*. The bird's activity can be monitored by using microswitches mounted beneath its perch. The bird whose activity is shown in Figure 9.11 was maintained in the laboratory at a constant temperature and with an unvarying length of day (12 hours of light alternating with 12 hours of darkness), so that it was deprived of the most obvious cues for the onset of winter or of spring. Notice that during the summer and winter months, its activity was limited to the daylight hours. However, during the autumn and spring, when it would be migrating in nature, the caged bird also became somewhat active at night. This nocturnal activity, called Zugunruhe, or migratory restlessness, serves as an important trigger for the onset of migration.

This timing function of the annual clock is particularly important for birds that winter close to the equator, where there are few cues to the changing season. At the equator, the photoperiod is constant throughout the year, just as it is in the laboratory, and rainfall and food abundance are too variable from year to year to serve as reliable cues signaling the appropriate time to begin migrating.

An annual clock also physiologically readies birds for migration and reproduction. The bird gets fatter (indicated by body mass) during the winter, which helps provide fuel for the spring migration; it molts during the winter; and its testes enlarge for summer reproductive activity. These cycles continue for many years in constant conditions, and the length of the cycle is generally slightly longer or shorter than a year (Gwinner 1996).

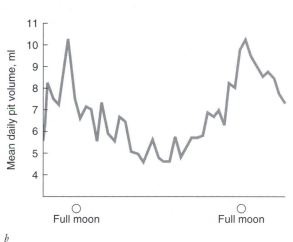

a

b

FIGURE 9.10   **A monthly rhythm in the pit size of the predatory ant lion (*Myrmeleon obscurus*). (*a*) The ant lion waits at the bottom of a self-constructed pit with only its pincers exposed. When a small arthropod, such as this ant, slips into the pit, the ant lion sucks the prey's body juices. (*b*) Monthly rhythms in the pit size of 50 ant lions maintained under normal daylight conditions in the laboratory. Each of the predators was fed one ant a day. Larger pits were constructed at full moon than at new moon. (From Youthed and Moran 1969.)**

The annual clocks of other species help them avoid the harsh conditions of winter by hibernating. It is especially useful for animals that hibernate in burrows or other locations where environmental cues are not available. For example, when the golden-mantled ground squirrel, *Citellus lateralis*, is maintained in the laboratory at a constant temperature and with an unvarying length of day (12 hours of light alternating with 12 hours of darkness), it will still enter a period of hibernation at approximately yearly intervals. This was true even when laboratory-born animals were subject to constant cold (3°C) and darkness; such animals still showed an annual rhythm of alternating activity and hibernation. Some of them even continued the pattern for three years.

Associated with hibernation but separate from it is a seasonal cycle in feeding. If a ground squirrel is subjected to a relentless "summer" by maintaining the laboratory temperature at an unvarying and torrid 35°C (95°F), it cannot enter hibernation. However, it will still reduce its food and water consumption during the assumed winter and begin to eat again and gain weight in the spring (Pengelley 1975).

One must be cautious in describing a behavior or physiological process that fluctuates annually as one that is controlled by an annual clock. In animals with short life spans, many of the seasonal changes in behavior are controlled by the changing photoperiod, the shortening of days during the winter months and the increasing daylight of the spring and summer. A daily

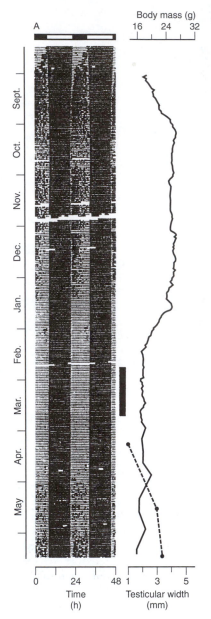

FIGURE 9.11   Annual cycles in migratory restlessness, body weight, testis size, and molting in a garden warbler held in a constant light-dark cycle (12 hours of light alternating with 12 hours of darkness) and at a constant temperature. Activity was measured with a microswitch mounted under the perch. Successive days are mounted underneath each other. The original record (0 – 24) is repeated on the right (24 – 48). Most of the bird's perch-hopping activity occurred during the day. When birds in nature are migrating, in the autumn and spring, the caged bird showed increased activity (migratory restlessness) during the night. The body weight changes throughout the year such that the bird fattens during the winter. These energy stores will increase the chances of successful spring migration. The testes enlarge during the spring in preparation for summer breeding. The molt (indicated by the vertical black bar) occurred in late February to March. (Data from Gwinner 1996.)

clock appears to be involved in measuring the interval of darkness to regulate photoperiodically controlled behaviors and processes. Unlike a response governed by photoperiod alone, a rhythm that is controlled by an annual clock will continue to be rhythmic even in the absence of changing day length.

THE CLOCK VERSUS THE HANDS OF THE CLOCK

When we study biological rhythms, we actually look at the rhythmic processes and make inferences about the clock itself. However, it is important to remember that the biological clock is separate from the processes it drives. Perhaps an analogy to a more familiar timepiece will emphasize this important point. The clock mechanism of an alarm clock is distinct from the hands of the clock, although it is responsible for their movements. If you are particularly vindictive after being awakened from a dream by the clamor of your alarm and tear the hands from the face of your clock, the internal gears will continue to run undaunted. And so it is with internal clocks. You can alter the rhythmic process without affecting the mechanism.

David Welsh and his colleagues (1995) performed the biological equivalent of tearing the hands from the clock. The suprachiasmatic nuclei (SCN) in the brain of a mammal are a "master" biological clock that drives rhythms in other processes. Welsh removed neurons from the SCN of a newborn rat and grew them in tissue culture. The spontaneous rate of firing of these single neurons varies regularly during each day, even in tissue culture, and so we can assume that the rhythm is driven by an internal cellular clock. This nerve firing was completely stopped by the addition of tetrodotoxin, a chemical that prevents action potentials that require sodium (Na^+) ions. However, 2.5 days later, when the tetrodotoxin was washed out of the cells, the rhythm reappeared with a phase predicted by the initial cycles (Figure 9.12). This suggests that although nerve firing had been halted, the clock was running accurately the entire time. Therefore, like the hands of a clock, the rhythmic process—nerve firing in this case—is separate from the clock mechanism. Processes are made to be rhythmic because they are coupled to and driven by a biological clock.

CLOCK PROPERTIES

Like any good clock, biological clocks measure time at the same rate under nearly all conditions, and they have mechanisms that reset them as needed to keep them synchronized with environmental cycles.

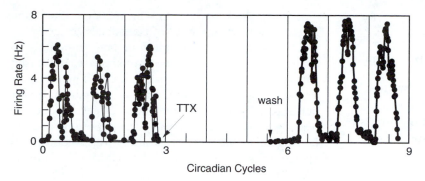

FIGURE 9.12 The rhythm in spontaneous electrical activity in isolated neurons from rat SCN. Sodium-dependent action potentials are blocked for 2.5 cycles by tetrodotoxin (TTX). When the inhibitor is washed out, the rhythm in spontaneous nerve firing returns with a phase predicted by the initial cycles. This demonstrates that the biological clock is separate from the rhythmic process it drives. (From Welsh et al. 1995.)

PERSISTENCE IN CONSTANT CONDITIONS

A defining property of clock-controlled rhythms is that cycles continue in the absence of environmental cues such as light-dark and temperature cycles. This means that the external day-night cycles in light or temperature are not causing the rhythms. Instead we attribute the ability to keep time without external cues to an internal, biological clock.

However, in the constancy of the laboratory, the period (the interval between two identical points in the cycle) of the rhythm is rarely exactly what it was in nature; that is, it becomes slightly longer or shorter. This change in the period is described with the prefix *circa*. So, a daily rhythm, one that is 24 hours in nature, is described as being circadian—*circa*, "about"; *diem*, "a day" (Figure 9.13). A lunar day (tidal) rhythm is described as being circalunidian; a monthly rhythm, circamonthly; and an annual rhythm, circannual. In other words, a laboratory hamster kept in constant conditions may begin to run a little later every night. If it starts to run 10 minutes later in each cycle, after two weeks its activity will be about 2.5 hours out of phase with the actual daily cycle.

When an animal is kept in constant conditions, the period length of its rhythms generally deviates from that observed in nature. An assumption is made that the period length is a reflection of the rate at which the clock is running. Sometimes this point is emphasized by describing the circadian period length in constant conditions as free-running, implying that it is no longer manipulated by environmental cycles.

STABILITY OF PERIOD LENGTH

If any clock is to be useful, it must be precise, and the biological clock is no exception. When an animal is cloistered in unvarying conditions and the free-running period length of its activity rhythm is determined on successive days of several months, the measurements are usually found to be extremely consistent. For some animals the precision is astounding. For example, the biological clock of the flying squirrel, *Glaucomys volans*, measures a day to within minutes without external time cues. In fact, the precision of the clock in most animals is greater than one might expect. The daily variability in the free-running period length is frequently no more than 15 minutes and is almost always less than 1 hour (Saunders 1977).

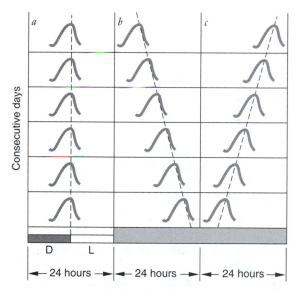

FIGURE 9.13 Diagram illustrating a biological rhythm in the entrained and free-running state. (*a*) The clock is entrained to the light-dark cycle indicated by the bars at the bottom of the column. Entrainment is the establishment of a stable phase relationship between the rhythm and the light-dark cycle, thus ensuring that the activities programmed by the clock occur at the appropriate times. When an organism is placed in constant conditions, the period length of its rhythms is seldom exactly 24 hours. Depending on the organism, the light intensity, and the temperature, the period length may be slightly longer than 24 hours (*b*) or slightly shorter than 24 hours (*c*). The adjective *circadian* is used to describe this change in period. (From Brown, Hastings, and Palmer 1970.)

ENTRAINMENT BY ENVIRONMENTAL CYCLES

The properties we have described so far are generally true for biological clocks of all cycle lengths. Although it is also true that tidal, fortnightly, monthly, and annual rhythms are set by environmental cycles, the nature of the environmental cycle varies for clocks of different period lengths. So, here we will focus on the environmental cycles that set circadian clocks.

Period Control: Daily Adjustment of the Free-running Period to the Natural Day-Night Cycle

Isn't it annoying when your watch runs slowly? If you did not reset it every day, your schedule would soon become a shambles. And so it is with biological clocks.

As we have just seen, the clocks themselves do not run on a precise 24-hour cycle. In nature, however, the period length of biological rhythms is strictly 24 hours. This is possible because the clock is entrained by (locked onto) the natural light-dark cycle. The clock is reset during each day by changes in light intensity as the sun rises and sets. For example, if the free-running period length of a mouse housed in constant darkness were 24 hours and 15 minutes, its clock would have to be reset by a quarter of an hour every day, or it would awaken progressively later each evening. These adjustments must be made so that the onset of activity, or any event programmed by the clock, will occur at the appropriate time each day.

Phase Control: Adjustment to a New Light-Dark Cycle

We have seen that one function of the biological clock is to time certain activities so that they occur at the best point of some predictable cycle in the environment. To be useful, then, they must be set to local time. But some species travel great distances during their lives. Clearly there must be a way to set the clock during long-distance treks, or those activities would occur at inappropriate times. In other words, biological clocks, like any clock, cannot be useful unless there is some way to set them (adjust the phase). If you were to fly from Cape Cod, Massachusetts, across three time zones to Big Sur, California, the first thing that you would want to do is to set your watch to the local time. It is obvious that if your biological clock is to gear your activities to the appropriate time of day in the new locale, it too must be reset.

When you first step off the plane after any flight across time zones, your biological clock is still set to the local time of your home. The clock will gradually adjust to the day-night cycle in the new locale (Figure 9.14). However, this shift cannot occur immediately; it may take several days. The length of time required for the

biological clock to be reset to the new local time increases with the number of time zones traversed. To make matters worse, not all your body functions readjust at the same rate, so the normal phase relationship among physiological processes is upset. Therefore, for a few days after longitudinal travel, your body time is out of phase with local time and your rhythms may be peaking at inappropriate times relative to one another. During this time you often suffer psychological and physiological disturbances. The syndrome of effects, which frequently includes a decrease in mental alertness and an increase in gastric distress, is referred to as jet lag.

For circadian rhythms the most powerful phase-setting agent is a light-dark cycle, although temperature cycles are also effective in plants and poikilothermic animals (those whose body temperature tends to be near that of their surroundings, commonly called cold-blooded). With a few exceptions, temperature cycles are generally not very effective in setting the clocks of birds or mammals (Hastings, Rusak, and Boulos 1991). When a rhythm becomes locked onto (synchronized with) an environmental cycle such as light-dark changes, we say that it is entrained to the cycle.

In the laboratory the biological clock can be reset at will by manipulating the light-dark cycle. If a hamster is kept in a cycle of 12 hours of light alternating with 12 hours of darkness such that the light is turned off at 6 P.M. real time, its activity begins shortly after 6 each evening. The light-dark cycle might then be changed so that darkness begins at midnight. Over the next few days the clock would be gradually reset so that at the end of about five days, the activity would begin shortly after midnight real time. If, after several weeks of this lighting regime, the hamster is returned to constant conditions, its activity rhythm would have a period that approximates 24 hours, and more important, the rhythm would initially be in phase with the second light-dark cycle.

Phase resetting occurs because a cue, a change from dark to light, for instance, affects the clock differently depending on when in the clock's cycle it occurs. Although the rhythms are separate from the clock itself, we assume that they indicate what time the clock is signaling. If an organism is kept in constant dark, its rhythms will free-run and we refer to points in the cycle as circadian time. For example, a hamster is active at night. So, when the hamster is kept in constant darkness, we refer to the time when it begins activity as early circadian nighttime. If a brief light pulse interrupts the darkness during early circadian night, it causes a phase delay. In other words, it resets the clock such that the hamster will become active later than expected on the next day. On the other hand, if a brief light pulse interrupts darkness during the late circadian night, it causes a phase advance: The animal becomes active sooner than expected in the next cycle. In most

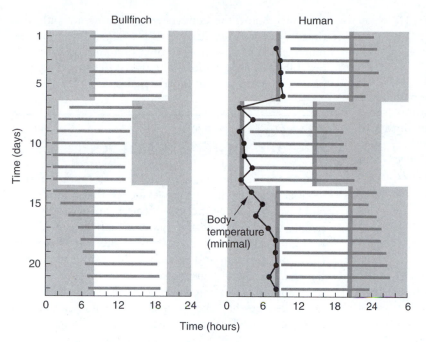

Bullfinch

Human

Time (days)

Body-temperature (minimal)

Time (hours)

FIGURE 9.14 The resetting of the biological clock by light-dark cycles. Immediately following a trip across time zones, the biological clock is still set to home time rather than that of the vacation locale. It may take several days for the clock to be reset so that it has the proper phase relationship with the new light-dark cycle. In addition, various rhythms (clocks) may rephase at different rates. During the interval of readjustment, the individual suffers from jet lag and may not feel well. (From Aschoff 1967.)

animals, a brief light pulse during circadian daytime has little or no effect. In nature, the clock is reset by light at dawn and dusk each day so that it keeps accurate time and is set to local time.

TEMPERATURE COMPENSATION

The biological clock remains accurate in spite of large changes in environmental temperature. This is somewhat surprising if one assumes that the timing mechanism is rooted in the cell's biochemistry. As a general rule, chemical reactions double or triple in rate for each 10°C change in temperature. However, the effect of an equal temperature rise on the rate at which the clock runs is usually minor, rarely as large as 20%. This effect may be described by a temperature coefficient, or Q_{10} value (calculated as the period at T° divided by the period at T° + 10°).* Typically, the Q_{10} values for the effect of temperature on the clock fall between 0.8 and 1.04. In contrast, the temperature coefficients for most chemical reactions are typically in the range of 2.0 to 4.0. This insensitivity to the effects of temperature suggests that the clock somehow compensates for them. It should be apparent that if the clock were as sensitive to temperature changes as most other chemical reactions are, it would function as a thermometer, indicating the ambient temperature by its rate of running, rather than as a timepiece.

---

*Remember that if the clock runs faster, it measures a cycle in a shorter amount of time. Thus the period length decreases as the rate increases.

ADVANTAGES OF BIOLOGICAL CLOCKS

We have seen that there are many behavioral rhythms that match the prominent geophysical cycles—a day, a lunar day, a lunar month, and a year. The geophysical cycles generate rhythmic changes in environmental conditions. One might wonder, then, why biological clocks exist at all. If the clocks cause changes that are correlated with environmental cues, why not just respond to the cues themselves?

ANTICIPATION OF ENVIRONMENTAL CHANGE

One reason for timing an event with a biological clock rather than responding directly to periodic environmental fluctuations is that it lets an animal anticipate the change and allow adequate time for the preparation of the behavior. For example, in nature, adult fruit flies (*Drosophila*) emerge from their pupal cases during a short interval around dawn. At this time the atmosphere is cool and moist, allowing the flies an opportunity to expand their wings with a minimal loss of water through the still permeable cuticle. This procedure takes several hours to complete. However, the relative humidity drops rapidly after the sun rises. If the flies waited until there was a change in light intensity, temperature, or relative humidity before beginning the preparations for emergence, they would emerge later in the day, when the water loss to the arid air could prevent the wings from expanding properly.

SYNCHRONIZATION OF BEHAVIOR WITH AN EVENT THAT CANNOT BE SENSED DIRECTLY

Another advantage of the clock's control of an event is that it allows a behavior to be synchronized with a factor in the environment that the animal cannot sense directly. An example is the timing of bee flights to patches of flowers that the bees have learned are open only during restricted times of the day (Figure 9.15). The flowers visited for nectar may be far away from the hive, and so the bee could not use vision or olfaction to determine whether the flowers were open.

CONTINUOUS MEASUREMENT OF TIME

Sometimes an animal may consult its clock to determine what time it is. As we have seen, this information is necessary to anticipate periodic environmental changes or to synchronize behavior with other events. However, at other times a clock is consulted to measure an interval of time. This, then, is a third benefit of a biological clock.

The ability to measure the passage of time continuously is crucial to an animal's time-compensated orientation. For example, a worker honeybee (*Apis* spp.) indicates the direction to a nectar source to recruit bees through a dance that tells them of the proper flight bearing relative to the sun. Since the sun is a moving

FIGURE 9.15 **Honeybees can use their biological clock to time their visits to distant patches of flowers so that they arrive when the flowers are open and nectar is available.**

reference point, the honeybee must know not only the time of day when it discovered the nectar but also how much time has passed since then. The biological clock provides this information. The use of the sun as a compass will be explored in more detail in Chapter 10.

ADAPTIVENESS OF BIOLOGICAL CLOCKS

Keeping these advantages of the clock in mind, we may wonder whether there is any evidence that a biological clock actually does increase fitness. Surprisingly few people have addressed this question. There is evidence that the clock enhances fitness in cyanobacteria (Johnson, Golden, and Kondo 1998; Yan et al. 1998), but that may be of marginal interest to those interested in animal behavior. However, Patricia DeCoursey and her colleagues have begun gathering evidence that the clock is indeed adaptive for certain rodents. They destroyed the SCN of some animals and compared their survival rate outdoors to that of intact animals. Since the SCN is the master biological clock in mammals, this procedure allowed them to compare the survival of animals with and without clocks. In a preliminary study, 12 intact control animals and 10 SCN-lesioned antelope ground squirrels (*Ammospermophilus leucurus*) were monitored in a desert enclosure. Their activity was continuously monitored in several ways, including a motion detector and a video camera. All the ground squirrels were primarily active during daylight.

However, an important difference in the behavior of the two groups is that the SCN-lesioned animals were more likely to be active on the ground surface of the enclosure during the nighttime than were intact animals. Whereas the amount of activity occurring during the night in SCN-lesioned animals ranged from 16% to 52.1%, nighttime activity represented no more than 1.3% of the activity of intact animals. Nine of the 12 control animals were active only in the day. This difference in nighttime activity had unfortunate consequences for the SCN-lesioned animals. One night, when 7 control animals and 5 SCN-lesioned animals had been introduced to the enclosure, a feral cat treated the enclosure as a kitty-convenience store. The videotape recorded the cat picking off ground squirrels that were active that night. As a result, the cat killed 60% of the SCN-lesioned animals, but only 29% of the intact controls (DeCoursey et al. 1997). Thus, it seems that an important function of the clock for these ground squirrels may be to reduce activity at dangerous times of the day.

DeCoursey then asked whether a biological clock enhances survival in eastern chipmunks, *Tamias striatus*. Animals were captured in the wild and taken to the laboratory for surgery. The SCN was destroyed in ten animals, and five others were given sham lesions; that is, they were anesthetized but not lesioned. The sham-lesioned animals served as surgical controls because they were removed from their habitat and suffered the consequences of removal, such as the possible takeover of their dens by other animals, but did not undergo surgery. The survival of these two groups of animals was compared with that of 13 intact controls. Survival was only 60% for SCN-lesioned animals but was 100% for surgical controls and 84.6% for intact controls. Thus, there was not a significant difference in survival between surgical control animals and intact controls. However, the survival of these two control groups was significantly greater than that of the SCN-lesioned chipmunks. The lack of a biological clock did not seem to affect the chipmunks' ability to obtain food because the annual body weight patterns for all the animals were similar. It would be interesting to know whether the clock enhances reproductive success, but those data are difficult to obtain. Most chipmunks in all groups were reproductively active during the fall and spring. However, the genetic studies necessary to determine the parents of surviving young have not been done (DeCoursey and Krulas 1998).

ORGANIZATION OF CIRCADIAN CLOCKS

Single cells may contain the necessary equipment for biological timing. Unicellular organisms have biological clocks, and the cells that make up tissues and organs often have their own independent clocks. Thus, a complex nervous system or endocrine system is not an essential component of the biological clock.

MULTIPLE CLOCKS

Since the clock can exist in single cells, does this mean that every cell in a multicellular organism has its personal timepiece? Not necessarily. One way to demonstrate that a group of cells has its own clock is to remove the tissue, grow it in culture, and see whether the rhythm persists. For example, when the adrenal glands of a hamster are grown in tissue culture, they continue to secrete their hormone, corticosterone, rhythmically for ten days (Shiotsuka, Jovonovich, and Jovonovich 1974). Likewise, pineal glands of chickens that are organ-cultured in continuous darkness display a circadian rhythm in the production of the hormone melatonin (Takahashi, Hamm, and Menaker 1980) and in N-acetyl transferase, the enzyme that controls the melatonin rhythm (Deguchi 1979; Kasal, Menaker, and Perez-Polo 1979). In fact, the chick pineal cells can be separated from one another and not only will the rhythm in melatonin continue but also each can be entrained by a light-dark cycle (Robertson and Takahashi 1988a, b). If different glands or organs have clocks that continue running even when removed from the body, it follows that a multicellular organism must have several clocks.

Another way to demonstrate that an organism may have more than one clock is to somehow get different clocks in one individual to run independently, each at its own rate. Occasionally this happens when an animal is kept in constant conditions. When humans lived in underground shelters without any time cues, about 15% of them had a sleep-wakefulness rhythm with a different period length than their body temperature rhythm (Aschoff 1965). Notice in Figure 9.16 that one person's body temperature rhythm had a period length of 24.7 hours, whereas the sleep-wakefulness rhythm had a period of 32.6 hours. If we follow the reasoning that the period length of a rhythm in constant conditions is a reflection of the rate at which the clock is running, this may be taken as evidence that these processes are controlled by different clocks.

More recently, it has been demonstrated that fruit flies (*Drosophila*) have a multitude of independent clocks through their bodies and that these clocks respond to

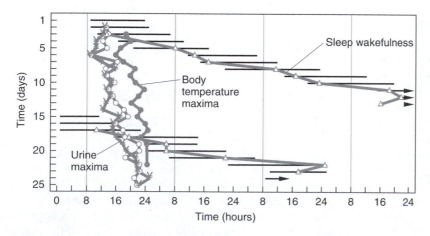

FIGURE 9.16   Desynchronization of rhythms of a man living in constant conditions. The maxima of body temperature (closed circles), urine volume (open circles), and potassium excretion (crosses) occurred at approximately 24.7-hour intervals, whereas the sleep-wakefulness rhythm (lines) and calcium excretion (open triangles) had a period of 32.6 hours. The fact that these rhythms had different period lengths is evidence that the processes are controlled by different clocks. (From Aschoff 1965.)

changes in light-dark cycles without any help from the head (Plautz, Kaneko, Hall, and Kay 1997; Plautz et al. 1997). This was shown by using an interesting technique that has since proved to be a valuable tool; it has advanced the study of rhythms because it allows researchers to observe the molecular activity of important clock genes in a single, living, intact animal. Before this, clock gene activity had to be studied by synchronizing the clocks of members of a large population of fruit flies with a light-dark or temperature cycle and then periodically selecting a group of flies from the population, grinding them up, and testing for gene activity.

Many researchers investigating the fruit fly's clock focus their efforts on the *period* (*per*) gene, which is an integral part of the clock mechanism (as discussed shortly). To monitor the clock's activity, the research groups headed by Jeffrey Hall and by Steve Kay genetically engineered fruit flies to contain the firefly luciferase gene. Luciferase is an enzyme that acts on luciferin to produce light, allowing the firefly to glow. The firefly luciferase gene was placed in the fruit fly's chromosome in the promoter region that turns on the clock gene, *per*. Then, whenever the *per* promoter turned on the *per* gene, it also switched on the luciferase gene, causing luciferase to be produced. Because the flies' diets were laced with luciferin, they glowed whenever luciferase was present. Thus, whenever the fly glowed, it meant that the *per* gene was turned on. Special cameras and video equipment measured the glow, and computers traced and recorded the glow pattern.

These glow rhythms will synchronize with light-dark cycles and will continue in constant darkness with a free-running period length. Not only do intact flies glow rhythmically, but so will cultures of head, thorax, or abdomen. Furthermore, separate cultures of body parts exposed to the same light-dark cycle will glow in unison, showing that each piece of cultured tissue has its own independent clocks and that these clocks have their own photoreceptors. Moreover, this raises the possibility that the insect's brain is not required as a master clock to synchronize rhythms throughout the body.

The glow rhythm of a fly kept in constant darkness gradually decreases in amplitude because the clocks in different cells run at slightly different rates without a light-dark cycle; thus the independent clocks gradually become asynchronous. It is interesting to note that the head is the only body part in which the clocks remain synchronized in the prolonged absence of light. But, when exposed to a new light-dark cycle, the clocks throughout the fly entrain within one cycle and the glow becomes rhythmic again. In nature, asynchrony among peripheral clocks is not a problem: Fruit flies almost always have an environmental light cycle that is able to synchronize their many independent clocks because each has its own photoreceptor, as we will see shortly (Figure 9.17).

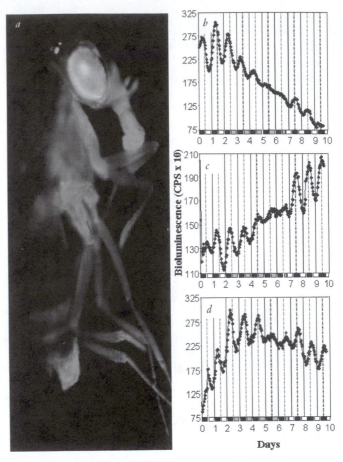

FIGURE 9.17   Biological clocks are found throughout fruit flies, not just in the brain. The *period* (*per*) gene is thought be an integral part of the clock's mechanism. To measure *per* activity, the firefly gene for luciferase was linked to the *per* promoter, which turns on the gene. Luciferase is the enzyme that causes a firefly to glow in the presence of luciferin. As a result, the fruit fly glowed with an eerie green color whenever the *per* gene was turned on (*a*). Computers measured and recorded the pattern of glow. The glow rhythm persisted in constant darkness for several cycles and could be synchronized with light-dark cycles. When parts of the fly were cultured separately, the cultured segments continued to glow rhythmically and could still be set by light-dark cycles. Rhythmic glow can be seen in separately cultured heads (*b*), thoraxes (*c*), and abdomens (*d*). Thus, these peripheral clocks do not require input from the brain. (From Plautz, Kaneko, Hall, and Kay 1997.)

A HIERARCHY OF CLOCKS

Most rhythmic organisms have a multitude of independent peripheral clocks in cells throughout the body. However, information about environmental cycles may not reach each clock directly. How, then, are an individual's many clocks synchronized so that all the rhythmic processes occur at the appropriate time relative to one another and the environment's cycles? It appears that there are one or more "master" clocks in the brain that are entrained by the light-dark cycle and regulate other clocks through the nervous and/or endocrine system. Therefore, we can consider three questions: (1) What photoreceptors are responsible for entrainment? (2) Where is the master clock? (3) How does the master clock regulate the other clocks in the body? Table 9.1 summarizes what is known about the location of circadian clocks in selected animals. We will approach these questions by focusing on the "clockshops" of silk moths, cockroaches, birds, and rodents.

Circadian Organization in Silk Moths

In silk moths, the photoreceptor for entrainment and the clock that controls eclosion (emergence of the adult from the pupal case) are located in the brain, which controls eclosion by producing a hormone. A series of elegant experiments revealed the details of the neuroendocrine control of the circadian eclosion rhythm (Truman and Riddiford 1970). The moths (*Hyalophora cecropia*) emerged at the usual time of day, just after sunrise, even if the nerves that connect the eyes and the brain were cut and certain parts of the nervous system (such as the subesophageal ganglion, corpora cardiaca, and corpora allata) or the developing compound eyes were removed. Apparently this treatment did not stop the clock or blind it to the light-dark cycle. However, when the brains were

TABLE 9.1   Known Circadian Clocks in Selected Invertebrates and Vertebrates

	SCN	Pineal	Eyes	Optic Lobe	Brain
Silk moths					X
Fruit flies					X
Cockroaches				X	
Crickets				X	
Sea hares			X		
Lampreys		X	Some		
Fish		X	Some		
Lizards		X	Some		
House sparrows	X	X			
Java sparrows	X	X			
Quail		X	X		
Pigeons		X	X		
Chickens		X	X		
Hamsters	X		X		
Rats	X				
Ground squirrels	X				
Humans	X				

removed from pupae, the adults emerged at random times throughout the day. When a brain was implanted in the abdomen of a pupa whose brain had been removed, the adult emerged at the customary time of day. So, the brain is needed if the clock is to tick; but how does it "know" when the sun rises or sets?

The search for the photoreceptor for entrainment involved removing the brains from 20 pupae and implanting a brain into the head end of 10 of these individuals and into the abdomen of the remaining 10. The pupae were then inserted through holes in an opaque board so that the two ends of each pupa could be presented with light-dark cycles that were 12 hours out of phase with one another. In other words, both ends of the pupae experienced 12 hours of light alternating with 12 hours of darkness, but the anterior end experienced light (day) while the posterior end was exposed to dark (night) and vice versa. The time of eclosion was determined by the light-dark cycle that the brain "saw," whether it had been implanted into the head or abdomen. So we see that the brain is the photoreceptor for the clock, and since there were no longer any nerves connected to the brains in these pupae, the brain must be hormonally linked to the eclosion process.

Although the previous experiments demonstrated that the brain is essential for rhythmicity, they did not conclusively show that it was the clock. An alternative explanation for these results is that the brain is responding to a signal from the clock by releasing a hormone necessary to initiate eclosion. However, if implantation of a brain transferred some clock property, such as a characteristic period or phase, this would be evidence that the brain is the clock. This was done by exchanging the brains of two species of moths, *H. cecropia*, which ecloses just after dawn, and *Antheraea pernyi*, which emerges just before sunset. The moths emerged at the time of day appropriate for the species whose brain they possessed. Since it is the clock that determines the appropriate time for eclosion, the transplantation of this phase information indicates that the brain contains the clock. Later experiments point to the cerebral lobes of the brain as the more specific site of the clock and the medial neurosecretory cell cluster as the source of the hormone that initiates eclosion (Truman 1972).

Circadian Organization in Cockroaches

The circadian system in cockroaches is somewhat different from that in moths. Unlike moths, in which the brain is the photoreceptor for entrainment, in cockroaches, such as *Leucophaea maderae* and *Periplaneta americana*, light information reaches the clock through the compound eyes (S. Roberts 1965). The location of the biological clock has been pinpointed to a region of the optic lobes of the brain. It is perhaps more precise to say that the cockroach has two clocks since there are two optic lobes and each can keep time independently (Page 1985; S. Roberts 1974). The clock regulates activity in the cockroach via nerves, not hormones as in the moths. We know this because cutting the nerves leading from the optic lobes causes the cockroach's activity to become arrhythmic. However, the neural connections will re-form within about 40 days if they have been cut or if optic lobes have been transplanted into a cockroach, and activity once again becomes rhythmic (Page 1983).

Circadian Organization in Birds

There is a good deal of redundancy in the circadian system of vertebrates. In birds, for example, the eyes, the pineal, and as yet unidentified photoreceptors in the brain can provide information about lighting conditions to the clock. Furthermore, birds have three interacting clocks: the pineal gland, the suprachiasmatic nuclei of the hypothalamus (SCN), and the eyes. The importance of each of these clocks varies among species. This system of interacting clocks controls other clocks within the body through the pineal's rhythmic output of the hormone melatonin (Binkley 1993).

Although the eyes can provide lighting information to a bird's clock, there are also photoreceptors in the brain of a bird that can cause entrainment (McMillan, Keatts, and Menaker 1975). Light, it seems, can penetrate through skin, skull, and brain tissue to reach photoreceptors within the brain itself (Figure 9.18). In one experiment, a blinded house sparrow (*Passer domesticus*) was exposed to cycles of 12 hours of dim light (0.2 lux) alternating with 12 hours of darkness. The light was so

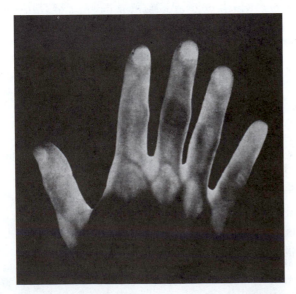

FIGURE 9.18   Flesh is amazingly transparent to light. The brain of some species contains photoreceptors that help coordinate the biological rhythms of some species of birds with the day-night cycle.

faint that most birds did not entrain to this cycle. However, when the feathers were plucked from a bird's head so that the amount of light reaching the brain was increased by several orders of magnitude, the bald bird entrained to the light-dark cycle. As the feathers grew back, entrainment failed. Replucking restored entrainment until India ink was injected beneath the skin of the head. The ink allowed only one-tenth of the light to reach the brain, and entrainment was lost. Sensitivity to the light cycle was regained when some of the scalp and the underlying India ink were scraped off. From the results of this experiment (shown in Figure 9.19), one can conclude that direct illumination of the brain can cause entrainment (Menaker 1968).

Where in the brain are the photoreceptors? We know that the pineal gland of at least some birds is sensitive to light because the rhythm in the hormone melatonin from pineals removed from chickens and kept alive in tissue culture will still entrain to light-dark cycles (Takahashi and Menaker 1979). But there is more to the story than this. The perch-hopping and body temperature rhythms of house sparrows (*P. domesticus*) whose pineals and eyes have been removed still entrain to light-dark cycles (Menaker 1968). So, there must be photoreceptors outside the pineal. Studies have tried to find these photoreceptors, but most have met with frustration because even the most focused light beam spreads outward within the brain tissue. Nonetheless, there do seem to be receptors within the hypothalamus and nearby regions (Silver et al. 1988). We still do not know whether the eyes, pineal, and other brain photoreceptors are doing the same job or are playing different roles in conveying light information to the clock.

We now know that there are several interacting clocks in the circadian system of birds: the pineal, the SCN, and the eyes. The search for the site of the circadian pacemaker began with the discovery that the pineal gland is an important master clock in some species. When the pineal is surgically removed from the brain of a house sparrow or a white-crowned sparrow (*Zonotrichia leucophrys*), the bird's perch-hopping activity and body temperature are no longer rhythmic in constant darkness (Gaston and Menaker 1968). This demonstrates only that the pineal is necessary for the expression of rhythmicity. But a later experiment indicated that the pineal is a circadian pacemaker. Two groups of donor house sparrows were entrained to light-dark cycles that were ten hours out of phase. The pineal glands of these sparrows were transplanted into the anterior chamber of the eye of arrhythmic pinealectomized sparrows that had been maintained in continuous darkness. Within a few days of a successful transplant, the recipients, still housed in constant darkness, became rhythmic. The phase of the newly instilled rhythm was that of the donor bird. This strongly sug-

Passer domesticus

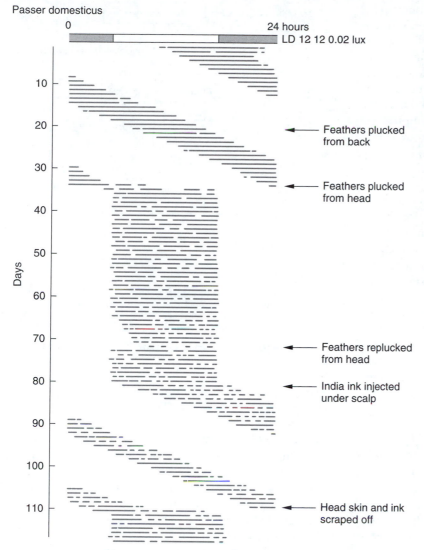

FIGURE 9.19 Activity rhythm of a blind house sparrow kept in a 24-hour light-dark cycle with 12 hours of dim (0.2 lux) light. The time of darkness is indicated by the shaded bars at the top of the figure. Activity is represented by the solid black horizontal bars. When the feathers were plucked from the bird's back, there was no effect on the rhythm. The bird was not entrained to the light-dark cycle. Then the feathers were plucked from the bird's head, thereby increasing the light that reached the brain. This resulted in entrainment until the feathers began to grow back. When the feathers were plucked once again, entrainment was reestablished. Next, India ink was injected under the skin of the scalp to decrease the light intensity that reached the brain. Entrainment was lost until the ink and the skin of the scalp were scraped off. (Modified from Menaker 1968.)

gests that the pineal gland of house sparrows contains a clock essential for the persistence of rhythmicity in the absence of external time cues (Zimmerman and Menaker 1979).

The importance of the pineal gland in the circadian system depends on the species of bird. We have seen its importance as a major clock in house sparrows, and the same is true in European starlings, *Sturnus vulgaris* (Gwinner 1978). However, it is not as important in the circadian system of Japanese quail, *Coturnix coturnix japonica* (Simpson and Follett 1981; Underwood and Siopes 1984).

In some species of birds, the SCN has been shown to be an important biological clock. The perch-hopping activity of the house sparrow, the Java sparrow (*Padda oryzivora*), and Japanese quail is severely disrupted following the destruction of the SCN (Ebihara and Kawamura 1981; Simpson and Follett 1981; Takahashi and Menaker 1979). The birds in which at

least 80% of the SCN was destroyed were arrhythmic in constant conditions. Nonetheless, they did entrain to light-dark cycles. Lesions that spared the SCN, that only partially destroyed them, or that were unilateral did not abolish the free-running activity rhythm.

The eyes may also be involved in the circadian system in some species. Quail, for instance, remain rhythmic after pinealectomy, unless they are also blinded (Underwood and Siopes 1984). Blinding also affects rhythmicity in pigeons, *Columba livia* (Ebihara, Uchiyama, and Oshima 1984), but not in sparrows or chickens (Hastings, Rusak, and Boulos 1991). In Japanese quail, the eye has been shown to be a clock in its own right (Underwood, Barrett, and Siopes 1990b).

How do the clocks in a bird's circadian system interact? The SCN is thought to communicate with the pineal gland over a neural pathway. The pineal may "talk back" to the SCN through its rhythmic production of melatonin. The SCN is known to bind melatonin, which is a good indication that it responds to the hormone (Vaněček, Pavlik, and Illnerová 1987). It may be that the eyes also play a role in the circadian system by rhythmically secreting melatonin, much in the same way as the pineal gland. The eyes of chickens, quail, and sparrows have rhythms in melatonin production (Binkley 1988). The eyes of quail produce about a third of the melatonin in the blood, and the pineal produces the rest. In spite of this, the link between a quail's eyes and the rest of its circadian system is a neural pathway, not the rhythmic melatonin secretion. Continuous administration of melatonin does not affect the free-running period of activity (Simpson and Follett 1981). Furthermore, cutting the optic nerve does not disrupt the rhythmic production of melatonin, but it does have a dramatic effect on the activity rhythm. In fact, the effect of cutting the optic nerve on the activity rhythm is the same as blinding the bird (Underwood, Barrett, and Siopes 1990a).

How does this multiclock system exert its control over the rest of the circadian system? The experiment in which the pineal placed into the anterior chamber of the eye restored rhythmicity to a pinealectomized host demonstrated that the pineal must exert its influence via hormones because the nerve connections of the gland were obviously eliminated.

Melatonin, a hormone produced by the pineal in greater amounts during the night than during the day, appears to be the hormone through which the pineal exerts its effect. If a capsule that continually releases melatonin is implanted in an intact house sparrow, either the bird becomes continuously active or the period length of its rhythm is shortened. In other words, when melatonin release is continuous rather than periodic (as it normally is), circadian rhythmicity is disturbed (Turek, McMillan, and Menaker 1976).

Furthermore, periodic release of melatonin can entrain the activity rhythm of a bird. This has been demonstrated by pinealectomizing a group of European starlings, *S. vulgaris*, and giving half the birds a daily injection of sesame oil (to ensure that any effect on the clock was caused by the hormone and not the injection). Almost all, 21 out of 22, of the starlings receiving melatonin injections entrained, whereas only 1 of the 10 control birds did. The daily injections seem to have mimicked the effect of the rhythmic release of melatonin by the pineal gland (Gwinner and Benzinger 1978).

Circadian Organization in Mammals

The circadian organization in rodents differs from that in birds in several ways: Mammals seem to lack extraretinal photoreceptors for entrainment, the pineal gland is not an independent clock, and the master clock seems to exert its influence through nerves and chemicals (Figure 9.20). However, as in birds, the SCN and the retina, the light-sensitive portion of the eye, are

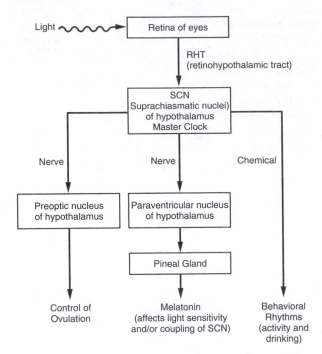

FIGURE 9.20   The circadian organization of mammals.

independent, self-sustaining clocks (Hastings 1997; Tosini and Menaker 1996). These two clocks interact to produce the observed characteristics of the circadian system (Lupi et al. 1999).

The eyes contain the photoreceptors for light entrainment in mammals. However, these are in a different part of the retina than the photoreceptors involved in vision (as discussed shortly). The information about the lighting conditions reaches the clock through the retinohypothalamic tract (RHT), a bundle of nerve fibers connecting the retina with the hypothalamus (R. Y. Moore 1979).

There is evidence that the SCN is a master clock in rodents and holds several "slave" clocks in the proper phase relationship. The initial studies in which the SCN was destroyed showed that the SCN is necessary for most rhythms in the rat (Moore and Eichler 1972; Raisman and Brown-Grant 1977) and the hamster (Stetson and Watson-Whitmyre 1976). Comparable destruction elsewhere in the brain has not so far destroyed rhythmicity.

The next step was to show that the SCN contains an autonomous clock. In one experiment, the SCN was isolated from neural input while still in place in the brain, and its activity remained rhythmic. Neural activity in the SCN and in one of several other brain locations was recorded simultaneously to be certain that the observations were specific to the SCN. In an intact animal, neural activity in all regions of the brain was always rhythmic. Then, a knife cut was made around the SCN, creating an isolated island of hypothalamic tissue that included the SCN. Following this treatment, the neural activity in the hypothalamic island remained

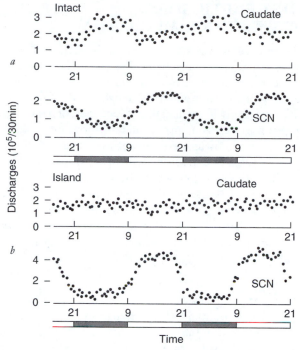

FIGURE 9.21 **Nerve activity in specific brain regions (the caudate and the suprachiasmatic nuclei) of a rat before and after isolating an "island" of brain tissue containing the suprachiasmatic nuclei. (*a*) The activity in a normal, intact rat is rhythmic in both regions of the brain. (*b*) After a region of the hypothalamus that contains the suprachiasmatic nuclei was isolated as an island, the nerve activity was rhythmic within the SCN but not outside the island in the caudate region. This supports the idea that the suprachiasmatic nuclei are self-sustaining oscillators. (Courtesy of S. T. Inouye.)**

rhythmic in constant darkness but activity in other brain regions was continuous (Figure 9.21). This strongly suggests that the SCN is a self-sustaining oscillator that instills rhythmicity in other brain regions through neural connections (Inouye and Kawamura 1979).

The activity of the SCN remains rhythmic in tissue culture, confirming that it is an independent clock (Inouye and Shibata 1994; Shibata and Moore 1988). Indeed, individual neurons within the SCN have their own clock. When cultured SCN neurons are separated, the spontaneous electrical firing of individual neurons is rhythmic in constant conditions, each of them with a slightly different period (Welsh et al. 1995).

The SCN was finally established as the primary clock in mammals by transplantation studies. When an SCN is transplanted into the brains of rats or hamsters that have been made arrhythmic by destroying their own SCN, their activity becomes rhythmic once again. The transplanted tissue is from fetal or newborn animals, so new neural connections may form with the host's brain. In those individuals whose activity rhythms

were restored after they received a new SCN, some neural connections had formed. As a result, the grafted SCN received input from the eyes and could communicate with other regions of the brain (Lehman et al. 1987; Sawaki, Nihonmatsu, and Kawamura 1984). It is important that the period length of the restored activity rhythm matches that of the transplanted SCN rather than the period length previously displayed by the recipient (Ralph et al. 1990). This result is what one would expect if the SCN were the clock that was providing timing information and not just a component needed to make the host's clock function.

As we have learned more about the SCN, we have begun to see that it contains many clocks, and this has raised new questions about how these clocks are linked together, or coupled, so that they function as a single unit. First, there is a suprachiasmatic nucleus in each half of the brain. There is evidence supporting the hypothesis that each of these centers is an independent oscillator (or contains a population of oscillators). In a normal hamster the two nuclei are mutually coupled and produce an activity rhythm with a single, well-defined band of activity during each cycle. When a nocturnal animal such as the hamster is maintained in constant light for an extended period, its activity pattern occasionally splits into two components that have different free-running periods until they become stabilized approximately 180° out of phase with each other. Some researchers have interpreted the splitting phenomenon as evidence for two circadian oscillators that are controlling activity and have become uncoupled as a result of the continuous illumination. There is some experimental support for this interpretation. When one of the suprachiasmatic nuclei in hamsters whose activity pattern had become split was destroyed, the split in the activity was eliminated and replaced by a new, single activity rhythm, presumably the one controlled by the remaining SCN (Pickard and Turek 1982).

There are other reasons for thinking of the SCN as a multiclock system. We now know, for instance, that each SCN has two structural components—an inner core and an outer shell. These differences in structure suggest that the core and shell may have different functions, but we don't know yet what these might be (Moore and Silver 1998; Weaver 1998). Recall also that individual neurons within the SCN have their own clocks. What synchronizes them? We have ideas but no answers (J. D. Miller 1998).

We have seen that the SCN is the primary clock in mammals that drives rhythms in behavior and physiology. Whereas some signals from the SCN are sent over neural pathways, others are sent by small, diffusible molecules (LeSauter and Silver 1998).

The SCN has at least two neural output pathways that affect rhythms. One of these is a pathway to the preoptic nucleus of the hypothalamus, and this seems to control the rhythm in ovulation but does not affect the activity rhythm. The second neural pathway leads to the paraventricular nucleus in the hypothalamus and then to the pineal. This pathway controls the pineal's production of melatonin by affecting transcription (Foulkes et al. 1997). Melatonin synthesis occurs only in the dark, and it is rhythmic.

In contrast to the effects of melatonin in birds, it has a subtle effect on the circadian system of mammals. Melatonin affects the mammalian SCN in some way, perhaps by decreasing the sensitivity of the circadian system to light or by altering the coupling between the SCN and its output (Cassone 1998). There are two other roles for melatonin in mammals. First, it serves as an entraining agent during development. Maternal melatonin cycles synchronize the clocks of fetuses with one another and with the mother. Second, it has become important in timing the seasonal reproduction of some species (Menaker 1997).

Activity rhythms, on the other hand, are thought to be caused by the SCN's release of a chemical signal to other parts of the brain without the help of neural connections. Rae Silver and her colleagues (1996) demonstrated this through transplant experiments similar to those described earlier, but with one important difference. The donor SCN tissue was enclosed within a capsule that allowed nutrients and diffusible molecules to flow between the host and graft tissue but did not allow neural processes to grow (Figure 9.22a). As in previous transplant experiments, both SCN of the host were destroyed prior to transplant, making the animal arrhythmic. The transplants were made between hamsters whose clocks ran at different rates because some carried the *tau* mutation, which alters the period length observed in constant conditions. The implanted SCN was the wild type, which has a period length of 24 hours in constant dark. Before SCN destruction, the activity rhythms of the *tau* mutant hosts were either 22 hours or 20 hours, depending on whether they were heterozygous or homozygous for the *tau* mutation. The encapsulated grafts restored the activity rhythm with the period length characteristic of the donor SCN (Figure 9.22b).

It is interesting that the chemical control of rhythms by the SCN seems to be unique to behavioral rhythms, such as activity and drinking. Reproductive responses to the length of day, which depend on melatonin from the pineal, and endocrine rhythms are not restored. These require neural connections.

THE MOLECULAR BASIS OF SELECTED CIRCADIAN CLOCKS

In recent years, we have learned a great deal about the molecular gears that make the clock tick and how it entrains to environmental light cycles.

THE GENETIC BASIS OF THE CLOCK MECHANISM

The molecular bases of the clocks investigated so far involve rhythmic gene activity. The products of one gene or set of genes activate or inhibit the activity of other genes, which in turn affect the activity of the first genes. This creates a self-regulated feedback loop of gene activity that measures an approximately 24-hour interval.

If your knowledge of genetics is rusty, you may want to refer to Chapter 3. In brief, however, genes consist of DNA. When a gene becomes active it is transcribed to mRNA, which is then translated to protein. Besides providing a more thorough discussion of gene activity, Chapter 3 discusses additional aspects of the *period* (*per*) gene that is known to be part of a gear in the clock of the fruit fly (*Drosophila*).

Fruit Fly

The starting point in the circadian cycle is arbitrary. We'll begin around noon, circadian time (Figure 9.23). Two clock genes, *period* (*per*) and *timeless* (*tim*), begin to be transcribed to mRNA (C. B. Green 1998; M. W. Young 2000). These genes have been turned on by a complex that forms between the protein products of two other clock genes, *clock* (*clk*) and *cycle* (*cyc*). The CLOCK-CYCLE complex binds to the promoter region of *per* and *tim* and turns on gene activity (Allada et al. 1998; Darlington et al. 1998). The *per* and *tim* mRNA slowly accumulate, and shortly after sunset their translation to proteins begins (C. B. Green 1998; M. W. Young 2000).

The PER protein forms a complex with the TIM protein. The level of the PER-TIM complex peaks around four hours before dawn. As a complex, PER-TIM is able to move into the nucleus, where it prevents the CLOCK-CYCLE complex from turning on the transcription of *per* and *tim*. In this way, PER and TIM shut down their own production (Darlington et al. 1998).

A time delay is built into the system because the PER-TIM complex can form only when there are high levels of *per* and *tim* mRNA, and these take time to accumulate (M W. Young 1998). In addition, the product of another gene, called *double-time* (*dbt*), adds a phosphate group to PER, which causes it to be destroyed. Thus, phosphorylation of PER protein controls its level. After awhile, however, the level of TIM is high enough that TIM binds to PER before a phosphate can be added. In this way, *dbt* adds a time lag between the transcription of *per* and *tim* and the accumulation of the PER-TIM complex (Price et al. 1998).

The TIM protein is broken down by light; so with the light of dawn, TIM degradation begins (M. W. Young 1998). PER is then released from the complex, and because of phosphorylation, it too is broken down.

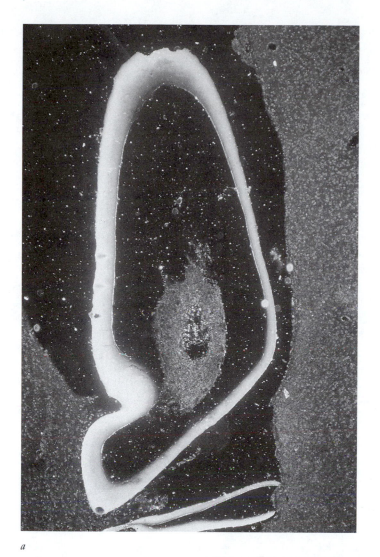

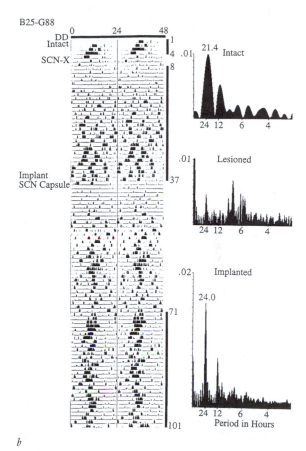

a   *b*

FIGURE 9.22   **Encapsulated donor SCN can reinstill rhythmic activity in SCN-lesioned hamsters. (*a*) The capsule, shown here in white, surrounds the donor SCN. Although the capsule allowed nutrients and small molecules to diffuse between the host and graft tissue, it prevented the formation of neural connections. (*b*) The activity rhythm of the intact host hamster (wild-type/tau) had a period length of 21.4 hours in constant darkness. After the SCN was destroyed, the animal was arrhythmic. Implantation of a capsule containing wild-type SCN (period length of 24 hours) restored the activity rhythm with a period characteristic of the implant. (Data from Silver et al. 1996)**

When the level of the PER-TIM complex falls, it can no longer prevent CLOCK-CYCLE from activating *per* and *tim*. The inhibition of transcription of *per* and *tim* is lifted, and they become active once again (Darlington et al. 1998).

Mammals

Less is known about the genetic basis of the clock of mammals than about the clock of fruit flies. But, based on what is known, it seems that the mechanisms of both clocks are based on self-regulated feedback loops in which the products of certain genes regulate the activity of other genes (Figure 9.24). The cast of molecular players in the clock of mammals is similar to those in the fruit fly. The names of some of these genes are familiar—*per*, *tim*, and *clk*. Their names are often distinguished from the *Drosophila* forms of the genes by the prefix *m*, for "mouse." The mouse has three different *per* genes, *mPer1*, *mPer2*, and *mPer3*. The proteins produced by these genes have regions that will allow them to form complexes with other specific proteins, but they do not have regions for binding to DNA. The protein product of another gene, *bmal1*, forms a com-

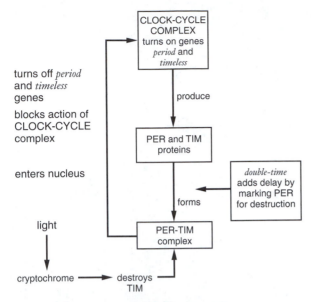

FIGURE 9.23   A model of gene activity thought to underlie circadian timing in the fruit fly.

plex with the CLOCK protein in much the same manner as CYCLE does in the fruit fly.

Two additional proteins—mCRYPTOCHROME1 (mCRY1) and mCRYPTOCHROME2 (mCRY2)—are also known to be involved in the circadian feedback loop of mammals. When mice are genetically engineered so that the *mCry1* and *mCry2* genes are knocked out (made nonfunctional), the animals are totally arrhythmic in constant darkness. However, if only one of these genes is disabled, the clock still functions in constant darkness but runs at a different rate, either faster or slower than

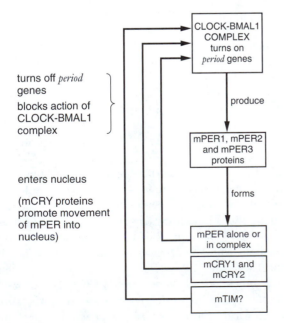

FIGURE 9.24   A model of gene activity thought to underlie circadian timing in mammals.

usual. Thus the *mCry* genes are necessary for clock functioning (van der Horst et al. 1999).

A cycle begins when the mCLOCK/BMAL1 complex binds the promoter region of *mPer1* and turns on gene activity, producing *mPer1* mRNA (Gekakis and Witz 1998). Soon afterward, *mPer3* and then *mPer2* become active. It is thought that the difference in the timing of gene activity might mean that factors in addition to the CLOCK/BMAL1 complex are affecting gene activity. Based on their role in the fruit fly's clock, it is suspected that the mPER proteins alone or in a complex with another clock protein enter the nucleus and turn off the activity of their own genes. The CRY proteins play two important roles in this feedback loop. First, the movement of the mPER proteins into the nucleus is promoted by CRY1 and CRY2. Second, once in the nucleus, CRY1 and CRY2 prevent the CLOCK/BMAL1 complex from turning on the *per* genes. The inhibition of the CLOCK/BMAL1 caused by the CRY proteins is stronger than that caused by any of the mPER proteins or by mTIM (Griffin, Staknis, and Weitz 1999; Kume et al. 1999). The proteins in the mPER complex then become phosphorylated, which causes them to break down. CLOCK and BMAL1 are still present, so as the level of mPER falls, they will once again turn on the *mPer* gene.

The role of mTIM in the mouse clock is unclear (Dunlap 1999). It might function along with the mPER proteins and the mCRY proteins in inhibiting the CLOCK/BMAL1 complex, thereby preventing it from turning on the activity of the *per* genes (Hardin and Glossop 1999).

ENTRAINMENT

If the clock is to be adaptive, it must be synchronized to environmental cycles, such as the light-dark cycles caused by the rising and setting of the sun. We are beginning to understand how this synchronization might occur.

Fruit Fly

We will describe the events involved in entrainment in the fruit fly. Recall that the TIM protein is broken down by light. (Cryptochrome is the pigment that absorbs light and causes TIM destruction, as we will see shortly.) The destruction of TIM is responsible for the changes in the phase of rhythms in response to light (Suri et al. 1998).

A light pulse can cause a phase delay or a phase advance, depending on the part of the cycle at which it occurs. In the early part of the evening, a light pulse causes a delay in the cycle. At this point in the cycle, the PER-TIM complex is still in the cytoplasm. The destruction of TIM by light will cause a delay until TIM can be replenished and the complex restored.

Only then can it move into the nucleus and shut off the transcription of *per* and *tim*. So, *per* and *tim* remain active longer than usual, causing a delay in the cycle.

A light pulse in the late night causes a phase advance. Late at night the PER-TIM complex is in the nucleus, preventing the CLOCK-CYCLE complex from activating the *per* and *tim* genes. At this point in the cycle, the light-induced destruction of TIM destroys the PER-TIM complex, allowing the CLOCK-CYCLE complex to turn on the *per* and *tim* genes sooner than usual. The early activation of *per* and *tim* advances the cycle. During the daytime portion of the cycle, light pulses have no effect because there is so little TIM present that further breakdown by light has no effect (Lee et al. 1996).

How does light information reach the clock? Before it can have biological effects, light must first be absorbed by a pigment. The pigments involved in vision are called opsins. Although the opsins play some role in the clock's response to light, we now know that they are not the most important way in which the clock senses light.

In fruit flies, the clock's primary light-sensing pigments are cryptochromes (CRY), which are blue light receptors. When cryptochrome absorbs light, it triggers a cascade of events, including the destruction of TIM, that synchronizes the biological clock to the environmental light-dark cycle.

The importance of cryptochrome in entrainment in fruit flies has been examined by looking at the effect of a mutation, *cry(b)*, that prevents cryptochrome production. As we describe the effects of this mutation, keep two things in mind. First, fruit flies have clocks throughout their bodies, and these clocks respond to changes in light-dark cycles without any help from the head. Second, cryptochrome is found in nearly all tissues (Plautz, Kaneko, Hall, and Kay 1997; Plautz et al. 1997).

Cryptochrome is the only photoreceptor for the peripheral clocks. This was demonstrated by observing the glow rhythm in fruit flies with the *cry(b)* mutation. (Recall that the glow rhythm can be observed when the *per* promoter is fused with the firefly luciferase gene. This procedure creates a fly that glows whenever the *per* gene is active.) The glow of mutant *cry(b)* flies is not rhythmic in light-dark cycles or in constant darkness. However, if *cry(b)* flies are exposed to a temperature cycle while they are in constant darkness, PER and TIM proteins are produced rhythmically. The *cry(b)* flies cannot produce cryptochrome, and without it the independent clocks cannot be synchronized to a light-dark cycle.

However, the clocks in a fruit fly's brain can receive information about light from visual pigments, as well as from cryptochrome. These clocks, specifically a group of cells on each side of the brain called the lateral neurons, drive the locomotor activity rhythm (Kaneko 1998). The locomotor activity rhythm of *cry(b)* flies is nearly normal in both light-dark cycles and constant darkness. We know that the clock in the lateral neurons is still running in *cry(b)* flies and that it can be set by light because the level of PER and TIM proteins cycle rhythmically in these neurons in a light-dark cycle.

Light can reset the brain clocks because light information reaches them by two pathways: one involving visual pigments and the other involving cryptochrome. In *cry(b)* flies, the visual pigments supply light information to the lateral neurons. However, the light responses aren't completely normal without cryptochrome. The clock cannot be reset by brief pulses of light, and light cannot cause large phase shifts in *cry(b)* flies (Stanewsky et al. 1998).

Light causes a change in CRY that allows it to form a complex with TIM both in the nucleus and in the cytoplasm. As a result of this interaction, the PER-TIM complex is unable to block the action of the CLOCK-CYCLE complex. Thus the *per* and *tim* genes remain turned on. Light enables CRY to block the action of the PER-TIM complex even if TIM is not broken down. Thus, TIM destruction must be a later consequence of the interaction between light and CRY. It is thought, then, that CRY acts as the photoreceptor for the circadian clock by interacting directly with a component of the clock mechanism, TIM. However, CRY is not thought to function as a component of the clock mechanism itself (Ceriani et al. 1999).

Mammals

In mice, information about the presence of light is sent to the SCN, where it turns on a set of immediate early genes (IEGs). You may recall that IEGs produce factors that turn on the activity of other genes. We know that light also turns on two clock genes, *mPer1* and *mPer2*, but we don't know whether IEGs play a role in this. It is interesting that the two clock genes are not activated simultaneously: *mPer1* is activated within 30 minutes, but *mPer2* is activated after about 3 hours (Shearman et al. 1997; Shigeyoshi et al. 1997). The activation of *mPer1* by light is essential for phase-shifting the clock (Akiyama et al. 1999). A phase delay results when light is present during the early night because it causes the levels of *mPer1* and *mPer2* mRNAs to remain high for a prolonged length of time. However, when light is present during the late evening, *mPer1* and *mPer2* are transcribed prematurely, causing a phase advance (Hardin and Glossop 1999).

What is the photoreceptor for the mammalian circadian clock? When cryptochrome was shown to be the photoreceptor for the clock in fruit flies, there was wide speculation that it would play a similar role in mammals. It does not.

Although the cryptochromes are central clock components in mammals, it seems that they are not involved in photoreception for the clock. Recall that light shifts the phase of the clock by turning on the *mPer* genes. Light still has this effect in mice that have been genetically engineered to lack both CRY1 and CRY2. Since the CRY proteins are not needed for light-induced phase shifting, light information must reach the clock through a different photoreceptor (Okamura et al. 1999). We know that the photoreceptor is present in the eyes, but it is not one of the visual pigments in the rods or cones that are important in vision (Freedman et al. 1999).

SUMMARY

Life evolved in a cyclic environment caused by the relative movements of the earth, sun, and moon. Often it is advantageous to gear an activity to occur at a specific time relative to some rhythmic aspect of the environment. Thus clocks evolved as adaptations to these environmental cycles.

There are many examples of rhythmic processes in animals that match the basic geophysical periods: a day (24 hours), the tides (12.4 hours), a lunar day (24.8 hours), a fortnight (14 days), a lunar month (29.5 days), and a year (365 days). Many of these processes remain rhythmic when the individual is isolated from the obvious environmental cycles that might be thought to provide time cues. For instance, many daily rhythms persist when the individual is kept in the laboratory without light-dark or temperature cycles. Therefore, we say that the rhythms are caused by an internal biological clock.

The biological clock is separate from the rhythms it drives. Processes become rhythmic when they are coupled to the biological clock.

In the constancy of the laboratory, the period length of biological rhythms may deviate slightly from the one displayed in nature. For this reason, periods are described with the prefix *circa*, meaning "about," and are called circadian, circalunidian, or circannual. The period length in constant conditions is described as the free-running period and is assumed to reflect the rate at which the clock is running. The free-running period is generally kept constant, which indicates that the biological clock is very accurate.

Although the period length of a biological rhythm is "circa" in the constancy of the laboratory, in nature it matches that of the geophysical cycle exactly because the clock is entrained to (locked onto) an environmental cycle. Entrainment adjusts both the period length and the phase of the rhythm. Daily rhythms can be entrained to light-dark cycles, and in some species, to temperature cycles.

Environmental temperature has only a slight effect on the rate at which the clock runs. This property is called temperature compensation.

There are several reasons why it may be advantageous to have a biological clock to measure time rather than responding directly to environmental changes: (1) anticipation of the environmental changes with enough time to prepare for the behavior, (2) synchronization of the behavior with some event that cannot be sensed directly, and (3) continuous measurement of time so that time-compensated orientation is possible. There is now some evidence that a functional clock enhances survival.

Biological clocks exist in single cells, and there may be many clocks in a single individual. It seems that there is a hierarchy of clocks, with one or more master clocks regulating the activities of other, slave clocks.

We are beginning to understand the circadian system in several species. A silk moth has photoreceptors for entrainment and a clock located in its brain, which then regulates eclosion by producing a hormone. In cockroaches, light information reaches the clock through the compound eyes. Two biological clocks, one in each optic lobe, regulate activity through nerves.

Birds have photoreceptors in their brains capable of providing information to the clock about lighting conditions. The circadian system in birds seems to contain three interacting clocks: the pineal gland, the suprachiasmatic nuclei of the hypothalamus, and the eyes. The importance of each of these clocks varies among species. This system of interacting clocks controls other clocks in the body through the pineal's rhythmic output of the hormone melatonin.

The photoreceptors for entrainment in mammals are the eyes. The information reaches the suprachiasmatic nucleus via nerves (the retinohypothalamic tract). The master clock control for certain rhythms is in the SCN, which regulates activity through nerves and chemicals.

The molecular basis of circadian clocks involves self-regulated feedback loops in which the products of one or more genes affect the activity of other genes. The pigment cryptochrome is important is bringing information about light to the clock of fruit flies but not to that of mammals.

10

Mechanisms of Orientation

Many of us have been moved on a crisp autumn day, enveloped in the reds, yellows, and browns of the season and watching formations of ducks or geese fly against a steely sky. We might have noticed that if it is early in the day, the flocks may be heading almost due south. If it is nearing dusk or if fields of grain are nearby, they may be temporarily diverted to resting or feeding areas. But when they resume their flight, they will head southward again.

In the following spring we may stand beside a swift-moving river in the Pacific Northwest and watch salmon below a dam or a fish ladder. As they lie in deeper pools, resting before the next powerful surge that will carry them one step nearer the spawning ground, they all face one way—upstream.

Both the birds and the fish are responding to a complex and changing environment by positioning themselves correctly in it and by moving from one particular part of it to another. Although the feats of migration are astounding, they are no more crucial to survival than are mundane daily activities such as seeking a suitable habitat, looking for food and returning home again, searching for a mate, or identifying offspring. These actions also depend on the proper orientation to key aspects of the environment. Indeed, an animal's life depends on oriented movements both within and between habitats. In this chapter we will explore some of the mechanisms by which animals orient themselves in space.

TYPES OF ORIENTATION

SIMPLE ORIENTATION RESPONSES

If you look into your aquarium and see a fish swimming belly up, you know something is wrong. This is not the normal posture for fish. Furthermore, if you found one of them on your desk, you would know for certain that something had gone awry, even if the fish were somehow propped into its proper swimming position. Not only does each species have a customary stance, but also each is usually found in a particular type of habitat. Orientation refers to all the reactions that guide an animal into its correct posture, into the proper habitat, or on long-distance travel such as migration and homing (Schöne 1984). We will consider some of the simpler responses first.

Kinesis

Perhaps the simplest response through which animals end up in a suitable location is a kinesis, a response in which an animal's rate of movement or the amount of turning in its path varies with the strength of the stimulus at each step of the movement (Benhamou 1989; Benhamou and Bovet 1989). Although a kinesis is not directed toward or away from the stimulus, it nonetheless results in the individuals settling in a suitable location. To illustrate, consider the humidity kinesis of wood lice. These small isopods are commonly found under rocks or flowerpots or in leaf litter (Figure 10.1). It is adaptive that they cluster in moist areas such as these because they will die if they are exposed to dry air for very long. However, they are not found in damp areas only because those that stray into dry areas die. This can be demonstrated experimentally by distribut-

ing wood lice over the floor of a chamber that has both humid and dry areas; the wood lice will soon congregate in the humid part of the floor. The explanation is rather straightforward: They are more active in the drier areas and they scramble about until, quite by accident, they encounter a more humid part, where activity slows down until many of them become motionless (Edney 1954; Gunn and Kennedy 1936). The effect of humidity on activity has also been demonstrated by determining the percentage of wood lice (*Porcellio scaber*) that were motionless under various constant humidity conditions: More than 70% of the animals were motionless at 90% humidity, but an average of only 20% were motionless at 30% humidity (Gunn 1937). Thus, the wood lice aggregate in the moist area because they move more slowly there, just as cars accumulate when drivers slow down to rubberneck at the scene of an accident. Furthermore, just as most drivers do not seek out traffic jams, wood lice do not specifically orient their movements toward dampness. Since a kinesis is a nondirectional response, the receptors must only be able to register variations in the intensity of the stimulus.

Kineses may also be used by certain vertebrates when locating an appropriate habitat. For a larva of the brook lamprey, *Lampetra planeri*, a suitable position is to be buried head downward in the muddy bottom of a stream or lake. A kinetic response to light helps them settle into this environment. These larvae have a light receptor near the tip of the tail. When the larva burrows deeply enough into the mud to cover the light receptor, it becomes motionless. However, if the light receptor is exposed, the larva becomes agitated and squirms about, with its head pointed downward, until it is once again covered. In one experiment the larvae were placed in an aquarium with a bottom that pre-

FIGURE 10.1   Wood lice such as these (*Oniscus asellus*) tend to accumulate under rocks. They are guided to favorable locations by a humidity kinesis because they are more active in dry areas than in moist ones.

vented burrowing. The aquarium was illuminated on one end and graded into a darkened portion. The lamprey became very active at the lighted end. The activity increased with the intensity of light to which the animals were exposed, but the activity was independent of the direction of the light. Since the larvae were not able to burrow as they randomly moved about, they eventually ended up in the darkened end, where activity was reduced (Jones 1955).

Taxis

Of much more interest biologically are movements that are oriented to the stimulus. The simplest directed orientation mechanism is a taxis, which generally moves the animal toward or away from the stimulus (Song and Poff 1989). A taxis, then, is based on a determination of the direction of the stimulus gradient (Benhamou and Bovet 1992).

A taxis can be described in different ways. One denotes the stimulus to which the animal is responding. Examples include a phototaxis to light, a geotaxis to gravity, a chemotaxis to a chemical, a rheotaxis to a current, or a phonotaxis to a sound. Another describes whether an animal moves toward or away from the stimulus. Thus, an animal that moves toward light and away from gravity is positively phototactic and negatively geotactic.

Different types of animals may use different methods to determine the direction of stimulation, depending on their perceptual abilities and the nature of the stimulus. An animal with a single receptor may determine the direction of stimulation by moving the receptor around and sampling in various positions. The closer to the source, the more intense is the stimulus. For example, a house fly maggot (*Musca domestica*) moves into a dark place when it is time to pupate by moving its "head" from side to side and sampling the light intensity (Figure 10.2). If the light is brighter on one side, it moves to the other (Mast 1938).

Some animals with two receptors make a simultaneous comparison of the intensity of the stimulus on each side and then orient themselves so that the receptors are stimulated equally. If such an animal displays a positive phototaxis and is blinded in one eye, it will continuously circle to the side of the good eye.

Some species have receptors that can determine both the direction to and the intensity of the stimulus. In this case, if an animal that showed a positive phototaxis were blinded in one eye, it would still be able to move directly toward the light source.

More Complicated Light Reactions

Many aquatic animals orient to light in a manner that helps them swim horizontally. Some of them, fish for instance, may show a dorsal light reaction, one in which the dorsal (back) side is kept toward the light. Others,

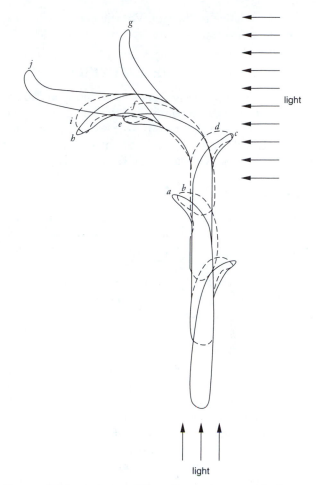

FIGURE 10.2   A taxis guides a fly maggot to a dark site for pupation. The maggot turns its body to measure the light intensity in various locations with its single photoreceptor. It moves in the direction of lowest light intensity. (From Fraenkel and Gunn 1961.)

such as the brine shrimp *Artemia*, show a ventral light reaction in which the ventral (belly) side is directed toward the light.

We all know that animals do not always move directly toward or away from a stimulus. Animals that show the light compass reaction assume an angle with a light source. An animal that makes short excursions from the nest may assume a fixed angle with the sun on its outward journey. This orientation angle would be reversed during the return trip. An ant (*Lasius niger*), for instance, may use the sun as a reference point during its foraging foray. The ant's bearing may be at some angle relative to the sun during its outward trip, say 45° to the right. If the trip was short, a path with the sun 135° to the left would bring the ant back to its nest (Santschi 1911). However, if it is important that the same compass bearing be maintained for a long time, the animal must change its orientation to the light as the sun moves across the sky. Using its biological clock (discussed in Chapter 9), the animal measures the passage of time and appropriately adjusts the angle of its movement relative to the sun. We will consider the use of the sun as a compass again later in the chapter.

LONG-DISTANCE ORIENTATION

Animals often travel long distances to and from home, but they do not all accomplish this feat in the same manner. Strategies for finding the way during long-distance travel are often grouped into three levels of ability (Able 1980; Griffin 1955; Terrill 1991).

Piloting

One level is piloting, the ability to find a goal by referring to familiar landmarks. The animal may search either randomly or systematically for the relevant landmarks. Although we usually think of landmarks as visual, the guidepost may be in any sensory modality. As we will see shortly, olfactory cues guide salmon during their upstream migration.

Compass Orientation

A second level, called compass, or directional, orientation is the ability to head in a geographical direction without the use of landmarks. As we will see shortly, the sun, the stars, and even the earth's magnetic field may be used as compasses by many different species. Most migrating birds use compass orientation (S. T. Emlen 1975). They simply fly in a particular compass direction and stop when they have flown a preprogrammed distance. Similarly, the white butterfly (*Pieris rapae*), known as the cabbage butterfly in the United States, heads in one compass direction while migrating during some seasons. In other seasons, it flies in roughly the opposite compass direction and ends up near its starting point (Baker 1978).

One way to demonstrate that an animal is using compass orientation is to move it to a distant location and determine whether it continues in the same direction or compensates for the movement. If it does not compensate for the relocation, compass orientation is indicated (Figure 10.3). When immature birds of certain migratory species, such as European starlings, were displaced experimentally, they flew in the same direction as the parent group that had not been moved, and they flew for the same distance (Perdeck 1967). In other words, they migrated in a path parallel to their original migratory direction. However, because they had been experimentally displaced before beginning their migration, they did not reach their normal destination. In some cases, this meant that they ended up in ecologically unsatisfactory places (Figure 10.4). In another experiment, storks were taken from nests in the Baltic region, reared in West Germany, and released after the local storks had left. The experimental group had a strong tendency to head south-southeast, the direction

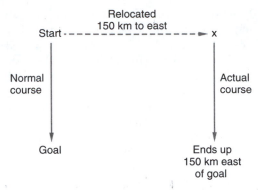

FIGURE 10.3 Experimental relocation of an animal that is using compass orientation causes it to miss the goal by the amount of its displacement.

of their parent population, even though the local West German storks were heading southwest (Schüz 1949).

There is growing evidence that the migratory direction of birds is genetically controlled (reviewed in Berthold 1991). Many bird species, in addition to the storks we just discussed, find their way without help from others of their species. Observations such as these suggest that migratory direction may be inherited. Individual birds held in the laboratory flutter in the direction in which they would be flying if they were free. When their cousins in nature have completed their migratory journey, the captive birds also cease their directional activity. Furthermore, many species, particularly those that fly from central Europe to Africa, change compass bearing during their flight. Garden warblers (*Sylvia borin*) and blackcaps (*S. atricapilla*) held in the laboratory change the direction in which they flutter in their cages at the time that free-flying members of their population change direction (Gwinner and Wiltschko 1978; Helbig, Berthold, and Wiltschko 1989). Crossbreeding studies have also shown the inheritance of migratory direction. Andreas Helbig (1991) crossbred members of two populations of blackcaps that had very different migratory directions. The orientation of the offspring was intermediate between those of the parents.

An animal may also use compasses in another way, one that may be used to find its way home after carrying out daily activities, such as foraging. In this type of navigation, called dead reckoning, the animal appears to know its position relative to home through the sequence of distance and direction traveled during each leg of the outward journey (Figure 10.5). The estimates of distance and direction are often adjusted for any displacement due to current or wind. Then, knowing its location relative to home, the animal can head directly there, using its compass(es). A compass may also be used to determine the direction traveled on each leg of the outward journey, or the direction may be estimated from the twists and turns taken, sounds, smells, or even

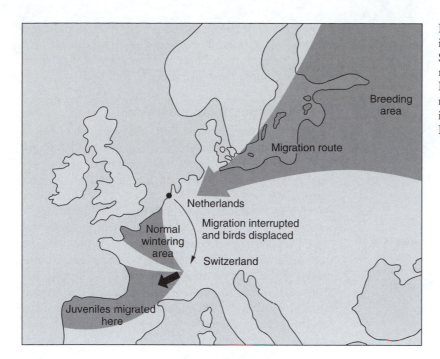

FIGURE 10.4   Immature starlings captured
in the Netherlands and released in
Switzerland did not compensate for the
relocation during their autumn migration.
Instead they traveled southwest, their normal migratory direction, and ended up in
incorrect wintering areas. (Modified from
Perdeck 1958.)

the earth's magnetic field. Once close to home, where landmarks are familiar, piloting may be used to pinpoint the exact location of home.

There is evidence that many types of animals use dead reckoning to find their way around. Consider, for example, the desert ant (*Cataglyphis bicolor*). During its foraging forays, this insect wanders far from its nest over almost featureless terrain. After prey is located, sometimes 100 meters away from the nest, roughly the distance of a football field, the ant turns and heads directly toward home. It appears that the ant knows its position relative to its nest by taking into account each turn and the distance traveled on each leg of its outward trip. To obtain this information, an ant must make the trip itself. If an ant that is leaving its nest is moved to a new spot by an experimenter, it does not seem to know where it is and searches randomly for its nest (Wehner and Flatt 1972). Furthermore, when an ant is captured as it is leaving a feeding station headed for home and is

relocated to a distant site, its path is in a direction that would have led it home if it had not been experimentally moved (Wehner and Srinivasan 1981).

True Navigation

A third level of orientation, sometimes called true navigation,* is the ability to maintain or establish reference to a goal, regardless of its location, without the use of landmarks. Generally, this implies that the animal cannot directly sense its goal and that if it is displaced while en route, it compensates by changing direction, thereby heading once again toward the goal (Figure 10.6).

We see, then, that true navigation requires the animal to have both a compass and a map that indicates position relative to a goal. Imagine yourself abandoned in an unfamiliar place with only a compass to guide your homeward journey. Before you could head home, you would also need a map so that you could know where you were relative to home. Only then could you use your compass and orient yourself correctly.

Only a few species, most notably the homing pigeon (*Columba livia*), have been shown to have true navigational ability. Certain other groups of birds, including oceanic seabirds and swallows, are also

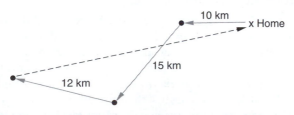

FIGURE 10.5   **Navigation by dead reckoning. This involves determining one's position by using the direction and distance of each successive leg of the outward trip. A compass can then be used to steer a course directly toward home.**

True navigation is an unfortunate term since it carries with it the implication that other means of finding one's way from place to place are not real methods of navigating. This is certainly not true. Nonetheless, we will use the term simply to distinguish this method of maintaining a course from the others.

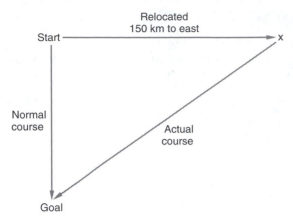

FIGURE 10.6   An animal that finds its way by using true navigation can compensate for experimental relocation and travel toward the goal. This implies that the animal cannot directly sense its goal and that it is not using familiar landmarks to direct its journey.

known to home with great accuracy (Able 1980; S. T. Emlen 1975).

MULTIPLICITY OF ORIENTATION CUES

The feats of migration are indeed astounding—an arctic tern circumnavigating the globe, a monarch butterfly fluttering thousands of miles to winter in Mexico, a salmon returning to the stream in which it hatched after years in the open sea. How do they do it? There is no simple answer. Different species may use different mechanisms, and any given species usually has several navigational mechanisms available. Indeed, common themes in orientation systems are the use of multiple cues, a hierarchy of systems, and transfer of information among various systems (Able 1991, 1996). When one mechanism becomes temporarily inoperative, a backup is used. Furthermore, a navigational system may involve more than one sensory system. These interactions can be quite complex, but we will simplify matters by considering each sensory mechanism separately.

VISUAL CUES

Visual mechanisms of orientation include the use of landmarks and celestial cues such as the sun, stars, and polarized light.

LANDMARKS

Landmark recognition is perhaps the most obvious way that vision may be used for orientation or navigation. Humans use landmarks frequently when giving direc-

tions: "Turn left before the bank" or "Make a right just after the gas station." Because the use of landmarks is so familiar to us, it is probably not too surprising to learn that many animals also use them to find their way.

Demonstrating Landmark Use

There are various ways to show that landmarks play a role in orientation. One way is to move the landmark and see whether this alters the orientation of the animal. Niko Tinbergen demonstrated that the digger wasp, *Philanthus triangulum*, relies on landmarks to relocate its nest after a foraging flight. While a female wasp was inside the nest, a ring of 20 pine cones was placed around the opening. When she left the nest, she flew around the area, apparently noting local landmarks, and then flew off in search of prey. During her absence, the ring of pine cones was moved a short distance (1 foot) away. On each of 13 observed trips, the returning wasp searched the middle of the pine cone ring for the nest opening. However, she did not find it until the pine cones were returned to their original position (Tinbergen and Kruyt 1938).

Animals can also be prevented from using landmarks by clouding their vision. Consider, for example, the ingenious way that Klaus Schmidt-Koenig and Hans Schlichte (1972) demonstrated that homing pigeons do not require landmarks to return to the vicinity of their home loft: They created frosted contact lenses for the pigeons (Figure 10.7). Through these lenses, pigeons could only vaguely see nearby objects and distant ones not at all. Nonetheless, the flight paths of these pigeons were oriented toward home just as accurately as those of control pigeons. Thus, the pigeons cannot be depending on familiar landmarks to guide their journey home. Note that this does not mean that they do not use landmarks when they are available, just that they can determine the homeward direction without them. Also, although pigeons with frosted lenses get to the general area of their home loft, they often cannot find the loft itself. Landmarks, then, may be important in pinpointing the exact loft location but are not necessary for determining the direction of home.

Models of Landmark Use

Knowing that an animal uses landmarks to find its way does not tell us *how* those landmarks are used. Do other animals use landmarks as humans do, as part of a mental map of the area? Perhaps some species do, but others might use landmarks in different ways. A simple model of landmark use is that the animal stores the image of a group of landmarks in its memory, almost like a photograph of the scene. Then it moves about the environment until its view of nearby objects matches the remembered "snapshot" (Srinivasan 1998). Rüdiger Wehner (1981) suggested that a whole series of mem-

FIGURE 10.7 Homing pigeons that are wearing frosted contact lenses are unable to use landmarks for navigation. However, these pigeons head home just as accurately as those with normal vision. Therefore, although pigeons may use landmarks if they are available, they do not require them to home.

ory snapshots might be filed in the order in which they are encountered. He added that invertebrates might be able to use landmarks by comparing the successive images of surrounding objects with a series of memory snapshots of the landmarks along a familiar route.

One animal that appears to use memory snapshots of landmarks is the desert ant (Figure 10.8). As previously mentioned, desert ants are able to plot a course back to the nest by dead reckoning; that is, they integrate the directions and distances traveled on all legs of the journey away from the nest to plot a direct course back. However, they also use landmarks, especially

FIGURE 10.8 The desert ant uses a remembered sequence of landmark images to find its way home in a familiar area.

when they have almost reached the nest on their return from the foraging site (Wehner, Michel, and Antonsen 1996). Desert ants tend to follow familiar routes. They follow one path during the outbound journey and another for the return, using a different set of memory images on the two legs of their trip (Wehner and Flatt 1972). The use of two sets of memory images would be predicted by the model since the ant could never match the memorized images of landmarks viewed from a particular spot along its outward path with the image of the same landmarks viewed while returning. If an ant that has nearly reached its nest is experimentally displaced to a new spot within a familiar area, it resorts to landmarks to guide the rest of its journey (Wehner 1983; Wehner and Raber 1979). In fact, desert ants often use landmarks instead of dead reckoning. If the most direct path is an unfamiliar route, it could lead over rocks or be blocked by scrub, and so landmarks are favored. Nonetheless, if the ant comes across a clearing, it can use dead reckoning to take the most direct course home (Collett et al. 1998).

Alternatively, animals might use landmarks by learning the relative positions among the landmarks, thereby forming a "cognitive map" of the area for reference. In this case, the animal could return directly home from a site *over a route it has never traveled* because it has a mental map of the area based on remembered relationships among familiar landmarks. This is a technique often used by humans. Even if you always walked to a friend's house along the same path, if you were dropped off in an unfamiliar spot but could see a familiar landmark, you would remember the position of your destination relative to that landmark and take a new trail to the house.

The idea that animals can form cognitive maps has become quite controversial. At least part of the controversy is caused by different definitions of the term because the definition determines what constitutes evidence of a cognitive map (Bennett 1996). Notice, for example, how slightly different definitions of cognitive maps, relating to the way in which landmarks are used for orientation, have created a controversy in studies of bee orientation.

The controversy began to brew when James Gould (1986) questioned whether the honeybee (*Apis mellifera ligustica*) forms cognitive maps. He trained bees to follow a specific route to a feeding station. Foragers regularly visiting the feeding station were captured at the hive entrance and transported in the dark to a new location the same distance from the hive as the feeding station but in a different direction. When released, the bees headed in the direction of the feeding station, exactly as would be predicted if they had mental maps of the landmarks in the area (Figure 10.9). Furthermore, the flight times of the bees indicated that they flew directly to the feeding station. This seems to provide evidence of a cognitive map as defined as a memory of landmarks that allows novel shortcutting to occur.

However, we should not be too quick to conclude that bees form mental maps. Experiments similar to

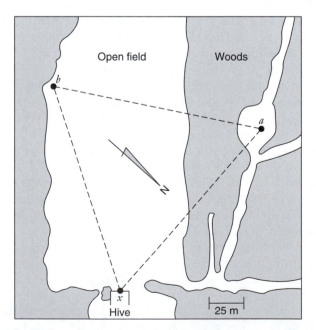

FIGURE 10.9 Bees may form a cognitive map of an area in which the relationship among the landmarks is known. To explore this possibility, bees were first trained to feeding site *a*. Later, they were captured as they left the hive and transported in the dark to site *b*. When released, the foragers flew directly to site *a*, even though it was hidden from view. This has been interpreted as an indication that bees have a mental map of the landmarks in a familiar area. (From Gould 1986.)

Gould's, but with attention paid to the landmarks available to bees, led Fred Dyer (1991) to conclude that the bees use landmarks associated with routes previously traveled. Gould's bees were able to head toward the feeding station by seeing landmarks associated with the goal. And, when Randolf Menzel (1989; Menzel et al. 1990) performed experiments similar to Gould's but without prominent landmarks associated with the goal, he found that the bees used the sun for orientation. Furthermore, the dances that bees perform to inform their nestmates of the location of a food source are based on the distance and direction flown while en route to the food, not on a cognitive map (Kirchner and Braun 1994).

Today, definitions still muddy the issue. After a careful review of the literature on cognitive maps, Andrew Bennett (1996) maintained that there was no conclusive demonstration that any animal, even a primate, had a cognitive map when it is defined as a memory of landmarks that permits the animal to take novel shortcuts. Simon Benhamou (1996) uses a different definition, and he reasons that a cognitive map allows an animal to know the relationships among familiar landmarks. Consequently, even if some of these were out of sight, the animal could use this mental map to determine its current location relative to a goal. Thus, an animal with a cognitive map should be able to find a goal without actually seeing any landmarks near the goal. To test this idea, rats were given an opportunity to build a cognitive map by exploring an area and learning the location of landmarks. Then the rats had to find a hidden goal (an escape platform) when the landmarks in view did not include any of those near the goal. Under these conditions, rats searched randomly for the goal. Although these results are not firm evidence against cognitive maps, they certainly do not support the idea. However, there is some evidence that Clark's nutcracker (*Nucifraga columbiana*), a bird that can remember the location of stored food for months, can form a cognitive map. In this case, however, a cognitive map is defined as an animal's mental image of geometrical relationships among landmarks (Kamil and Jones 1997). And so we see that in science, as in life, the answer to a question often depends on how it is asked.

THE SUN

Many animals use the sun as a celestial compass. In other words, these animals can determine compass direction from the position of the sun. Because of the earth's rotation, the sun appears to move through the sky at an average rate of 15° per hour. In the temperate latitudes of the Northern Hemisphere, for example, the sun rises in the east and moves across the sky to set on the western horizon. The specific course that the sun appears to take varies with the latitude of the observer and the season of the year, but it is predictable (Figure

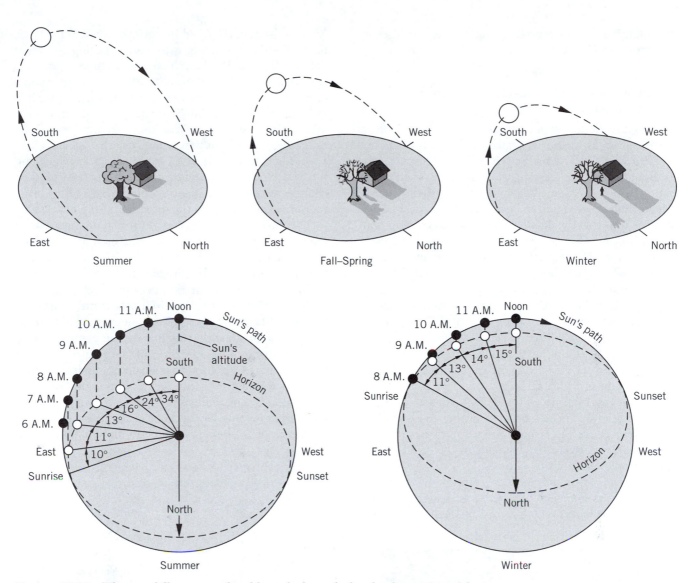

FIGURE 10.10 **The sun follows a predictable path through the sky that varies with latitude and season. If the sun's course and the time of day are known, the sun's bearing (azimuth) provides a compass bearing. The sun appears to move across the sky at an average rate of 15° an hour. Therefore, if the sun is to be used as a compass for a long time, the animal must compensate for its movement.**

10.10). Therefore, if the path and the time of day are known, the sun can be used as a compass.

Knowledge of one compass bearing is all that is necessary for orientation in any direction. Consider this simplified example. Suppose you decided to camp in the woods a short distance north of your home. As you headed for your campsite at 9 A.M., the sun would be in the east, so you would keep the sun on your right to travel north. However, during your homeward trek the next morning, you would keep the sun on your left to travel south.

The use of the sun for orientation is complicated by its apparent motion through the sky. The sun appears to move at an average rate of 15° an hour. Therefore, an animal heading straight for its goal and

navigating by keeping a constant angle between its path and the sun would, after one hour, be following a path that would be off by 15°. Some species take only short trips, so errors due to the sun's apparent motion are inconsequential. These species do not adjust their course with the sun's. But if the sun is to be used as an orientation cue for a prolonged period, the animal must compensate for the sun's movement. To do so it must be able to measure the passage of time and correctly adjust its angle with the position of the sun. At 9 A.M. an animal wishing to travel south might keep the sun at an angle of 45° to its left. By 3 P.M., however, the sun will have moved approximately 90° at an average rate of 15° an hour. To maintain the same southward bearing, the animal must now assume a 45° angle, with the sun on its right. Time is measured by using a biological clock (discussed in Chapter 9).

Evidence for the Time-Compensated Sun Compass

The first work on sun compass orientation was done on birds and bees in the laboratories of Gustav Kramer (1950) and Karl von Frisch (1950), respectively. Although these two investigative groups worked at the same time, neither knew of the other's work. Nevertheless, they often used similar experimental designs to reveal the details of sun compass orientation. We will take a closer look at the experiments of Gustav Kramer here, but if you want to compare these studies to those of von Frisch, consult von Frisch's (1967) fascinating book, *The Dance Language and Orientation of Bees.*

Gustav Kramer (1949) began his studies by trapping migrant birds and caring for them in cages. He then noticed that they became restless during their normal migration season. Furthermore, most of their activity took place on the side of the cage corresponding to the direction in which the birds would be flying if they were free to migrate. This activity has been aptly named migratory restlessness. In noting these tendencies, Kramer set the stage for a series of experiments that would yield valuable evidence in the quest for the navigational mechanisms of birds.

The indication that birds migrating during the day use the sun as a navigational cue was that the orientation (directionality) of migratory restlessness was lost when the sun was blocked from view. Kramer (1951) set up outdoor experiments with caged starlings, *Sturnus vulgaris* (Figure 10.11), which are daytime migrators, and found that they oriented in the normal migratory direction unless the sky was overcast, in which case they lost their directional ability and moved about randomly. When the sun reappeared, they oriented correctly again, suggesting that they were using the sun as a compass. Then Kramer devised experiments in which the sun was blocked from view and a mirror was used to change the *apparent* position of the sun. The birds reoriented according to the direction of the new "sun."

Because migration occurs during limited periods in the fall and spring, experiments using migratory restlessness to study orientation mechanisms are limited to two brief intervals a year. To eliminate this problem, Kramer (1951) devised an orientation cage in which there were 12 identical food boxes encircling a central birdcage (Figure 10.12). Kramer and his students trained birds to expect food in a box that lay in a certain compass direction. This ring of food boxes could be rotated so that a bird trained to get food in a given compass direction would not always be going to the same food box. This eliminated the possibility that the bird might learn to recognize the food dish by some characteristic, such as a dent. As long as the birds could see the sun, they would

FIGURE 10.11   Starlings are daytime migrators and were the subject of Gustav Kramer's pioneering work on bird navigation.

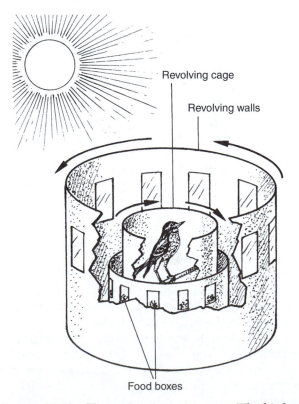

Revolving cage

Revolving walls

Food boxes

FIGURE 10.12   Kramer's orientation cage. The bird can see the sky through the glass roof but is prevented from seeing the surrounding landscape. It is trained to look for food in a food box that is placed in a particular compass direction.

approach the proper food box. However, on overcast days the birds were often disoriented, as would be expected if they were using a sun compass.

The results of experiments with birds in Kramer's orientation cages not only confirm those on migratory restlessness (Kramer 1951), but also indicate that the birds compensate for the movement of the sun. Actually, the idea of time-compensated sun compass orientation began when Kramer noticed that the birds in his orientation cages were able to orient in the proper direction even as the sun moved across the sky. When the real sun was replaced with a stationary light source, the birds continually adjusted their orientation with the stationary sun as though it were moving. The orientation with the artificial sun changed at a rate of about 15° an hour, just as it would to maintain a constant compass bearing using the real sun.

The birds are able to compensate for the sun's apparent movement; therefore, they must possess some sort of independent timing mechanism. As we saw in Chapter 9, the biological clock that allows birds to compensate for the movement of the sun can be reset by artificially altering the light-dark regime. Initially the birds are placed in an artificial light-dark cycle that corresponds to the natural lighting conditions outside; the lights are on from 6 A.M. to 6 P.M. The light period is then shifted so that it begins earlier or later than the actual time of dawn. For example, if the animal is exposed to a light-dark cycle that is shifted so that the lights come on at noon instead of 6 A.M., the animal's biological clock is gradually reset. In this case, the animal's body time would be set six hours later than real time. Therefore, if the biological clock is used to compensate for the movement of the sun, orientation should be off by the amount that the sun had moved during that interval. In this example, orientation should be shifted 90° (6 × 15°) clockwise, for example, west instead of south (Figure 10.13).

One of Kramer's students, Klaus Hoffmann (1954), was the first to use the clock-shift experiment to demonstrate the involvement of the biological clock in sun compass orientation. After resetting the internal clock of starlings by keeping them in an artificial light-dark cycle for several days, the birds' orientation was shifted by the predicted amount.

Development of the Sun Compass

How does a bird learn that the sun rises in the east? What reference system is used to link the position of the sun with geographic direction so that it can be used as a compass? We do not know the answer yet, but there are suggestions. One idea is that the pattern of polarization of light in the sky is used to calibrate the sun compass. As will be described in more detail later in this chapter, this pattern of polarization depends on the position of the sun and, therefore, seems to move through the sky as the sun does. The centers of rotation of the polarization pattern are at the north and south poles, and therefore, these centers could serve as reliable indicators of geographic direction (Brines 1980; Phillips and Waldvogel 1988). Another suggestion is that the earth's magnetic field is the reference system by which the sun compass is set. As we shall see later in

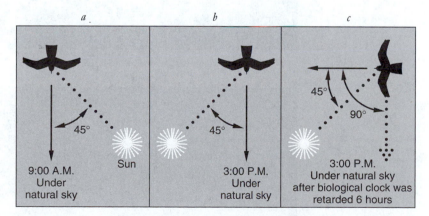

FIGURE 10.13   A clock-shift experiment demonstrates time-compensated sun compass orientation. (*a*) The flight path of a bird flying south at 9 A.M. might be at an angle of 45° to the right of the sun. (*b*) By 3 P.M., the sun would have moved roughly 90°, so to continue flying in the same direction, the bird's flight path might be at an angle of 45° to the left of the sun. (*c*) If the bird's biological clock were delayed by six hours and the bird's orientation tested at 3 P.M. (when the bird's body time was 9 A.M.), it would orient to the west. The flight path of the bird would be determined by the bird's biological clock. The flight path would, therefore, be appropriate for 9 A.M., and orientation would be shifted by 90° clockwise. (From Palmer 1966.)

this chapter, there are several potential ways in which geographic direction could be obtained from the earth's magnetic field. Wolfgang Wiltschko and his coworkers (1983) have provided some support for the idea that the earth's magnetic field is involved in the learning process that sets the sun compass. They raised homing pigeons in a magnetic field that was rotated either 60° or 120° clockwise. The young birds could see the sun only in an abnormal relationship to the magnetic field. When the pigeons were released for their first flight in sunshine, their orientation was shifted slightly clockwise.

The Sun and Nocturnal Orientation

We have seen how the sun is used for orientation during the daytime, but it is interesting to note that some nocturnal migrants select their flight direction by the point of sunset. They then associate this point with other cues that will be available to guide their flight throughout the night, such as the wind direction, stars, or the earth's magnetic field (reviewed in F. R. Moore 1987).

Although Kramer (1950) noticed that the nocturnal restlessness of his birds was oriented more precisely if they were put into their cages before sunset, the first experiments designed to determine the role of the setting sun were performed by Frank Moore (1978, 1980) on the Savannah sparrow (*Passerculus sandwichensis*). He found that the nocturnal restlessness of these birds was well oriented if they could see the sunset, even if they were prevented from viewing the stars. However, orientation deteriorated when they could see the stars but not the sunset. The activity of several other nocturnal migrants is also more oriented when they are permitted to see the sunset—white-throated sparrows, *Zonotrichia*

FIGURE 10.14 (*a*) Star compass orientation was explored by exposing nocturnal migrants, indigo buntings, to a planetarium sky. During the normal time of migration, caged birds will flutter in the proper migratory direction if the stars are visible. (*b*) In some studies a record of activity was created on the sides of a funnel-shaped cage by the bird's inky feet.

a

b

albicollis (Bingman and Able 1979; Lucia and Osborne 1983); tree sparrows, *Spizella arborea* (Cherry 1984); and the European robin, *Erithacus rubecula* (Katz 1985).

STAR COMPASS

Many species of bird migrants travel at night. Even if they set their bearings by the position of the setting sun, how do they steer their course throughout the night? One important cue is the stars. This was first demonstrated by Franz and Eleonore Sauer (Sauer 1957, 1961; Sauer and Sauer 1960). Using several species of sylviid warblers, they performed a series of experiments aimed at discovering just which objects in the nighttime sky that the birds use as cues. The Sauers kept their caged warblers inside a planetarium so that the nighttime sky could be controlled. They first lined up the planetarium sky with the sky outside and found that the birds oriented themselves in the proper migratory direction for that time of year. Then the lights were turned out and the star pattern of the sky was rotated. The birds continued to orient according to the new direction of the planetarium sky. When the dome was diffusely lit, the birds were disoriented and moved about randomly. In some experiments, even though the moon and planets were not projected, the birds oriented correctly, apparently taking their bearings from the stars.

We know the most about the mechanism of star compass orientation in the indigo bunting (*Passerina cyanea*). Our knowledge has been gained primarily through Stephen Emlen's systematic planetarium studies. These indicate that the indigo bunting relies on the region of the sky within 35° of Polaris (Figure 10.14). Since Polaris is the pole star, it shows little apparent movement and, therefore, provides the most stationary reference point in the northern sky. The other constellations rotate around this point (Figure 10.15). The stars nearer Polaris move through smaller arcs than do those farther away, closer to the celestial equator. The

birds learn that the center of rotation of the stars is in the north, information that is used to guide their migration either northward or southward. The major constellations in this region are the Big Dipper, the Little Dipper, Draco, Cepheus, and Cassiopeia. Experiments have shown that it is not necessary for all these constellations to be visible at once. If one constellation is blocked by cloud cover, the bird simply relies on an alternative constellation (S. T. Emlen 1967a, b).

Young birds learn that the center of rotation of stars is north. The axis of rotation then gives directional meaning to the configuration of constellations. Once their star compass has been set in this way, the birds do not need to see the constellations rotate. Simply viewing certain constellations is sufficient for orientation. This was first demonstrated by exposing groups of young indigo buntings to normal star patterns in a planetarium sky. One group saw a normal pattern of rotation, one that rotated around Polaris. The other group viewed the normal pattern of stars, but instead of rotating around Polaris, these stars rotated around Betelgeuse, a bright star closer to the equator. When the birds came into a migratory condition, their orientation was tested under a stationary sky. Although each group was headed in a different geographic direction, both groups were well oriented in the appropriate migratory direction relative to the center of rotation they had experienced, either Betelgeuse or Polaris (Figure 10.16). In other words, in the autumn, when the birds would be heading south for the winter, those that had experienced Betelgeuse as the center of rotation interpreted the position of that star as north and headed away from it (S. T. Emlen 1969, 1970, 1972).

The development of the star compass has been studied in only a few species other than the indigo bunting. Garden warblers (W. Wiltschko 1982; W. Wiltschko et al. 1987) and pied flycatchers, *Ficedula hypoleuca* (Bingman 1984), also learn that the center of celestial rotation indicates north.

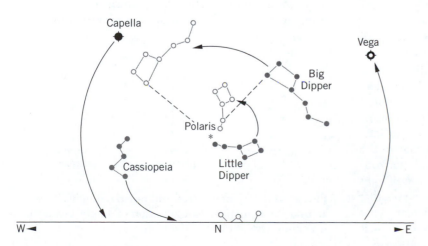

FIGURE 10.15   The stars rotate around Polaris, the North Star. The center of rotation of the stars tells birds which way is north. The positions of stars in the northern sky during the spring are shown here. The closed circles indicate star positions during the early evening, and the open circles indicate the positions of the same stars six hours later.

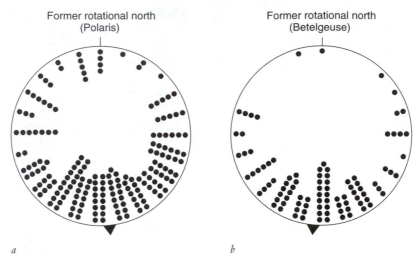

Former rotational north
(Polaris)

Former rotational north
(Betelgeuse)

a

b

FIGURE 10.16   The orientation of indigo buntings to a stationary planetarium sky after exposure to different celestial rotations. During their
first summer, indigo buntings learn that the center of celestial rotation
is north. This was demonstrated by exposing a group of young birds to
a planetarium sky that (*a*) rotated around Polaris (the North Star) or (*b*)
around Betelgeuse. During their first autumn, when they would be
migrating south, they were exposed to a stationary planetarium sky.
Each dot is the mean direction of activity for a single test. The arrow on
the periphery of the circle is the overall mean direction of activity. Each
group oriented away from the star that had been the center of rotation.
(Modified from data of Able and Bingman 1987; Emlen 1970.)

POLARIZED LIGHT AND ORIENTATION

One of the puzzling facets of sun compass orientation is
that many animals continue to orient correctly even
when their view of most of the sky is blocked. How is
this possible? For at least some of these animals,
another celestial orientation cue is available in patches
of blue sky—polarized light. Before considering orientation to polarized light, it might be useful to become
familiar with the nature of polarized light and how the
pattern of skylight polarization depends on the position
of the sun.

The Nature of Polarized Light

Light consists of many electromagnetic waves, all
vibrating perpendicularly to the direction of propagation (Figure 10.17). As a crude analogy, think of a rope
held loosely between two people as a light beam. The
rope itself would define the direction of propagation of
the light beam. If one person repeatedly flicked his or
her wrist, the rope would begin to wave or oscillate.
These oscillations would also be perpendicular to the
length of the rope, but they could be vertical, horizontal, or any angle in between. The same is true of light
waves. Most light consists of a great many waves that
are vibrating in all possible planes perpendicular to the
direction in which the wave is traveling. Such light is

described as unpolarized. In fully polarized light, however, all waves vibrate in only one plane. Our rope light
beam, for instance, would become vertically polarized if
the person's wrist was flicked only up and down. In this
case, the rope might oscillate vertically in the spaces
between the boards of a picket fence.

As sunlight passes through the atmosphere, it
becomes polarized by air molecules and particles in the
air, but the degree and direction of polarization in a
given region of the sky depend on the position of the
sun. In other words, there is a pattern of polarized light
in the sky that is directly related to the sun's position
(Figure 10.18). One aspect of this pattern is the degree
of polarization. To picture the pattern of polarization,
think of the sky as a celestial sphere with the sun at one

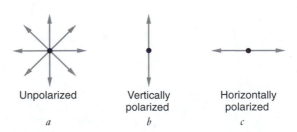

Unpolarized

Vertically
polarized

Horizontally
polarized

a

b

c

FIGURE 10.17   Unpolarized and polarized light. The
arrows show the planes of vibration of a light beam that
is coming straight out of the page.

FIGURE 10.18   The sky viewed through a polarizing filter to show the pattern of skylight polarization at (*a*) 9 A.M., (*b*) noon, and (*c*) 3 P.M. The darker region of the sky is the band of maximum polarization. The diagrams below show the pattern of polarization (*d*) with the sun on the horizon, (*e*) at 45° elevation, and (*f*) at zenith. The arrows indicate the direction of the plane of polarization. The small circle denotes the position of the sun. The pattern of polarization depends on the position of the sun. The blue sky provides an orientation cue for animals that can perceive the plane of polarization.

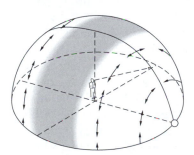

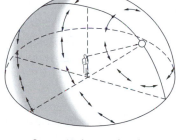

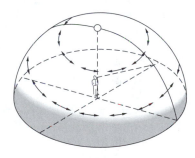

d Sun at horizon (dawn)

e Sun at 45 degree elevation
(perhaps 9 am)

f Sun at zenith (noon)

pole and an "antisun" at the other. The light at the poles is unpolarized, but it becomes gradually more strongly polarized with increasing distance from the poles. Thus, between the sun and the antisun, there is a band where the light in the sky is more highly polarized than in other regions. This region is described as the band of maximum polarization. But there is more to the pattern than this: The direction of the plane of polarization (called the e-vector) also varies according to the position of the sun. The plane of polarization of sunlight is always perpendicular to the direction in which the light beam is traveling. If you were to draw imaginary lines of latitude on the celestial sphere so that they formed concentric circles around the sun and antisun, these lines would indicate the plane of polarization at any point in the sky. Since the entire pattern of polar-

ization of light in the sky is determined by the sun's position, the pattern moves westward as the sun moves through the sky (Waterman 1989).

Uses of Polarized Light in Orientation

Polarized light reflected from shiny surfaces, such as water or a moist substrate, is used by certain aquatic insects to detect suitable habitat. Indeed, polarized light may actually attract them (Schwind 1991). For the backswimmer, *Notonecta glauca* (Figure 10.19), not only is the horizontally polarized light that is reflected from the surface of a pond a beacon that helps the insect locate a new body of water during dispersal, but it also triggers a plunge reaction that brings the insect closer to a new home (Schwind 1983).

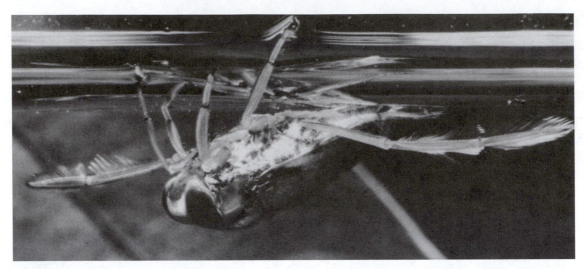

FIGURE 10.19   Many aquatic insects, such as this backswimmer, use polarized light reflected from water or a moist surface to locate an appropriate habitat. A backswimmer spends almost its entire life underwater. These insects are commonly seen in ponds, suspended beneath the water surface, as this one is. Adults can fly, however, and may disperse to a new pond before laying the second batch of eggs of the season.

There are two possible ways in which the plane of polarization of the light in the sky is used as an orientation cue. First, polarized light is used as an axis for orientation. In other words, an animal might move at some angle with respect to the plane of polarization. Many animals use polarized light in this way. Salamanders living near a shoreline, for instance, can use the plane of polarization to direct their movements toward land or water (Adler 1976). When a predator approaches from the shoreline, the grass shrimp (*Palaemonetes vulgaris*) darts quickly away from the shore toward deeper water. This escape response is at least partially guided by the celestial pattern of polarization (Goddard and Forward 1991). Polarized light may also be used by young sockeye salmon to maintain their proper migratory direction while traveling through rivers or lakes (Dill 1971). Indeed, many species of fish are able to detect polarized light, although we do not yet know how they use this ability (Hawryshyn 1992).

Second, the pattern of polarization of sunlight might be used to determine the sun's position when it is blocked from view. This method is possible because the pattern of polarization depends on the position of the sun. Honeybees (*A. mellifera*) and desert ants do this in an amazingly simple manner. The receptors for polarized light are located in a small region near the top of the rim of the compound eye. The response of a given receptor depends on its alignment with the plane of polarization. The receptors are arranged so that if the insect is headed directly toward or away from the sun, the combined responses of these receptors are greater than when the insect is at any other angle to the sun. Therefore, to be headed toward or away from the sun,

all the insect must do is move around until the summed output of its receptors is at its maximum. The next step, determining which end of its body is facing the sun, is usually fairly easy. Having located the sun, the insect can then orient accordingly (Wehner 1987).

The polarization of light in the sky could also provide an orientation cue at dawn and dusk, when the sun is below the horizon. Many birds that migrate at night set their bearings at sunset. Apparently, the pattern of skylight polarization at sunset (Able 1982) and at sunrise (F. R. Moore 1986) assists the orientation of birds migrating at these times because experiments have shown that the birds' directional tendencies are altered when the plane of polarized light to which they are exposed is experimentally shifted by rotating polarizing filters. Indeed, when a bird is setting its bearings for the night, polarized light is a more important orientation cue than the sun's position along the horizon at dusk or the geomagnetic field (Able 1993; Able and Able 1996).

MAGNETIC CUES

CUES FROM THE EARTH'S MAGNETIC FIELD

Our earth acts like a giant magnet. To picture the geomagnetic field around the earth, imagine an immense bar magnet through the earth's core from north to south. However, this bar magnet is tilted slightly from the geographic north-south axis, and the magnetic poles are shifted slightly from the geographic, or rotational, poles (Figure 10.20). The difference between the magnetic pole and the geographic pole is called the

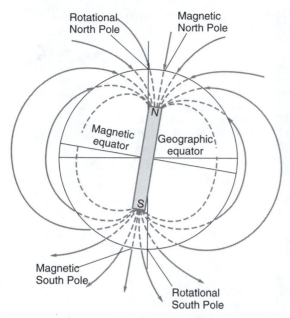

Rotational North Pole

Magnetic North Pole

N

Magnetic equator

Geographic equator

S

Magnetic South Pole

Rotational South Pole

FIGURE 10.20 **The earth's magnetic field. The lines of force leave magnetic south vertically; curve around the earth's surface; and enter magnetic north, heading straight down. The geomagnetic field provides several possible cues for navigation: polarity, the north-south axis of the lines of force, and the inclination of the lines of force. The magnetic compass of most animals appears to be an inclination compass. They determine the north-south axis from the orientation of the lines of force but assign direction to this by the inclination of the force lines. In the Northern Hemisphere, north is the direction in which the force lines dip toward the earth.**

cally with latitude, providing three potential orientation cues. Which of these are used? Our own experience with compasses immediately brings polarity to mind. When the needle on a compass points north, it is responding to the polarity of the earth's field. Indeed, some species of animals seem to respond to polarity. This list includes invertebrates, the lobster, for instance (Lohmann et al. 1995), as well as vertebrates, including newts (Phillips 1986; Phillips and Borland 1994); sockeye salmon (Quinn 1980; Quinn, Merrill, and Brannon 1981); bobolinks (Beason and Nichols 1984); and the mole-rat (*Cryptomys* spp.), a rodent that lives underground (Marhold, Wiltschko, and Burda 1997). We know that an animal responds to polarity when its orientation changes in response to an experimental shift in the direction of magnetic north.

However, most other animals, including birds and sea turtles, appear to use the magnetic field inclination. They distinguish between "poleward," where the lines of force are steepest, and "equatorward," where the lines of force are parallel to the earth's surface. Although the horizontal component of the earth's field, which runs between magnetic north and magnetic south, indicates to the animal the north-south axis, the inclination of the field is the cue that tells the animal whether it is going toward the pole or toward the equator.

The classic experiments on birds that demonstrate the use of an inclination compass are typical of those that have been done on other species. For example, the migratory restlessness of European robins remains oriented in the proper direction even when the birds have no visual cues. When the magnetic world that the birds experienced was reversed by switching the polarity (north and south poles) of an experimental field, there was no effect on their orientation. However, the birds reoriented if the inclination in the experimental field was altered. It is interesting that these birds were not able to orient according to horizontal field lines, that is, field lines that were horizontal to the earth's surface. Horizontal field lines occur around the equator. A bird could determine the north-south axis in a horizontal field, but without the inclination it would not know which direction is toward the equator (W. Wiltschko and Wiltschko 1972). Just how a migrant deals with this problem while in the equatorial regions is not known. It is assumed that the birds use factors other than the geomagnetic field to maintain their orientation (W. Wiltschko and Wiltschko 1996).

The results of an experiment on free-flying homing pigeons are also consistent with the idea that a bird's magnetic compass is based on the inclination of the magnetic lines of force. Small Helmholtz coil hats were fitted onto the heads of homing pigeons (Figure 10.21a). A Helmholtz coil is a device that generates a magnetic field when an electric current runs through it. The magnetic field experienced by the birds can be altered by reversing the direction of current flow

declination of the earth's magnetic field. Because the declination is small in most places, usually less than 20°, magnetic north is usually a good indicator of geographic north. The declination is, of course, greatest near the poles.

Several aspects of the earth's magnetic field vary in a predictable manner and could, therefore, provide directional cues. One aspect is polarity. The magnetic north pole is called the positive pole, and the magnetic south, the negative pole. The second aspect is the course of the lines of force. These leave the magnetic south pole vertically; curve around the surface of the earth; become level with the surface at the magnetic equator; and reenter the magnetic north pole, going straight down. The angle of inclination, or dip, of the magnetic field is the angle that the line of force makes with the horizon. The angle of inclination is steepest (vertical) near the poles and near zero (horizontal) near the equator. The third aspect that varies predictably is the intensity (or strength) of the geomagnetic field. It is greatest at the poles and least at the equator.

Thus we see that the polarity, inclination, and intensity of the earth's magnetic field vary systemati-

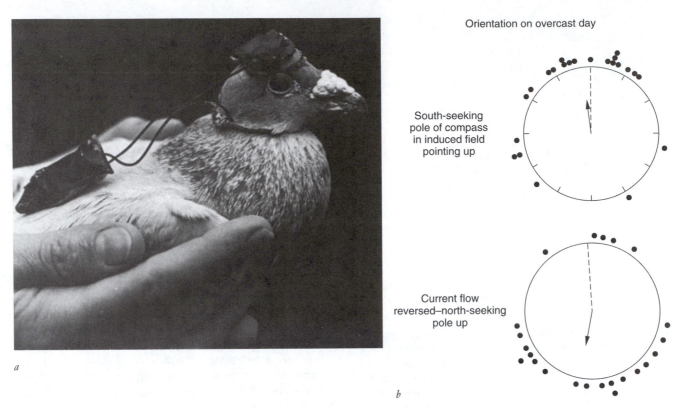

Orientation on overcast day

South-seeking
pole of compass
in induced field
pointing up

Current flow
reversed–north-seeking
pole up

a

b

FIGURE 10.21   (*a*) A pigeon with a Helmholtz coil, a device that generates a magnetic field, on its head. (*b*) The magnetic field experienced by the pigeon can be altered by changing the direction in which the electric current runs through the coil. On overcast days, when the birds could not use the sun as a compass, the magnetic field influenced their orientation. They oriented as if they interpreted north as the direction in which the magnetic lines of force dip toward the earth. Each dot indicates the direction in which a bird vanished from sight after being released. The arrow in the center indicates the mean vanishing bearing. (Modified from data of Walcott and Green 1974.)

through the coil. On cloudy days, when the pigeons relied on magnetic cues rather than their sun compass, they oriented as if they considered north to be the direction in which the magnetic lines of force dip into the earth. Those birds that experienced the greatest dip in the magnetic field in the north, as it is in the normal geomagnetic field, headed home. In contrast, the birds that experienced the greatest dip in the magnetic field in the south were misdirected by the reversed magnetic information and headed directly away from home (Figure 10.21b) (Visalberghi and Alleva 1979; Walcott and Green 1974).

There are also some indications that certain animals respond to the small differences in the intensity of the geomagnetic field. Among these animals are bees (Kirschvink et al. 1997; Walker and Bitterman 1989), homing pigeons (Keeton, Larkin, and Windsor 1974; Kowalski, Wiltschko, and Fuller 1988), sea turtles (Lohmann and Lohmann 1996a), and the American alligator (Rodda 1984). If changes in magnetic intensity can be sensed, the gradual increase in strength between the equator and the poles could also serve as a crude compass.

THE MAGNETIC COMPASS

If we keep in mind that orientation is essential to the survival of migrating or homing animals, it should not come as a surprise that orientation is affected by the interaction of many cues, as well as many factors, including experience, species differences, and amount of stored energy. We will separate some of these interacting variables to try to understand just how animals remain oriented when faced with the real problems of navigating.

The Magnetic Compass and Bird Navigation

As we have seen, there is evidence that birds use the earth's magnetic field as a compass. They seem to use the horizontal component of the field to identify the north-south axis and the angle of inclination to determine whether they are headed toward the pole or the equator.

Migratory birds inherit a program that tells them to travel in a certain geographical direction for a certain amount of time. Birds follow the migratory program with the help of two types of orientation cues: (1) the earth's magnetic field and (2) celestial cues, such as the sun, stars, or the pattern of polarized light at sunset (W. Wiltschko, Wiltschko, Munro, and Ford 1998). These cues interact in some interesting, flexible ways.

Development of the Magnetic Compass
The primary magnetic compass is innate. In regions where magnetic information is adequate, in Germany, for instance, a magnetic compass can develop in birds that have never seen the sky. Groups of young garden warblers (W. Wiltschko and Gwinner 1974), pied flycatchers (Beck and Wiltschko 1982), and blackcaps (Bletz et al. 1996) that were hand-reared without access to celestial cues were able to choose their correct migratory direction.

Although the magnetic compass is innate, during development the preferred magnetic direction may be set by cues from celestial rotation—the stars during the night and the patterns of skylight polarization during the day (reviewed in Able and Able 1993; W. Wiltschko and Wiltschko 1991). Why is calibration necessary? Celestial cues may be more accurate indications of geographic north, but they are not always available. Magnetic cues are always available, but the difference between magnetic north and geographic north varies with latitude. A young migrant is likely to live in regions fairly close to the poles. The closer one is to the poles, the greater the distance between the magnetic north pole and the geographic north pole. Thus, near the poles, the magnetic and celestial compasses indicate north in slightly different directions. The center of celestial rotation always indicates geographic north, and it is used to calibrate the magnetic direction for a specific locale.

Growing up near the poles poses a special problem for the development of a magnetic compass in some species. In extreme polar regions, the magnetic field lines are nearly vertical and, therefore, provide scanty directional information. Both celestial cues and a magnetic compass may have to be present simultaneously for the magnetic compass of birds raised in polar regions to determine the correct migratory direction. As you may recall, experiments in Germany, where the inclination of the magnetic field is 66°, showed that the magnetic compass of young blackcaps and pied flycatchers could develop normally in the absence of celestial cues. But this was not the case for the same species in Latvia, a more northern region, where the inclination of the magnetic field is 73°. In Latvia both magnetic and celestial cues were necessary. Pied flycatchers that were hand-raised in the normal geomagnetic field but without celestial cues showed a bimodal orientation during the autumn migration, in this case along the southwest-northeast axis. Birds that were raised with a view of the daytime sky, but without magnetic compass information, oriented in the same bimodal pattern. The bimodal pattern resulted because a bird would choose southwest on one night and switch to northeast on another night. Apparently, the birds could determine the correct migratory axis from the horizontal component of the geomagnetic field, but they didn't know which end was southwest. Only those birds that saw the celestial rotation during the daytime and, at the same time, were exposed to correct magnetic information could orient in the appropriate migratory direction (Weindler et al. 1998).

Other studies have suggested that when young birds leave on their first migration, celestial rotation triggers a tendency to move away from its center, which is south in the northern hemisphere and vice versa (Weindler, Wiltschko, and Wiltschko 1996). Presumably, that is enough to point a bird to the correct end of its migratory axis.

The difference in the role of celestial cues in the development of the magnetic compasses of birds growing up in Germany versus those raised in Latvia shows an effect of early experience with the geomagnetic field due to location. There are also species differences. For example, snow buntings (*Plectrophenax nivalis*) begin life even closer to the magnetic north pole than do blackcaps or pied flycatchers. One might expect that the near vertical (89°) lines of force at the snow buntings' breeding areas would render a magnetic inclination compass useless. However, snow buntings can orient there without any celestial cues (Sandberg, Bäckman, and Ottosson 1998).

We see a somewhat different interaction of magnetic compass and celestial cues in the homing ability of pigeons. The magnetic compass develops when they are very young. In fact, on their first flight, homing pigeons seem to rely on it exclusively. As the birds gain flying experience, however, they learn to use other cues. When they are about three months old, they begin to use the sun compass, provided they have been able to see the sun (R. Wiltschko 1983). Once established, however, the sun compass becomes the primary orientation mechanism (R. Wiltschko and Wiltschko 1981), and the magnetic compass serves mainly as a backup system for cloudy days, when the sun cannot be seen (Keeton 1971). Even on sunny days, however, the magnetic compass is not turned off completely. Instead there is some interaction between the two systems (Visalberghi and Alleva 1979; Walcott 1977). For instance, the shift in orientation following a clock-shift experiment is often much smaller than would be expected if the birds were using only the sun compass (R. Wiltschko, Kumpfmiller, Muth, and Wiltschko 1994).

Use of the Magnetic Compass During Migration
We have seen that the migratory route is genetically coded with respect to the earth's magnetic field. Because the magnetic compass of birds is an inclination compass, migrants from either the northern or

the southern hemisphere might use the same migratory program—toward the equator (where the lines of force are more horizontal) in the fall and toward the pole (where the lines of force are more vertical) in the spring (R. Wiltschko and Wiltschko 1999b; W. Wiltschko and Wiltschko 1996).

We might wonder, then, how a bird from northern regions that crosses the equator during migration can continue to fly south in the southern hemisphere. To continue flying in the same geographical direction when the equator is crossed, the birds must reverse their migratory direction with respect to the inclination compass: They must now fly "poleward" instead of "equatorward." It turns out that experience with the horizontal magnetic field around the equator is the switch that causes the birds to begin flying "poleward" (W. Wiltschko and Wiltschko 1996).

The interactions between cues from the earth's magnetic field and celestial rotation may continue into adulthood. Birds migrate from higher latitudes—where "true" geographic north, as determined by the center of celestial rotation, and magnetic north point in different directions—to lower latitudes, where the two cues are in greater agreement. During the migration of at least one species, the Savannah sparrow, it seems that the magnetic compass is continuously recalibrated to true north by information from celestial rotation (Able and Able 1996). Furthermore, the appearance of the sky changes as the bird progresses in its migration, and the new stars that come into view are calibrated using the earth's magnetic field (R. Wiltschko and Wiltschko 1999b).

It is interesting to note that the sensitivity of the magnetic compass of birds corresponds to the strength of the earth's magnetic field. A bird generally does not respond to magnetic fields that are much stronger or weaker than that which is typical in the area where it has been living. In fact, the range of intensities to which a bird may respond on a given day is usually narrower than those that it might experience during migration. However, it seems that the range of sensitivity may be adjusted by exposure to a field of a new strength for a period of time. Thus, responsiveness may be fine-tuned during migration (W. Wiltschko 1978; R.Wiltschko and Wiltschko 1999a).

The Magnetic Compass of Sea Turtles

Some sea turtles travel tens of thousands of kilometers during their lifetimes, a feat that can require continuous swimming for periods of several weeks, with no land in sight. As a loggerhead sea turtle, *Caretta caretta* (Figure 10.22), makes its way across the featureless Atlantic Ocean from the coast of Florida (perhaps to the Sargasso Sea and back), it is guided by the earth's magnetic field, in addition to the pattern of waves (Lohmann 1992). Laboratory experiments have

FIGURE 10.22   A hatchling loggerhead sea turtle. These turtles may use the earth's magnetic field to guide their travels through the open ocean.

revealed that loggerhead hatchlings use the earth's magnetic field for orientation. The hatchlings swim toward *magnetic* northeast in the normal geomagnetic field and continue to do so when the field is experimentally reversed (Figure 10.23) (Lohmann 1991). And, similar to a bird's magnetic compass, a sea turtle's is based on the inclination of the magnetic lines of force (Light, Salmon, and Lohmann 1993).

How does a sea turtle use this magnetic compass to maintain its correct heading across the featureless ocean? When sea turtles first enter the ocean, they simply swim into the waves to maintain an offshore heading. Near the shore, the waves come directly toward land, so swimming into the waves takes the turtles out to sea. The course that is initiated by swimming into the waves is later transferred to the magnetic compass. In the open ocean, waves can no longer serve as a cue because they can come from any direction. Here, sea turtles maintain the same angle with the magnetic field that they assumed while swimming into the waves. In this way, they stay on course. Simultaneous experience with both cues seems to be important. This was revealed in a recent experiment in which hatchling loggerhead sea turtles swam into surface waves in tanks for either 15 minutes or for 30 minutes. Their orientation was then tested in still water and in a magnetic field. Only those hatchlings with 30 minutes of experience swimming into waves in a magnetic field were able to maintain their orientation in still water (Goff, Salmon, and Lohmann 1998).

IS THERE A MAGNETIC MAP?

As we've seen, true navigation requires not only a compass but also a map. The map is necessary to know one's

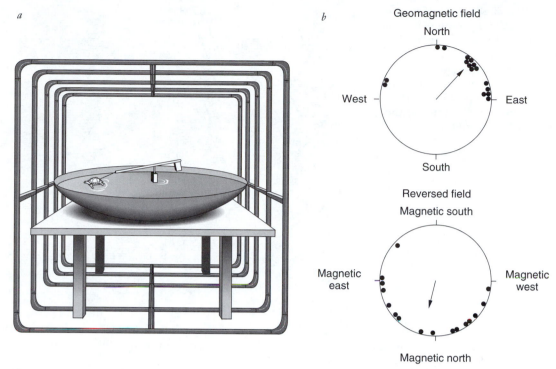

FIGURE 10.23 A demonstration of the ability of loggerhead sea turtle hatchlings to orient to magnetic fields. (*a*) A sea turtle is harnessed in a small tank so that its swimming direction can be determined. A coil that can alter the magnetic field experienced by the turtles surrounds the tank. (*b*) In the earth's magnetic field the turtles orient toward magnetic northeast. When the field is reversed, the hatchlings still orient to magnetic northeast, even though this is in the opposite geographic direction. (From Lohmann 1991.)

location relative to home, and then a compass is needed to guide the journey in a homeward direction. So far, we have discussed reasons for thinking that the earth's magnetic field serves as a compass. But a bird such as the homing pigeon also needs a map. Although the nature of the pigeon's map is unknown, it has been popular to speculate on the possibility that the earth's magnetic field might be its basis (Gould 1980, 1982; Lednor 1982; B. R. Moore 1980; Walcott 1980, 1982). Indeed, there are characteristics of the geomagnetic field that could serve as a map.

Some of the magnetic effects on pigeon homing seem to be more than interference with the magnetic compass and, therefore, may support the idea of a magnetic map. One example is the disorientation of pigeons released in magnetic anomalies, places where the earth's magnetic field is extremely irregular. We might expect that a pigeon relying on the predictable changes in the geomagnetic field would become confused in areas where the field is abnormal. We do find that some magnetic anomalies disorient pigeons even under sunny skies, when presumably they would be using the sun as a compass (Frei 1982; Frei and Wagner 1976; Kiepenheuer 1982; Wagner 1976; Walcott 1978). A perfect compass (the sun) cannot help if the map is

messed up. This suggests that the geomagnetic field may be more than just a compass. As you can see in Figure 10.24, some birds released at magnetic anomalies appear to follow the magnetic topography, usually preferring the magnetic valleys, where the lower field strength is closer to home values.

Another observation consistent with the idea of a geomagnetic map is the shift in the initial bearings of pigeons that occurs when the geomagnetic field increases during magnetic storms (Frei 1982; Keeton, Larkin, and Windsor 1974). If the regular increase in the earth's magnetic field from the equator to the poles somehow told a bird its latitude, we would expect a magnetic storm would cause a bird to read its magnetic map incorrectly and use its compass to head off in the wrong direction.

Additional evidence for a magnetic map in migratory birds comes from experiments in which Australian silvereyes (*Zosterops l. lateralis*) were exposed to a strong magnetic pulse to determine its effect on orientation. As discussed in the Appendix, two general categories of mechanisms have been proposed for magnetoreception—light-dependent and magnetite-based mechanisms. Magnetite may play a role in both compass orientation and map sense. Birds have deposits of

FIGURE 10.24   The flight paths of pigeons in magnetic anomalies. In some places the geomagnetic field is highly irregular. Pigeons released in these areas may be completely disoriented, even on sunny days. The paths of these pigeons seem to follow the magnetic valleys, where the field strength is closer to the value at the home loft. (From Gould 1980.)

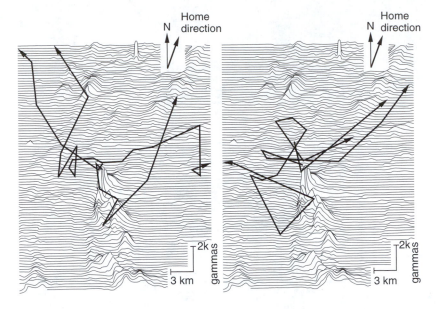

magnetite in their heads, especially in the ethmoid region and in the beak (Beason and Nichols 1984; Holtkamp-Rötzler et al. 1997; Walcott, Gould, and Kirschvink 1979). A strong magnetic field will alter the magnetization of the magnetite deposits. When adult silvereyes that were flying northward from Tasmania during their autumn migration were exposed to a strong magnetic pulse, their orientation was shifted clockwise by about 90° toward the east (W. Wiltschko, Munro, Beason, and Wiltschko 1994). Similar results were obtained when adult silvereyes were exposed to a strong magnetic pulse during the spring migration (W. Wiltschko, Munro, Ford, and Wiltschko 1998). These observations support the idea that a magnetite-based receptor plays a role in orientation, but they don't indicate whether it is involved in the compass sense or the map sense. However, when juvenile silvereyes were exposed to a magnetic pulse shortly after fledging, before they started to migrate, it had little effect on their orientation. They continued to orient in their normal autumnal migratory direction. Unlike adult migrants, which have established a navigational map during previous migrations, the juveniles rely on an innate migratory program that heads them in the appropriate compass direction for their first migration. Because a magnetic pulse disrupts orientation in adults but not in juveniles, it is thought that the earth's magnetic field provides at least one coordinate in the navigational map of adults (Munro et al. 1997).

It is also possible that the sea turtles use a magnetic map to locate their position. We do know that they can detect both the inclination and the intensity of the earth's magnetic field. Both these features vary across the earth's surface, and they vary in different directions. Thus, they form a grid that could be used for position finding. It is possible that adult sea turtles use such a map to locate an isolated target such as a nesting beach (Lohmann and Lohmann 1996a, 1996b).

CHEMICAL CUES

In this section we will focus on the use of olfactory cues for orientation during homing. We will discover that salmon are guided to the stream where they hatched by chemical landmarks, and we will examine the more recent suggestion that pigeons may also use olfactory cues when homing.

OLFACTION AND SALMON HOMING

One of the most remarkable stories in the annals of animal behavior concerns the travels of the salmon. Salmon hatch in the cold, clear fresh water of rivers or lakes and then descend from the streams that flow from those areas and swim to sea, fanning out in all directions. Once they reach the ocean, depending on the species, they may spend one to five years there until they reach their breeding condition. Now large, glistening, beautifully colored creatures, they head from their feeding grounds back through the trackless sea to the very river from which they came. When they reach the river, they swim upstream, always turning up the correct tributary until they reach the very one where they spent their youth.

Although navigation in the open seas appears to depend on the integration of several sensory cues, including magnetism, sun compass, polarized light, and perhaps even odors, navigation up the rivers is primarily based on olfactory cues (reviewed in Dittman and Quinn 1996). During their upstream migration, the salmon follow a chemical trail back to the tributary where they hatched. When they come to a fork in the river, they may swim back and forth across the two branches. If they mistakenly swim up the wrong branch and lose the scent of the home stream, they drift downstream until the scent is encountered again. Then, they usually take the correct route.

Sensory deprivation experiments have demonstrated the importance of olfaction in salmon homing. Blinding the fish had no effect, but plugging their nasal cavities impaired their ability to home correctly. Coho salmon (*Oncorhynchus kisutch*) were trapped shortly after they had made their choice of forks in a Y-shaped stream. The nasal cavities of half of those caught in each branch were plugged. The other half were untreated. All the fish were then released downstream from the fork and allowed to repeat their upstream migration. Whereas 89% of the control fish returned to the branch where they were originally captured, only 60% of the fish with nose plugs made the correct choice (Wisby and Hasler 1954). In another study, a fish with its nose plugged swam with others of its kind to the opening of its home pond. However, unable to smell the special characteristics of its home waters, it did not enter the pond (Cooper et al. 1976).

The salmon learn the home-stream odors during a single sensitive period of development, the parr-smolt transformation. Andrew Dittman and his colleagues exposed groups of hatchery-reared salmon to an artificial odorant, β-phenylethyl alcohol (PEA), during different, specific developmental stages. When the salmon were mature, their behavioral responses to these odors were tested. One test involved a two-choice arena. In the absence of PEA, significantly more salmon entered arm A of the arena, regardless of previous exposure to PEA or the timing of PEA exposure. Next, PEA was metered into arm B, the less preferred arm of the arena. As you can see in Figure 10.25, the responses of control fish that had no previous exposure to PEA and fish that had been exposed to PEA during the alevin or parr stages of development were indistinguishable from their responses in the absence of PEA. However, the behavior of mature salmon that had been exposed to PEA during the smolt stage was different from that of other salmon. Significantly more of the exposed salmon entered the PEA-scented arm when PEA was present than when it was absent. The parr-smolt transformation was also shown to be the critical time period for olfactory imprinting when the fish were tested in a natural stream in which PEA was added to one fork of a river (Dittman, Quinn, and Nevitt 1996). At least part of the cellular basis of olfactory imprinting in salmon is due to odorant-induced changes in the olfactory receptor cells (Nevitt et al. 1994).

Although it seems certain that salmon use olfactory cues when migrating upstream, the source of the specific stream odor is still being debated. Two hypotheses have been suggested. One is that fish are guided to their

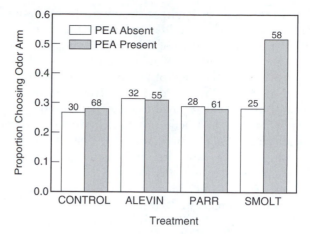

FIGURE 10.25   The sensitive period for olfactory imprinting in coho salmon is during the parr-smolt transition. Groups of young salmon were exposed to an artificial odorant, PEA, during different, specific developmental stages. The mature salmon were tested in a two-arm arena. Control fish that had never been exposed to PEA and fish that had been exposed to PEA during the alevin or the parr developmental stages responded in the same way whether or not PEA was present in one arm of the test arena. In contrast, significantly more of the salmon that had been exposed to PEA during the smolt stage chose the arm containing PEA when it was present than when it was absent. (From Dittman, Quinn, and Nevitt 1996.)

natal stream, the one in which they were hatched, by its characteristic fragrance from the rocks, soil, and plants. The fish imprint on (learn) the unique combination of stream odors when they are young and follow the scent to their natal stream (Hasler and Scholz 1983; Hasler, Scholz, and Horrall 1978; Hasler and Wisby 1951). This hypothesis was later extended by the suggestion that salmon actually learn the sequence of stream odors encountered during their outward migration (Brannon and Quinn 1990; Harden-Jones 1968). An alternative hypothesis is that the odors are pheromones (communicatory chemicals, discussed in Chapter 18), mucus, or fecal material from other fish of the same species. In this case, the fish would not learn the characteristic odor of the home stream but rather that of their colleagues (Nordeng 1971, 1977).

In a series of experiments designed to test the imprinting hypothesis, Arthur Hasler and his coworkers demonstrated that salmon are able to imprint on a chemical not usually found in natural waters and that they use this cue to locate their home stream. Groups of young salmon (smolts) were exposed to minute quantities of either morpholine or PEA. A third group, the controls, were not exposed to either chemical. The fish were marked and released in a lake that was equidistant from the two streams that would later contain the test chemicals. Eighteen months later, the mature fish were ready to begin their upstream migration. Morpholine was then added to one of the rivers and PEA to the other. All 19 of the streams along 200 kilometers of shoreline were monitored for marked fish. Only a small percentage of any marked fish returned, but those that did returned to the correct stream. Over 90% of those that returned to a morpholine- or PEA-scented stream had been previously exposed to that chemical. In contrast, the number of control fish visiting one of the scented streams was never more than 31% and was usually much lower (Scholz et al. 1976). In other studies, individual fish were followed during migration (Johnson and Hasler 1980; Madison et al. 1973; Scholz et al. 1975). Fish that had been imprinted on morpholine stopped and milled about in an area that had been treated with morpholine, but they continued to migrate through the same area when morpholine was absent. Fish that had not been previously exposed to morpholine continued to swim through the area even when morpholine was present. The same was true when PEA was used as the chemical label.

Other studies, however, make the pheromone hypothesis quite plausible. Coho salmon can discriminate among their population (Quinn and Tolson 1986) and even their siblings (Quinn and Busak 1985; Quinn and Hara 1986) by using odor cues, and salmon are attracted to water that bears the scent of their own species (Hoglund and Astrand 1973) or their own population (Selset and Doving 1980). Water that contains odors from spawning conspecifics is particularly attractive (Newcombe and Hartman 1973). Furthermore, breeding Atlantic adult salmon (*Salmo salar*) suddenly appeared in a previously barren stream shortly after fry were added (Solomon 1973). These studies suggest that odors from conspecifics may be one part of the symphony of odors characterizing the home stream, but they do not demonstrate that odors from other fish are important or necessary in guiding the upstream migration.

When both cues are present, salmon tend to return to the site they experienced as juveniles instead of to an area containing the odor of conspecifics. In an experiment, groups of juvenile coho salmon were released as smolts. As adults, they returned to their release site. To reach this site, they had to swim within about 100 meters of a hatchery that contained their siblings, as well as other conspecifics. Although the hatchery's effluent contained odors from its captive fish, the migrating salmon ignored these conspecific odors and swam past the hatchery to their release site. Likewise, Atlantic char could not be lured into returning to a nonnatal stream by the presence of adults (Black and Dempsey 1986). Thus, site-specific odors seem to be more important to homing salmon than odors from conspecifics.

Rather, population recognition odors may have functions other than homing. For example, they may be involved in sibling recognition and serve to reduce aggression, increasing inclusive fitness by allowing a sibling to establish a neighboring territory and raise offspring. Alternatively, they might function in mate choice and reduce outbreeding (Courtenay et al. 1997).

OLFACTION AND PIGEON HOMING

No one denies that olfactory cues are of paramount importance during the upstream migration of salmon or that they play some role in pigeon homing. However, not everyone agrees that olfactory cues form the map used by homing pigeons (Able 1996; Papi 1986, 1995; Schmidt-Koenig 1987; Wallraff 1996; R. Wiltschko 1996).

Models of Olfactory Navigation

Two models for olfactory navigation have been suggested. According to Floriani Papi's "mosaic" model, pigeons form a mosaic map of environmental odors within a radius of 70–100 kilometers of their home loft. Some of this map would take shape as the young birds experienced odors at specific locations during exercise and training flights. More distant features of the map would be filled in as wind carried faraway odors to the loft. One odor might be brought by wind from the north and another by wind from the east. The bird would associate each odor with the direction of the wind carrying it. When the wind shifted direction, the odors that arrived first would be closer than those that took longer to arrive (Papi et al. 1972). For instance, a hypothetical pigeon might learn that the sea is to the west, an evergreen forest is south, a large city is north, and a garbage dump is east. If the bird in this example smelled pine needles at its release site, it would assume that it was in the forest south of its loft and would use one of its compasses, perhaps the sun or the earth's magnetic field, to fly north.

Hans G. Wallraff (1980, 1981) has suggested a "gradient" model of olfactory navigation that assumes that there are stable gradients in the intensity of one or more environmental odors. Then, wherever it was, the bird would determine the strength of the odor and compare it to the remembered intensity at the home loft. Unlike the mosaic model, which requires only that the bird make qualitative discriminations among odors, the gradient model demands that the bird make both qualitative and quantitative discriminations. Reconsider the previous example. The smell of the ocean might form an east-west gradient, and the fragrance of the evergreen forest might generate a north-south gradient. If the bird in the previous example smelled the air at a release site and determined that the scent of the sea was stronger but the smell of the forest was weaker than at the home loft, it would determine that its current position was northwest of home.

Tests of the Models

These models of olfactory navigation have stimulated intensive research, but it is still not clear how important odors are in the navigation of homing pigeons. Let us see how various researchers have approached the question.

Depriving Birds of Their Sense of Smell One approach in testing olfactory hypotheses is to deprive the pigeon of its sense of smell and observe the effect on its orientation and homing success. These smell-blind (anosmic) pigeons are less accurate in their initial orientation, and fewer return home from an unfamiliar, but not from a familiar, release site. Regardless of its effect on orientation, olfactory deprivation always delays the bird's departure from the release site (Able 1996). These results are consistent with the idea that olfaction plays an important role in pigeon homing.

Besides its effect on the pigeon's sense of smell, perhaps olfactory deprivation affects another behavior, one not primarily controlled by olfaction, and this other behavior alters homing performance. Suppose the procedures that impair the sense of smell also affect the pigeons' motivation or their ability to process information. Although possible, the evidence does not support these possibilities. Anosmic pigeons home as well as control pigeons when they are released from familiar sites. Thus the procedures do not seem to affect the birds' motivation to return home. Furthermore, pigeons whose sense of smell is temporarily blocked by an application of zinc sulfate to the olfactory epithelium have problems in returning home from unfamiliar locations, but they perform as well as controls in a spatial memory task that does not involve homing (Budzynski, Strasser, and Bingman 1998). Could it be that some other sense, say sensitivity to magnetism, is blocked along with olfaction? The discovery of cells thought to be magnetoreceptors in the olfactory epithelium of rainbow trout makes this an intriguing possibility (Walker et al. 1997). So, let us consider some of the other ways in which the idea of an olfactory map has been tested.

Distorting the Olfactory Map Another way of testing olfactory hypotheses is to manipulate olfactory information to distort the bird's olfactory map. This has been done by deflecting the natural winds to make it seem that odors are coming from another direction. The deflector lofts used in these experiments typically have wooden baffles that shift wind flow in a predictable manner (Figure 10.26). For instance, wind from the south might be deflected so that it seemed to come from the east. A pigeon in this loft would form an olfactory map that was shifted counterclockwise by 45°. When it was released south of its loft, we would expect it to interpret the local odors as being east of its loft and fly west to get home.

Unlike the olfactory deprivation studies previously discussed, deflector loft experiments have shown consistent shifts in the orientation of homing pigeons in every country in which tests have been conducted (Baldaccini et al. 1975; Kiepenheuer 1978; Waldvogel et al. 1978; Waldvogel and Phillips 1982). However, there are reasons to believe that the shift in orientation observed in pigeons from deflector lofts might be due to something other than a distorted olfactory map. We would expect pigeons that were temporarily prevented from smelling at the time of their release to be unable to read their olfactory map and to orient randomly. But this is not the case: The orientation of anosmic pigeons from deflector lofts is still shifted (Kiepenheuer 1979). In light of this, it has been suggested that the baffles in these lofts also deflect sunlight and that the consistent shift in pigeon orientation is caused by an alteration in the sun compass (Phillips and Waldvogel 1982; Waldvogel and Phillips 1982; Waldvogel, Phillips, and Brown 1988).

Manipulating Olfactory Information Although the interpretation of olfactory deprivation and deflector loft experiments is quite controversial, the experiments in which olfactory information predictably alters the orientation of pigeons remain as unshaken support for an olfactory hypothesis. For example, the orientation of pigeons was influenced by their experience with an unnatural odor, benzaldehyde (Figure 10.27). Pigeons were kept in lofts where they were fully exposed to the wind. The experimental birds were exposed to an air current coming from a specific direction and carrying the odor of benzaldehyde in addition to the natural breezes. We would expect these pigeons to incorporate this information into their olfactory maps. The control birds were exposed to only the natural winds, so they would not have an area with the odor of benzaldehyde in their olfactory map. All the birds were exposed to benzaldehyde while they were transported to the release site and at that site. The experimental birds took off in a direction opposite to that from which they had experienced benzaldehyde at the loft. In other words, they oriented as if they used an olfactory map that contained an area scented with benzaldehyde. If the release site did not smell of benzaldehyde, the experimental birds were homeward oriented. The control birds were not confused by the smell of benzaldehyde at the release site and flew home. Since benzaldehyde was not part of their olfactory map, they did not associate it with a particular direction. They used other cues to guide them home (Ioalè, Nozzolini, and Papi 1990).

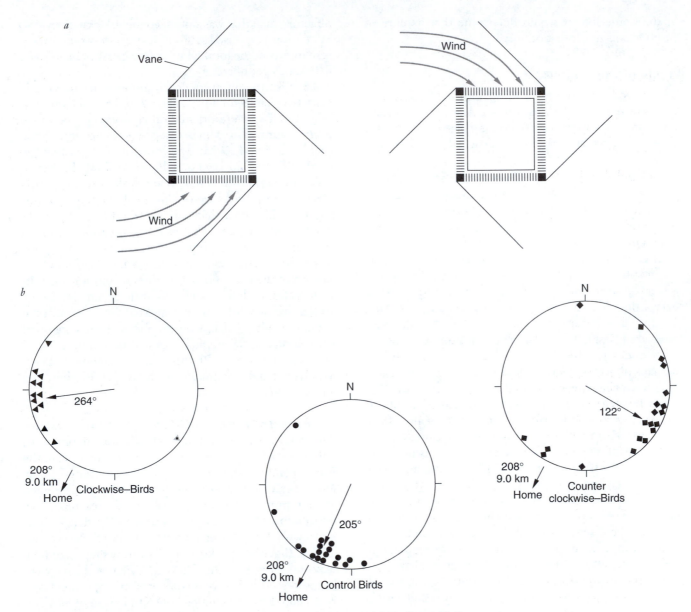

FIGURE 10.26   Deflector lofts shift the orientation of pigeons. (*a*) Deflector lofts have baffles that shift the apparent direction of the wind by 90°. Pigeons living in deflector lofts should form shifted olfactory maps. (*b*) The vanishing directions of these pigeons is shifted by about 90°. The dots at the periphery of the circle denote the direction in which the pigeon flew out of sight. The arrow within the circle indicates the mean bearing of all birds. The light and dark data points refer to the responses of birds released on different dates. Although a shift in orientation is reported in all deflector loft experiments, it may be due to the deflection of light rather than a shift in the olfactory map. (Data from Baldaccini et al. 1975.)

SOUND CUES AND ECHOLOCATION

The world is a noisy place, and most animals take advantage of some sounds for orientation. Many, in fact, intentionally produce sounds that help others pinpoint the sender's position. Others unintentionally make noises as they move about and thereby reveal their location. For instance, the sounds of a scampering mouse guide the deadly strike of a barn owl. Sounds from nonbiological sources may also convey directional information. Waves crashing on the shore or wind whistling through a mountain range might help tell an animal where it is.

Most animals must be fairly close to the source to be able to use these sound cues. However, such sources

Geographic orientation of apparatus

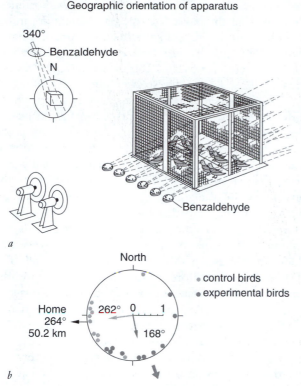

a

b

North

control birds
experimental birds

Home
264°
50.2 km

262° 0 1

168°

FIGURE 10.27 The results of an experiment that manipulated a pigeon's olfactory information. (*a*) The experimental pigeons were kept in a loft that was exposed to natural odors, as well as to a breeze carrying the odor of benzaldehyde from a source northwest of the loft. Control birds were exposed to only natural odors. While they were transported to the release site, all birds were exposed to the odor of benzaldehyde. (*b*) The orientation of the experimental birds, but not the control birds, was altered by exposure to benzaldehyde. The initial orientation of control birds was homeward. However, the initial orientation of experimental birds was toward the southeast, as would be expected if they had interpreted the odor of benzaldehyde as an indication that the release site was northwest of the loft. The experimental birds oriented as if they formed an olfactory map containing an area with the odor of benzaldehyde. (Data from Ioalè, Nozzolini, and Papi 1990.)

also generate low-frequency sound, infrasound, that could theoretically carry information on distant features of the landscape over thousands of kilometers to the animals that can hear these deep tones. If a migrant bird can hear infrasound, it could put this ability to good use. While flying high above the Mississippi Valley, for example, it could conceivably hear a thunderstorm in the Rockies or the waves pounding the Atlantic coast. We know that pigeons are among those birds that can hear infrasound (Kreithen and Quine 1979; Schermuly and Klinke 1990), but we do not know whether they use this information for orientation.

Perhaps the most amazing use of sound for orientation is echolocation—a process by which the animal

FIGURE 10.28 An echolocating bat produces loud pulses of sound (indicated here as semicircles radiating outward from the bat's mouth) and then analyzes the returning echoes (indicated here as semicircles radiating from the insect toward the bat) to get an acoustic picture of its surroundings.

makes sounds and analyzes the returning echoes to create an acoustic picture of its surroundings (Figure 10.28). Because it is among the most fascinating uses of sound for orientation, we will focus on it in more detail.

ECHOLOCATION IN BATS

When making a list of things that go bump in the night, do not include bats. Many of us have marveled at bats that are darting between trees at daredevil speeds, catching tiny insects on the wing while cloaked in the darkness of night (Figure 10.29). Although not all species of bats use echolocation, most of those in the United States, Canada, and Europe do.

As a bat cruises through the air in search of a meal, it emits a series of high-frequency (ultrasonic) pulses of sound. We cannot hear sounds at such a high pitch, so the hunting bat seems silent to us. A bat's cries generally fall somewhere in the frequency range of 10–200 kilohertz. These very short wavelengths are ideally suited for detecting small- or medium-size insects because the wavelengths are not much longer than the prey. However, the use of high-frequency sounds is an evolutionary tradeoff. They allow the bat to locate small prey but limit the range over which the prey can be detected because high-frequency sounds do not carry as well as those with low frequencies.

Although we cannot hear them, the cries of most species of bats are intensely loud. The softest (least intense) sound that most humans can hear is about 0.0002 dyne (a unit of energy) per centimeter squared.

FIGURE 10.29 **Bats dominate the night sky. In spite of the darkness, insectivorous bats perform intricate maneuvers to catch small insects in flight. They detect their prey and avoid colliding with obstacles by echolocation.**

Among the quietest of bats is the whispering bat, whose cry is about 1 dyne per centimeter squared. This is roughly as loud as someone whispering at a distance of 10 centimeters. In contrast, the cries of other bats may be as loud as a nearby jet engine, 200 dynes per centimeter squared (Fenton 1983). So our quietest night might be a cacophony for other species.

A bat alters the rate of its cries during a hunt (Griffin, Webster, and Michael 1960). When it is searching for prey, it produces about 20 sound pulses per second. This rate of pulsing allows the bat to get a general picture of its surroundings. But once an object is detected, the rate of pulsing is increased to roughly

50 to 80 cries per second and each pulse is shortened. By increasing the pulse rate, the bat gains more information about its prey and can track it more accurately (Suga 1990). Finally, just before the prey is captured, the pulse rate increases to 100 or even 200 pulses per second, depending on the species (Figure 10.30)

There is great variety in the structure of cries from different species of bats, but the pulses fall into three general categories (Figure 10.31). One category is the frequency-modulated (FM) calls. These sweep downward through a broad range of frequencies, usually covering at least one octave. The more frequencies in the outgoing pulse, the greater is the number of altered frequencies in the echo that can be analyzed to provide more information about the target. In other words, the amount of information about the target conveyed by the echo increases with the bandwidth of the call. The FM calls are, therefore, well suited to determine the size, shape, and surface texture of the target. Constant-frequency (CF) calls, which contain a single frequency, are the second category of signals. Although this type of call does not reveal as many details about the target, it allows the bat to detect an object and determine whether it is moving toward or away from the bat and at what rate. CF calls allow a bat to detect an insect at greater distances than do FM calls. In addition, they allow bats to identify objects as insects by their fluttering wings. The third category includes calls that combine a CF component with an FM component. Typically these have a constant tone followed by a downward sweep (Neuweiler 1984).

These types of calls are analyzed differently. The interval between the emission of an FM pulse and the return of the echo tells the bat the distance to the target. If the interval gets shorter, the bat knows it is getting closer (Simmons 1973, 1979). In contrast, when a bat produces a CF pulse, it determines whether it is

FIGURE 10.30 **The increasing rate of a bat's echolocation calls at different stages of the hunt. During the search phase, the bat scans the area, producing about 20 sound pulses per second. When prey is detected, the pulse rate more than doubles so the bat can get a better picture and track the prey more accurately. During the final stages, the pulse rate increases dramatically to about 100–200 cries per second. (From Simmons, Fenton, and O'Farrell 1979.)**

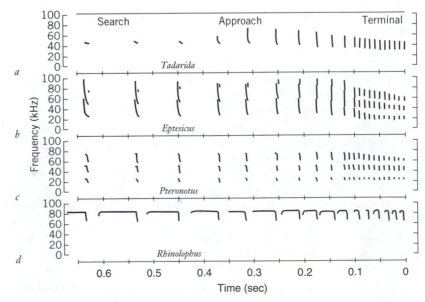

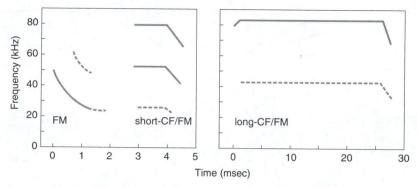

FIGURE 10.31   Three categories of echolocation sounds produced by bats. The sounds emitted by bats differ according to the species and the situation. Frequency-modulated (FM) calls sweep downward through a broad range of frequencies, usually covering at least one octave. Constant-frequency (CF) calls contain a single frequency. Some calls begin with a constant frequency and end with a modulated frequency. (From Simmons, Howell, and Suga, 1975.)

closing in on its prey by measuring the Doppler shift. The Doppler shift is the apparent change in frequency of a sound source that is moving relative to the listener (or vice versa). You have undoubtedly experienced the Doppler shift when listening to the whistle of a passing train or the whine of a race car at the track. Although the actual pitch of the sound remains constant, it appears to become higher as the source moves toward you and to drop as the source moves away from you. This is because the pitch of the sound you hear depends on the number of sound waves per second that strike the eardrum. You can imagine the sound waves bumping into one another and becoming compressed as the source moves toward you. For this reason you hear a higher frequency (pitch) sound as the source moves closer. As the source moves away, the sound waves are stretched out and the apparent pitch is lower. Bats that are using CF calls usually adjust the pitch of their cry so that the Doppler-shifted echo is in the range of frequencies to which their ears are most sensitive (Schuller, Beuter, and Rübsamen 1975; Simmons 1974).

Most species of bats specialize in producing a single type of pulse, one suited to the style of foraging typical of the species, but some can adjust their echolocation strategy to the situation. The Mexican free-tail bat (*Tadarida brasiliensis*) forages in open spaces, where insects can be detected against an uncluttered background and chased with little danger of bumping into obstacles. When searching for prey, this bat uses a CF pulse that is well suited for simply detecting the prey. The pulse becomes increasingly frequency modulated once the prey is detected. The FM signal provides information about the insect that helps the bat decide whether to attack, and if the answer is yes, it guides the pursuit. Species that forage within vegetation must distinguish the echo of an insect fluttering by from the clutter of echoes caused by twigs and leaves. These bats

typically use a signal made up of a long CF pulse with one or two FM components. With a pulse of this structure, the bat can identity its prey as a fluttering target amid a background of stationary objects and track its movements. Other bats are gleaners that pick up prey from a surface. They must distinguish a slowly moving or stationary object against a highly reflective background. The fishing bat, *Noctilio leporinus*, detects ripples in the water caused by small surface-feeding fish. Its echolocation signals switch from CF to CF/FM pulses while searching for prey to an FM cry once a fish has been detected. This echolocation strategy helps it notice a disturbance against a uniform background and then determine whether that disturbance is caused by a fish (Neuweiler 1984, 1990; Simmons, Fenton, and O'Farrell 1979; Simmons, Howell, and Suga 1975).

We are most familiar with echolocation in bats, but several other groups of animals share this ability (reviewed in Vaughan 1986)—the toothed whales and dolphins, several species of shrews, and a few species of birds (oilbirds and cave swiftlets).

ELECTRICAL CUES AND ELECTROLOCATION

Electrical cues have a variety of potential uses for those organisms that can sense them. As we will see in Chapter 12, certain predators use the electrical cues given off by living organisms to detect their prey. In addition, electrical fields generated by nonliving sources, such as the motion of great ocean currents, waves and tides, and rivers, could provide cues for navigation. Although there is currently no evidence that migrating fish such as salmon, shad, herring, or tuna are electroreceptive, there is some evidence that electrical features of the ocean floor may help guide the

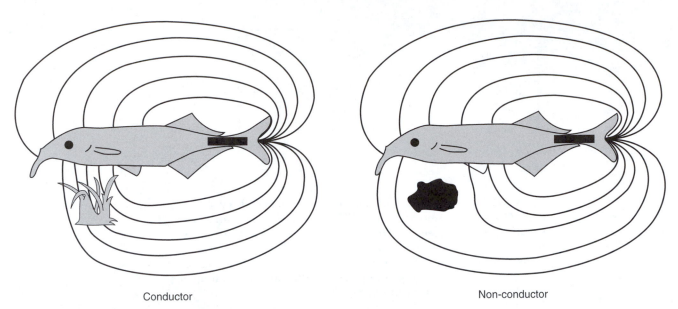

Conductor   Non-conductor

FIGURE 10.32   Electroreception. The electrical field generated by this fish is distorted by nearby objects. A good conductor, such as another living organism, draws the lines of force together. A nonconductor, such as a rock, spreads them out. Using electroreceptors distributed over its body surface, the fish senses the changes in the electrical field to "picture" its environment. (From von der Emde 1999.)

movements of bottom-feeding species such as the dogfish shark (Waterman 1989).

Although most living organisms generate weak electrical fields in water, only a few species have electric organs that generate pulses, creating electrical fields that can be used in communication (discussed in Chapter 18) and orientation. The electric organs of weak electric fish (mormyrids and gymnotids), located near their tail, for instance, generate a continuous stream of brief electrical pulses. The result is an electrical field around the fish in which the head acts as the positive pole and the tail as the negative pole. Nearby objects distort the field, and the distortions are detected by numerous electroreceptors in the lateral lines along the sides of the fish. A weakly electric fish generally keeps its body rigid, a posture that simplifies the analysis of the electrical signals.

These fish examine their surroundings by using their electrical sense. Since they live in muddy water, where vision is limited, and are active at night, electrolocation is quite useful. Objects whose electrical conductivity differs from that of water disturb this electrical field. An object with greater conductivity than that of water—another animal, for instance—directs current toward itself. Objects that are less conductive than water, such as a rock jutting into its path, deflect the current away (Figure 10.32). Thus, the fish can distinguish between living and nonliving objects in its environment.

The distortions in the electrical field create an electrical image of objects that can tell the fish a great deal about the features of its environment. The distortion varies according to the location of the object relative to

the fish, so the location of the image on its skin tells the fish where in relation to its own body the object is located. If the distortion is greatest on the right, the object is located on the right. An object near the fish's head creates the greatest distortion near the head (Caputi et al. 1998). The degree to which the electrical field is distorted by an object (the amplitude of the

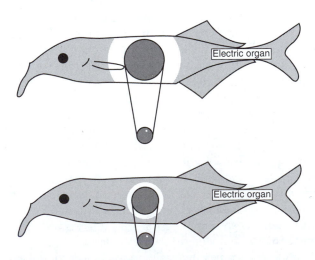

FIGURE 10.33   The electrical image of a metal sphere at different distances from the fish. The size (width) of the electrical image increases with distance. The amplitude differences in the degree of distortion of the electrical field between the center and the periphery of the electrical image decrease with increasing distance. (From von der Emde 1999.)

image) is greater in the center of the image than at the periphery. The fish often performs a series of movements close to the object under investigation. These actions might provide sensory input that helps the fish determine the object's size or shape (von der Emde 1999).

Electric fish can even measure the distance of most objects accurately, regardless of its size, its shape, or the material of which it is composed. In contrast to a visual image, the size (width) of an electrical image increases with distance. In addition, the amplitude differences between the center and the edges of an electrical image become smaller with the increasing distance of the object (Figure 10.33). The fish uses both of these features—size and amplitude—together to determine the distance of an object. A large, nearby object might cast the same sized image as a smaller, distant object, but the more distant object would have smaller amplitude dif-

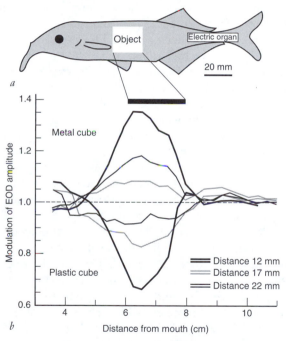

FIGURE 10.34 (*a*) **A weakly electric fish,** *Gnathonemus petersii*, **with a 2-cm cube positioned for electrical image measurement. (*b*) The electrical images of a metal or plastic cube at three distances from the fish's surface, measured at the midline. The electrical image of the metal cube is shown as a peak and the image of the plastic cube is shown as a trough because metal (a conductor) pulls the lines of force together and plastic (a nonconductor) spreads them out. Regardless of the composition of the cube, the width of the electrical image increases with increasing distance. The difference in amplitude between the center and the periphery of the image gets smaller with increasing distance. The fish uses the ratio of two features of the image—size and the amplitude differences between the core and the rim—to determine the distance of an object (From von der Emde 1999.)**

ferences between the central and outer areas of the image. The electrical images of a 2-cm cube of metal or plastic presented at different distances and measured along the midline of an electric fish are shown in Figure 10.34 (von der Emde 1999).

SUMMARY

Orientation to key aspects of the environment is critical to an animal's survival. Orientation refers to all the reactions that guide an animal into its correct posture, into its proper environment, or during migration and homing.

In some species, simple orientation responses maintain the individual's correct posture or guide it to suitable surroundings. A kinesis, such as the humidity kinesis of wood lice, is an undirected response that brings the individual into a suitable environment. The response is proportional to the intensity of the stimulus. A taxis is a response in which the animal moves directly toward or away from the stimulus. The animal moves in the direction of most favorable stimulus intensity.

A dorsal light reaction keeps an animal's dorsal (back) side toward the light. A ventral light reaction orients an animal so that its ventral (belly) side is toward the light. In a light compass reaction, the animal assumes a certain angle to a light source.

Navigation describes an animal's ability to orient itself over long distances. One level of orientation, called piloting, is the ability to locate a goal by referring to landmarks. A second level is compass orientation, in which an animal orients in a particular compass direction without referring to landmarks. This is the type of navigation used by most bird migrants. A third level describes an animal's ability to locate the goal without the use of landmarks, even if it is released in an unfamiliar location. True navigation requires a map to determine location and a compass to guide the journey.

Animals have access to and use many different cues for orientation and navigation. The sensory modality of the primary cue varies among species, and many species have a hierarchy of cues. Although the interactions among cues can be complex, we have considered each sensory basis separately.

Visual cues include landmarks; the sun, stars, or moon; and the pattern of skylight polarization. Methods of demonstrating that landmarks are used in navigation include moving the landmark to see whether the animal reorients or becomes disoriented and impairing the animal's vision so that landmarks cannot be used. Some species use landmarks by matching the objects viewed with the remembered image of the array of landmarks. When landmarks are used in this way, the animal must always follow the familiar path. A controversial idea is that certain species are able to form a mental (cognitive) map of an area by remembering the relationships among landmarks.

The sun may be used as a point of reference by assuming some angle relative to it during the journey and then reversing the angle to get home. Alternatively, since the sun follows a predictable path through the sky, if the time of day is known, the sun's position provides a compass bearing. If the sun is used as an orientation cue over a long interval of time, the animal must compensate for the sun's movement. Animals must learn to use the sun as a compass. The point of sunset is also an orientation cue that some nocturnal migrants use to select their flight direction, which is then maintained throughout the night by using other cues.

The stars provide an orientation cue for some nocturnal avian migrants. Birds such as the indigo bunting learn that the center of celestial rotation is north. This gives directional meaning to the constellations in the circumpolar area. Since the spatial relationship among these constellations is constant, if one is blocked by cloud cover, the birds can use the others to determine the direction of north.

Sunlight becomes polarized as it passes through the atmosphere. The pattern of polarization of light in the sky varies with the position of the sun. Polarized light may provide an axis for orientation, or it may allow animals to locate the sun from a patch of blue sky even when their view of the sun is blocked.

The earth's magnetic field provides several cues that could be used for orientation. Most animals appear to have an inclination compass. They distinguish between equatorward (where the magnetic lines of force are horizontal) and poleward (where the lines of force dip toward the earth's surface). The primary magnetic compass is innate, but the preferred magnetic direction may be calibrated by cues from celestial rotation during development. The magnetic compass is influenced by early experience and by species and individual differences.

Birds migrating from either the northern or the southern hemisphere can use the same magnetic migratory program—travel equatorward in the fall and poleward in the spring. As a migrant crosses the equator, the horizontal fields experienced there reset its magnetic inclination compass so that it then travels poleward. The magnetic compass is also recalibrated during migration.

The earth's magnetic field is one of the first orientation cues used by homing pigeons and many species of bird migrants. The sun compass later becomes the primary orientation cue of homing pigeons. Some researchers have suggested that in addition to serving as a compass, the geomagnetic field may serve as the homing pigeon's map, revealing its location relative to home.

Sea turtles use a magnetic compass while migrating across the Atlantic Ocean. They calibrate their magnetic compass relative to the direction of the surface waves that they experienced as they initially swam offshore.

Certain observations suggest that homing pigeons and sea turtles may have a magnetic map.

Chemical cues can also be used for orientation. Salmon are guided to their natal stream by chemical cues. One hypothesis is that young salmon learn (imprint on) the characteristic odors of their natal stream and then follow the odor trail back to that place. The alternative hypothesis is that fish are innately attracted to pheromones, communicatory chemicals, produced by fish of the same species. When both cues are present, salmon tend to return to the site they experienced as juveniles instead of an area containing other fish of their kind.

It has also been suggested that homing pigeons rely on olfactory navigation, but the results of olfactory deprivation studies are inconsistent. Although the results of deflector lofts are consistent, they may not be due to a shifted olfactory map. However, the results of experiments in which olfactory information is manipulated are consistent with an olfactory basis for pigeons' navigation. The role of olfaction in their homing remains controversial.

There are many ways in which sound can be used for orientation. Most species orient to sound by listening and then locating the source. Some species, the best known of which are bats, echolocate. They produce bursts of sound and then analyze the echoes to develop a picture of the environment. The echolocation cries of bats are very high-pitched, and those of most species are intensely loud. The rate of pulsing increases during a hunt, helping a bat characterize its prey and track it more accurately. There are three categories of bats' cries: frequency modulated (FM), constant frequency (CF), and constant frequency with a frequency sweep at one end. The echoes from FM pulses provide the most information about the target. A bat analyzes an FM call by measuring the interval between the pulse and the return of the echo to determine the distance to the target. CF pulses allow a bat to detect prey from a greater distance than do FM calls and to determine whether the target is approaching or moving away. A CF pulse is analyzed by the Doppler shift. Most bat species produce only one type of call, and its structure is suited to their style of foraging.

Some aquatic species can detect electrical fields. These could be of use in navigation. A few species have electric organs that can generate electrical fields, which can be used in communication and navigation. The weak electric fish generate a stream of electrical pulses and then sense objects by the disturbance created in this symmetrical field.

11

The Ecology and Evolution of Spatial Distribution

Beginning in April, the flutelike phrases of the wood thrush's (*Hylocichla mustelina*) song can be heard at dawn and dusk in the deciduous forests from Manitoba, Ontario, and Nova Scotia and south to Florida (Figure 11.1). Slightly smaller than a robin, this bird is considered to be one of the best songsters. The first melodies heard in the spring are a sure sign that the male has arrived on his territory—generally the same territory he held the previous year. The males usually arrive a few days before the females and begin defending territories. When an interloping male is within about 10 meters of the territory holder, a vocal duel may begin, and the males will sing alternately. As the interloper approaches, the defending male may flick his tail and/or wings repeatedly, indicating that he is perturbed. When

he becomes more excited, he spreads his breast feathers and compresses his head feathers. The intruding male is generally chased off the territory.

When a female arrives, she, too, is chased by the territorial male. But, she does not flee. After several days, the male's aggression subsides, and they mate. The female builds a cuplike nest of grass and twigs and lines it with grass. In this she lays four or five greenish-blue eggs. While the female is incubating the eggs, she spends 80% to 90% of her time on the nest. When she leaves the nest, the male usually perches near it and sings, unless he is busy tending to a previous brood. After the eggs have hatched, both parents feed the nestlings. In two to three weeks, the young gradually leave the territory.

In October, groups of wood thrushes form as they begin their southward migration to Mexico, Central America, and sometimes the Caribbean Islands. Here, they will spend the winter. While on the winter grounds, the group splits up and birds live singly. They will form groups again in March, when the spring migration begins. In mid-May, they reach the most northern parts of their range, and the cycle begins again.

All animals, not just wood thrushes, must make many "decisions" about where to live. One decision is whether to remain on or to return to a breeding site or to disperse. Male wood thrushes return to their previ-

FIGURE 11.1 All animals must make decisions about where to live. Within two to three weeks after fledging, the young wood thrushes shown here will disperse. The male wood thrush, however, will return to the same territory year after year. In contrast, females do not return to the same breeding area in subsequent years. Wood thrushes migrate between the Great Lakes or southern Canada to Mexico and Central America.

ous territories, and females disperse (Clarke, Sæther, and Røskaft 1997). Another decision is whether to defend a territory. Wood thrushes do. Finally, animals must "choose" whether to migrate. A wood thrush's migration is more modest than those of some other species—several thousand kilometers between the Great Lakes or southern Canada to Mexico and Central America.

These are not conscious choices. They are determined by the bottom line on natural selection's accounting ledger. Each choice has costs and benefits, and these may differ between the sexes. The payoff is often influenced by specific ecological factors. In this chapter, we will consider some of the factors that affect the costs and benefits of each choice and how they may be influenced by ecological conditions.

NATAL PHILOPATRY AND NATAL DISPERSAL

Some animals are born in one place and then move to another place, where they breed, never to return to their birthplace. This general behavior has been called natal dispersal, and it has come to be defined in many ways. Here, we will use Walter E. Howard's (1960) definition of natal dispersal as "the movement an individual makes from its point of origin to the place where it reproduces or would have reproduced had it survived and found a mate." This definition refers to the permanent movement of individuals away from their birth site.

Natal dispersal can be contrasted with natal philopatry, in which offspring remain at their natal area and share the home range or territory with their parents (Waser and Jones 1983).

COSTS AND BENEFITS OF NATAL PHILOPATRY VERSUS NATAL DISPERSAL

What determines whether a juvenile should remain in the area of its birth or disperse? There is no simple answer to that question. Indeed, there are probably multiple influences on dispersal (Dobson and Jones 1985). Furthermore, such factors differ in their importance between species, sexes, and individuals. Here we will consider some of those factors as they relate to the costs and benefits of philopatry and dispersal. (We will discuss dispersal and philopatry again in Chapter 17 when we explore factors that lead to helping.)

One potential cost of remaining in the birth area is inbreeding—mating between relatives. Extreme inbreeding involves mating between parents and offspring or between siblings. Inbreeding is costly both because it reduces variation among offspring and because it increases the risk of producing offspring that are homozygous for harmful or lethal recessive alleles. All organisms probably carry some harmful alleles that are recessive and therefore not expressed. Close relatives are more likely to have inherited the same versions of these alleles than are other members of the population. If close relatives mate, their offspring are at increased risk of inheriting a copy of the deleterious allele from each parent. These offspring will then display the harmful trait (Shields 1982).

It can be difficult to determine the fitness costs of inbreeding because the frequency of inbreeding is usually low. However, a long-term study of the Mexican jay (*Aphelocoma ultramarina*) did find fitness costs associated with inbreeding. The brood sizes of inbred pairs were smaller than those of outbred pairs, which suggests hatching failure. Furthermore, compared with outbred nestlings, significantly fewer of the inbred nestlings survived to the next year (Brown and Brown 1998).

A second possible cost of philopatry is increased competition. If conditions are crowded at home, young that remain in their birth area may have to compete with relatives for food, nest sites, or mates. Limited access to critical resources could result in lower survival and/or less reproductive success than could be achieved by dispersing (Shields 1987). In other words, sometimes the grass really is greener on the other side of the street, and so juveniles are better off leaving home than competing with others for limited resources.

What benefits might be associated with philopatry? Genetically speaking, inbreeding may not be all bad. A certain level of inbreeding, for example, could maintain complexes of genes that are particularly well adapted to local conditions. If animals that dispersed from the birth site settled a short distance away and mated with nonrelatives, such complexes of genes would be diluted or perhaps weeded out of the population. Thus, we see that there may be an optimal level of inbreeding (Shields 1982).

Another benefit of remaining near the birthplace is familiarity with the local physical and social setting. Such familiarity may enable philopatric young to be efficient not only at finding and controlling food but also in escaping from predators. Familiarity with family and neighbors is also likely to reduce the levels of aggression and stress associated with social interactions. In short, philopatric young may live longer and leave more offspring because of the relatively low risks and energy use associated with living in familiar surroundings (Shields 1982).

Dispersing also has costs and benefits. Dispersers are thought to face high energy costs and risks of predation as a result of increased movement and lack of familiarity with the terrain (e.g., Ambrose 1972; Metzgar 1967). In some cases, dispersers may encounter high levels of aggression when attempting to establish residence in an existing population (e.g., Joule and Cameron 1975). Despite such costs, recall that dispersers may avoid crowded conditions at the natal area and thereby benefit through increased access to mates and critical resources.

NATAL PHILOPATRY AND NATAL DISPERSAL IN BIRDS AND MAMMALS—SOME HYPOTHESES

Males and females of a particular species often differ in whether or not they disperse from their birthplace. Even more striking is the observation that the direction of the sex bias in natal dispersal differs between birds and mammals (Table 11.1). In the majority of bird species that have been studied, females are more likely to disperse than males. In mammals, however, just the

TABLE 11.1   Number of Species of Mammals and Birds in Which Natal Dispersal Is Male-Biased, Is Female-Biased, or in Which Offspring of Both Sexes Disperse

	Predominant Dispersing Sex		
	Male	Female	Both
Mammals	45	5	15
Birds	3	21	6

Source: Data from P. J. Greenwood (1980).

reverse is true—males are more likely to disperse than females.

We can ask why this is so. There are several hypotheses. Those that we will consider are that sex differences in natal dispersal reduce the (1) probability of inbreeding, (2) competition for mates, and (3) reproductive competition between parents and offspring.

Reduction in Inbreeding

One hypothesis is that a sex bias in dispersal evolved as a way to avoid the genetic costs of inbreeding while enjoying the benefits of familiarity with local physical and social conditions. A sex bias in dispersal seems to be the perfect compromise: Extreme inbreeding is prevented because members of one sex disperse, and individuals of the other sex experience the benefits of philopatry (P. J. Greenwood 1980).

Although this hypothesis explains why sex biases in dispersal tendencies occur, it does not explain which sex leaves home and why the direction of the bias differs in birds and mammals. There are two suggested explanations for the direction of the sex bias.

The first idea is that sex differences in dispersal are related to differences in the ways in which male birds and mammals compete for mates (P. J. Greenwood 1980). Most birds are monogamous, and rather than competing for females directly, male birds usually compete for territories that attract females. This is called a resource-defense mating system. Under conditions such as these, familiarity with a particular area might be more important to males than to females. Female birds might disperse to avoid the genetic costs of close inbreeding and to choose territories with the best resources.

In contrast, most mammals are polygynous; that is, a single male defends a group of females. Males directly compete with one another for females in this mate-defense mating system. Young or subordinate males, unable to compete successfully for access to females, may disperse to increase their chances of mating. Also, female mammals often live in matrilineal social groups (groups of mothers, daughters, and granddaughters) in which the benefits of living with kin may be quite high. Because of this social system, males may also disperse to avoid the genetic costs of extreme inbreeding. Thus, female-biased dispersal in birds seems to be linked to resource-defense mating systems, and male-biased dispersal in mammals is linked to mate-defense mating systems (reviewed in Clarke, Sæther, and Røskaft 1997).

Although most birds are monogamous and defend territories, and most mammals are polygamous and directly defend females, there are exceptions. For example, some species of mammals are territorial. Based on the inbreeding avoidance hypothesis, we would predict that territorial mammals would show the same pattern of dispersal observed in territorial birds—females should be more likely than males to disperse. However, this pattern was observed in only 1 of 13 species of territorial mammals examined (Liberg and von Schantz 1985).

The second idea to explain the direction of the sex bias is that the sex that gets first choice of breeding sites is the one that remains in the natal area; the other sex disperses. This model was first developed to explain sex-biased dispersal patterns in mammals (Clutton-Brock 1989), but it has been recently extended to birds (Wolff and Plissner 1998). In either case, the model assumes that philopatry is more desirable than dispersal. According to this model, mating systems affect dispersal patterns of mammals indirectly by influencing whether the father will be present when his daughters are old enough to breed. If he is not, females have first choice of the breeding site, and they choose to stay at home. On the other hand, if the father is still around when his daughters reach sexual maturity, he has first choice of the breeding site, and so females disperse to avoid inbreeding.

Because female mammals nurse their young, in most species males show little parental care. This allows males to avoid long-term pair bonds, and they are free to wander over large areas. When competition over mates is intense, as among elephant seals or red deer, a male's opportunity to breed may be limited. As a result, he is likely to be gone before his daughters are old enough to reproduce. Thus, daughters don't have to disperse to avoid inbreeding. However, in those species in which a male's reproductive life span is long and he is present when his daughters are old enough to breed, as in chimpanzees, the females usually disperse (Clutton-Brock 1989; Wolff 1994).

Reduction in Mate Competition

A second hypothesis is that differences between males and females in levels of competition for mates might be involved in sex differences in the dispersal tendency in mammals (Dobson 1982). Because most mammals are polygynous, competition for mates would be more intense among males than among females, and thus dispersal should be more common in males. Furthermore, in species with monogamous mating systems, levels of competition for mates would be more equal between the sexes, and males and females should disperse in similar proportions.

When dispersal data for species of mammals with different mating systems were examined, they revealed remarkable agreement with this hypothesis (Table 11.2). However, avoidance of inbreeding is probably a significant influence on natal dispersal in many species because reduction in competition for mates can't explain all sex differences in natal dispersal. (For instance, it doesn't explain why females are more likely than males to disperse in monogamous birds and in polygamous mammals that defend territories.)

Reduction of Competition with Parents

A third hypothesis is that juveniles are thrown out of their homes, instead of leaving on their own volition. Why would parents force their young to leave? According to the Oedipus hypothesis, sex differences in natal dispersion in both birds and mammals reduce reproductive competition between parents and offspring (Liberg and von Schantz 1985). This differs from the previous hypotheses in that the question of dispersal is viewed from the standpoint of the parents, not the offspring. This model assumes that it is usually in the offspring's best interest to stay at home. Because parents occupy the superior position in the parent-offspring relationship, it is they that make the "decisions" about those that can stay and those that should leave. Furthermore, if parents benefit by letting some offspring stay at home but cannot afford to have all offspring stay, then they should force those young that exact the highest toll to leave.

Following this reasoning, we see that levels of reproductive competition between parents and male and female offspring are related to mating systems and modes of reproduction in birds and mammals. In species with polygynous or promiscuous mating systems, male offspring that remain at home are predicted to compete

TABLE 11.2   Number of Species of Mammals in Which Natal Dispersal Is Male-Biased, Is Female-Biased, or in Which Offspring of Both Sexes Disperse, as a Function of Type of Mating System

Mating System	Predominant Dispersing Sex		
	Male	Female	Both
Monogamous	0	1	11
Polygynous or promiscuous	46	2	9

Source: Data from Dobson (1982).

with their fathers for mates, but female offspring do not compete with either parent. In monogamous mating systems, neither sons nor daughters pose a competitive threat to either parent. Furthermore, differences between mammals and birds may be based on their different modes of reproduction—specifically, egg laying versus gestation and birth. In birds, a daughter that is allowed to stay at home can cheat her mother, or both parents in the case of monogamous species, by laying eggs in the family nest and thereby imposing her own breeding costs on her parents. (Nest parasitism is discussed further in Chapter 15.) In mammals, however, daughters cannot hide their reproductive efforts because pregnancy and birth are obvious events, and parents cannot be fooled into caring for their daughters' offspring. There are no corresponding ways in which sons can cheat their parents in either birds or mammals.

These ideas lead to the following predictions for dispersal and philopatry: In monogamous birds, sons cannot steal copulations from their fathers, but daughters can cheat parents by laying eggs in the family nest. In such cases, parents should force daughters to disperse. In polygynous and promiscuous birds, sons can steal copulations from their fathers and daughters can cheat parents, so both sexes should be driven away from the nest. In the case of monogamous mammals, sons cannot steal matings from their fathers and daughters cannot trick parents into caring for offspring, so the prediction is that both sexes should be allowed to remain at home. Finally, in polygynous and promiscuous mammals, sons can cheat their fathers out of copulation but daughters cannot fool parents into caring for offspring, so daughters should be allowed to stay home and sons should be forced to disperse. The predictions of the Oedipus hypothesis are summarized in Table 11.3.

The Oedipus hypothesis (Liberg and von Schantz 1985) provides explanations for some of the areas of disagreement between the hypotheses of inbreeding avoidance (P. J. Greenwood 1980) and of reduction in mate competition (Dobson, 1982). For example, the latter hypothesis predicts that competition for mates should be fairly equal between the sexes in monogamous mammals, and so there should be no sex bias in dispersal. However, most bird species are monogamous and most are characterized by female-biased dispersal. The Oedipus hypothesis provides the explanation that in any mating system, female birds should be forced to disperse because they can cheat their parents by laying eggs in the family nest. The hypothesis does not apply to situations in which young animals disperse on their own initiative, perhaps to avoid inbreeding or to gain access to better resources.

We see, then, that none of the three hypotheses alone can explain patterns of dispersal in all species. So, when we want to explain sex differences in dispersal in a particular species, we should explore the factors that may cause the observed dispersal pattern to avoid the cost of inbreeding, reduce competition for mates, and reduce reproductive competition between parents and offspring. We should also examine the relative roles that parents and offspring play in dispersal.

TERRITORIALITY

In much the same way that patterns of dispersal or philopatry can be examined through an analysis of costs and benefits, we can also consider patterns of exclusivity of space use by animals. For example, we can ask whether an animal should maintain exclusive use of an area or share it with others. An area that is maintained for exclusive use by defending it from intruders is called a territory. As we will see, decisions about territoriality, like those involving dispersal, depend to a large extent on access to resources and mates. We will look more closely in Chapter 16 at the costs and benefits of territoriality as they relate to animals that are competing for a limited resource. Here, however, we will consider some of the ecological factors that affect those costs and benefits.

THE ECOLOGY OF TERRITORIALITY IN MICROTINE RODENTS

Microtine rodents, a group that includes the voles, lemmings, and muskrats, are small mammals that exhibit

TABLE 11.3 **Predictions of the Oedipus Hypothesis**

Mating System	Birds (Daughter Can Cheat Parents by Laying Eggs in the Family Nest)	Mammals (Daughter Cannot Trick Parents into Caring for Her Own Young)
Monogamous (son cannot steal matings from father)	Daughters forced to disperse	Equal proportions of both sexes disperse
Polygynous or promiscuous (son can steal matings from father)	Daughters and sons forced to disperse	Daughters allowed to stay; sons forced to disperse

Source: Modified from Liberg and von Schantz (1985).

amazing variation in the degree of territoriality, both among species and between sexes within species. In some species, only females are territorial; in other species, only males are territorial; and in still others, territoriality is exhibited by both male and female members of a breeding pair. Hypotheses put forth to explain the different patterns of territoriality suggest that whether or not females are territorial depends on characteristics of the food supply, and whether or not males are territorial depends on the availability of reproductive females (reviewed by Ostfeld 1990). The important point here is that males and females differ in terms of which resource is most critical to their reproductive success, that is, which resource is worth defending—food for females and mates for males. The two major hypotheses that we will discuss—the "females in space" hypothesis (Ostfeld 1985) and the "females in space and time" hypothesis (Cockburn 1988; Ims 1987a)—are similar in their explanations for female territoriality but differ somewhat in their explanations for male territoriality.

The Females in Space Hypothesis

As previously mentioned, for females, food appears to be the critical resource in reproductive success. Rick Ostfeld (1985) proposed that characteristics of the food supply, specifically abundance, distribution, and renewal rate, ultimately determine the occurrence of territoriality in microtine rodent females. Some species of voles feed primarily on seeds, fruits, and forbs. Such foods are sparse, patchy in distribution, and slowly renewed. Ostfeld thus predicted that females of species feeding primarily on seeds, fruits, and forbs would be territorial. Other species of voles feed predominantly on grasses, sedges, and horsetails. Because these plants

are abundant, widely distributed, and rapidly renewed, Ostfeld predicted that females of species feeding on grasses, sedges, and horsetails would be nonterritorial. Although there are some exceptions, comparisons across species generally support these predictions (Table 11.4).

In addition, two single-species studies in which the amount of available food was either changed or experimentally manipulated also support the idea that characteristics of the food supply determine spacing patterns of females. Bank voles (*Clethrionomys glareolus*) and gray-sided voles (*C. rufocanus*) feed mostly on fruits and seeds, and as predicted by Ostfeld (1985), females are territorial. However, when food was unusually abundant for a population of bank voles (Ylonen, Kojola, and Vitala 1988) or when supplemental food in the form of oats, corn, and sunflower seeds was provided to a population of gray-sided voles (Ims 1987b), the degree of territoriality exhibited by females decreased. In the study with gray-sided voles, females that exploited the same supplemental food store became more clumped in distribution, and the degree of overlap in their home ranges increased considerably (Figure 11.2).

Most male mammals do not help in the rearing of offspring, and as a result access to mates is thought to be a more critical determinant of male reproductive success than is access to food. Ostfeld (1985) predicted that the spatial distribution of females determines whether or not males are territorial. Thus, food resources determine the spatial organization of females, which in turn affects the spacing pattern of males (Figure 11.3). According to this idea, when females are nonterritorial, they occur in clumps and can be more easily defended by males (i.e., males can be territorial). When females are territorial, however, they are widely

TABLE 11.4 **Type of Diet and the Occurrence of Territoriality in Microtine Rodents: Testing the Females in Space Hypothesis**

Species	Diet*	Female Territoriality	Male Territoriality
Clethrionomys gapperi	SFF	yes	no
Clethrionomys glareolus	SFF	yes	no
Clethrionomys rufocanus	SFF	yes	no
Microtus arvalis	SFF	yes	no
Microtus ochrogaster	SFF	yes	yes
Microtus pinetorum	SFF	yes	yes
Microtus pennsylvanicus	SFF	yes	no
Microtus agrestis	GSH	no	yes
Microtus californicus	GSH	no	yes
Microtus xanthognathus	GSH	no	yes

*SFF = seeds, fruits, and forbs; such foods are scarce, patchy in distribution, and slowly renewed. GSH = grasses, sedges, and horsetails; such foods are abundant, widely distributed, and rapidly renewed.

Source: Modified from Ostfeld (1985).

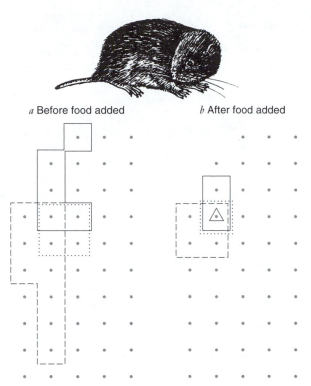

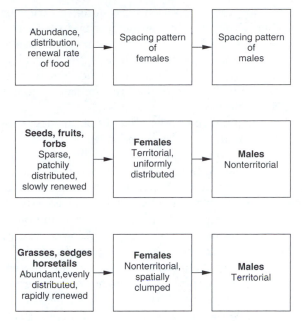

FIGURE 11.3   The females in space hypothesis for the occurrence of territoriality among microtine rodents. (Modified from Ostfeld 1990.)

FIGURE 11.2   In microtine rodents, females are predicted to be territorial when food resources are scarce. In this field experiment with gray-sided voles, the home ranges of specific females were measured both before and after supplemental food was added (the triangle denotes location of supplemental food). The addition of food resulted in smaller home ranges and a greater degree of home range overlap. In other words, an abundance of food resulted in a relaxation of territorial behavior among neighboring females. (Modified from Ims 1987b.)

distributed, making defense by males much more difficult. Under these latter conditions, males are not expected to be territorial but are predicted to roam widely in search of receptive females. An inevitable result of such wandering is that the ranges of males will overlap. According to the females in space hypothesis, then, territoriality in males is expected to be associated with lack of territoriality in females, and conversely, lack of territoriality (home range overlap) in males is expected when females are territorial. Interspecific comparisons generally support these predictions (refer again to Table 11.4).

Is there evidence from single-species studies to support the idea that the spatial distribution of females determines the spatial distribution of males? In the study previously described, in which supplemental food was given to populations of gray-sided voles, a reduction in the overlap of male home ranges (i.e., an apparent increase in territorial behavior) coincided with the decreased territoriality and spatial clumping of females (Ims 1987b). Although consistent with the females in

space hypothesis, these results might also be explained by other factors, such as breeding synchrony among females (see later discussion). Finally, results contrary to the predictions of the females in space hypothesis were obtained by Rolf Ims (1988). He placed female gray-sided voles in wire-mesh cages and arranged them in either a clumped or a dispersed distribution, and over a ten-day period he recorded patterns of space use by males. The home ranges of males overlapped more (i.e., males were less territorial) when females were clumped than when they were widely dispersed. Ims suggested that the clumped arrangement of females induced intense competition among the males, making territorial behavior too costly.

It is important to note that the predictions of the females in space hypothesis pertain only to the breeding season, the time when food is especially important to females because of the high costs of reproduction and the time when males compete for copulations. During the nonbreeding season, Ostfeld (1985) predicted that the patterns of sex-specific territoriality exhibited by microtine rodents during the breeding season should break down and mixed-sex aggregations should form. In support of this prediction, communal groups of males and females of different ages have been found during late fall and winter in several species of voles.

The Females in Space and Time Hypothesis

Andrew Cockburn (1988) and Rolf Ims (Ims 1987a), although agreeing with Ostfeld (1985) on the causes of female territoriality in microtine rodents, proposed a modification of factors responsible for male territoriality. They suggested, based on ideas outlined earlier by Emlen and Oring (1977), that the distribution of fertilizable females has not only a spatial component but also a temporal component. In other words, the availability of females varies in both space and time. Not surprisingly, then, this has been called the "females in space and time" hypothesis.

The temporal component focuses on the degree of breeding synchrony in the population. Female voles and lemmings typically produce several litters within a breeding season. Although the first period of estrus in a female's life appears to be induced by exposure to a male (this first estrus is called male-induced estrus), subsequent estrus periods occur just after giving birth. Within hours of parturition, a female becomes sexually receptive for about one day. This period of receptivity is known as postpartum estrus. A female that is impregnated during postpartum estrus does not become receptive again until she gives birth to her litter, about three weeks later. Most microtine females are thus sexually receptive for fairly short periods of time (about one day) at regular intervals (every three weeks) throughout the breeding season.

Do most females within a population become sexually receptive at the same time (i.e., synchronously) or at widely dispersed points in time (i.e., asynchronously)? According to Ims (1987a), when females become receptive asynchronously, males should search for females, stay with them only during their brief period of sexual receptivity, and then move on in search of other mates. Regardless of the spatial distribution of females, the time taken for a male to search for new mates is probably less than the three weeks it would take for a particular female to enter estrus again. When breeding is asynchronous, then, males should not be territorial but should roam widely in search of receptive females (Figure 11.4). Ims further predicts that these patterns of movement and mating will most likely result in a promiscuous mating system. Because males have highly overlapping home ranges, a single female will probably mate with more than one male over the course of the breeding season, and males, constantly on the move in search of receptive females, will probably mate with more than one female.

The male spacing strategy is predicted to change dramatically when breeding occurs synchronously. Because resources, in general, are more easily defended through territorial behavior when they have a clumped distribution, males can monopolize females that become receptive at the same time. If such females have a clumped distribution or fairly small home ranges, a single male can monopolize a group of females, and the mating system is likely to be polygamous (see Figure 11.4). If, however, such females have a dispersed distribution or extremely large home ranges, a single male may monopolize only a single female, and the mating

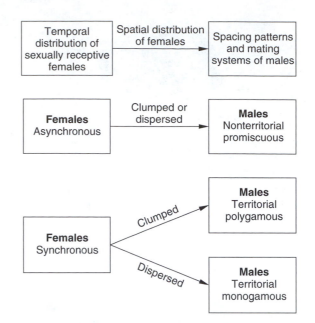

FIGURE 11.4   The females in space and time hypothesis for the occurrence of territoriality among microtine rodents. (Drawn from Ims 1987a.)

system is likely to be monogamous (again see Figure 11.4). Thus, although the type of mating system depends on the spatial distribution of females, the spacing system of males can be predicted by the temporal component of female distribution alone.

It turns out, unfortunately, that it is extremely difficult to disentangle the effects on male territorial behavior of spatial and temporal clumping of females because the likelihood of synchronous breeding is probably tied to the spatial distribution of females. Because social interactions, olfactory cues, and shared food resources may be involved in achieving breeding synchrony, such synchrony is probably more common among females that are clumped. Thus, clumping by females may result in territoriality by males, but the question remains, is male territorial behavior due to the spatial distribution of females, breeding synchrony among females, or both?

At least one study has supported the females in space and time hypothesis, and one has not. When a population of meadow voles (*Microtus pennsylvanicus*) was monitored, females were found to breed asynchronously, and as predicted by the hypothesis, males were not territorial (McShea 1989). However, in another study, this time involving the gray-sided vole, females in the population were found to breed synchronously, but contrary to the predictions of the hypothesis males were nonterritorial (Kawata 1985).

Unanswered Questions

We see, then, that there are studies that support and studies that refute both the females in space hypothesis

FIGURE 11.5 A pair of prairie voles. Understanding the spacing patterns of these small, apparently monogamous rodents has proven to be a challenge for those who study territoriality.

and the females in space and time hypothesis. There are also certain territorial systems that cannot be completely explained by either hypothesis. Pine voles (*M. pinetorum*) and prairie voles (*M. ochrogaster*; Figure 11.5), for example, eat seeds, fruits, and forbs. According to the females in space hypothesis, because these foods are sparse, patchy, and slowly renewed, females should be territorial—and they are. However, contrary to the predictions of the females in space hypothesis, males of both species are also territorial. Rather than roaming widely to come in contact with as many females as possible, male pine voles and prairie voles appear to mate monogamously (FitzGerald and Madison 1983; Getz et al. 1993). Although there is some variation within populations, in both species a given male and female share a nest and territory. The females in space and time hypothesis predicts a monogamous mating system only when breeding occurs synchronously, and females have widely dispersed or have exceptionally large home ranges. To date, the degree of breeding synchrony has not been studied in natural populations of pine voles or prairie voles, but female home ranges do not appear to be particularly large or widely spaced.

MIGRATION

In some species, spatial distribution varies over time, usually with the seasons. Indeed, there may be dramatic mass movements of animals, and some are migrations. We will consider migration to be the movement of animals away from an area and the subsequent return to that area. Animals usually migrate between breeding areas and overwintering, or feeding, areas (Figure 11.6). The distance may be short. For example, a salamander or newt may travel less than a kilometer from its woodland home to the pond where it breeds. But, in other cases, the distances are truly astounding. Northern elephant seals (*Mirounga angustirostris*) migrate twice a year—once to breed and again mostly to eat—from beaches in southern California and Baja California, Mexico, to northern feeding grounds in the Aleutian Islands. Thus they migrate about 8000 kilometers each year, and that is just the horizontal distance. They make frequent, deep dives that can add another 3000 kilometers of vertical distance to their journey (Tennesen 1999). The arctic tern (*Sterna paradisea*) migrates even further, about 20,000 kilometers, between its southern wintering area and northern breeding area (Baker 1980).

Why should an animal bother to travel hundreds or thousands of kilometers to one location only to return to its starting point half a year later? There are probably many answers to this question, and no single one could apply to the diverse species of migrators. However, the simplicity of one explanation hides its profundity: Those animals that migrate do so because they produce more offspring this way. Not all species will have greater reproductive success by migrating, and so not all species migrate.

Although the actual costs and benefits of migration vary among species, the benefits must result in the production of more offspring than would be possible if the migrator stayed put. To explore this question further, we will consider some of the possible costs and benefits that might accompany migration.

COSTS OF MIGRATION

Migration takes a tremendous toll. Only half of the songbirds that leave the coast of Massachusetts each year ever return, and less than half of the waterfowl in North America that migrate south each fall return to their breeding grounds (A. C. Fisher 1979).

One reason for these enormous losses is that traveling such long distances requires a great deal of energy. For instance, a bird uses about six to eight times more energy when flying than when resting. Imagine yourself running 4-minute miles continuously for 80 hours. This would require roughly the same amount of

FIGURE 11.6   Migrating caribou marching through Alaska in July. During the spring, many caribou breed in the tundra. Beginning in July, they migrate south, where food will be available through the winter.

energy per kilogram of body mass as a blackpoll's non-stop transoceanic flight of 105 to 155 hours (Williams and Williams 1978). It is a good thing the blackpoll is not walking, though. It turns out that the energy costs of traveling a certain distance vary with the mode of locomotion. If each obtained the same amount of energy from each gram of fat, then using 1 gram, a mammal could walk 15 kilometers, a bird could fly 54 kilometers, and a fish could swim 154 kilometers (Aidley 1981). However, even if it is easier for birds to travel great distances than it is for you, migration is still metabolically demanding. It is not unusual for ships at sea to have migrants land on the deck, exhausted and sometimes dying.

Natural selection favors behaviors that reduce the risk of starvation during migration. One way to do this is to store fat before the journey begins. Gram for gram, fat provides more than twice the energy of carbohydrate or protein. Thus it is an extremely important energy source during migration. It should not be surprising, then, that migratory animals as diverse as insects, fish, birds, and mammals put on fat reserves prior to migration (Berthold 1993). Indeed, the body mass of birds that migrate over long distances may more than double before migrating (Klaassen 1996).

Another way to reduce the risk of starvation during migration is to head to refueling areas if fat reserves are low. This is observed in certain migratory birds. For instance, snow buntings (*Plectrophenax nivalis*) that have stored enough energy for migration orient in the correct migratory direction, southwest. In contrast, lean birds select northeast headings (Sandberg, Bäckman, and Ottosson 1998). Similar differences in magnetic orientation between fat and lean birds have been noted in other species (Bäckman, Pettersson, and Sandberg 1997; Lindström and Allerstam 1986; Sandberg 1994; Sandberg and Moore 1996). This difference in behav-

ior according to energy reserves is adaptive. An individual who doesn't have enough body fat to fuel a long migratory journey is better off backtracking in search of better refueling areas than running out of energy during the trip.

Many weary migrants fall to predators. For example, songbirds are "fast food" for the Eleanora's falcon. The songbirds, worn out by their flight across the Mediterranean, land in the nesting area of this falcon, which times its breeding so that its young will hatch when the songbirds arrive (Walter 1979). Furthermore, birds such as yellow-rumped warblers (*Dendroica coronata*) that are in migratory condition take greater risks of predation than do nonmigratory members of their species. Following exposure to a predator, the migratory birds resume feeding sooner and handle food items faster than birds not in migratory condition (F. R. Moore 1994). Other migratory species experience heavy predation because their predators follow their seasonal movements. Lions, cheetahs, and hyenas often track the movements of African ungulates (Schaller 1972); wolves follow North American caribou (Sinclair 1983); and water pythons migrate seasonally to exploit their migratory prey, the dusky rat (Madsen and Shine 1996).

The cost of migration is high in many cases because large areas of inhospitable terrain must be passed. Not only do terrestrial birds pay a large cost for crossing the sea, but when a land area is unfamiliar, an animal may also have problems in finding food or water. And then there is the matter of obstacles. Birds often crash into tall structures such as lighthouses, skyscrapers, and TV towers. In a single night, seven towers in Illinois felled 3200 birds (A. C. Fisher 1979).

Migration generally occurs during the spring and fall, times of notoriously unstable weather, which can drastically raise the cost of migration. Severe rainstorms and snowstorms kill millions of migrating monarch butterflies. It is not uncommon to see thousands of dead or dying monarchs on the shores of Lake Ontario or Lake Erie following a severe storm (Urquhart 1987). A flock of Lapland longspurs that were migrating through Minnesota encountered a sudden snowstorm one night. The next morning, 750,000 of these small birds were found dead within 1 square mile (S. T. Emlen 1972). In Manitoba, Canada, a spring storm killed so many warblers that dead birds were found about 8 inches apart (A. C. Fisher 1979). A tornado and storm at Grand Isle Louisiana occurred when birds were arriving at the coast after flying all night across the Gulf of Mexico. That storm killed an estimated 40,000 birds of 45 species (Wiedenfeld and Wiedenfeld 1995). Butterflies are even more sensitive to freezing temperatures. Although the monarch's Mexican wintering grounds usually remain above freezing, one frigid night killed over 2 million monarchs (Calvert and Brower 1986).

In addition to such dangers, territorial animals, such as birds, must relinquish the rights to a hard-won territory each year and compete vigorously to become reestablished the following year. You may recall that this was the case for the wood thrush, described in the opening example of this chapter.

So what rewards could possibly override such disadvantages?

BENEFITS OF MIGRATION

Energy Profit

We can intuitively understand the advantages of moving from approaching arctic winters to the sunny tropics. We can even see why animals might move shorter distances, for example, from Nebraska to Oklahoma, for the winter. Even a cursory familiarity with the elements of nature could also convince us of the advantages of simply moving from a mountaintop to a valley every winter. In each case, the animals are trading a less hospitable habitat for a more hospitable one.

The severe weather during the northern winters has favored migration. Each fall, millions of monarch butterflies (*Danaus plexippus*) migrate southward from the central and eastern regions of Canada and the United States to fir forests in central Mexico—sites with particular characteristics that enable the survival of the butterflies (Figure 11.7a). These forests are found about 3000 meters (nearly 2 miles) above sea level on the southwest slopes of a very small area of mountaintops. An important characteristic of these forests is that their temperature is cool but not freezing. Exposure to freezing temperatures would kill the monarchs. However, warmer temperatures would unnecessarily elevate their metabolic rates and waste energy reserves (Calvert and Brower 1986).

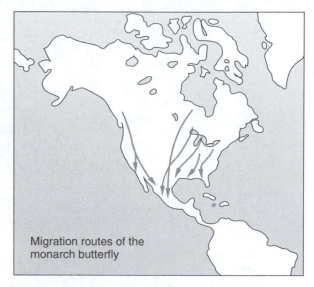

Migration routes of the monarch butterfly

a

b

FIGURE 11.7 (*a*) The monarchs travel in large groups from northeastern North America to overwinter in Mexico or from central California to the coast. (*b*) Monarch butterflies roost on trees in their overwintering sites in the mountains west of Mexico City.

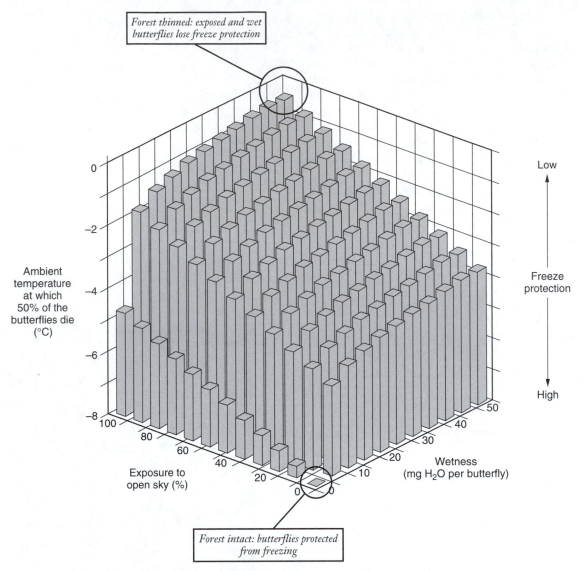

FIGURE 11.8   The intact forest canopy of the monarch butterfly's overwintering site normally serves as both an umbrella and a blanket, which keeps the monarchs from freezing. The vertical axis indicates the ambient air temperature at which 50% of a butterfly sample will freeze to death. A wet butterfly freezes at a warmer ambient air temperature than a dry one. Because of radiational cooling, exposure to open sky also allows a butterfly to freeze at a higher ambient temperature than if it were protected by the forest canopy. When the forest is thinned because local farmers need the wood, monarchs are vulnerable to freezing, especially on clear nights after wet winter storms. Thus, thinning of the forests in the monarch's overwintering area in Mexico is endangering the monarch population. (Data from Anderson and Brower 1996.)

Another characteristic of these overwintering sites that enhances the survival of monarchs is the tall trees, primarily oyamel firs. Besides providing branches on which the butterflies can roost, the trees form a thick, protective canopy over the butterflies (Figure 11.7b). The canopy serves as an umbrella, shielding the butterflies from rain, snow, or hail. This increases survival because a dry monarch can withstand colder temperatures than can one with water on its surface (Figure 11.8). If monarch butterflies are wet, 50% of the population will freeze at –4.2°C. If they are dry, however, the temperature can dip to –7.7°C before 50% of the population freezes. The canopy also serves as a blanket that keeps the butterflies warm. Openings in the forest canopy increase radiational cooling, which can lower body temperatures to as much as 4°C below the ambient air temperature. The body temperature of monarchs under a dense canopy is approximately the same as the air temperature. However, body temperature drops in proportion to the degree of exposure, increasing the chances that the butterflies will freeze to death (Anderson and Brower 1996).

Unfortunately, we may soon see just how important the forest canopy is to the survival of the monarch butterflies. The forest is being thinned because the local farmers who live near the monarchs' overwintering sites need the wood for firewood and construction. Such thinning allows both wetting and exposure of monarchs, increasing winter mortality. Conservationists are planting trees in an effort to save the monarchs, but the trees are not intended to provide new roosts for monarch clusters. Instead, it is hoped that the trees will provide local farmers with the wood they need for survival.

Seasonal changes in climate also affect food supply. In some species, migration is an adaptation that permits the exploitation of temporary or moving resources. The larvae of monarch butterflies feed only on milkweed. In regions of the eastern United States, however, the milkweed plants grow only during the spring and summer months (Urquhart 1987). Certain species of insectivorous bats may migrate in response to the size of the insect supply. Mexican free-tailed bats (*Tadarida brasiliensis*), for instance, leave the southwestern United States as the harsh winter climate causes the insect supply to dwindle. They migrate to regions of Mexico where insects are available throughout the winter (Fenton 1983). Wildebeest on the Serengeti also migrate, following temporary resources. During the wet season, they graze in the open grasslands. The short grasses found there are more easily digested and have higher concentrations of calcium and protein than taller grasses found elsewhere. However, the wildebeest must have drinking water, and the water holes of the open grasslands evaporate during the dry season. At this time, then, the wildebeest migrate to the wooded grasslands, where water is available (Kreulen 1975; Maddock 1979).

As winter approaches, increasing the animals' energy needs, the food supply drops and forces any resident species into more severe competition for such commodities. So, in spite of the energy required for migration, it may result in an overall energetic savings. For example, a study of the dickcissel (*Spiza americana*) revealed that despite the energy costs of migration, this bird enjoys an energetic advantage from both its southward autumn migration and its northward spring migration. Studies of the junco (*Junco hyemalis*), white-throated sparrow (*Zonotrichia albicolis*), and American tree sparrow (*Spizella arborea*) show that by avoiding the temperature stresses of northern winters, these species compensate for at least some of the energy spent on migration (reviewed in Dingle 1980).

The question arises, then, if there is so much food in the warmer winter habitats, why do species migrate from such areas? Why do they return to their summer homes at all?

Reproductive Benefits

One answer might be that there are important advantages in rearing broods in the summer habitats. For example, days in the far north are long, and the birds' working day can be extended—they can bring more food to their offspring in a given period of time and perhaps rear the brood faster. Another result of long days is that more food is available for offspring and more young can be raised (Figure 11.9). Although factors other than food availability may also play a role, generally the farther north from the tropics a species breeds, the larger is its brood (Welty 1962).

In some species, migrations might enhance reproductive success by bringing members of the opposite sex together, increasing the chance of mating. For instance, this is a partial explanation for the return of salmon from the sea to the freshwater streams where they hatched.

Other species migrate to areas that provide the necessary conditions for breeding or that offer some protection from predators. Gray and humpback whales, for example, breed in coastal bays and lagoons that provide the warmer temperatures needed for calving and help protect the calves from predation. The need for protected rookery sites may prompt seal, sea lion, and walrus migrations, as they come ashore on their traditional beaches after months at sea. Sea turtles also regularly migrate thousands of kilometers between feeding grounds and breeding areas, usually the same beaches on which they hatched. For example, female green tur-

FIGURE 11.9 The arctic tern, a champion migrator, travels 20,000 kilometers each year in migratory flights. It breeds in northern regions, where the days are long, allowing a parent to gather more food for its young.

FIGURE 11.10   The green turtle. One population migrates 1800 kilometers from its feeding ground off the coast of Brazil to Ascension Island to breed on the sheltered beaches, where it is safe from predators.

tles, *Chelonia mydas* (Figure 11.10), feed in the warm marine pastures off the coast of Brazil and then swim roughly 1800 kilometers to the sandy shores of Ascension Island, where they lay about 100 eggs, each the size of a golf ball (Lohmann 1992). The beaches where sea turtles lay their eggs are on isolated stretches of continental shores or small remote islands (Lohmann and Lohmann 1996). Because of their isolation, these beaches might have fewer predators.

Reduction in Competition

Another advantage in returning to the temperate zone is that of escaping the high level of competition that exists in a warmer, more densely populated area. The annual flush of life in the temperate zones provides a predictable supply of food that can be exploited readily by certain species without competition from the large number of nonmigrants that inhabit the tropics (Lack 1968).

Reduction in Predation

A third advantage in returning to temperate zones to breed lies in escaping predation. If predators are unable to follow herds of migratory ungulates, for instance, each individual's chance of survival is enhanced. Thus, escape from predation has been suggested as the reason that the number of migratory ungulates is so much greater than the number of nonmigratory ungulates (Fryxell, Greever, and Sinclair 1988).

In the far north, breeding periods are very short because of the weather cycles. This can be an advantage to nesting birds, which are in danger from predators. The short season results in a great number of birds nesting simultaneously, and thus, the likelihood of any single individual being taken by a predator is reduced. Also, since there is not an extended period of food availability for predators, their numbers are kept low. By leaving certain geographical areas each year, migratory species deprive many parasites and microorganisms of permanent hosts to which they can closely adapt. Long, harsh winters in the frozen north further reduce the numbers of these threats.

SUMMARY

Natal dispersal may be defined as "the movement an individual makes from its point of origin to the place where it reproduces or would have reproduced had it survived and found a mate." Thus, natal dispersal involves permanent movement away from the birth site. Natal philopatry occurs when offspring remain at home until death.

Costs and benefits are associated with philopatry and natal dispersal. Potential costs of philopatry include those of inbreeding. Inbreeding increases homozygosity and thus reduces variation among progeny and increases the risk of producing offspring that are homozygous for deleterious or lethal recessive alleles. Competition with relatives is another cost of philopatry. If conditions are sufficiently crowded at home, philopatric young may face intense competition from relatives for food, nest sites, or mates. Limited access to critical resources could result in reduced survival or fertility of philopatric young relative to those offspring that disperse.

Since a certain level of inbreeding may actually be beneficial because of the maintenance of gene complexes that are particularly well suited to local conditions, genetic benefits are also associated with philopatry. Familiarity with the local physical and social setting is another benefit because animals can find food and escape predators more easily in a familiar area. Dispersers, on the other hand, are thought to face high energy costs and risks of predation as a result of increased movement and lack of familiarity with the physical and social environment.

The direction of sex differences in dispersal differs between birds and mammals. In most birds, females are more likely to disperse than males. In most mammals, however, males are more likely to disperse than females. At least three hypotheses have attempted to explain patterns of dispersal in birds and mammals—inbreeding avoidance, reduction in mate competition, and reduction in reproductive competition with parents. The inbreeding avoidance hypothesis is that one sex disperses to avoid inbreeding and the other is philopatric and enjoys the benefits of familiarity with the locality. There are also two suggestions regarding the direction of the sex bias in dispersal. One focuses on differences in the ways in which male and female birds and mammals compete for critical resources (in birds, mating systems are largely resource-defense, and in mammals mate-defense). The other suggests that the sex that gets first choice of the breeding site is philopatric; the other sex disperses. A second hypothesis for sex-biased dispersal is that it reduces competition for mates. Most mammals are polygamous, and competition among males for females can be intense. Subordinate males may

disperse to increase their chances of mating. Most birds are monogamous, and competition for mates is equal between the sexes. Thus, males and females should disperse in equal proportions. The Oedipus hypothesis focuses on the abilities of male and female offspring to cheat their parents.

Consideration of the costs and benefits of each choice can also help us understand the patterns of space use among animals. A territory is a defended area that generally contains a critical resource. The microtine rodents show tremendous variation in the degree of territoriality among species and between sexes within species. The hypotheses to explain the differences in territoriality emphasize that males and females differ in which resource is critical for their reproductive success and is, therefore, worth defending. For females, food is critical; for males, it is mates.

Migration is repeated movement between two locations, usually a feeding or wintering site and a breeding area. Migratory habits exist if the individuals that migrate leave more offspring than those that do not. Migration also has costs and benefits. Among the costs are energy expenditure, predation, dangers of crossing unfamiliar and inhospitable terrain, and risk of exposure to severe weather. The advantages may include a favorable net energy balance. In spite of the energy cost of migration, an animal may gain by escaping the metabolically draining harsh temperatures and by avoiding the increase in competition that accompanies a reduction in food supplies. It may also gain by a reduction in predation or parasitism.

12

Foraging

Atop a rock in a meadow in California sits a western fence lizard (*Sceloporus occidentalis*), basking in the sun and surveying the area for grasshoppers, a favorite prey. Nearby, a common raven stalks the lizard, but the lizard darts to safety in a crevice beneath the rock. The raven does not remain hungry very long, however. A neighboring fence lizard venturing far from its protective fence post in search of an insect meal is snatched and quickly eaten by the raven.

A careful observer of this scene will note that the distribution of plants in the vicinity of a lizard-occupied rock varies from that surrounding a similar rock with no lizard. These lizards don't eat plants, so how can this be? The answer lies in the food web interactions of this community. The western fence lizard is a central place forager,

which means that it always returns to the same place after each foraging bout. Although these lizards will eat many types of insects, they consume primarily grasshoppers. The grasshoppers eat plants. Grasshoppers are found throughout the meadow, but a lizard doesn't travel far from its refuge—a rock, tree stump, or brush pile. It hunts close to home. As a result, the density of grasshoppers increases with distance from lizard-occupied structures. And the abundance of plants (the food of grasshoppers) varies inversely with the density of grasshoppers. Plant biomass is high near lizard refuges and decreases with distance. The types of plants that are present also varies with distance from lizard-occupied structures. The proportion of herbaceous forbs, the favorite food of grasshoppers, is highest closer to lizard refuges, and the proportion of edible forbs is higher at distances from these structures (Chase 1998). We see, then, that although the western fence lizard doesn't eat plants, its presence can change the distribution of plants, as well as animals, in its vicinity.

Indeed, life on this earth is tied together in an intricate web in which energy is exchanged between those that eat and those that are eaten. Although the names of the players may differ, the deadly pageant that paradoxically sustains life is played out on all the earth's various stages—marsh and river, tundra and forest, desert and steamy jungle alike.

OBTAINING FOOD

One of the hallmarks of life is the transference of energy. Once life has captured energy in molecular bonds—whether as a tiny flowering plant on an alpine hillside or as some little-known life form on the ocean floor—that energy is destined to be transferred from one kind of organism to another as one is eaten by the next.

Essentially, what we will consider here is the transference of the energy in molecular bonds from any organism to some animal. Therefore, we will consider the foraging, or food-getting, behavior of both herbivorous (plant-eating) and carnivorous (flesh-eating) animals, keeping in mind that most species are omnivorous and eat a variety of food types.

FILTER FEEDING

Many aquatic animals feed by straining food from the surrounding water. For example, bivalve mollusks, such as clams, are able to filter food particles from their water environment by pumping water across feathery gill structures and mucous membranes. Although filter feeding is more common in invertebrates than in vertebrates, there are some interesting vertebrate examples. One is the flamingo. Its beak is modified for filtering crustaceans and small mollusks from the mud of shallow ponds and lakes. Another significant example is one of the largest known life forms: the baleen whale. Epidermal plates in the roof of its mouth have become specialized into filter structures that strain small crustaceans and other zooplankton from the water.

PLANT EATING

Plants have a variety of parts—leaves, stems, fruits, and flowers, to name a few—each of which may be used by different species of animals. Giraffes may browse on leaves and wildebeest graze on grasses. The larvae of many insects are leaf-miners, and they chew tunnels through the nutritious photosynthetic tissue of the leaf that is found between the upper and lower surfaces. Other species, bumblebees, for instance, collect nectar from flowers for a living. Others, including the fruit bat, eat fruit, and yet others gather seeds.

A variety of animals cultivate some or all of the food they need. For example, leaf cutter ants (*Acromyrmex* spp.) cut fresh leaves and carry the pieces back to the nest under the ground (Figure 12.1). There they alter the plant material, encouraging the growth of a special fungus on the leaves. This fungus, whose existence is unknown outside the ant nests, is thought to be the primary food source for these ants (Weber 1972). The ants actively prepare their fungus gardens. Pieces of suitable leaves are first licked on both sides, a process that removes the waxy layer covering the leaf and reduces the population of microorganisms that might compete with the desired fungus. The leaf fragments are then chewed to a pulp, placed in the fungus garden, and inoculated with hyphae of fungus. This preparation makes the leaves a richer source of nourishment for the fungus. The ants are, indeed, good farmers. Fungus gardens in abandoned nests degenerate quickly as they become overrun with other microbes (Quinlan and Cherrett 1977).

Certain mammals also cultivate plants. Prairie dogs pull out the grasses that they do not like to eat, resulting in the proliferation of the grasses they do like

FIGURE 12.1   Leaf cutter ants are transporting leaves to fertilize their fungus gardens. These ants cultivate a special fungus that serves as their primary source for food.

around their burrows. Gorillas are herbivorous and are known to rip down large plants, the result being a surge in the growth of young, fast-growing vegetation, which is encouraged by the new light.

HUNTING

Traps

Some predators are able to trap their prey, that is, to manipulate objects or alter their environment in such a way as to capture, or at least restrain, prey. For example, humpback whales (*Megaptera novae-angliae*) build bubble "nets" to trap, or more accurately to corral, their prey. Beginning about 15 meters deep, the whale begins to blow bubbles by forcing bursts of air from its blowhole while swimming in an upward spiral. The big bubbles, along with a mist of small ones, form a cylindrical net that concentrates krill and small fish (Figure 12.2). The whale then swims upward through the center of the bubble net with an open mouth and devours the prey in a single gulp (Earle 1979).

More familiar examples of animals that trap their prey are spiders. In the summertime you may notice the traps of orb-weaving spiders in the doors and windows of old houses and barns, in grasses or trees, or in any area where their sticky filaments can intercept flying insects. The resident spider usually positions itself so it will not be obvious to the prey. Then, when the web vibrates at a given frequency, the spider rushes out. The vibration is the signal that prey is in the web and must be secured with additional silk before it can escape.

Certain spiders have evolved hunting techniques that take advantage of the ways in which insects perceive and process visual information. An example is the golden orb-weaving spider, *Nephila clavipes*, which is found in a variety of habitats throughout the tropics and subtropics. A golden orb-weaver spins webs of different colors and adjusts the color to the light characteristics of the habitat in which it forages. In shady environments, such as forests, this spider spins webs that appear to be white (unpigmented) to approaching prey. However, in sunny, nonforest or edge habitats, it spins conspicuous yellow webs. Yellow is the color of new leaf material and flowers and is therefore associated with positive rewards for the herbivorous and pollinating insects that the golden orb-weaver preys on. Although the stingless bee, *Trigona fluviventris*, can see and learn to avoid spider webs, it is attracted to the yellow webs and has difficulty learning to avoid them. It is thought that the bee has evolved to associate yellow, the color of many flowers, with food, and so it has difficulty associating the web with danger. The ability to adjust the color of its silk to the lighting conditions in a variety of habitats may allow the golden orb-weaver to forage wherever insects are locally abundant. This, in turn, might explain why the spider is so widely distributed (C. L. Craig 1994a; C. L. Craig, Weber, and Bernard 1996).

Spiders of the genus *Argiope* lure certain flying insects to their traps by decorating their webs with bars and crosses that reflect ultraviolet light (Figure 12.3). These web adornments actually seem to attract prey, especially pollinating insects, such as stingless bees, that forage on flowers. Studies have shown that webs with ultraviolet reflecting decorations catch more insects per hour than do undecorated ones. Since many flowers have regions that reflect ultraviolet light to attract pollinators, it is thought that the web decorations attract insects by sending the same visual cues as a flower (C. L. Craig and Bernard 1990). Furthermore, the spiders change their decorations daily, and so the webs look dif-

FIGURE 12.2 The humpback whale traps krill in cylindrical bubble "nets" that it constructs by forcing air out of its blowhole. In this photo the bubble net is seen as a ring of bubbles on the surface of the water. Using a bubble net, a whale can corral the fast-swimming krill that would otherwise scatter and consume them in a single gulp.

FIGURE 12.3   Ultraviolet decorations on the webs built by *Argiope* spiders lure certain flying insects into the trap.

ferent to foraging insects that return to the areas where the spiders are found. The changing visual cues make it difficult for insects, such as stingless bees, to learn to avoid the webs (C. L. Craig 1994b).

It is interesting that the spiders adjust the web size, design, and degree of decoration to the abundance of prey. When there are few prey, the spiders build larger webs, using more silk, but with few decorations. This web-weaving strategy is used as long as the availability of prey is constant. However, when the prey supply is variable, the spiders increase the number of decorative bands but not the amount of silk invested in their webs (Herberstein, Craig, and Elgar 2000).

Aggressive Mimicry

In aggressive mimicry, the predator gets close to its prey because it mimics a signal that is not avoided by the prey and may even be attractive to it. Some predators have specialized structures that are used as lures. The alligator snapping turtle holds its mouth wide open while it lies on the bottom of muddy streams and lakes. The tongue has a wormlike outgrowth that is wiggled, attracting fish that the turtle then snaps up. It is interesting that lures can be used where vision is normally not possible. The deep-sea angler fish lives at extreme ocean depths, where virtually total darkness prevails. Nonetheless, the female has a long, fleshy appendage attached to the top of the head that is luminous and acts as a lure, bringing curious fish within the reach of its jaws (Wickler 1972).

Some predatory species draw within striking distance of their unsuspecting prey by mimicking beneficial species. For example, certain blennid fish look and

act like beneficial cleaner fish to lure their prey. The cleaning wrasse (*Labroides dimidiatus*) removes external parasites, diseased tissue, fungi, and bacteria from other fish (Figure 12.4a). Indeed, parasite-laden fish line up at coral reef "cleaning stations" much like cars at a car wash to avail themselves of the services of the cleaning wrasse. Some cleaning stations may have hundreds of patrons each day. The cleaning wrasses generally advertise their services with distinctive swimming motions performed above the cleaning station. Customers show their willingness to be cleaned by postures that are characteristic of the species. The blennid fish (*Aspidontus taeniatus*), a phony cleaner, looks and behaves like the cleaning wrasse (Figure 12.4b), but when it is invited to approach, it takes a bite out of the customer (Wickler 1972).

Many predators attract their prey by sending signals that mimic the mate of the prey species. For instance, a certain jumping spider of Queensland, Australia, *Portia fimbriata*, mimics the vibratory courtship display of another type of jumping spider (*Euryattus* spp.) on which it feeds. Unlike most other jumping spiders, the *Euryattus* spiders nest within rolled-up leaves that dangle from rocks or vegetation by heavy guylines. A courting *Euryattus* male stands on the suspension nest of a female and performs a vibratory display called shuddering. This causes the female to come out of her nest. *Portia* predators shudder in a manner similar to that of a courting *Euryattus* male (Figure 12.5), causing the female to emerge from her leaf nest. If she retreats before capture, the hunter waits and repeats the action. In experimental tests in which *Portia* females encountered suspension nests of female *Euryattus*, the predator successfully attacked and killed the prey 40% of the time (Jackson and Wilcox 1990).

The young of a certain fish, *Erythrinus erythrinus*, use a variation of this tactic to attract their prey, which is another fish, *Rivulus agilae*. In this case, the color pattern of the young predator mimics that of the female of its prey. Males of the prey are attracted to the predator and begin to court it. While the male is courting the "female," the predator catches the suitor by the tail and swallows him (Brosset 1997).

Antidetection Adaptations

In some cases the ability of the predator to draw near is enhanced by camouflage, rendering the predator more difficult to detect. Jumping spiders of the genus *Portia* use environmental disturbances as a smokescreen to camouflage their approach to the prey. *Portia* preys on other spiders, which often involves crossing the prey's web. *Portia*'s movement shakes the web slightly, and this can alert the prey to the predator's approach. But, *Portia* often masks its movements by timing the hunt to occur when the prey's web is shaking because of some

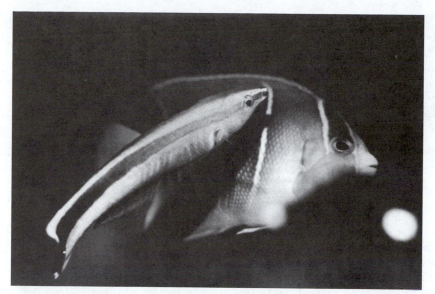

(a)

FIGURE 12.4 (a) A cleaner wrasse is removing parasites from another fish. The service of cleaner fish is beneficial because it removes external parasites, diseased tissue, fungi, and bacteria. The cleaner wrasse has distinctive markings and behavior patterns that advertise its services. (b) *Aspidontus* is disguised as a cleaning wrasse and mimics its behavior. As a result, fish in need of the cleaning service allow the phony cleaner to approach. *Aspidontus* then takes a bite out of the would-be customer.

(b)

environmental disturbance, such as wind blowing through the web (Wilcox, Jackson, and Gentile 1996).

Adaptations for Detection of Prey

Most foragers must search for food items. It should not be surprising, then, to find that they have sensory specializations that increase the chances of finding an edible item. As we look around the animal kingdom, we find many sensory specializations for prey detection that may seem unusual to us because the sense is poorly developed or lacking among humans.

The snakes in two large families, the Crotalidae, or pit vipers (e.g., rattlesnake, water moccasin, and copperhead), and the Boidae (e.g., boa constrictor, python, and anaconda), use their prey's body heat to help guide their hunt. They have special receptors that are so sensitive to infrared radiation (heat) that these snakes can locate their warm-blooded prey even in the darkness of night (Figure 12.6). A rattlesnake, for instance, can detect a mouse whose body temperature is at least 10°C above air temperature within a range of 40 centimeters. Furthermore, the detection is swift, within half a second, as it must be if the snake is to strike successfully at a mouse scampering by, hidden in the night's blackness.

These snakes can locate the source of infrared radiation with amazing accuracy. Laboratory studies have shown that even with both eyes covered, a rattlesnake can strike a warm object, such as a soldering iron, within 5° of dead center of the source (Figure 12.7). This accuracy, which may be deadly for a mouse, can be traced to the structure of the pit organs that house the

FIGURE 12.5   The predatory jumping spider *Portia* performs a vibratory display that mimics the courtship display of another species of jumping spider, *Euryattus,* on a female's leaf nest. This action lures her out of her nest, where she can be attacked.

infrared sensory endings. A small, warm object within half a meter of the pit opening stimulates only a small region of the epithelium, and the location of the stimulated portion will be interpreted by the brain to reveal the location of the source. Furthermore, the great density of heat-sensitive endings in the membrane permits accurate determination of the position of the stimulated patch. By turning its head from side to side, the snake can locate the boundaries of the heat source, thereby determining the size of the warm object. In addition to allowing the snake to select prey of an appropriate size, a mouse and not a moose, for example, knowledge of the boundaries of the potential meal allows the snake to strike with great accuracy (Gamow and Harris 1973; Newman and Hartline 1982).

The sand scorpion (*Paruoctonus mesaensis*), a nocturnal predator of the Mojave Desert, detects and locates prey by the vibrations they generate in the sand.

FIGURE 12.6   (*a*) The heat sensors of the black-tailed rattlesnake, located in pit organs between each eye and nostril (the dark circular structures in this photo), allow it to detect its warm-blooded prey in complete darkness. Snakes of the families Crotalidae and Boidae have infrared heat sensors that help them find their prey. (*b*) A rattlesnake's head-on infrared picture of a mouse.

(*a*)

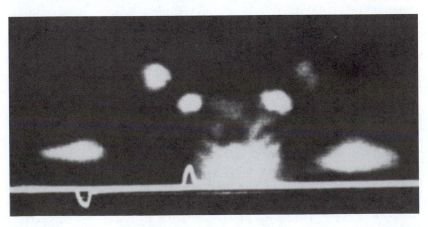

(*b*)

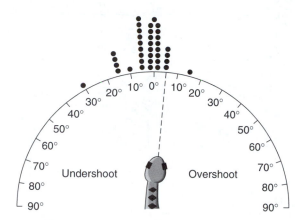

FIGURE 12.7   **A rattlesnake's heat sensors can direct an amazingly accurate strike. Each circle indicates the angular error of a strike at a warm soldering iron by a blindfolded rattlesnake. The average error of the strikes was less than 5°. (Redrawn from Newman and Hartline 1982.)**

The scorpions can detect disturbances up to 30 centimeters away and can locate the distance and direction of the source of vibration almost perfectly at distances of 10 centimeters or less (Brownell 1984).

The star-nosed mole (*Condylura cristata*) finds prey by using its sense of touch. Critical to that sense is its unusual star-shaped nose, consisting of 22 fleshy, mobile appendages. Aside from its snout, this mole looks like most moles. It is about 6 inches long and has powerful digging claws and poorly developed eyes. Rarely coming to the surface, the star-nosed mole lives in the extensive tunnel system it digs in the wetlands. It searches these tunnels for earthworms and other small prey by touching the walls with its nose as it moves forward. As the nose touches the surface, its fleshy appendages sweep forward; they sweep backward as the nose is lifted. Each second, the nose samples ten or more spots in the tunnel. The movements are controlled by muscles that are attached to the skull with tendons that attach to the base of each appendage. Each appendage is sheathed with touch receptors, called Eimer's organs, that communicate with the somatosensory area of the cortex of the mole's brain. When the nose touches prey, the mole grabs the animal in its mouth and quickly eats it (Catania and Kaas 1996).

Certain seabirds seem to sniff out seafood patches scattered over vast expanses of open ocean (Nevitt 1999a). Procellariiform seabirds, including the petrels and albatrosses, are sometimes called the tube-nose seabirds because of their long tubular nostrils, a feature that suggests that they have a well-developed sense of smell. Indeed, they do, and this sense seems to help them locate patches of food such as squid, fish, and krill. Krill are small shrimplike animals that are the principal herbivores of the Antarctic food web. All animals in the region depend on krill for food—directly or

indirectly. Thus, areas with high densities of krill are likely to attract many other animals in the food web. So, if these seabirds find a patch of krill, they are in the right general area for finding food. Once in the foraging hot spot, the birds can use other cues to guide their hunt.

Odors associated with krill seem to be "peaks" in the olfactory landscape that help the seabirds pinpoint good foraging areas. Some such odors may come from the krill themselves. As predators eat krill, the macerated krill bodies may release volatile compounds that seabirds can smell. Another odorous gas, dimethyl sulfide (DMS), is released from phytoplankton. The krill graze on phytoplankton, and so DMS can also lead the seabirds to krill.

Gabrielle Nevitt and her colleagues have demonstrated that the odors of macerated krill and of DMS attract seabirds and that the attractiveness is species-specific. They created vegetable-oil slicks, some of which were perfumed with krill extract; plain vegetable-oil slicks served as controls. Observers who did not know which of the slicks were krill-scented counted the number of seabirds attracted to the slicks and measured how quickly they appeared. Cape petrels, southern giant petrels, and black-browed albatrosses showed up within minutes. Five times more of these birds appeared at the krill-scented slicks than at the control slicks. The odor did not attract storm petrels or Antarctic fulmars. Equal numbers of these birds showed up at control and krill-scented slicks (Nevitt 1999b). Similar experiments compared the attractiveness of DMS-scented slicks with plain vegetable-oil slicks. Prions, white-chinned petrels, and two species of storm petrels showed up at DMS-scented slicks twice as often as at control slicks. But, three species of albatrosses and Cape petrels were just as attracted to the control slicks as to the scented ones (Nevitt, Veit, and Kareiva 1995). Thus, it seems that the odors of krill or DMS can provide an immediate and direct way to assess the potential productivity of an area but that different species of seabirds use these cues in different ways.

Sharks have a well-deserved reputation as consummate predators. Many senses are involved in the shark's astounding ability to detect and track down prey. It can hear prey from a distance of over nine football fields and will turn and swim toward it. When the shark is within a few hundred meters of its prey, its nose can help direct its search. As the shark nears its prey, its lateral line organs detect small disturbances in the water caused by the swimming motion of prey, and so the shark knows the location of the prey even if it cannot see it. Within close range, the shark can see its prey. Perhaps the most amazing of all the shark's senses is its ability to detect small electrical fields.

Nearly all living organisms generate electrical fields when they are in seawater that could reveal their presence, type, and perhaps even their physiological condition to an interested predator. The slightest movement, even the beating of the heart, could betray the quarry to an electrosensitive predator, such as a shark. Even if the prey could remain perfectly motionless and will its heart to stop beating, the voltage difference between its body fluids and seawater or between different parts of its body would send an electrical message, revealing its location to its predator. Wounded prey are an easy mark because even a minor scratch can double this voltage difference and send an even stronger signal (Bullock 1973).

Adrianus Kalmijn (1966, 1971) was the first to demonstrate that sharks (*Scyliorhinus canicula*) are sensitive to weak electrical fields, and he later showed that sharks can use this sense to locate prey by the small electrical fields caused by action potentials within the victim's body. More recently, Michael Watt and his colleagues (1999) have shown that the Australian lungfish (*Neoceratodus forsteri*) also has an electrical sense that is used in locating prey. First the lungfish was observed, and its behavior while locating food was described. Then stimuli from a prey animal, a crayfish, were manipulated to determine which ones were most important. The crayfish was housed in a small chamber hidden beneath the substrate. The chamber blocked visual, mechanical, and chemical cues from the crayfish, although it did allow the transmission of bioelectric fields. The lungfish located the hidden chamber containing the crayfish (the target) and showed sustained feeding movements above the chamber (Figure 12.8). When the chamber was empty or when the crayfish was dead, the activity of the lungfish was not in the vicinity of the chamber and no feeding movements were observed. We see, then, that the lungfish is capable of detecting hidden prey by passive electroreception alone.

Next, the relative importance of bioelectric and chemical cues was determined. A length of flexible tubing led from the chamber to another area of the aquarium (the distractor). The experimenters then added chemicals of crayfish odor to the tube so that the chemical stimuli would be displaced from the bioelectric stimuli. The lungfish continued to show foraging activity above the target chamber, not in the area containing crayfish odor. However, if a dead crayfish was in the chamber, or if the chamber was shielded to block bioelectric cues and crayfish odor was released in the distractor region of the aquarium, the lungfish did respond to the chemical stimuli and foraged in the distractor region. So, the lungfish can and does use chemical cues to locate prey; but when both cues are present, bioelectric cues of the prey's location are more important to the lungfish than chemical cues.

Other electroreceptive fish and some amphibians also locate live prey by using bioelectric cues. It is inter-

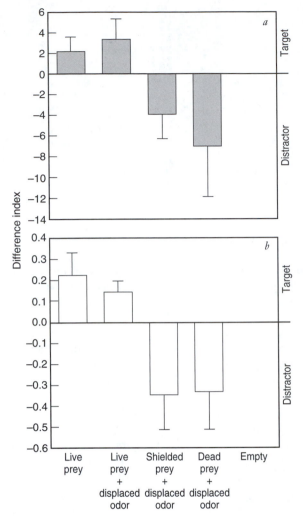

FIGURE 12.8 **The results of experiments demonstrating that Australian lungfish can locate their prey by using the electrical field generated by any living organism in seawater. The results are shown as a difference index, which is the amount of foraging activity above the target, the hidden chamber where the prey was located, and the distractor, the region of the aquarium to which the prey's odor was displaced. A positive score indicates that the fish were foraging above the target; a negative score indicates that the fish were foraging over the distractor. The distractor indexes are shown for (*a*) the foraging intensity and (*b*) the foraging accuracy of each treatment. Although the lungfish can and do use chemical cues to locate prey, bioelectric cues are more important. (From Watt, Evans, and Joss 1999.)**

esting that the platypus (*Ornithorhynchus anatinus*), an egg-laying Australian mammal, has both electroreceptors and mechanoreceptors on its ducklike bill that help in locating prey (Figure 12.9) (Griffiths 1988; Scheich et al. 1986).

The platypus hunts at night in murky streams. It swims slowly, patrolling for food with its eyes, ears, and

FIGURE 12.9 **The duck-billed platypus is among the predators that can use electrical fields generated by its prey to locate them.**

nostrils closed. As a platypus swims, it makes intermittent head sweeps, scanning the area for electrical signals. When an electrical stimulus is detected, it swims two to three times faster and makes larger, regular head sweeps at an average rate of two sweeps per second. The platypus is most sensitive to electrical stimuli that originate slightly below it and almost completely to its side (80° lateral from the tip of the bill). A hunting platypus will often dive to the bottom of the body of water and dig in the silt with its paws and bill. As it probes, it stirs up small animals, including crayfish, insect larvae, and freshwater shrimp. When disturbed in this way, the prey generally darts away. These movements cause electrical fields that are picked up by the electroreceptors, as well as mechanical waves in the water that are sensed by the mechanoreceptors on the platypus's bill. Together the electroreceptors and mechanoreceptors give the platypus a complete, three-dimensional fix on the position of its underwater prey (Manger and Pettigrew 1995; Pettigrew, Manger, and Fine 1998).

The lungfish and the platypus use electrical cues given off by animals to locate prey, but other species locate prey by generating their own electrical field and sensing disturbances in that field created by the prey. Consider, for example, the black ghost knifefish (*Apteronotus albifrons*). This nocturnal predator uses its electrical sense to feed on small prey such as the insect larvae and small crustaceans in the freshwater rivers of South America. This fish has two types of electroreceptor organs distributed over its entire body. One type, the tuberous electroreceptor organ, is specialized for detecting alterations in the electrical field that the fish generates itself and uses for orientation (electrolocation is discussed further in Chapter 10). The ampullary organs, which are specialized to detect electrical fields generated by external sources, such as prey animals,

make up the second type of electroreceptor organs. These receptors, along with the mechanosensory lateral line system, help the fish hunt for prey at night or in muddy water. An agile swimmer, this fish can swim backward as easily as forward. It can even swim upside down or horizontally while making sweeping searches for prey, and it can hover. By changing its position, velocity, and orientation, these fish can influence the pattern of incoming electrical signals to assist a hunt for prey.

Mark Nelson and Malcolm Maciver (1999) have studied prey capture by the black ghost knifefish by analyzing infrared video recordings of it feeding on water fleas (*Daphnia*) in total darkness. While searching for prey, the black ghost knifefish usually swims with its dorsal side leading, which is accomplished by swimming forward with its head pitched downward, or backward with its head pitched upward. This body position increases the volume of space that is searched. When the fish is swimming with its body in a horizontal position, the head and tail receptors pass through the same volume of space, whether it is swimming forward or backward. The dorsal surface has the greatest density of electroreceptors. This swimming posture also makes it easier for the fish to approach the prey. Once the prey is detected, the fish performs a reverse scan of the area, during which the distance between the prey and the fish is shortened. Then the fish lunges to capture the prey.

Hunting Cryptic Prey

Prey animals have developed several means of reducing the probability that they will be detected by their predators, as we will see in the next chapter. One such technique is crypsis (or camouflage)—blending in with the background.

FIGURE 12.10   (*a*) A blue jay is "foraging" for cryptic prey in a Skinner box. The blue jay was shown slides, some of which contained a cryptic moth. Among the cryptic moths shown to the blue jays on slides were (*b*) *Catocala retecta* and (*c*) *C. relicta*. If the bird was shown a slide that contained a moth and pecked the appropriate key after spotting it, the bird received a mealworm reward.

a

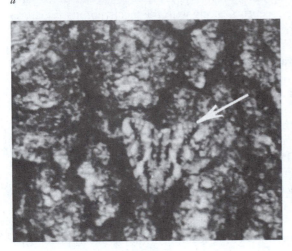

b

c

Clearly any method that a predator can use to increase its ability to detect its cryptic prey will be beneficial. One method might be the formation of a "search image," in which the predator learns characteristic features of its prey and searches for these when hunting. This concept is a familiar one. If we are looking for our car keys, we do not look for something large, round, and blue; instead, we keep our eyes peeled for something small and shiny. Our search image, then, is centered on the characteristics of the keys. Similarly, an animal's search image is thought to be a mechanism, such as filtering out unimportant stimuli or focusing attention on the important stimuli, that increases a predator's ability to find cryptic prey.

Luuk Tinbergen (1960) employed the concept of search image formation to explain the change in diet of the birds he observed feeding their young in the Dutch pinewoods. He noted that the abundance of particular species of insect varied throughout the season. A

species might be scarce at first, become plentiful later, and then disappear. However, the birds did not consume all perfectly palatable prey types in direct proportion to their abundance. There was usually a delay between the appearance of a given insect species in the environment and the time when the birds began to sample it. But once they did, there was often a sharp increase in the rate of predation on the new prey. Eventually, abundant prey were selected more than would be expected by their relative frequency. Tinbergen suggested that the sudden change in diet might have occurred because the predators learned to see their cryptic prey through the use of search images as they learned the prey's traits.

We do, indeed, find fairly sudden changes in the diet of several kinds of predators, and that acceptable prey species are not always consumed in proportion to their abundance. But are these changes in menu due to the formation of a search image? To find out, one would need to devise an experiment in which several types of prey were equally acceptable and all equally easy to catch and to handle.

There have been several approaches to testing the hypothesis that a search image is responsible for the change in a predator's diet, but we will discuss only two. One uses operant conditioning. Birds respond to images projected on a screen by pecking at keys in a Skinner box (see Chapter 5). The birds do not eat the prey they detect. Instead they receive a standard food reward for a correct response. Thus, palatability and ease of capture and handling are constant. The other approach involves "free-feeding" under controlled conditions. The type of food item is kept the same, but its crypticity is altered by varying its color or that of the background so that they match or contrast.

Alexandra Pietrewicz and Alan Kamil (1981) used the operant-conditioning approach to simulate predator-prey interactions between blue jays (*Cyanocitta cristata*) and one of their normal prey types, underwing moths (*Catocala* spp.). Hungry blue jays were given a slide show in a Skinner box theater (Figure 12.10). Some slides contained a moth; some did not. If a moth was present and the bird saw it, it could peck at the slide and receive a mealworm reward. After a short interval, the next trial began. When no moth was present, the blue jay could abandon its search of that area and look elsewhere by pecking at a key that would then advance the projector to the next slide. There are two types of mistakes the bird could make. One is a false alarm—incorrectly responding as if a moth were present. The other is a miss, in which the bird failed to see the cryptic moth and pecked at the advance key. Either of these mistakes resulted in a delay before the hungry predator could search for prey in the next slide.

Pietrewicz and Kamil (1979, 1981) used this operant-conditioning procedure to determine whether or not the birds formed search images. The search image hypothesis would predict that recent experience with a prey type would allow the birds to learn its key characteristics and prime them to look for others of that type. It was, therefore, reasoned that encounters with one species of cryptic moth would help jays find that type more accurately and quickly, but experience with another species would interfere. The jays were shown a series of slides, half of which contained a moth. They were tested under three conditions: (1) The moth-containing slides were of *C. retecta* only, (2) the moth-containing slides showed only *C. relicta*, and (3) the two moth species were shown in random order. The results, shown in Figure 12.11, are exactly what would be predicted by the search image hypothesis. The jays' ability to detect one prey type improved with consecutive encounters. However, although both prey types were cryptic, prey detection did not get better when the two species were encountered in random order. Thus, we see that experience with one type of cryptic prey improved the predator's ability to find that type of prey but not other kinds of cryptic prey.

Marion Dawkins's (1971) investigations on search image used domestic chickens freely feeding on colored rice grains. In some trials, these rice grains were made conspicuous by placing them on a contrasting background. In others, they were made cryptic by presenting them on a matching background. Each chick was observed feeding under both conditions. The chicks quickly spotted the conspicuous grains and consumed them at a high rate. Although the chicks took several minutes to detect the cryptic grains, by the end of a test period they were eating them at the same rate as they did the conspicuous grains. Thus, the results of this experiment are also consistent with the idea of search image formation.

Although it is clear that predators get better at finding cryptic prey with experience, not everyone agrees that this is accomplished by forming a search image. It has been suggested, for instance, that changes in search rate may account for the increased ability to find cryptic prey (Gendron and Staddon 1983, 1984). The general idea of the search rate hypothesis is that after a predator first finds a cryptic prey item, it may scan the area more slowly and look more carefully for others. If more time is spent viewing an area, the predator is more likely to notice camouflaged prey than if it only glances at the spot. However, if there is no prey hidden in the area, the time spent looking for it is wasted. Therefore, the predator faces a speed/accuracy tradeoff. As a result, it might optimize its foraging by adjusting its search rate to the degree of crypticity and abundance of the prey. If conspicuous food items are abundant, it is best to inspect the area quickly. But if

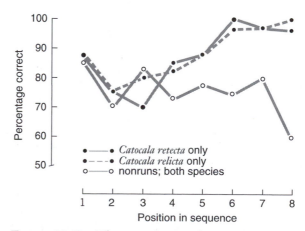

FIGURE 12.11   The percentage of correct responses by blue jays when shown moth slides in sequences of the same species (runs) or a random sequence of both species. After experience with one cryptic species of moth, blue jays became better able to detect that species. However, if the species were shown in random order, the jays' performances did not improve. This is consistent with the search image hypothesis. (Data from Pietrewicz and Kamil 1981.)

there are many cryptic prey to be found, a slower search may actually increase the rate at which food items are discovered because fewer will be missed.

It seems that birds do modify their search rate when the degree of crypticity of their prey changes. Observers of birds foraging in nature often notice that birds appear to hunt more slowly when their prey is cryptic (Goss-Custard 1977; J. N. M. Smith 1974). However, it is difficult to measure crypticity and search time simply by watching birds hunt. Quantification is easier under controlled conditions and in studies in which quantification has been possible the results confirm that the search rate varies with the degree to which the prey is camouflaged. For instance, measurements were made of the search rate of bobwhite quail (*Colinus virginianus*) that were hunting artificial prey in an aviary. The degree of camouflage of the prey, wormlike pellets made from dough, was manipulated by changing their color. The birds reduced their hunting speed as their prey became more cryptic (Gendron 1986).

When a predator becomes more efficient at discovering cryptic prey, we might wonder whether it is forming a search image or adjusting its search rate. The increased efficiency would be predicted by either hypothesis. Similarities in predictions of the two hypotheses have triggered an argument over whether the results of the studies on search image previously described, as well as others, are equally consistent with the search rate hypothesis (Guilford and Dawkins 1987). If the simpler search rate hypothesis can explain the change in the accuracy of prey detection, should we toss aside the idea of search image as an unnecessary complication? This question led to a reexamination of the search image hypothesis.

Evidence for search image formation is provided by studies using operant techniques. Consider the work of Patricia Blough (1989, 1991), who used a rather abstract foraging situation. In her experiments, the "prey" were letters of the alphabet displayed on a screen. The predators were pigeons (*Columba livia*), trained to look for a particular target letter of the alphabet in a display of several letters. When the pigeons pecked at the correct letter, they received a food reward. The degree of crypticity could be enhanced by increasing the number of nontarget letters in the display. Blough reasoned that if a predator knows that cryptic prey are available, its attention can be focused on finding them, and it will find them more efficiently. This priming of attention might be analogous to a search image. She tested this idea by providing a cue, for instance, vertical screen borders, that would prime the pigeon to expect a target letter in the display that followed. The results show that pigeons that were cued to expect a target did find it more quickly (Blough 1989).

The best way to determine whether predators use search images is to observe their behavior in situations in which the search image and search rate hypotheses predict different responses. One such situation is when there are several types of cryptic prey hidden in the hunting area. The search rate hypothesis predicts that a predator in this situation will find all types of cryptic prey with equal ease. If recent experience with one type of cryptic prey causes a predator to slow its search for others, it will be more likely to find cryptic prey of any type. However, a search image is specific for one type of prey and would, therefore, be less likely to help a predator detect other prey types. The search image hypothesis assumes that because of experience with one type of cryptic prey, a predator forms a search image based on the key characteristics of that particular prey. Therefore, a search image for one type of cryptic prey might even interfere with the detection of other types.

Blough (1991) simulated a situation in which there were two forms of cryptic prey by making the target either one of two letters. She also provided two types of cues—one signaled that either of the target letters would be present; the other signaled a specific letter. She found that a pigeon primed to expect a particular letter found it more quickly than if the cue simply informed the bird that one of the two letters would appear in the display. Furthermore, when the pigeon was misinformed about the identity of the expected target, it took longer to find it, even if it was conspicuous, than to find a cryptic target that was correctly cued. This specificity is just what would be expected if a search image were formed.

Another study that used operant techniques distinguished between the search image and search rate hypotheses by varying the frequency of the cryptic prey (Plaisted and Mackintosh 1995). In this case, the prey were checkerboard stimuli hidden in a larger checkerboard display. There were two target stimuli, which were previously determined to be equally cryptic to pigeons. During the first half of the experiment, one of the stimuli was presented on 80% of the trials and the other stimulus on the other 20%. These frequencies were reversed during the second half of the experiment. If the pigeons slowed their search rate when faced with cryptic stimuli, we would expect to see improved detection of both stimuli. This did not occur, however. Instead, as you can see in Figure 12.12, the pigeons' ability to detect a stimulus improved only during the part of the session in which that stimulus appeared more frequently than the other. Thus, these results are consistent with the search image hypothesis but not the search rate hypothesis.

Another prediction that would distinguish between the search rate and search image hypotheses involves changes in the search rate as a result of experience with cryptic and conspicuous prey. The search rate is most easily measured as response time since the animal is presumed to respond as soon as prey is detected. The search rate hypothesis is that predators learn to adjust their search rate to the degree of crypticity of common prey. Therefore, it would predict that response time

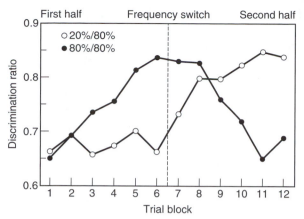

First half   Frequency switch   Second half

○ 20%/80%
● 80%/80%

FIGURE 12.12 **Pigeons presented with two cryptic stimuli in different frequencies show improved detection of the stimulus that is presented more often, a result that is consistent with the search image hypothesis. The cryptic stimuli were squares of checkerboard patterns hidden in a larger checkerboard display that had been previously shown to be equally cryptic. The stimuli were presented randomly. During the first half of the experiment, stimulus A was presented 80% of the time and stimulus B 20% of the time. The frequencies of presentation were reversed during the second half of the experiment. The discrimination ratio is a calculated measure of the pigeons' ability to detect the stimulus. The pigeons' ability to detect the more frequently presented stimulus improved. If the pigeons were slowly to look for cryptic stimuli, we would have expected improved detection of both stimuli. (Data from Plaisted and Mackintosh 1995.)**

would increase as a result of experience with cryptic prey. It should take longer to detect prey and to recognize that there is no prey in that patch of the environment. The search image hypothesis would predict no change or a slight improvement in response time.

These predictions were tested by measuring the response times of pigeons that were detecting the presence of wheat or beans on backgrounds that altered the degree of crypticity. On a sandy background, beans were more cryptic than wheat. The search rate hypothesis, therefore, would predict increases in response time during and following successive presentations of cryptic beans but decreases during and following successive presentations of the more conspicuous wheat. This was not observed. Instead, as the search image hypothesis predicts, the birds detected the beans more quickly as they gained experience in detecting beans through successive encounters (Bond and Riley 1991).

So we see that both ideas seem to have merit. There is evidence that predators form search images, at least in some circumstances, and they may also adjust their search rate. Perhaps this should not be surprising. A forager's life depends on finding food, and so we might expect it to use any technique that will enhance its success.

OPTIMAL FORAGING

A foraging animal may have a variety of potential food items available. Some are easier to find, some are easier to handle and digest, some are easier to capture, and some have a higher nutritive value than others. So, how does an animal decide which to eat?

As we discussed in Chapter 4, optimality theory assumes that natural selection will favor the optimal behavioral alternative. In other words, when an animal is faced with several possible courses of action, the one whose benefits outweigh its costs by the greatest margin will be favored by natural selection.

Foraging has been a favorite subject for testing optimality theory for several reasons. One is that the costs and benefits are easily translated into energy units. This is important because optimal foraging theory attempts to predict the foraging patterns of animals on the basis of net energy gained per unit of time. Some of the factors considered include the caloric value of the food, the amount of energy required to digest it, the amount of time it takes to find it, and the amount of time it takes to handle it. Another reason that foraging is often used to test optimality theory is that it is relatively easy to think of foraging as a series of decisions and then to focus on one decision at a time. Examples of some of these decisions are what to eat, where to look for food, how long to search one area before moving on, and what sort of path to take through the area. Finally, foraging behavior is suited to tests of optimality theory because once an aspect of foraging has been chosen for study, it is relatively easy to identify the behavioral alternatives the individual has available.

According to theory, individuals that forage optimally are likely to be fitter, that is, to leave more offspring, than those that do not. It seems intuitively logical that success in getting food is going to be translated into other kinds of successes because food provides the energy for other activities. But as long as an individual's stomach is full, does it really matter whether it foraged in the most efficient manner? To assume that foraging success need only be adequate would ignore the fact that foraging competes with other activities, even in species that have less stringent energy demands than do shrews or titmice. An individual's "spare" time can be spent on other activities related to survival and reproduction.

We should keep in mind that the link between fitness and optimal foraging is difficult to verify because fitness is measured as lifetime reproductive success and foraging success is usually measured in much more immediate terms. Nonetheless, several studies have suggested a relationship between foraging success and fitness. For example, Peter Sherman (1994) monitored the web building, foraging success, and reproductive efforts of the orb-weaving spider *Larinoides cornutus*. He experimentally supplemented some of the webs with prey. Throughout the summer, only 14 of 38 adult female spiders produced a total of 21 egg sacs. Seventeen of these, nearly 81%, were produced following heavy experimental or heavy natural prey consumption. Another example is provided by a songbird, the water pipit (*Anthus spinoletta*). The number of fledglings produced increased with the biomass of food delivered to them by the parents (Frey-Roos, Brodmann, and Reyer 1995). Thus, the premise that foraging and fitness are related seems sound. Although optimal foraging theory seems reasonable, we should nonetheless examine it with a critical eye.

FOOD SELECTION

The first step in acquiring a meal generally is deciding what to eat. As we will see, this may involve several types of decisions.

Distinguishing Between Profitable and Nonprofitable Prey

An optimal forager must be able to distinguish between food that is worth eating and food that requires more energy to process than it provides. So, we might begin our consideration of optimal foraging by looking for examples of animals that are able to distinguish between profitable and nonprofitable food items. Some evidence comes from crows that feed on shellfish along Canada's western seacoast.

Northwestern crows (*Corvus caurinus*) search along the waterline during low tide for whelks, which are large mollusks. When a suitable whelk is found, the crow carries it in its beak toward the shore, flies almost vertically upward, and then drops the whelk. If the shellfish breaks open, the crow eats it. If not, the bird retrieves the whelk, flies upward, and drops it again, repeating the procedure until it does break.

Reto Zach wondered whether the crows distinguish profitable from nonprofitable whelks. He reasoned that foraging crows must make an energetic profit or they would not do it. In other words, they must gain more calories from the whelk than they use obtaining it. He thought about factors that might influence the profit margin. He asked first whether the crows were selective in their choice of whelks. He wondered whether there were differences in the ease of handling whelks of different sizes. One might expect that flying upward would be the most energetically costly part of eating whelks. Then he asked whether crows adjust their cost according to the expected caloric value of the meal. How many times would a crow continue to drop a whelk that was difficult to crack open?

Zach (1978) found that the crows preferentially prey on large whelks. He demonstrated this in several ways. First, he collected broken pieces of whelk shells,

and by comparing the size of the base of the shells to those of living whelks, he estimated the size of the whelks that had been eaten. The whelks selected by the crows appeared to be among the largest and heaviest on the beach. Larger whelks provide more energy, so this observation was consistent with the hypothesis that the crows could distinguish profitable prey. However, Zach was cautious in his interpretation. Perhaps, he reasoned, the pieces of smaller shells were more easily washed out to sea by wave action, leaving the larger pieces over-represented in this sample. So, Zach offered each of three pairs of crows equal numbers of small, medium, and large whelks and, at hourly intervals, recorded the number of each size taken. Crows selected large whelks (Table 12.1). Even after the crows had eaten most of the large whelks, they continued to ignore smaller ones. Was this because large whelks are more palatable? No. By removing whelks from their shells and presenting equal numbers of each size class to crows, Zach demonstrated that all size classes were equally palatable.

Why, then, do crows accept only the largest and heaviest whelks on the beach? One reason seemed simple enough—larger whelks have a higher caloric value. However, Zach wondered whether size affects the ease of breaking the shell as well. To answer this question, he collected whelks of different sizes and dropped them onto rocks from different heights. He found that large whelks were more likely to break than the medium and small whelks; the large ones required fewer drops from any given height.

We see in Figure 12.13 the total height to which a crow would have to fly, on the average, to break whelks of the three different sizes at different heights. Notice that the probability that a whelk will break increases with the height of the drop. The lower the dropping

TABLE 12.1   Numbers of Small, Medium, and Large Whelks Laid Out and Cumulative Numbers of Whelks Taken Over the Subsequent 5 Hours*

	Type of Whelk		
	Small	Medium	Large
Laid out	75	75	75
Taken after 1 hour	0	2	28
2 hours	0	3	56
3 hours	0	4	65
4 hours	0	4	68
5 hours	0	6	71

*Results from three pairs of crows were homogeneous (replicated goodness of fit test) and therefore combined. Right from the start whelks were taken nonrandomly ($p<.005$; single classification goodness of fit test).

Source: Zach (1978).

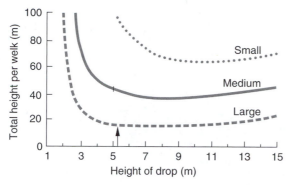

FIGURE 12.13 **The total height of the drop required to break the shells of different-sized whelks. Northwestern crows choose large whelks and drop them from a height of about 5 meters (indicated by the arrow). This minimizes the energy used in obtaining food and yields an energy profit. The crows would use more energy than they obtain by feeding on small or medium whelks. (From Zach 1979.)**

height, the more drops are required to break it. The total vertical height needed to break the shell can be determined by multiplying the number of drops by the height of the drop. There is a height—slightly more than 5 meters—at which the total height required to break a whelk is minimized. Zach (1979) found that the crows dropped whelks from a mean height of 5.23 meters (±0.07 m).

Given this information, Zach (1979) was able to calculate a crow's expected energy profit from whelks of different sizes. He found that crows are likely to gain 2.04 kilocalories from a large whelk, but they use 0.55 kilocalories to obtain it. Thus, the net energy gain for a large whelk is 1.49 kilocalories per whelk. Medium and small whelks, on the other hand, require more energy to handle because they are harder to break. They also contain fewer calories. Zach calculated that the net energy gain for a medium whelk would be –0.30 kilocalories. We see, then, that the bird would use more energy in attempting to eat a medium whelk than it would get in return. Obviously, small whelks, which are even harder to break and contain fewer calories, would be even more costly to eat. Crows are able to distinguish between profitable and nonprofitable food items.

Optimal Diet

The previous example shows us that animals can distinguish between prey that will give them a net increase in energy and prey that will actually cost them energy. This is not surprising since individuals that do not make an energy profit while foraging would soon starve to death, leaving fewer offspring.

However, when faced with a variety of types of food, all of which would yield a net energy gain, should the individual eat some of each? Optimal foraging the-

ory answers no. There exists an optimal diet, one that maximizes energy gain (Schoener 1987).

To determine the optimal diet, we must first calculate the profitability of each food item. Profitability, in this case, is calculated by dividing the net energy gain by the handling time:

Profitability = net energy gain/handling time

The net energy gain represents the number of calories the animal can get from the food minus the number of calories required to process that food, because of handling or digestion, for instance. Handling time includes both the amount of time the animal needs to pursue the food item and the amount of time needed to prepare it so that it can be swallowed.

Obviously, the food item that has the greatest profitability should be eaten whenever it is encountered. But if a forager finds the second–most profitable food item, should it continue searching for the best item or accept this one? It turns out that this decision depends on the abundance of the most profitable prey, not on the abundance of the second-best food item. If the best type of food is rare, it will take a long time to find. In this case, the animal will increase its overall energy intake by accepting any food that provides an overall energy gain. But as the abundance of the best food increases, the animal loses valuable time in attending to other, less profitable items when it could be dealing with only the best ones. Consider the following analogy. Assume for the moment that lobster is your favorite food. Although hot dogs may stop your hunger pangs, you are not fond of them. Obviously, if you can get enough lobster to satisfy your needs, you should not bother with hot dogs, no matter how plentiful they are.

We see, then, that optimal diet theory leads to at least two testable predictions (Pyke, Pulliam, and Charnov 1977):

1. The acceptability of a food item depends on the abundance of more profitable food items and is independent of its own abundance.

2. As high-ranking items become more common, less profitable items will be eliminated from the diet.

Observations that are qualitatively consistent with these predictions have been obtained in a field study of redshanks (*Tringa totanus*), shorebirds that feed on worms (Figure 12.14). Although large worms provide more energy than small ones, they are difficult to find in some locations. The rates at which redshanks preyed on large and small worms at various study sites with different abundances of prey sizes were compared. As the abundance of the larger, more profitable prey increased, the birds became more selective. Whenever large worms were common, they were eaten instead of the small worms, even if the small ones were most plentiful (Goss-Custard, Cayford, and Lea 1998).

FIGURE 12.14   The redshank is a shorebird that feeds on worms. Large worms are more profitable than small ones. When redshanks forage in areas where large worms are abundant, they are more selective and eat more large worms than they do in areas where large worms are rare.

FIGURE 12.15   American crows adjust the manner in which they handle food so that energy gain is maximized. The crows eat both English and black walnuts, which they break open by flying up to 30 meters upward and dropping them onto the ground. The energetic cost is the upward flight to drop the walnut. American crows adjust the drop height to account for the hardness of both the walnut and the ground. Since they feed in flocks, there is a risk that a dropped walnut will be stolen by another bird. The crows lower the drop height when the threat that the walnut would be stolen exceeded a certain threshold. (Photograph courtesy of Paul Switzer.)

In field studies such as this, the rate at which the subject encounters different prey types is largely out of the control of the experimenter. To eliminate this difficulty, John Krebs and his coworkers (1977) designed an apparatus that would allow them to control the abundance of two different food items presented to great tits (*Parus major*). The birds were placed in cages in which they obtained food from a conveyor belt that ran past the cage. This allowed Krebs to control both the type of prey encountered and the abundance of the different prey. The prey items were small and large pieces of mealworms. The energy value of large mealworms was greater than that of small ones, but the handling time was equal for the two prey types. Krebs found that when there were few mealworms of any size, large or small, the birds ate both kinds of prey. They showed no preferences. However, when the large mealworms were made more plentiful, the birds became selective and tended to ignore the small ones. As long as the abundance of the large mealworms was kept at this level, the birds selected them, even if the small mealworms became more abundant than the large ones.

The way food is handled can also influence its profitability. Thus, it should not be surprising to learn that American crows (*C. brachyrhynchos*) handle their food in a manner that maximizes the energy gained from each item. These urban-dwelling crows eat two types of walnuts, which they break open by dropping them from

heights in much the same way that the northwestern crow drops whelks (Figure 12.15). Recall that the upward flight to drop the food item is energetically costly. Daniel Cristol and Paul Switzer (1999) have shown that American crows adjust the height from which they drop a walnut according to the circumstances. English walnuts break more easily than black walnuts, and the crows drop the English walnuts from lower heights than black ones. Any nut will crack more easily on a hard surface than on a soft surface. The crows adjust the height of the drop to account for substrate hardness. Moreover, walnuts are more likely to crack with each successive drop. Remarkably, the crows lower the height with each drop when repeatedly dropping the same walnut. (It is interesting to note that whelks are not more likely to break open with each successive drop, and the crows in Zach's (1979) study always dropped a whelk from approximately the same height.) American crows feed in large flocks, and so there is always the threat that a dropped nut will be stolen by another bird. The crows also adjust the height to minimize the chances of theft. The risk of theft was determined by using an index that combined the number and proximity of conspecifics. When the risk exceeded a certain threshold, the crows lowered the drop height so that they could recover the nut before it was snatched by another bird. Thus, American crows adjust the way in which they handle food to maximize profitability by taking into account the walnut's and the substrate's hardness, as well as the risk of theft.

Nutrient Constraints

So far we have assumed that the only thing a predator has to consider is maximizing its energy intake. But if obtained calories were the only nutritional concern, we could eat nothing but our favorite food—say, hot fudge sundaes or Spam. There would be no need for parents to insist that children eat their vegetables.

The commonly touted nutritional guideline for humans, "Eat a variety of foods," may also apply to other animals. It has been shown, for instance, that nestlings of the European bee-eater (*Merops apiaster*) convert food to body weight more efficiently if they are fed a mixture of bees and dragonflies than if they receive only bees or only dragonflies (Krebs and Avery 1984).

Animals may have specific nutritional requirements, and they may have to balance their diet so that the essential dietary component is supplied along with sufficient energy. The moose (*Alces alces*) that live on Isle Royale in Lake Superior are an example (Figure 12.16). They must obtain enough energy for the growth and maintenance of their huge bodies, but they also have a minimum daily requirement for sodium. The leaves of deciduous trees on the shore contain more calories than aquatic plants, so to maximize energy intake the moose should eat only land plants. However, land plants have a low sodium content. In contrast, low-calorie aquatic plants have a higher concentration of sodium. A moose balances these needs by eating a mixture of plants so that energy intake is as great as it can be while sufficient sodium is still obtained (Belovsky 1978).

SEARCHING FOR FOOD

So far we have used optimality theory to help us understand how an animal chooses a given type of food. Optimality theory can also help explain how animals actively search for food.

Choosing a Site

Once a forager has decided what to eat, it must then decide where to look for it. Since food is typically found in patches rather than evenly spread across the environment, an animal that is searching for food is likely to encounter areas where food is easier to obtain than in other locations.

Assessing Site Quality Optimal foraging theory would predict that animals will choose the best area (patch) for foraging. Operant-conditioning techniques have been useful in testing this prediction. In one experiment, patch choice of caged rats was explored by using bar pressing for a food reward to simulate foraging. The abundance of food could be varied by altering the size of the food-pellet reward, and foraging costs could be changed by increasing or decreasing the number of bar presses required to obtain a reward. When permitted to choose between levers that differed in food abundance or in the ease of obtaining food, rats fed more frequently and ate larger meals at more profitable patches (levers). This was true whether the difference in profitability was in the size of the pellet or in

FIGURE 12.16 Moose must obtain enough energy for growth and maintenance, and the leaves of deciduous trees contain more energy than aquatic plants. However, moose must also obtain a minimal amount of sodium. So, although sodium-containing aquatic plants provide less energy, moose must eat some low-calorie plants. This is an example of how nutrient requirements can constrain optimal foraging.

the work required to obtain it (Johnson and Collier 1989). So, as predicted, the animals chose the best patch in which to forage.

To determine which of several possible foraging sites is the most profitable, a forager must sample them periodically. It has been demonstrated that great tits sample feeding patches and use this information to choose the most profitable patch. In this experiment, the food patches were grids that contained different numbers of mealworms. The birds quickly learned to feed from the one with the highest density of mealworms, but they continued to sample less profitable patches. The value of sampling became apparent when the density of mealworms in the most profitable patch was reduced suddenly. The tits then switched to the grid with the second-highest prey density (Smith and Sweatman 1974).

The astute reader has probably realized that sampling is not without cost. The time spent in sampling different patches before concentrating on the most profitable one lowers the average rate of food intake. Compromises must be made. Optimality theory has permitted predictions about the conditions under which a forager should explore patches other than the one currently being exploited. Consider, for example, an individual with a choice of two foraging sites. One patch reliably provides a moderate amount of food. The other fluctuates in richness, reaching extremes far above and below the stable one. It is clearly beneficial to exploit the fluctuating site during times of peak abundance, but it is detrimental to waste time there when food is less plentiful than at the stable site. To determine the level of food in the fluctuating site, an individual must sample—but how often?

One model predicts that the fluctuating site should be sampled more often when the value of the stable one falls below the average value of the fluctuating one (Stephens 1987). Chipmunks are well suited to test this prediction because much of their natural food supply, seeds from deciduous trees, is found in patches of fluctuating abundance (Figure 12.17). The supply of seeds below a particular tree may be here today and gone tomorrow, varying with the schedule of ripening, amount of wind, and activities of other animals. As a result, chipmunks must decide how often to check the seed supply at other trees to determine whether it might be beneficial to switch foraging locations. The amount of time individual chipmunks spent foraging and exploring alternative sites was monitored to determine what factors influence how frequently alternative locations were sampled. In one experiment, the chipmunks (*Tamias striatus*) foraged from artificial patches—trays of sunflower seeds. The value of a given patch was manipulated by varying the number of seeds on the tray. It was found that the chipmunks spent more time sampling the food density at other locations as the quality of the patch being exploited decreased (Kramer and Weary 1991).

As might be expected, behaviors that reduce the time or energy spent on sampling the patch's quality have evolved. For example, after sampling a buried patch of seeds and finding that the patch has been depleted, least chipmunks (*T. minimus*) sometimes mark the patch with urine. The urine mark deters the marker, as well as other chipmunks, from sampling that patch again. In this way, it improves the foraging efficiency of the marker by eliminating unnecessary patch sampling (Devenport, Devenport, and Kokesh 1999).

Animals that forage in groups can minimize their own sampling effort by paying attention to the sampling activity of others in the same patch. Starlings (*Sturnus vulgaris*) are group foragers that sample a food

FIGURE 12.17 Chipmunks generally forage on seeds beneath deciduous trees. Thus, their food is found in patches of fluctuating abundance. As the quality of their current feeding location declines, they spend more time sampling the food abundance at other locations.

patch by probing their bills into the ground. Jennifer Templeton and Luc-Alain Giraldeau have shown that "public information" provided by the activity of some starlings can improve the foraging efficiency of other starlings. A field experiment demonstrated that starlings use the successful foraging activities of others to assess patch quality (Templeton and Giraldeau 1995). Starlings that observe the unsuccessful sampling activities of others can also use that information to supplement the information from their own sampling when deciding when to leave a depleted patch. This was shown by allowing the birds to forage in artificial patches that offered 30 probe sites. Because the patches were nearly or completely empty during training, several sites had to be probed before deciding that the patch really was empty. During the test sessions, all probe sites were empty so that differences in the amount of sampling could not be attributed to differences in the rate of food depletion. A starling was first tested alone to see how many probes it would make before leaving the patch. Next, the starling was tested with a partner whose foraging activity could be easily observed. There were two types of partners. The high-information partner had previously learned that only one hole contained a food pellet. As a result, this partner probed many holes during the test session. The low-information partner had been trained to probe only a few holes. The results are shown in Figure 12.18. The number of empty holes probed by the starling before leaving the patch was highest when it was tested alone. Apparently, the bird supplemented its own sampling information with that gained from observing the other bird because it probed fewer holes when foraging with a partner. When the information provided by the partner was low, the test starling probed more holes than when the partner provided a great deal of information (Templeton and Giraldeau 1996).

Sensitivity to Variability in Food Abundance

Some species live in an environment in which the food supply is unpredictable. As a result, an individual may have to choose between a site that reliably supplies a moderate amount of food and one that fluctuates between a rich and poor food supply. The individual that chooses a variable site could get lucky and find plenty of food quite easily, but there is always the risk that food will be scarce. In the jargon of foraging theory, the term *risk* refers to variability in food abundance. Some species seem to consider the degree of variability in food supply when deciding where to forage. These species are described as being risk-sensitive. Our discussion so far has assumed that animals are indifferent to risk, that they behave as if all they care about is maximizing energy gain. It has been presumed that animals will always choose the patch that provides the highest average yield despite differences in the fluc-

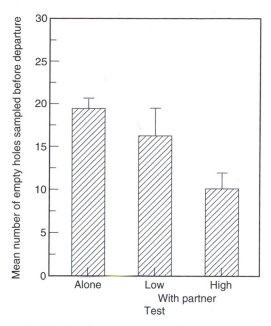

FIGURE 12.18　Starlings are group foragers that use public information gained by observing the sampling activity of others, in addition to that gained by their own sampling activities, to decide when a patch is depleted. The mean number of empty probe sites sampled before leaving the patch is highest when a starling is foraging alone. Furthermore, when the starling was foraging with a partner, the number of empty holes sampled by the test starling before leaving was higher when the partner's sampling activities provided little information about patch quality than when the partner's sampling activities provided a great deal of information. (Data from Templeton and Giraldeau 1996.)

tuation in food supply among sites. This view is too simple, however. Studies have shown that many animals do discriminate between reliable and risky sites, even when the average amount of food is the same at both locations. Some animals are gamblers and choose the variable site. They are called risk-prone. Others, those who are risk-averse, tend to choose reliable sites where they are more or less guaranteed of finding at least a mediocre amount of food (Stephens and Krebs 1986).

Why might it be advantageous to consider the variability of food supply when deciding where to look for food? One hypothesis for the function of risk sensitivity is that it reduces the chances of starvation (Stephens and Charnov 1982). The reasoning is as follows: An animal that fails to find a certain minimal amount of energy each day will die of starvation. If enough food can be found at the site that provides a stable food supply, there is no benefit in gambling on finding sufficient food at the variable location. However, if the stable site does not provide enough food to prevent starvation, the only chance for survival is to forage in the location where the food supply is variable and hope for the best.

The hypothesis that risk sensitivity minimizes the risk of starvation predicts that individuals will consider the variability in food supply in addition to the average amount of food available at two locations. A preference for stable sites should change to one for risky patches as the individual gets closer to starvation. This prediction has been met in studies of several species of birds, including juncos (Caraco 1981), white-crowned sparrows (Caraco 1983), hummingbirds (Stephens and Paton 1986), and warblers (Moore and Simm 1986), as well as common shrews (Barnard and Brown 1985). In contrast to common shrews, hungry round-eared elephant shrews do not show risk-prone behavior (Lawes and Perin 1995).

Risk sensitivity in bumblebees (*Bombus* spp.) has been the basis of several experiments and has led to disagreement over the function of risk sensitivity (Figure 12.19). In one experiment, Leslie Real (1981) permitted bumblebees to choose between two types of artificial flowers, one that provided a variable supply but both yielding the same average amount of nectar. Eighty-five percent of the bees' visits to flowers were to those that provided a constant source of nectar. Thus, they were risk-averse. When the average volume of nectar supplied by variable flowers was greater than in the constant sources, the bees would visit variable sites more frequently (Real, Ott, and Silverfine 1982).

A reanalysis of the data from these experiments led to the suggestion of an alternative hypothesis for risk sensitivity in bumblebees. It has been proposed that

FIGURE 12.19 Bumblebees are risk-sensitive foragers; that is, they behave as if they consider whether the foraging location provides a stable food supply or one that varies in richness. When the colony has sufficient energy reserves, the bees chose to forage at flowers that provide a stable, but mediocre, amount of nectar. However, when the colony's energy reserves are low, they prefer to gather nectar from flowers that fluctuate in nectar abundance.

when the mean reward of reliable and variable sources is similar, bumblebees prefer reliable sources of nectar so that their net rate of energy intake is maximized (Harder and Real 1987). It turns out that when the time needed to handle flowers and to fly between them is taken into account, the expected rate of net energy intake is lower at variable sites.

Ralph Cartar and Lawrence Dill (1990) distinguished between these hypotheses by keeping the expected net energy gain the same in variable and constant flowers and observing the bees' preference as the colony's energy reserves were manipulated. A bumblebee colony's energy source is nectar stored in open-topped honey pots, so the level of nectar in the honey pots could be altered to change the colony's energy reserves. If risk sensitivity makes it more likely that bees will obtain some minimal amount of energy, their behavior should change with changes in the amount of nectar in the honey pot. However, if risk sensitivity in bumblebees is a result of lower rewards at variable flowers, there should be no change when the energy reserves are altered. It was found that when the reward rate was kept constant, energy reserves did affect the bees' flower choice. Risky, variable flowers were preferred when the honey pots were experimentally drained of nectar, but constant flowers were preferred when the honey pots were experimentally supplemented with nectar. Thus, the hypothesis that risk sensitivity increases the likelihood of obtaining some minimal amount of energy is supported.

In the case of bumblebees, however, it seems that this energy requirement is necessary to minimize the probability of reproductive failure rather than to prevent starvation. An individual forager could easily obtain enough energy to sustain itself with the first few flower visits, but if the colony does not have enough energy, it may be unable to repel parasites and will have a longer period of brood development* (Cartar and Dill 1990). Thus, the function of risk sensitivity may depend on how the forager uses the energy it obtains. Different models of risk-sensitive foraging can be developed for cases in which the energy is used for reproduction and those in which the energy is used to minimize the risk of starvation (McNamara, Mermad, and Houston 1991).

Deciding When to Leave a Patch— Marginal Value Theorem

As an animal forages within a particular patch, food may become more difficult to obtain. The supply becomes depleted as it is eaten because the best sources were utilized first, leaving only poorer quality sources,

---

* Worker bees are sterile. Their reproductive gains are in the sisters they help to raise. See Chapter 17 for a discussion of how this might have evolved.

or because prey begin to take evasive action. Regardless of the cause, at some point it may become advantageous for the animal to move to a new patch, where food will be easier to find.

According to optimality theory, there are several factors for an animal to consider when deciding whether to leave a patch. One is the abundance of food at the current site relative to other locations. Clearly, it would be wise to move on if one could do better somewhere else. This could be determined by comparing the food abundance at other feeding locations to the expected food abundance at the current location. To maximize energy gain, the individual should look for a new patch when its rate of intake at the current feeding location drops below the average value for other areas. If the food availability at the current location is below the average value of other locations, the animal has a better than even chance of moving to a better patch.

In deciding whether to leave its present site, the animal must also consider the difficulty in getting from one patch to another. The more difficult it is to get to a new location, measured in time or effort, the longer the individual should continue feeding at its current spot. The time and effort spent getting to the next patch must be subtracted, in a sense, from the value of that patch.

The marginal value theorem, which describes behavior that maximizes the long-term rate of return for resources that are patchily distributed, has often been applied to foraging situations (Charnov 1976). Put simply, this model predicts that as travel time between patches increases or as the average quality of the overall habitat decreases, the animal should spend relatively more time in any given patch. If the animal leaves a patch too soon, its overall rate of energy intake will decrease because it will be wasting time by traveling, and if the animal remains in the patch too long, it will waste time searching for food in a depleted area.

We might ask, then, what rule of thumb a forager uses when deciding whether to leave a patch. The decision rules most frequently suggested are (1) remain in the patch for a certain length of time, (2) remain in the patch until a certain number of food items have been obtained, and (3) remain in the patch until the rate of food capture declines to some threshold level. As might be expected, different species appear to follow different rules.

In laboratory experiments, the blue-gill sunfish (*Lepomis macrochirus*) seems to use the rate of prey capture as a measure of the quality of the patch (DeVries, Stein, and Chesson 1989). However, these fish remain in a patch longer than predicted by the marginal value theorem (Crowley, DeVries, and Sih 1990). Although there are several other possible explanations for why blue-gills linger at a depleted site, one is that the costs of traveling to a new area make it worthwhile to linger at the current location.

Travel time does appear to be a factor in determining how long the individuals of some species remain in a foraging patch. One such species is the eastern chipmunk. Chipmunks gather food to bring back to their nests and, therefore, provide an example of central place foraging. In an experiment similar to one previously described, the food patches were trays of sunflower seeds. By varying the distance of the tray from the burrows, the travel time of the chipmunks was altered. It was found that the amount of time spent at a tray (patch time) increased with the distance between the seed tray and the burrow (Giraldeau and Kramer 1982).

CONSTRAINTS ON OPTIMAL FORAGING— PREDATION AND COMPETITION

In our discussion so far, we have considered examples of individuals who are foraging in a manner that allows them to make the largest possible energy profits. However, this is not always true. When we observe animals that are not foraging optimally, should we cast out the whole hypothesis? No, we should look for other factors, such as the presence of predators or competitors, that require the forager to compromise.

The ultimate challenge is to eat without being eaten. Therefore, a forager may have to sacrifice some energy gain for safety. This tradeoff may be made in a variety of ways. One way is to minimize the time spent foraging. Indeed, whirligig beetles (*Dineutes* spp.) do just that. During the daytime, groups of these insects can be seen swimming on the surface of streams, ponds, and lakes. It is easiest to catch prey, insects that have become trapped in the surface tension, from the periphery of the group. However, central positions within the group are safest from predators. When food levels were experimentally manipulated, it was found that hungry beetles were more likely to be found at the edge of the group, where they were more likely to encounter prey. In contrast, well-fed beetles moved to the center of the group (Romey 1995).

Another way to make the tradeoff between predation risk and energy gain is to feed in safer locations, even if food is not plentiful. This is the strategy adopted by a desert population of chacma baboons, *Papio cynocephalus ursinus* (Figure 12.20a), studied by Guy Cowlishaw (1997). The site was Tsaobis Leopard Park, Namibia, located in southwest Africa. The park's environment is rugged, with mountains and ravines, as well as gravel and alluvial plains. The Swakop River to the north dries up at some times of the year, but its water table does support patches of woodlands.

Four different habitats can be identified in the reserve—river bed, woodland, plains, and hills—each differing in food availability and predation risk (Figure 12.20b; Table 12.2). These baboons have a simple diet.

FIGURE 12.20 (*a*) Chacma baboons make a tradeoff between predation risk and energy gain. They feed on leaves, flowers, fruits, and pods. They sleep in trees or on cliffs, but they spend most of the day on the ground, where they are exposed to predators. The primary predators are leopards and lions. (*b*) The four habitats available to the baboons in the study are riverbed, woodland, plains, and hills. Both the risk of predation and the abundance of food differ in each habitat. The baboons do not spend much time in foraging in the woodland, where food is most abundant. Instead, they feed mostly in the bed habitat. Although less food is available, this habitat is safer from predators. Thus, optimal foraging is constrained by predators. (Diagram from Cowlishaw 1997.)

a

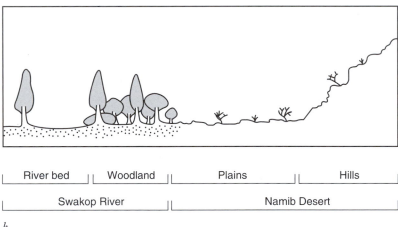

b

During the time span of this study (late winter), 92%–97% of the baboons' feeding time was spent gathering leaves, flowers, fruits, and pods from only five plant species. Over 90% of the food energy was found in the woodland and almost all of the rest in the bed. The most common predators of baboons are leopards and lions, but only leopards were present during this study. Both predators hunt by stalk-and-ambush methods. They have better luck capturing prey from shorter ambush distances and are, therefore, more successful in habitats with cover to hide their approach. Based on the

hunting strategies of these predators, Cowlishaw estimated the predation risk in each habitat. The woodland had the highest estimated risk of predation, the river bed and the plains had a modest risk, and the hills had the lowest risk.

Cowlishaw wondered how the baboons would balance the ease of foraging with the threat of predation, so he measured the amount of time spent in each habitat and recorded what the animals were doing. There are four groups of baboons in this population, and there were some differences in where they spent their time. Nonetheless, it was clear that they were making a tradeoff. Although the woodland had more food, it also had the highest predation risk. When foraging, the baboons spent more time in the safer, bed area even though much less food could be found there. Furthermore, other activities, such as resting and grooming, were usually done in the hills, where the predation risk was lowest.

We could expect that the balance between safety and energy gain might be influenced by increases in either the value of the food or the degree of danger. Indeed, several studies have confirmed these expecta-

TABLE 12.2 **Relative Food Abundance and Predation Risk in Four Habitats Available to Baboons in Tsaobis Leopard Park**

	Food Abundance	Predation Risk
Bed	Modest	Modest
Woodland	High	High
Plains	Negligible	Modest
Hills	Negligible	Lowest

tions. One examined the importance of these factors in foraging young coho salmon, *Oncorhynchus kisutch*. During the first two years of their lives, coho salmon live in streams, where they feed on small invertebrates that drift by. They wait for suitable prey in a central location, where they blend with the gravel background and are difficult for predators to spot. When a prey item appears, the fish swims to gulp it. While moving, the fish is much more visible to predators, so the risk increases with the distance traveled.

The value of food was manipulated by varying the size of the prey, the degree of hunger, and the amount of competition for the food item. Juvenile coho salmon will take greater risks when the importance of the food is greater. For example, they will travel longer distances to obtain larger prey (Dunbrack and Dill 1983). Furthermore, a hungrier fish will travel farther for food than a satiated fish. The fish also weighs the importance of obtaining the food item against the chance of losing it to a competitor. In the field, the prey that is drifting along will be gulped down by the first fish to reach it. So, a fish that does not risk swimming toward its prey may lose it to a neighboring territory holder. In one laboratory study, mirrors reflected the fish's own image, thereby increasing the apparent density of competitors. Again, the fish increased the distance it traveled from its territory to obtain prey.

However, if the fish sees a predator and, therefore, perceives that the potential danger is high, its attack distance decreases. This was demonstrated by using a pulley system to threaten the fish periodically with a model of a natural predator, a photograph of a rainbow trout. The more frequently the model was presented, the shorter the attack distance (Figure 12.21) (Dill and Fraser 1984).

It is often the case with humans that values change with maturity, so it should not be surprising that among animals the relative weights given to foraging success and predation risk may change with age. This has proven to be true among certain colonial web-building spiders (*Metepeira incrassata*). The spider's position within the colony influences both foraging success and the risk of predation or parasitism. Individuals on the edges of the colony are more successful at capturing prey, but they are also at greater risk. Prey capture rates are 24%–42% higher for individuals on the periphery of the colony. Unfortunately, spiders in these locations are also subject to three times as much predation as individuals in core locations. Thus, there is an inevitable tradeoff between foraging success and predation risk. The outcome of this tradeoff will affect the number of offspring an individual is likely to leave. Reproductive success is greater in the central locations, where the risk of predation and egg parasitism is lower. Predators always prefer large reproductive females over smaller spiders, and since predators approach the web from the edges, the large females living there are at par-

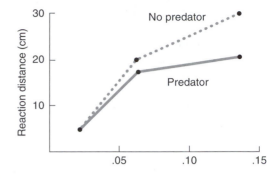

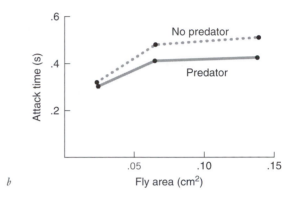

FIGURE 12.21 The risk of predation constrains optimal foraging in young coho salmon. When the young salmon swim away from their covering vegetation to attack prey, they are more visible to predators. Larger flies, which are more valuable prey, cover a larger area. Notice that the fish will risk swimming farther from its protective resting place for larger flies. However, it will not swim as far in the presence of a predator, regardless of the size of the fly. (Data from Dill and Fraser 1984.)

ticularly high risk. Furthermore, if the eggs are left unguarded at the periphery, there is an increased chance of cocoon parasitism. As a result, most individuals hatch and begin life in the central regions of the colony. Here, prey is scarce, so larger and older spiders have an advantage in competition for prey. Therefore, younger, smaller spiders may take a chance on living on the edges of the colony, where they can obtain more food and grow faster, thereby increasing the odds of reaching sexual maturity. But as the spiderlings mature, values change and safety becomes more important than foraging success. So, the larger spiders that have reached sexual maturity prefer the core positions (Raynor and Uetz 1990).

SUMMARY

There are many ways in which an animal can obtain food. Many aquatic animals filter food from the surrounding water. Certain species work as farmers and cultivate their specialized or preferable food. The leaf cutter ant, for instance, maintains fungus gardens and ensures a supply of a special fungus that is its primary food source.

Animals that hunt for prey employ special techniques that enhance their success. Some build traps. Others lure their prey with specialized structures or behaviors that mimic the courtship displays of the prey species. Still other species employ a hide-and-wait strategy and ambush their prey.

Certain hunters also have adaptations that reduce the likelihood that they will be spotted by their prey, thereby allowing them to come within striking distance. Camouflage, for instance, makes the predators more difficult to see because they blend in with the environment. Certain other predators are disguised as species that are beneficial to the prey.

Many sensory specializations enhance a predator's ability to detect its prey. Some animals, such as mosquitoes and certain snakes (the Crotalidae and the Boidae), detect the body heat that emanates from their warm-blooded victims. The sand scorpion, on the other hand, is exceedingly sensitive to vibrations in the sand that are created by its prey. From a human perspective, perhaps the most unusual sensory specialization is the ability of certain animals, including the Australian lungfish, to detect the electrical fields that living organisms generate when they are in seawater. The electrical sense was shown to exist in lungfish by eliminating other sensory cues that could be used for prey localization and determining whether the lungfish could still locate its prey. Although visual and olfactory cues may be used in nature, they are not necessary for prey detection.

We often find that predators become more adept at finding cryptic prey following experience with them. Two hypotheses have been presented to explain this observation. The search image hypothesis is that the predator learns key features of the cryptic prey and then focuses on them while hunting. The search rate hypothesis is that after a predator finds a few cryptic prey, it slows the rate at which it searches for prey and, as a result, misses fewer cryptic food items. Studies have shown that predators use both techniques to locate cryptic prey.

According to optimality theory, natural selection favors the behavioral alternative whose benefits outweigh its costs by the greatest amount. The optimal foraging pattern, then, is the one that maximizes net energy gain.

One of the first decisions a forager must make is what to eat. Studies have shown that animals such as northwestern crows choose food items that provide a net energy gain (large whelks) and not those that require more energy to obtain than they provide (medium or small whelks).

When faced with a variety of foods, all of which would yield an energy gain, the forager should choose the optimal diet, the one that maximizes energy gain. The profitability of a food item is defined as the energy it provides divided by the time it takes to find, capture, prepare, and digest that item. According to theory, the acceptability of any particular food item depends on the abundance of the most profitable food item. When the most profitable items are abundant, less profitable items will be eliminated from the diet. The results of field studies on redshanks feeding on worms and laboratory studies of great tits feeding on mealworms are consistent with the predictions of optimal diet theory.

Animals may choose less profitable food items because of specific nutritional requirements. The individual may have to eat some less energetically profitable items to obtain an essential dietary component.

Food is often easier to find in some places than in others. Optimal foraging theory predicts that animals will choose to forage where food is most easily obtained. To determine which area is most profitable, many locations must be sampled. But sampling different patches before concentrating on the best one reduces the average rate of food intake. Optimal foraging theory predicts that the time spent in sampling should increase as the quality of the patch being exploited decreases.

As an animal forages within a patch, the food may become more difficult to obtain. The marginal value theorem predicts that (1) an individual should leave its current feeding location when the abundance of food there drops below the expected average for the environment, and (2) an individual should remain at its current feeding location longer the more difficult it is to reach a new location.

Individuals of some species may have to choose between a feeding location that supplies a moderate, but constant, supply of food and one that fluctuates between a rich and a poor food supply. Species that seem to prefer or avoid variable feeding sites are called risk-sensitive. Risk sensitivity may lessen the chances of failing to obtain enough energy to stay alive or for reproduction.

Animals may not be foraging optimally because they are sacrificing energy gain to avoid predation or because they are competing with other individuals for food.

13

Antipredator Behavior

Animals are never safe from predators, and monarch butterflies (*Danaus plexippus*) are no exception, falling prey to birds and small mammals. In light of the constant threat of predation, these butterflies have developed an impressive array of devices to outsmart their enemies, and their protective strategies appear to work, at least some of the time.

Like many animals, monarch butterflies may use a combination of color pattern and behavior to avoid being eaten. Their boldly patterned, orange, black, and white wings warn potential predators that they taste bad, their unpalatability being due to their assimilation of noxious chemicals from food plants. In particular, monarch larvae feed on milkweed plants (Asclepiadaceae) and incorporate toxins, called cardiac glycosides, into their own tissues (Brower et al. 1968). Predatory birds that eat one of these insects, even as adults, have severe vomiting and tend to avoid butterflies of similar appearance in the future. However, one might wonder what good it is to be filled with toxins if the individual must be eaten before the poisons will work. The advantage is that many predators release unharmed prey that are brightly colored and bad tasting. Poisons, stolen from plants, may thus deter some predators. However, not all milkweed plants contain cardiac glycosides. Butterflies reared on plants that do not contain the poison are quite palatable, although they may still be avoided by predators who have had experience with noxious members of the species. Even butterflies of other species try to cash in on the defense system of *D. plexippus*. By resembling the sometimes unpalatable monarch, individuals of other species deceive predators into avoiding them as well. Such deception is a common component of antipredator strategies.

No defense system works all the time. Even for the monarch butterfly, the effectiveness of protective strategies often varies with the species of the predator, the season, and the context of the predator-prey encounter. Each fall, monarch butterflies, some from as far away as the northeastern corner of the United States, migrate to the mountains of central Mexico, where they spend the winter in densely packed aggregations of tens of millions of individuals. The months spent in Mexico, however, are far from a winter vacation (Figure 13.1). Birds of two species, the black-backed oriole (*Icterus galbula abeillei*) and the black-headed grosbeak (*Pheucticus melanocephalus*), have penetrated the monarch's chemical defense system; these two species eat an estimated 4,550 to 34,300 butterflies per day in some overwintering colonies (Brower and Calvert 1985). The oriole selectively strips off relatively palatable portions of the butterflies' bodies (e.g., the thoracic muscle and abdominal contents), and the grosbeak appears insensitive to the cardiac glycosides (Fink and Brower 1981). However, all is not lost for the monarch. Facing each winter with the prospects of an avian feeding frenzy, the butterflies reinforce their antipredator system by converging in great numbers (Calvert, Hedrick, and Brower 1979). By forming dense aggregations, they dramatically dilute the predation risk to any one individual. Also, because predation is most intense at the periphery of the colony, central positions are highly sought after and are quickly assumed by the first individuals to arrive. In the life of a monarch butterfly, it does not pay to be fashionably late in arriving at the overwintering site.

Since predation is a pervasive theme in the pageant of life, we can ask how other animals cope with its constant threat. Which devices aid in escaping detection by a predator (primary defense mechanisms) and which come into play once a prey animal has been detected and capture seems all too imminent (secondary defense mechanisms)? Does membership in a group always confer antipredator privileges?

CRYPSIS

Animals that are camouflaged to blend with their environment are said to be cryptic; "I am not here" is the message of cryptic species. Crypticity may range from simple markings that break the body contour to devices that render the animal almost invisible. There are many well-known examples of crypticity, including reed-nesting marsh birds that sit immobile with their bills pointed upward, moths that seem to disappear on the bark of a tree, and frogs that are almost invisible against the leaves of the forest floor (Figure 13.2). By blending with their background, these animals may escape the notice of visually hunting predators.

FIGURE 13.1   Avian predators penetrate the chemical defense system of monarch butterflies at their overwintering sites in Mexico.

FIGURE 13.2 **Frogs of Malaysia, among dead leaves.**

AVOIDING DETECTION THROUGH COLOR AND MARKING

Breaking the Contour

Many animals avoid being seen by simply matching their background color, but sometimes such coloration is not enough because visually hunting predators may recognize prey by their body contour. Some animals break their outline by developing bizarre projections; other species do so with bold markings. This latter antipredator device, disruptive coloration, is perhaps best illustrated by the black-and-white vertical stripes of the zebra (Figure 13.3).

Countershading

Countershading is another option for avoiding the unwanted attentions of predators. Because light normally comes from above, the ventral surface of the body is typically in shadow and predators may cue in on darkened bellies. Many animals appear to obscure the ventral shadow by being paler ventrally and darker dorsally (Figure 13.4). For animals that rest upside down, we would expect just the opposite color pattern; that is, the dorsal surface would be pale and the ventral surface would be dark.

FIGURE 13.3 **The principle of disruptive coloration. To see how it works, place the book in an upright position and then back away. See which disappears first, the zebra or the pseudozebra. The bold stripes of the zebra should obscure the body's outline best. (After Cott 1940.)**

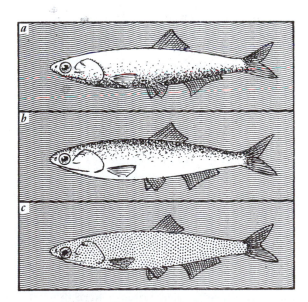

FIGURE 13.4 **The effects of countershading on conspicuousness. Light normally comes from above, and under these circumstances fish that are (*a*) uniformly colored have a conspicuous ventral outline. A countershaded fish (*b*) is darker dorsally than ventrally (as shown here, illuminated from all sides), and thus its body outline is obscured (*c*) when light comes only from above. (After Cott 1940.)**

Direct evidence for countershading as a concealing adaptation is embarrassingly meager. Until such evidence is obtained, we cannot discount the possibility that the combination of dark backs and light bellies is totally unrelated to crypsis and is involved instead in biological functions such as thermoregulation or protection against ultraviolet radiation (Kiltie 1988).

Transparency

Some animals are cryptic simply by being transparent. Although no animal is completely transparent, organisms such as cnidarians (e.g., hydroids and jellyfish), ctenophores (e.g., comb jellies), and the pelagic (open ocean) larval stages of many fish achieve near transparency by such means as high water content of tissues, small size, and reduced number of light-absorbing molecules or pigments (McFall-Ngal 1990).

Frequently neglected in discussions of cryptic mechanisms, transparency is probably the dominant form of crypsis in aquatic environments. It is more common in aquatic than in terrestrial habitats for two reasons. The first is based on differences in the refractive indexes (the angle at which light bends when passing from one medium into another) of water and air. Because animals' bodies are largely water, when light travels from the surrounding water into the tissues of an aquatic animal, the angle of light is virtually unchanged; and in the absence of light-scattering or light-absorbing elements, the animal appears to be transparent (light, then, is basically passing from water into water). In contrast, in a terrestrial environment, light must pass from air into the water-filled tissues of an animal. The difference in the refractive indexes of air and the terrestrial animal's tissues creates an obvious body outline, greatly diminishing transparency. The second reason transparency is rarely used as a camouflaging mechanism by terrestrial animals has to do with the deleterious effect of ultraviolet radiation on land. In aquatic habitats, much of the ultraviolet radiation is filtered out within a few meters of the water's surface, and thus animals living beyond this distance are not subject to the same radiation damage as terrestrial organisms.

Color Change

Usually, cryptic animals are camouflaged in some habitats but not in others, and thus their occurrence is often restricted to those particular areas where they are best concealed. One way some species get around this restriction is by changing color as they change backgrounds. Although the chameleon is perhaps the most familiar example, the cuttlefish (*Sepia officinalis*) is the true master of color change. The cuttlefish is a cephalopod mollusc related to such creatures as squid, octopus, and nautilus. (Chances are that you have actually seen part of the white, internal shell support of this animal hanging in a birdcage. Called cuttlebone, it is not bone at all, but it helps parakeets and other birds keep their beaks sharp.) Among naturalists, the cuttlefish is best known for the speed with which it changes color to match its background, each pattern rendering it virtually invisible to both its predators and its prey. William Holmes (1940), working in a laboratory in England, documented the magnificent changes in its color and pattern (Figure 13.5). According to Holmes, the swimming cuttlefish adopts disruptive coloration in the form of a brown-and-white zebra pattern. When resting on the bottom, however, *Sepia* adjusts its color to that of the substrate at hand. Within a matter of seconds, the dorsal color can change from dark brown to sandy brown to almost white as the cuttlefish settles on one background after another. Not surprisingly, the cuttlefish also demonstrates countershading. Typically darker dorsally than ventrally, the cuttlefish thus obscures the ventral shadow created by light from above. When turned on its back in the water, however, its ventral surface darkens and its dorsal surface pales.

Color Polymorphism

Many insects have the ability to change color. Usually the transition from one color to the next does not occur instantaneously as the individual moves from one background to another (as with cuttlefish) but rather at specific times in their natural history, such as molting. Joy Grayson and Malcolm Edmunds (1989) examined the causes of color and color change in caterpillars of the poplar hawkmoth (*Laothoe populi*), a species that normally proceeds through four larval instars (stages before pupation and metamorphosis to the adult moth). Final instar caterpillars of this species can be yellow-green, dull green, or white. The dull green color is genetically determined, and yellow-green and white are environmentally induced polymorphisms (i.e., different forms, or "morphs," whose color depends on the immediate surroundings). Apparently, the main factor that determines whether a caterpillar becomes white or yellow-green is the surface on which it rests and feeds during its first two or three instars. It is neither the color (wavelength of light) nor the nutritive qualities of the leaf that are critical. Instead, it is the intensity of the light reflected from the leaf and perceived by the young caterpillar: If a larva sees white, then it becomes white; but if it should see green, grey, or black, then it becomes yellow-green.

Environmentally induced polymorphisms in body color also occur in grasshoppers that inhabit the savanna of Africa (Hocking 1964). Fires, either natural or set by humans, sweep through these areas annually or sometimes every second or third year. Following a fire, some grasshoppers change color in a matter of days to blend with the blackened ground. Other grasshop-

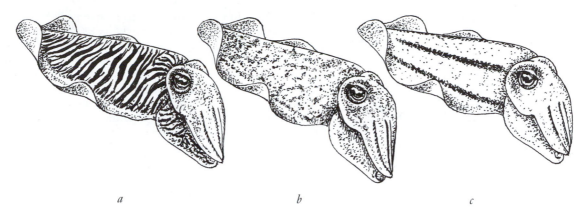

a   *b*   *c*

FIGURE 13.5   Many cryptic animals are restricted to portions of their habitat in which they are well concealed. The cuttlefish gets around this restriction by changing its color to match the particular background on which it rests. When free-swimming, the cuttlefish has disruptive marks reminiscent of the zebra pattern (*a*), but within a matter of seconds of settling on a particular substrate, its dorsal pallor can change to either dark brown, white, or a light mottle pattern characteristic of sand (*b*). Pattern (*c*) occurs just before the cuttlefish ejects a cloud of ink and disappears from view completely. (After Holmes 1940.)

pers cope with periodic burning of the vegetation by using two color forms, one green and one black, each of which seeks its appropriate burnt or unburnt background.

A similar situation has been reported for fox squirrels (*Sciurus niger*) in the eastern United States. Fox squirrels have been described as the most variable in color of all mammals in North America (e.g., Cahalane 1961). Color varies both among and within populations. Dorsal coloration may range from gray or tan to black, and coloration on the head and ear region is often distinctive (Figure 13.6). Even within a single litter, both melanistic (black) and nonmelanistic young can be found. Intrigued by the variation in coat color of fox squirrels, Richard Kiltie (1989) examined close to 2000 museum specimens of this species. He determined

the percentage of dorsal black for each skin and compiled information on the occurrence of wildfires in the eastern United States. Taken together, his data on coat color and fires show that the incidence of melanistic individuals is correlated with the frequency of wildfires over the total range of the squirrels. Both wildfires and melanistic squirrels are more common in the southeastern United States (Figure 13.7).

In fox squirrels, the melanistic polymorphism in coat colors may thus be maintained by the periodic blackening of the ground and lower portions of tree trunks by wildfires. One would imagine that dark squirrels are less conspicuous to hawks than are light or variably colored individuals against a blackened background. The advantage, however, does not remain with the black squirrels for long. As rainfall and new plant growth convert a charred area into a less uniformly black substrate, fox squirrels with variable amounts and patterns of black dorsal coloration would be more cryptic than uniformly black individuals against the patches of light and dark underground. Finally, when the period of regrowth of the pine and oak forest is almost complete, the advantage may shift to squirrels that are uniformly light in coloration. Thus, variable coat color in fox squirrels may result from the alternating cryptic superiority of light and dark individuals in an environment that periodically burns and regenerates.

AVOIDING DETECTION THROUGH BEHAVIOR

The vast majority of animals lack the ability to rapidly adjust to the color and pattern of their immediate surroundings. Since individuals that cannot change color

FIGURE 13.6   Some color morphs of the fox squirrel. Note the variation in the percentages of dorsal black and the pattern around the head and ear region. (After Kiltie 1989.)

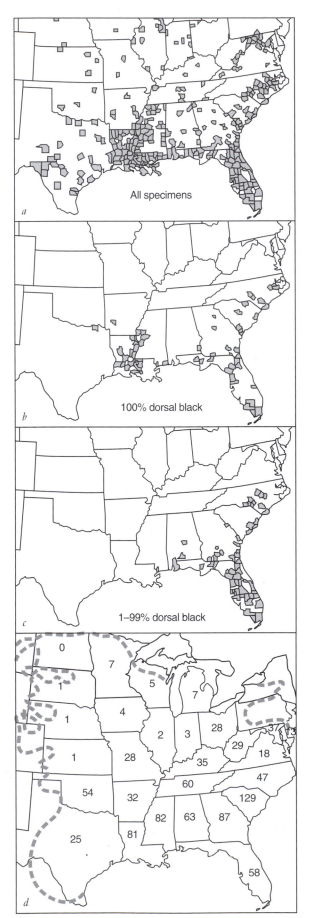

FIGURE 13.7   In the fox squirrel, the incidence of melanism is correlated with the frequency of wildfires. (*a*) Most of the counties in the eastern United States from which Kiltie (1989) examined museum specimens of fox squirrels. (*b*) Counties from which specimens with 100% dorsal blacks were noted. (*c*) Counties from which specimens with intermediate levels of dorsal black (1–99%) were recorded. (*d*) Average number of wildfires per state in protected forestlands during the years 1978–1982; values have been normalized to take into account the area of land under wildfire surveillance (dashed lines depict limits of the fox squirrel's range). Note that melanistic fox squirrels occur primarily in the southern portions of the species range (*b*) and that individuals with intermediate levels of dark coloration are limited to the eastern Gulf and Atlantic coastal plains (*c*). These areas in the southeastern United States are also the areas in which wildfires are most common (*d*). (From Kiltie 1989.)

will be conspicuous if they rest in the wrong place, selection of the appropriate background and correct orientation on that background are critical.

The Peppered Moth: A Case in Point?

The peppered moth (*Biston betularia*) is usually touted as a clear example of crypsis, the choice of appropriate background providing a selective advantage in predator evasion. This moth comes in both typical and melanic forms (Figure 13.8). The typical form (known as *typica*) is whitish gray with a sprinkling of black dots, and the most extreme melanic form, known as *carbonaria*, is almost completely black. Whereas the paler, typical form would be cryptic against lichen-covered trees, the melanic form would be well concealed against the bark of dark trees.

Dramatic changes in the frequencies of these color morphs have been correlated with the degree of environmental pollution. Before 1850 and the industrialization of England, less than 1% of the peppered moths around Manchester were melanic, but by 1895 melanic moths made up 98% of the Manchester population (Howlett and Majerus 1987). In North America, the frequency of the melanic form began to rise around 1900 and reached 90% by 1959. Around that time, laws were passed to curb industrial pollution. The air became cleaner, and the frequency of melanic forms in Britain and North America began to drop (Figure 13.9) (Grant, Owen, and Clarke 1996).

The "classical" explanation for these changes in the frequency of melanic forms of peppered moths goes as follows: Before industrialization, the trees in England were lichen-covered. When resting on this background, the paler, typical form is cryptic and would be difficult for visually guided predators, such as birds, to detect. In

FIGURE 13.8   Two forms of the peppered moth. Above is the typical form and below is the melanic form. Typical moths are more difficult to see against lichen-covered trees, but melanic moths are more difficult to see on soot-covered trees.

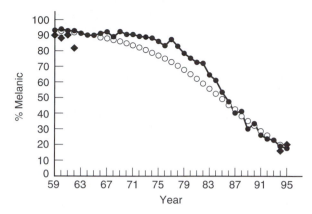

FIGURE 13.9   The decline in the frequency of the dark, melanic form of the peppered moth in America (solid diamonds) and Britain (solid circles). This moth comes in a pale, speckled form and a dark melanic form. Although the frequency of the melanic form rose dramatically during the first half of the twentieth century, it declined during the second half. These changes in frequency of the melanic form mirror changes in air pollution, particularly sulfur dioxide, that accompanied industrialization and then the passage of laws designed to improve air quality. The open circles show the theoretical expectations based on a constant selection coefficient of $s = 0.153$ against the dominant allele. (Data from Grant, Owen, and Clarke 1996).

the mid-1880s, England, converting to an industrial economy, started burning soft coal in its factories. The lichens that covered the trunks of trees were destroyed, and the greenery of the English countryside quietly submitted to its new layer of soot. The tables had turned. Now, the melanic moths were hidden against the soot-covered vegetation and the typical form was conspicuous to predators. As a result, the frequency of melanic forms increased (summarized in Kettlewell 1973). About 50 years later, laws designed to improve air quality were passed. As the air quality improved, the paler forms increased in frequency.

But did the observed changes in air quality lead to changes in the frequency of melanic forms *through effects on crypsis and predation*? The answer isn't as clear as one might think and has become quite controversial (Hagen 1999; Rudge 1999; Sargent, Millar, and Lambert 1998). Let's take a closer look.

If the classical explanation for the changes in frequency of the melanic forms is correct, we would predict that predators would be less likely to detect a melanic moth on a dark background (or a paler moth

on a light background). One way to test this prediction is to mark moths of both forms and release them into a polluted, dark environment and into an unpolluted environment and then to recapture as many as possible to see if the relative numbers of each form remain the same. This series of experiments was conducted by H. B. D. Kettlewell (1955, 1956). At one heavily polluted site and one less polluted site, he marked, released onto tree trunks, and then recaptured light and dark individuals, each time recording the number of survivors of each morph. Kettlewell also observed the behavior of insectivorous birds that were feeding on the trees where the moths had been released. The results, summarized in Table 13.1, demonstrate that the dark forms survived better at the heavily polluted site and the light forms fared better at the less polluted location. Apparently, when placed on the bark of soot-covered trees, the dark form is less conspicuous to birds than is the light form and suffers less predation. The reverse is true in relatively pollution-free areas, where the light form is almost invisible against the lichen-covered trees.

However, the difference in the predation rate on each form may have occurred because the moths, released during the daytime, hastily settled for a nearby resting spot. Thus, on the first day, they might not have been resting in a well-chosen location. However, if the first day's recapture figures are removed from the

Table 13.1   Industrial Melanism in Peppered Moths

Location	Typicals		Melanics	
	Number Released	Number (Percentage) Recaptured	Number Released	Number (Percentage) Recaptured
Dorset (unpolluted)	496	62 (12.5)	473	30 (6.3)
Birmingham (polluted)	137	18 (13.1)	447	123 (27.5)

Source: Data from Kettlewell (1955, 1956).

recapture figures, those that remain are insufficient for statistical analysis (Mikkola 1984).

In other studies, frozen moths of each form have been glued onto dark- or light-colored tree trunks in "lifelike" positions and the predation rate on each form determined. Some studies do demonstrate the differences in predation that would be expected based on the degree of crypticity of each moth form. However, other studies don't show any differences in predation rates (Sargent, Millar, and Lambert 1998).

Even if there is differential predation, based on the degree of crypticity when the moths are resting on tree trunks, that doesn't tell us much about what is happening in nature. We really don't know where these moths rest during the daytime, but we do know that it is not on the vertical trunks of trees. There is some evidence that they may spend the day resting on the undersides of small horizontal branches in the treetops (Mikkola 1979). And, there is also some experimental evidence that moths that do rest on the shaded areas just below limb joints are more likely to be overlooked by predators (Howlett and Majerus 1987).

However, some observations are inconsistent with the classical explanation. Over the years, as England's countryside has returned to a less polluted state, many researchers have attempted to predict changes in the frequencies of dark and light forms by using computer models. In most instances, the observed frequencies of moths have not agreed with those predicted, and dark forms were more common in unpolluted areas than the models projected.

There are also observations that don't fit the story at all. In the forests near Detroit, Michigan, the frequency of melanic forms rose and fell, as they did in Britain, but the tree trunks remained lichen-covered the entire time. No soot ever covered the tree trunks. Thus, in Michigan the changes in the frequency of melanic forms could not have been caused by changes in the degree of crypticity (Grant et al. 1998).

It has been suggested that "the intuitive appeal of the 'crypsis' explanation may have blinded us to the role that other selective factors might be playing in the melanism story" (Sargent, Millar, and Lambert 1998). Perhaps the larvae of the different color forms differ in tolerance of a specific pollutant. With that in mind, it is interesting to note that the decline in melanic forms during the last 30 years seems to be correlated with a decrease in atmospheric sulfur dioxide (Grant et al. 1998). Alternatively, the parasites or predators of the moths might be the individuals most directly affected by pollution. How many other hypotheses can you think of? Even if the classical explanation for the rise and fall of melanic moths turns out to be correct, the recent controversy reminds us that alternative hypotheses should always be proposed and tested.

Selection of the Appropriate Background

Many animals appear to select "correct" backgrounds, and once there they exhibit behavior that maximizes their crypsis. The California yellow-legged frog (*Rana muscosa*) inhabits swift-flowing streams in the woodlands of southern California. The light gray granite boulders that line the streams seem conspicuous resting spots for the yellow-brown frog. Below the water, however, these same boulders are covered by a yellow-brown layer of algae. At a moment's notice, *R. muscosa* leaps into the water and lies motionless against a background to which it is perfectly matched (Norris and Lowe 1964).

Although the mechanisms for color matching among most cryptic animals have gone largely unexplored, it appears from experiments with insects that choice of a background in some species results from inherited preferences for particular backgrounds (Sargent 1968). Remember that cryptic animals not only select appropriately colored backgrounds but also assume positions that maximize the effectiveness of their crypsis. For example, moths that exhibit barklike crypsis orient themselves on tree trunks in such a way that their markings are aligned with the direction of the lines and ridges of the bark (Figure 13.10). Although some moth species appear to use tactile cues associated with the immediate substrate to determine the proper

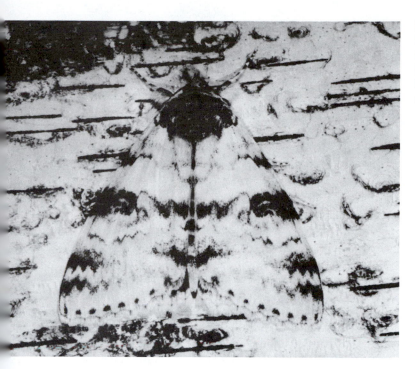

FIGURE 13.10 Many moths that exhibit barklike crypsis orient in such a manner that their markings match up with the lines and ridges of the bark. This moth has oriented to be virtually indistinguishable from the bark on which it rests.

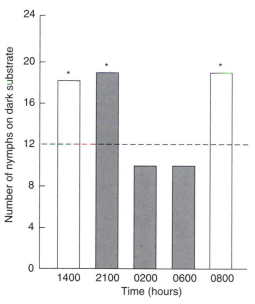

FIGURE 13.11 Substrate selection in stonefly nymphs at various times of the day (white bars represent data when lights are on in the laboratory and dark bars when lights are off) when given the option of resting on dark brown or light gray tiles. During the lights-on period (0800 and 1400 hours) and shortly after the lights go off (2100 hours), a larger number of nymphs were observed on dark tiles, a background on which they were cryptic, than on light tiles. Substrate selection was not apparent during the two remaining dark observations (0200 and 0600 hours). The dashed line represents expected results if no selection occurred; the asterisks indicate selection for dark brown tiles. (From Feltmate and Dudley Williams 1989.)

resting position, other species adopt resting attitudes on the basis of cues, such as gravity, that are independent of the bark (Sargent 1969). Now that we have some ideas about how background selection occurs in at least a few species, we will turn our attention to a test of why it occurs.

Is the combination of cryptic coloration and background selection adaptive? If it is, prey should experience less predation when sitting on the substrates that they tend to select as resting spots than they do when sitting on other surfaces. Blair Feltmate and D. Dudley Williams (1989) tested this idea by using rainbow trout (*Salmo gairdneri*) as predators and stonefly nymphs (*Paragnetina media*) as prey. Background color preferences of stoneflies, stream insects that are dark brown to black in color, was first tested by placing each of 24 nymphs into its own aquarium along with one dark brown and one light gray commercial tile on the bottom of the aquarium. Nymphs were left to settle for 24 hours, and then at 1400 hours (2 P.M.) the researchers recorded whether they rested on the dark brown or light gray substrate. The experiment was then repeated with recordings of nymphal position at 0200, 0600, 0800, and 2100 hours to test whether selection of substrate varied as a function of time of day (lights in the laboratory were on timers and were off from 1900 to 0700 hours). Thus, independent replicates of the experiment were run at five different times of the day, three in the dark (2100, 0200, and 0600 hours) and two in the light (0800 and 1400 hours). The results, depicted in Figure 13.11, demonstrate that stoneflies selected the dark brown substrate rather than the light gray one at

0800, 1400, and 2100 hours; no selection was observed at 0200 or 0600 hours. Although stonefly nymphs selected the dark over the light substrate, this selection ceased approximately two hours after the lights in the laboratory went off and resumed within one hour of their being turned on.

In the next experiment, Feltmate and Williams (1989) examined whether stoneflies resting on the light substrate were more vulnerable to predation by rainbow trout. As before, each stonefly was introduced into its own aquarium. This time, however, the tank contained either light or dark tiles (not both, as in the first experiment). A trout was released into each tank after the nymphs had two hours to adjust to their new surroundings. Twenty-four hours after releasing the nymphs, the authors recorded the number of stoneflies consumed in tanks containing either the light or dark substrate. The consumption of nymphs by trout was lower in tanks that contained the dark substrate (3 of 24 nymphs eaten) than in tanks that contained the light substrate (19 of 24 nymphs eaten). These data suggest that the selection of dark resting spots by stoneflies has been naturally selected, at least in part, because it reduces the risk of being found and eaten by visually hunting fish. The breakdown in substrate color selection during the hours of darkness also links visual predation to the distribution of nymphs. After all, animals need to be cryptic only when they are most vulnerable to predation by visual hunters (Endler 1978). The choice of substrate by stoneflies may also conceal them from their own prey, as has been shown for other aquatic insects (Moum and Baker 1990).

Movement and Absence of Movement

Movement, and in some cases absence of movement, is an important component of crypticity. Once oriented correctly on the appropriate background, many cryptic animals remain immobile most of the time, and when they do move, they move so slowly that they attract as little attention as possible. In other cases, rapid motion followed by abrupt cessation of movement contributes to crypsis. The escape behavior of newborn northern water snakes (*Natrix sipedon*) illustrates this point. Although adults of this species are uniformly colored, the young have a cross-banded pattern and appear cryptic when lying motionless in vegetation. When disturbed, however, newborns zip along the ground and then come to a complete stop. The rapid movement blurs the cross bands, making the snake appear to be unicolored (Figure 13.12). This visual illusion and the abrupt transition from motion to stillness combine to lead the observer (and presumably a predator) to search for the snake ahead of its actual position, for one's eyes tend to follow the path of a moving object even after that object has stopped moving (Pough 1976). Here, then, behavioral responses combined with cryptic coloration maximize camouflage.

Modification of the Environment

Some animals improve their chances of survival by modifying the background on which they normally rest. The spider *Tetragnatha foliferens* hides under a leaf that it folds into a tube and fastens to the center of its web (Hingston 1927b). From this vantage point, the spider is not only inconspicuous to predators but also in the best position to gain quick access to prey entangled in its web.

Insectivorous birds appear to use leaf damage as a cue to the location of cryptic caterpillars (Heinrich and Collins 1983). Thus, it should come as no surprise that some species of palatable caterpillars avoid attracting birds by snipping off partially eaten leaves at the end of a feeding bout (Heinrich 1979). Caterpillars of the underwing moth *Catocala cerogama* feed on the leaves of basswood trees, and their dorsal coloration resembles the bark of basswood twigs. At the end of a nocturnal foraging bout, an individual of this species backs off the leaf on which it has been feeding; chews through the stalk of the leaf; turns around; and as the chewed leaf falls to the ground, crawls onto the twig, where it will remain for the daylight hours. The branch on which the caterpillar was feeding now appears to be ungrazed. Although *C. cerogama* caterpillars are cryptically colored, they further enhance their chances of survival by removing evidence of their presence. (In contrast, unpalatable caterpillars—those with "hair," spines, or toxic glycosides—are quite blatant in their feeding behavior, foraging both day and night, and show no patterns of behavior associated with hiding leaf damage.) Finally, some animals achieve crypsis by carrying parts of their environment around with them. Many species of spider crabs (family Majidae) decorate themselves with algae, barnacles, sponges, and bottom debris and become virtually indistinguishable from the ocean floor (Figure 13.13). According to Mary Wicksten (1980), who has studied camouflage in these crabs, even an experienced underwater naturalist might sit on a crab festooned with bottom debris before realizing that it is there.

OTHER FUNCTIONS OF COLOR

Evasion of predators is not the only function of color pattern in animals. Color affects heat balance and thus plays a role in thermoregulation. Color and pattern are also important in many aspects of communication, including mate recognition, courtship, male-male competition, and territorial defense.

a   *b*

FIGURE 13.12   Newborn water snakes use a combination of cryptic coloration and movement to fool visually hunting predators. (*a*) A young snake with a cross-banded pattern appears cryptic when lying still in vegetation. (*b*) When startled, the snake moves rapidly across the substrate, causing the cross bands to blur, and then comes to an immediate halt. A predator that is tracking the snake's movement ends up searching for the snake ahead of its actual position.

The various functions of animal color and pattern may act in concert or in opposition. Let us consider a case in which they act in opposition. If color and pattern are adjusted for thermoregulation, how can animals communicate effectively with mates and competitors and at the same time be inconspicuous to visually hunting predators? Although some cryptically colored animals have evolved alternative means of exchanging information (e.g., relying on auditory or olfactory signals to communicate with conspecifics), many still rely on visual cues. As we will see, the color pattern displayed by a particular animal may be a compromise between factors that favor crypsis and those that favor conspicuousness.

John Endler's (1978) work with wild populations of guppies (*Poecilia reticulata*) in northeastern Venezuela and Trinidad provides an excellent example of how color patterns may represent a balance between mate acquisition and crypsis. Whereas a female's choice of mate and competition among males favor brighter colors and more visible patterns in guppies, selection by diurnal visual predators (at least six species of fish and one freshwater prawn) favors less colorful and less conspicuous patterns. It is interesting that as predation risk increases across communities, the colors and patterns of guppies become less obvious because of (1) shifts to less conspicuous colors, (2) reductions in the number of spots, (3) reductions in the size of spots, and (4) slight reductions in the diversity of colors and patterns (Figure 13.14). In areas in which guppies encounter low predation pressure, however, the balance shifts toward attracting mates, and colors and patterns become more conspicuous.

FIGURE 13.13   Camouflage through gaudy decoration. A crab has decorated its carapace, rostrum, and legs with strands of algae, strings of hydroids, and bottom debris. Such ornaments are picked up by the chelae or pincers, carried to the mouth for roughening of the edges, and then impaled on the shell. (Redrawn from Wicksten 1980.)

POLYMORPHISM AS DEFENSE

Like most things, cryptic coloration is not foolproof. Although individuals of a cryptically colored species blend with their background, predators in a given area

FIGURE 13.14   An animal's color is often a compromise solution to the problem of selection for conspicuousness in courtship displays and selection for inconspicuousness to visually hunting predators. Changes in the color and pattern of guppies as a function of predation pressure reflect the fine balance between these two selective forces. As predation risk increases, the number of spots (*a*), size of color patches (*b*), and diversity of patterns (*c*) decrease. (Modified from Endler 1978.)

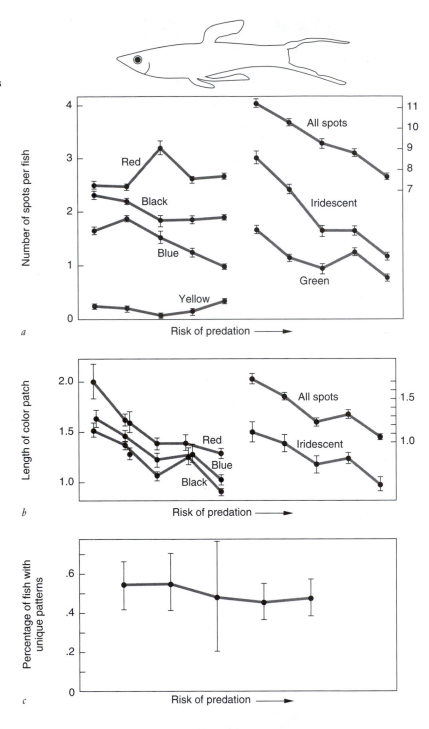

may develop a search image for that particular species and systematically search out and consume remaining individuals. If individuals of the prey species are widely spaced, however, predators will rarely encounter them and will soon forget the search image. Indeed, individuals of many cryptic species occur at widely spaced locations throughout their environment.

Other cryptic species get around the problem of search images by occurring in several different shapes and/or color forms, that is, by being polymorphic. We

have already discussed some examples of polymorphism with respect to industrial melanism in peppered moths and fire melanism in fox squirrels and grasshoppers. In some cases, however, polymorphic species are not cryptically colored at all and rely solely on their diverse appearance to evade detection by predators. Whether cryptic or not, by being different, individuals of prey species can occur at higher densities without suffering increased mortality from predators searching for individuals with a specific appearance. Some species that

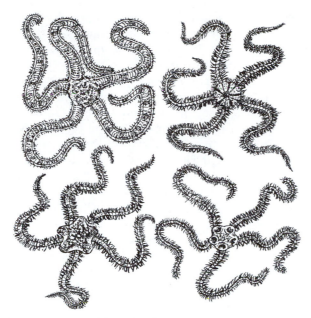

FIGURE 13.15 **Although all the same species, these four brittle stars are dramatically different in appearance, thereby inhibiting the formation of search images in predators. (Drawn from photograph in Moment 1962.)**

occur at very high densities exhibit extreme polymorphism, making it almost impossible to find two individuals that look alike (Figure 13.15).

Gairdner Moment (1962) described the phenomenon in which members of a population look as little like one another as possible. In such populations the probability of an individual's having a certain appearance is inversely related to the number of other individuals in the population that have that appearance. If one morph in a polymorphic population is much more common than another morph, predators are likely to develop a search image for the more common, rather than the rare, morph. The end result is that predators take more of the common form relative to its frequency in the population. Thus, when two morphs are equally cryptic and are exposed to predators that use search images when hunting, the rare morph will have a selective advantage over the common morph. This form of frequency-dependent selection has been called apostatic selection (Clarke 1969) or reflexive selection (Moment 1962). Its strength varies as a function of factors such as density, palatability, and conspicuousness of prey (Allen 1988). Furthermore, Jeremy Greenwood (1984) indicates that predators need not hunt by search image to cause apostatic selection in prey. Some predators, for example, may simply have an aversion to prey that are rare or unfamiliar to them.

What experimental evidence do we have that being different pays off? Croze (1979), working on a sandy peninsula in England, placed 27 painted mussel shells with pieces of meat under them on the ground and

exposed them to predation by carrion crows (*Corvus corone*). In some of the 14 trials, the shells were monomorphic (i.e., all the same color), whereas in others they were trimorphic (9 red, 9 yellow, and 9 black). The results, summarized in Table 13.2, show that the crows took fewer of the trimorphic than the monomorphic prey. The percentage of survival for each of the three morphs in a trimorphic population was two to three times higher than in monomorphic populations. Thus, a morph had a twofold to threefold selective advantage when occurring as part of a trimorphic population. Croze's results demonstrate that when prey populations occur at the same density, individuals in polymorphic populations experience less predation than those in monomorphic populations.

Before leaving the topic of polymorphism, we should make two points. First, throughout our discussions of cryptic and diverse coloration, we have focused on visually hunting predators and have ignored predators that detect prey through other means. Many animals detect prey by using their sense of smell, and the formation of olfactory search images seems quite reasonable. In the case of prey hunted on the basis of olfactory cues, one could imagine animals that are cryptic by being either odorless or similar in odor to their background, or even populations that have odor polymorphisms (Edmunds 1974).

Second, although being different to avoid being eaten may be the primary explanation of polymorphism in a population, it may be less important or totally unimportant in other populations. Populations of the banded snail *Cepaea nemoralis* are notoriously polymorphic in color and pattern, and explanations for their polymorphism have often focused on protection from avian predators through cryptic and diverse coloration. However, in some locations there are physiological differences in heat resistance among the various morphs, and such differences appear more important than predation in determining morph frequencies—just a reminder that there may be more to an animal's color and pattern than meets the eye.

Table 13.2 **Survival of Painted Mussel Shells in Either Monomorphic or Trimorphic Populations When Exposed to Predation by Carrion Crows (in Percentages)**

	Type of Population	
Shell Color	Monomorphic	Trimorphic
Yellow	10	31
Black	12	40
Red	19	45

Source: Data from Croze (1970).

WARNING COLORATION

Many animals that have dangerous or unpleasant attributes advertise this fact with bright colors and contrasting patterns. Bold markings, typically in black, white, red, or yellow, warn the predator of the prey's secondary defense mechanism, and through this warning discourage an attack. The phenomenon is called aposematism, and there are many familiar examples. The dramatic black and white markings of spotted and striped skunks (*Spilogale* and *Mephitis*) are truly exceptional amid an array of brown coat colors in mammals. The markings may serve, in part, to warn predators of the foul-smelling repellent that may, upon further harassment, be released from the skunks' anal scent glands. Many insects, such as the social wasps (*Vespula*), have a boldly patterned yellow and black body to warn of their painful sting, and the bright colors of several species of butterfly advertise their unpalatability. Frogs of the genus *Dendrobates*, and especially those of the genus *Phyllobates*, have toxic skin secretions. A single individual of the species *P. terribilis* has enough toxin in its skin to kill about 20,000 house mice or, in more familiar currency, several adult humans. The Choco Indians of western Colombia make deadly weapons by simply wiping their blowgun darts across the back of one of these frogs (Myers and Daly 1983). Not surprisingly, frogs that have toxic skin secretions are aposematically colored to warn predators that they are best left alone. In addition to conspicuous colors, characteristic noises (e.g., buzzes) and strong smells may also warn a predator. *Sternotherus odoratus*, indelicately but accurately called the stinkpot, is a musk turtle of the eastern United States that ejects an odorous secretion when disturbed. The stink is thought to be an aposematic signal that warns predators of the turtle's bad-tasting flesh, pugnacious disposition, and unhesitating bite (Eisner et al. 1977).

Animals that are colored in this way often enhance their conspicuousness behaviorally. Many are active during the daytime, and individuals of some species form dense, obvious aggregations. Although rare forms in aposematic animals are typically selected against (predators will not be as familiar with the rare form as they are with the common form and may attack), they are at less of a disadvantage when they occur in clusters (Greenwood, Cotton, and Wilson 1989). Thus, dense aggregations of aposematic prey not only emphasize the warning but also function as areas in which rare forms may arise and survive.

The response of predators to aposematic coloration may be learned or innate. In the first case, predators sample some of the prey, discover their unpleasantness, and learn to avoid animals of similar appearance when searching for subsequent meals. For example, garter snakes (*Thamnophis radix*) develop a much stronger avoidance of conspicuously colored noxious prey than of nonaposematic prey, even though olfaction plays an important role in detection and ingestion of prey. Two different types of prey, earthworms and fish, were offered to garter snakes on forceps that were aposematically colored (yellow and black) or nonaposematically colored (green). After they had consumed fish that was presented on forceps of either type of coloration, snakes in the experimental group were injected with lithium chloride to induce illness. Following the induced illness, the garter snakes avoided all fish, regardless of the color of the forceps. However, the garter snakes that had been offered fish on aposematically colored forceps had a much stronger aversion to fish than those who had been offered fish on green forceps (Figure 13.16). Seven days after the induced illness, only one of five snakes in the aposematic treatment group ingested any fish during the 120-second test interval. In contrast, all five of the snakes in the nonaposematic treatment group eventually attacked fish. Thus we see that predators learn to avoid unpalatable prey more readily if the prey are con-

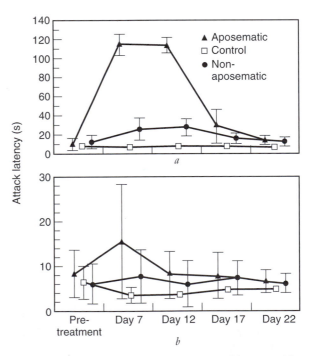

FIGURE 13.16 Predators learn more quickly to avoid distasteful prey that are conspicuous. Here, garter snakes were first offered pieces of fish on aposematic (yellow and black) or on nonaposematic (green) forceps. The snakes were then made ill by an injection of lithium chloride. The post-treatment attack latencies to pieces of fish (*a*) or earthworms (*b*) indicate that snakes in the aposematic treatment group had a stronger aversion to fish than did snakes in the nonaposematic treatment group. (Data from Terrick, Mumme, and Burghardt 1995.)

spicuously colored (Terrick, Mumme, and Burghardt 1995).

Sometimes two warningly colored species look alike. Apparently, two noxious species can benefit from a shared pattern because predators consume fewer of each species in the process of learning to avoid all animals of that general appearance. This phenomenon is called Mullerian mimicry. Although some predators learn through memorable experiences to avoid aposematic prey, others display innate avoidance. An innate response to warning coloration might be favored over a learned response when the secondary defense of the prey has the potential of being fatal to the predator. Learning, at the moment of death, is of little value.

Sometimes, like advice, warning coloration is ignored. A predator that is starving might tackle a noxious prey that it would normally pass up during better times. Wolves will attack both skunks and porcupines when other prey is scarce. In addition, some predators are specialists and are able to eat certain aposematic animals, or the least noxious parts of them, as we saw for the black-headed grosbeak and black-backed oriole that were preying on the unpalatable monarch butterfly. Others (such as arthropods) are attracted to the bold patterns and movements of warningly colored animals but are unable to make the connection between color and inedibility, and they repeatedly attack. However, as long as an antipredator device confers a net advantage in survival and reproduction, it will continue in the population.

BATESIAN MIMICRY

Batesian mimicry is named after the nineteenth-century English naturalist Henry Walter Bates, and it refers to a palatable species that has adopted the warning characteristics of a noxious or harmful species. The harmless species is called the mimic and the noxious one, the model. By resembling a noxious species, the mimic gains protection from predators. The evolution of mimicry has recently been the center of lively discussion (e.g., Holmgren and Enquist 1999; Joron and Mallet 1998; Mappes and Alatalo 1997).

The degree of protection experienced by the mimic varies as a function of numerous factors. The ratio of models to mimic is important, for example. The mimic does better when it is rare and therefore less likely to be detected by the predator than the noxious model (Turner 1977). And, the more distasteful the model, the better the mimic fares. Recently, these predictions have been experimentally tested, using Great tits (*Parus major*) as the predators. The birds were offered model and mimic prey one at a time, as prey would be encountered in nature. Both prey types survived better when there were fewer mimics. The models survived signifi-

cantly better the more distasteful they were, and the degree of unpalatability also affected the survival of the mimics (Lindstrom, Alatalo, and Mappes 1997). The memory of predators, availability of alternate prey, and whether mimics and models are encountered simultaneously or separately may also play a role (Speed and Turner 1999). As a result, a mimic may gain most if its habits and daily activity overlap those of its model species.

This increase in benefits was demonstrated by using naive birds, brown-eared bulbuls (*Hypsipetes amaurotis pryeri*), that were trained to take food from two feeders in captivity. The model prey was *Pachlopta aristolochiae*, a butterfly that sequesters alkaloids as a larva and is, therefore, distasteful. A noxious model butterfly was placed at one of the feeders. After an unpleasant experience eating the model, the bird took less of the palatable food from the feeder. This suggests that the bird associated the unpleasantness not just with the model but also with the place where it was experienced. The birds were then offered female swallowtail butterflies (*Papilio polytes*), which come in mimic and non-mimic forms. The birds avoided the mimetic forms of swallowtails (Uesugi 1996).

Although in some instances the resemblance between model and mimic seems almost exact, the likeness usually does not have to be perfect because predators appear to generalize conspicuous features of noxious prey. In some cases, poor mimics may exist because they exploit some constraint in the predator's visual or learning mechanisms. An example is provided by hoverflies, which mimic certain wasps. Studies have shown that pigeons rank hoverflies according to their similarity to the wasp model. To human eyes, the two most common types of hoverflies show the least resemblance to wasps, yet the pigeons rank them as being very similar to wasps. It is thought that these wasps have some key feature that is used by pigeons in pattern recognition (Dittrich et al. 1993).

Most known examples of mimicry are visual, probably reflecting the fact that we humans are visually oriented creatures. Other animals rely on smell and hearing more than vision, and thus olfactory and auditory mimicries may be quite common.

Many fascinating examples of Batesian mimicry can be found among insects and spiders. Some perfectly harmless flies mimic the yellow and black bands or buzzing sounds characteristic of bees and wasps. Predators all too familiar with the painful stings of bees and wasps may leave these flies alone. Ants are typically avoided by insectivorous predators because of their sting or bite and unpleasant taste (formic acid lends ants their bad taste). It should come as no surprise, then, that ants have many mimics, and the likeness can include features of color, morphology, and behavior. Major R. W. G. Hingston (1927a) recorded several instances of spiders mimicking species of ants in India. One species closely resembled the large Indian black ant, *Camponotus compressus*. The spider was the size of a *Camponotus* worker and shared the uniform black color, elongated shape, and slender legs of the worker physique. Because ants have three pairs of legs and spiders four, the spider used its front pair of legs to simulate the antennae of the ant. These legs were thrust forward and the tips kept in continual motion to mimic the methodical movement characteristic of the ant's antennae. In a second species of ant studied by Hingston, individuals dragged workers of other species back to their nests, decapitated their victims, and threw the heads in a refuse heap. Hingston (1927b) reported that one species of spider curled into a tight motionless ball, with its head and legs tucked underneath its pear-shaped abdomen (Figure 13.17), and sat amid the discarded heads. By mimicking a fragment of a noxious species (and associating with an even more formidable species), these spiders appear to gain protection from predators and may also be cryptic to their own prey that may wander by. Thus, mimicry can serve both defensive and foraging functions.

Despite the diverse array of mimetic resemblances recounted in the literature, only a few studies demonstrate that the purported mimics actually gain protection from their natural enemies. T. E. Reimchen (1989) first described a system of Batesian mimicry involving the juvenile stage of a snail (the mimic) and the tubes of a polychaete worm (the model) and then provided evidence that the resemblance actually conferred some degree of protection to the young snails. The snail, *Littorina mariae*, lives in the intertidal zone of the North Atlantic. The shells of some juveniles have a conspicuous white spiral, and the shells of others are yellow or brown. When adult, the snails are either yellow or brown, and the white spiral possessed by some is visible only as a white apex on the shell. Egg masses of the snail are deposited directly on algal fronds, and once the juveniles hatch they disperse on the fronds.

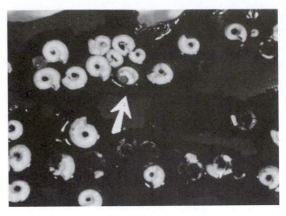

FIGURE 13.18   The white phase juvenile of a snail (shown by the arrow) is a Batesian mimic of the tubes of certain polychaetes. By resembling the tubes, young snails may gain protection from fish that are searching for food on algal fronds.

Snails with white-spiral coloration were observed only in habitats where the polychaete *Spirobis* was present. In these habitats, white-spiral phase juveniles are virtually indistinguishable from the tubes of *Spirobis* that are cemented to the fronds (Figure 13.18).

Reimchen collected the intertidal fish *Blennius pholis*, an important predator on juvenile snails, and conducted predation experiments in aquaria in the laboratory. Although the polychaete tubes are not noxious to the fish, they represent a substantial investment in time and energy because they are difficult to remove from the substrate, and once removed, they may prove to be unoccupied. In the experiments, blennies were housed alone in an aquarium and were presented with juvenile snails on either an algal frond with polychaete tubes or an algal frond without tubes. At each presentation, three juvenile snails (one white-spiral, one yellow, and one brown) were randomly positioned on the frond and the frond was lowered to the bottom of the tank. Once blennies detected a snail, they plucked it off the frond and swallowed the shell whole. Reimchen recorded the first snail taken at each trial. Overall, white-spiral snails suffered the lowest number of attacks and the reduction in attacks was greater on fronds with polychaete tubes (9.4%) than on fronds devoid of tubes (22.9%). Thus, in this unusual system of snail-polychaete mimicry, resemblance to the model does appear to confer a protective advantage to the mimic.

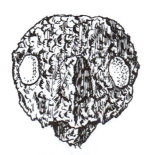

Ant's head   Spider head

FIGURE 13.17   On the left is the head of an ant discarded by a worker of another ant species. On the right is a spider from India that mimics the discarded remains. (Redrawn from Hingston 1927a.)

DIVERTING COLORATION, STRUCTURES, AND BEHAVIOR

A great many animals have evolved colors, structures, and patterns of behavior that seem to divert a predator's attention, while they, and in some cases their offspring, escape with little or no damage. Whereas crypsis, diverse coloration, warning coloration, and Batesian mimicry help prey avoid an encounter with a predator, distraction devices come into play once a prey animal has been discovered or when discovery seems all too imminent. Thus, we will now begin our discussion of secondary defense mechanisms, that is, those mechanisms that operate during an encounter with a predator.

EYESPOTS

Eyespots can serve two defensive functions (Owen 1980). First, if the spots are large, few in number, brightly colored, and suddenly flashed, they may startle or frighten the predator. Second, spots that are small and less gaudy may instead serve as targets to misdirect the predator. Such eyespots are typically located on nonvital portions of the body, and thus prey can often escape with less than fatal damage.

FALSE HEADS

Many predators direct their initial attack at the head of the prey. Some prey species have taken advantage of this tendency by evolving false heads that are located at their posterior end, a safe distance from their true heads. Lycaenid butterflies (Lepidoptera: Lycaenidae) display patterns of color, structure, and behavior that are consistent with deflecting predator attacks toward a false head (e.g., Robbins 1981). Individuals of the species *Thecla togarna*, for example, have a false head, complete with dummy antennae, at the tips of their hind wings (Figure 13.19). These butterflies enhance the structural illusion of a head at their hind end by performing two rather convincing behavioral displays. First, upon landing, the butterfly jerks its hind wings, thereby moving the dummy antennae up and down while keeping the true antennae motionless. The second ploy occurs at the instant of landing, when the butterfly quickly turns so that its false head points in the direction of previous flight. An approaching predator is thus confronted with a prey that flutters off in the direction opposite to that expected. Experimental tests have demonstrated that markings associated with false heads misdirect the attacks of avian predators and, in particular, increase the possibility of escape if the prey is caught to begin with (Wourms and Wasserman 1985).

AUTOTOMY

Rather than simply diverting a predator's attack toward a nonvital portion of the anatomy, some prey actually hand over a "disposable" body part to their attacker, almost as a consolation prize. Autotomy, the ability to

FIGURE 13.19 **The false head of a butterfly. Note the pattern of markings that tends to focus attention on the posterior end of the butterfly and the dummy antennae and eyes at the tips of the hind wings. Markings and structure combine with behavior (e.g., movement of dummy, rather than true, antennae) to divert the attention of a predator away from the true head. (Redrawn from Wickler 1968.)**

break off a body part when attacked, has evolved as a defense mechanism in both vertebrates and invertebrates. Tail autotomy in lizards, for example, is commonly reported, as well as in some salamanders, a few snakes, and even some rodents. A more dramatic autotomy, however, is seen in sea cucumbers (members of the phylum Echinodermata), which, upon being attacked, forcefully expel their visceral organs (guts, in the vernacular) through a rupture in the cloacal region or body wall. The predator may then begin to feed on the sea cucumber's offering as it makes its slow escape. In most autotomy cases, the disposable body part is subsequently regenerated. As an example of the phenomenon, we will focus on tail autotomy in lizards.

Tail autotomy benefits the lizard in two ways: First, it allows the lizard to break away from its attacker, and, second, if the detached tail continues to thrash and writhe, the attacker is distracted as the lizard runs away (E. N. Arnold 1988). Although the vigor and duration of postautotomy tail movement varies among species, in some lizards the tail may thrash for as long as five minutes. The effectiveness of tail autotomy is underscored by the presence of tails in the stomachs of predators, as well as the occurrence of tailless lizards and lizards with regenerated tails in natural populations.

Direct experimental evidence for the importance of tail autotomy as an antipredator device comes from a laboratory story by Benjamin Dial and Lloyd Fitzpatrick (1983). These researchers tested the effec-

Table 13.3   Responses of a Feral Cat to Simultaneous Presentation of Autotomized Tails (Either Thrashing or Exhausted) and Live Tailless Bodies of Two Species of Lizards

	Number of Responses		
Tail	Attack to Tail	Attack to Body	Escape of Lizard
Anolis carolinensis			
Exhausted	0	8	3
Thrashing	0	6	1
Scincella lateralis			
Exhausted	0	6	0
Thrashing	7	0	7

Source: Data from Dial and Fitzpatrick (1983).

tiveness of tail autotomy and postautotomy tail movements in permitting the escape of lizards from mammalian and snake predators. In the first study, staged encounters were conducted between a feral cat (*Felis catus*) and two species of lizards, *Scincella lateralis* (a species with vigorous postautotomy tail thrashing) and *Anolis carolinensis* (a species with less vigorous thrashing). Dial and Fitzpatrick recorded the cat's reaction to lizards of both species under two conditions: (1) thrashing tail trials—lizards and their autotomized tails were placed in front of the cat immediately after autotomy, and (2) exhausted tail trials—tails were allowed to thrash to exhaustion and then lizards and their autotomized tails were placed in front of the cat. In both types of trials, autotomy was induced by the experimenters. They gripped the lizards' tails at the caudal fracture plane with forceps (in many species of lizards, tail breakage takes place at preformed areas of weakness). The results, summarized in Table 13.3, show that the dramatic postautotomy tail thrashing of *S. lateralis* is an effective escape tactic, whereas the more subdued tail movement of *A. carolinensis* is not. Note that in all of the thrashing-tail trials with *S. lateralis*, the cat attacked the tail, rather than the lizard, and in all cases the lizard escaped. In 100% of the exhausted-tail trials with this species, however, the cat attacked and captured the lizards. The results for *A. carolinensis* were quite different: The cat attacked the lizards and ignored the tails in all trials.

In the second experiment, Dial and Fitzpatrick (1983) examined whether postautotomy tail movement influenced the predator's handling time. The authors staged encounters between *S. lateralis* and the snake *Lampropeltis triangulum*, again using autotomized tails that were either thrashing or exhausted. On the average, the snakes required 37 seconds longer to handle thrashing tails than exhausted tails, providing the tail-

less lizard with more time to escape. Thus, for the lizard *S. lateralis*, postautotomy tail movement supplements the simple mechanism of breaking away from the predator's grasp and, depending on the type of predator, may either attract the predator's attention (as in the case of the cat) or increase the time required to handle the autotomized tail (as in the case of the snake). Either way, postautotomy tail movement enhances the opportunity for the lizard to escape.

Until this point we have focused on the benefits of tail autotomy without mentioning potential costs. Depending on the species of lizard, tail loss may lead to reductions in speed, balance, swimming, or climbing ability, and when the tail is used as a display, even to declines in social status (Fox and Rostker 1982). Furthermore, regeneration of the tail must certainly entail costs in energy and materials. Many lizards, after all, have substantial fat deposits in their tails that are also lost with the tail. One possible recourse to the cost of leaving behind energy reserves was suggested by Donald R. Clark (1971) after he observed that postautotomy tail movement in *S. lateralis* pushes the tail through leaf litter. He suggested that such movements propel the tail out of sight of the predator and facilitate later retrieval and eating by the original owner. There are some reports of lizards ingesting their own autotomized tails. W. W. Judd (1955), in describing the capture of a lizard that had escaped in his laboratory stated, "When an attempt was made to capture it, the skink snapped off the terminal one inch of its tail and wriggled free. However, it immediately turned around, and grasping the severed tail by its narrow end, gulped the whole thing down." However, Dial and Fitzpatrick (1983) found that when snakes grabbed the tail of *S. lateralis* they never subsequently lost their hold. Thus, the question of whether lizards routinely lose their tails and eat them too remains unanswered.

FEIGNING INJURY OR DEATH

In ground-nesting birds such as the killdeer (*Charadrius vociferus*), a parent may feign injury in an elaborate effort to divert the attention of an approaching predator away from its nest and young, particularly soon after hatching, when offspring are most vulnerable (Brunton 1990). Upon spying a predator, an adult may suddenly begin dragging its wing, as it flutters away from its nest. The predator follows, and as it closes in the killdeer suddenly recovers and flies away, giving a loud call. If all goes as planned, the predator will continue to wander off.

Some animals rely not on diverting the attention of a predator but on causing the predator to lose interest. Because many predators kill only when their prey is moving, an animal that feigns death may fail to release the predator's killing behavior, and with any luck the predator will lose interest and move along in search of a more lively victim. Perhaps the most familiar feigner of death is the opossum, *Didelphis virginiana* (Figure 13.20). Hence the phrase "playing possum" has come to be synonymous with "playing dead." Although their performance is less well publicized than that of opossums, juvenile caimans (*Caiman crocodilus*) react aggressively toward humans when approached on land but feign death when handled in water (Gorzula 1978). The response of an individual to a particular predator may thus vary as a function of context, and prey animals typically have several antipredator devices at their disposal.

Hognose snakes (*Heterodon platirhinos*) have a complex repertoire of antipredator mechanisms, and feigning death is one option. These fairly large nonvenomous or slightly venomous snakes occur in sandy habitats in the eastern United States. When first disturbed, the hognose opts for bluffing—it flattens and expands the front third of its body and head, forming a hood and causing it to look larger. It then curls into an exaggerated S-coil and hisses, occasionally making false strikes at its tormentor. When further provoked, however, it drops the bluff and begins to writhe violently and to defecate. Then it rolls over, belly up, with its mouth open and tongue lolling. If the predator loses interest in the "corpse" and moves away, the snake slowly rights itself and crawls off.

The complete repertoire of antipredator mechanisms occurs in young hognose snakes, and Gordon Burghardt and Harry Greene (1988) have shown that newborn snakes are capable of making very subtle assessments of the degree of threat posed by a particular predator. The researchers conducted two experiments in which they monitored the recovery from feigning death (i.e., crawling away) of newly hatched snakes under various conditions. In experiment 1, the recovery of snakes was monitored in the presence or absence of a stuffed screech owl (*Otus asio*) mounted on a tripod 1 meter from the belly-up snake. In experiment 2, the snake recovered (1) in the presence of a human who was staring at the snake from a distance of 1 meter, (2) in the presence of the same person in the same loca-

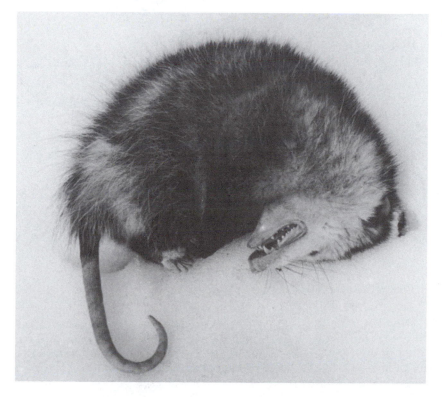

FIGURE 13.20 An opossum is playing dead.

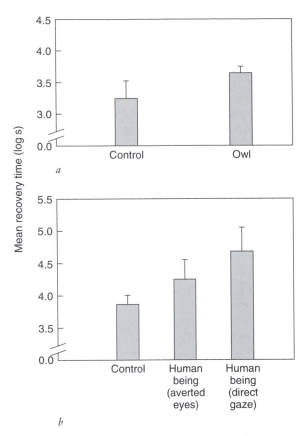

FIGURE 13.21   **Mean time to recovery from feigning death in neonatal hognose snakes exposed to various recovery conditions. (*a*) In experiment 1, snakes recovered in the presence of a stuffed owl or in the absence of a stuffed owl (control condition). (*b*) In experiment 2, snakes recovered in the presence of a human being (with eyes staring at the snake or with eyes averted) or in the absence of a human being (control condition). Because the recovery times were skewed, the data were transformed into logarithms of seconds. (From Burghardt and Greene 1988.)**

tion but whose eyes were averted, and (3) in a control condition in which no human being was visible. Both the presence of the owl (experiment 1) and the direct human gaze (experiment 2) resulted in longer recovery times than the respective control conditions (Figure 13.21). When the human being averted his or her eyes, the recovery time was intermediate. Thus, young snakes are capable of using rather subtle cues to make adjustments in their antipredator behavior.

INTIMIDATION AND FIGHTING BACK

Prey animals have many ways of communicating "I am formidable" to a predator. Presumably, when a predator encounters a large, threatening, well-armed prey, it will continue on its way, searching for a less challenging meal.

ENHANCEMENT OF BODY SIZE AND DISPLAY OF WEAPONRY

When dealing with potential predators, some animals employ the size-maximization principle. A cat hunches its back and erects its fur in the presence of a dog. Some toads and fishes inflate themselves when disturbed. In each case, the animal increases its size and appears more formidable or unswallowable. Several displays of intimidation through an increase in size are shown in Figure 13.22, but threat maximization need not always be visual. Loud calls, hisses, or growls may also cause a predator to look elsewhere for its next meal.

Sometimes eyes can be threatening. Several species of animals have utilized the relationship between eyes and threat and have developed eyespots as a means of repelling predators. Such eyes usually appear to be large, wide open, and staring straight ahead, although they may in actuality be sightless spots on the wings of a harmless insect or on the backside of a toad (Figure 13.23). Remember that eyespots may also startle a predator or misdirect its attack. Finally, some animals display their weapons when confronting a predator. Ungulates often display their horns and paw at the ground, perhaps to draw attention to their dangerous

FIGURE 13.22   **Intimidation displays in several species of animals. The displays make the animal appear larger and more formidable to predators. The animals shown here are (counterclockwise) the frilled lizard, cat, short-eared owl, and spotted skunk. (Modified from Johnsgard 1967.)**

FIGURE 13.23 A toad directs its backside toward an attacker, revealing a pair of frightening eyespots.

hooves. Porcupines erect their spines and cats display their teeth. All these postures are probably meant to intimidate a predator.

CHEMICAL REPELLENTS

A wide variety of insects can discharge noxious chemicals when they are captured. Some of these chemicals are powerful toxins or irritants, and in some species they can be shot with considerable accuracy in several directions. The assassin bug (*Platymeris rhadamantus*) reacts to a disturbance by spitting copious amounts of fluid in the direction of the attacker. The saliva is rich in enzymes and causes intense local pain when it comes in contact with membranes of the eyes or nose.

Other masters of chemical warfare are the bombardier beetles, which deter predators by emitting a defensive spray that contains substances stored in two glands that open at the tip of the abdomen (Dean et al.

1990; Eisner 1958). Because the tip of the abdomen acts as a revolvable turret, the spray can be aimed in all directions (Figure 13.24). The chemical reactants from the two glands are mixed just before they are discharged, producing a sudden increase in temperature of the mixture. The hot spray is ejected, accompanied by audible pops, in quick pulses. The effect has been likened to that of the German V-1 "buzz" bomb of World War II (Dean et al. 1990).

Chemical deterrents are by no means limited to arthropods, as anyone who has had the misfortune of surprising a skunk or who owns a dog that has enjoyed the same experience must surely know. Although the defensive response of the horned toad (*Phrynosoma cornutum*) is perhaps less well known than that of the skunk, it is certainly no less spectacular. When disturbed, this small, spine-covered lizard of the southwestern United States may spatter its attacker with a stream of blood ejected from its eyes (Lambert and Ferguson 1985). At the turn of the century, Charles Holder (1901) examined this behavior and suggested, on the basis of trials in which his fox terrier posed as a predator, that the ejected blood contained noxious components. Apparently, contact between the ejected blood and nasal membranes of the dog was particularly irritating, and only a single encounter was required to produce "a wholesome dread" in the lizard's canine tormentor. Whether the discharged blood actually contains noxious components and just what these components might be remain to be determined.

STARTLE MECHANISMS

Sometimes even an extra second or two is enough time for an animal to escape from what appears to be certain

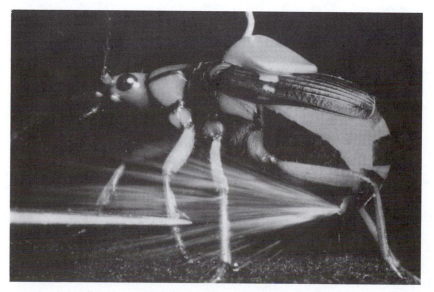

Figure 13.24 The bombardier beetle ejects a hot, irritating spray at its attackers. This beetle, tethered to a wire fastened to its back with wax, responds with excellent aim to a forceps pinch on its left foreleg.

death. In some cases prey can escape if it can startle the predator into delaying for only an instant. Called deimatic displays by Edmunds (1974), startle mechanisms involve sudden and conspicuous changes in the appearance or behavior of prey that can produce confusion or alarm in a predator. The sudden presentation of a visual stimulus (such as large eyespots) or an auditory stimulus (such as squeaks, rattles, or screams) may startle the predator to the extent that it withdraws or hesitates just long enough for the prey to escape.

Many insects have deimatic displays that involve the sudden exposure of bold colors or patterns that are concealed when resting. Moths of the genus *Catocala* are palatable to avian predators and have cryptic barklike forewings that cover flamboyantly patterned hind wings. Apparently, the word *Catocala* is derived from the Greek words *kato* and *kalos*, meaning "beautiful behinds" (Sargent 1976). When crypsis fails and the moths are disturbed by predators, they suddenly flash their striking hind wings. Do the hind wings of *Catocala* serve as startle devices? Debra Schlenoff (1985) investigated this possibility by examining the response of blue jays (*Cyanocitta cristata*) maintained in indoor-outdoor aviaries to models of *Catocala* moths. She constructed artificial moths of gray cardboard, gave them pinyon nut bodies (delicacies for blue jays), and attached brightly colored hind wings that popped into view when the model was removed by the bird from the presentation board (Figure 13.25). During the training phase, blue jays were taught to capture and eat artificial moths with uniform gray hind wings. During the test phase of the experiment, jays were presented with seven of the moths with gray hind wings and one randomly placed moth having a boldly patterned hind wing, that is, a *Catocala* hind wing. When jays picked up the moth with the *Catocala* pattern, they raised their crests and gave alarm calls, sometimes dropping or flying away from the model.

Auditory stimuli can also be startling to predators. Some species of arctiid moths click in response to touch or sound. These moths are usually distasteful and when harassed emit a repellent froth from their thorax. The disturbance clicks of arctiid moths often cause bats to abort their predatory attacks. Are the sudden clicks of these moths startle devices or simply warnings of bad taste? Big brown bats (*Eptesicus fuscus*) that were trained to fly to a platform where they sometimes received a mealworm reward, veered away from the platform when arctiid clicks were broadcast (Bates and Fenton 1990). Individual bats, however, quickly habituated to the clicks and soon did not respond to the sound by avoiding the platform. In a second experiment, when broadcast clicks were paired with mealworms injected with quinine sulfate, bats rapidly learned to associate moth clicks with bad-tasting mealworms. Thus, in the case of bats and arctiid moths, the clicks may serve both

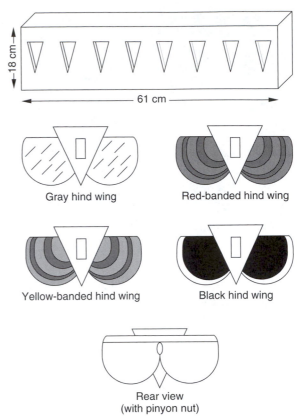

FIGURE 13.25   Presentation board and moth models. When the models were correctly positioned in the board, the hind wings were folded back into the open triangles of the board and only the gray forewings were visible. Blue jays trained to capture models with gray hind wings exhibited startle responses when they captured randomly placed models with brightly colored hind wings. (From Schlenoff 1985.)

as startle and warning devices, depending on the experience of the predator—inexperienced bats are startled by clicks, but experienced bats recognize clicks as a warning of distasteful prey.

PRONOUNCEMENT OF VIGILANCE

Some prey appear to inform predators that they have been spotted. The advantage might be in discouraging the predator by confrontation with an alert and aware prey. Stalking predators, for example, may abandon the hunt once they receive the signal that they have been detected. Stotting, a stiff-legged jumping display performed by many species of deer, pronghorn, and antelope, in which all four feet are off the ground simultaneously (Figure 13.26), appears to be just such a signal. The stotting display has attracted the attention of a number of investigators, and arrival at prey detection as

FIGURE 13.26 **Stotting by a Thomson's gazelle.**

Table 13.4 Hypotheses to Explain the Function of Stotting

Benefits to the individual

Signaling to the predator
1. Pursuit invitation
2. Predator detection
3. Pursuit deterrence
4. Prey is healthy
5. Startle
6. Confusion effect

Signaling to conspecifics
7. Social cohesion
8. Attract mother's attention

Signaling not involved
9. Antiambush behavior
10. Play

Benefits to other individuals
11. Warn conspecifics

Source: After Caro (1986a).

a plausible function has involved testing predictions from a diverse array of hypotheses.

At least 11 hypotheses have been proposed for the function of stotting (Caro 1986a, Table 13.4). Although not mutually exclusive, the hypotheses range from the interpretation of stotting as a signal given by a hunted animal to either a predator or a conspecific to the interpretation that stotting has no signal value at all and is simply a form of play or, alternatively, a means to visually survey the flight path away from a predator. In the first true effort to distinguish among the hypotheses, Tim Caro (1986b) recorded the response of Thomson's gazelles (*Gazella thomsoni*) to naturally occurring predators, usually cheetahs (*Acinonyx jubatus*), in the Serengeti National Park of Tanzania. He analyzed prey behavior, cheetah behavior, and the outcome of hunts and found that cheetahs were more likely to abandon hunts when their prey stotted than when they did not (Table 13.5). These results, combined with other data that refuted many of the remaining hypotheses, suggested that stotting typically functioned to inform the predator that it had been detected. Two other functions for stotting were supported by Caro's observations. First, mothers may stott to distract a predator from their fawn, a function much like the broken wing displays described for killdeer. Second, fawns appear to stott to inform their mother that they have been disturbed at their hiding place.

A more recent study suggests that the context of the cheetah-gazelle encounter and age of the performing gazelle are not the only factors to influence the function of stotting. The type of predator is another consideration. When hunted by coursing predators that rely on stamina to outrun their prey, gazelles appear to use stotting as an honest signal of their ability to outrun predators (FitzGibbon and Fanshawe 1988). Coursing

Table 13.5 Outcome of 31 Cheetah Hunts Involving Thomson's Gazelles That Did or Did Not Stott

	Chase Occurred			
	Chase Successful	Chase Unsuccessful	Hunt Abandoned	Total
Gazelle stotts	0	2	5	7
Gazelle does not stott	5	7	12	24

Source: Modified from Caro (1986b).

predators such as African wild dogs (*Lycaon pictus*) concentrate their chases on those individuals within a group that stott at lower rates, and thus they appear to use information conveyed in stotting to select their prey. In the study by FitzGibbon and Fanshawe, the mean rate of stotting by gazelles that were chased was 1.64 stotts per second, and for those not chased, 1.86 stotts per second. By signaling their ability to escape at the start of a hunt, those gazelles with high stamina and/or running speeds may not have to prove their physical prowess by outrunning wild dogs in long, exhausting, and potentially dangerous chases. If the function of stotting varies with the species of predator, we should not be surprised if future studies reveal that the function varies with the species of prey as well. Finally, although often performed in the presence of predators, stotting also occurs during intraspecific encounters, and we can only guess what its function is under these circumstances.

GROUP DEFENSE

Until now we have focused almost exclusively on strategies employed by individual animals to avoid being eaten. Some animals, however, are social, and membership in a group makes accessible a host of antipredator tactics that are not available to solitary individuals. Generally, predators experience less success when hunting grouped rather than single prey because of the superior ability of groups to detect, confuse, and repel predators. In addition, an individual within a group has a lower probability of being selected during any given attack. We will now consider some examples of how social animals cope with predators. Keep in mind that group living has many advantages, including those totally unrelated to protection from predators (see Chapter 16).

ALARM SIGNALS

When a predator approaches a group of prey, one or more individuals within the aggregation may give a signal that alerts other members of the group to the predator's presence. Alarm signals may be visual, auditory, or chemical, and they often either enlist support in confronting the attacker or inspire retreat to a safe location. In some cases the alarm may aid the signaler or its relatives; in other instances the alarm appears to benefit all those exposed to the signal by permitting members of the group to escape in a coordinated fashion. The proposed selective advantages of signaling alarm are covered in more detail in Chapter 17. We will focus our discussion here on the chemical alarm systems of some fish and amphibians.

Some species of fish show escape responses to chemical stimuli from injured conspecifics. For example, if the skin of a minnow is broken, an alarm substance, called Schreckstoff, is released from skin cells. Conspecifics that smell the chemical respond by rapid dashes, followed by hiding and reduced activity. Although once thought to be peculiar to the minnows and their relatives, an analogous alarm system has been reported for other groups of fish, including the darters and gobies (R. J. F. Smith 1982, 1989). In most cases, the alarm response is displayed by fish that form schools.

Although the presumed function of releasing Schreckstoff is to warn other fish within the school of the danger of attack, there is little experimental evidence for its effectiveness as an antipredator mechanism. Such evidence is available, however, for the alarm substance produced by injured tadpoles of the western toad, *Bufo boreas*. Individuals of this species live in the ponds and lakes of western North America, where the tadpoles form dense aggregations. Diana Hews (1988) first documented the response of toad tadpoles to release of the alarm substance and then tested whether tadpoles alerted by the substance had higher survival rates than those not exposed. Two natural predators of western toad tadpoles, giant water bugs (*Lethocerus americanus*) and dragonfly naiads (*Aeshna umbrosa*), were used in the experiments.

When tested in aquaria, western toad tadpoles increased their activity and avoided the side of the tank that contained a giant water bug that was feeding on a conspecific tadpole (in a visually isolated but interconnected container). Tadpoles did not increase their activity or avoid the predator's side of the tank when the water bug was feeding on a tadpole of another species. It is important to note that toad tadpoles alerted by the conspecific alarm substance were less vulnerable to predation. Dragonfly naiads had fewer captures per attack in tests with tadpoles exposed to the toad extract containing the alarm substance than with tadpoles exposed to the control extract, water (Figure 13.27). In addition to warning conspecifics, the alarm substance of *B. boreas* may function directly in deterring predators. Many larval and adult toads are distasteful to predators because of a toxin in their skin, and this "bufotoxin" is a likely component of the alarm substance. Again a given defensive mechanism can have more than one protective function.

IMPROVED DETECTION

Early detection of a predator can often translate into escape for prey, and groups are typically superior to lone animals in their ability to spot predators. Increases in the number of group members (and hence the number of eyes, ears, noses, etc.) often result in increases in

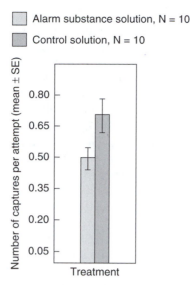

☐ Alarm substance solution, N = 10

☐ Control solution, N = 10

FIGURE 13.27   **Effects of alarm substance of toad tadpoles on the attack success rate of dragonfly naiads. Tadpoles exposed to the alarm substance were less vulnerable to predation by naiads than those exposed to the control substance, water. (Modified from Hews 1988.)**

the immediacy with which approaching predators are detected. Furthermore, as a result of the enhanced vigilance of groups, a given group member can often spend more time foraging and less time on the lookout for predators.

The benefits of increased predator-detecting ability may accrue to members of single-species or mixed-species groups. Florida scrub jays (*Aphelocoma coerulescens*) form single-species groupings, usually of from two to eight family members. Because these birds live in small, permanent groups of stable composition, it is possible for individuals to coordinate their vigilance in a highly structured sentinel system (McGowan and Woolfenden 1989). At any given time, only one family member typically sits on an exposed perch and continually scans the surroundings for predators. If a predator is spotted, the sentinel sounds the alarm, and family members respond by either mobbing a ground predator or by fleeing or monitoring the movements of an aerial attacker. Periodic exchanges among family members occur to relieve the sentinel bird of its duties. Sentinel systems have also been reported for mammals such as the dwarf mongoose, *Helogale undulata rufula* (Rasa 1986), and the meerkat, *Suricata suricatta* (Moran 1984), two species that live in family-based social groups.

The benefit of improved vigilance will apply to members of mixed-species groupings, providing that they are on the lookout for the same species of predators and that they communicate detection to other group members. Also, some members of heterospecific groupings benefit if predators display a preference for

individuals of the other prey species in their group. For example, Thomson's gazelles (*G. thomsoni*), a species familiar from our discussion of stotting, and Grant's gazelles (*G. granti*) often form mixed-species groups in the Serengeti National Park of Tanzania (FitzGibbon 1990). Thomson's gazelles that joined Grant's gazelles to form larger mixed-species flocks were less vulnerable to cheetahs than those gazelles that remained as smaller groups of conspecifics. Grant's gazelles, on the other hand, benefited from the association because of the cheetahs' preference for the smaller Thomson's gazelles.

DILUTION EFFECT

Individuals in groups are safer not only because of their enhanced ability to detect predators but also because each individual has a smaller chance of becoming the next victim. Called the dilution effect, this advantage for grouped prey operates if predators encounter single individuals or small groups as often as large groups and if there is a limit to the number of prey killed per encounter. As group size increases, the dilution effect becomes more effective, and improved vigilance appears to provide relatively less benefit (Dehn 1990).

Although this notion of safety in numbers has intuitive appeal, in some cases, predators aggregate in areas where their prey are abundant. As a result, some grouped prey may actually suffer higher predation rates. In an examination of the balance between the forces of the dilution effect and the aggregating response of predators, Turchin and Kareiva (1989) studied grouping in aphids (*Aphis varians*). These small insects form dense clusters on the flowerheads of fireweed, and it is here that they are preyed on by ladybird beetles, typically *Hippodamia convergens*. In one experiment, the researchers quantified per capita population growth rates (a measure of individual survivorship) for aphids living singly and for those living in colonies of over 1000 individuals and found that individual aphids benefited by forming groups. Grouping was only advantageous, however, in the presence of predators; when ladybird beetles were excluded from fireweed plants, the individual survivorship of aphids did not increase with the colony's size.

The next question, then, is how do ladybird beetles respond to the grouping of their prey? Turchin and Kareiva found that beetles exhibited a strong aggregation response: More than four times as many beetles were found at aphid colonies of over 1000 individuals than at small and medium-sized colonies (Figure 13.28). In addition to gathering at large colonies, ladybird beetles also increased their feeding rate as the aphids' density increased. On average, beetles consumed 0.9 aphids per ten minutes in colonies of 10 individuals and 2.4 aphids per ten minutes in colonies

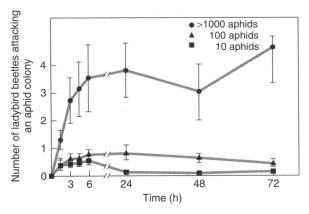

FIGURE 13.28   The tendency for prey individuals to form large groups and thereby dilute their chances of becoming the next victim may be countered by the tendency by their predators to aggregate where prey are most common. Here, predatory ladybird beetles gather in larger numbers at aphid colonies that contain the most individuals. (Modified from Turchin and Kareiva 1989.)

of 1000 individuals. Thus, the group size of aphids appears to affect the per capita growth rate of the aphid colony, the number of predators attracted to the colony, and the rate at which predators feed. Given all these factors, does grouping reduce predation risk for aphids? Apparently so. When the researchers calculated the instantaneous risk of predation to an individual aphid in a ten-minute period, they obtained values of 0.05 for colonies of 10 and 0.008 for colonies of 1000 or more. Thus, in the aphid-ladybird system, the dilution effect still occurs despite the strong tendency of predators to aggregate at large colonies of prey. Turchin and Kareiva are quick to point out, however, that predators are not the only enemies of aphids. Parasitoids and pathogens may increase rapidly in large groups of aphids and may profoundly affect mortality, perhaps even eliminating the antipredator advantages of the dilution effect.

SELFISH HERD

In most groups, as we saw with the monarch butterfly, centrally located animals appear to be safer than those at the edges. By obtaining a central position, animals can decrease their chances of being attacked and increase the probability that one of their more peripheral colleagues will be eaten instead. This antipredator mechanism, often referred to as the "selfish herd" (Hamilton 1971), emphasizes that although a given group appears to consist of members that coordinate their escape efforts, it is actually composed of selfish individuals, each trying to position as many others as possible between itself and the predator.

Consistent with the selfish herd hypothesis, individuals of some species do, indeed, aggregate when they

are alarmed. Tadpoles of the common toad, *Bufo bufo*, provide an interesting example. The tadpoles tended to form more cohesive aggregations in the presence of chemical cues of fish predators than in their absence. Thus, aggregating is an antipredator response. Furthermore, group formation is beneficial to the individual. Tadpoles were placed in a floating arena where fish predators could see them and strike at them but could not capture them. The strike rate per individual decreased. But, the strike rate at the group as a whole increased with increasing group size (Watt, Nottingham, and Young 1997).

Another study examined the way in which alarmed or unalarmed individuals would place themselves in a group. As you may recall Schreckstoff is an alarm chemical produced by some species of fish. Jens Krause (1993) habituated 14 dace (*Leuciscus leuciscus*) to the odor of Schreckstoff, and so they no longer reacted to its presence. A lone minnow (*Phoxinus phoxinus*), which was still responsive to Schreckstoff, was added to the group. Before Schreckstoff was added to the water, the minnow randomly intermingled in the shoal of dace. In repeated tests, when Schreckstoff was added, however, the minnow moved closer to the other fish and positioned itself so that it was surrounded. Only the alarmed fish chose a central location in the group.

One might ask, then, are central locations within the group *always* the best? The answer is no. In fact, a study on the antipredator advantages of schooling in fish suggests that the center is sometimes the most dangerous place to be. When in the company of a predatory seabass (*Centropristis striata*), silversides (*Menidia menidia*) at the center of a school suffered the most attacks (Parrish 1989). Rather than assaulting the margins, seabass swim toward the center of the school, split the school into two groups, and then strike at the tail end of one of the groups, where individuals that were in the center now find themselves. The relative safety of a location within a group thus depends on the predator's method of attack. Because schools of fish undoubtedly cope with a number of predators, each possibly using a different attack strategy, the relative advantage of central versus peripheral locations may change. In addition, factors such as foraging efficiency (those in the front see the food first) and the energetics of locomotion (fish in the front of a school may experience more "drag" than those at the back) probably also influence optimal positions within the school.

CONFUSION EFFECT

Predators that direct their attacks at a single animal in a group may hesitate or become confused when confronted with several potential meals at once. No matter how brief, any delay in the attack will operate in favor of the prey. The so-called confusion effect was first described by Robert Miller (1922) for flocks of small birds in the presence of a hawk. He noted that upon detecting an approaching hawk, individual birds within the flock sat motionless in the foliage and all produced a shrill, quavering note, the rendition of which was known as the "confusion chorus." This particular call was hard to locate, and Miller thought that it might function to distract attention from any particular individual in the group. Hawks apparently experienced difficulty in selecting a victim and were less successful in attacks on such groups than on a solitary bird. Miller described this dilemma of the hawk as follows: ". . . the more attention is divided, the greater is the possibility of failure."

The confusion effect is thought to be one of the primary antipredator advantages of schooling in fish. When the fish in the school scatter, it makes it difficult for the predator to focus on a single one. Neill and Cullen (1974) examined the effects of the size of the school on the hunting success of two cephalopod predators (squid, *Loligo vulgaris*, and cuttlefish, *S. officinalis*) and two fish predators (pike, *Esox lucius*, and perch, *Perca fluviatilis*). Whereas squid, cuttlefish, and pike are ambush predators, perch typically chase their intended victims. In most cases, predators were tested with fish of their natural prey species in schools of 1, 6, and 20 individuals. For all four predators, attack success per encounter decreased as the size of the school of prey increased (Figure 13.29). In the three ambush predators, the increased size of the prey's school appeared to produce hesitation and behavior characteristic of conflict (such as alternating between approach and avoidance). Perch, on the other hand, switched targets more frequently as the size of the school increased and, with each switch, reverted to an earlier stage of the hunting sequence. Under natural conditions, predators of fish may achieve hunting success by restricting their attacks to individuals that have either strayed from the school or have a conspicuous appearance; in both cases, the predator can concentrate on the odd target. In some prey species, individuals within schools appear to segregate by size to reduce their conspicuousness to predators (Theodorakis 1989).

MOBBING

Sometimes prey attack predators. Approaching and harassing one's enemies is called mobbing, and this antipredator strategy typically involves visual and vocal displays, as well as frequent changes in position that culminate in swoops, runs, and direct hits on the predator. Mobbing is usually initiated by a single individual, and then conspecifics, or members of another species, join in the fracas. When in hot pursuit of a predator, mobbers appear to have a greater chance of being preyed on than nonmobbers, although there is some disagreement over whether mobbing actually entails a deadly risk to those that participate (Curio and Regelmann 1986; Hennessy 1986). The possible functions of mobbing include, but are not limited to, (1) confusing the predator; (2) discouraging the predator either through harassment or by the announcement that it has been spotted early in its hunting sequence; (3) alerting others, particularly relatives, of the danger; and (4) providing an opportunity for others, again particularly relatives, to learn to recognize and fear the object that is being mobbed (Curio 1978). Most evidence suggests that mobbing is not an act performed by a cooperative group of individuals that is attempting to protect the group as a whole, but rather it is the selfish

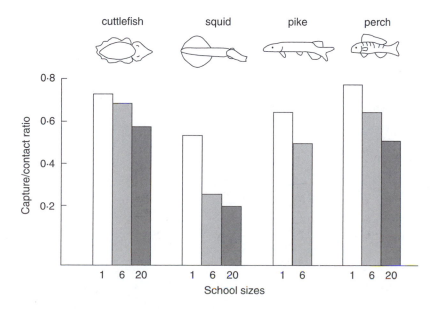

FIGURE 13.29 The confusion effect. As the size of the school of prey increases, four predators have reduced hunting success because of either hesitation and conflict behavior (in the case of the cuttlefish, squid, and pike) or frequent switching of targets (perch). (Modified from Neill and Cullen 1974.)

act of individuals that are attempting to protect only those that will benefit them directly, that is, themselves and their mates, offspring, and relatives (Shields 1984; Tamura 1989).

SUMMARY

Antipredator mechanisms may be classified as primary or secondary. Primary defenses, which operate whether a predator is present or not, decrease the probability of an encounter with a potential predator. The prey may go undetected if it blends with its background (crypsis) or occurs in a wide variety of colors (diverse coloration). Alternatively, the prey may be detected by a predator and either be recognized as inedible (warning coloration) or go unrecognized as a potentially tasty meal (Batesian mimicry). Although many primary defenses involve colors and patterns, the behavior of a prey animal is critical to the success of these mechanisms.

Secondary defenses operate during an encounter with a predator and increase an animal's chances of surviving the encounter. Amid the many options available, an individual may divert the predator's attention, inform the predator that it has been spotted early in the hunt, make the predator hesitate, or turn the tables and fight back.

Membership in a group makes available a number of protective devices that often combine primary and secondary defense mechanisms. Generally, predators have less success when hunting grouped rather than solitary prey because of the superior ability of groups to detect, confuse, and discourage predators. In addition, during any given attack, an individual in a large group has a lower probability of being the one selected by the predator (dilution effect) and may use other group members as a shield between itself and the enemy (selfish herd).

Although for convenience we have discussed antipredator behavior as several distinct defensive mechanisms, our intent is not to imply that a given individual or species is characterized by a single protective strategy. Indeed, most animals face a large number of potential predators with a diverse array of methods for detecting and capturing prey, and thus a variety of deceptive and defensive tactics is crucial. The use of any particular device probably reflects the relative risks and energetic demands of the predator-prey encounter, and almost no defense works all the time. Finally, the behavior and color patterns of animals must be interpreted in the context of several selective forces; after all, animals must not only avoid being eaten but also must feed and reproduce.

14

Sexual Selection

Sex is a powerful force in nature. Males of some species compete with one another, sometimes in titanic battles, just for the opportunity to mate. Others have exaggerated traits, or ornaments, to attract females. It is easy to see that producing offspring is adaptive, but it isn't always easy to understand male competition or female choice.

Why did the elaborate plumage of peacocks evolve? Surely the long, shimmering train is energetically costly to produce. Furthermore, the iridescent splendor is heavy, as well as conspicuous to predators. Equally striking are the enormous horns on some male

scarab beetles (Figure 14.1). At what costs did they evolve? How can we reconcile the production of such bizarre structures with natural selection?

Darwin (1871) was the first to suggest that spectacular structures such as the plumage of peacocks and the horns of male beetles could arise and be maintained through the process of sexual selection. According to Darwin, sexual selection can occur through two mechanisms: male competition for access to mates and female choice of mates. In the first, called intrasexual selection, individuals of one sex, usually males, gain a competitive edge by fighting with each other, and winners could claim the spoils of victory—females. Intense fighting and competition for mates could lead to selection for increased size and elaborate weapons such as the horns of male beetles. Sexual selection can also involve selective mate choice, in which individuals of the sex in demand, usually females, choose mates with certain preferred characteristics. Thus, males not only fight with each other for access to females but also compete to attract females through the elaboration of structures or behavior patterns.

Throughout the animal kingdom, males typically compete for females and females actively choose their mates. Why? Many people think that these differences in mating strategies between the sexes are related to differences in investment in gamete production or parental care.

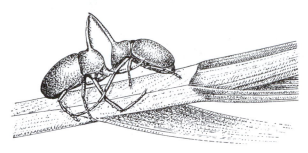

FIGURE 14.1   Males of the scarab beetle *Golofa porteri* have a head horn and enormously elongated front legs. Weapons such as these may have evolved in the context of male-male competition. (Redrawn from Eberhard 1980).

Whereas females produce a small number of large, energetically expensive eggs, males produce millions of small, relatively inexpensive sperm. As a result of differences in the number of gametes produced by the sexes, females (or more correctly, their eggs) become a limited resource for which males compete. A female is likely to maximize her reproductive success by finding the best-quality male to fertilize the limited number of eggs she produces. In contrast, a male's reproductive success will probably increase with the number of mates he has. This difference in gametes may help to explain the "undiscriminating eagerness" of males and the "discriminating passivity" of females (Bateman 1948).

Robert Trivers (1972) suggested that differential investment by the sexes in offspring, rather than in gametes, was responsible for competition and mate choice. In Trivers's view, the sex that provides more parental investment for offspring (usually the female) becomes a limiting resource for which the sex that invests less (usually the male) competes. As a result, males compete for access to females and females have the luxury of choosing among available suitors. Trivers's theory is perhaps best tested by examining the select group of nontraditional species—those in which males invest more than females in the care of offspring. If Trivers is correct, we would predict that females of these species would compete for males, who in turn would be quite discriminating in their choice of mates. As we will see in the following chapter on parental care and mating systems, Trivers's hypothesis is often supported by these exceptions to the rule.

Within-species variation in the courtship roles of some insects also supports the idea that the degree of parental investment is important in determining which sex competes for access to individuals of the other sex. For example, courtship roles are far from fixed in certain species of katydids (relatives of the grasshoppers). Indeed, in one species,* whether males compete for

---

* The genus and species of this particular katydid are undescribed, and thus scientific names cannot be provided.

females or females compete for males varies with the relative importance of male parental investment. In katydids, male parental investment consists of a nuptial meal. That is, at the time of mating, a male katydid transfers his spermatophore (a packet of sperm and fluids) to the female. Following separation of the couple, the female bends and eats part of the spermatophore (Figure 14.2). This protein-rich meal is important in successful reproduction because both the number and fitness of her offspring are enhanced by the male's gift. The relative importance of the male's gift, however, varies with food availability. The gift is especially important to females when food is scarce, and it is during these times that females compete for males. In contrast, when food is plentiful, the relative value of the gift declines, and males compete for females (Gwynne and Simmons 1990). We see, then, that a change in courtship roles within a species of katydid results from variation in the relative importance of male parental investment.

Despite the continuing accumulation of evidence in favor of Trivers's (1972) ideas, scientists have begun to challenge the notion that patterns of parental investment are prime determinants of the nature and strength of sexual selection. Indeed, there may be several factors, independent of parental care, that affect mating effort and success. In other words, there are more conflicts between the sexes than simply who cares for the kids. Such conflicts might include which sex searches for mates and, once the sexes actually meet, which sex is more accepting of potential partners (Hammerstein and Parker 1987).

MALE MATING COSTS REVISITED

Some researchers have questioned the long-held notion that males incur trivial costs when producing gametes (Dewsbury 1982b; L. W. Simmons 1988). They are quick to point out that although, gamete for gamete, sperm are vastly smaller and cheaper to produce than eggs, sperm are probably never passed along one at a time. Instead, millions of sperm are transferred in groups along with accessory fluids in ejaculates (unpackaged fluids) or spermatophores (packaged fluids). Although the cost of producing a single gamete may be minuscule, the costs of producing sperm groups and accessory fluids may limit the reproductive potentials of males. In field crickets (*Gryllus bimaculatus*), for example, the costs of spermatophore production (calculated by determining the percentage of the donor male's body weight made up by his spermatophore) are relatively greater for small males than for larger ones. Small males appear to cope with these higher costs not by producing smaller spermatophores but by increasing the refractory period between matings (time from when

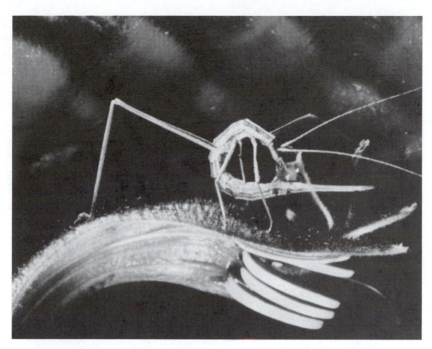

FIGURE 14.2　A female katydid is eating part of the spermatophore deposited by her mate. This protein-rich gift constitutes the male katydid's parental investment. When food is scarce, the gift has relatively greater value, and rather than females choosing mates, males choose females.

the male attaches a spermatophore to a female until the onset of the next attempt at courtship, when another spermatophore is ready). Thus, at least for small male field crickets, the costs of spermatophore production appear to limit their number of matings (L. W. Simmons 1988).

Understanding male mating costs is essential to our understanding of the evolution of reproductive patterns, and males may, in fact, have higher costs than previously believed. In most cases, however, we can still expect females to be more selective than males as a result of their greater gametic and parental investment. With this introduction to some of the historical and theoretical background for sexual selection, we will now consider intrasexual selection, and then intersexual selection, in more detail.

INTRASEXUAL SELECTION— COMPETITION FOR MATES

Intrasexual selection has led males to evolve a broad spectrum of attributes related to intense competition for mates. Some mechanisms operate prior to copulation, whereas others have their effect once mating has occurred. We will now consider a few of the many tactics males use to gain a competitive edge in the mating game.

ADAPTATIONS THAT HELP A MALE SECURE COPULATIONS

Dominance Behavior

Males in some species may secure copulations by dominating other males and thereby excluding competitors from females. This behavior, of course, places a certain premium on greater male strength and more effective weaponry. Since the female makes no such investments, males tend to diverge from females in both appearance and aggressiveness.

In many birds and mammals, males are larger than females, presumably because large body size improves the fighting ability and hence the reproductive success of males. It can be expected that the greatest levels of sexual dimorphism (difference in the appearance of the sexes) will be found in species in which the competition between males is most intense.

This is, in fact, often the case, but sexual dimorphism can result from other factors as well (Arak 1988b; Hedrick and Temeles 1989). Apparently, in many organisms the effect of female body size on female reproductive success is equally important in determining which sex is more "built to last" than the other. Also, although environmental factors usually affect males and females in a similar manner, sexual dimorphism in body size sometimes results from sex differences in food preferences (Selander 1966) or predation pressures (Bergmann 1965).

TABLE 14.1   Patterns of Copulation in a Captive Colony of Stumptail Macaques

Male Rank	Type of Copulation	Total Copulations[a]
Alpha male	Visible	424
	Surreptitious	—
14 other adult	Visible	69
males[b]	Surreptitious	393

[a] Per 1050 hours of observation.
[b] Of 18 males in the colony, 3 failed to secure a copulation.
Source: Estep et al. (1988).

In species in which males form dominance relationships, high-ranking males typically engage in more sexual activity than low-ranking males (Dewsbury, 1982a). In their attempts to monopolize sexual access to females, dominant males sometimes interfere directly with the copulations of subordinate males. Daniel Estep and his colleagues (1988) examined interference by dominant males and the consequent inhibition of sexual behavior among subordinate males of the stumptail macaque (*Macaca arctoides*). Long-term observations of a captive colony of this Old World primate revealed the existence of a stable, linear dominance hierarchy among adult males. Dominant males monopolize access to females by inhibiting subordinate males through aggressive threats and disruption of mating attempts. Within the dominance hierarchy, copulation frequencies were rank-related; the alpha (highest-ranking) male had almost as many copulations as the other 17 adult males combined (Table 14.1). Within this group, the three lowest-ranking males failed to mate even once during the 34-week study. If low-ranking males mated at all, they usually did so surreptitiously, out of view of dominant males. When tested in the absence of dominant males, low-ranking males copulated far more frequently than they did in their presence. (Dominance relationships are discussed further in Chapter 16.)

Sexual Enthusiasm

Despite costs associated with mating, males of most species appear to follow a strategy of copulating with as many females as possible. To this end, males employ a variety of precopulatory mechanisms to ensure that few, if any, mating opportunities slip by. If we assume that males have a lower investment in gamete production and parental care, it is apparent that they would usually pay a small price for mating mistakes. Thus it behooves a male to take advantage of each mating opportunity that he encounters. The result is that males can be expected to have a low threshold for sexual arousal. Males may be so easily triggered, in fact, that they end up directing their sexual attentions to biologically inappropriate stimuli such as the wrong species, inanimate objects, and individuals of the same sex (Figure 14.3). For example, on a warm, rainy night in March, upward of 4000 wood frogs (*Rana sylvatica*) may make their way toward a woodland pond. Once a female has been located, a male will try to clasp her in amplexus, the position that will allow him to fertilize her eggs. The movement of one male wood frog toward a female is sufficient to set off a chain reaction among neighboring males. The end result may be a pileup in which males, more often than not, end up clasping each other (Noble and Farris 1929).

The Coolidge effect is a good example of the seemingly boundless sexual enthusiasm of the males of many species. As the story goes (Bermant 1976), President and Mrs. Coolidge were visiting a farm and Mrs. Coolidge was the first to tour the premises. She was

FIGURE 14.3   Most males have a low threshold for sexual arousal and may direct their ardor to biologically inappropriate objects. This male Rio Grande leopard frog (on top) has mistaken both the species and sex of his partner, a male bull frog.

TABLE 14.2   Sexual Performance of 16 Male Rats

Test Condition	Mean Number of Intromissions	Mean Number of Ejaculations	Mean Time to Satiation (Min)
Same female	43.0	6.9	93.8
Different female	85.5	12.4	257.7

Source: A. E. Fisher (1962).

shown a yard with a number of hens and one rooster. When she was told that the one rooster was sufficient because he could copulate many times each day, she said, "Please tell that to the President." When the President came along, he was told the story, to which he asked, "Same hen every time?" No, he was told. After nodding slowly, the President said, "Tell *that* to Mrs. Coolidge."

In many species of rodents, sexually satiated males immediately resume copulation when they encounter a new female. Alan Fisher (1962) found that a male Norway rat (*Rattus norvegicus*) will copulate more or less continuously for many hours as long as novel females are presented (Table 14.2). Whereas male rats left with the same female generally reached sexual satiation in approximately 1.5 hours, some males could be maintained in a state of cyclic sexual arousal for up to 8 hours simply by introducing new females at appropriate intervals. Furthermore, it is not simply the case that "absence makes the heart grow fonder," for removal and replacement of the same female did not reactivate sexual behavior.

Female Mimicry and Satellite Behavior

In keeping with the old adage "out of sight, out of mind," subordinate stumptail macaques cope with dominant males by copulating on the sly. This question arises, then: How do males of other species deal with dominant males? Two common tactics employed by subordinate males are female mimicry and satellite behavior. In the former, subordinate males mimic females and slip inside the territory of a dominant male. Once there, these pseudofemales are able to fertilize eggs on the sly. We see this, for example, in bluegill sunfish (*Lepomis macrochirus*). Small males mimic the coloration and behavior of females and thereby succeed in entering the nests of larger males. Once inside the nest, the small male positions himself between a female and the larger male, all the time engaging in behavior typical of an egg-bearing female (e.g., slow dipping movements and exaggerated rubbing of the side of the male). At the time of spawning, both the larger male and the female mimic release sperm (Gross 1982).

The second strategy, satellite behavior, is characteristic of many amphibians in which males employ loud (and energetically expensive) calls to attract females. Although all males may be sexually mature and capable of calling, some remain silent and associate closely with a calling male, ready to intercept females attracted to the calls of the other male.

An example of satellite behavior is found in natterjack toads, *Bufo calamita* (Arak 1988a). On spring evenings, males of this species gather at the edges of temporary ponds and call loudly to attract females. Callers typically adopt a head-up posture and energetically belt out their call (Figure 14.4). Satellites, on the other hand, keep a low profile and remain stationary and silent in a crouched position next to a calling male. If a female is attracted to the caller, a satellite male will make every effort to intercept and clasp her. In natterjack toads, body size and call intensity are highly correlated with mating success; in general, small males with weak calls are at a reproductive disadvantage. Not surprisingly, the consistency with which males in a population adopt the calling or satellite tactic is correlated with their body length: Small males tend to be satellites; large males tend to be callers; and intermediate males are switch hitters, alternating opportunistically between the two strategies.

FIGURE 14.4   A male natterjack toad calls for females at the edge of a pond. Calling males compete with other males, called satellite males, that crouch silently nearby and intercept females on their way to the pond.

ADAPTATIONS THAT FAVOR THE USE OF A MALE'S SPERM

Mate Guarding

In addition to the option of adopting alternative reproductive strategies, males display a host of adaptations that increase the probability that their sperm, and not the sperm of a competitor, will fertilize the eggs of a particular female. An obvious strategy is to be present when eggs can be fertilized and prevent other males from gaining access to the female during this time. Depending on whether the female stores sperm between copulation and fertilization, mate guarding can occur before or after copulation or both.

In some species, including most crustaceans, mating is often restricted to a short period of time after the female molts, and females don't store sperm. Consequently, the first male present fertilizes all the eggs. In such species, we generally find precopulatory mate guarding (Birkhead and Parker 1997). The tiny amphipod *Gammarus lawrencianus* (Figure 14.5), for example, lives in estuaries along the coast of North America. Typically, a free-swimming male grabs a passing female before she molts and draws her to his ventral surface. Once the male has a firm hold on the female, he moves her in such a way that the long axis of her body is at right angles to his. In this posture the male palpates the female with various appendages, and after a minute or so, rotates her into the precopula position in which her body is now underneath and parallel to his. It is during the period of palpation and inspection that mate-guarding decisions are made. One factor in the decision is whether the female has already mated. Males are more likely to guard females with an empty brood pouch than those with a brood pouch full of juveniles or recently fertilized eggs (Dunham 1986). A second

factor is the length of time the male will have to guard the female before mating. Although females early in their reproductive cycle are also typically rejected (presumably because of the greater time investment in guarding), males that have gone a long time without female social contact will accept them (Dunham and Hurshman 1990). Once in the precopula position, the two swim as a single unit until the female molts and sperm transfer is complete. Thus, the male has ensured that his sperm, rather than the sperm of a competitor, are used in fertilization.

In some amphibians, males maintain the vigil for long periods. *Atelopus oxyrhyncus*, a bright yellow frog of the Venezuelan cloud forest, exhibits a particularly prolonged amplexus (Dole and Durant 1974). During the nonbreeding season, males and females of this species maintain individual home ranges on the leaf-covered forest floor. As the reproductive season approaches, males clasp neighboring females and the two move as a unit to nearby streams to spawn. Although most males initiate amplexus approximately one month before arrival at the spawning site, one notorious male clasped a female for a full 125 days prior to breeding. Prolonged mate guarding in *A. oxyrhyncus* is costly not only to females, which are faced with carrying males on their backs, but also to males, which are reduced to feeding only when insects chance to land on the head of their mate.

In some cases, a male may protect his reproductive interests by guarding his mate after copulation as well. If a female were to mate with more than one male, the first male to copulate with her would presumably have a reduced chance of fertilizing her eggs as a result of sperm competition with the second male (see later discussion). A male usually guards his mate during the postcopulatory period either by staying close to her and chasing away other males or by maintaining direct physical contact with her, thereby impeding access to her.

Male blue crabs (*Callinectes sapidus*) guard females before and after mating. A male guards a female before she molts so that he will be present when her eggs can be fertilized. But, about 12% of the females mate again within a few days of the final molt. A female that mates more than once stores all the sperm from both males. Thus, sperm from both males have equal access to her eggs. So, to protect his interests, a male blue crab also guards his mate after copulation. If many other males are present, a male guards a female for a longer time after copulation than he does if few males are present (Jivoff 1997).

Rove beetles (*Leistotrophus versicolor*) employ a unique variation on the theme of postcopulatory mate guarding: Males chase *females* away after mating with them (Alcock and Forsyth 1988). Males and females of this species aggregate at dung and carrion because these substances attract flies that serve as their food. Male bee-

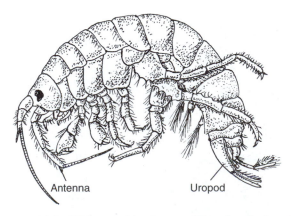

FIGURE 14.5   Male amphipods guard females during the precopulatory period, just before the female molts. By maintaining constant contact with a premolt female, the male ensures that his sperm rather than the sperm of a rival are used in fertilization.

tles compete for territories on the dung and thereby gain access to food and mates. Rather than guarding females after copulation, males attack and chase them. Such ungallant behavior probably serves to move still-receptive females away from dung and rival males. Males then quickly return to the tasks at hand, namely, catching flies and courting new females. Traditional postcopulatory mate guarding might be ineffective in rove beetles because several males are often present at a given patch of dung, and while one male was chasing off another male, a second rival could gain access to his mate.

Mechanisms to Displace or Inactivate Rival Sperm

Although physical struggles among males for females are often conspicuous, perhaps the fiercest clashes related to mating occur more quietly—in the female reproductive tract. If the female has mated with more than one male, their sperm must compete for the opportunity to fertilize her eggs. For example, the quality of a male's sperm—measured, perhaps, by their motility or the number of viable sperm per ejaculate—may determine the proportion of young that he sires. Another factor that may come into play is his position in the line of suitors to mate with a specific female. In some species, the advantage goes to the first male to mate with the female, in others the last male, and in still others mating order does not seem to be an important factor in determining patterns of paternity. G. A. Parker (1970, 1984) pioneered research in the area of sperm competition and described the phenomenon as a "push-pull" relationship between two evolutionary forces—one that acts on males to displace previous ejaculates left by rivals, and one that acts on early males to prevent such displacement.

In some cases, the interference is rather crude—the male simply removes rival sperm. For example, males of the damselfly *Calopteryx maculata* use their penis not only to transfer sperm but also to remove sperm previously deposited by competitors (Waage 1979). Backward-pointing hairs on the horns of the damselfly's penis appear to aid in scooping out clumps of entangled sperm left by earlier rivals (Figure 14.6). Some male crustaceans employ equally subtle tactics. Rather than scooping out their rival's sperm, male spider crabs (*Inachus phalangium*) push the ejaculates of earlier males to the top of the female's sperm storage receptacle; seal them off in this new location with a gel that hardens; and then place their own sperm near the female's oviduct, the prime location for fertilization to occur (Diesel 1990).

Males of other species stimulate the female into ejecting the sperm of another male. During copulation, a male damselfly (*C. haemorrhoidalis asturica*) stimulates the female sensory system that controls egg laying and fertilization, causing her to release sperm from her sperm storage structure. This allows him to gain access to rival sperm that would otherwise be unreachable (Cordoba-Aguilar 1999). A similar strategy is used by male dunnocks, which are small songbirds. If the male sees his mate near another male, he pecks at the female's cloaca, causing her to eject sperm-containing fluid (Davies, 1983).

Sperm competition in moths and butterflies (Lepidoptera) takes a somewhat different form. Male Lepidoptera produce two different types of sperm. One type (eupyrene sperm) contains genetic material and can fertilize eggs. The second type (apyrene sperm) might be considered to be "dud" sperm because it lacks genetic material and cannot fertilize eggs. This nonfertilizing sperm usually makes up at least half of the sperm complement in any given ejaculate.

The function of nonfertilizing sperm in moths and butterflies has been a mystery since its discovery in the early 1900s. One idea is that this kind of sperm embarks on a "seek and destroy" mission, the ultimate goal of which is to displace or inactivate sperm from previous matings by other males, sperm that would compete for access to eggs. A second idea is that nonfertilizing sperm may be the functional equivalent of cheap filler, preventing or delaying further matings by the female. Distention of a female's sperm storage organ by these energetically low-cost duds could signal a successful insemination and thereby reduce the female's receptivity to further matings (Silberglied, Shepherd, and Dickinson 1984).

Mechanisms to Avoid Sperm Displacement

Since males have evolved ways to give their own sperm an advantage over that of other males, obviously natural selection would have evolved ways for these other males to keep this from happening. So, we will now consider some tactics utilized by male mammals to avoid or reduce the effects of matings by other males.

Male mammals may increase the proportion of their own sperm in the female reproductive tract through their mating behavior. For example, despite the fact that female Norway rats, golden hamsters (*Mesocricetus auratus*), and deer mice (*Peromyscus maniculatus*) can become pregnant after only one complete ejaculation series, males of these species typically attain multiple sets of ejaculations (Dewsbury 1984). In this way, not only do they leave more ejaculate in the female tract, but they also tend to tie up receptive females longer, rendering them less available to competitors. Multiple and prolonged copulation may also increase fertility or accelerate pregnancy initiation. In some species, then, males that attain multiple ejaculations sire a greater proportion of pups in resulting litters than do males that have just one ejaculation.

FIGURE 14.6 (*a*)   A copulating pair of damselflies. (*b*) The penis of a male damselfly serves not only to transfer sperm to the female but also to remove sperm previously left by competitors; backward-pointing hairs on the horns of the penis remove clumps of rival sperm.

a

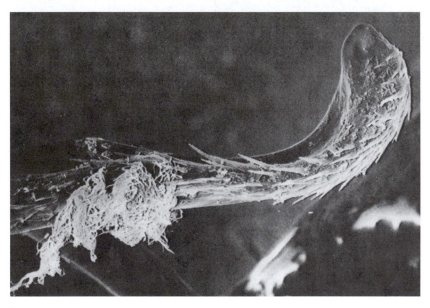

b

Like those of some promiscuous rodent species, females of the Rocky Mountain bighorn sheep (*Ovis canadensis canadensis*) usually mate with several males during a single period of estrus. Sperm competition in this species is intense because subordinate males (called coursing rams) are remarkably successful in forcing copulations on ewes guarded by defending rams (Hogg 1988). Defending and coursing rams copulate with estrous ewes at extremely high rates (0.83 and 0.90 copulations per hour of estrus, respectively; estrus typically lasts for two days). Similar to multiple ejaculations of rodents, frequent copulations by individual males of these bighorns presumably function to increase the proportion of their own sperm in the female reproduc-

tive tract. Defending rams copulate at especially high rates immediately after successful copulation by a coursing ram. Such "retaliatory" copulations by dominant males (Figure 14.7) may serve a function akin to that of the damselfly penis—mechanical displacement of rival sperm. Alternatively, retaliation may be advantageous to the dominant male because sperm from the last matings are more likely to fertilize eggs. A phenomenon similar to retaliatory copulation has been reported for roof or brown rats (*R. rattus*). In this species, males exhibit reduced postejaculatory intervals (time from ejaculation to next intromission) if a second male has copulated with the female in the interim (Estep 1988).

FIGURE 14.7 Competition for mates is intense among males of Rocky Mountain bighorn sheep. Although dominance and access to mates are established by fierce physical clashes, male competition continues in the reproductive tracts of females. Dominant males copulate at high rates immediately after successful copulation by a subordinate male. Such retaliatory copulations may displace the sperm of the subordinate male.

FIGURE 14.8 In deer mice, copulatory plugs deposited by males probably retain sperm in the female reproductive tract rather than prevent inseminations by rival males.

Repellents and Copulatory Plugs

In an effort to reduce the likelihood of future matings by competitors, males of many taxa apply a repellent odor to their mates. In the case of the neotropical butterfly, *Heliconius erato,* for example, a male transfers an "antiaphrodisiac" pheromone to his mate during copulation, which makes her repulsive to future suitors (Gilbert 1988).

In other species, males may deposit a copulatory plug, a thick, viscous material that tends to clog the female's reproductive tract. The males of many species, from garter snakes (Devine 1977) to small mammals, deposit copulatory plugs. It has been suggested that the copulatory plugs of rodents evolved in the context of "chastity enforcement," although with the exception of guinea pigs (Martan and Shepherd 1976) copulatory plugs do not appear to block subsequent inseminations.

Donald Dewsbury (1988) examined the function of copulatory plugs in deer mice, *P. maniculatus* (Figure 14.8), by allowing females to mate with two males in succession. In the control group of females, the copulatory plug produced by the first male was left in place. In the experimental group, the first male's copulatory plug was removed before introduction to the second male. It was found that the percentages of pups sired by the first male did not differ between groups of females with or without copulatory plugs. Dewsbury suggests that rather than protecting against subsequent inseminations, copulatory plugs in deer mice ensure that sperm are retained within the female reproductive tract.

Although for the vast majority of species the precise function of copulatory plugs remains a mystery, R.

Robin Baker and Mark A. Bellis (1988, 1989) suggest an interesting hypothesis. They note that in mammalian ejaculates, there is a large proportion of abnormal sperm. In healthy humans, for example, up to 40% of the sperm are irregular in size or shape, perhaps having two heads or two tails. Baker and Bellis suggest that such irregular sperm assume a "kamikaze" role, sometimes staying behind and forming a kind of plug or sperm aggregation (Figure 14.9) that acts as a roadblock or barrier in the lower region of the female reproductive tract (e.g., the vagina or cervix). In this manner, kamikaze sperm may thus protect "egg-getters" (i.e., normal sperm) from being ousted by a second ejaculate. Perhaps, as Baker and Bellis suggest, the proportions of normal and kamikaze sperm vary according to whether plugging or fertilization is of prime importance; when a male copulates with the same female several times during a single mating session, it would be advantageous to have a preponderance of normal sperm in early ejaculates and larger numbers of kamikaze sperm in the last ejaculate. The possibility that male mammals somehow manipulate ratios of kamikaze and egg-getter sperm in their ejaculates has exciting implications for future interpretation of the patterns of reproductive behavior.

So far in our discussion of sperm competition, we have focused on interactions between males. Do females benefit by mating with more than one male? Multiple matings may, in fact, be to a female's advantage. Mating with several males may (1) increase the probability of fertilization, (2) increase the genetic diversity of offspring, (3) result in the accumulation of material benefits if males provide nutritional gifts at copulation, or (4) ensure that a female's sons are good at the game of sperm competition, if the trait is heritable.

FIGURE 14.9   Intermeshed sperm, digested out of the copulatory plug of a rat, form the framework on which seminal fluids coagulate. Baker and Bellis (1988) have suggested that the large numbers of apparently deformed sperm in mammalian ejaculates assume a kamikaze role and stay behind in a plug or sperm aggregation. By forming the framework of barriers in the female reproductive tract, kamikaze sperm could protect egg-getter sperm (i.e., sperm that are physiologically capable of fertilizing eggs) from ousting by a second ejaculate.

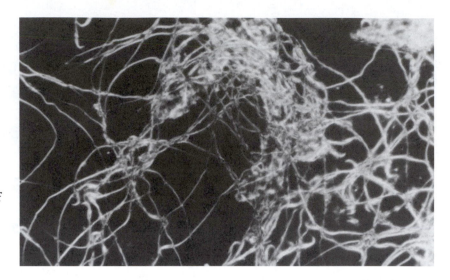

These and other suggestions are reviewed in more detail by Møller and Birkhead (1989).

SEXUAL INTERFERENCE: DECREASING THE REPRODUCTIVE SUCCESS OF RIVAL MALES

As we have seen, mating success is often measured by how well one advances one's own reproductive efforts and how effectively one interferes with a competitor's efforts. We will now concentrate on the latter. Any behavior that reduces a rival's fitness by decreasing his mating success is called sexual interference (S. J. Arnold 1976).

Perhaps the most effective animals in sexual interference are male salamanders. Adrianne Massey (1988) studied sexual interactions in field populations of red-spotted newts (*Notophthalmus viridescens*) and noted three tactics of sexual interference: (1) spermatophore transfer interference, (2) pseudofemale behavior, and (3) amplexus interference. These are all methods through which male newts decrease the reproductive success of competing males.

The first method, spermatophore transfer interference, occurs when a rival male inserts himself between a female and the courting male that has just dismounted. The rival male not only induces spermatophore deposition by the first male but also slips his own spermatophore into position so that the female picks that up rather than the spermatophore of the first male. This form of sexual interference depletes the first male's supply of spermatophores for future inseminations and prevents him from inseminating the female (as well as permitting the intruding male to inseminate her). The second method, pseudofemale behavior, also causes courting males to waste spermatophores, but this time in the context of male-male pairings (Figure 14.10). Male red-spotted newts often clasp other males, although such pairings are usually brief because clasped males give a head-down display that elicits release. In some cases, however, the clasped male does not signal his maleness to the clasping male. Furthermore, once the clasping male dismounts, the clasped male may nudge him, in the manner of females, and get him to uselessly deposit his spermatophore. Males of the red-spotted newt occasionally engage in a third method of interference, amplexus interference. Here, an intruder simply inspects a pair in amplexus at close range, usually just before the mating male dismounts to deposit a spermatophore. The presence of a voyeur leads the amplexing male to pause. When he resumes his mating behavior, he usually picks it up at an earlier stage. The interference, then, causes the mating male to waste time and energy without increasing his probability of fertilizing the eggs.

Other species exhibit more dramatic forms of sexual interference, such as infanticide, the killing of a competitor's offspring (Figure 14.11) (Hausfater and Hrdy 1984). In some mammals, sexually selected infanticide often occurs when one or more males from outside the social group usurp the resident male. In Hanuman langurs (*Presbytis entellus*), the primate species in which infanticidal male intruders were first reported, infanticide occurs when individuals from all-male bands invade harems (Sugiyama 1965). With her offspring gone, the female quickly returns to estrus; in Yukimaru Sugiyama's (1984) words, infants are attacked because they are "little more than an obstacle to activation of the mother's receptivity."

Although most primatologists accept the hypothesis that male infanticide in Hanuman langurs is sexually selected, acceptance is not universal (discussed in K. S. Brown 1996; Dixson 1998). Nonetheless, supporting evidence is slowly accumulating. One criticism was that the acts of killing were not actually observed and were only assumed to have been committed by the males that

a

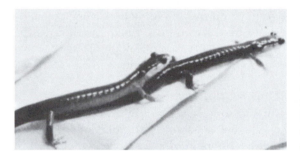

b

FIGURE 14.10 There are few differences in heterosexual and homosexual courtship in salamanders. (*a*) A receptive female has straddled the tail of a male to stimulate the male to deposit a spermatophore. (*b*) A male mimics female behavior to cause the courting male to deposit a spermatophore. Pseudofemale behavior is one form of sexual interference used by male salamanders to decrease the reproductive success of competitors.

took over the groups. However, many instances have now been observed (Sommer 1993). Other critics argue that infanticide is a result of crowded living conditions, but it turns out that male infanticide is equally common in low-density and high-density populations (Dixson 1998; Newton 1986). Furthermore, two predictions of the sexual selection hypothesis for male infanticide— that the infanticidal males will kill only infants that are unrelated to them and that infanticide will increase a male's chance of siring the next infants—have recently been supported by DNA analysis (Borries et al. 1999).

For male tree swallows (*Tachycineta bicolor*), nest sites are essential for acquiring a mate (Figure 14.12). Males of this species compete intensely among them-

FIGURE 14.11 Although infanticide may serve several functions, it is a dramatic example of sexual interference. Here, a lion has killed the cub of a rival male.

FIGURE 14.12 Nest sites are often in short supply for male tree swallows. In the absence of resident males, male "floaters" may kill unrelated young to gain access to nest sites and mating opportunities.

selves for access to limited numbers of nest sites. Many individuals do not acquire nests, and thus there is a large population of male and female "floaters." When resident male tree swallows were experimentally removed from their nests shortly after the eggs hatched, some males from the floating population killed the nestlings and thereby acquired nest sites (Roberston and Stutchbury 1988). In some cases, infanticidal males nested with the resident female, whereas in others they recruited females from the floating population. In tree swallows as in langurs, then, killing of unrelated young not only damages the male whose offspring are killed but also increases chances for mating with "widows" or other available females.

INTERSEXUAL SELECTION— MATE CHOICE

Intersexual selection occurs when one sex is in the position of choosing individuals of the other sex as mates. This puts members of the sex being chosen in the position of competing among themselves for that privilege. Usually, as we have said, females do the choosing and males compete among themselves to be chosen. Females generally have more to lose by choosing a poor-quality mate than do males. There are several reasons for this. First, eggs are usually larger and contain more energy than sperm. Second, females often must care for the developing embryos in some way, generally by incubating them or through some form of pregnancy. Third, females often must play the larger role in bringing up the offspring. So, it is in a female's best interest to mix her genes with the best male possible. Thus, females are choosy.

CRITERIA BY WHICH FEMALES CHOOSE MATES

Characteristics used by females to select a mate should affect female fitness, be assessable, and vary among males (Searcy 1979). Given these criteria, females are thought to choose mates on the basis of their ability to provide sufficient sperm, useful resources, parental care, or good genes. Although some females may evaluate potential mates on only one characteristic, others may base their choice on multiple criteria. Some of these characteristics are material benefits and easy to assess; others are indirect cues of quality. Next we will consider the characteristics that females might use when searching for "Mr. Right."

Material Benefits

Females of some species may base their choice of a mate on the quality of resources provided by males. By so doing, they could receive either immediate gains from gifts presented during the courtship period or more long-term benefits from access to valuable resources, such as food or nest sites, that are controlled by males. Females that exchange mating for material goods could place themselves at an advantage. The increased commodities could obviously enhance reproductive output by enabling the female to live longer by being well fed or having access to a protected nest site. Not only would she have a competitive advantage over females without mates, but also high-quality resources would probably improve the survivorship and competitive ability of her offspring. A female, then, must be able to assess the quality of a male's provisions not just at the moment but also for the future.

Nuptial Gifts and Cannibalism   The males of many birds and some species of insects offer nutrition or other valuable substances to the female during courtship. These nuptial gifts may take various forms, including prey, seminal nutrients, glandular secretions, and the spermatophore. Indeed, a male may even offer parts or all of his body (Andersson 1994). The functions of the gifts generally fall into two categories, which are not mutually exclusive. First, they may increase the male's chances of mating by making him more attractive to the female, by making it easier to copulate, or by maximizing the amount of sperm transferred. Second, they may serve as paternal investments by increasing the number or fitness of his offspring (Vahed 1998).

In some species, for example the katydid mentioned previously, the spermatophore that the male presents as a nuptial gift provides nutrients for the female or the eggs. Depending on her own nutritional state, a female can either put a large portion of the male's contribution into the eggs or use the materials from the spermatophore for herself, increasing the chances that she will live long enough to breed again (Gwynne and Brown 1994).

The benefit obtained from the nuptial gift isn't always energy. There are many other valuable substances. Larvae of the arctiid moth *Utetheisa ornatrix* sequester alkaloids from the plants they feed on. These alkaloids make them quite distasteful to certain predatory spiders. Males transmit some of these alkaloids to the female with their spermatophores. The female bestows some of the alkaloids on her eggs, giving them protection from predators (Eisner and Meinwald 1995). The female is also protected by this gift. Almost immediately, she becomes unacceptable prey to these spiders (Gonzalez et al. 1999). In other species, limited micronutrients are passed to the female or her eggs in the nuptial gift. Certain male moths (*Gluphisia septentrionis*) endow the offspring with sodium, which is an ion important in many cellular processes. However, the plants that the larvae eat contain very little sodium. A male moth obtains sodium by drinking from mud puddles, extracting the sodium, and squirting the water out his anus. Eggs sired by "puddling" males have two to four times as much sodium as those of other males (Smedley and Eisner 1996).

In other species, the male presents a gift of prey that the female feeds on during copulation, buying him time to transfer his sperm. Male hangingflies (Bittacidae) usually offer a large prey item, one that will take at least 20 minutes to consume. It takes the male about 20 minutes to completely transfer enough sperm and fluids to make the female unreceptive to other males and begin laying her eggs (Thornhill 1976). Males of the hunting spider *Psaura mirabilis* gift wrap the prey in silk before presenting it to a female. The more silk used in the wrapping, the longer it takes for a female to digest it. Although a larger prey doesn't require more silk to wrap it, a male who uses more silk might prolong copulation during mating (Lang 1996).

In some insects, the male's nuptial gift is a part or all of his body. During copulation, a female sagebrush cricket (*Cyphoderris strepitans*) feeds on the male's fleshy hind wings and the haemolymph that oozes from the wounds. This courtship feeding keeps the female mounted while the male transfers his sperm. Males whose hind wings have been surgically removed transfer significantly less sperm than do intact males (Figure 14.13) (Eggert and Sakaluk 1994).

In some species of spiders, scorpions, mantids, and diptera, the male makes the ultimate sacrifice during copulation. He gives his body to the cannibalistic female. Oddly enough, there are some circumstances in which this ultimate sacrifice is adaptive. The male redback spider (*Latrodectus hasselti*) stores his sperm in a tightly coiled structure on his head, called a palp. To transfer sperm, he scrapes the palp across the female's genital opening, which causes the coil to unspring. The open coil is then inserted into the genital opening and sperm is transferred. A few seconds later the male does a somersault, and so his abdomen is placed directly above the female's jaws. About 65% of the time, the female eats him while he's in the somersault position. His chances of being consumed are much greater if the female is hungry. The males gain two paternity advantages from this suicidal behavior. One advantage is that a cannibalized male fertilizes about twice as many eggs as a male who survives copulation because he copulates about twice as long. Another advantage is that the female is about 17 times less likely to mate again after consuming the first mate (Andrade 1996).

Territory A male's territory may provide many valuable resources, including food and nest sites. In many species, a male's mating success depends on the quality of his territory. A territory with oviposition sites is attractive to female dragonflies (*Plathemis lydia*), as well as to certain frogs (*R. clamitans* and *R. catesbeiana*). Among fish, female wrasse prefer to mate with males whose territories are in deep water, where egg predation is lower. Territory size or quality is also important in many species of birds. Among mammals, female pronghorn antelopes (*Antilocapra americana*) choose males with the best forage on their territories (Andersson 1994).

Sufficient Sperm Given the vast number of sperm produced by most males, is it possible that females in the market for a mate would take into consideration a male's sperm-producing ability? Indeed, females of some species appear to choose "fresh" males over those that are "spent." Multiple copulations by male fruit flies (*Drosophila melanogaster*) cause depletion of accessory gland substances and a temporary reduction in fertility. Not surprisingly, when given a choice, female fruit flies choose virgin males over sexually exhausted ones (Markow, Quaid, and Kerr 1978). In species in which males can quickly regenerate sperm, such as some of the larger mammals, including humans, discrimination does not seem to be based on how many sperm a male is likely to be carrying at the time. (Thus we maintain the delicate integrity of our social fabric.)

Parental Ability Females may also assess a male's parental abilities. This seems even more challenging than estimating the value of future material resources, but it appears to occur, nonetheless. For example, females may use physical or behavioral features of

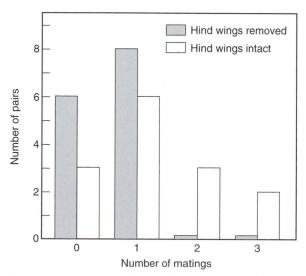

Figure 14.13 Male sagebrush crickets with intact hind wings transfer significantly more sperm packets to females than do males whose hind wings have been surgically removed. A female sagebrush cricket feeds on the male's fleshy hind wings during copulation. This helps to keep her mounted while the male transfers his sperm. (Data from Eggert and Sakaluk 1994.)

FIGURE 14.14   In red-winged blackbirds, females may use the size of the male's epaulette and his courtship intensity to estimate the quality of care that he is likely to provide to her offspring.

FIGURE 14.15   A male European crossbill passes regurgitated seeds to a female. Female birds may judge the parental ability of a male by the quality and quantity of gifts provided during courtship feeding.

males to predict parental quality. In fish such as the mottled sculpin (*Cottus bairdi*), large males are better able to guard eggs from predators, and thus females probably use size to assess parental ability (Downhower and Armitage 1971). Several reliable predictors of male parental ability have also been identified for red-winged blackbirds, *Agelaius phoeniceus* (Figure 14.14). In this species, the effort devoted to nest defense is correlated with the size of the epaulette, the patch of red feathers on the shoulder (Eckert and Weatherhead 1987), and feeding effort is correlated with male courtship intensity (Yasukawa 1981). Therefore, females could use epaulette size and intensity of courtship to estimate the quality of care that a male is likely to provide. In still other species of birds, females may judge male parental ability on the basis of the quality of nutritional gifts provided during the period of courtship (Figure 14.15). A large number of high-quality gifts may signal a male's superb foraging skill and willingness to feed his mate and offspring during incubation and post-hatching stages. Finally, some females may use the success of previous nesting attempts to judge the parental ability of males. Despite striking mate fidelity between the members of breeding pairs of some species of birds, bonds are often severed after unsuccessful breeding attempts (Coulson 1966).

Male Traits or Courtship Displays That Indicate Quality

Barring direct examination of a male's genotype, how could a female evaluate variation in genetic quality among suitors? Females might judge genetic quality by examining a male's (1) general physical well-being, (2) capacity to dominate rival males, or (3) capacity for prolonged survival. Often, certain male traits or aspects of their courtship displays are correlated with desirable qualities in a mate and father of one's offspring, and females prefer mates with these characteristics.

In many species, females chose larger males as mates. For example, the nocturnal desert beetle, *Parastizopus armaticeps*, can breed only after a heavy rainfall, which usually occurs twice a year. Because of the dry desert conditions and the scarcity of food, few offspring are produced in a brood, in spite of the efforts of both parents. The female forages on the surface at night for high-quality detritus. The male's parental responsibilities include digging a breeding burrow and maintaining the moisture level within it. The larvae require 100% relative humidity to survive, so if moisture levels in the burrow drop, the larvae die. The sand dries out rapidly after a rainfall. Thus, the deeper the burrow, the longer the time available for larval development and the greater the number of larvae that reach the pupal stage (Figure 14.16). Since larger males can dig deeper burrows than smaller ones, a female will increase her reproductive success by choosing a larger male. In choice tests, females preferred larger males, apparently estimated by body mass. When a weight was experimentally attached to the back of a smaller male, he became the chosen one. It is thought that the female estimates the body mass of a male during courtship, probably with the female behavior pattern called "push under" (Rasa, Bisch, and Teichner 1998).

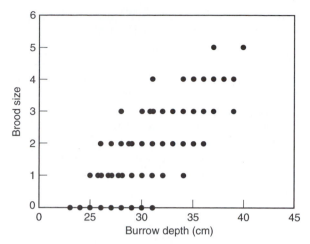

FIGURE 14.16   In the desert beetle, *Parastizopus armaticeps*, **increased burrow depth increases the number of larvae that survive to the pupal stage. Larger males dig deeper burrows. Thus a female enhances her reproductive success by choosing a larger male as a mate. The size of the male is assessed during courtship (Data from Rasa, Bisch, and Teichner 1998).**

Females also often prefer flashy, brightly colored males. In many species the preferred colors are red, orange, or yellow, which come from carotenoid pigments in food and cannot be synthesized. In regions where red food items are rare, the brightness of coloration may be a reliable indicator of the male's foraging ability or nutritional status. Among house finches (*Carpodacus mexicanus*) there is a great deal of variation in male plumage coloration, and females prefer males with red plumage as opposed to yellow or orange (Hill 1990). This choice may bestow direct resource benefits to the female because males with brightly colored plumage tend to be more attentive to the nests (Hill 1991). The female that chooses a bright male may also gain indirect benefits, good genes for her offspring. A male obtains these pigments from fruits and seeds consumed at the time of molt, and so the brightness of his plumage may reflect his nutritional condition (Hill 1995). The brightness of the male's coloration may also signal his overall health because parasites dull a male's plumage (Thompson et al. 1997).

This female preference for brightly colored males is well known in certain fish, including the guppy and the stickleback. Again we find correlations with male quality. Among guppies, redness signals male health. The offspring of redder males grow faster and have higher survival rates (Reynolds and Gross 1992). And in sticklebacks (*Gasterosteus aculeatus*), redder males are less likely to have parasites (Milinski and Bakker 1990a).

In other species, certain behaviors of the courtship display indicate a male's quality. For example, some courting male fiddler crabs build a small pillar next to their burrow, making them more attractive to females. Not all males build pillars, and those that do don't build them every day that they are active on the surface. Although the pillars don't require a lot of energy to erect, they do take time to build—time that could be spent foraging. The hypothesis that pillar building is an honest, condition-dependent signal of a male's quality was tested by providing supplemental food and observing the effect on pillar building. The additional food did not increase the number of males that built pillars. However, the males that did build pillars erected more than twice as many as the pillar-building males in the control group, which did not receive additional food (Backwell et al. 1995).

A display feature favored by certain female songbirds is a large song repertoire (Searcy and Yasukawa 1996). Male great reed warblers (*Acrocephalus arundinaceus*) defend territories on which several females can breed at the same time. They also help care for the young by feeding them and protecting them from predators. The number of offspring a male produces that survive to breeding age increases with the size of his repertoire. Females might be using this feature as an indication of his age or the quality of his territory. The size of a male's repertoire does, in fact, increase with age, and older males generally do have more attractive territories. But, that is not the whole story: DNA fingerprinting has shown that female reed warblers seek copulations with males that have larger song repertoires than their neighbors. The females did not gain material benefits, such as being able to raise young on a high-quality territory or having a mate that will care for the young, by mating with the other male. Instead, they may have been seeking high-quality genes for their offspring. This idea is supported by the observation that the relative post-fledgling survival of the offspring was related to the size of the song repertoire of the genetic father (Hasselquist 1998; Hasselquist, Bensch, and vonSchantz 1996).

Male satin bowerbirds (*Ptilonorhyncus violaceus*) build unique structures called bowers in which they display to females in attempts to secure copulations (Figure 14.17). Males decorate their bowers with flowers and feathers and display strong preferences for inflorescences of certain colors; blue and purple flowers are relatively rare in the environment of satin bowerbirds and are preferred over yellow and white flowers, whereas orange, red, and pink inflorescences are completely unacceptable (Borgia, Kaatz, and Condit 1987). The number of decorations on a bower is an important determinant of a male's mating success, and males go to great lengths to steal rare decorations from the bowers of competitors. Bowers and their decorations appear to have no intrinsic value to either sex outside the context of sexual display and thus probably serve primarily as indicators of male quality. Females may favor males who exhibit exotic decorations because the ability to

FIGURE 14.17   Males of the satin bowerbird decorate their bowers with flowers and feathers. The number of decorations on a bower is an important determinant of mating success, and males steal rare decorations from rivals with impunity. Female bowerbirds may judge the genetic quality of a male by the attractiveness of his bower.

accumulate and hold these decorations indicates that a male is in top physical condition. Because the number of inflorescences on bowers is correlated with age, the number of flowers, and specifically the number of rare decorations, they could be used by females to assess the experience of a male. Thus, although female bowerbirds cannot directly examine the genotype of potential mates, they may evaluate a male's genetic quality on the basis of the attractiveness of his bower.

ORIGIN AND MAINTENANCE OF MATE CHOICE PREFERENCES

We have seen that females of many species are attracted to the flashiest, most ostentatious males. It isn't too difficult to see why a female should be choosy or why she would choose the male that offers her the most in material benefits. But, how did female mate preferences originate when there are no material benefits? And, how are they maintained, especially if the traits increase male mortality? We'll consider several possible answers to these questions: runaway selection, choosing males with good genes, sensory bias, and mate-choice copying.

Runaway Selection

One hypothesis for the evolution of female preferences for elaborate male characteristics is called runaway selection. The basic idea is that female preference and

the male trait evolve together because the same genes control them. It begins when females evolve a preference for a particular male characteristic, perhaps because it is associated with a characteristic that increases fitness. For example, when females are present to see the heroics, male guppies (*Poecilia reticulata*) often approach and inspect a nearby predatory fish. It has been suggested that this boldness advertises to females a male's vigor. The boldest fish are also the most colorful. So, female preference for bright coloration may have begun because it indicated increased male fitness (Dugatkin and Godin 1998; Godin and Dugatkin 1996).

This preference can set the stage for runaway selection (R. A. Fisher 1930). A female that mates with a male that has an attractive characteristic will have sons with the trait, providing that the attractive character of the male is inherited, and she will have daughters who show a preference for that trait. The attractive sons will acquire more mates with a preference for that characteristic than other males and thus will leave more progeny. In this way, runaway selection can produce increasingly exaggerated male traits and a stronger female preference for them.

Lande (1981) and Kirkpatrick (1982) have developed mathematical models of Fisher's (1930) ideas for the evolution of female choice. These models demonstrate that runaway selection can indeed result in mate choice for characteristics that are arbitrary or even disadvantageous to the health and survival of individuals, providing that females prefer to mate with males that possess them.

Experimental support for the runaway selection hypothesis comes from studies on several species that show that female preference and exaggerated male traits evolve together. Studies on guppies (*P. reticulata*), for instance, show the same pattern of geographic variation in the brightness of male orange coloration and the strength of the female's preference for orange (Endler and Houde 1995). Selection experiments have also demonstrated that a male trait and female preference can coevolve. Studies on stalked-eyed flies (*Cyrtodiopsis dalmanni*) provide an example (Figure 14.18). As their name suggests, the eyes of these flies are located at the end of long stalks. The eyestalks of males are much longer than those of females, and females usually prefer to mate with the male with the longest eyestalks. After 13 generations of selective breeding, Gerald Wilkinson and Paul Reillo (1994) created two lines of stalked-eyed flies—one with long stalks and the other with short stalks. Female preference for eyestalk length changed accordingly. In the line of flies in which the males sported long eyestalks, females chose to mate with the longest stalked male. However, in the line of short-stalked males, females preferred shorter-stalked males.

FIGURE 14.18   The eyes of stalk-eyed flies are located at the end of long eyestalks. In males, the eyespan can be greater than body length. Females usually prefer to mate with the longest-stalked male. By selectively breeding either the longest-stalked males or the shortest-stalked males, two lines of flies were created. Female preference for eyestalk length coevolved with male's eyestalk length. Females from the long-stalked line chose the longest-stalked males as mates. However, females in the short-stalked line chose shorter-stalked males as mates. This genetic correlation between a male trait and a female's preference is consistent with the runaway selection hypothesis.

Runaway selection would favor ever more exaggerated male characteristics and females that find them attractive. So when does it all stop? When does a peacock tail become too long? The process will be stabilized only when natural selection balances sexual selection (R. A. Fisher 1930). In other words, when the peacock's tail becomes too energetically costly to produce or when it is so long and heavy that he cannot escape from predators, selection will no longer favor its increased length.

Choosing Males with Good Genes

Amotz Zahavi (1975) suggested an alternative to the runaway selection model for the evolution of female choice. Zahavi's alternative, called the handicap principle, states that females prefer a male with a trait that reduces his chances of survival but announces his superior genetic quality precisely because he has managed to survive despite his "handicap." In short, male secondary sexual characteristics act as honest signals, indicating high fitness, and females choose males with the greatest handicaps because their superior genes may help produce viable offspring.

The hypothesis that females choose males with good genes predicts that the mating preferences of females should increase the viability of their offspring. Supporting evidence can be classified into four categories (Ryan 1997): (1) studies that show a relationship between exaggerated male traits and offspring viability, (2) studies that show female choice for parasite-resistance genes, (3) female choice for trait symmetry as an indicator of overall genetic quality, and (4) female choice based on the degree of genetic similarity.

Exaggerated Male Traits and Offspring Viability

One of the best known examples of an exaggerated male trait is the ostentatious tail of the male peacock, and this is also a species in which there is good evidence that the females get genetic benefits that increase sur-

vival of their offspring through mate-choice preferences. Marion Petrie has studied a free-ranging population of peacocks *(Pavo cristatus)* in Whipsnade Park in England. During the early spring, adult male peacocks gather to court females on a communal breeding ground called a lek. An adult male peacock is a thing of beauty (Figure 14.19). His chief glory is a train of long, beautifully marked feathers, each tipped with an iridescent "eye." A courting male lifts his tail, which lies under the train. This elevates the train and spreads it out like a fan. He then struts around on his small territory on the lek, vibrating his tail rapidly. In turn, the tail

FIGURE 14.19   A peacock displays his train. Perhaps the elaborate plumage evolved to attract mates.

vibrations cause the plumes in his train to rattle audibly. He attempts to copulate with a "hoot-dash," in which he begins to lower his train and rushes toward the female while giving the "hoot" call. If his copulation attempt is successful, the female will squat in front of him, allowing him to mount her.

It turns out that peahens are choosy, and they prefer males with elaborate trains. When it comes to peacock mating success, the "eyes" have it; that is, mating success is significantly correlated with the number of eye-spots in the male's train. On one lek, which consisted of ten courting males, the most successful male copulated 12 times, but the least successful males never did. A female never accepted the first suitor she saw. On average, she visited about three males before copulating. In 10 out of 11 observed courtship displays that ended with successful copulation, the female chose the male with the highest number of eye-spots in his train of those she visited (Petrie, Halliday, and Sanders 1991). When 20 eye-spots were experimentally cut out of the trains of some peacocks, the male's mating success was significantly less than his success during the previous year. The attractiveness of control males, whose tails were left intact after being captured and handled, remained the same (Petrie and Halliday 1994).

Besides boosting a male peacock's mating success, the extent of sexual ornamentation appears to be correlated with survivorship. In the spring of 1990, of 33 displaying males in the same population of peacocks, 22 copulated successfully at least once and 11 were never successful. During the following winter, two foxes managed to enter the park, and they killed five peacocks. Four of the birds that were killed were among those that had been unsuccessful in gaining any copulations in the previous season. The fifth bird had copulated only twice. The males that were killed also had shorter trains with fewer eye-spots than surviving males had. This observation is consistent with the hypothesis that only the most healthy males can develop long, elaborate trains. Thus females are choosing males with good genes (Petrie 1992).

It also seems that the offspring of females who choose mates with elaborate trains benefit from their father's good genes. Petrie (1994) demonstrated this by pairing males with females chosen at random. In each large cage, she placed one male and four females. The mated males varied in attractiveness to females, as measured by the mean area of eye-spots in the train. All the offspring were raised under common conditions. When they were 84 days old, the offspring were weighed. The offspring of males with more elaborate trains weighed more than those with less showy fathers. After two years, more of the offspring of the highly ornamented males were still alive than were those of less attractive males. Since the matings were arranged by Petrie, the differences in offspring viability cannot be due to differences in the quality of females. By ruling out maternal effects, Petrie has provided strong evidence that a female can enhance the survival chances of her offspring through mate-choice preferences.

Parasite Load   In some species, brighter, showier males have fewer parasites than their dull competitors. It makes intuitive sense that females should prefer to mate with a healthy male, but how did a relationship between female preference for a male secondary sex characteristic and parasite load come about? There are three common hypotheses for why females prefer showy males (Møller, Christie, and Lux 1999):

1. A showy secondary sex characteristic is a reliable, honest indicator of a male's health, so a female will obtain good genes (for parasite resistance) for her offspring by choosing showy males. (Honest signals are discussed further in Chapter 19.)

2. Showy males make better parents because healthy males will have more time to devote to paternal duties.

3. A female is less likely to acquire contagious parasites from a showy male than from a dull, more heavily parasitized male.

William D. Hamilton and Marlene Zuk (1982) were the first to propose that the elaborate ornaments of males represent reliable signals of health and nutritional status. They examined plumage coloration in North American birds and suggested that only males in top physical condition would be able to maintain bright, showy plumage. Because bird species vary in their susceptibility to parasitic infection, Hamilton and Zuk predicted that the degree of male brightness would be correlated with the risk of attack by parasites. Accordingly, if males of a species vary substantially in their parasite load, it would behoove a female to choose a male that honestly signals his good health, for the offspring of this male may experience increased viability if they inherit his resistance to parasites. Because the brightness of plumage is closely tied to a male's general health, it is a reliable signal that cannot be faked by a parasite-laden male.

Based on this hypothesis, Hamilton and Zuk reasoned that in species in which the risk of infection by parasites is minimal, information regarding parasite load has little value to females in the market for a mate, and thus females should not display a preference for males with showy features. Hamilton and Zuk therefore predicted that in species with low risk of parasite infection, males would not be brightly colored.

In testing their hypothesis, Hamilton and Zuk surveyed the literature on avian parasites and determined the risk of infection for each bird species. They then ranked each species from 1 (very dull) to 6 (very striking) on a plumage showiness scale. In support of their ideas, there was a significant association between showiness and the risk of parasitic infection: Those species with the highest risk of infection from blood parasites had the most showy males.

Although consistent with the good genes hypothesis, Hamilton and Zuk's (1982) results could also be interpreted as arising from the runaway selection process. A stronger test would entail demonstrating that females that choose males with super-bright plumage produce offspring of higher viability than do other females.

In the last decade, the influence of parasites on sexual selection has been examined in many species of birds (e.g., Borgia 1986; Møller 1991), as well as in a host of other animals including gray treefrogs, *Hyla versicolor* (Hausfater, Gerhardt, and Klump 1990); three-spined sticklebacks, *G. aculeatus*, (Milinski and Bakker 1990); and humans, *Homo sapiens* (Low 1990). Some studies support the idea that females choose mates by using cues that will lead to more viable offspring; others do not.

More recently, all the available evidence was reanalyzed. This analysis confirmed that showy males have fewer parasites. However, this was true in species in which males show parental care and in those that do not. Thus, this observation sheds doubt on the hypothesis that females choose showy males because they will provide more help in raising the offspring. Furthermore, the relationship between showiness and parasite load held true even for parasites that are not directly transferred from one individual to another. When parasites cannot be transferred directly, a female would not avoid being parasitized herself by choosing a showy male. However, the analysis did reveal that showy males are better able than dull males to mount a stronger immune response against a wide variety of parasites. This observation supports the first hypothesis—that females choose showy males to obtain good, parasite-resistance genes for their offspring (Møller, Christie, and Lux 1999).

Trait Symmetry and Genetic Quality

The extent of asymmetry, called fluctuating asymmetry, in otherwise bilaterally symmetrical traits might indicate overall genetic quality. Because the growth and development of both sides of the body are controlled by the same set of genes, we would expect the right and left sides to be identical. This is not always the case, however. Environmental insults or genetic defects can cause one side of the body to develop in a slightly different manner from the other, causing the traits to be asymmetrical. Females in many species prefer males with symmetrical traits, presumably because symmetry signals a healthy condition and good genes (Møller and Pomiankowski 1993; Møller and Swaddle 1997).

Consider the sailfin molly *Peocilia latipinna*, a small tropical fish commonly found in brackish coastal marshes. It is so named because the males have a large dorsal fin that resembles a sail. The fin is displayed to the female during courtship. Males are brightly colored and may have vertical bars on the sides of their bodies. The number of bars may differ between sides. In nature males differ in the presence of bars, the number of bars, and symmetry in the number of bars. In laboratory choice tests, female sailfin mollies prefer males with vertical bars to those without bars. Furthermore, when bars are present, females prefer symmetrical males—those with the same number of bars on each side of the body (Schluter, Parzefall, and Schlupp 1998).

Studies on barn swallows suggest a way in which trait symmetry might reflect good genes. Female barn swallows (*Hirundo rusticus*) prefer to mate with males with long tails (Møller 1988b). Mite infestations stunt the growth of tail feathers, and so tail length advertises whether a male has been previously infected with mites. Mite infestations also increase asymmetry in tail length. Besides preferring males with longer tails, female swallows prefer males with tails that are symmetrical in length (Møller 1990). So, by choosing males with long, symmetrical tails, females are choosing males with genes for parasite resistance (Møller 1992).

Degree of Genetic Similarity

In certain species, female mate choice for good genes involves genes of the major histocompatibility complex (MHC). This is a large chromosomal region that varies tremendously among individuals. It is important in the immune responses that protect us from disease-causing organisms. (The role of the MHC in kin discrimination is discussed in Chapter 17.)

Several studies have shown that house mice (*Mus musculus*) prefer mates who differ from themselves in the makeup of the MHC region (reviewed in Penn and Potts 1999). The mice use odor cues to determine the degree of similarity of the potential mate's MHC region to their own. They can even distinguish individuals that are nearly identical to themselves in every gene except those of the MHC region (reviewed in Penn and Potts 1998).

We still aren't sure *why* MHC-dependent mating preferences exist, but two hypotheses have been suggested. The first is that they evaluate the degree of genetic relatedness of potential mates to avoid inbreeding. Inbreeding caused by mating with close relatives increases homozygosity and hence the risk of producing offspring that are homozygous for deleterious or lethal recessive alleles. On the other hand, extreme outbreeding may cause the breakup of successful parental complexes of genes. Given the potential costs associated with mating with either very close relatives or complete strangers, females may strike a balance between extreme inbreeding and outbreeding when choosing a mate. The second hypothesis is that MHC-dependent mating preferences allow a female to increase her offspring's resistance to disease. Recall that the genes of the MHC region are important in protecting against disease-causing organisms. By choosing a male whose MHC alleles differ from hers, she increases the variability of the MHC region in her offspring. This may make them resistant to a wider variety of disease-causing organisms (Penn and Potts 1999).

Sensory Bias

According to the sensory exploitation hypothesis (discussed further in Chapter 19), mating preferences evolved before the appearance of the sexually selected trait because of a sensory bias that was selected because it is (or was) adaptive in another context. The male's signals evolved to take advantage of the female's sensory bias (Ryan 1997, 1998).

An example of a female preference for a male trait that is due to a sensory bias is the female swordtail's preference for a male with a sword. The sword is an elongated, colored structure that extends from the lower part of the caudal fin. Although the genus *Xiphophorus* consists of the platyfish and the swordtails, only swordtails have swords. Nonetheless, females of two species of platyfish prefer males to which swords have been surgically added over normal, swordless males of their own species. The same is true for females of a closely related swordless genus, *Priapella*. Figure 14.20 shows a graphic representation of the evolutionary relationships among the species of *Priapella* and *Xiphophorus*. If this diagram is correct, the ancestors of swordtails lacked swords, and yet the females of these swordless species prefer males with swords. Thus, the female preference evolved before the male trait. It seems, then, that the male sword evolved by sensory exploitation of this preexisting sensory bias of females (Basolo 1990, 1995a, b).

It has been suggested that this female preference for males who sport swords may be a result of a more general preference for large body size. Video sequences of a swordless courting male were computer-altered so that the total body length of the swordless male equaled that of a sworded male. This modification eliminated the female preference for swords (Rosenthal and Evans 1998).

However, female preferences for sword length and body length may actually reflect two separate biases. The strength of these preferences were examined in female swordtails (*X. helleri*) and female *P. olmecae*, closely related species. Females of both species preferred males with longer swords. But, the strength of the preference in *P. olmecae* females was much stronger than in swordtail females, suggesting that the strength of this preference changed in one or both lineages.

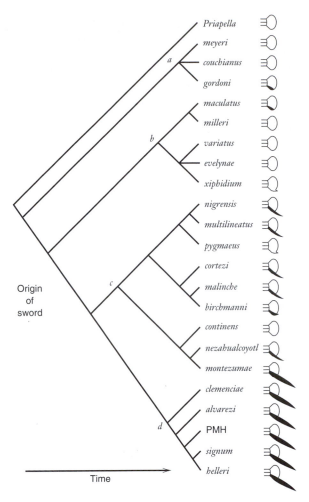

FIGURE 14.20 The evolutionary relationships among fish in the genus *Xiphophorus* (the platyfish and the swordtails) and the closely related genus *Priapella* suggest that the female preference for swords evolved before the male's sword. (From Basolo 1995a.)

Furthermore, although the preference for sword length and body size are equally strong in female swordtails, the preference for sword length is significantly stronger than for body size in *P. olmecae* females. So, whereas an ancestral bias for large body size might explain a female swordtail's mating preferences, the size bias alone cannot explain the preexisting bias for sword length in *P. olmecae* females (Basolo 1998).

Mate-Choice Copying

It's an old story—the male, perhaps even previously ignored by the female, becomes much more attractive to her because another female finds him attractive. Mate-choice copying occurs when one female's choice of a mate affects another female's choice (Pruett-Jones 1992).

The guppy, *P. reticulata*, provides an example. An aquarium system was arranged so that a female guppy (the focal female) was allowed to watch another female (the model) make a choice between two males. Then the focal female was given the opportunity to make her own choice between the same two males. On significantly more trials, she spent more time near the male that the model female had chosen (Dugatkin 1992). In fact, this tendency to copy can be strong enough to reverse the fickle focal female's previous mating choices. In another study, the focal female was first allowed to make her own choice between two males. Then she observed the previously unpreferred male consorting with a model female. When the focal female was offered the chance to choose between the males again, there was a significant change in her preference (Dugatkin and Godin 1992).

Why would a female be influenced by someone else's mate choice? There are several possible advantages. One advantage might be that mate-choice copying teaches a female what to look for in a mate. In this case, we would expect to find young females who are not yet skilled in discriminating differences among males to copy the choice of older, more experienced females (Nordell and Valone 1998; Stöhr 1998). This expectation is borne out among guppies. Although young females are likely to be influenced by the mate choice of older females, older individuals don't copy the choices of younger ones (Dugatkin and Godin 1993). Mate-choice copying may also be an adaptive strategy when direct assessment is costly. These costs might include reduced time for foraging, exposure to predators, and delayed breeding.

Both social factors (copying) and genetic factors (an inherited preference for orange body color) play a role in mate choice in guppies. The relative contributions of each of these factors was explored in a clever series of experiments. As in previous studies, a female guppy was offered a choice between two males. This time, however, the males differed in the amount of orange color on their bodies. The difference in coloration varied by 12%, 24%, or 40%. In nearly all trials, the focal female chose the more orange male. Then she observed a staged encounter between another female and the same two males in which the female apparently chose the duller of the two males. When the focal female was tested again, her choice of males depended on the amount of difference in their coloration. Social factors were more important when the differences in coloration were small or moderate. The female was swayed by the choice of the other female and chose the duller male when the paired males differed by 12% or 24%. However, when there was a great (40%) difference in coloration, the genetic preference for orange overrode the social cues. Thus, there is a color threshold, above which genetic factors will be most important and below which social factors will predominate (Dugakin 1996). That threshold can be shifted by the amount of apparent interest other females show toward the duller male of the pair. The test female is more likely to chose the duller of two males that differ dramatically in coloration if she sees *two* model females independently make that choice or if one model female spends a longer time (twice as long as in previous studies) near the dull male (Dugatkin 1998).

CRYPTIC FEMALE CHOICE

The observation that "It ain't over 'till it's over" may ring true for female mate choice. It is thought that females of certain species can choose the sperm that will fertilize her eggs after copulating with several males. This is called cryptic female choice because it is a hidden, internal decision made after copulation (Eberhard 1996). In the sperm wars, this choice is the female equivalent of male sperm competition.

So far, evidence for sperm choice is limited because it is difficult to distinguish from sperm competition. As a result, its importance as a force in sexual selection has been controversial. In those studies in which sperm choice has been demonstrated, the females are not choosing the sperm of the most attractive males. Instead, sperm is chosen to avoid incompatible genetic combinations (Birkhead 1998).

CONSEQUENCES OF MATE CHOICE FOR FEMALE FITNESS

Having explored those characteristics females might be assessing in potential mates, we come to an important question: Do choosy females produce fitter offspring? A laboratory study of female choice in fruit flies (*D. melanogaster*) provided evidence that mate choice by females can indeed result in the production of fitter offspring (Partridge 1980). Larvae produced by females that could choose their mates and females that could not were allowed to compete in laboratory vials with larvae from a standardized strain of flies. The proportion of offspring emerging as adults was slightly but significantly higher when females could choose their mates. These results suggest that in fruit flies, female choice improves the competitive ability of offspring. Similar results have been obtained for seaweed flies (*Coelopa frigida*). In these experiments, females that were allowed to exercise mate choice not only were more likely to mate but also produced offspring with higher survival rates than did females that were given a single, randomly chosen male (Crocker and Day 1987).

STRATEGIES OF FEMALE MATE CHOICE

Until now we have focused on criteria used by females to choose mates, and we have said nothing about how females actually go about the process of selecting the best possible male. How much time should they devote to searching, and once located should prospective mates be judged against a fixed standard or by some relative criterion?

Ideally, a female would have the opportunity to inspect every available suitor and choose the best from among them. In reality, however, females in search of mates probably operate under constraints of time, mobility, and memory (Janetos 1980). A female whose reproduction is restricted to a particular season does not have all the time in the world to locate a mate and initiate breeding. Ecological conditions may limit her mobility and thereby prevent complete inspection of all males in the surrounding area. Also, the number of males that a female can remember at one time will certainly shape strategies of mate choice. In a theoretical analysis of female choice, Anthony Janetos compared five different strategies of mate selection and concluded that a strategy in which females rank and choose among available males on a relative basis (comparison to an external standard—i.e., the choice is based on relative differences among males that are sampled) is better than random mating or alternative strategies based on a fixed threshold criterion (comparison to some internal standard). Although mate choice based on relative differences among available males may be the more common strategy, mate choice based on comparison to some internal standard has been reported for cockroaches, *Nauphoeta cinerea* (Moore and Moore 1988), and scorpionflies, *Hylobittacus apicalis* (Thornhill 1980). Different tactics of mate choice are likely to evolve under different circumstances, and knowledge of *how* females choose mates is just as essential to our understanding of intersexual selection as knowledge of *why* females choose mates.

MATE CHOICE BY MALES

In spite of our emphasis on female choosiness, some males may also be selective in picking of mating partners. The more a male has to offer in parental care or material resources, the more selective he is expected to be. Two other factors that are important in the evolution of male choice are a limited ability to fertilize all available females and variation in female quality. As an example of this latter factor, consider the case of thirteen-lined ground squirrels (*Spermophilus tridecemlineatus*). Despite the fact that male ground squirrels provide neither parental care nor material resources, they often reject females willing to mate with them. As it turns out, male choice is favored in this species because of predictable variation in the quality of females—variation that can be attributed to sperm competition (Schwagmeyer and Parker 1990). Each female, on average, mates with two males, and the first male sires about 75% of the litter. This percentage, however, is not fixed. Factors such as the time between matings of the first and second males (the second males wait until the first males depart to begin mating, and longer delays favor the first males) and the duration of the second male's longest copulation (increases favor second males), affect the fertilization rate. Using measurements such as these, the researchers developed a model to determine whether male choice in ground squirrels is consistent with what would be expected from the known effects of sperm competition on the fertilization rate. They then examined whether field data on the male response to previously mated females conformed to the model. Consistent with their predictions, male ground squirrels were more likely to reject previously mated females the longer the time since the female's copulation with another male.

METHODOLOGICAL PROBLEMS IN THE STUDY OF MATE CHOICE

Several methodological problems are commonly encountered when studying mate choice (Halliday 1983). Aside from the fact that mate choice may be very subtle, it is often masked or confounded by extraneous effects, such as motivation of the female at the time of assessment. If a female does not display a preference for a male during a choice test, is it because she has rejected him as a potential mate, or is it simply because she is not receptive? Studies of female choice face the difficult task of unraveling the ties between a motivational state and the choice of a mating partner. Also, female choice in the laboratory is often measured by orientation toward or time spent near a particular male; yet in some cases, these behavioral responses may not relate to actual mating inclination.

Perhaps the most difficult problem for students of mate choice is determining the precise roles of male competition and female choice within a single mating system. In the plethodontid salamander *Desmognathus ochrophaeus*, staged laboratory encounters between one female and two males (one large and one small) always resulted in courtship between the female and large male. Although one might be tempted to claim that females of this species display a clear preference for large males, Lynn Houck (1988) has shown that pairing between females and large males results mostly from the fact that large males attack and threaten small males and effectively exclude them from courtship activity. Further evidence against female choice for large males comes from observations that in the presence of only one potential mate, females mated with small males just as often as with large males. Houck is careful to point out, however, that female choice of mates cannot be ruled out completely for this species; female salamanders in the field are not restricted physically or socially and probably have the option of examining several males before engaging in courtship behavior.

Pregnancy block, also called the Bruce effect, is an example of a phenomenon that has been interpreted both as a product of male-male competition and as a mechanism of female choice. In the Bruce effect, the pregnancy of a recently inseminated female is terminated upon exposure to an unfamiliar male. Although first reported in house mice, *M. musculus* (Bruce 1959), pregnancy block has also been documented in a variety of rodents, including deer mice and voles. Unlike house mice, in which pregnancy terminations occur only during the first few days following insemination, female voles that are exposed to unfamiliar males may resorb or abort fetuses up to day 17 of a 21-day gestation period (Stehn and Richmond 1975).

Initial interpretations of pregnancy block focused on advantages to unfamiliar males (e.g., R. L. Trivers 1972). A male, new to the neighborhood, that encountered a pregnant female could not only disrupt the previous insemination of a rival male but also mate with the female once she became receptive again. The incidence of pregnancy termination in mice, however, depends on such factors as the presence of the first mate or other females, age of the female, and length of exposure to the unfamiliar male (Bruce 1961, 1963; Chipman and Fox 1966; Parkes and Bruce 1961), suggesting that females have some control over whether or not termination of pregnancy occurs. Perhaps pregnancy block is not simply a product of male-male competition, but rather a mechanism by which females can select better mates or better mating environments (Schwagmeyer, 1979).

The ability to terminate a pregnancy might be beneficial under conditions in which a female, deserted by her initial mate, has little chance of successfully rearing the upcoming litter (Dawkins 1976). Upon encountering a new male, her best option might be to terminate investment in the current pregnancy and mate with the newcomer to improve her chances of receiving parental assistance. Moreover, if the female carried the litter to term, there is always the possibility that the new male would commit infanticide; thus, it may be in the best interest of the female to cut her losses and resorb the doomed youngsters (Labov 1981). Indeed, in a more recent study, newly pregnant house mice were more likely to exhibit pregnancy block when exposed to infanticidal males than when exposed to noninfanticidal males (Elwood and Kennedy 1990). These data suggest that females can somehow assess the risk to their future offspring and adjust their reproductive response accordingly. Although the functional significance of pregnancy block is far from resolved, the possibility that female rodents exercise mate choice even after copulation has exciting implications.

In summary, then, intrasexual and intersexual selection may operate within a single mating system. We should thus be very cautious about attributing any single characteristic of males to one form of sexual selection or the other. Perhaps the elaborate train of peacocks, often thought of as an example of a character that probably resulted from female choice, results instead from male-male competition or some combination of the two processes. Although a recent study supports a role for female choice in the peafowl's mating system (Petrie, Halliday, and Sanders 1991), male-male competition cannot be excluded as an important factor in mating success in this species. From this classic example we see that determining the roles of intersexual and intrasexual selection in the lives of animals remains an enormous challenge.

SUMMARY

Sexual selection results from (1) competition within one sex for mates (intrasexual selection) and (2) preferences exhibited by one sex for certain traits in the opposite sex (intersexual selection). Because females invest more in gametes and parental care, they are usually a limited resource for which males compete. Competition among males for access to females appears responsible for the evolution in males of a broad spectrum of physical and behavioral attributes that enhance fighting prowess. Weapons, large size, and physical strength have evolved in males of many species. In addition, males devote much time and energy to ensuring that their sperm and not the sperm of a competitor are used to fertilize a female's eggs. In some species this may take the form of dominance behavior, mate guarding, or alternative reproductive strategies; in others, repellents or copulatory plugs are used. Because reproductive success is measured in relative terms, males may also enhance their position by employing tactics of sexual interference that decrease the sexual success of other males.

As a result of their greater investment in gametes and parental care, females are usually more discriminating than males in their choice of mates. In some species, female mate choice is based on the ability of a male to provide material benefits, either for immediate gains or for long-term gains such as access to resources. Nuptial gifts, including prey, seminal fluids, glandular secretions, and spermatophores, provide immediate benefits. Females may also choose mates according to the quality of the male's territory, his ability to provide sperm, or his parental abilities. In addition, a male's courtship display or certain other traits may be assessed by females to judge the suitor's overall quality.

Several hypotheses explain the origin and maintenance of female mate-choice preferences when the male does not provide material benefits. According to one hypothesis, called runaway selection, female preference and the preferred male trait evolve together because they are controlled by the same genes. In this case, a female mates with a male who bears the preferred trait, producing sons with the trait and daughters with a preference for the trait. The attractive sons will acquire more mates with a preference for the trait and will, therefore, leave more offspring. Runaway selection, then, can result in increasingly exaggerated male traits and stronger female preferences for them. The preferred traits can be arbitrary or even disadvantageous, providing that females prefer to mate with males that have them.

A second hypothesis is that females choose mates who have good genes. Supporting this hypothesis are studies showing that males with exaggerated male traits leave more viable offspring than do other males, studies showing female choice for parasite-resistance genes, female choice for trait symmetry, and female choice based on the degree of genetic similarity. A third hypothesis is that the evolution of a male's attractive trait took advantage of preexisting sensory biases in females. Mate-choice copying is a fourth hypothesis. In this case, one female's mate choice affects the choice made by another female.

When selecting a mate, females probably operate under constraints of time, mobility, and energy.

15

Parental Care and Mating Systems

PARENTAL CARE

Inside silken nest chambers *Stegodyphus mimosarum* spiderlings sit on the abdomen of an adult female and begin to devour her. Chances are the victim is their mother. The *S. mimosarum*, a social spider found in Africa, is only one of about ten species of spiders in which the young eat their mother at the end of the brood care period (Seibt and Wickler 1987). Unlike species of solitary spiders, in which gerontophagy (the consumption of elderly individuals) is confined to the mother and her offspring, social *Stegodyphus* have a more varied menu in that they can choose between eating their own mother or someone else's. However, in the interests of seeing her genes passed on through the survival and reproduction of her relatives, the designated victim should "prefer" to be eaten by her own offspring (particularly if she cannot produce another brood) rather than by a nonrelative. Because spiderlings

do not cannibalize one another or individuals of different spider species, gerontophagy in *S. mimosarum* appears to represent a mother's final act of parental care as she sends her offspring into the world on a full stomach. This bizarre example illustrates the often unexpected adaptive relationships between parents and offspring. In this chapter, we will focus first on parent-offspring relationships and then on mating systems in an effort to better understand these reproduction-enabling mechanisms across the spectrum of animal life.

ALLOCATION OF PARENTAL REPRODUCTIVE RESOURCES

There are very few parents as self-sacrificing as *S. mimosarum* mothers. As a rule, parental efforts serve to maximize an individual adult's lifetime reproductive success and not necessarily each reproductive event. Thus, all parents must make two important "decisions." First, they must decide how much of their own resources to devote to reproduction instead of to their own growth and survival. Second, they must decide how to allocate the available resources among their offspring (Clutton-Brock and Godfray 1991). It is easy to see how these decisions can lead to conflicts of interest both between the parents and their offspring and among siblings. As we will see, many factors are involved.

An obvious factor influencing how much effort parents should invest in the current offspring is the likelihood that they will have future opportunities to breed. This, in turn, will be affected by the parent's age (Trivers 1974) and the life span of individuals in that species (Linden and Møller 1989). It might be expected that in short-lived species with little hope of producing additional young in the future, parents would invest more heavily in the present young. On the other hand, parents of long-lived species might spend more of their resources on their continued growth and survival because they might have the opportunity to breed again.

This hypothesis has been tested experimentally on several bird species by handicapping the parents so that more parental effort was required to raise the young and then determining whether the parents would bear the increased costs of reproduction themselves or pass the costs on to their young. Leach's storm-petrel (*Oceanodroma leucorhoa*) is a relatively long-lived seabird. Adult petrels make long journeys to ephemeral food patches to gather planktonic crustaceans, drops of oil, and small fish to feed their chicks. A foraging trip usually lasts two to three days. About 30% of that time is spent airborne, and so the cost of flight for a parent that is provisioning chicks is significant. Shortening the wing span by clipping feathers increases the energetic cost of flight, raising the cost of reproduction. When parent petrels were handicapped in this way, they passed the increased reproductive costs to their offspring and maintained their own nutritional condition. Feather growth, which is a measure of a bird's nutritional state, did not differ between adult birds whose wings were clipped and untreated control birds. However, the chicks whose parents' wings had been clipped grew more slowly and spent more nights without food than chicks with untreated parents (Mauck and Grubb 1995). In contrast, when parents of short-lived species, such as starlings (Wright and Cuthill 1990), flycatchers (Slagsvold and Lifjeld 1988), and tits (Slagsvold and Lifjeld 1990), were handicapped in a way that increased their reproductive costs, they bore at least part of the increased costs themselves and continued to allocate nearly the same amount of resources to their chicks. Whereas the single chick raised each year by long-lived petrels represents only a small part of the parent's lifetime reproductive success, in these short-lived species, each clutch represents a large proportion of the parent's lifetime reproductive success. Thus, in these studies, we do find that expected life span influences a parent's allocation of resources in a way that maximizes its lifetime reproductive success.

When resources are limited, parents must decide which of their offspring to feed. How is that decision made? Do parents feed each youngster according to its need, or do they give more to the one most likely to survive? The answers to these questions often depend on the species and sex of the parent, as well as size differences among the offspring. In many species of birds, the young hatch over a several-day period. The larger, first-hatched young are generally fed more than other nestlings. As a result, the first-hatched young grow faster and are more likely to survive. An Australian parrot, the crimson rosella (*Platycerus elegans*), is an interesting exception to this general trend. The pattern of food distribution is complex. For one thing, the parents differ in their strategies of food delivery. The chicks hatch over a period of 1.5 to 7 days, and so the first-hatched chicks are up to seven times larger than the last-hatched chicks. Parents feed the chicks by direct regurgitation, and males delivered larger loads and had higher feeding rates. Females distribute food equally to all chicks in the brood. In contrast, males feed the first-hatched chicks more than the last-hatched ones. Food distribution also depends on the sex of the chick. Large male chicks receive the most food. All female nestlings, regardless of size, are fed equally. Consequently, unlike many other species in which the young hatch asynchronously, the last-hatched crimson rosellas do not always have reduced growth and survival (E. A. Krebs 1999; Krebs, Cunnigham, and Donnely 1999). In other bird species such as bluethroats (*Luscinia s. svecica*), parents distribute food differently, depending on the size differences among the hatchlings. When the size differences are large, male and female parents feed the largest nestlings nearly twice as often as small ones. But when size differences are small, food is more equally distributed (Smiseth, Amundsen, and Hansen 1998).

It is easy to see how differences in the distribution of resources by parents can lead to sibling rivalry. In some species, this rivalry leads to siblicide, in which one offspring attacks and kills its brother or sister. Siblicide is most common in species in which resources are limited and parents deposit eggs or young in a "nursery" with limited space (Mock and Parker 1998). The nursery can take various forms, including a uterus, a brood pouch, a parent's back, a nest, or a den. Although it seems odd at first, siblicide may be advantageous to the parents of some species. For example, when more young are produced than can be raised successfully, siblicide can save the parents time and energy by eliminating the young that are least likely to reach adulthood.

Why would a parent produce more young than it can raise successfully? Several answers to this question have been suggested. One idea is that overproduction of young is insurance in case some eggs or offspring fail to develop. In some species of eagles, for instance, the female typically lays two eggs, but only one chick reaches fledgling age. The eggs are usually laid a few days apart. If both eggs hatch, the older, stronger chick generally kills its younger nestmate (Mock, Drummond, and Stinson 1990). Another possible explanation is that the overproduction of young is an adaptation to a variable food supply. In years when food is plentiful, all the young may reach adulthood. However, in years of food scarcity, when sibling competition for resources is severe, the weaker siblings will be killed (Clutton-Brock and Godfray 1991). In other cases, extra offspring may be produced to benefit the stronger siblings, either by helping them raise offspring when they become adults or by serving as critical meals to provide nourishment when conditions are particularly harsh (Mock and Parker 1998).

Siblicide is common among spotted hyenas (*Crocuta crocuta*). The female generally gives birth to twins. In nature, the newborns spend most of their time underground. They don't appear at the surface until they are about two weeks old. At this time, most of the litters consist of one male and one female. Many same-sex litters are probably produced, but these are not observed because one sibling killed the other during the first few weeks of life (Frank, Glickman, and Lichte 1991). It was hypothesized that the function of siblicide of a same-sex twin is to eliminate future reproductive competition (Frank, Holekamp, and Smale 1995). It turns out, however, that siblicide among same-sex twins is not as routine as was once thought. Instead, it seems to occur when resources are insufficient to sustain two cubs. Thus same-sex siblicide may be a response to harsh environmental conditions (Smale, Holekamp, and White 1999).

PROVIDING THE CARE

Maternal and Paternal Care

Male Versus Female Care and Mode of Fertilization In species with parental care, why do parental duties fall to the female in some species and to the male in others? The mode of fertilization (internal versus external) appears to be an important variable in determining which parent cares for the offspring (Ridley 1978). Specifically, female care is associated with internal fertilization and male care with external fertilization. Three hypotheses—the certainty of paternity hypothesis, the gamete order hypothesis, and the association hypothesis—have been proposed to explain the evolution of male versus female care as it relates to the mode of fertilization. We will consider each hypothesis in turn.

The certainty of paternity hypothesis is based on the idea that parental solicitude toward the young is correlated with the likelihood of genetic relatedness (Trivers 1972). Females, regardless of the mode of fertilization, can be absolutely certain that they are related to their offspring. Certainty of maternity guarantees that 50% of a mother's genes are present in each of her

progeny. Males, especially of species with internal fertilization, cannot be so confident. Even though a male copulates with a female, he has no guarantee that his sperm, rather than the sperm of a competitor, will fertilize her eggs. In short, because males of internally fertilizing species run the risk of investing time and energy in raising another male's offspring, the odds run against the evolution of paternal behavior. The reliability of paternity is assumed to be greater when eggs are fertilized externally instead of inside the female, and thus external fertilization opens the way for male parental care.

Although appealing to our sense of intrigue, the certainty of paternity hypothesis has not held up under scrutiny. Theoretical models developed by John Maynard Smith (1977) and Werren, Gross, and Shine (1980) raise questions about the usefulness of the paternity hypothesis as a general explanation for the evolution of patterns of parental care. Briefly stated, these authors argue that if paternity is similar for all matings regardless of whether or not there is paternal care (i.e., within a population, the reliability of paternity does not differ between males that are parental and those that are nonparental), paternity itself cannot influence selection for parental behavior by males. Paternal care may still be selected if a male leaves more surviving offspring by caring for them than he would if he did not care for them. This might occur, for example, if the predation rate on unprotected offspring is much higher than on offspring that are protected by a male. If paternal care is genetically controlled, the paternal male should spread his genes because his offspring would have a greater chance of surviving than would the offspring of a nonpaternal male.

The gamete order hypothesis suggests that patterns of parental care result from differences in opportunities of males and females to desert offspring. Natural selection should favor desertion by whichever parent has the earliest opportunity, thereby forcing the remaining partner to provide care (Dawkins and Carlisle 1976). According to this hypothesis, the partner that releases gametes last gets saddled with parental responsibilities. In species with internal fertilization, females are likely to care for the young because they are unable to desert the embryo(s). By contrast, in externally fertilizing species, females usually release their eggs before males release their sperm, and thus females, rather than males, have the first opportunity to leave the scene.

The association hypothesis, perhaps not as flashy as the first two hypotheses, relies only on the proximity of adults and offspring to explain the evolution of male versus female care (G. C. Williams 1975). With internal fertilization, it is usually the female that carries the embryos, and thus she is in the best position to care for the young when they enter the world. Paternal behavior is less feasible because fathers may not be in the vicinity when the eggs are laid or the young are born. According to this hypothesis, external fertilization, particularly when it occurs in a territory defended by a male, would be associated with parental care by males. Because territorial males are almost always in the neighborhood of the eggs they fertilize, paternal behavior is likely to evolve.

These three hypotheses, of course, cannot cover all the cases. The evolution of male versus female care in some species remains somewhat of a mystery. In an attempt to test the hypotheses in certain species, Gross and Shine (1981) examined reams of data for amphibians (35 families) and teleost fish (182 families), two groups that show both male and female care and internal and external fertilization. On the basis of their survey, the authors rejected the gamete order hypothesis, discovered some instances in which the certainty of paternity hypothesis failed to predict the pattern of parental care, and concluded that the association hypothesis was most consistent with the available data.

Paternal investment in fish and amphibians usually takes the form of solitary male care rather than shared male and female responsibilities. In contrast, paternal investment in birds and mammals almost always occurs in addition to maternal care. Can the association hypothesis explain this basic dichotomy in patterns of parental behavior? As Gross and Shine (1981) point out, the association hypothesis should not be invoked to explain this basic difference in the parental care of ectothermic (fish and amphibians) and endothermic vertebrates (birds and mammals), but it is valuable in interpreting the distribution of male versus female care within each group. The higher frequency of biparental care in birds and mammals with paternal investment probably reflects the fact that parental care in these groups usually involves both the feeding and the guarding of young. Two parents are better than one for feeding offspring, and thus biparental care evolves (Maynard Smith 1977). Fish and amphibian parents, on the other hand, rarely feed their offspring, and parental duties consist largely of guarding, a task that may be performed almost as well by one parent as by two (J. M. Emlen 1973).

Having stated the broad generalization of the adequacy of solitary male care in fish and amphibians, we should point out that there are many exceptions to this rule, including some cases of biparental care in fish. For example, cichlids (family Cichlidae; approximately 1800 species worldwide) are an important lineage of freshwater fish in Africa, South and Central America, and India. Most cichlids breed in areas rich in potential predators, and unlike many fish, they display elaborate parental behavior that includes solitary male care, maternal care, and biparental care (Keenleyside 1979).

What is it about some species of cichlids that led to the evolution of biparental care? In studying several species of biparental cichlids in Lake Jiloa, Nicaragua, McKaye (1977) discovered that less than 10% of breeding pairs were able to raise their offspring to the stage at which they left their natal territory. If two parents have such dismal luck in protecting their offspring, how successful could a single parent be? Experimental evidence verifies the severe handicap suffered by single parents of biparental cichlid species: In the presence of predators, removal of either member of a brooding pair resulted in increased loss of young compared to control situations in which both parents remained with the brood (Keenleyside 1979). Thus, under certain ecological conditions—in this case, extremely intense predator pressure—two parents may be necessary to successfully rear the young.

For a second example of biparental care among fish, consider the elaborate parental behavior of another cichlid, the discus (*Symphysodon discus*). Two days after fertilization, the fry of this species hatch, aided by both parents who chew open the egg cases. The two parents then deposit their youngsters on aquatic vegetation, where the wrigglers dangle from threads as they live off their yolk supply. Within two or three more days, however, the brood can swim freely, and they attach themselves to their parents and begin feeding on parental skin secretions (Figure 15.1) (Skipper and Skipper 1957). Quite possibly, young that can feed off two parents may grow and develop more rapidly than those that have only a single parent. In some fish, whether or not they live in predator-rich waters or their pattern of care involves the feeding of young, biparental care may be necessary to ensure the survival and health of the offspring.

Patterns of Parental Care and Phylogenetic History   What role might phylogenetic history play in predicting patterns of parental care? The relationship between phylogeny and strategies of reproduction is perhaps best illustrated by a broad comparison between patterns of parental care in birds and mammals. Whereas biparental care occurs in only about 3% of mammalian species, it is found in roughly 70% of bird species. In mammals, internal gestation and lactation necessitate a major parental role for the female and restrict the ability of the male to help offspring during early development. Male mammals cannot take over the duties of pregnancy and lactation for their mates, and rather than hang around during the period of early development, males, in most cases, seek mating opportunities elsewhere. However, in some mammalian species the male remains with the female during the rearing of offspring. This raises the interesting question of why male lactation has not evolved in these species, such as incubation, feeding, and guarding are usually divided somewhat equally between the sexes. We see from this general comparison of birds and mammals that the basic biological attributes of a lineage constrain evolutionary possibilities: Phylogenetic history, indeed, is important in determining patterns of parental care.

Sex Role Reversals

In the vast majority of internally fertilizing species, sperm is the male's only contribution to the survival of his offspring. As previously mentioned, Trivers (1972) argued that differential parental investment by the sexes governs the operation of sexual selection: The sex with greater parental investment becomes a limiting resource, essentially an object of competition among individuals of the sex that is investing less. The end result of the traditional asymmetry in patterns of parental investment is that males compete among themselves for access to females, and females in turn are selective in their choice of mates. (Recall, however, from our discussion of sexual selection in Chapter 14, that patterns of parental investment are probably not the sole determinants of sexual selection.) Are there exceptions to the general rule of predominantly female care in internally fertilizing species? Although in the minority, there are some species in which the burden of parental care falls largely or entirely on the male. These nontraditional species provide an excellent test of the importance of relative parental investment in controlling sexual selection. Let's consider two well-known examples of species that exhibit the phenomenon of sex role reversal in parental care—the giant water bug and the northern jacana.

Female giant water bugs (*Abedus herberti*) lay their eggs on the backs of males. Oviposition represents the end of female investment, and male water bugs are left with the task of ensuring the development of their offspring. Males aerate the eggs by exposing them to the air-water interface and through gentle rocking motions just below the surface. They assist nymphs during hatching and avoid eating their carefully nurtured investment by simply not feeding while the nymphs on their backs are emerging (Figure 15.2). Trivers's (1972) hypothesis predicts that because male water bugs show greater parental investment than females, they should be a limited resource for which females compete. We would also expect females to actively pursue males during courtship, and males in turn should be quite discriminating. Robert Smith (1979) tested these predictions in the field and in the laboratory by carrying a large kitchen strainer with which to catch these insects while stalking Arizona streams. Smith's data indicated that space on the backs of males was a limiting resource during periods of peak egg production (from April to

Figure 15.1 **Young cichlids of the biparental species** *Symphysodon discus* **graze on parental skin secretions.**

given that the physiological hurdles to such an occurrence seem somewhat surmountable (Daly 1979). Although we do not have the answer to this question, two possibilities have been suggested. First, the reproductive success of male-female pairs may not be limited by the lactational abilities of the female at all but rather by other factors, such as food availability (Daly 1979). According to this argument, then, males have evolved other forms of parental investment, and the addition of lactation to the list of paternal activities may simply not be useful. The second possibility concerns potential limitations on a male's time and energy. If a male remains with a female to prevent other males from mating with her, then perhaps the amount of time and energy that the male devotes to guarding either his mate or their territory makes the evolution of lactation in males impossible (Clutton-Brock 1991).

In contrast to the extended period of internal development in mammals, birds develop outside the mother's body. Embryos, along with food in the form of yolk, are packed in eggs that develop largely in the external environment. Because male birds are just as capable as their mates at providing care, parental duties

FIGURE 15.2   A male giant water bug broods eggs glued to his back. A single nymph has just hatched, and the male, as part of this complete takeover of parental duties, suppresses his predatory urges while the young assume independence.

August). In staged encounters in the laboratory, females always approached males first, and although males usually responded with a display called "pumping" (a rapid rocking movement that sometimes reached a crescendo of 300 pumps per minute), they occasionally opted not to display and thereby gracefully declined the female's invitation to mate. Trivers's predictions were further met by Smith's observations that males exercised considerable control over a female's success in courtship, mating, and egg laying. At the conclusion of these male-orchestrated activities, females departed, the weight of parental care now resting squarely on the male.

The northern jacana (*Jacana spinosa*), previously called the American jacana, a small bird of tropical marshes best identified by its spindly greenish legs and toes, provides a second example of sex role reversal in parental care. Donald Jenni and Gerald Collier (1972) conducted an extensive study of the population dynamics, behavior, and social organization of individually marked jacanas in Costa Rica. They found that jacanas have a polyandrous mating system in which a given female is simultaneously paired with several males. At their study site, northern jacanas breed year-round and a female defends a large territory that may encompass from one to four male territories. In addition to helping her mates defend their individual territories, the female independently repels intruders from her entire territory. A female's critical role in territorial defense is matched by her dominant role in courtship. Because female jacanas keep harems of males, many females are excluded from breeding. The result is heightened competition among females for males. A female may attack the eggs or chicks of a neighbor. The male may then

mate with that female and raise a new clutch of eggs (Emlen, Demong, and Emlen 1989).

Parental activities, however, are the province of the male (Jenni and Betts 1978). Nest building and incubation are male duties exclusively, a division of labor underscored by the fact that only male jacanas have incubation patches (highly vascularized bare patches of skin on the belly). Once the precocial young have hatched, males brood and defend their chicks (Figure 15.3). A female aids her mate in caring for the brood only in the sense that she helps repel intruders from his territory, a task that she performs at all times regardless of whether or not young are present.

Morphological correlates appear to accompany the female jacana's malelike role in territorial defense and competition for males during courtship: Breeding females weigh approximately 145 grams; adult males typically weigh in at a trim 89 grams.

The high predation rate probably favored sex-role reversal among jacanas. More than half of all egg clutches are lost to predators. A female lays only four eggs per clutch. Thus, she has an incentive to spend less time parenting and more time producing eggs (Weisner 1992).

Giant water bugs and northern jacanas are not the only species in which males assume primary responsibility for the care of offspring. With the notable exception of mammalian species, some level of sex role reversal has been reported in other species of insects and birds, as well as in crustaceans, fish, and amphibians. The courtship behavior of males and females in species that exhibit sex role reversal in patterns of parental care is strong support for Trivers's hypothesis which relates differential parental investment to sexual selection. As

FIGURE 15.3   Male northern jacana and his two chicks confront an intruder, a purple gallinule. In this polyandrous species, males assume all parental responsibilities and females defend areas that encompass the smaller territories of their mates.

is often the case in biology, exceptions to the rule provide the clearest insight.

Brood Parasitism

Intraspecific Brood Parasitism
Waterfowl exhibit two patterns of behavior in which an individual other than the genetic parent provides care for conspecific young (Eadie, Kehoe, and Nudds 1988). In several species, some individuals lay their eggs in the nests of other individuals of their species, and from that point on the foster parent incubates these eggs along with her own and then cares for all the young once they hatch (although especially common in ducks, geese, and swans, this phenomenon has also been reported in other species of birds, including house sparrows, swallows, and starlings). In another situation, the young of some waterfowl individuals mix with the young of other conspecifics soon after hatching, and then adults provide all subsequent care for these mixed broods. Rather than staying with its offspring and simply receiving assistance from another parent, the genetic parent permanently leaves its offspring in the care of another parent.

Does the genetic parent (the "donor") profit from leaving its young in the care of another individual (the "recipient")? Do the recipients of foreign young suffer reduced reproductive success? In some instances, the repercussions for both the donors and recipients of eggs are known. In bar-headed geese (*Anser indicus*), for example, some females lay their eggs in the nests of other females in their colony, and although the hatching success of their donated eggs may be only 5% to 6%, the gain in reproductive success is higher than if such females had not produced any eggs (Weigmann

and Lambrect 1991). In other words, donors benefit. The hatching success of the eggs of recipients, on the other hand, is substantially less (29%) than that of females that incubate only their own eggs (67%). Recipients are thus harmed by the presence of eggs in their nests that are not their own. In bar-headed geese, then, the phenomenon of laying eggs in the nests of other conspecifics is an example of intraspecific brood parasitism because there is an advantage for the donor and a disadvantage for the recipient. In many other cases, however, the benefits to donors and costs to recipients have not been documented in sufficient detail to justify calling it parasitism. Until the precise costs and benefits to participants have been quantified, it may be more appropriate to refer to these phenomena as brood amalgamation (amalgamation meaning mixing or commingling) (Eadie, Kehoe, and Nudds 1988).

Interspecific Brood Parasitism
Some animals exploit the parental behavior of other species. Approximately 1% of all species of birds lay their eggs in the nests of another species, thus leaving host parents with the task of raising foster young. The phenomenon is called interspecific brood parasitism because of the apparent benefit experienced by the true parents, who dispense with all parental responsibilities, and the harm that befalls the heterospecific host parents, who typically experience a reduction in reproductive success (see later discussion). Interspecific brood parasites include several species of honeyguides, cuckoos, finches, and cowbirds, as well as at least one species of ducks (Payne 1977).

Damage to the host or its young may be directly inflicted by either the parasitic adult or its offspring

FIGURE 15.4   A nestling parasitic cuckoo is evicting a host's egg from the nest.

(reviewed by Payne 1977). The adult, for example, must place its egg in the host's nest when the host is beginning to incubate its own eggs. Upon discovering a nest after incubation or hatching has begun, a female cuckoo may eat the eggs or kill the young, causing the potential host to nest again. If the cuckoo is on time, however, she may throw out a host's egg before laying her own in the nest. To add insult to injury, cuckoos often lay their eggs from a perch above the nest, and their thick-shelled eggs break the host's eggs when they strike them. Nestling cuckoos are renowned for methodically evicting eggs or young from the nest of their foster parents. By positioning itself under a nearby egg or nestling, a young cuckoo may lift a nestmate onto its back, slowly work its way to the edge of the nest, and

nudge the host's egg or nestling to its death below (Figure 15.4). Nestling honeyguides employ an even more gruesome tactic to ensure full attention from their foster parents. At hatching, young honeyguides use their hooked bills to kill their nestmates. An alternative strategy to the outright killing of foster siblings is simply to monopolize parental care. Brood parasites usually mature more rapidly than a host's young, thus gaining a critical head start in growth and development. Also, their huge mouths, brightly colored gapes, and persistent begging often elicit preferential and prolonged feeding by foster parents (Figure 15.5). In England, a young cuckoo was so successful at begging that adults of three different species were observed feeding the youngster (Hardcastle 1925). The host's young are usually no match for their larger, more aggressive foster sibling and often die from starvation, crowding, or trampling.

In response to the devastating effects of brood parasitism, host species have developed ways to avoid being parasitized. The most common defenses are those used to reduce predation—the host species simply conceal their nests and defend them when they are discovered. Some hosts identify and remove the eggs or young of parasites from mixed clutches or broods. In response, many aspects of the laying behavior of brood parasites are designed for better deception of hosts. Cuckoos (*Cuculus canorus*) time their laying for the late afternoon, when hosts are less attentive (e.g., Davies and Brooke 1988), and whereas some passerines spend at least 20 minutes laying an egg in their own nests, female cuckoos can deposit an egg in a host's nest in less than 10 seconds (Wyllie 1988). Parasitic eggs or young often resemble those of the host species. Some species of parasitic finches mimic the mouth color and pattern of host nestlings, and others go so far as to imitate the

Figure 15.5   Parasitic young are often larger than their foster parents. Here a European cuckoo begs for food from its smaller foster parent, a hedge sparrow.

begging calls and head-waving behavior of their nest-mates. Such coevolution between the brood parasite and host may produce adaptations and counteradaptations of increasing complexity (Davies and Brooke 1988).

Before leaving the topic of interspecific brood parasitism, we should note that a somewhat similar phenomenon has been reported for three species of mouth-brooding fish in Africa (Ribbink 1977). Like other mouth-brooding cichlids, *Haplochromis polystigma*, *H. macrostoma*, and *Serranochromis robustus* females pick up their eggs soon after laying and brood them in their mouth until the fry are ready to emerge. At the time of emergence, females release their offspring at a suitable site and remain to guard them. In the event of danger, the young retreat to the safety of their mother's mouth. Females of these three predaceous cichlids have been observed tending mixed broods of their own fry and the fry of *H. chrysonatus*, a plankton-feeding cichlid. Although *H. chrysonatus* apparently broods its own eggs, it turns over the duties of guarding the fry to foster parents of the other cichlid species. Is the phenomenon in which some parental duties are carried out by foster parents of other fish species analogous to that of interspecific brood parasitism in birds?

In reviewing the evidence for what had initially been described as "cuckoo-like" behavior on the part of *H. chrysonatus*, Kenneth McKaye (1981) noted several differences between the phenomena of mixed-species broods in fish and interspecific brood parasitism in birds. For example, in contrast to the situation in birds, in fish (1) the energetic cost of rearing foster young is minimal; (2) brood sizes are large, and when a predator attacks, it takes only a portion of the brood; (3) foster young or their parents pose no threat to the host's young; and most important, (4) unlike the devastating effects of brood parasitism on the reproductive fitness of avian hosts, caring for the young of another fish species may actually increase the reproductive success of foster parents. In fish, the host may actually "adopt" the young of other species. Such adoptions might be advantageous to the foster parent if the addition of foreign fry to a brood reduces the probability that the parent's own young will be taken by a predator (see Chapter 13 for a discussion of the benefits of being part of a group when predators are nearby).

Although this explanation for the occurrence of mixed-species broods in fish is possible and indeed intriguing, given that the three cichlid species seen guarding broods that contained their own young and those of *H. chrysonatus* inhabit waters rich in potential predators, more data are needed on the benefits and costs to foster parents. Clearly, however, if guarding the young of other species is beneficial to these cichlid parents, it isn't truly parasitism. If both parties benefit, the relationship may actually be mutualistic. Until the pre-

TABLE 15.1   **Comparison of Interspecific Brood Care in Birds and Fish**

Characteristic	Birds	Fish
Size of brood	Small	Large
Effect of predation on brood	Total loss	Partial loss
Parental feeding of young	Yes	No
Energy devoted by foster parents to feed alien young	High	None or low
Competition among members of brood	High	Probably low
Behavior of alien young or their natural parent toward host young	May kill or remove	No removal or replacement
Active recruitment of alien young into brood	No	Yes
Apparent nature of relationship	Parasitic	Mutualistic
Ultimate "adaptive" response of host to alien young	Rejection	Acceptance and search for foster young

Source: Modified from McKaye (1981).

cise details of the relationship are known, we will describe the formation of mixed-species broods in cichlids as interspecific brood amalgamation. Table 15.1 contrasts the characteristics of interspecific brood parasitism in birds and brood amalgamation in fish.

MATING SYSTEMS

Earlier in this chapter we described how parents and offspring often disagree over the details of their interactions. As we will see in the pages that follow, such conflict is certainly not restricted to the parent-offspring relationship. In fact, adult males and females are often at odds over what constitutes an ideal mating relationship. Although the ultimate goal of reproduction for both sexes is to maximize fitness (the relative number of offspring that survive and reproduce), the reproductive success of males and females is constrained by different factors. In most species, a male's success is limited by access to females, and a female's is limited by access to resources (Davies 1991). Consequently, a male can often boost his reproductive success by mating with more than one female. In contrast, a female increases her reproductive success by gathering more resources, which can be accomplished in a number of ways, including male parental care and access to a high-quality territory. Generally speaking, then, males focus on mating effort, and females tend to emphasize parental effort. Therefore, we should not expect perfect parental harmony in the production of offspring. We would predict instead that each parent will attempt to maximize its own reproductive success, even if this is costly to the other (Trivers 1972).

A CLASSIFICATION OF MATING SYSTEMS

We will begin our consideration of mating systems with some definitions. As is often the case, there isn't universal agreement on the characteristics that should be used in classification. Randy Thornhill and John Alcock (1983) use only the number of copulatory partners per individual (Table 15.2). According to them, polygyny is a mating system in which some males copulate with more than one female during the breeding season; monogamy is a mating system in which a male and female have only a single mating partner per breeding season; and polyandry, the least common mating system, is one in which some females mate with more than one male during the breeding season. Other people believe that besides the number of mates, additional characteristics—the formation of a pair bond between the male and female, the duration of the pair bond, and cooperation in the care of the young—should be considered (e.g., Table 15.3; Wittenberger 1981).

We tend to follow the simpler classification system, which considers only the number of mates during a breeding system. And, as is common throughout much of the literature on mating systems, we group polygyny, in which one male mates with more than one female, and polyandry, in which one female mates with more than one male, under the general category of polygamy.

It is important to keep in mind, however, that regardless of the mating system, sexual fidelity is hard to find. Genetic analyses of parents and offspring often reveal that apparent social partners mate with other individuals: The social father is not always the genetic

TABLE 15.2   Classification of Mating Systems Based Only on the Number of Copulatory Partners Per Individual

Monogamy:	An individual male or female mates with only one partner per breeding season
Polygyny:	Individual males may mate with more than one female per breeding season
Polyandry:	Individual females may mate with more than one male per breeding season

Source: Thornhill and Alcock (1983).

TABLE 15.3   Classification of Mating Systems Based on the Number of Copulatory Partners Per Individual and Nature of the Pair Bond

Monogamy:	Prolonged association and essentially exclusive mating relationship between one male and one female
Polygyny:	Prolonged association and essentially exclusive mating relationship between one male and two or more females at a time
Polyandry:	Prolonged association and essentially exclusive mating relationship between one female and two or more males at a time
Promiscuity:	No prolonged association between the sexes and multiple matings by members of at least one sex

Source: Wittenberger (1979).

father of all the offspring, even in social monogamy. For example, DNA studies of the offspring of 180 species of socially monogamous* songbirds showed that only 10% of these species were genetically monogamous (Morell 1998). Cuckoldry can be a problem for polygamous males as well. Indeed, extra-pair matings seem to be the rule rather than the exception. The likelihood of extra-pair matings may even play a role in determining which mating system evolves.

Extra-pair matings have costs and benefits, especially if they result in fertilizations. A male's costs include the time and energy used in searching for receptive females other than his mate. While he's away, his primary mate may mate with another male, which reduces his reproductive success. However, if he is successful in inseminating mates of other males, he can boost his reproductive success substantially.

Hypotheses for female benefits of extra-pair matings can be grouped into two categories: material or genetic. By mating with several males, a female may gain added assistance in raising her offspring. In some species, such as red-winged blackbirds (*Agelaius phoeniceus*), the extra-pair males helped defend the female's nest from predators, improving her fledging success (Gray 1997). Females may also exchange copulations for a valuable resource, food, for instance. (In Chapter 14, the material benefits of nuptial gifts presented to the female at the time of copulation were discussed.) Extra-pair matings may also provide genetic benefits to the female, for example, the assurance of ample sperm to ensure the fertilization of all the eggs (Birkhead 1996). We see this in red-winged blackbirds. In addition to higher fledging success, females that

---

* Social monogamy is an association between one male and one female. It makes no assumptions about mating exclusivity or biparental care.

engaged in extra-pair matings hatched a significantly greater proportion of eggs than did females who were faithful to their primary mate (Gray 1997).

There is also evidence that a female may seek extra-pair matings to obtain "good genes" for her offspring. You may recall the discussion of female reed warblers in Chapter 14. These females seek extra-pair matings with males with a larger song repertoire than their mate's. Although they don't get material benefits from the extra-pair mate, they apparently do get good genes for their offspring. The post-fledgling survivorship of the young was related to the genetic father (Hasselquist 1998; Hasselquist, Bensch, and vonSchantz 1996). Socially monogamous female alpine marmots (*Marmota marmota*) also seem to copulate with males whose genetic quality is higher than that of their mate. Alpine marmots are burrowing rodents that live in mountain pastures. They live in family groups, made up of the resident pair, juveniles, yearlings, and subordinates. Satellite males, who don't have family groups, live around the periphery. The resident male may be cuckolded by his own subordinate males, satellite males, and males of neighboring groups. Cuckoldry is, indeed, a problem for some resident males. Genetic analyses have shown that some males, presumably those of lower genetic quality, are frequently cuckolded, whereas others are the genetic fathers of 100% of the offspring of their social mates (Figure 15.6) (Goossens et al. 1998). This variability in reproductive success among resident males would be predicted by the hypothesis that females engage in extra-pair matings to obtain high-quality genes for their offspring.

THE EVOLUTION OF MATING SYSTEMS

Two important factors in the evolutionary choice of mating systems are the need for parental care and the distribution of females. When each male and female can leave more descendants by sharing the work of raising the young, monogamy is favored. This helps explain why 90% of all bird species are socially monogamous (Lack 1968). Female mammals nurse their young, a trait that often lessens the requirement for male help in raising them. Thus, only 3% of mammalian species are socially monogamous (Kleiman 1977). The distribution of females throughout the habitat also plays a role in determining the mating system. Males can more easily monopolize females when females have small ranges or when they form small stable groups, and so polygyny is favored (Clutton-Brock 1989b).

Ecological factors, such as predation, resource quality and distribution, and the availability of receptive mates, affect the need for paternal care, the ability of males to monopolize females, and the ability of females to choose among potential suitors (Emlen and Oring 1977). Because such ecological conditions often vary both within and between locations, considerable flexibility is usually associated with the mating patterns of a given species.

Monogamy

Monogamy results when a male and female have only a single mating partner per breeding season. Because sperm from one male is often sufficient to fertilize a female's limited number of eggs, monogamy is sufficient from the female perspective, but for males, confining copulation to a single female is a rather conservative means of ensuring genetic representation in the next generation. We might wonder, then, why a male would be monogamous and what ecological circumstances might favor monogamy over polygyny?

Inability to Monopolize More Than One Female
Circumstances that make it difficult for a male to monopolize multiple mates will favor monogamy. For example, when receptive females are scarce or widely distributed, it might be more beneficial for a male to remain with a given female rather than to search endlessly for additional mates. In other cases, male monogamy may be enhanced by female reproductive synchrony (Knowlton 1979). When females within a population are receptive at the same time, a male's chances of copulating with several females in succession

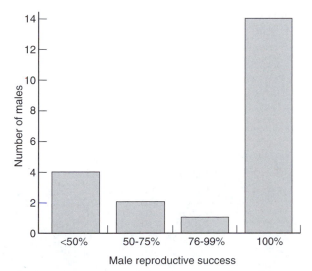

FIGURE 15.6 **The distribution of the percentage of young sired by the resident male in socially monogamous Alpine marmots. In some cases a male, presumably one of higher quality, is the genetic father of all of the offspring in his group. Other males were occasionally cuckolded to varying degrees. This is consistent with the hypothesis that the females engage in extra-pair matings for genetic benefits—high-quality genes—for their offspring. (Data from Goossens et al. 1998).**

FIGURE 15.7   A male termite is closely pursuing a female he has met on the nuptial flight. Synchrony in the availability of receptive females forces monogamy on male termites.

are exceedingly small. For example, pair formation in male termites takes place during the nuptial flight, when vast numbers of individuals from neighboring colonies become airborne simultaneously (E. O. Wilson 1971). During the flight, each male attempts to sequester a single female for himself. The short duration of the flight, the intense competition for mates, and each female's steadfast refusal to actually mate until well established in a safe burrow all favor a male termite's act of guarding the sole female he has managed to sequester. In fact, males of most termite species remain with their mates for life (Figure 15.7).

Necessity of Biparental Care   Monogamous males are often in the position of providing useful materials to their mate and offspring. Hornbills (Bucerotidae) are large-bodied monogamous birds of the Old World tropics and subtropics. All 45 species display the peculiar habit of sealing their nestholes. After a female enters a nesthole to begin laying eggs, she carefully plasters mud around the edge of the hole until only a small crack remains to connect her to the outside world. Once sealed within her nesthole, a female relies entirely on her mate to bring food for herself and her nestlings (Figure 15.8). Although variable across species, the period of time in which a female remains ensconced in her nest may extend to over four months. Nesthole sealing, apparently an adaptation to reduce predation on females and nestlings (Kemp 1971, 1976), obviously means that a female will be monogamous, and the need for male care ensures that he, too, will not take another mate. A male hornbill, constrained by the enormous task of feeding his mate and offspring, simply cannot care for a second nest (Leighton 1986). Thus, monogamy in hornbills is linked to their adaptation to reduce nest predation, which requires that the male be able to provide sufficient food for the female and the young.

As we've seen, monogamy is common among birds but rare among mammals. This striking phylogenetic difference is said to result from the difference between avian and mammalian parents discussed earlier in the chapter, namely, that whereas male birds are equally adept as females at most aspects of parental care, male mammals typically can provide little useful help during the female-dominated activities of gestation and lactation (Trivers 1972).

In some mammalian species, however, males do have important parental responsibilities, and the fitness of both mates depends on the male's parental investment. For example, silver-backed jackals (*Canis mesome-*

FIGURE 15.8   A male red-billed hornbill is feeding his mate, which is sealed inside the nest hole. Male hornbills are monogamous because they cannot simultaneously provision two nests.

las; also called black-backed jackals) are monogamous canids that inhabit the brush woodland of the Serengeti Plain of Tanzania. For over a decade, Patricia Moehlman (1986) has studied this species and yet has never observed a single change of mate among breeding pairs. Adult members of a breeding pair appear to mate exclusively with their partners and exhibit a high degree of behavioral synchrony that includes cooperative hunting and tandem marking of territory boundaries. Male silver-backed jackals participate in parental activities such as guarding the den and regurgitating food for the young. In this species, male parental investment seems critical to the reproductive success of the female; in the two instances in which the male of a pair disappeared during the whelping season, the female and pups all died within a week.

The necessity of male parental care has also played a role in the evolution of monogamy in the California mouse (*Peromyscus californicus*), a rodent that is perhaps the only truly monogamous mammal in California. Genetic analyses have shown that once paired, these mice never stray (Ribble 1991). The pups are born at the coldest time of the year and need their parents' body heat for survival. Both parents take turns huddling over the pups to keep them warm. When environmental temperatures are warm and there is plenty of food and water, male attention is not important to the survival of the young and may even slightly lower their survival rate. However, in the laboratory or in nature, when temperatures are low and food is scarce, pup survival was enhanced because the father huddled over them (Gubernick, Wright, and Brown 1993).

However, even in monogamous species that show male parental care, this care doesn't always boost fitness. Patricia Gowaty (1983) removed males from pairs of the eastern bluebird (*Sialia sialis*) to examine the effects of male parental care on female reproductive success. This species is socially monogamous, but we now know from DNA fingerprinting that it is not genetically monogamous. As a result of extra-pair fertilizations, 15% to 20% of the nestlings are not fathered by the male in the social partnership (Morell 1998). Males were removed from the territories of breeding females after eggs had been laid. These experimentally deserted females were compared to paired females with respect to several measures of reproductive success (e.g., clutch size, number of eggs hatched, and number of nestlings fledged) during a season. Contrary to general expectation, lone and paired females did not differ for any measure of reproductive success considered, indicating that female fitness is not always dependent on male parental care, even in monogamous species. Gowaty suggests that rather than male parental care, dispersion of nest sites is the important variable for maintaining monogamy in eastern bluebirds. More recently, she has suggested that males are monogamous to prevent extra-pair matings by the female (Gowaty

1996). So, if parental care by monogamous males does have positive effects on female fitness, these effects may be subtle and much more difficult to observe in some species than originally expected. It is also apparent that long-term field studies, which permit the determination of lifetime reproductive success, are critical in evaluating the role of paternal behavior in monogamous mating systems.

Mate Guarding

Kirk's dik-dik (*Madoqua kirkii*) is a small antelope (only about 15 inches tall) that lives in dry scrub in many parts of Africa. Dik-diks form monogamous bonds that last for several years, if not for an entire lifetime. Unlike many species, dik-diks seem to be faithful to their mates: Genetic analyses revealed no evidence of extra-pair paternity (Brotherton et al. 1997). With monogamy rare among mammals, we may wonder why such devotion has evolved in dik-diks.

Several hypotheses can be ruled out. The need for biparental care can't be important in the evolution of monogamy, because paternal care is absent—the male did not defend resources or reduce predation risk, and there isn't a risk of infanticide (Brotherton and Rhodes 1996). Furthermore, at least some monogamous males defend territories that could support more than one female. In a rare population in which some males were polygynous, the territories of the polygynous males were not of higher quality than those of monogamous males. And, it seems unlikely that monogamy evolved because constant male attention is necessary to keep females from wandering off the territory and mating with another male. Pair mates spend only about 64% of their time together, and so there would be ample opportunity to wander away.

Instead, it seems that monogamy evolved because the costs of guarding more than one female from intruding males would be too great. The male guards his female by preventing other males from knowing when she is in estrus. He does this by covering up the scent of his female's territorial markers—piles of dung—by scratching dirt over them and then defecating on top of them. It is also essential that a male advertise his territorial ownership because rivals bent on filling vacant territories are never far off. This is accomplished by marking his territorial borders with the scent from glands under his eyes (Komers 1997). If a male were to try to overmark the scent of two females on separate territories, he might fail to mark his territory sufficiently. As a result, he could lose it in the intense competition for vacant territories. In the few instances in which a male dik-dik succeeded in polygyny, the females shared a territory. The female accepts being guarded because an extra-pair mating might incite a fight between her male and the rival that could prove harmful to her or her offspring (Brotherton and Manser 1997).

Mate guarding also seems to explain monogamy in the coral-reef fish *Valenciennea strigata*. These fish are socially and genetically monogamous (Reavis 1997). In this species, a male and female share a territory that has several burrows, which are used both as nests and as refuge. Only the males care for the eggs. Because there is an abundance of resources, nearly all males breed. Those that don't are generally juveniles or small adults, unable to attract an already-paired mate or find an unpaired mate. The size of a mate is important to females because females paired with large males fed more than those paired with small males. Thus, a female benefits from guarding her current mate because chances are good that any unmated males will be small. The cost of mate guarding for a female is minimized by a one-to-one sex ratio. Males guard their mates to guarantee paternity. Widowed fish remated quickly, and so the pairs don't remain monogamous because potential mates are rare. When a mate is lost, the new mate is usually smaller. We see, then, that males and females of this coral-reef fish remain monogamous because, respectively, mate guarding prevents polygyny and current mates are larger and more valuable than new mates (Reavis and Barlow 1998).

Polygyny

Natural selection's ledger sheet shows that polygyny has both costs and benefits for males and females. Because a male can generally fertilize more eggs than a female can produce, a male usually benefits from polygyny by producing more offspring (if paternal care is not required). As males maximize their reproductive output through multiple matings, polygyny results, and we find that this is the most common mating system among vertebrates, including humans (Flinn and Low 1986). Possible costs to a male include an increased chance of cuckoldry, because he is not guarding each female from other males, and the costs of achieving dominance or defending a territory.

Polygyny has several significant costs for females. For example, males in such cases usually do not help to rear the young. In species in which the males do provide some parental care, it is divided among the offspring from more than one female. Females must also share essential resources, such as nest sites or territories. Also, the activity around these areas may attract predators, and other receptive females may increase the competition for commodities such as food.

Why would a female accept the costs of polygyny? There is no simple answer to this question. Indeed, the answers may vary among species and even within a species. At least part of the variability is due to ecological factors that influence whether a female will reap a net benefit by mating with an already mated male. We'll consider some of these ecological factors, along with several hypotheses for females' acceptance of polygyny, in the discussion that follows.

Hypotheses for Females' Acceptance of Polygyny

There are several hypotheses to explain why females accept the costs of polygyny. We will consider three—polygyny threshold hypothesis, "sexy son" hypothesis, and best alternative hypothesis.

1. Polygyny threshold hypothesis. Polygynous matings will be advantageous to females when the benefits achieved by mating with a high-quality male and gaining access to his resources more than compensate for these costs. In other words, a female may reproduce more successfully as a secondary mate on a high-quality territory than as a monogamous mate on a low-quality territory. Jared Verner and Mary Willson (1966) coined the phrase *polygyny threshold* to describe the difference in a territory's quality needed to make secondary status a better reproductive option for females than primary status.

Gordon Orians (1969) elaborated on Verner and Willson's ideas and reasoned that if polygyny is always to a male's advantage and yet does not always occur, the circumstances under which it does occur must be those in which there is some advantage to the female. He reasoned that females should join a harem when this decision confers greater reproductive success than monogamous alternatives. According to this argument, then, the average reproductive success of females should not decrease as the harem's size increases. Figure 15.9 illus-

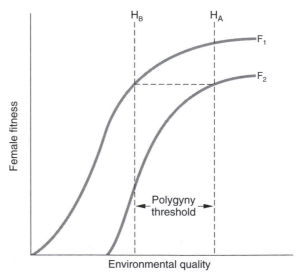

FIGURE 15.9 The polygyny threshold model. The polygyny threshold is the difference in quality between H_a and H_b that would favor polygynous matings. F_1 = fitness curve for monogamous females F_2 = fitness curve for secondary females H_a = highest quality breeding habitat H_b = marginal breeding habitat (From Wittenberger 1979; modified from Orians 1969.)

FIGURE 15.10   A male elephant seal. Males of this species compete for access to, and defend, large numbers of females at birthing sites. This is an example of female defense polygyny.

trates this model's way of relating differences in a territory's quality to a female's choice of already mated versus unmated males.

The polygyny threshold model has been extremely important in shaping our views of mating systems, especially the mating relationships of birds. Predictions of the model have been subjected to numerous tests in the field, and most studies confirm the expected relationship between the number of mates and environmental factors believed to reflect a territory's quality. However, the critical question is whether unmated females do better by accepting secondary rather than primary status. Although this point can be tested by comparing the success of monogamous and secondary females, procuring the necessary data on reproductive success is no easy task.

2. "Sexy son" hypothesis. According to this hypothesis, access to good genes for her offspring compensates a female for the costs of polygyny. A female may benefit from mating with an already mated male if her sons inherit the genes that made that male attractive. Her sexy sons will presumably provide her with many grandchildren. Thus, the female's lifetime reproductive success may be enhanced by choosing to mate with a male that is attractive to many females (Weatherhead and Roberston 1979, 1981).

In other words, a female that chooses an already-mated male may benefit *indirectly* if the good genes she acquires for her offspring boost their survival and reproductive success. In Chapter 14 we saw examples of female mate choice for attractive males whose good genes did enhance the survival of the offspring (e.g., peacocks with more eyespots in their trains; Petrie 1994). The issues that remain to be demonstrated are the extent to which the reduction in the direct breeding success of a female can be compensated for by the indirect benefits of enhanced reproduction of offspring and extra-pair fertilizations by genetically high-quality males (Alatalo and Rätti 1995).

3. Best alternative hypothesis. In some species, females may accept polygyny because it is the best of the available options. For example, when a fraction of the males in the population control all the suitable breeding habitat, a female that did not accept the costs of polygyny might have to forego breeding. In species such as lions, females accept polygyny because the only way to reproduce is to remain in groups so that they can acquire and defend territories on which to hunt and

breed (Packer, Scheel, and Pusey 1990). In other species, it might be less costly for a female to accept polygyny than it would be to drive off other females. This may help explain polygyny among yellow-bellied marmots (discussed shortly).

Evolution of Female Defense Polygyny   In female defense polygyny, a male defends a harem of females. This type of polygyny occurs when females live in groups that a male can easily defend. In some species, female gregariousness may be related to cooperative hunting or increased predator detection; in other species, clumping is more directly related to reproduction. For example, female elephant seals (*Mirounga angustirostris*) become sexually receptive less than one month after giving birth. Each year pregnant females haul themselves onto remote beaches of Año Nuevo Island, California. Female gregariousness, a shortage of suitable birth sites, and a tendency to return annually to traditional locations result in the formation of dense aggregations of receptive females. Under these crowded conditions, a single dominant male, weighing in the vicinity of 8000 pounds and sporting an enormous overhanging proboscis (Figure 15.10), can monopolize sexual access to 40 or more females (LeBoeuf 1974). This male defends his harem against all other male intruders in bloody, and sometimes lethal, fighting.

The importance of the distribution of females in the evolution of female defense polygyny is seen when the harem size on densely populated areas such as Año Nuevo Island is compared to those on the much less populated beaches of Patagonia. In Patagonia, females

FIGURE 15.11   A yellow-bellied marmot. Results from studies on the social behavior of this polygynous mammal do not support the polygyny threshold model.

are not confined to small areas along the coast because the beaches are long and continuous. When females are dispersed over a large area, a harem is more difficult to defend. As a result, the median harem size in Patagonia is 11 females. Thus, a greater percentage of males in the population has an opportunity to breed, and the variability in reproductive success among males is much lower than on Año Nuevo Island (Baldi et al. 1996).

Yellow bellied marmots (*Marmota flaviventris*) also practice female defense polygyny. As in elephant seals, the distribution of females allows a male to economically defend a harem. These rodents inhabit forest clearings and alpine meadows of the western United States (Figure 15.11). Active from early May to mid-September, they spend the rest of the year in hibernation. For over 20 years, Kenneth Armitage has studied the social behavior of marmots in Colorado. Through a combination of live trapping, direct observation, and genetic analysis of blood samples, he and his colleagues have discovered that although some individuals are solitary, others live in social groups made up of one adult

male, one or more adult females, and a number of yearlings and juveniles. Matrilines, made up of from one to five related females, often occupy a common home range, and a male's harem may contain one or more matrilines.

Not surprisingly, a male marmot's reproductive output, measured by the number of yearlings produced, increases as harem size increases. However, in apparent contradiction to the notion that polygyny occurs when it is advantageous to females, the reproductive success of females decreases with increases in harem size (Figure 15.12) (Downhower and Armitage 1971).

Several explanations have been offered for the finding that female reproductive success varies with harem size. One idea is that the typical harem size of two in yellow-bellied marmots is a compromise between the polygynous inclinations of the male and the preference for monogamy by the female (Downhower and Armitage 1971). However, Elliott (1975) and Wittenberger (1979) criticized this interpretation, suggesting that loss in the annual reproductive output of females living in large harems could be compensated for by increased survival and hence greater *lifetime* reproductive success. According to their argument, female marmots in large harems may lose a little on an annual basis but surely would make up for this loss by longer survival and increased opportunity for reproduction. The problem with this argument is that survivorship of females does not vary as a function of harem size.

Although Armitage (1986) revised the earlier Downhower and Armitage model (1971), the polygyny threshold model still does not apply to yellow-bellied marmots. The model is based on the assumption that polygyny evolves when already mated males attract additional females to their territory. In yellow-bellied mar-

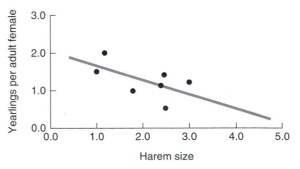

FIGURE 15.12   Contrary to predictions of the polygyny threshold model, female reproductive success, measured in terms of the number of yearlings produced per adult female, decreases as harem size increases in yellow-bellied marmots. (Modified from Downhower and Armitage 1971.)

FIGURE 15.13   A male scorpionfly. Scorpionflies display resource defense polygyny in that a male will defend the area around a dead insect and females must mate with him to gain access to the meal.

mots, however, males are attracted to females that reside in a suitable habitat. In marmots then, females appear to determine the spacing patterns that make polygyny possible. Furthermore, the organization of marmots into matrilines does not meet the implicit assumption of the polygyny threshold model that females act independently of other females. Results from the study of the behavior and ecology of yellow-bellied marmots not only raised questions about the universality of the polygyny threshold model but also, perhaps more importantly, have emphasized the conflicting nature of male and female reproductive interests.

Evolution of Resource Defense Polygyny

Resource defense polygyny occurs when males defend resources essential to females rather than defending females themselves. Those resources can be nest sites or food. If such resources are unevenly distributed in an area, it is possible to defend areas that have more or better resources than others (Emlen and Oring 1977).

In some cases there is good evidence that a female's choice is often based on the quality of resources controlled by a male. Thus, even though receptive females do not live together, a male can monopolize a number of mates by controlling critical resources. An example is provided by male scorpionflies (Mecoptera: Panorpidae; Figure 15.13), which fiercely defend the area around a dead arthropod (Thornhill 1981). Standing next to his nutritious find, a male will disperse a sex attractant and display with rapid wing movements and abdominal vibrations. If a female approaches, copulation is the fee that she must pay to gain access to this food. Thus, while males enjoy the benefits of mating with several females, females enjoy a hearty meal that

may reduce the amount of time that they need to spend foraging.

As we saw in Chapter 14, females of many species choose males on the basis of the quality of their territory. In these species, males must cope with the uneven distribution of resources in nature. However, males of the African cichlid fish, *Lamprologus callipterus*, create their own defensible distribution of essential resources—a collection of snail shells in which females can lay their eggs. Males gather shells and even steal them from rivals. The females, which are minuscule compared to their mates, live inside the shell with their offspring until the young are old enough to venture out on their own. Males don't care for the young, but they do defend the shells. Larger males collect more shells and, therefore, attract more females (Sato 1994).

Unlike female yellow-bellied marmots, the reproductive success of females of several species of birds doesn't seem to be reduced by polygyny. Among great reed warblers (*Acrocephalus arundinaceus* L.), the reproductive success of females who chose a mated male was only marginally lower than that of females who chose an unmated male—0.51 as opposed to 0.53 offspring that survived to the next breeding season (Bensch 1996). In lapwings (*Vanellus vanellus*), polygyny definitely boosts male reproductive success, as expected. Polygynous males raised 58% to 100% more young than monogamous males. Even when cuckoldry due to extra-pair fertilizations is taken into account, polygynous males raised 67% to 97% more young than monogamous males. The reproductive success of polygynous females was slightly lower than monogamous females, but the difference was not significant (Parish and Coulson 1998). In these species, then, females seem to be compensated for any costs associated with polygyny, as predicted by the polygyny threshold model.

Evolution of Lek Polygyny

The third category of polygyny, lek polygyny, occurs when males defend "symbolic" territories that are often located at traditional display sites called leks. Males of the lek species do not provide parental care and defend only their small territory on the lek, not groups of females that happen to be living together nor resources associated with specific areas. Females visit these display arenas, select a mate, copulate, and leave. This extreme form of polygyny occurs when environmental factors make it difficult for males to monopolize females either directly, as in female defense polygyny, or indirectly, as in resource defense polygyny.

A well-known example of a species that displays lek polygyny is the hammer-headed bat, *Hypsignathus monstrous* (Figure 15.14), an inhabitant of lowland rainforests of western and central Africa. Adult males, nearly twice the weight of adult females, have not only an enormous muzzle that terminates in flaring lip flaps but also a larynx that fills more than one-half of their body cavity. Mature females have a much less dramatic muzzle and a larynx approximately one-third the size. Jack Bradbury (1977), intrigued by the extreme sexual

FIGURE 15.14 A male hammer-headed bat is at his calling site on the lek.

dimorphism, modifications of the male larynx, and reports of aggregations of calling males, observed the bats from platforms in the forest canopy, netted animals at the leks, and tracked individual males and females by radiotelemetry.

As with other lek species, male hammer-headed bats show no parental care and have no significant resources within their display territories. Adult males select sites that are at least 10 meters apart along streams and rivers. These linear arrays of displaying males may contain over 100 individuals and occur at traditional sites used for at least 60 years. Individual males are quite particular about their choice of calling branch, selecting the exact same place night after night and often making from five to ten passes over the location to ensure that they land at just the right spot. Once in position, they begin wing flapping and loud calling (a noise described by Bradbury as sounding like "a glass being rapped hard on a porcelain sink"). These displays often continue without interruption for hours. Interested females fly along the calling assembly and make short, hovering inspections of the males. The mere pause of a female near a male inspires him to accelerate his calling rate, and should she move closer, he pulls his wings tightly against his body and emits several long buzz notes. If the female turns and leaves, the male will extend his wings and resume his display. However, if the female lands on his calling branch, signaling that he has been chosen to donate his sperm, he copulates rapidly and typically resumes calling within one minute of mating. Copulation appears to occur exclusively on these display territories, and some males have far greater success than others. In fact, at the major assembly site studied in 1974, 6% of the males achieved 79% of the copulations.

How might lek behavior such as that seen in the hammer-headed bat have evolved? Several hypotheses have been suggested. These have been divided into two groups, based on potential benefits for males or females.

From the male perspective, group, rather than individual, displays may increase the signal range or the amount of time that signals are emitted (e.g., Lack 1939; D. W. Snow 1963). Or, males may aggregate because they require specific display habitats that are limited and patchily distributed (e.g., B. K. Snow 1974). Leks may provide protection from predators through increased vigilance—with more eyes watching, a predator would have a harder time sneaking up on them (e.g., Wiley 1974). Leks may also serve as information centers, where males exchange the latest news on good foraging sites (e.g., Vos 1979). Males may gather near "hot spots," areas through which a large number of females are likely to pass (e.g., Bradbury, Gibson, and Andersson 1986; Payne and Payne 1977). Finally, leks

FIGURE 15.15   Polyandry is practiced by spotted sandpipers. Here two females fight over a prime nesting habitat on which the winner will probably pair with several males in succession.

may arise because less successful males generally have better mating chances in the vicinity of highly successful males. Because certain males are extremely successful at attracting females, other, less successful males gather around these "hotshots" and obtain more copulations than they would have had they displayed by themselves (Beehler and Foster 1988).

From the female perspective, large groups of males may facilitate mate choice (Alexander 1975; Bradbury 1981). After all, it might be easier to distinguish between superior and inferior males when comparison shopping is possible. Mating within a group of males may reduce the vulnerability of females to predation since any predator might be distracted by so many displaying individuals (Wittenberger 1978); or if males aggregate in a less desirable habitat, lek mating may reduce competition between the sexes for resources (Wrangham 1980). All of these hypotheses could be useful as explanations for the development of leks but for now are best regarded as only possible explanations.

Polyandry

Polyandry, defined as the mating relationship that occurs when some females have more than one mate during the breeding season, is the least common mating system. Because most males offer only sperm in return for copulation and because the sperm from one male is often sufficient to fertilize all the eggs of a given female, most females have little to gain by mating with more than one male. However, if copulation opens the door to critical resources or male parental assistance, then mating with several males may result in reproductive benefits for females. From a male perspective, however, polyandry seems to be the most unsatisfactory mating option. If, in exchange for copulation, a male must commit himself to the provision of resources or parental care, the last thing he should want is to share his mate with other males. The rarity of polyandry results from the basic fact that under most circumstances, polyandrous matings are disadvantageous to both males and females.

Polyandry occurs to varying degrees in different species. The females of some species mate with more than one male to supplement sperm received from previous matings (e.g., fruit flies; Pyle and Gromko 1978), whereas others copulate with several males to gain

access to resources that males exchange for copulation (e.g., hummingbirds; Wolf 1975). These mating relationships do not involve sex role reversal (i.e., females competing for males that provide the bulk of parental care and are selective in their choice of mate) and are best described as combinations of female polyandry and male polygyny. "True" polyandry is always associated with sex role reversal, such as that described for the northern jacana in the section on parental care. Thus, in its most striking form, true polyandry is characterized by males assuming most, if not all, parental responsibilities, thereby freeing females to pursue mating opportunities.

Two hypotheses have been proposed to account for the evolution of sex role reversal and polyandry. Graul, Derrickson, and Mock (1977) suggested that sex role reversal and polyandry evolved in response to dramatic fluctuations in food availability. During periods of scarcity, a male's takeover of parental duties may be necessary if the energy reserves of his mate have been drastically depleted by egg production. Faced with an exhausted mate, the best option for a male may be to increase parental investment. Should food availability increase, the female can lay additional eggs in a second nest that she might tend (double clutching) or leave to another male. The second hypothesis suggests that high predation pressure, rather than food availability, is the ecological factor of importance in the evolution of role reversal and polyandry (Maxson and Oring 1980). Intense predation on nests may force males into parental duties so that females can lay a second clutch, thereby increasing the chances for both partners of fledging young.

The spotted sandpiper, *Actitis macularia* (Figure 15.15), is an example of a polyandrous species subjected to intense nest predation and at the same time demonstrates the flexibility associated with strategies of mating and parental care (Oring and Lank 1986). For several years, Lewis Oring and his colleagues have studied a population of spotted sandpipers on Little Pelican Island in Leech Lake, Minnesota (a name that fails to conjure up images of a swimmer's paradise). Female spotted sandpipers lay separate clutches with one or more males on all-purpose territories that they defend against other females. Although females typically acquire mates sequentially, an occasional female will have two males on her territory at the same time. Males engage in territorial defense and provide most of the parental care. Female spotted sandpipers engage in a strategy of bet hedging, vigorously pursuing mating opportunities when they arise and, in their absence, investing in parental care. In essence, if a new male appears, she will mate with him; if no new males appear, she will tend to her young. Variations in social behavior result from a combination of changing environmental circumstances and social conditions. As is characteristic of other species that display sex role reversal, female spotted sandpipers are larger and more striking in their coloration (i.e., they have a denser spot pattern) than males.

Nest loss plays an important role in determining the incidence of polyandry in spotted sandpiper populations. Intense predation on nests leads to concentrations of individuals in more protected areas and thereby increases the potential for females to acquire multiple mates. However, even in relatively safe sites such as Little Pelican Island (the small size of the island makes it unsuitable for year-round occupation by most mam-

malian predators), nest losses to predation are quite high. For example, despite the removal of some mammalian predators from the island, only 37% of the 1289 eggs laid between 1973 and 1982 hatched. Although deer mice (*Peromyscus maniculatus*) are the most consistent source of egg destruction, a single mink on the island in 1975 caused total breeding failure. Females counter high predation rates with a remarkable ability to rapidly lay replacement clutches. Interclutch intervals may be as short as 3 days, and one female laid as many as 20 eggs (totaling more than 400% of her body weight) over a period of 42 days (unpublished data cited in Emlen and Oring 1977). In summary, intense predation on spotted sandpiper nests leads to the clumping of individuals in more protected areas and increases the potential benefits of laying several clutches (Figure 15.16).

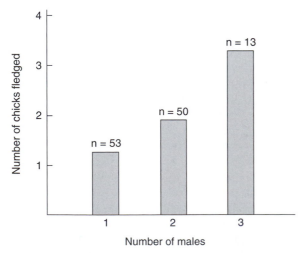

FIGURE 15.16 In spotted sandpipers, female reproductive success, measured in terms of the number of chicks fledged, increases with the number of males monopolized by each female. (From Oring and Lank 1986.)

SUMMARY

Parental care is one component of the overall life history of a species. Because animals have limited time, energy, and resources to devote to reproduction, evolutionary "decisions" must be made about the amount of care and who should assume parental responsibilities. The issue of who provides the care appears to be related to factors such as mode of fertilization (male care being associated with external fertilization and female care with internal fertilization) and phylogenetic history (in mammals parental care usually rests with the female, whereas in birds parental duties are typically partitioned somewhat equally between the sexes).

Conflicts of interest are not restricted to parent-young interactions, and in fact such conflicts characterize most social behavior. In the case of mating relationships, males usually produce more offspring by seeking additional mates. In contrast, females tend to emphasize parental effort rather than mating effort and can usually produce more offspring by gaining male parental investment. The disparity between the sexes in parental investment interacts with ecological factors such as predation, resource quality and distribution, and the availability of receptive mates to shape the mating system of a species.

Mating systems are usually defined on the basis of one or more of the following factors: (1) the number of individuals that a male or female copulates with, (2) whether or not males and females form a bond and cooperate in the care of the offspring, and (3) the duration of the pair bond. We favor a system of classification based solely on the number of copulatory partners per individual. According to this system, polygyny occurs when some males copulate with more than one female during the breeding season; monogamy is a mating system in which a male and female have only a single mating partner per breeding season; and polyandry, the least common mating system, occurs when some females mate with more than one male during the breeding season.

Extra-pair matings are common regardless of the mating system. These have different costs and benefits for males and females and will affect the evolution of mating systems. Potential benefits to males include an increased number of offspring. Females gain material benefits, such as help from the extra-pair male in raising offspring or food, or genetic benefits, such as fertility insurance or high-quality genes.

Two important factors influence the evolution of mating systems, the need for biparental care and the distribution of females. When females are widely distributed, it is difficult for a male to monopolize more than one mate, and so monogamy is favored. Monogamy is also favored when the male and female can raise more offspring when both help to raise the young.

PART THREE

Behavior of Groups— Social Behavior

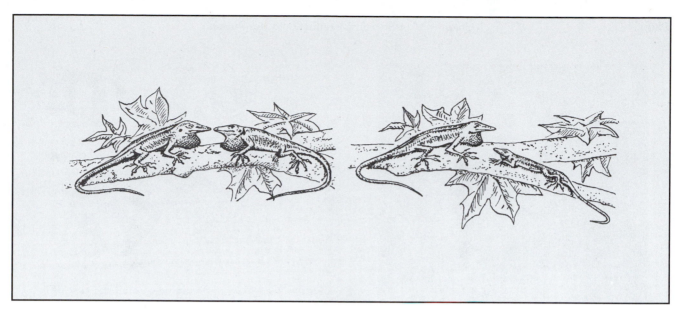

16

Sociality, Conflict, and Resolution

Tiny *Holocnemus pluchei* spiderlings hatch from the eggs that their mother has held close to her body for some time (Figure 16.1). For the next four or five days, the spiderlings remain on the web of their mother, but at the end of this grace period, each spiderling is faced with a choice—to build its own web or to move onto the web of a conspecific, typically that of a larger individual. (Although up to 15 conspecifics may share a single web, most groups contain just two individuals.) The choice is by no means an easy one—there are costs as well as benefits to group living in *H. pluchei* (Jakob

1991). Spiderlings that move onto a web that is already occupied catch fewer prey items than do spiderlings that opt for a solitary existence because the larger occupants of the web outcompete them. Some group-living spiderlings may actually be eaten by their "hosts."

Why do some spiderlings choose to share a web, given that they lose food (and possibly their lives), rather than constructing their own web? It seems that webs are quite costly structures to build. One estimate suggests that the average spiderling must forage for five days to recoup the energy used in web building (in the life of a spiderling, five days is a substantial amount of time). When spiderlings move onto a web that is already occupied, they make few or no home improvements. Thus, because they add little or no silk, they essentially dispense with the costs of web building. In short, group living in this species entails losing food but saving silk (Jakob 1991).

This brief sketch of the options available to *H. pluchei* spiderlings reveals that there are both pros and cons of living socially. In the section that follows we will consider the variety of costs and benefits experienced by group-living animals.

LIVING IN GROUPS

Jane Goodall, the famed African researcher, is reported to have said, "One chimpanzee is no chimpanzee at all."

FIGURE 16.1   *Holocnemus pluchei* female with eggs. Soon after hatching, spiderlings have the option of building their own web or moving onto the web of a conspecific. Although living socially saves the spiderling the time and energy needed to build a web, it also results in decreased food because of competition from the original owner.

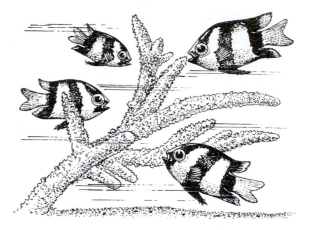

FIGURE 16.2   Group living may entail the cost of competing with other group members for food. Humbug damselfish, for example, live in groups amid staghorn corals off the Great Barrier Reef. In this species, the larger, more dominant group members gain first access to the best food floating by in the current. As a result of strategic positioning by larger group members, the growth of smaller individuals is often suppressed.

In her efforts to encourage research institutes to offer chimpanzees a more natural social environment, Goodall illustrates the fact that many animals are highly social, and into the very fabric of their lives is woven the need to be with others of their kind. In fact, if we think about animals in general, we are likely to come up with case after case of animals that live in groups.

We should quickly point out that not all group-living animals are technically social. For example, some arthropods tend to gather in places that have a certain moistness and temperature. They therefore incidentally end up in the same vicinity through what is called passive grouping. Other species, though, are actually social; that is, they seek out each other's company. Being social, however, has both costs and benefits (Alexander 1974).

COSTS OF SOCIALITY

The costs to individuals that live in groups may include increased competition for mates, nest sites, or food. Consider the case of the humbug damselfish, *Dascyllus aruanus* (Figure 16.2). This small fish lives in fairly stable social groups of about 5 to 35 individuals amid branching corals in lagoons off the Great Barrier Reef of Australia. The groups are organized into strict dominance hierarchies in which each individual has a rank. Each hierarchy is organized according to a simple rule—larger fish are always dominant over smaller fish—and this hierarchical organization affects patterns of feeding within groups. Humbugs feed on plankton

by orienting into the current and intercepting prey items that flow past their home coral. As it turns out, the larger (more dominant) fish within a group feed farther upstream than do the smaller fish and thereby gain first access to food arriving in the current. Smaller, lower-quality prey eventually reach the smaller group members farther back. This method of feeding within the group probably explains the suppressed growth observed for juveniles that live in groups with many large adults (Forrester 1991). We see, then, an example in which social living entails the cost of competing with other group members for food.

Other costs of social living include increased exposure to parasites or disease, increased conspicuousness due to the increased level of activity in groups, and a greater possibility of providing parental care to young that are not one's own or spending time trying to find one's own young from amid the clamoring throng. Mexican free-tailed bats (*Tadarida brasiliensis mexicana*), for example, live in aggregations, sometimes with millions of individuals, where mismatches between nursing mothers and their offspring sometimes occur (McCracken 1984). Although it may seem surprising that about 17% of mothers suckle young that are not their own, imagine the task of a mother when she returns to the cave after foraging for insects and tries to locate her own pup amid a dense mass of wriggling youngsters (Figure 16.3). Social living may also provide opportunities for exploitation of some individuals by others. In large groups, a male may have difficulty in guarding his mate from other males, and hence he may end up caring for another male's offspring. Females

FIGURE 16.3 A crèche of Mexican free-tailed bat young. Mothers returning from a night's foraging sometimes mistakenly nurse the offspring of another female. In this species, then, group living sometimes entails the cost of providing parental care to another group member's young.

may also be duped into caring for young that are not their own. Amid the seemingly frantic activity of a nesting bird colony, a female may fail to notice the arrival of another female at her nest and the quick deposit of an egg by the intruder. (For more extensive discussion of intraspecific brood parasitism, see Chapter 15.)

BENEFITS OF SOCIALITY

Obviously, since so many species are social, there must be certain distinct advantages in group living. Recall from our discussion of antipredator behavior in Chapter 13 that with membership in a group comes a variety of antipredator tactics that are not available to solitary individuals. Predators are usually less successful when hunting grouped rather than solitary prey because of the superior ability of groups to detect, confuse, and repel predators. Furthermore, during an attack by a predator, an individual within a group has a smaller chance of becoming the next victim.

Individuals living in groups may also have greater foraging success. In some species, animals in groups cooperatively hunt together, with increased success, on the average, for each individual. Cooperative hunting sometimes enables predators to kill animals larger than themselves. This seems to be the case for Harris's hawk, *Parabuteo unicinctus* (Bednarz 1988), a bird of prey that lives (and hunts) in family groups in the southwestern United States. During the early morning hours, family members typically gather at one perch site, from which the group then splits into smaller subgroups of from one to three individuals, each subgroup making alternate short flights throughout the family's home area. This type of movement has been described as "leapfrogging." Upon discovering a rabbit, one of three hunting tactics, or more typically a mixture of the

three, may be employed by the hawks. The most common, the surprise pounce, occurs when several hawks arrive from different directions and converge on a cottontail or jackrabbit unfortunate enough to be out in the open. Even if the rabbit escapes under vegetation, however, the safety may only be temporary. At this point, the hawks employ their flush-and-ambush tactic, a strategy in which one or two hawks flush the rabbit from the cover, and then family members perching nearby pounce on it. Relay attack is the third, and least common, hunting tactic used by family members. Here, they engage in a constant chase of a rabbit, with the position of the lead bird changing each time the lead attempts a kill and misses. Once killed, the prey is shared by all members of the hunting party.

Using this combination of hunting techniques, Harris's hawks probably confuse and exhaust their prey. In fact, success in hunting rabbits is correlated with group size—hunting parties of five to six individuals do better than smaller parties (Figure 16.4*a*). Thus the average energy intake available per individual from rabbit kills is also higher in groups of five or six members than in smaller groups (Figure 16.4*b*). As noted previously, cooperative hunting makes possible the killing of prey that may be much larger than the hawks themselves. Jackrabbits, for instance, are two to three times heavier than female and male hawks, respectively.

Unlike the situation for Harris's hawk, in which members of hunting parties participate in all aspects of foraging, from searching for prey to eating it, ants of the species *Formica schaufussi* forage individually but form groups to bring oversized prey back to the nest. Once a scout has found a relatively large food item (typically a dead insect), it then lays a chemical trail back to the nest. Recruits eventually assemble at the item and then either drag or carry it back. In this way,

FIGURE 16.4   In Harris's hawks, members of family groups engage in cooperative hunting. Groups of five or six individuals are most successful at killing cottontails and jackrabbits (*a*) and individuals in such groups have a higher average energy intake than individuals in smaller groups (*b*). (From Bednarz 1988.)

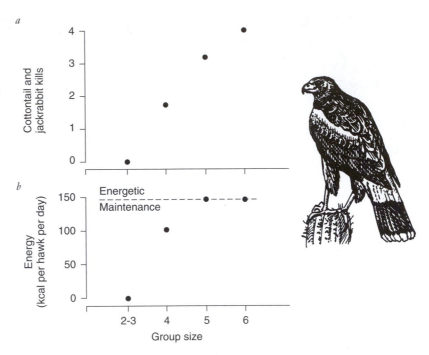

F. schaufussi can include a larger range of prey items in its diet. The retrieval groups are also likely to be able to defend the prey against other ant species in the area. In staged contests in the field, larger retrieval groups of *F. schaufussi* were more successful in securing prey from their competitor *Myrmica americana* than were smaller groups (Figure 16.5). Here, then, groups are effective not only in prey retrieval but also in defense of the prey (Traniello and Beshers 1991).

Another species in which more than one factor influences the formation of groups is the lion (*Panthera leo*). Although ideas on patterns of grouping in lions initially focused on hunting success (Figure 16.6), more recent data suggest that this factor alone cannot explain the formation of prides in this species (Packer 1986; Packer, Scheel, and Pusey 1990). Scavenging may be a more efficient means of obtaining food than hunting, and under conditions such as these, groups may be necessary to defend carcasses against lions from other prides. Group living also appears to be important in protecting cubs from nomadic males that commit infanticide and in the defense of the pride's home area against intrusion by neighboring prides. Thus, for lions, the benefits of group living include the defense of food, young, and living areas against conspecifics.

Since there are costs, as well as benefits, in group living, we might expect that some social interactions might involve conflict. Individuals would seek to maximize their benefits and minimize their costs. Conflicts are resolved through agonistic behavior, which involves aggressive behaviors such as threats, chasing, and fighting, as well as submissive behaviors. In this chapter, we will consider conflict among animals and its resolution. Two ways in which conflict in a stable group is minimized while animals compete for limited resources are dominance hierarchies and territoriality.

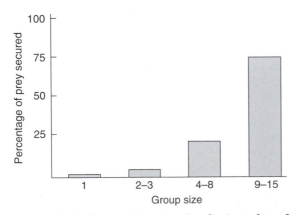

FIGURE 16.5   In the ant *Formica schaufussi*, workers forage individually but retrieve heavy prey items in groups. Such group retrieval is important in the defense of prey against workers of other ant species. Here, in staged encounters with a competitor, *Myrmica americana*, larger retrieval groups of *F. schaufussi* are more successful than smaller groups at securing prey. (Data from Traniello and Beshers, 1991.)

DOMINANCE HIERARCHIES

Dominance refers to the ability of one member of a group to assert itself over others in acquiring access to a resource, such as food, a mate, a display or nesting site, or any other limited factor that would increase fitness (E. O. Wilson 1975). When the social group is stable, the same two individuals are likely to repeatedly

FIGURE 16.6 A group of lions is sharing a wildebeest kill. Studies of group living in lions have often focused on hunting success. As it turns out, some of the major functions of grouping relate to the defense of cubs, space, and scavenged food. (Photo from Willy Bemis.)

encounter one another in a competitive situation. In such cases, the animals don't usually fight each time they meet. Instead, relationships develop among them so that a submissive animal predictably yields to the dominant one (Kaufmann 1983). Confrontations often involve nothing more than display.

Once stable dominance relationships have formed in a group, its members minimize the costs of aggression by limiting interactions to threats, as opposed to fights. By reducing the chance of injury during fighting, a dominance relationship may benefit both the dominant and the submissive animal. When a subordinate learns not to challenge a dominant animal for a resource, it no longer risks injury in contests that would probably be lost anyway, and a dominant individual doesn't have to waste energy and risk injury by repeatedly establishing superiority each time it encounters a subordinate (Bernstein 1981; T. E. Rowell 1974). For example, in one chimpanzee colony, damaging fights occurred five times more often when male ranks were unstable. Group stability can translate into enhanced reproductive success for each individual. This can be observed in flocks of domestic chickens. When the dominance hierarchy is stable, hens fight less and lay more eggs than when dominance is being established (Pusey and Packer 1997).

The form of the dominance hierarchy can vary among species, and within a species it can vary with conditions and over time. The simplest form of a dominance hierarchy is despotism, in which one individual rules over all others in the group and the subordinates are equal in rank. Triangular relationships may form in other species. For example, A may dominate both B and C, but B may also dominate C. Hierarchies may also be

linear. In this case, A is dominant over all other animals, B is dominant over all but A, and C is subordinate to A and B but dominant over the rest. Such a dominance hierarchy might be represented as follows:

$$A \longrightarrow B \longrightarrow C \longrightarrow D \longrightarrow E \longrightarrow F \longrightarrow G \longrightarrow H \longrightarrow I$$

This is often called a pecking order because it was first described in chickens, a species in which dominance is commonly shown by pecking lower-ranking animals.

In some species, dominance may be earned through individual efforts. Dominance or submissiveness is often learned from the outcome of one or a few initial interactions. For example, if several unfamiliar domestic hens are housed together, there will at first be frequent fights over resources such as food. As time passes, however, the amount of fighting decreases. Each hen remembers the outcome of encounters with every other hen. Based on the results of previous interactions, a hen will either defer or be assertive during a subsequent encounter with a particular individual. Gradually, a linear dominance hierarchy develops (Schjelderup-Ebbe 1922).

There are also instances in which dominance is attained through an association with a high-ranking individual. For example, when two flocks of dark-eyed juncos merge, all the birds of one flock tend to rank above those of the other. It is as if the subordinate birds ride the coattails of the highest-ranking bird to achieve dominance in the combined flock. A possible explanation for the subordinates' rise in dominance is that the highest-ranking individual behaves differently toward birds that are familiar, because they were members of its original flock, than toward unfamiliar flockmates (Cristol 1995).

In other species, dominance may be a birthright based on the status of one's parents. This is the case among rhesus monkeys (*Macaca mulatta*). Adult females have a linear dominance hierarchy, and offspring assume a dominance position just below their mother (de Waal 1991).

DETERMINANTS OF DOMINANCE

When dominance relationships are settled through individual efforts rather than parental rank, several factors can affect the outcome of the contests. An important factor is each competitor's ability to win a fight—its resource holding potential (discussed shortly)—which is determined by characteristics including size, strength, weapons, and fighting ability. As we will see later in this chapter, the battle may never be fought or may be short-lived. If the weaker contestant assesses its competitor to be superior, it usually concedes without risking injury by continuing a fight in vain.

After the initial contest, however, the experience of either winning or losing can influence the outcome of subsequent encounters. Winners usually win again, and losers generally lose. These effects have been shown across a wide variety of species, including crickets (Alexander 1961), molluscs, fish, birds, and rodents (Chase, Bartolomeo, and Dugatkin 1994).

Both winner and loser effects can be seen among nestling blue-footed boobies, *Sula nebouxii* (Figure 16.7). Each brood typically has two eggs. The chicks hatch over a four-day period from eggs of similar size and mass. The older one becomes more aggressive than the younger one and threatens or attacks it every day during the three- to four-month nestling period. The younger chick responds with submissive gestures and

rarely challenges its older sibling. These observations led to the hypotheses that the younger nestmate's early experience of losing increases its tendency to be submissive and that the older nestmate's experience of winning increases its tendency to be aggressive. The hypotheses were tested by separating the siblings when they were two to three weeks old and permanently pairing each with a singleton, a chick that had been raised naturally without a nestmate. The singleton had no previous experience with conflict, so it had never won or lost. During the first four hours of pairing, the results were as expected. The dominant chicks were six times more aggressive than their new, inexperienced nestmates. Submissive chicks were seven times less aggressive than their new nestmates.

Although all of the previously subordinate chicks remained so over the next six days, half of the previously dominant chicks lost their dominant roles. It is thought that during this six-day period, the initially submissive chicks continued to behave submissively toward their new nestmates. The newer nestmates apparently responded to their submissive gestures and behaved aggressively. The previously dominant chicks experienced stimuli quite opposite to those experienced in their home nests: the new nestmates behaved aggressively instead of submissively. Thus, the previously dominant chicks' aggressiveness was eroded by the hostility of the new nestmates (Drummond and Canales 1998).

Loser effects can be quite long-lasting, which Gordon Schuett (1997) demonstrated in male copperheads (*Agkistrodon contortrix*). In the initial experiment, a female snake and two males with no fighting experience during the previous 6 to 12 months were placed in a larger arena. One of the males was 8% to 10% longer than the other, measured as the distance between the snout and the vent, and had greater body mass. In all

FIGURE 16.7 Nestling blue-footed boobies exhibit both winner and loser effects in the contests for dominance. An animal's prior experience in winning or losing can affect the outcome of future contests. Winners generally continue to win, and losers usually remain losers. Loser effects often last longer than winner effects.

cases, the larger male won the fight and the female. When the pairs were rematched 24 hours later, prior losers gave up without even challenging the competitor. The next day, prior losers were paired with unknown males that closely matched their own length. This time the losers did give some challenge displays but significantly fewer than their opponent. No fighting occurred. Prior losers just gave up and were chased away. It seems, then, that for male copperheads, the phrase "once a loser, always a loser" rings true.

It is likely that an individual will have multiple fights, and it might win some and lose some. In several species, the loser effect lasts longer than the winner effect. One explanation for this might be that whereas a poor fighter always benefits by learning this and adjusting its fighting strategy accordingly, the winner of one conflict does not always gain from aggressively challenging its next competitor. But, regardless of the reasons for winner and loser effects, their existence means that an individual's position in the dominance hierarchy depends on the order of the opponents it encounters, as well as its fighting ability (Hsu and Wolf 1999).

BENEFITS OF DOMINANCE

We will focus on two benefits of dominance: enhanced reproductive success and increased access to food.

Enhanced Reproductive Success

In many social groups with dominance hierarchies, the dominant female (and in some cases the dominant male) suppresses reproduction by other members of the group. The most dramatic examples of reproductive suppression occur in the eusocial species, such as the social insects and certain mole rats, in which a single female, the queen, reproduces. All other females are workers and generally do little, if any, reproduction. (Eusociality and cooperative breeding are discussed further in Chapter 17.)

More commonly, however, the dominant animals have a clear reproductive advantage, but they don't completely suppress reproduction by subordinates. For example, each pack of African wild dogs (*Lycaon pictus*) has a clear dominance hierarchy in each sex. In one study, it was found that the top-ranking alpha females in the Selous Game Reserve produced 76% of all litters, and those in Kruger National Park produced 81% of the litters. Furthermore, although 82% of the dominant females gave birth each year, only 6% to 17% of the subordinate females did so (Creel et al. 1997). A model for the evolution of the reproductive suppression in mammals is presented in Wolff (1997), and models for the evolution of the reproductive advantage enjoyed by dominant animals can be found in Reeve and Keller (1995) and in Cant (1998).

In many social species, dominant males mate more frequently than subordinates. For example, among red deer (*Cervus elaphus*), a single master stag holds a harem of females, and all other males are subordinate. The deer in one herd were observed daily, and records were kept of the date of each mating and the individuals involved. Subordinate males managed to mate before and after the main rutting season, but dominant males didn't waste time or energy mating during the off season. During prime time, however, the dominant males did most of the mating. When calves were born, the mean gestational time for red deer was used to calculate the most likely date of conception. No calves were conceived during the out-of-rutting season. The father of the calf was identified as the stag who mated with that female on the date closest to the calculated conception date. The master stag did, indeed, sire most of the calves born to the females in his harem (Bartos and Perner 1998).

A reproductive advantage to dominance has been experimentally demonstrated in laboratory mice. Each group in the study was made up of three males and three females. The males were permitted to fight for a day or so, forming rigid hierarchies. The members of each group were then permitted to mate. Paternity could be determined because each male could be identified by genetic markers. In 18 of the 22 groups, the dominant male fathered all the litters. In 3 groups, a subordinate male managed to father one litter, and a subordinate male in 1 group managed to sire two litters. Although dominant males represented only a third of the population, they managed to father 92% of the offspring (DeFries and McClearn 1970).

Although there are many examples in which high rank is associated with increased reproductive success, this is not always the case. Alternative reproductive strategies employed by subordinate males can reduce the reproductive success of the dominant male. Among Rocky Mountain bighorn sheep, for instance, subordinate rams occasionally fight a dominant ram for a few seconds of access to a female—all that is necessary to copulate. When lambs were born, the male most likely to be the father was identified by using a combination of genetic and behavioral data, as well as a model of bighorn reproductive competition. It was estimated that challenging males fathered 44% of the 142 lambs born in two natural populations (Hogg and Forbes 1997).

Other examples of alternative mating strategies, satellites and sneaks, are not as risky as that of the subordinate bighorn rams. Sneaky males generally mimic female characteristics and sneak past a dominant male to copulate with his female. Satellite males generally position themselves so that they can intercept females who are attracted to a dominant male and copulate with them. We describe examples of these alternative reproductive strategies elsewhere in this book: sneaky male side-blotched lizards (Chapter 4), sneaky male plainfin midshipman fish (Chapter 7), and satellite male natterjack toads (Chapter 14). In all cases, the underlying strategy is to avoid the costs of achieving and maintaining dominance and still enjoy some reproductive success. Consequently, the reproductive success of the dominant male may be diminished.

The reproductive success of a dominant male may also be lessened by coalitions. In this case, two or more low-ranking males may form a coalition and successfully challenge a higher-ranking male for possession of a female. For example, in a troop of savanna baboons with eight adult males, the three lowest-ranking males regularly formed alliances to oppose a single higher-ranking male. The alliances won the female on 18 of 28 attempts (Noë and Sluijter 1990). Other examples of cooperation in acquiring a mate are described in Chapter 17.

Increased Access to Food

Dominants often get more food than subordinates. For example, this occurs in captive rhesus monkeys, female woodland caribou (*Rangifer tarandus*), female bison (*Bison bison*), and female mountain goats (*Oreamnos americanus*) (Fournier and Festa-Bianchet 1995). Besides increased reproductive success, food is an important benefit of dominance among brown hyenas (*Hyaena brunnea*) of the central Kalahari (Figure 16.8). Each sex has a clear linear dominance hierarchy, and the male and female at the top have equal rank. Although brown hyenas live in a clan, they forage alone. During the rainy season, the primary component of their diet is the remains of kills, such as giraffe, gemsbok, and wildebeest, made by other predators. As many as six hyenas may arrive at a carcass together, but only one or two will feed together. The top-ranking animals of the clan have more feeding time at carcasses than subordinates. In addition, subordinate males and females are significantly more likely to leave the carcass without feeding if a dominant animal is present (Owens and Owens 1996).

In wood pigeons (*Columba palumbus*), the dominant birds feed in the center of the flock, where it is safer. As a result, the dominant animals eat more food than do subordinates. In fact, subordinates often gather only enough food to last them the night. They could starve if energy demands increase because of a drop in the temperature or if they are prevented from foraging the next day by bad weather (Murton, Isaacson, and Westwood 1966).

WHY BE SUBORDINATE?

With the benefits of a dominance hierarchy clearly stacked on the side of the high-ranking members, we may wonder why subordinates stay in the group. The answer is that they stay because it is the best option available—it minimizes their losses. If they persisted in contesting their rank or competing for every resource, they would risk injury in conflicts they would most likely lose (Fournier and Festa-Bianchet 1995).

Subordinates could leave and join another group. Indeed, some do, although this, too, is a risky option. For example, although subordinate red foxes (*Vulpes vulpes*) have little hope of living long enough to become dominant in their natal group, the mortality rate of those who disperse is also very high (Baker et al. 1998). So, although the benefits of group living may not be as great for a subordinate animal as they are for a dominant one, they may still outweigh the costs of leaving the group. Besides, things could get better. The domi-

FIGURE 16.8 Top-ranking brown hyenas enjoy two important benefits of dominance—enhanced reproductive success and increased access to food. Dominant hyenas have more feeding time at carcasses than subordinates.

nant animal could die or become weak enough to be displaced by a subordinate. In the meantime, since many groups consist of family members, a subordinate animal may gain some fitness through kin selection by helping to raise its siblings. (A more detailed discussion of helping is found in Chapter 17.)

TERRITORIALITY

Another way in which animals compete for resources is by defending the area that contains the resource. This area is called a territory. We will now consider territoriality and its evolution.

HOME RANGES, CORE AREAS, AND TERRITORIES

We begin with some definitions. The home range of an individual animal is that area in which it carries out its normal activities. Within the home range there is often an area in which most activities are concentrated—the core area. In some cases, the core area may be the area immediately surrounding the nest site or perhaps a food or water source. Although the home ranges of individuals may overlap, this is less likely in core areas. If overlap of the core area, or even the entire home range, is prevented by agonistic behavior by the resident(s)—individuals, pairs, or groups may actively defend a given area—then the defended area is called a territory (Noble 1939). Although this definition of territory emphasizes active defense of an area, other definitions downplay defense and emphasize instead the exclusive use of space (Schoener 1968). This latter definition is often more practical for describing the space use patterns of animals whose territorial behavior is difficult to observe in the field. Because of the secretive habits of many small mammals, for example, it is virtually impossible to state with any certainty that the exclusive use of an area is maintained by active defense (Ostfeld 1990). A survey of the literature on territoriality revealed no fewer than 48 different definitions. The most common one, "a defended area," was used in only 50% of the papers. Since the definition affects the type of data collected, the lack of consistent terminology can make comparisons among studies difficult (Maher and Lott 1995). Keeping this problem in mind, we will forge ahead.

Territories may have different uses, depending on the resource being contested. They may be used solely for feeding, mating (recall our discussion of leks in Chapter 15), or raising young, or they may be used for a variety of purposes, in which case they are called multipurpose territories.

COSTS AND BENEFITS OF TERRITORIALITY

Both costs and benefits are associated with the defense of an area (J. L. Brown 1964). Depending on the type of territory maintained, some benefits might include (1) a longer-lasting food supply, given that fewer individuals have access to it; (2) increased access to mates, if individuals of the opposite sex prefer to mate with territory holders; or (3) a greater opportunity to rear offspring at a high-quality site. On the other hand, acquiring or defending a territory can be downright costly when one considers the energy needed to patrol boundaries, to display to potential intruders, and to forcibly evict trespassers or a previous owner. Territory acquisition and defense also take time away from other essential activities such as foraging.

Because feeding and territorial behavior cannot be performed simultaneously, great tits (*Parus major*), small perching birds that live in England and feed on seeds and insects, face a dilemma. In this species, activities related to feeding and territorial defense occur in different microhabitats and are completely incompatible (Ydenberg and Krebs 1987). Whereas feeding occurs within 3 meters of the ground, territory defense, particularly the singing component, takes place high in trees, about 10 meters above the ground. Obviously, then, tradeoffs are involved in territorial behavior: Time spent in defense cannot be spent foraging.

How do great tits "decide" how to allocate time between territorial defense and feeding? An important factor seems to be food availability. In one field experiment, supplemental food was placed on tables and provided to some males in their territories; these males were designated "owners." Owners, however, were not the only birds to make use of the extra food. Males in neighboring territories also visited the food tables and were called "visitors." A third group of males—control males—did not visit the feeding tables. After five days of supplemental feeding in certain territories, each male was exposed to a single simulated intrusion that consisted of exposure to a stuffed male great tit mounted on a stake. At the same time that the stuffed male was placed on each male's territory, one minute of a taped great tit song was played in the background. The responses of the three types of males were scored. Males that had access to supplemental food (both owners and visitors) responded more strongly to the stuffed intruder than did control males (Figure 16.9). Owners and visitors exhibited shorter latencies in approaching the stuffed intruder, spent more time in its vicinity, and were more likely to attack. Control males, on the other hand, were more likely to respond to the intruder with a song, a less energetically expensive display than outright approach and attack. These results suggest that access to the supplementary food allowed owners and

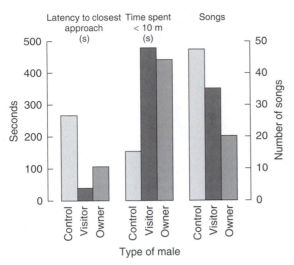

FIGURE 16.9   Great tits, like many animals, cannot feed and defend their territory at the same time. Thus, there are tradeoffs between foraging and territorial defense. One factor that seems to influence the allocation of time between the two activities is food availability. Here, male great tits that had access to supplemental food (owners and visitors; see text for details) responded more strongly to a stuffed "intruder" than did control males that did not make use of the extra food. Although owners and visitors engaged in energetically expensive methods of territorial defense (approach and attack), control males engaged in less expensive displays such as singing. Thus, access to high levels of food appeared to allow males to devote more time and energy to territory defense. (From Ydenberg and Krebs 1987.)

visitors to meet their food requirements rapidly and thereby to spend more time in territorial activity, particularly in energetically expensive activity (Ydenberg 1984). That visitors were similar to owners in their level of response to the intruder suggests that the response was not simply a result of birds defending a rich food resource on their own territories.

ECONOMICS AND TERRITORIALITY

We would predict, then, that territoriality will occur only if the benefits from enhanced access to the resource are greater than the cost of defending the resource—that is, when the territory is economically defendable (J. L. Brown 1964). A net benefit of territoriality translates into increased fitness for the individual territory holder. Factors such as abundance, spatial distribution, and renewability are thought to affect whether or not a particular resource is economically defendable (Davies and Houston 1984). Generally speaking, territoriality is favored when resources are moderately abundant. If the resource is scarce, an individual may not gain enough to pay the defense bill, and it may be economically wiser to look for greener pas-

tures. Accordingly, the golden-winged sunbird (*Nectarinia reichenowi*) will abandon a territory when it no longer contains enough food to meet the energy costs of daily activities, as well as defense (Gill and Wolf 1975). On the other hand, if there is more than enough of the resource to go around, it would be energetically wasteful and economically unsound to defend it. Water striders (*Gerris remiges*) are among the species that will cease to defend territories if supplied with abundant food (Wilcox and Ruckdeschel 1982). Also, female marine iguanas do not bother defending territories with nest sites on most of the Galápagos Islands. They defend territories only on Hood Island, the only Galápagos island where nest sites are in short supply (Eibl-Eibesfeldt 1966).

An important factor in determining whether it is economically sensible to defend a resource is its distribution in space and time. Generally, animals are territorial if the resource occurs in patches or clumps. A pile of food, for instance, is easier to defend than food that is spread thinly over a large expanse, as long as the number of competitors anxious to contest ownership is not too great. Thus we find that male Everglades' pygmy sunfish defend territories near a clump of prey in an aquarium but are not territorial when prey are evenly dispersed (Rubenstein 1981).

The degree of competition for the resource also influences its economic defendability. The more competitors, the greater the cost of defense. Male fruit flies (*Drosophila melanogaster*) defend small areas of food that are suitable for oviposition, particularly if there are females in the vicinity. As would be predicted, however, the incidence of territoriality is reduced at higher densities of males (Hoffmann and Cacoyianni 1990).

Another example is the ayu (*Plecoglossus altivelis*). These are fish that graze on algae attached to the substrate. Some fish defend territories, and others, called floaters, intrude into the school to feed on the algae. When the number of floaters was experimentally increased in a tank in which the natural habitat was simulated, the size of the territories decreased (Iguchi and Hino 1996).

If animals are territorial when the benefits exceed costs, individuals with high-quality territories should enjoy greater fitness than those with poor-quality territories or no territory at all. This prediction is confirmed among male tree-dwelling iguanas (*Liolaemus tenuis*). Larger males defend larger trees, which are occupied by more females than are small trees. Thus, small males are usually monogamous and larger ones are polygamous. In other words, the size of a male's harem, and therefore his fitness, depends on his size and the size of his tree (Manzure and Fuentes 1979).

It is generally assumed that both the benefits and the costs increase with territory size. A larger territory might be expected to provide more food, for instance,

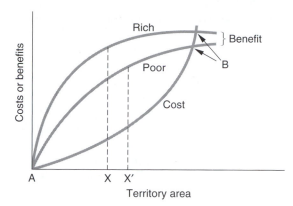

FIGURE 16.10   The hypothetical relationship between the costs and benefits of territoriality. Both costs and benefits increase with territory size. The shape of the benefits curve varies with the quality of the territory. It is profitable to defend the territory as long as the benefits exceed the costs, between points A and B. If the animal is to maximize its net gain, the optimal territory size is at point X or X′, depending on the exact placement of the cost and benefit curves.

but an animal would have to use more energy to patrol the increased border. Therefore, we might expect animals to adopt a territory size that would give them the greatest profit. However, there is no universally perfect territory size. The exact shape of the benefit and cost curves and their relative placement will vary among species and environments (Figure 16.10). The precise placements of these curves may alter or even reverse predictions of optimal territory size (Schoener 1983).

Many studies that apply optimality theory to territoriality concentrate on feeding territories and measure costs and benefits as energy gains and expenditures. We might imagine at least two fundamental strategies. The first would be to spend the least amount of time possible to gain the bare minimum of necessary energy. This would be optimal, for example, if the risk of predation were high. In this case, it would be better to gather just enough food to get by and return quickly to a safe location. However, some other activity, courting or nest building, for example, might be more important than eating. The second strategy would be to eat to maximize energy gains. For instance, it may be important to put on weight for migration or before hibernation.

We might predict, therefore, that in some species the size of an individual's territory would be adjusted to maximize energy gains. At least some individual rufous hummingbirds (*Selasphorus rufus*) appear to do this. During their southward migration, these birds pause for a few days in the mountain meadows of California to build the fat reserves needed to fuel the next leg of the journey. During this interval they feed on the nectar of the flowers of the Indian paintbrush. Each bird defends a group of flowers as a territory. The territory

size and weight gain for a single individual are shown in Figure 16.11. As you can see, this bird adjusted the size of its territory so that it could gain weight as quickly as possible. This individual began with a small territory that contained few flowers, so its weight gain was minimal. It increased the territory size greatly on the third day. This territory had more flowers and the bird could obtain more energy, but it had to invest more energy in defense. Nonetheless, it gained somewhat more weight than possible on the smaller territory. Next, however, it reduced the size of its territory slightly on the fourth and fifth days, thereby cutting the energy costs of defense and maximizing its weight gain (Carpenter, Paton, and Hixon 1983).

GAME THEORY AND SINGLE CONFLICT ENCOUNTERS

So far we have considered ways in which animals resolve conflicts over limited resources when there are likely to be repeated contests between the same individuals. Now we will consider the factors that influence the outcome of a single encounter between individuals that are competing for the same resource.

In conflicts, the reproductive success gained by each contestant by behaving in a certain way depends

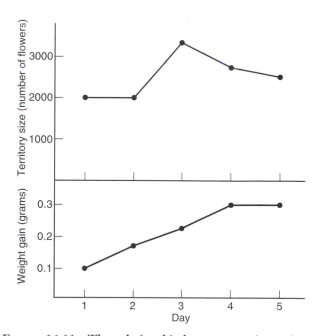

FIGURE 16.11   The relationship between territory size and weight gain for one rufous hummingbird. It is important for these birds to gain weight maximally during their stopovers along their migratory route. These data indicate the weight gained by a single territorial bird on five successive days. This individual adjusted its territory size so that its weight gain was maximal. (From Carpenter, Paton, and Hixon 1983.)

largely on its competitors' actions. Therefore, we need more information than just the costs and benefits to predict the evolutionarily stable strategy. Recall from Chapter 4 that an evolutionarily stable strategy (ESS) is one that cannot be bettered (in terms of fitness) by any other strategy when most members of the population have adopted it.

Game theory is a tool for analyzing such situations. It treats members of a population as opponents in a game. (The mathematical basis of game theory is similar to that underlying chess or checkers.) Game theory models require that we take into account (1) all the existing strategies, (2) the frequency of each strategy in the population, and (3) the fitness payoff for each possible combination of strategies. This information is often presented in a matrix that indicates the payoff for each possible interaction. In this matrix, the payoff is +1 if you win, −1 if you lose, and 0 if you tie. Similar payoff matrices can be created to predict ESSs.

HAWKS AND DOVES

We begin by considering aggressive games in which only two strategies might be employed. The "hawk" strategy is to fight to win. A hawk will continue to fight until either it is seriously injured or its opponent retreats. The "dove" strategy (originally called "mouse") is to display but never engage in a serious battle. If a dove is attacked, it retreats before it can be seriously injured.

We can calculate the costs and benefits of each strategy, keeping in mind that the realized payoff will depend on the opponent's strategy. The expected payoffs are shown in Table 16.1. The possible benefit in all matches is the value of the resource to the victor. In a contest between two hawks, the cost is the injury that might result. The net benefit would be the value of the resource less the cost of injury. It is assumed that all hawks have an equal chance of winning. A hawk that is fighting another hawk will win half of its battles and be injured in the other half. Thus, when two hawks square off, the payoff is one-half of the net benefit. If a hawk meets a dove, the hawk always wins. Since a dove

immediately retreats, it gains nothing but it loses nothing. On the other hand, the hawk in a hawk-dove contest gets the full value of the resource at no cost. In a contest between two doves, the cost is the time spent displaying. Although the price of display is generally considerably less than the cost of a wound, both the winning and the losing doves must make this investment. It is assumed that all doves have an equal chance of winning. The benefit, therefore, is one-half the value of the resource. The payoff in a dove-dove contest is one-half the value of the resource minus the cost of displaying.

Dove-Dove Interactions— The War of Attrition

A dove, we have seen, displays, but it retreats if attacked. What happens, then, if the opponent is another dove? There is no attack, and no one retreats. When two equally matched doves confront each other, the winner of this so-called war of attrition is the one that displays longer. It seems obvious that this would result in a mixed ESS of varying display times because a pure ESS of a fixed display time could always be bettered by the strategy of a slightly longer display. The stable strategy is to be unpredictable, to display for a randomly chosen length of time. Since the cost of display increases with duration, shorter displays should be more common than longer ones (Maynard Smith 1974).

Conflicts Involving Hawks

When the cost of injury is greater than the value of the resource, as is usually the case, the ESS will always be a mixture of hawk and dove strategies. This is because the success of either strategy depends not only on its payoff but also on the chances that an opponent will be playing the same or the alternate strategy. The greatest possible payoff, the full value of the resource without cost, goes to a hawk when it meets a dove. Therefore, as long as the population is mostly doves, a hawk fares quite well. As a result, hawks become more common and spread through a population of doves, making

TABLE 16.1   Payoff Matrix for Hawk-Dove Game

Payoff Received by	Opponent	
	Hawk	Dove
Hawk	$\dfrac{\text{Value of resource} - \text{Cost of injury}}{2}$	Value of resource
Dove	0 (Gains nothing, loses nothing)	$\dfrac{\text{Value of resource}}{2} - \text{Cost of displaying}$

encounters between two hawks more likely. However, since the cost of injury is greater than the value of the resource, hawks experience a decrease in fitness in hawk-hawk encounters. The fitness gained by a dove that is fighting another dove is meager compared to the fruits of victory enjoyed by a hawk that is conquering a dove, but at least there is no chance of *losing* fitness. Doves could, therefore, invade and spread through a population of hawks. Thus, we see that neither the dove nor the hawk strategy is a pure ESS. Since the fitness gain of each strategy is frequency-dependent, the ratio of strategies would fluctuate until a point was reached at which the fitness of the hawk and of the dove was equal. This mixture would be evolutionarily stable. The precise mixture of hawk and dove strategies that would confer equal fitness to each would depend on the relative value of costs and benefits (Maynard Smith and Price 1973).

Factors That Affect Strategy Success

Cost　As the cost of fighting increases, the dove strategy should be favored. The potential cost of injury is greater in species that have weapons, such as horns or sharp teeth. Therefore, the hawk-dove model predicts that contests between members of well-armed species should rarely escalate from display to battle (Figure 16.12). Conversely, the hawk strategy should be more common when the risk of injury is low. Accordingly, animals without weapons, such as toads (*Bufo bufo*), fight fiercely (Figure 16.13) (Davies and Halliday 1978). Although some individuals are injured, the risk of injury is much less than it would be if they had the prowess of lions.

Resource Value　It seems reasonable that the intensity of fighting should increase with the value of the resource. After all, an animal that fights harder to win a richer prize will enjoy the same net benefit (value of resources minus cost of injury). This prediction seems intuitively obvious from human experience. You would probably put up more of a struggle if a mugger were stealing your life savings than you would to save a pocketful of change.

We see this prediction fulfilled in many animal encounters as well. Offspring, a direct measure of fitness, are obviously of great value. It is not surprising, then, that a mother will fight fiercely to defend her young. It would also be expected that the most intense fights among males for access to females should take place when females are most likely to conceive. Indeed, red deer stags (*C. elaphus*) fight most fiercely and, as a result, are wounded most frequently during the period when most calves are conceived (Figure 16.14) (Clutton-Brock, Guiness, and Albon 1982). Territories also differ in value, and it would be expected that animals would

FIGURE 16.12　Male rattlesnakes that are fighting. Game theory predicts that fights are not likely to escalate when the cost of injury is great. Each of these males could kill the other, but they do not bite. The males press against one another, belly to belly. Finally, the weaker individual yields, and his head is pushed to the ground by the stronger animal.

FIGURE 16.13 Two bullfrogs that are fighting. Game theory predicts that fights are more likely to escalate when the costs (risk of injury) are low.

fight more intensely when the stakes are high. For example, among grassland populations of funnel-web spiders (*Agelenopsis aperta*), females expend more energy and risk greater injury in disputes over high-quality web sites (Riechert 1986b).

As the value of the resource increases, we would expect more hawklike behavior. The hawk strategy can even become a pure ESS if the value of the resource exceeds the cost of injury. Furthermore, when the value

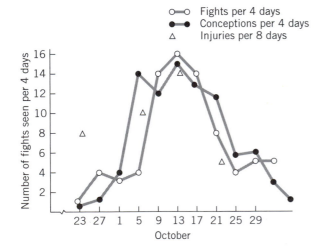

FIGURE 16.14 The number of fights among male red deer (open circles) are compared to the number of conceptions during the rutting season. The number of fights and conceptions are indicated for four-day intervals. The number of injuries per eight-day interval during the rut is indicated by triangles. The number of fights and injuries peak during the interval when conception is most likely. As predicted by game theory, male red deer fight harder when the value of the resource is greater. (From Clutton-Brock, Guiness, and Albon 1982.)

of victory is especially high, fatal fighting may evolve. To understand more completely why an animal might fight to the death over a resource, we must consider the lifetime consequences of a battle. What can be gained by retreating? The answer is one's future. In a sense, natural selection weighs the value of the resource against the value of the future. If the value of the future is close to or less than the value of the resource, the opponents have little to gain by giving up the fight. Thus, extreme hawklike strategies may evolve. That is, when the value of the future is close to zero, the contestants have nothing to gain by retreating and will fight to the death. It should not be surprising, then, that fighting that leads to severe injury or death is usually seen when a major chunk of the contestants' lifetime reproductive success is at stake (Enquist and Leimar 1990).

Male elephant seals (*Mirounga angustirostris* and *M. leonina*), for instance, fight brutal and often bloody battles for the right to mate (Figure 16.15). Adult males bear many scars that serve as silent testimony to the intensity of combat. After a fight, when an adversary enters the sea, the water is often reddened with blood. The rivalry is intensified by the females. A female protests loudly when a male tries to copulate (Cox and LeBoeuf 1977). This attracts the attention of other males, who attempt to interfere. As a result, only the largest and strongest males are able to mate. (The benefit to the female is obvious.) All matings are performed by a few dominant males who defend harems of females. The duels between males are so strenuous that a male can usually be harem master for only a year or two before he dies. Thus, battles are titanic because a male's entire reproductive success is at stake (LeBoeuf 1974; McCann 1981).

FIGURE 16.15 Male elephant seals are scarred and bloody from battle. Fights between bulls are brutal and often result in injury. Such titanic battles are predicted by game theory when the value of the resource is very high. Only the most dominant bull will be harem master and leave offspring.

ASYMMETRIES AND CONDITIONAL STRATEGIES

In the hawk-dove model, the strategies are played randomly with some frequency. It assumes that all hawks are equally matched competitors, as are all doves. But how often is this actually the case? In real life, rivals are rarely true equals. Instead, they generally vary in some quality. Contests are, therefore, usually asymmetric.

Given that life is not fair and inequalities do exist, it seems reasonable that a competitor might be more successful if its strategy were adjusted to the situation—for instance, if a male played hawk to a puny opponent and dove to a mammoth rival. When a strategy is adjusted to the circumstances, it is called a conditional strategy. As might be expected, conditional strategies are highly successful and may invade and replace the mixed ESS of hawk-dove, in which the strategies are played randomly.

The inequalities, or asymmetries, among rivals can be grouped into three categories: (1) differences in the ability of each contestant to defend the resource, (2) differences in the value of the resource to each contestant, and (3) arbitrary differences between the opponents that affect the outcome even though they are unrelated to the resource value or the ability to defend the resource (Maynard Smith and Parker 1976). We will look at each type of asymmetry more closely to see how each might affect the outcome of a contest.

Asymmetry in the Ability to Defend the Resource
In many instances, the contestants differ in their ability to defend the resource. One may be larger or heavier, have bigger weapons, or be a more skilled fighter. Characteristics such as these that bear on an opponent's ability to defend a resource describe its resource holding potential (RHP). It seems intuitively obvious that

contestants would increase their fitness by assessing their opponent's RHP and adjusting their fighting strategy to suit the occasion (Archer 1988; Parker 1974). It would be a bad idea, for instance, for a chihuahua to play hawk against a rottweiler.

Since fighting is costly, contestants usually assess one another's RHP, often by using displays. If a contestant judges that its opponent is likely to win a fight should one develop, it is likely to stop competing for the resource before a fight ensues (Whitehouse 1997). When the contestants know from the start the ability of their opponent to defend the resource, another fighting strategy, assessor, becomes unbeatable (Maynard Smith 1982). Assessor is a conditional strategy that follows this rule: If one's RHP is greater than the opponent's, play hawk; if one's RHP is smaller, play dove. In other words, if one opponent is assessed to be worthier, the other would gain more in fitness by playing dove and retreating than it would by playing hawk and risking injury in a fight it is likely to lose. However, assessor is only an ESS when the cost of assessing the opponent's RHP is less than the cost of losing a fight.

Size is one of the most important qualities in an individual's ability to defend a resource. Larger animals are more likely to start a fight, persist longer, and eventually win the dispute than are smaller competitors. For example, female funnel-web spiders defend their territories from floaters, who lack a territory, and from neighboring territory holders. The contestants assess the weight of the opponent from movements on the web before any contact is made. The strategies played by each female depend on their relative weights. If the territory holder is larger, the intruder plays dove and immediately retreats. However, if the intruder is larger, she escalates the confrontation. The smaller web owner plays a strategy called retaliator—display unless attacked; if attacked, strike back (Riechert 1984).

Weapons are also a component of RHP in those species that bear them. A male mountain sheep with small horns will defer to a competitor with larger horns (Geist 1971). Likewise, male red deer judge the size of their competitor's antlers and retreat when they are outclassed (Clutton-Brock, Guiness, and Albon 1982). Indeed, the size of the chela (claw) that one male shore crab (*Carcinus maenus*) presents to his opponent during an agonistic contest is more important than his overall body size in determining the outcome of the conflict (Sneddon, Huntingford, and Taylor 1997).

Asymmetry in the Resource Value

It is not difficult to imagine situations in which the resource is more valuable to one of the contestants. Food, for instance, would be more precious to a starving animal than to a well-fed one. A female might vary in importance, depending on whether she had already been inseminated. A territory might be more important to the resident because it knows the location of refuges or food sources there. As we have seen, game theory predicts that an animal should be willing to accept greater cost as the value of the resource increases, just as you might pay more for a more valuable item. The animal that considers the resource to be more valuable would be expected to fight harder in a hawklike battle or display longer in a war of attrition.

Food is a resource that often has different values to the contestants. A hungry animal may fight hard for its dinner, but after eating it may be less willing to compete for a second course (Houston and McNamara 1988). This prediction is borne out in northern harriers (*Circus cyaneus*), for instance. These small hawks catch small prey that requires some time to digest before they can consume another one. Once the owners of a territory have eaten, the value of the prey temporarily decreases. During this interval, harriers are not as aggressive toward competitors that might steal food (Temeles 1989).

Asymmetries in the value of food might also allow a ravenous, yet small, subordinate individual to win a contest over food from one that was larger but well fed (Popp 1987). This has been shown to be true for bluethroats (*Luscinia s. svecica*) before migration. These small birds molt at their breeding site before beginning their long-distance flight. At this time the birds are generally quite lean and have little fat reserves. Body fat is important because during the flight food may be scarce and its availability is unpredictable. A fatter bird has a better chance of survival, and every bit of added fat is an extra safeguard. When food was provided at the molting sites, the birds would fight frequently, with the winner chasing the loser from the food bowl. Individuals that had no body fat won more frequently than those that were close to their maximum recorded weight. Leaner birds were able to chase away larger birds that were already positioned at the feeding bowl, presumably because the food was more important to the lean birds, and so they were more highly motivated to win (Lindstrom et al. 1990).

There are also examples in which a particular territory may have different values to the contestants. The worth of a territory may increase as the owner learns the location of good feeding and nesting sites. We would predict the winner in a territorial dispute to be the individual that valued the property more. This expectation is supported in removal studies of male red-winged blackbirds (*Agelaius phoeniceus*). Territory owners were removed until replacement pairs moved in. When the original owners were released, they fought to reclaim their territory. We might assume that the newcomer would require a few days to become familiar with the territory. As he did, the property would become gradually more valuable to him. The original owner is already familiar with the benefits of the territory and places a high, constant value on it. As expected, when the original territory holders were released after being held in captivity for up to 49 hours, they nearly always won back their territories (Beletsky and Orians 1987). However, when the original owners were retained for up to a week, the new residents usually defeated the former owners. The released owners were just as persistent in their attempts to recover their territories after they had been removed for seven days as they were after two days. The difference in the outcome of the contests, therefore, seems to be due to a change in the behavior of the new residents. They were more willing to escalate contests as the territory became more valuable to them (Beletsky and Orians 1989).

Arbitrary Asymmetry

The differences between contestants that we have discussed so far have a fairly straightforward relationship to the fighting strategy of each contestant. We have seen that a competitor might follow a conditional strategy such as "play hawk if larger; dove, if smaller" or "persist longer when the resource is more valuable." The value of the resource and resource holding potential are, therefore, called correlated asymmetries. The animal maximizes its fitness by responding to differences in resource value or RHP because the differences are likely to influence the outcome of the battle.

Other differences between contestants are not logically connected to the fighting strategy of the opponents but do affect the outcome of the dispute. These arbitrary asymmetries (also called uncorrelated asymmetries) are, in essence, rules or conventions used to settle conflicts. Examples of arbitrary rules for settling differences among humans are flipping a coin or pulling straws. There is no reason that "heads" should win or that the "short straw" should lose. These are simply rules that are mutually agreed upon.

Prior ownership (or residency) is a common arbitrary asymmetry. Animals often appear to adhere to the principle that "possession is nine-tenths of the law." The bourgeois strategy sets rules for dealing with prior ownership: Play hawk if you had it first; otherwise, play dove. If the bourgeois strategy is added to hawk and dove strategies in a population, it does better than either. Bourgeois is, in fact, the only ESS in a population initially using hawk, dove, and bourgeois strategies.

In the bourgeois strategy, the owner always wins, with barely a squabble, and the outcome of any dispute can be reversed by switching ownership. Studies of hamadryas baboons (*Papio hamadryas*) fit this bill. A male permitted to associate with a female for as little as 20 minutes was perceived as the "owner" by a second, newly introduced male. That the second male was deferring ownership was revealed when, several weeks later, he was permitted to associate with the female. When the tables were turned, the first male did not challenge the second male's ownership (Kummer, Gotz, and Angst 1974). Another study demonstrated that prior ownership is more important than dominance in deciding the outcome of disputes. Generally, a dominant individual has an automatic right to the resource. However, when a subordinate hamadryas baboon has possession of a piece of food, a dominant individual will not attempt to take it. However, if food is thrown between two individuals, the dominant one does not permit the subordinate to take it (Sigg and Falett 1985).

Although it may seem at first that the bourgeois strategy would always be used to settle territorial disputes, it is not. The territory holder may have won the property because of better fighting skills. In this case, the dispute is likely to be settled by asymmetries in RHP. In other cases, as we have seen, the territory may have different resource value to the competitors. The owner may know the location of abundant food, safe refuges, or nesting sites. This information would make the property more valuable to the resident, so the owner would be expected to fight harder than an intruder.

Because there are usually several possible explanations for a particular behavior, it is important that each alternative be tested. For example, male speckled wood butterflies (*Parage aegeria*) seem to follow a bourgeois strategy when defending spots of sunlight that serve as mating territories. When an intruder approaches, a resident flies upward in a spiral pattern. An intruder never challenges. When a resident is experimentally removed, he is almost instantly replaced by another male. Within minutes, the new male's ownership is established so firmly that he will win in an encounter with the former resident. However, if two males are experimentally tricked into joint "ownership," an escalated contest involving spiral flight occurs (Davies and Halliday 1978).

Although it is true that the resident always wins, the victory may actually be due to differences in the RHP of a territory holder and an intruder. At first it was thought that the spiral flight of the territory holder was a signal of ownership that was respected by the intruder. But the possibility that it was a display of RHP was not ruled out. It turns out that the length of time a butterfly can fly increases with its body temperature. Thus, a butterfly that has raised its body temperature by basking in a sun spot may have a greater RHP than one that has just flown in from a shady location.

This possibility was tested by Alastair Stutt and Pat Willmer (1998) by experimentally manipulating the body temperature of two butterflies. Two butterflies were caught and marked, and each was placed into one of two clear plastic boxes. One box was insulated with polystyrene; the other was covered in black plastic. A 30-mm area at the top of each box was left transparent, allowing light to enter. The boxes were then placed next to each other and moved into a sun spot. The temperature in each box was monitored, and it quickly climbed higher in the black box than in the insulated one. Within five minutes the temperature in the black box rose to that previously measured in sun spots. However, the temperature in the insulated box still approximated those in shady locations. Then the butterflies were released, each considering itself to be the resident owner of the sun spot. The duration of the escalated flight was recorded, and the winner was noted as the one who returned to the sun spot where the boxes had been placed.

The two hypotheses lead to different predictions. If the spiral flight were a signal of ownership, since both males considered themselves to be owners we would predict that cool males would win as frequently as warm males. However, if the spiral flight were a display of RHP, then we would predict the warm males to win more contests because they could fly longer.

When escalated flight occurred, the warmer male won significantly more often. This is consistent with the idea that spiral flights are true contests, not just signals of ownership, and that the winner will be the male with the higher RHP as measured by body temperature. The conclusion is the same as before—the resident wins—although the reasons for the conclusion are different, serving as a reminder of the need to test alternative explanations for every observation.

SUMMARY

Group living has many benefits, particularly those relating to eating and not being eaten. Some predators, for example, engage in cooperative hunting and as a result have higher capture success and average energy intake per individual than do solitary hunters.

Cooperative hunting may also allow predators to kill animals much larger than themselves. Because groups are usually more successful than solitary individuals at detecting, confusing, and repelling predators, there are also antipredator advantages in grouping. In addition, during an attack by a predator, an individual in a group has a smaller chance of becoming the next meal. Finally, living in groups may aid in defending young, food, or space against conspecifics.

Group living also has costs. In some cases, group members face increased competition for mates, nest sites, or food resources. Other costs include increased exposure to parasites or disease, increased conspicuousness to predators or prey, and a greater possibility of providing care to young that are not one's own.

Because group living has both costs and benefits, we would expect that some interactions would involve conflict. Two ways in which animals compete for limited resources in a stable group are dominance hierarchies and territoriality.

Dominance refers to the ability of one member of a group to assert itself over others in acquiring access to a limited resource. Members of a social group benefit from stable dominance relationships because the cost of aggression is minimized by limiting interactions to threats instead of fights. Dominance relationships can have many forms. In some species, a single despot rules over all other members of the group. Other dominance hierarchies are linear.

In some species dominance is earned by winning contests, and in others it is based on the status of one's parents. In contests to establish dominance, an individual's fighting ability (resource holding potential) and its prior experience of winning or losing can influence the outcome of the contest.

Dominant animals enjoy enhanced reproductive success and increased access to food. In some species, dominants leave more offspring because they suppress reproduction by subordinates. In others, dominants mate more often than subordinates.

Subordinates reduce their risk of injury by avoiding fights they are likely to lose. They may also stay with the group because it is the best option available. Leaving and joining another group is risky. The subordinate's status could increase if the top-ranking animal dies.

The establishment of territories is another way in which animals compete for resources. The home range of an animal is the area in which it carries out most of its normal activities. Some animals may defend all or a certain portion of their home range. The defended area, or area maintained for their exclusive use, is called a territory. In some instances, territories are used solely for feeding, mating, or raising young, whereas in others they serve a variety of purposes.

Some benefits associated with territoriality include a longer-lasting food supply because fewer individuals have access to it, increased access to mates, and a greater opportunity to rear young on a high-quality site. Territoriality can be quite costly, however, given the energy needed to advertise ownership and to keep out and evict intruders. Territory acquisition and defense also take time away from other essential activities such as foraging.

From the standpoint of economics, territoriality is expected to occur only when the benefits outweigh the costs. Factors such as abundance, spatial distribution, and renewability are thought to influence whether a particular resource is economically defendable. Although it is usually individuals of the same species that exploit the same resources, and thus territories are usually defended against conspecifics, sometimes individuals of different species use the same resources and interspecific territoriality results.

The outcome of a single contest over limited resources is a situation in which the fitness payoff for a particular action depends on what the other members of the population are doing. The mathematics underlying game theory is helpful in predicting behavior in such situations. Game theory models require that we take into account (1) all the existing strategies, (2) the frequency of each strategy in the population, and (3) the fitness payoff for each possible combination of strategies.

Game theory has been used extensively to analyze animal conflict. A simple model involves the hawk strategy (fight until seriously injured or until your opponent retreats) and the dove strategy (display but retreat if attacked). The benefit is the value of the resource. The cost is the risk of injury. A strategy's fitness payoff depends on the opponent's strategy. The precise mixture of strategies that would confer equal fitness to each depends on the relative values of costs and benefits.

The hawk-dove model assumes that the contestants are equally matched. In reality, this is rare, and contests are usually asymmetric. Conditional strategies, which are adjusted to the circumstances, are more successful in asymmetric contests. Asymmetries can exist in resource holding potential (fighting ability), resource value, or some arbitrary factor such as prior ownership that influences the outcome of the contest.

17

Cooperation and Altruism

WHAT IS ALTRUISM?

Nice guys finish last—or do they? In recent years the concept of "nature, red in tooth and claw" has been changing with the realization that some animals act kindly to other members of their species. For example, an individual Belding's ground squirrel (*Spermophilus beldingi*) risks being spotted by the approaching predator when it sounds an alarm. Nevertheless, it barks at the sight of a badger, and all those in the area scurry to safety. In other species, such as wild turkeys (*Meleagris gallopavo*), a male attracts a mate by displaying in com-

mon mating areas, where both dominant and subordinate males aggregate (Figure 17.1). These communal displays surely increase the chance of being seen by a predator, yet the subordinate has little hope of immediate fatherhood. Members of still other species cooperate in rearing offspring that do not belong to them. One can see why it is adaptive for a parent to care for its own offspring until they can survive on their own, but in some species, for example, the dwarf mongoose (*Helogale parvula*), even unrelated immigrants may forgo reproducing themselves and assist another female in raising offspring by bringing food to the young and by guarding the den from predators. Perhaps the best example of altruism among animals involves the social insects (ants, termites, social wasps, and social bees). In many species, the workers toil tirelessly to care for their colony. They may even die in defense of the nest. However, the young they help rear are not their own; the workers are sterile.

When we observe animals cooperating with or helping one another, such as in these examples, we often describe their behavior as altruistic. We might identify an action as altruistic when it appears to be costly or detrimental to the altruist but is beneficial to another member of its species. Although it is more difficult to measure, altruism is more precisely defined in terms of fitness: It is the behavior of an individual that raises the fitness (number of offspring produced that

FIGURE 17.1   In communal displaying by wild turkeys, groups of brothers strut together to attract mates. In some populations, only the most dominant male in the most dominant sibling group will have the opportunity to copulate. Since the cooperating males are closely related, kin selection is a reasonable hypothesis for the evolution of this behavior.

live to breed) of another individual at the expense of the altruist's personal fitness (Hamilton 1964).

Our discussion of cooperative and altruistic behavior will begin with some of the hypotheses for how giving aid may have evolved. This should set the stage for later sections in which the various hypotheses will be evaluated as possible explanations for the evolution of specific types of aid-giving behavior. However, as we will see, questions about the evolution of altruism are only part of the fascinating puzzle.

HYPOTHESES FOR THE EVOLUTION OF ALTRUISM

At first, it may not be obvious just why the evolution of altruism has been difficult to explain. Part of the reason seems to lie in the philosophy and values that many of us have been taught. We may believe that good or altruism *should* triumph over evil or selfishness. However, as morally satisfying as it may seem to humans, a good deed costs the altruist. The cost may be in time, energy, or increased danger—factors that may reduce the altruist's reproductive potential. Unless the benefits to the altruist outweigh these costs, it is difficult to see how the behavior would have been selected because natural selection does not give points for being nice. Natural selection gives points for reproductive success, period. Evolution occurs when individuals that carry certain inherited traits leave more reproductively capable offspring than do individuals that lack the alleles* for those

characteristics. By this reasoning, a population of altruists would be vulnerable to a selfish mutant, one that would accept the altruists' services and never reciprocate. A selfish mutant would gain at the altruists' expense. As a result, alleles that promote selfish behavior would multiply more quickly in the population, always pressing out alleles for altruism. It would be predicted, therefore, that selfish individuals would inherit the earth. However, we do see examples of helpful or cooperative behavior in many species of animals. So, let's look at some of the explanations offered to resolve this enigma.

INDIVIDUAL SELECTION

As we saw in Chapter 4, natural selection responds to the fitness costs and benefits of a behavior and selects the behavioral alternative that shows the most favorable balance sheet. We would expect, then, that if animals tend to perform some behavior, even a cooperative one, the benefits of the deed must outweigh the costs. When we see helpful or cooperative behavior, we might expect the donor to be gaining more than it is losing.

When the behavior seems to benefit others, at some cost to the actor, it is not always easy to see what the benefit to the actor may be. However, there are several ways in which an individual may gain through an action that appears to be costly. Consider alarm calling by birds. It seems at first that a bird that sounds an alarm at the sight of the predator risks drawing the predator's attention to itself. It has been suggested, however, that the caller may, in fact, benefit by the act. The call may cause the predator to leave since its presence has obviously been discovered and the flock is on the alert. What about animals that help others raise young? In some species of cichlid fish, adults adopt unrelated young into their own brood, caring for them and defending them from predators as if they were fam-

---

*You may recall from the detailed discussion in Chapter 3 that an allele is one of the alternative forms of a gene. For instance, fruit flies have several alleles that influence the length of their wings: one for normal wings, one for short wings (apterous), and one for vestigial wings.

FIGURE 17.2 **A cichlid caring for fry. Some cichlids adopt fry of other pairs and care for them as if they were their own. This seemingly altruistic behavior is actually quite selfish. By adopting unrelated young, the parents reduce predation on their own offspring.**

ily (Figure 17.2). Although this may at first seem to be quite a generous act, on closer inspection it seems that the parents gain because the adopted young reduce the predation risk of their own young (McKaye and McKaye 1977). So, there may be personal gains that outweigh the costs of a helpful action.

KIN SELECTION

We have said that fitness is a measure of reproductive success. So far we have defined individual fitness, which is measured by the number of offspring an individual leaves that survive and reproduce. Offspring carry copies of some of an individual's alleles into future generations, and if a particular allele survives, so does the trait for which it codes. Now we can expand the definition a bit. We can remind ourselves that family members other than offspring also possess copies of some of that individual's alleles. An allele is an allele, regardless of its bearer. Therefore, if *family members* are assisted in a way that increases their reproductive success, the alleles that the altruist has in common with them are also duplicated, just as they would be if the altruist reproduced personally. In other words, because common descent makes it likely that a certain percentage of the alleles of family members are identical, helping kin is another way to perpetuate one's own alleles. Kin selection, then, includes fitness gained not just by helping one's offspring (direct selection) but also fitness gained by helping one's nondescendant kin (indirect selection).

The fraction of alleles shared between two individuals as a result of common descent is called their coefficient of relatedness, or *r*. You can see that *r* also indicates the probability that two individuals will have inherited a particular allele from a common ancestor. In

fact, it is possible to calculate the percentage of alleles one is likely to share with particular relatives. You may recall from our discussions in Chapter 3 that with few exceptions, animals have two alleles for each gene and that these separate during the formation of gametes (eggs or sperm). There is, therefore, a 50–50 chance (a probability of 0.5) that any particular allele will be found in an egg or sperm produced by the parent. Thus, an individual is likely to have 50% of its alleles in common with a parent, sibling, or offspring ($r = 0.5$) and 25% in common with a half sibling or a grandparent ($r = 0.25$), but only 12.5% (one-eighth) in common with a first cousin ($r = 0.125$). In contrast to individual selection, which considers only the survival of alleles through one's own offspring, kin selection also includes the survival of alleles through the offspring of relatives (Figure 17.3).

Kin selection is based on the idea of inclusive fitness, which considers *all* adult offspring, an individual's own or those of its relatives, that are alive because of the actions of that individual. Since an individual shares more alleles by common descent with certain relatives, the possibility of genetic gain increases with the closeness of the relationship. Helping a cousin, which has an average of only one-eighth of the same alleles, is less productive than assisting a brother or sister, which is likely to have half of its alleles in common with the altruist. In other words, the fitness gained through family members must be devalued in proportion to their genetic distance (diminished relatedness); the allele for an altruistic act would survive if a small number of close relatives or a large number of distant relatives were assisted. We can calculate inclusive fitness by counting offspring, but not all offspring are counted, and a relative's offspring do not count as much as one's own.

Which offspring are included when determining inclusive fitness? According to W. D. Hamilton (1964), when calculating inclusive fitness we count the offspring that would have existed without help or harm from others and some fraction (determined by *r*) of the offspring of relatives that the individual helped or harmed. Simply put, to calculate an individual's inclusive fitness we add (1) the individual's offspring and (2) the extra offspring that a relative was able to raise because of that individual's efforts, devalued by the genetic distance between the individual and the relative who was helped (e.g., multiplied by *r*), and we subtract any of the individual's offspring that resulted from the help or harm of others (Grafen 1982, 1984).

What does all this have to do with the evolution of altruism? It gives us a broader view of the costs and benefits of a particular helpful or cooperative behavior so that we see how the allele that promotes the action may have spread through the population. We will find that an allele that promotes a helpful behavior will spread if the fitness benefits are greater than the fitness

FIGURE 17.3   **Relatives share a portion of their alleles by common descent. Parents and offspring share 50% of their alleles, as do siblings. Half siblings share only 25%.**

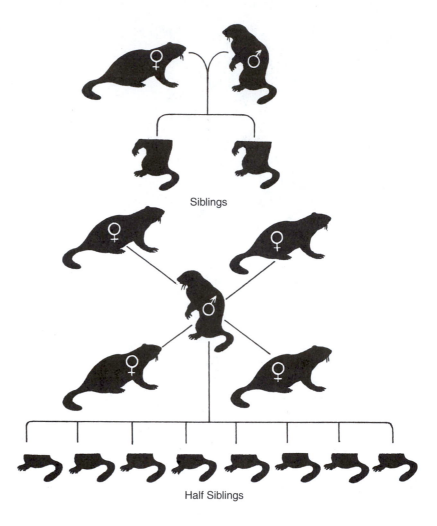

Siblings

Half Siblings

costs. (Notice that fitness benefits and costs also tell us the probability that the allele that favors the helpful behavior will be shared by the offspring. So, phrased slightly differently, the allele that favors the helpful behavior will spread if helping results in leaving more copies of the allele that promotes helping than would be left if the individual did not help.) The benefits, the fitness gains to the altruist, are calculated as the *extra* offspring that were raised by relatives because of the actions of the altruist, multiplied by r, to take into account the genetic distance between the altruist and the offspring. The cost, or the fitness loss to the altruist, is the difference in the animal's lifetime number of offspring that results from choosing to be helpful rather than behaving in some other way. It is, of course, difficult to know "what would have been." However, we might estimate by comparing the number of offspring produced by animals that followed one course of action with the number produced by those that did not. There are reasons that this estimate may be inaccurate (Grafen 1984), but it is a beginning.

In the end we would expect the allele to spread if the benefit exceeds the cost. If, for example, a male animal helped his brother raise four more nieces than he could

have without assistance, the altruist's genetic gain would be 1 ($r \times$ number of extra nieces = 0.25 × 4 = 1). If he raised one fewer offspring of his own by helping his brother raise nieces than he could have if he had devoted himself entirely to raising his own young, then his genetic cost would have been 0.5 ($r \times$ number of fewer offspring = 0.5 × 1 = 0.5). In this case, because the benefit (1.0) exceeds the cost (0.5), the allele that promotes helping brothers would spread through the population. So we see that altruism, a behavior that lowers one's individual fitness, could evolve if it increased the fitness of enough relatives that also carry alleles for altruism.

Kin-Biased Behavior

If altruism evolves by kin selection, we might expect animals to assist relatives more often than they assist others. Kin-biased behavior may be based on a variety of mechanisms, some of which might involve the ability to discriminate kin from nonkin. We might wonder, then, how an animal might determine who its relatives are. We will consider four possible mechanisms of kin discrimination (Blaustein, Bekoff, and Daniels 1987; Holmes and Sherman 1983).

Location   When relatives are distributed throughout the habitat in a predictable way, kin selection can work if the altruistic deeds are directed toward those individuals in areas where relatives are most likely to be found. For instance, we might see location as a mechanism underlying kin-biased behavior in species in which there is limited dispersal from the nest or burrow. As you may recall from Chapter 11, in many mammal species, the males disperse and the females remain in their natal area. Among many bird species, however, the females tend to disperse and the males remain at home. In any event, those animals that remain in the natal area tend to be related. Therefore, any good deed directed toward a neighbor that is performed by an individual of the nondispersing sex will usually be received by a relative.

The most obvious example of how location might work as a mechanism for kin-biased behavior occurs when a parent identifies its offspring as the young in its nest or burrow. In many species of birds, parents will feed any young they find in their nest. We see this among bank swallows (*Riparia riparia*), for instance. The parent bank swallow learns its nest hole locations and feeds any chicks in these holes, including any neighbor's chicks placed inside by experimenters. However, a parent will ignore its own chicks if they are moved to a nearby nest hole. After about two weeks, at the time the young begin to leave the nest and fledglings unrelated to the parent might enter the nest, the parent begins to recognize its young by its distinctive calls and also begins to reject foreign young experimentally placed in the nest (Beecher, Beecher, and Hahn 1981).

Spadefoot tadpoles (*Scaphiopus bombifrons*) also use location cues to identify kin. In this case, though, the good deed is for a cannibalistic tadpole to refrain from eating its relatives. Spadefoot toad tadpoles develop in temporary ponds in the desert. Cannibalism provides extra nourishment, which hastens their growth and allows them to metamorphose to a toad before the pond dries up. All spadefoot tadpoles begin life as omnivores. However, if a tadpole happens to eat another one or a fairy shrimp, it is transformed into a cannibal. Kin discrimination would benefit the cannibal because eating only nonkin saves family members, who are likely to bear copies of its own alleles. After a tadpole becomes cannibalistic, it avoids eating relatives by staying away from the areas of the pond where relatives are found. If the cannibal should nip at another tadpole and find that it tastes like a sibling, the meal is immediately released unharmed. Noncannibalistic tadpoles, on the other hand, congregate with their siblings in the pond (Pfennig, Reeve, and Sherman 1993).

Familiarity   Kin may also be treated differently from nonkin because they are familiar. In some species, the young learn to recognize the individuals with which they are raised through their experiences during early development, and then, later in life, they treat familiar animals differently from unfamiliar ones. The ideal setting for this learning is a rearing environment such as a nest or burrow that excludes unrelated individuals.

Familiarity is apparently a mechanism used by a young spiny mouse (*Acomys cahirinus*) to distinguish its siblings. When weanling pups are released into a test arena, they often huddle together in pairs, and the members of the pair are generally siblings. However, the spiny mice do not identify their siblings per se, but rather they prefer to huddle with familiar pups, their littermates. We know this because siblings separated soon after birth and raised apart treat one another as nonsiblings. However, if unrelated young are raised together, they respond to one another as siblings. So we see that kin-biased behavior among spiny mice seems to develop as a result of familiarity (Porter, Tepper, and White 1981).

Strangers might be recognized as kin by this method if they are associated with a mutual relative such as a mother. For example, sisters born in different litters might identify each other as siblings because their common mother treats them both as offspring.

Phenotype Matching   Another mechanism for discriminating kin, called phenotype matching, allows animals to identify kin even if they have never met before (Alexander 1979; Holmes and Sherman 1982; Lacy and Sherman 1983). Phenotype, as you may recall from Chapter 3, is the physical, behavioral, and physiological appearance of an animal. Because one's genetic inheritance has so much to do with appearance, family members often resemble one another in one or more characteristics.

Phenotype matching can be thought of as a multistep process with three components (Sherman, Reeve, and Pfennig 1997): (1) It begins with a cue or label produced by the individual that can be used to identify it as a relative. (2) The cue is compared to a template, an internal representation of the way the characteristic is expressed among kin or nonkin. The degree of similarity between the cue and the template reflects the level of relatedness. (3) A response to the individual in question is made. The most appropriate response will depend on how closely it resembles kin, as well as the context of the encounter. We will consider the aspects of phenotype matching in more detail.

Cues of Kinship   We have already seen that the location of an animal might be an indirect cue of relatedness. In other species, the cue labeling an individual as kin is a physical feature directly associated with that individual. This characteristic may be determined by the organism's genes, or it may be acquired from the environment. It is only necessary that the physical characteristic reliably indicates kinship.

Recognition cues can have a genetic basis, or they can be acquired. Those with a genetic basis commonly involve alleles of the region of DNA, called the major histocompatibility complex (MHC), that code for molecules on the surface of cells that allow the body to distinguish between "self" and "nonself." For example, differences in the alleles of the MHC cause grafted tissue to be rejected and trigger protective immune responses when disease-causing organisms enter the body. These same genes serve as direct cues of relatedness, allowing individuals to identify their kin (reviewed in Brown and Eklund 1994; Penn and Potts 1999).

The larvae of the sea squirt *Botryllus schlosseri* use MHC to discriminate kin. These larvae, which superficially resemble a frog tadpole, are planktonic for a short time and then settle, attach to the sea bottom, and develop to the adult form. When groups of siblings settle, they tend to clump together, but groups of unrelated larvae settle randomly. If the siblings do not share an allele in the MHC region, they do not settle together. Unrelated larvae that happen to share an allele are just as likely to settle together as are siblings that share an allele (Grosberg and Quinn 1986).

Experiments in house mice (*Mus musculus*) also indicate that social preference may be influenced by a small genetic difference in the MHC. Mice were repeatedly inbred so that all their alleles were identical except for a single gene, the same region that was shown to be important in kin discrimination in sea squirts. In mating preference tests, these inbred male mice could choose among females that carried different alleles of the MHC. Surprisingly, these alleles made a difference in their preference. Males from most strains chose females whose alleles differed from theirs, but males from a few strains preferred mates whose alleles matched theirs (Yamazaki et al. 1980). The MHC alters mating choices through odor cues (Yamazaki, Singer, and Beauchamp 1998). Individuals of outbred populations of mice maintained in enclosures large enough to allow normal patterns of social competition and reproductive behavior also prefer mates whose histocompatibility genes differ from their own (Potts, Manning, and Wakeland 1991).

In other species, direct recognition cues are acquired rather than inherited. For example, paper wasps (*Polistes fuscatus*) discriminate between kin and nonkin by using the odor of hydrocarbons that become locked into the insect's cuticle (skin) before it hardens. The odor comes from the nest. Nest odor differs among colonies because it depends on the type of plant fiber used to build the nest, as well as on secretions produced by wasps that built the nest. Each colony uses a unique combination of plants to construct the nest, so each has a distinctive odor. The odoriferous hydrocar-

bons are transferred from the nest to the workers as they emerge from the pupal case. Since a colony consists of a queen and her worker daughters, the nest odor labels colony members as relatives (Breed 1998).

THE TEMPLATE   Once the recognition cue has been produced, the next step in phenotype matching is to assess relatedness by comparing the cue to a template of the way in which the trait is expressed among kin. In most, perhaps all, cases, the template is learned. The referent for the template can come from the environment, the family, or the individual itself. As we've seen, a feature of a paper wasp's environment, specifically its nest odor, correlates with kinship. Thus, a paper wasp's recognition template is learned from experience in the nest, not from nestmates or from itself (Pfennig et al. 1983).

In other species, kin recognition templates are learned from family members. As we will see shortly, Belding's ground squirrels (*Spermophilus beldingi*) give alarm calls to warn their mothers, daughters, and sisters of an approaching predator. Pups apparently identify their siblings because they learn one another's odors while still in the same nest burrow. In nature, unrelated pups are not usually found in the same burrow. When pups from different nests are experimentally switched, unrelated pups that are raised together will treat one another as siblings later in life (Sherman, Reeve, and Pfennig 1997).

Tadpoles of the Cascades frog (*Rana cascadae)* behave as if they match their *own* phenotype with that of a stranger to identify kin. This was shown in a series of experiments by Andrew Blaustein and his colleagues (summarized in Blaustein, Bekoff, and Daniels 1987; Blaustein and O'Hara 1986; Blaustein and Waldman 1992). All tadpoles that hatch from a single egg mass are siblings, and in nature, Cascades frog tadpoles remain near their hatching site and form small cohesive schools. At first, then, one might suspect that location or familiarity might be the basis of their preference for associating with kin. However, when these hypotheses were tested, location cues and interaction with siblings were shown to be unnecessary. In one series of experiments, tadpoles were reared for about four weeks in one of four conditions: (1) exposure only to their siblings, (2) exposure to both sibling and nonsibling tadpoles, (3) exposure only to nonsiblings, and (4) total isolation. Each tadpole was then given the choice of associating with siblings or with nonsiblings. The test aquarium was a long tank with mesh barriers, forming two end compartments. About two dozen siblings were placed in one compartment and the same number of nonsiblings in the other. The tadpole being tested swam freely within the central chamber. The amount of time a tadpole spent near each end was recorded and used to determine whether the tadpoles preferred to associate with kin. Indeed, most of the tadpoles for all four rearing conditions did spend more time swimming near their siblings. Even the tadpoles raised with non-siblings preferred to swim near their own siblings, which they had never met before. So we see that Cascades frog tadpoles can identify their brothers and sisters, even if they did not grow up with them (O'Hara and Blaustein 1981).

Since the siblings develop within a common jelly matrix, it was then suspected that the developing tadpoles may learn to identify some cue that comes from the jelly, but this, too, was shown to be unimportant. Tadpoles reared without any jelly mass or with the jelly mass of nonsiblings substituted for their own still preferred to associate with their siblings. Furthermore, tadpoles raised in isolation preferred to associate with full siblings over half siblings, maternal half siblings over paternal half siblings, and paternal half siblings over nonsiblings (Blaustein and O'Hara 1982). A possible explanation for this ability to recognize kin is that the tadpoles, even those raised alone, learn cues associated with themselves, and when they later make an association choice, they match their own phenotype with those of other tadpoles.

Golden hamsters (*Mesocricetus auratus*) also seem to use cues associated with themselves as a template for phenotype matching. In this case, the cue seems to be the quality of odor of flank gland secretions (Todrank, Heth, and Johnston 1998). Hamsters scent-mark with their flank glands by rubbing their sides against surfaces in a stereotyped manner. The scent marks communicate a number of things, including individual identity, social status, and agonistic motivation. Male hamsters will scent-mark more to odors of nonkin males than to those from male siblings. The degree of genetic relatedness is more important than familiarity in determining a male's response to another male. This was demonstrated by transferring week-old male pups from one litter to another and raising them as foster brothers for one month before housing each male individually. When the males were three to four months old, the odor of another male was introduced into the test male's home cage and the number of flank scent marks elicited by the odor was recorded. The odor donor was either a familiar brother, an unfamiliar brother who had been cross-fostered and reared in a different litter, a familiar foster brother, or a unfamiliar brother of a foster brother (unfamiliar nonkin). As you can see in Figure 17.4, males scent-marked less to the odors of siblings, even if the male had never met the sibling. Furthermore, they scent-marked more to nonsiblings, even when raised with the male as a foster brother. Since the males respond to unrelated foster brothers as nonkin even though they were raised together, they did not use the odors from all their nestmates to form a family template. Thus, either the template is formed

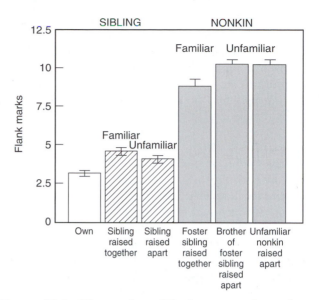

FIGURE 17.4　**The number of flank scent marks made by male golden hamsters in response to the flank gland odor of another male. Some of the week-old male pups in litters were transferred to other litters. The siblings and foster siblings were then raised together for a month before housing each male individually. Several months later, a male's flank scent-marking response to the odor of another male was recorded. Males scent-marked less to the odor of a brother, even if they had never met. On the other hand, males scent-marked frequently to the odors from all nonsiblings, even if the male had been raised with the donor of the odor. These results suggest that a male hamster uses the odor of his own flank gland secretions as a template and compares the odors of other males to his own to assess the closeness of their genetic relationship. (Data modified from Heth, Todrank, and Johnston 1998.)**

during the first week of life (before the pups were cross-fostered) or the male uses his own flank gland odor as the template to assess his genetic relationship with others (Heth, Todrank, and Johnston 1998).

APPROPRIATE RESPONSE　Once an animal has assessed how closely related it is to another individual, it should respond appropriately. If the match between the stranger's cues and the receiver's template is close, the stranger should be accepted as kin. If it isn't close, the stranger should be rejected. But the match is rarely exact, and the response may not be complete acceptance or rejection. Instead there may be an optimal acceptance threshold that depends on the consequences of mistakenly accepting an unrelated individual or rejecting kin (Reeve 1989). As a result, the response often depends on the context of the encounter. In paper wasps, for example, accepting an unrelated intruding wasp into the nest could be disastrous. Intruders sometimes steal eggs to feed to their own larvae. If an unrelated wasp is encountered away from the nest, however, the stranger doesn't pose a threat. Paper wasps are aggressive toward unrelated wasps in the presence of a nestmate or a familiar nest fragment, cues that indicate proximity to the nest. They are more tolerant of unrelated wasps when there are no cues that the nest is nearby (Starks et al. 1998).

Recognition Alleles It has also been suggested that animals might inherit the ability to recognize kin, instead of having to form a template of the important recognition cues (Hamilton 1964). This would be possible if an individual inherited a recognition allele or group of alleles that enabled it to recognize others with the same allele(s). The postulated allele would have three simultaneous effects: It would (1) endow its bearer with a recognizable label and (2) the sensory ability to perceive that label in others, and (3) it would cause the bearer to behave appropriately toward others with the label. This recognition system has been named the "green beard effect" to indicate that the label could be any conspicuous trait, such as a green beard, as long as the allele responsible for it also causes its owner to behave appropriately to other labeled individuals (R. Dawkins 1976, 1982; Hamilton 1964). Demonstrating the existence of recognition alleles has proven quite difficult, primarily because it is so hard to eliminate all the possible opportunities for learning recognition cues during an animal's lifetime.

Laurent Keller and Kenneth Ross (1998) may have identified a green beard allele in the red fire ant (*Solenopsis invincta*). Originally from South America, the fire ant is a recently introduced pest in the southern United States. This ant has two forms of social organization—the monogyne form, which has a single egg-laying queen per colony, and the polygyne form, which has several queens per colony. Furthermore, the social organization is controlled by the protein-encoding gene, *Gp-9*. This gene has two alleles, *B* and *b*. It turns out that queens in the monogyne subpopulation are all homozygous *BB*, but those in the polygyne subpopulation that live long enough to lay eggs are all *Bb*. Why? It's because the workers in polygyne colonies kill any *BB* queens, and *bb* queens die young of physiological causes. These workers act as if they follow the rule, "If *b* is present, let it be. If *b* is missing [e.g. a *BB* genotype], kill it." The workers that execute the *BB* queens are primarily *Bb* workers rather than *BB* workers. We see, then, that in this case, the *b* allele is a green beard allele (or is closely linked to one) that induces workers who bear the allele to kill all queens who do not bear it. Apparently, workers distinguish the *BB* queens from *Bb* queens on the basis of a chemical carried on the cuticle of the homozygous queens. We know this because workers that rub against a *BB* queen during an attack or are rubbed against a *BB* queen by an experimenter are subsequently attacked by other workers, but they are not attacked after contact with a *Bb* queen.

RECIPROCAL ALTRUISM

Reciprocal altruism is evolution's version of "you scratch my back and I'll scratch yours" or "one good deed deserves another." Reciprocal altruism can be selected if the favor is returned with a net gain to the altruist. Although evolution's tally sheet records gains and losses in terms of inclusive fitness, reciprocal altruism does not require that the cooperating animals be related (R.L. Trivers 1971).

At first it is difficult to understand how reciprocal altruism evolved when there is a time lag between the deed and repayment because delayed restitution opens the possibility of cheating. A cheater is an individual that received a service but failed to repay the altruist at a later time.

In reciprocal altruism, the costs and benefits to the altruist depend on whether the recipient returns the favor. You may recall from Chapter 16 that evolutionary game theory is designed to handle situations such as this, in which the best course of action depends on what others are doing.

The Prisoner's Dilemma is a mathematical game that has been applied to assist our understanding of the evolution of reciprocal altruism (Figure 17.5). The name of the game comes from an imaginary story in

Player B	Player A: Cooperate	Player A: Defect
Cooperate	R Reward for mutual cooperation (neither provides evidence)	S Sucker's payoff (sucker keeps quiet but the partner squealed)
Defect	T Temptation (Temptation to defect)	P Punishment for mutual defection (both partners defect)

FIGURE 17.5 Fitness payoffs for Player A in Prisoner's Dilemma. In this game, two crime suspects are being interrogated in separate rooms. There is enough evidence to guarantee a short jail sentence for each of them. But, each is offered a deal—complete freedom for squealing on the other one. If each informs on the other, however, they both serve an intermediate jail sentence. With these payoffs, reciprocal altruism can evolve only if the individuals have repeated encounters. In that case, the tit-for-tat strategy, which instructs each player to cooperate on the first move and then do whatever the opponent did on the previous move, is an evolutionarily stable strategy.

which two suspects are arrested for a crime and kept in separate jail cells to prevent them from communicating. Certain that one of them is guilty but lacking sufficient evidence for a conviction, the district attorney offers each a deal. Each prisoner is told that there is enough incriminating evidence to guarantee a short jail term, but freedom might be obtained by "squealing" and providing enough evidence to send the other to jail for a long time. However, if each informs on the other, they both go to jail for an intermediate length of time.

Given these payoffs for selfishness (informing) and altruism (concealing evidence), selfishness is the best strategy. A selfish prisoner cannot lose. If the partner is altruistic, the selfish prisoner goes free. Even if the partner turns out to be selfish, too, the first prisoner is better off being selfish than taking the rap for both of them. We see, then, that neither prisoner would be expected to behave altruistically. The cost, if the other does not reciprocate, would be too great. If we translate jail sentences into fitness payoffs, the solution to the Prisoner's Dilemma seems to imply that reciprocal altruism cannot evolve. Indeed, this may be true *if* the prisoners will never meet again.

In real life, however, individuals interact repeatedly, and when they do, reciprocal altruism might evolve. Assume, for instance, that the prisoners are Bonnie and Clyde, long-term partners in crime, who are repeatedly arrested. In this case, Bonnie would succeed if she altruistically concealed evidence against Clyde each time they were arrested, *as long as* Clyde reciprocated. However, if Clyde informed on Bonnie, she should do likewise at her first opportunity. If, in a subsequent encounter, he concealed evidence, she should also do so.

This strategy, called "tit-for-tat", can be a winner in the game of Prisoner's Dilemma if there are repeated encounters between the prisoners. The strategy of tit-for-tat might be paraphrased, "Start out nice and then do unto others as they did unto you." In this strategy, an individual begins by being cooperative and in all subsequent interactions matches the other party's previous action. This strategy, then, involves both retaliation (it follows defection, or cheating, with defection) and forgiveness (it "forgets" a defection and cooperates if the partner later cooperates). If a population of individuals adopts this strategy, it cannot be overrun by a selfish mutant (Axelrod 1984). So, when the individuals have repeated encounters, reciprocal altruism can be an evolutionarily stable strategy (ESS), one that cannot be invaded by another strategy (see Chapter 4).

We see, then, that game theory modeling of the evolutionary process informs us that a population that is genetically programmed to unfailingly perform some service for others can be invaded and overrun by any mutants that cheat because a swindler's gain is always greater than a sucker's. However, alleles creating altru-

ists that retaliate against cheaters by refusing to assist them on the next encounter will survive and prosper if there are enough individuals carrying the allele in the initial population to make meetings between them frequent enough. In other words, reciprocal altruism can evolve if there is discrimination against cheaters (Axelrod and Hamilton 1981; R. Dawkins 1976).

Besides discrimination against cheaters, another condition that is required for the evolution of reciprocal altruism is a fair return on the altruist's investment. Certain factors lead to a profitable balance sheet and will, therefore, favor the evolution of reciprocal altruism: (1) The opportunity for repayment is likely to occur, (2) the altruist and the recipient are able to recognize each other, and (3) the benefit of the act to the recipient is greater than the cost to the actor. These factors are most likely to occur in a highly social species with a good memory, long life span, and low dispersal rate (Trivers 1971).

If these characteristics bring our species to mind, it is not surprising. Robert Trivers (1971) points out that reciprocal altruism is particularly common among humans. Not only do humans help the infirm through various programs (with the expectation that they, too, might someday benefit from such a program), but they also help one another in times of danger and they share food, tools, and knowledge. Trivers even argues that our feelings of envy, guilt, gratitude, and sympathy have evolved to affect our ability to cheat, spot cheaters, or avoid being thought of as a cheater.

In real life, cooperation is rarely an all or nothing phenomenon. Instead, the level of cooperation can vary. For example, when a macaque grooms another, the investment varies with the amount of time spent on the activity. Variable investment in cooperation has been incorporated into a strategy called "raise the stake." In this strategy, an individual tests the water during the first encounter with a new partner by investing only a small amount. If that investment is matched, the individual will increase its investment in subsequent encounters. The early encounters are used to build trust (Roberts and Sherratt 1998). Computer simulations that employ this strategy suggest that altruism can spread more quickly than previously thought, especially if an individual has the freedom to refuse to cooperate with a potential partner because of past experiences with that individual (Sherratt and Roberts 1998).

The best-known example of reciprocal altruism among nonhuman animals is provided by vampire bats (*Desmodus rotundus*), who share food with needy familiar roostmates even if they are not related (Figure 17.6) (DeNault and McFarlane 1995; Wilkinson 1984, 1990). This generosity may make the difference between life and death for the recipient. If a vampire bat fails to find food on two successive nights, it will starve to death unless a bat that has successfully fed regurgitates part of

FIGURE 17.6 Vampire bats are reciprocal altruists. A vampire bat that has just fed will regurgitate blood to a hungry roostmate, thereby saving it from starvation. Only bats that have a prior association will regurgitate blood to one another.

its blood meal. The hungry bat begs for food by first grooming, which involves licking the roostmate under the wings, and then by licking the donor's lips. A receptive donor will then regurgitate blood (Figure 17.7). The regurgitated food must be enough to sustain the bat until the next night, when it may find its own meal. Although the benefit to the recipient is great, the cost to the donor is small. Since a bat's body weight decays exponentially after a meal, the recipient may gain 12 hours of life and, therefore, another chance to find food. However, the donor loses fewer than 12 hours of time until starvation and usually has about 36 hours, another two nights of hunting, before it would starve (Figure 17.8).

Vampire bats roost in somewhat stable groups of both related and unrelated members. A typical group consists of 8 to 12 adult females and their pups, a dominant male, and perhaps a few subordinate males. Males leave their mothers when they are about 12 to 18 months old, but females usually remain well past reproductive maturity. Thus, many of the individuals in a roost are related, but perhaps not so closely as might be expected. According to biochemical analysis of blood samples, only half of the individuals in a cluster share the same father. Furthermore, about once every two years, a new female will join the group.

Although the groups may change slightly over time, there are numerous opportunities to share food. Females may live as long as 18 years, and in one study, two tagged females shared the same roost for more than 12 years. Additionally, a blood meal is not always easy to obtain. On any given night, roughly 33% of the juveniles less than two years old and 7% of older bats fail to feed.

Generally, only individuals who have had a prior association share food. In an experiment, a group of bats was formed from two natural clusters in different areas and maintained in the laboratory. Aside from a grandmother and granddaughters, all the bats were unrelated. The bats were fed nightly from plastic measuring bottles so that the amount of blood consumed by each bat could be determined. Then, every night one bat was chosen at random, removed from the cage, and deprived of food. When it was reunited with its cagemates the following morning, the hungry bat would beg for food. In almost every instance, blood was shared by a bat that came from the starving bat's population in nature. Furthermore, there seemed to be pairs of unrelated bats that regurgitated almost exclusively to each other, suggesting a system of reciprocal exchange.

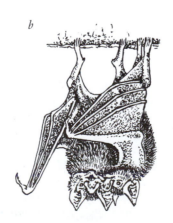

FIGURE 17.7 A vampire bat that was unsuccessful in obtaining a meal during a night's hunt begs for food from a roostmate. First it grooms the roostmate by licking it under the wings (*a*), then it licks it on the lips (*b*). If receptive, a well-fed roostmate will respond by regurgitating blood to the hungry partner (*c*). (From Wilkinson 1990.)

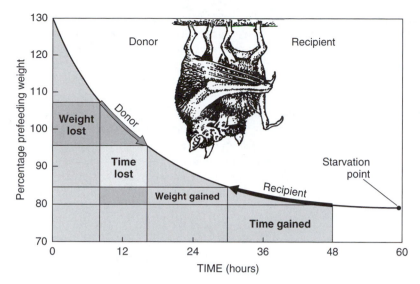

FIGURE 17.8   Blood sharing by vampire bats costs the donor less than the benefit to the recipient because the weight loss after a meal is exponential. By donating 5 milliliters of blood to a hungry roostmate, the donor loses only about 6 hours of the time it has left until starvation. However, the recipient may gain 18 hours, and therefore a chance to hunt again. (From Wilkinson 1990.)

PARENTAL MANIPULATION

Parental manipulation, as proposed by Richard Alexander (1974), is a means for the evolution of altruism in which parents are selected to produce offspring that behave altruistically toward their siblings. This selection occurs because the parents' fitness is increased in spite of a reduction in the personal fitness of a particular youngster.

Alexander's (1974) reasoning is as follows: If an allele were to cause an offspring to maximize its personal fitness by grabbing more than its fair share of parental favors from siblings, it would be beneficial to that individual as a juvenile. However, this behavior would decrease the parents' fitness because each and every offspring that parents produce, even the one that loses sibling disputes, bears copies of the parents' alleles. The greedy juvenile will mature and someday become a parent. It will then pass this allele to at least some of its own offspring, and their selfishness would reduce the parent's fitness. Thus, as a parent, the selfish sibling would become the victim of its genetic inheritance. In the end, an allele that favors selfish behavior toward siblings would reduce its bearer's overall reproductive success (i.e., there would be selection against such selfish behavior). The bottom line of Alexander's argument is that parents will always win when their fitness conflicts with that of their offspring.

Richard Dawkins (1989) later pointed out that this argument is flawed because it assumes a genetic asymmetry between the parent and offspring that does not exist. The genetic relationship between a parent and offspring is 50%, whether you look at it from the parent's or from the offspring's point of view. Thus, the argument could just as easily be made with the parent's and offspring's actions reversed, and then the opposite conclusion would be reached. If a parent had an allele

that improved parental fitness because one of the offspring was favored, that allele would be expected to have lowered its fitness when the parent was a juvenile. By this line of reasoning, the child should always win. We see, then, that there is no blanket answer to the question of whether the parent or the offspring will be favored by natural selection in the battle of the generations. In the end, we might expect the parent-offspring conflict to end in compromise.

EXAMPLES OF COOPERATION AMONG ANIMALS

As we consider various forms of cooperation among animals, we will note many similarities among distantly related groups. We will also notice that the selective forces leading to similar forms of cooperation may be quite different.

ALARM CALLING

Individuals of many social species emit an alarm call when a predator or other source of danger is discovered. As a result, others in the area are alerted to the danger, enabling them to take appropriate action to protect themselves.

Alarm Calls of Birds

Is Avian Alarm Calling Altruistic?   Does the bird who calls to alert its neighbors assume a risk? The answer has been controversial. If an acoustic engineer were to design a sound to be difficult for humans to localize, its structure would be similar to that of the alarm call of certain birds (Marler 1955). Because of this, it has sometimes been inferred that the caller is difficult to locate and may not be endangering itself. However, laboratory experiments on raptors such as barn owls, *Tyto alba* (Konishi 1973; Shalter and Schleidt 1977); pygmy owls, *Glaucidium perlatum* and *G. brasilianum;* and goshawks, *Accipiter gentilis* (Shalter 1978) showed that these predators are able to locate the source of alarm calls. Although avian alarm calls are usually almost pure tones (i.e., they consist of few frequencies) and pure tones are difficult for a barn owl to locate, the calls are usually centered between 6 and 8 kHz, wavelengths that are among those most easily located by barn owls (Figure 17.9). This suggests that the caller might indeed attract the attention of the predator and thus be behaving altruistically. Also consistent with this idea is the observation that willow tits (*Parus montanus*) keep silent when a hawk is within 10 meters and could hear an alarm call, but they emit an alarm call when a hawk is 40 meters away and out of earshot (Alatalo and Helle 1990).

Hypotheses for the Evolution of Avian Alarm Calling

Although alarm calling is quite common among birds, the selective forces that have led to it are still not clear.

INDIVIDUAL SELECTION   There are several suggestions for how alarm calling may have evolved through individual selection; a common theme among them is that calling reduces the caller's risk of predation. An alarm call informs a predator that it has been detected, and since many predators prefer to sneak up on unwary prey, the call may discourage an attack. For instance, in one study, although the goshawks and pygmy owls tested could orient to alarm calls, the predators were less likely to attack the prey if an alarm was sounded (Shalter 1978). In fact, the one goshawk tested that had been captured as a mature hunter failed to respond to

nine of the ten alarm calls. (It can be speculated that she was particularly unresponsive because she had had more unprofitable hunting experiences with alarm-calling birds than did the other birds tested.) In this way, calling may avert a predator's attack.

A slightly different variation on this theme is that the caller gains by warning others because they flee and no longer lure the predator to the area. A group of birds feeding in an open area is likely to be noticed by a predator. Once in the area, a predator will be a hazard to the would-be caller as well. However, a warning causes the entire flock to take flight. Thus, the caller's chances of survival are increased because its warning caused the others to clear out, thereby reducing the predator's interest in that area (R. Dawkins, 1976).

A corollary to this hypothesis is that alarm calls evolved because they indirectly reduce the chance that the predator will develop a preference for the caller's particular species. If a hunt is successful, the predator will be more likely to return to the area on future hunting expeditions. In time, then, the predator will become more familiar with the general area in which the prey was obtained and with the prey's habits. An individual bird may be safer if the predator never learns to prefer its species (Trivers 1971).

It has even been argued that the caller is completely selfish. It has seen the predator and knows its location. Calling informs others that a predator is near but does not tell them where it is. When a predator draws near, it is advantageous to get out of sight, but it is even better to take the flock with you as cover. Because only the caller knows the predator's location, it can place itself within the flock in such a way that it is less likely than the others to be the victim (Charnov and Krebs 1975).

We might expect the costs and benefits of alarm calling to vary among members of a flock, and if they do, members should differ in the frequency with which they emit alarm calls. For example, the frequency of alarm calling varies with the social status of willow tits—older, dominant males are more likely to emit alarm calls than are young, subordinate males. Older males may have more to gain by warning others because it may help keep their mate alive for the next breeding season (Alatalo and Helle 1990). Indeed, an adult male does produce more frequent alarm calls when he can see his mate than when she is out of view (Hogstad 1995).

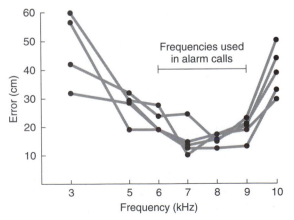

FIGURE 17.9   **The overlap between the frequencies employed in alarm calls and those that a barn owl locates most easily. (Modified from Konishi 1973.)**

KIN SELECTION   If kin selection were the basis for giving warning calls, the caller would be saving others that bear alleles in common with it. There may, in fact, be a risk in calling, but if the risk is outweighed by the benefit to family members, the allele for calling could be maintained in the population. Since this hinges on the presence of relatives in the rescued flock, John

Maynard Smith (1965), who originally made this suggestion, added that alarm calling could only have evolved during the breeding season, when it would enhance the survival of the caller's offspring.

Several lines of evidence suggest that kin selection is not the force that led to alarm calling among birds. For one thing, alarm calls *are* given throughout the year, rather than being restricted to the breeding season. In addition, since many flocks are made up of several species, one must wonder how many of the caller's relatives could be listening. Data for several bird species indicate that relatives usually disperse and thus simply would not be present to heed the warning (Trivers 1971).

RECIPROCAL ALTRUISM　It seems reasonable at first that reciprocal altruism could account for the evolution of warning calls; the risk assumed by one caller would be repaid when others return the favor. However, even Robert Trivers (1971), an originator of the concept of reciprocal altruism, believes that this is an unlikely explanation. Recall that altruism can evolve by this means only if cheaters are discriminated against. There is no evidence that warning calls are ever withheld because of the past behavior of a companion. In addition, the members of many flocks come and go, making it difficult to identify cheaters.

Alarm Calls of Ground Squirrels

Although the basis for the evolution of avian alarm calls is still disputed, Paul Sherman (1977) has built a convincing argument that some alarm calls of Belding's ground squirrels (*S. beldingi*) evolved through kin selection (Figure 17.10). These rodents are often victims of aerial predators, such as hawks, or of terrestrial predators, such as coyotes, long-tailed weasels, badgers, and pine martens. The alarm calls for these classes of predators are different. If the villain approaches on the ground, the alarm is a series of short sounds, whereas the warning of an attack from the air is broadcast as a high-pitched whistle. Thus, the selective forces behind the evolution of alarm calls in response to aerial predators appear to be different from those behind the evolution of alarm calls in response to terrestrial predators.

Individual Selection　Although the alarm calls in response to aerial predators appear to promote self-preservation, those in response to terrestrial predators do not. When a hawk is spotted overhead or when an alarm whistle is heard, near pandemonium breaks out in a Belding's ground squirrel colony. Following the first warning, others also whistle an alarm and all scurry to shelter. As a result, a hawk is rarely successful. However, when it is, the victim is most likely to be a noncaller. In one study, only 2% of the callers but 28%

FIGURE 17.10　A female Belding's ground squirrel is emitting an alarm call. When a terrestrial predator is spotted, females with living relatives nearby are more likely to call than are either females without kin as neighbors or males that rarely have kin in the vicinity. Observations such as these support the kin selection hypothesis for the evolution of alarm calling to warn of terrestrial predators.

of the noncallers were caught (Table 17.1). The most frequent callers were those that were in exposed positions and close to the hawk, regardless of their sex or relationship to those around them. Thus it seems that the alarm whistles given at the sight of a predatory bird directly benefit the caller by increasing its chances of escaping predation (P. W. Sherman 1985).

In contrast, individual selection does not seem to be behind the evolution of the ground squirrels' alarm trills, which are issued in response to terrestrial predators. In this case, the caller is truly assuming a risk; we know this because significantly more callers than noncallers are attacked. As can be seen in Table 17.1, 8% of the ground squirrels that called in response to terrestrial predators were captured, whereas only 4% of the noncallers were caught. The predators, even coyotes whose hunting success often relies on the element of

TABLE 17.1 Alarm Calling and Survival in Belding's Ground Squirrels at Tioga Pass, California.[a]

	Number of Ground Squirrels			
Category	Captured	Escaped	Percent Captured	$P(x^2$ Test)
Aerial predators				
Callers	1	41	2%	
Noncallers	11	28	28%	<0.01
Total	12	69	15%	
Terrestrial predators				
Callers	12	141	8%	
Noncallers	6	143	4%	<0.05
Total	18	284	6%	

[a]All data are from observations made during attacks by hawks (n = 58) and predatory mammals (n = 198) that occurred naturally during 1974–1982.

Source: P. W. Sherman (1985).

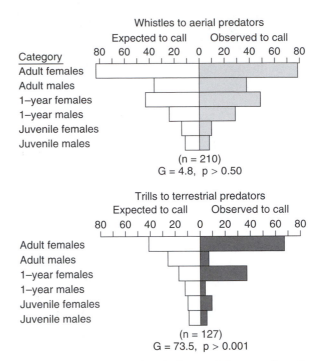

FIGURE 17.11 Expected and observed frequencies of alarm calls by Belding's ground squirrels in response to aerial and terrestrial predators. Expected frequencies are those that would be predicted if the animals called randomly. The calls in response to aerial predators are close to the expected frequencies. However, the calls in response to terrestrial predators are more likely to be given by females with relatives nearby than would be predicted if the animals called randomly. (From P. W. Sherman 1985.)

surprise, did not give up when an alarm call was sounded. Furthermore, the caller was not manipulating its neighbors to its own advantage. Generally, the reaction of other ground squirrels was to sit up and look in the direction of the predator or to run to a rock. Their reaction did not create the pandemonium that might confuse a predator. Nor did the caller seek safety in the midst of aggregating conspecifics (P. W. Sherman 1977).

Kin Selection The evidence that kin selection is the basis for alarm trills to terrestrial predators by ground squirrels is strengthened by information on the structure of their society. The squirrels in an area are usually closely related because of the stability of females in the population. Because daughters tend to settle and breed near their birthplace, the females within any small area are usually genetically related to one another. The sons, on the other hand, set off independently before the first winter hibernation, never to return to their natal burrow.

The population of Belding's ground squirrels studied by Paul Sherman (1977; 1980a, b; 1985) lives in Tioga Pass Meadow high in the Sierra Nevada Mountains of California. Because members of this population have been individually marked since 1969 and their genealogies are known, Sherman is able to keep records of which individuals called and when. His data, which are summarized in Figure 17.11, suggest that the ground squirrels practice nepotism, favoritism for family members. Notice in the figure that when a terrestrial predator appears, females are more likely than males to sound an alarm. This is consistent with kinship theory

because it is females that are more likely to have nearby relatives that would benefit from the warning. In addition, reproductive females are more likely than nonreproductive females to call (Figure 17.12). An even finer distinction can be made: Reproductive females with living relatives call more frequently than reproductive females with no living family.

Alarm calling in the round-tailed ground squirrel, *S. tereticaudus*, also appears to be a result of kin selection (Dunford 1977). The females tend to settle near their mothers and the males emigrate, just as Belding's ground squirrels do. It is the females, whose neighbors are likely to be family, that are most likely to call. An interesting difference between this species and Belding's ground squirrels is the incidence of male alarm calling. During the spring, when the males in an area are not likely to have kin nearby, they rarely warn others of impending danger. However, by July, a few juvenile males have settled near their mothers and sisters, and these males do issue warnings. If there were individual gain from calling, one would predict that males would emit alarm calls throughout the year. Instead, as predicted by the kin selection hypothesis,

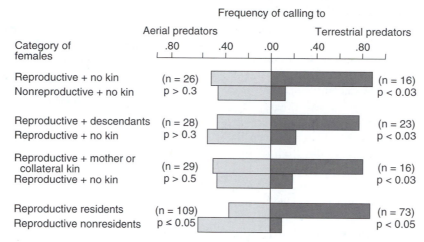

FIGURE 17.12   The effects of residency and genetic relatedness on the frequency of alarm calling for terrestrial and aerial predators in Belding's ground squirrels. Notice that when a terrestrial predator approached, reproductive females called more frequently than nonreproductive females. Furthermore, reproductive females with kin nearby called more than reproductive females with no kin, and residents called more frequently than nonresidents. Kinship and residency did not affect the frequency of calling when an aerial predator approached. (From P. W. Sherman 1985.)

they call only during the time of the year when kin are nearby.

COOPERATION IN ACQUIRING A MATE

Males of some species cooperate in attracting a mate. Some even relinquish the opportunity to pass their alleles into the future generation personally, at least temporarily. Instead, they concentrate their efforts on making another male more attractive to females. This presents a problem that should now be familiar: Since behaviors persist over time only when the alleles for them are perpetuated, how could this form of cooperation have evolved?

Kin Selection

When the cooperating males are related, kin selection seems to be a reasonable explanation for the evolution of this collaboration. Two examples, one from birds and the other from mammals, are presented to support this claim.

Wild Turkeys   Strangely, most male wild turkeys (*M. gallopavo*) in some Texas populations never mate. Toward the end of a young cock's first autumn, when he is about six to seven months old, he and his brothers forsake the others in their family and form a sibling group that will be an inseparable unit until death. This sibling group and all other juvenile male sibling units in the area flock together for the winter. During the first

winter, each male's status within this fraternity is determined. Only one of the males will mate, that individual being determined through competition.

Each male's reproductive fate is decided by the outcome of two contests. One competition is for dominance within the sibling group. Brothers battle by wrestling, spurring, striking with their wings, and pecking at heads and necks. Endurance is the key to success: The turkeys fight until they are exhausted. When only one is able to do battle, however weakly, he is the winner. The second contest is between rival sibling groups. The groups challenge and fight one another until a dominance hierarchy is established. The sibling group with the most members is usually victorious. Renegotiation of rank is rare; the dominance hierarchy within and between sibling groups is stable.

When the breeding season begins, females interested in mating visit the open meadows, where the males congregate. Although the male winter flocks have disbanded, individual sibling groups remain together. The brothers of each unit court the hens by strutting in unison, even though only the dominant male in the highest-ranking sibling group will mate. Of 170 tagged males displaying at four grounds, not more than 6 males accounted for all 59 observed matings. If a subordinate male is presumptuous enough to attempt a mating, the dominant male chases him away and then mates with the hen.

A subordinate male gains inclusive fitness by helping his brother to perpetuate his alleles. On the other hand, without his assistance, the brother could not be successful; the cooperative efforts of siblings are necessary for their unit to become dominant, and the synchronous strutting of siblings makes the dominant male more attractive to the hens. Thus, the subordinate brother reproduces by proxy (Watts and Stokes 1971).

Lions   Male lions (*Panthera leo*) also cooperate in attaining mates. When the group of males, called a coalition, is larger than three, the males are usually brothers, half brothers, and cousins that left their natal pride as a group (Packer et al. 1991). They remain together, and after one to three years of traveling nomadically, they challenge the males of other prides. The coalition may take over a pride by slowly driving out the resident males, or it may be a hostile takeover, involving serious fighting (Figure 17.13). In such contests, the larger coalition usually wins. The reward for the victors is a harem of lionesses. When the females come into reproductive condition, which is sometimes hastened if the new males kill any cubs that are present, they often do so simultaneously. During the two- to four-day period when a female is in reproductive condition, she mates about every 15 minutes around the clock. Any of the males in the coalition may be the first to find her, mate with her, and keep others away by his presence. A female may change mates during this period but generally not more than once a day (Bertram 1975, 1976). Thus the male who mates may gain fitness directly. When another male takes over, it is likely to be a relative. Then the first male still may gain fitness indirectly.

Individual Selection

As you must now be aware, similar behaviors can evolve through different mechanisms, and more than one mechanism may shape a behavior. Although kin selection may be responsible for the evolution of assistance in mate acquisition among wild turkeys in Texas, it is only part of the story among lions and is not involved in

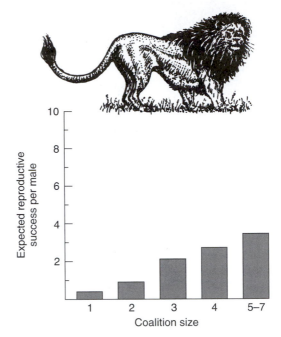

FIGURE 17.14   The size of male coalitions of lions is related to each male's reproductive success. As the coalition's size increases, so does reproductive success. (Data from Packer and Ruttan 1988.)

other cases of cooperative courting. As we will see, it may be to a male's advantage to assist another's reproductive efforts if this service increases his chances of personal reproduction, now or in the future.

Lions   It is now known that roughly half of male coalitions contain at least one unrelated male, and coalitions of two or three usually consist of unrelated males (Packer 1986; Packer et al. 1991; Packer and Pusey 1982). Immediately the question arises, then, of why an unrelated male would be accepted in a coalition. The answer turns out to be quite simple—the larger the coalition, the greater a male's reproductive success (Figure 17.14). Larger coalitions have a better chance of ousting the current coalition in a pride, maintaining control of that pride, and perhaps even gaining residence in a succession of prides. A solitary male has little chance of reproducing and, therefore, much to gain by joining another coalition. A small coalition may also gain by accepting an unrelated male because the extra member may help it take over prides. Indeed, coalitions accept unrelated companions only while they are not yet resident in a pride (Packer and Pusey 1987).

Larger coalitions also remain in control of a pride longer than smaller ones. A coalition of three to six males may remain in control as long as two to three years. A coalition of two might be in possession for over a year. If a lone male manages to gain control of a pride, which happens infrequently, his tenure generally lasts only a few months (Bertram 1975). As a result, the life-

FIGURE 17.13   Two male lions are fighting for control of a pride.

time success of a male lion increases by cooperating with other males in taking over a pride, even if all the males are not related (Packer et al. 1988).

Long-Tailed Manakins

Among male lions, cooperation in acquiring a mate may have evolved because of both direct and indirect fitness gains. However, cooperating long-tailed manakins (*Chiroxiphia linearis*) are not related. Therefore, there are no indirect fitness gains and kin selection cannot be playing a role. It takes two, or sometimes three, of these beautiful male birds to court a female. First, they must attract a female, and two males will perch together and call synchronously. The call, which sounds similar to the word *toledo*, may be emitted as many as 19 times a minute and 5000 times a day. If these vocalizations are successful in attracting a female, the males move to a display perch. The next step is for one male to stimulate the other to perform the acrobatic display that will induce the female to mate. The male is put "in the right mood" by the solicitation display executed by his masculine partner.

When the males are ready, they proceed to the jump display that will stimulate the female. This courtship display cannot be performed alone. In one of the most common variations of the jump display, called the up-down variant, the movements of the males are somewhat like those of two children on a seesaw, but the birds do not use any apparatus. One male jumps into the air, emitting a wheezy *buzzee* call and hangs there momentarily; just as he lands, the other male jumps up. In this way they alternately jump up and down. Another common variant of the jump display is the cartwheel variant, in which the males make a continuous moving circle around each other. The pattern of their movements resembles that of objects being juggled. They begin by perching next to each other on the display branch. The front male jumps up and backward just far enough to land on the spot where the second male had previously perched. Collision is avoided because while the first male is in midair, the partner moves anteriorly to occupy the spot where the other male had perched. Now this male takes a turn at jumping up and moving to the rear. The courtship sequence may be repeated only once or as many as a hundred times in succession. When the display bout is over, one male leaves the display branch and watches while the remaining male does a solo performance of the precopulatory display. If his gymnastics have impressed the female, she mates with him.

Although the males take turns jumping to court a female, it is always the same male who mates. The benefit to this male is obvious, but what is the advantage to the other male? It is not indirect fitness because it seems unlikely that the two males are related. A typical brood consists of only one or two offspring, and there is no reason to assume that the siblings are necessarily the same sex. Furthermore, just before and after each breeding season, the young, particularly the subadult males, disperse. It seems unlikely, therefore, that male relatives would stay in proximity for the three to four years it takes them to acquire adult plumage. If the partners are not genetically related, the nonbreeder is not increasing his indirect fitness. So why doesn't he compete in attracting females?

A possible answer to how the cooperative courting of the long-tailed manakin evolved emerges when the choices of the subordinate bird are considered. Although he will not mate while he is a member of his current male-male alliance, his chances of mating would not be increased by deserting his partner. Solitary males cannot mate. They cannot even perform the courtship display. If he cannot dominate this partner, his chances of becoming the dominant member of another pair would be low. However, if the subordinate male outlives his partner, it is likely that a younger male, one that can be dominated, will become his new associate. Then it will be his turn to mate and raise his personal fitness (W. A. Foster 1977; McDonald and Potts 1994).

Reciprocal Altruism

Cooperation in mate acquisition may also evolve through reciprocal altruism. An example of this is the coalitions formed among male olive baboons, *Papio anubis* (Figure 17.15). A male who lacks a female consort sometimes enlists the help of a friend to win another male's mate. The following scenario is typical of what often occurs: Male A is consorting with an estrous female. Male B covets this female since he has none of his own. He solicits the help of male C, and the two form an alliance and challenge male A. While the battle is in progress, male B gets away with the female. Male C has acted altruistically; he risked injury while assisting another to acquire a mate. However, at some time in the future, he will enlist the help of male B in winning a consort of his own (Packer 1977).

COOPERATIVE BREEDING AND HELPING

Cooperative breeding occurs when individuals (helpers) assist in the care and rearing of another's young rather than producing offspring of their own. It was first described among birds (Skutch 1935). Indeed, Alexander Skutch (1961) has defined a helper as "a bird which assists in the nesting of an individual other than its mate, or feeds or otherwise attends a bird of whatever age which is neither its mate nor its dependent offspring." Cooperative breeding has now been described in roughly 3% of bird and mammal species (S. T. Emlen 1997) and in several species of fish (Dugatkin

FIGURE 17.15 An alliance between two male olive baboons. The two males on the right are cooperating to challenge the male on the left. Alliances such as this one are formed to win the lone male's consort for one of the challengers. At a later time, the male that was assisted will have to reciprocate.

1997). In any species, though, the helper facilitates the genetic legacy of others.

Helpers' Duties

Basically, helpers give parental care to offspring that are not their own. Helping is more precisely called alloparental care (E. O. Wilson 1975). Most commonly, helpers assist in one of two ways—providing food for or protecting the offspring of others—but there are other ways to help as well.

Providing Food   Most helpers bring food to another individual's offspring. Florida scrub jay (*Aphelocoma coerulescens*) helpers, for example, deliver about 30% of the food consumed by the nestlings. Although this help does not increase the total amount of food brought to the nest, it does reduce the parents' share of the job and enables them to enjoy better health, as is suggested by the improvement in survivorship of breeders with helpers. In one study, 87% of the breeders with helpers lived to the next year, but only 80% of the breeders without helpers did so (Stallcup and Woolfenden 1978).

Among some mammals, the helper may bring food to the mothers, as well as to the offspring. Blackbacked jackals (*Canis mesomelas*), also known as silverbacked jackals, deliver food by regurgitation (Figure 17.16). Helpers contribute between 18% and 32% of all regurgitations to pups. In addition, they contribute to the nourishment of the lactating mother, which might remain with the pups while the others hunt (Moehlman 1979).

Protection of Offspring   One advantage of group living is increased vigilance and defense against predators. Therefore, it is not surprising that helpers provide extra protection for the young. Besides issuing alarm calls to warn the chicks, Florida scrub jay helpers actively defend them by mobbing predators such as snakes (Woolfenden 1975). Jackal families with helpers always have an adult on guard to drive away predators,

whereas groups lacking helpers may have to leave the pups unattended while the others hunt. Among fish, the helpers mainly contribute by offspring protection; they can contribute little to the nourishment of the young (Dugatkin 1997).

Other Activities   Depending on the species, however, helpers may engage in a variety of other activities. In certain bird species, for instance, helpers may build and clean nests or incubate and brood the nestlings

a

b

FIGURE 17.16 A jackal helper is about to regurgitate food to a pup (*a*) and is chasing away a predator (*b*).

(Skutch 1987). Although most primate helpers serve primarily as baby-sitters, certain ringtailed lemur helpers are an exception. These helpers may nurse the infants in addition to assisting in their care (Pereira and Izard 1989). In another primate, the saddle-backed tamarin (*Saguinus fuscicollis*), there are two types of helpers. One type consists of extra males whose job is to carry the heavy juveniles. At birth a tamarin is almost 20% of its adult weight, and typical litters consist of twins. Thus, carrying these youngsters is burdensome, and if the duty were not shared the mother might not be able to obtain enough nourishment for herself and to ensure a milk supply (Terborgh and Goldizen 1985).

Do Helpers Help?

Although there are exceptions, many studies do indicate that the helped reap fitness benefits from the helper's assistance. These benefits may take two forms—increased survival of the young and increased survival of the breeders. We have mentioned the Florida scrub jay, and this species is one of the most thoroughly studied to date (Woolfenden 1975; Woolfenden and Fitzpatrick 1990). These birds live in territories that contain one breeding pair and a varying number of helpers—from none to as many as six. The breeding success for pairs with helpers clearly exceeds that of pairs without helpers (Woolfenden 1975). Figure 17.17 shows the breeding success of experienced pairs with and without helpers during one five-season study. It can be seen that the presence of helpers has no effect on the number of eggs laid but does increase the chances that

the young will hatch, leave the nest, and become independent birds.

Similarly, breeding success increases with the presence of helpers in some mammalian and fish species. Blackbacked jackals form monogamous breeding pairs. Between one and three of the young from previous litters remain with their parents and help them rear the next pups. As seen in Figure 17.18, the breeding success of a pair of blackbacked jackals increases with the number of helpers (Moehlman 1979). A similar relationship is found in the Princess of Burundi cichlid fish (*Lamprologus brichardi*). More young are successfully raised when the young from one brood help their parents guard the eggs and larvae of subsequent broods from predators (Taborsky and Limberger 1981).

But increased breeding success in groups with helpers is not by itself sufficient to demonstrate that breeders are actually benefiting from the assistance of helpers because correlation does not show cause and effect. To see this, consider the following example: The number of bars is usually greater in towns with more churches. This is not because the fear of God causes people to drink but rather because the frequency of both enterprises increases as the population does.

In the case of helpers, some other factor, territory quality, for example, could be responsible for the increase in both reproductivity and group size. If reproductive success is higher among individuals on good territories and the young stay with their parents, we would expect groups on good territories to have more helpers and higher reproductive success, even if the helpers did nothing at all.

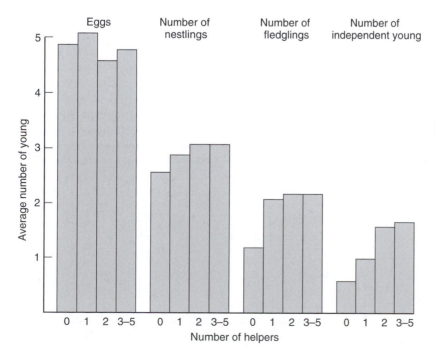

FIGURE 17.17 **The relationship between the number of Florida scrub jay helpers and the breeding success of the experienced parents. Helpers do not increase the number of eggs laid. They do, however, increase the chances that the eggs will hatch and that the young will survive to become independent. (Data from Woolfenden 1975.)**

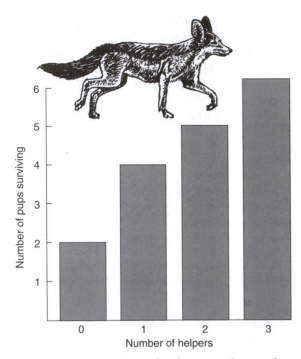

FIGURE 17.18   The relationship between the number of blackbacked jackal helpers and the number of the pups surviving. (Modified from Moehlman 1979.)

One way to examine whether territory quality, not the presence of helpers, is responsible for the enhanced reproductive success of groups with helpers is to remove the helpers and look at the effect, if any, on reproductive success. If helpers really do help, we would expect that their removal would lower reproductive success. On the other hand, if they do not help, their removal should have no effect. In one experiment, helpers were removed from the nests of gray-crowned babblers (*Pomatostomus temporalis*), a bird that lives in year-round territorial groups of 1 to 13 birds in the open woodland of Queensland, Australia. Parents are usually assisted by a variable number of their offspring from previous broods, but in this experiment, 9 of the breeding groups were reduced to a single helper. These groups then raised an average of 0.8 young, less than half the number of fledglings produced by the 11 control groups, which had more assistance. Therefore, the positive relationship between breeding success and the number of helpers found in gray-crowned babblers does seem to be a result of the presence of helpers (Brown et al. 1982).

However, not all data are consistent with the idea that helpers increase the fitness of the breeders; Amotz Zahavi (1974) reached the opposite conclusion from his studies of the Arabian babbler (*Turdoides squamiceps*). These birds nest communally in the deserts of Israel. As in the other examples, the young from previous broods remain to help their parents raise the next brood, but

Zahavi did not find that helpers increased the reproductive success of the group. Furthermore, he observed agonistic behavior within the breeding groups. One nestling died as the result of a wound probably caused by a peck. Zahavi suggests that helpers may do more harm than good by reducing the amount of food available for nestlings and by increasing the chances that a predator will be attracted by activity around the nest. He argues that the helpers are tolerated only because they are offspring of the breeders and because they help to defend the nest when a predator approaches. His conclusions remain controversial, however, because Jerram Brown's (1975) reanalysis of Zahavi's data did show a positive relationship between the breeding success and the presence of helpers.

When considering the balance of benefits and costs of harboring a helper from the breeder's point of view, it is interesting to note that in at least one species, the pied kingfisher (*Ceryle rudis*), helpers are tolerated only when their services are needed. These birds usually have primary helpers, which are older offspring, but may also have secondary helpers, which are unrelated. Heinz-Ulrich Reyer (1980) compared two colonies of pied kingfishers in East Africa. Breeding pairs at Lake Naivasha typically have only one primary helper. When males apply for a job as secondary helpers, they are persistently chased away by the male territory holder. In contrast, at Lake Victoria, secondary helpers are eventually tolerated and permitted to stay and feed the young. Why? The answer is that the services of secondary helpers are needed to raise offspring at Lake Victoria but not at Lake Naivasha. These birds fish for a living, and Lake Victoria is a harder lake to fish. Victoria's rougher waters increase the time it takes to catch a fish, and the fish are smaller. Furthermore, the fishing grounds are farther from the colony. With the additional fish provided by secondary helpers, the breeding pair can raise more offspring (Table 17.2).

TABLE 17.2   The Effect of Helpers on the Reproductive Success of Pairs of Pied Kingfishers

	Lake Victoria			Lake Naivasha		
	Mean	SD	*n*	Mean	SD	*n*
Clutch size	4.9	0.6	22	5.0	0.6	8
Young hatched	4.6	0.5	14	4.5	0.7	2
Young fledged						
No helpers	1.8	0.6	14	3.7	0.9	9
1 Helper	3.6	0.5	12	4.3	0.5	4
2 Helpers	4.7	1.0	6	—	—	—

Source: Reyer (1980).

Why Do Helpers Help?

Although helpers are not always related to the breeders, in most species helpers are older offspring who postpone departure from the natal territory and remain, at least for a while, within the parental group. In some species, the young may remain at home because they benefit so greatly from group living. Juveniles remaining with a group may enjoy access to some critical resource (either now or in the future), protection from predators, more food as a result of cooperative hunting, or more effective territorial defense (Stacey and Ligon 1987). The costs of dispersing may also favor juveniles that remain at home. At least four factors probably enter into the decision: (1) risk of dispersal, (2) chances of finding a suitable territory, (3) chances of finding a mate, and (4) chances of successful reproduction once established.

As we will see, the costs and benefits of helping are influenced by certain ecological conditions. Severe ecological conditions, such as the prohibitive costs of reproduction or shortages of territories or mates, limit the options of independent breeding and favor staying with the parents (S. T. Emlen 1982a, 1997; Emlen and Vehrencamp 1983).

Habitat Saturation in Stable Environments

A young male scrub jay probably stays with his parents because he has little choice. It is very unlikely that he could find a suitable area for a territory because the available space is filled. A bird that wins a territory keeps it for life (Woolfenden 1975). The most common way for a male to acquire a territory is by inheriting a portion of his parents' property, either by replacing his father after his death or by subdivision of his father's territory. If there is more than one son helping, the most dominant one is favored in the property settlement (Woolfenden and Fitzpatrick 1978).

The scrub jay's problem is typical of that encountered by animals living in stable environments. When ecological conditions are predictable, numbers increase and suitable habitat becomes saturated. The result is severe competition for territories. An individual has the option of leaving the parental nest and breeding independently only if a territory can be established by challenging and defeating a breeder, successfully competing for any vacancies that result from the death of nearby breeders, or inheriting or budding off a portion of the parental territory. If an individual must postpone breeding until a territory is available, it is best to wait at home because this is an area of proven quality, a factor that will increase the chances of surviving until the next year. Meanwhile, alliances can be formed that may enable the takeover of other territories, including a portion of the parents'. The helper's waiting time is best spent in increasing his inclusive fitness by helping to raise siblings (S. T. Emlen 1982a).

This contention is supported by a comparison of the extent of helping in populations of the acorn woodpecker (*Melanerpes formicivorus*) that are residing in habitats with varying degrees of saturation. These gregarious birds, best known for their meticulous habit of individually caching thousands of acorns in storage trees called granaries (Figure 17.19), have been studied in several locations in the United States. Study sites in California, New Mexico, and Arizona vary with respect to woodpecker density, territory turnover rate, and territory fidelity. The occurrence of helping at these three locations parallels the gradient in difficulty associated with territory establishment (S. T. Emlen 1982a).

Michael and Barbara MacRoberts (1976) studied acorn woodpeckers in coastal California and noted that not a single territory became vacant during their three-year study. In this extremely saturated habitat, birds were permanently territorial and lived in family groups that consisted, on average, of 5.1 adults plus the young of the year. Forty-nine percent of the juveniles remained at home, and 70% of the groups had helpers. Young acorn woodpeckers in the Magalena Mountains of New Mexico face somewhat better odds in their quest for suitable territories than do those on the West Coast: 19% of the territories in New Mexico became vacant over a three-year study by Peter Stacey (1979). The average group size at this locality was 3.0, and 29% of all the youngsters stayed at home. Helpers were present in only 59% of the breeding groups. However, acorn woodpeckers in the Huachuca Mountains of southeastern Arizona seem free of the shortage of available housing. In contrast to their California and New Mexico counterparts, those in Arizona tended to disperse or migrate between seasons (Stacey and Bock 1978). The average group size at the Arizona site was 2.2, and only 16% of the breeding units had helpers. In

FIGURE 17.19 An acorn woodpecker places an acorn in a previously drilled hole in a dead tree. Food stores must last the communal group through the winter.

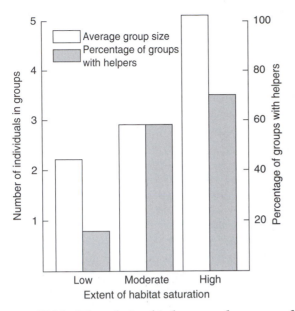

FIGURE 17.20 The relationship between the extent of habitat saturation and both the average group size and the frequency of helping behavior in acorn woodpeckers. In environments with few available territories, the group size and the percentage of groups with helpers is much greater than in areas of low habitat saturation. (Data from S. T. Emlen 1982b.)

sechellensis) disperse to fill vacant territories unless the vacancy is of equal or higher quality than its natal territory (S. T. Emlen 1997).

Biased Sex Ratios   Sexual partners, rather than territories, may be in short supply for some cooperatively breeding birds. Typically involving an excess of males, this demographic constraint limits the option of becoming established as an independent breeder (Emlen and Vehrencamp 1983). Male-biased sex ratios characterize the splendid fairy-wren (*Malurus splendens*), an inhabitant of heathlands and scrubs of western Australia. This small passerine, whose tail makes up nearly half of its body length (Figure 17.21), has been studied since 1973 by Ian Rowley and his colleagues. Females of this species suffer much greater annual mortality than do males (57% and 29% respectively; Rowley 1981) and thus are frequently in short supply. As would be predicted from the scarcity of females, helpers tend to be males that are awaiting an available mate. For some individuals, the wait can be as long as five years. As patterns of mortality and the resultant sex ratios vary, so does the percentage of groups with helpers: When females are scarce, male helpers are plentiful. Breeding females with helpers survived better

short, the frequency of helping in populations of acorn woodpeckers varies directly with the scarcity of open territories (Figure 17.20).

As usual, however, some parts of these stories do not fit as neatly into the interpretation of helping as a consequence of habitat saturation. In populations of both the Florida scrub jay and the acorn woodpecker, available territories occasionally remain vacant. These inconsistencies certainly deserve further attention, but we should not be too quick to abandon the hypothesis completely because of them. There is some evidence, for example, that the Florida scrub jay helpers that decline to establish their own territories are somehow biologically inferior and would be unable to establish and maintain a territory of their own. Instead, they survive as nonbreeders in the relative safety of their parents' territory and gain some indirect fitness by helping to raise siblings (Woolfenden and Fitzpatrick 1990).

It has also been suggested that the benefits of group living, rather than the inability to win and maintain a territory, may explain why territories remain vacant. A major benefit to acorn woodpeckers that remain on their natal territory is present and future access to the trees needed for acorn storage. A juvenile does better by remaining on its parents' high-quality territory as a nonbreeder and postponing breeding than it would if it attempted to breed on a low-quality territory (Koenig and Stacey 1990). Indeed, neither acorn woodpeckers nor Seychelles warblers (*Acrocephalus*

FIGURE 17.21 A male splendid wren carries an insect to feed its young. When adult females are in short supply in the population, this breeding male can count on his sons to help rear the next brood.

than those without helpers (Rowley and Russell 1990; Russell and Rowley 1988). Furthermore, helpers, combined with breeding experience, caused more females to renest after a first brood had been raised.

Cost of Reproduction in Unstable Environments

Some populations of cooperative breeders live in environments in which conditions fluctuate. Breeding attempts among novices are largely unsuccessful for species that inhabit variable, unpredictable habitats. Faced with erratic changes in environmental conditions, juveniles often remain at home and help rear their relatives.

Support for the idea that helping should increase with the degree of breeding difficulty comes from studies on white-fronted bee-eaters, *Merops bullockoides* (Figure 17.22). The life history of this bird makes it a good species in which to look for a relationship between helper frequency and environmental harshness. It lives in the savannahs and scrub grasslands of the Rift Valley in eastern and southern Africa. The breeding of these insectivores is most successful when it is timed to coincide with periods of insect abundance. However, synchronization of these events is difficult to achieve because there is extreme variation in both the timing and amount of rainfall and the response of insects to the rain. As a result, the ease of raising offspring and, therefore, the cost of breeding independently vary from year to year.

The frequency of helping does indeed increase during hard times. During periods of little rainfall and hence low food availability, bee-eaters suffer high losses from nestling starvation. It is during these periods of extremely harsh environmental conditions, when the chances of successful reproduction are slim, that older

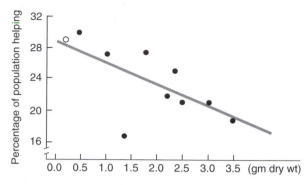

FIGURE 17.23   The incidence of helping in white-fronted bee-eaters varies as a function of food availability; when insect abundance increases, fewer young remain at home to help care for younger siblings. The closed circles indicate insect availability during the month before breeding. The open circle indicates the insect availability for one colony during the month after breeding. (From S. T. Emlen 1982b.)

offspring remain in their natal groups as helpers. Once conditions become more favorable, however, young bee-eaters are more likely to disperse and initiate breeding on their own. Notice in Figure 17.23 that helping increases with decreases in either rainfall or the availability of food, conditions that make it difficult to raise young (S. T. Emlen 1982a).

The Evolution of Cooperative Breeding

In this discussion and throughout most of this book, we have assumed, as most people do, that behavior is a result of natural selection. However, the view that helping is an adaptation shaped by natural selection has been challenged (Jamieson 1986, 1989, 1991; Jamieson and Craig 1987), and this has led to considerable debate (S. T. Emlen 1991; Ligon and Stacey 1991; White et al. 1991). Put simply, Ian Jamieson has suggested that helping is a byproduct of the evolution of parental care and communal breeding. As we have seen, for some species dispersal may be difficult. Also, there may be advantages in group living, including increased vigilance against predators and cooperative foraging. As a result, natural selection may favor group breeding in some species. Natural selection may also favor individuals that feed begging youngsters wherever they are found. In other words, helpers and parents may be responding to the stimuli that trigger parental care. Communal breeding brings nondispersing older animals, usually older siblings, in contact with begging youngsters. Helping, then, is seen as an unselected consequence of group living.

If helping is an unselected byproduct of communal breeding, we might expect all adults to follow the general rule, "If it begs, feed it." However, this is not always

FIGURE 17.22   When environmental conditions are harsh, young white-fronted bee-eaters forsake independent reproduction and remain at home to rear their younger siblings.

the case. Whether or not an immigrant acorn wood-pecker feeds nestlings depends on when it joined the group. If the immigrant arrived before the eggs were laid, it will feed the youngsters. But if it arrived after the eggs were laid, it will ignore the begging (Stacey and Ligon 1987). Furthermore, among white-fronted bee-eaters, only about half of the nonbreeding members of the group become helpers, even though all were exposed to the same begging stimuli (Emlen and Wrege 1989). Exceptions such as these are not expected if helping is triggered by the stimulus of begging young.

It has also been argued that helping, at least in some species, is costly to the helper, and thus we might anticipate natural selection to refine parental responses, causing helping to be weeded out (Emlen et al. 1991). Among white-winged choughs (*Corcorax melanorhamphos*), for instance, helpers have a variety of duties, one of which is incubating the eggs. Although all the helpers develop a brood patch, not all incubate the eggs. Those that do incubate lose weight in proportion to the time they spend on the nest (Figure 17.24) (Heinsohn, Cockburn, and Mulder 1990). It would be surprising that such a costly behavior would be maintained even if it originated as an unselected byproduct of selection for communal breeding and parental care.

Jamieson and Craig's (1987) suggestion that helping has not been selected has been useful in focusing attention on the untested assumptions that can pervade the study of any behavior. As a result, it has stimulated research in many new directions. Keeping in mind that any direct or indirect benefits that result from helping are consistent with the hypothesis that helping is a result of natural selection, but not proof of it, we will assume that natural selection has favored helping when the benefits gained by the helper exceed the costs of helping.

Individual Selection   As we consider the potential costs and benefits, we should keep in mind that each of these will vary not just among species but also within species (Heinsohn and Legge 1999). Andrew Cockburn (1998) has suggested that males and females may help for different reasons. He notes that among bird species in which the males help, their efforts don't usually increase the number of young produced. However, in species with female helpers, groups with helpers do raise more offspring than those without. Thus, whereas females may help for indirect fitness gains by increasing the number of nondescendent kin raised, males may help for direct fitness gains.

COST TO THE HELPER   Most parents will agree that raising young is energetically draining. Recall that white-winged choughs lose weight in proportion to the amount of time they spend on incubating the eggs. Mongoose (*Suricata suricatta*) helpers also experience an energetic cost from helping. In this case, a helper forgoes feeding and stays at the burrow to baby-sit for the young pups and guard them from predators for an entire day while the parents and others forage. During a 24-hour shift, the baby-sitter loses 1.3% of its body weight. In contrast, the foraging group members gain roughly 1.9% of their body weight (Heinsohn and Legge 1999). Helpers among Princess of Burundi cichlid fish grow more slowly than nonterritorial fish. In this case, however, the slow growth isn't due to the increased energy expenditure of helping. Instead, it seems to be caused by the helper's low rank in the dominance hierarchy (Taborsky 1984).

Helping may even reduce survival. For example, the helpers among stripe-backed wrens (*Campylorhynchus nuchalis*) that bring the most food to the young have reduced survival. Furthermore, many of them will never get a chance to breed themselves. Of course, it is possible that these enthusiastic helpers don't live as long as others simply because they are poor-quality birds (Rabenoid 1990).

BENEFITS OF HELPING   Although in many species older offspring remain at home during periods of either harsh environmental conditions or territory or mate shortages, why should they help? Furthermore, not all helpers are relatives. We have already discussed one example of unrelated helpers, the pied kingfisher. Others, including mammals such as the dwarf mongoose (*Helogale parvula*), can be listed. The inclusive fitness of helpers cannot be increased by the additional offspring they help to raise.

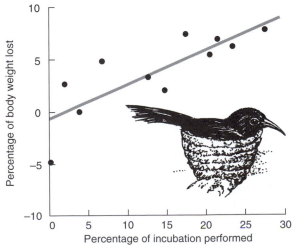

FIGURE 17.24   Some white-winged chough helpers incubate the eggs. Those that do lose weight steadily. It would be expected that such a costly behavior would be selected against if there were no benefits from helping that outweighed the costs. (From Heinsohn, Cockburn, and Mulder 1990.)

The question is, then, what might they be gaining? Many different hypotheses have been offered as answers (J. L. Brown 1987; S. T. Emlen 1991).

1. *Enhanced survivorship.* As we have already noted, some ecological conditions make it likely that an individual will live longer if it stays with a group. The Princess of Burundi cichlid fish is an example. When the fish are small, the chances of survival off the natal territory are slim because of predation. Even when breeders are experimentally removed from a territory, potential helpers choose to stay at home instead of moving to the unoccupied territory (Taborsky 1985).

Helping, then, may be "paying" the breeders for permission to stay on a territory of known quality until the helper can win one of its own. For example, when helpers of another cooperatively breeding cichlid fish (*Neolamprologus pulcher*) were temporarily removed, other helpers attacked them upon their return. After their return, the helpers increased their rate of territory maintenance, number of visits to the brood chamber, and defense of the brood chamber. In spite of the increase in their efforts, 29% of the returned helpers were eventually evicted. These observations are consistent with the idea that the helpers must pay with help to be allowed to remain on the family territory (Balshine-Earn et al. 1998).

2. *Increased reproduction in the future.* There are also many ways that helping might increase the chances of successfully raising offspring in the future. We have seen that helping may be a way of gaining a territory, as it is for the Florida scrub jay.

When mates are scarce, the helper may gain an opportunity to acquire the original breeder as a mate in a later year. Frequently, helpers who are unrelated to the breeder are lacking a mate of their own. Sometimes, this helper assists a female breeder one year and mates with her the following year. This is the case in the pied kingfisher. You may recall that the secondary helpers are unrelated to the breeders, and all secondary helpers are unmated. More than half of them return the following year, and of these half succeed in mating with the female they had assisted (Reyer 1980; 1984; 1986).

If the helper finally becomes a breeder, it may gain assistance in raising its own young from the offspring it helped to rear. The white-winged chough, a bird that cannot breed without helpers, is an excellent example. Groups with fewer than seven members cannot support even a single youngster through the first winter. The main reason seems to be that foraging is a skill that takes years for a chough to master. Their diet consists of small invertebrates that are found by digging in the soil. Finding sufficient food is a difficult and time-consuming activity. Juveniles must be fed for up to eight months, and even then they are poor foragers (Heinsohn, Cockburn, and Mulder 1990). Thus, helpers are a necessity and the group grows slowly, usu-

ally only through reproduction. The young remain in the group and help raise the next generation of choughs (Heinsohn 1991).

3. *Increased reproductive success as breeder.* In addition, the helper may benefit if its activities help it to learn or practice skills that will be necessary when it raises its own young. Although there may be examples of this among certain birds, the best illustration might be among primates. Immature female vervet monkeys (*Cercopithecus aethiops sabaeus*) often baby-sit for the infants of others. Juvenile females that initiate caretaking more often and have more experience in carrying infants are more successful in raising their own first-born when they become mature than are females with less baby-sitting experience (Fairbanks 1990).

4. *Social prestige.* It has also been suggested that the return on a helper's investment is a boost to its social prestige, which provides direct fitness benefits through increased dominance rank or additional opportunities to mate or form coalitions (Zahavi 1995). Because helping is costly, it provides a reliable signal of the helper's quality to others in the group. This explanation has been used to interpret certain observations of Arabian babblers. The dominant birds in the social group sometimes hinder the helping efforts of individuals ranked slightly below them but not those of individuals whose rank was greatly below their own. By preventing the subordinate from gaining social status through helping, the dominant bird reinforces its position over those who threaten it most. The idea that helpers gain fitness through social prestige remains controversial, especially in light of a more recent study in which dominants were not observed to interfere with subordinates and all group members, regardless of rank or sex, brought food to the chicks at the same rate (Wright 1997).

Kin Selection   Most helpers are previous offspring of a breeding pair that assist in the rearing of their siblings. Since an individual is as closely related to a sibling as to an offspring, it seems reasonable that kin selection might play a role in the evolution of cooperative breeding. Furthermore, most cooperatively breeding species live in habitats that either prevent the individual from breeding independently or make it difficult to raise young successfully without assistance.

There is still much controversy over the importance of kin selection in helping. The hypotheses we have discussed are not mutually exclusive. Any one of them or any combination might be valid for a particular species. This question then arises: How might we determine, for any given species, whether any of the proposed hypotheses for the evolution of helping by individual or kin selection are correct? The answer is simple—we make predictions that are consistent with each hypothesis and then evaluate them. The evaluation may be based on observation or experimentation.

TABLE 17.3   Predictions Concerning (1) Fitness Gains Accruing to Helpers and
(2) Personal Characteristics of Birds That Become Helpers

Hypothesis	Prediction 1	Prediction 2
Helping results in direct fitness gains through:		
Increased survival to next breeding season	Survival probability should increase with group size; birds that have served as helpers should have a higher survival than those that have not.	No prediction.
Increased future opportunity to breed	Birds that have served as helpers should have a greater probability of breeding than those that have not.	Birds that have not bred before should be more likely to help than individuals that have already achieved breeding status.
Increased reproductive success as breeder	Reproductive success for first-time breeders should increase as some function of the amount of previous helping experience. and/or The likelihood of first-time breeders having helpers of their own should increase with previous helping experience.	Birds that have no prior experience should be more likely to help than birds that have extensive prior experience.
Helping results in indirect fitness gains due to the extra related offspring reared	The presence of helpers should significantly increase the production of young at helped nests. and/or The presence of helpers should increase the survival of recipient breeders. and Helpers should, on average, be closely related to the nestling beneficiaries.	Birds should be more likely to help when the recipients are close kin than when they are distant kin or unrelated.

Source: Modified from Emlen and Wrege (1989).

Stephen Emlen and Peter Wrege (1989) used five years of data on white-fronted bee-eaters to test alternate hypotheses for helping behavior (Table 17.3). The birds in the study were individually marked, and their family relationships were known. Each of the hypotheses we have discussed makes certain predictions that can be tested by examining the fitness benefits and the characteristics of helpers.

If helping increases an animal's chances of living to the next breeding season, we would predict that survival would increase with group size and helpers should have higher survival than nonhelpers. These predictions were not fulfilled in bee-eaters. Individuals in small groups survived as well as those in large groups. Moreover, breeders, helpers, and nonhelpers all survived equally well. Thus, enhanced survival does not seem to be an advantage of helping in bee-eaters.

The hypothesis that helping increases future reproductive success predicts that helpers should have a better chance of breeding than nonhelpers and that animals that have never bred should be more likely to help than those that have offspring. Neither of these predic-

tions was met in bee-eaters. Forty percent of all helpers are former breeders, and half of the birds that hatched in one season became breeders the following year, without having helped others at all.

If the hypothesis that helping increases reproductive success applies, we would predict that animals that had been helpers would raise more offspring than those that had never helped. Again, this prediction is not fulfilled in bee-eaters. There were no significant differences between helpers and nonhelpers in the number of young fledged.

Finally, the hypothesis that helping increases the helper's inclusive fitness through the extra offspring that relatives produce makes several predictions possible. First, breeding units with helpers should produce more young than units without helpers, and/or the number of surviving young should be greater for units with helpers. Second, helpers should be closely related to those that are helped. Bee-eater helpers did, indeed, benefit by increasing the production of nondescendant kin. Although they did not help the breeders live longer, they did dramatically increase nestling survival.

Helpers and breeders are closely related, and so the extra nestlings that were raised because of the helper's efforts increased its inclusive fitness greatly. It was calculated, in fact, that most of the benefit a bee-eater derives from helping is through the extra nondescendant kin that survive. So we see that in bee-eaters, kin selection is the major factor in the evolution of helping.

EUSOCIALITY

Traditionally, a species is considered to be eusocial if it has three characteristics: reproductive division of labor, cooperation in the care of young, and overlap of at least two generations capable of sharing in the colony's labor (E. O. Wilson 1971). The astute reader may realize that these characteristics are also found among the cooperative breeders that we have just discussed, and this has led to some controversy about the definition of eusociality. One suggestion is to restrict the use of the term to those societies with castes, groups of individuals irreversibly specialized to perform a task (Crespi and Yanega 1995). Another is to expand the definition and consider cooperative breeding and eusociality as part of a continuum. Cooperatively breeding species could be placed along a eusociality continuum according to how evenly reproduction is shared among group members (Figure 17.25) (Lacey and Sherman 1997; Sherman et

al. 1995). The predicted location of several of the cooperatively breeding species we discussed in the previous section are indicated along this continuum. Most of them are in the middle or the low end, indicating that several or all of the group members breed. In this section we will focus on those species in which breeding is restricted to one or very few members of the group— the high end of the continuum. The species discussed here also have castes, which makes them eusocial by either definition.

Eusocial Species

Social Insects   Most eusocial insects are either hymenopterans (ants, bees, and wasps) or isopterans (termites). However, eusociality has also been reported in a few other groups, including aphids (Aoki 1972, 1979, 1982), some beetles (Kent and Simpson 1992), and thrips (Crespi 1992).

The members of a group of social insects labor together as if they are parts of one large organism. Reproduction is the sole responsibility of the queen. The sterile castes perform several altruistic services. One is defense. For example, some members of each ant colony, the soldiers, specialize in the colony's defense. They place themselves between the threat and the colony and fight intruders to the death. Even when

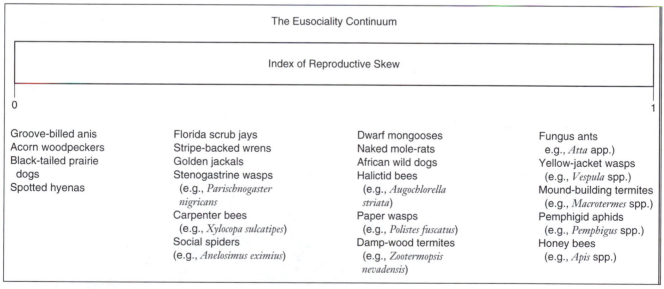

FIGURE 17.25   The eusociality continuum blurs the distinction between cooperative breeders and eusocial species. The traditional definition of a eusocial society is one with reproductive division of labor, cooperation in caring for the young, and an overlap of adult generations. Cooperative breeding species share these characteristics with eusocial species, and so it has been suggested that they form a continuum of social systems. The primary differences among the social systems is the degree to which reproduction is shared among group members. When reproduction is restricted to a single individual, the reproductive skew is 1. A reproductive skew of 0 indicates that the lifetime reproductive success of all group members is equal. This diagram shows predicted locations of certain cooperatively breeding species along the eusociality continuum. (From Lacey and Sherman 1997.)

FIGURE 17.26   One of the jobs of sterile worker honeybees is to guard the nest. They check the odors of all incoming bees. Here, a guard bee inspects an approaching male at the hive's entrance. If the intruder is not a nestmate, it is expelled from the hive.

they are injured, they appear to behave in the best interests of the colony. Wounded soldiers leave the nest, or if they are already outside they refuse to enter. This is adaptive because dead bodies in the colony would pose a sanitation problem. Honeybees also have nest guards (Figure 17.26). As honeybee workers defend the nest, they commit suicide. When a bee stings an intruder, the barbs on the stinger anchor it in the victim's body, causing it to be torn from the bee's body. The sterile workers do more than just defend the colony: They also provide food for nestmates. In almost all eusocial species, the workers indiscriminately share food with colony members. Finally, the most amazing aspect of eusocial behavior in many species of social insects is that the workers are sterile. They toil tirelessly, caring for offspring that are not their own, and without the hope of ever having any of their own.

Mole-Rats   Mole-rats are so named because they behave like moles but look like rats. These rodents are particularly intriguing because two species, the naked mole-rat, *Heterocephalus glaber* (Jarvis 1981), and the Damara mole-rat, *Cryptomys damarensis* (Bennett and Jarvis 1988), appear to be eusocial. The social structure of these burrowing animals fits the classical definition of eusociality (E. O. Wilson 1971) that we applied to social insects in the previous section. First, breeding is restricted to a single female, aptly named the queen, even in groups with almost 300 members (Figure 17.27). Other adult females are smaller than the queen and neither ovulate nor breed. One to three males breed with the queen, although most adult males do produce sperm. Second, mole-rat colonies contain overlapping generations of offspring. Third, there is differentiation of labor among individuals within the

FIGURE 17.27   A queen naked mole-rat, the only reproductive female of the colony, is resting on the workers that feed her and help care for the young.

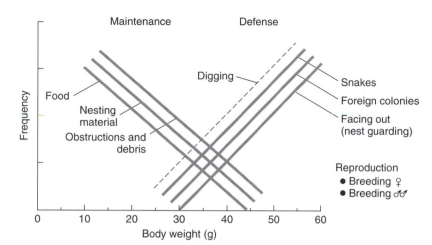

FIGURE 17.28 The duties assumed by mole-rats vary with their body size. Smaller individuals gather food and nesting materials. Slightly larger ones clear and dig tunnels. The largest members defend the colony from invasion by members of other colonies and from predatory snakes. (Data from Lacey and Sherman 1991.)

colony. This specialization is much less rigid than among the castes of social insects. The duties assumed by the nonbreeding members seem to depend more on their size and age (Figure 17.28). Smaller members' duties generally include gathering food and transporting nest material. As they grow, they begin to clear the elaborate tunnel system of obstructions and debris. Larger members dig tunnels and defend the colony (Honeycutt 1992; Lacey and Sherman 1991; Lovegrove 1991; Sherman, Jarvis, and Braude 1992).

The Evolution of Eusociality

The sterility of workers raises an interesting question. How can the alleles for the altruistic deeds be perpetuated when the individuals displaying the behaviors do not produce any offspring? This observation mystified even Charles Darwin (1859). In his *Origin of Species by Means of Natural Selection*, he wrote, " I can see no real difficulty in any character having become correlated with the sterile condition of certain members of insect communities: the difficulty lies in understanding how such correlated modifications could have been accumulated by natural selection."

As we consider the evolution of eusociality, keep in mind that it involves the interplay of factors at various levels (Figure 17.29). One group of factors includes those that cause the formation of stable groups in which generations can overlap—the benefits of group living and the costs of dispersal. Some of these were discussed in Chapter 16, and others will be mentioned here. Once the group has formed, we can consider the factors that favor caring for offspring that are not one's own, including the closeness of the genetic relationship among group members. Many of these factors were discussed in the previous section on helping. Finally, we can consider the factors that would favor the restriction of reproduction to one or a few individuals. In the discussion that follows, notice that it is impossible to keep

these groups of factors completely separated because of the interactions among them.

Kin Selection Members of eusocial colonies are almost always closely related to one another. Although genetic relatedness among colony members is not a requirement for the evolution of eusociality, it does boost indirect fitness gains and favors the evolution of eusociality.

THE PROPOSED ROLE OF HAPLODIPLOIDY The preponderance of eusociality among the hymenopterans provided a valuable clue to W. D. Hamilton (1964) as he puzzled over the evolution of the behavior. It drew

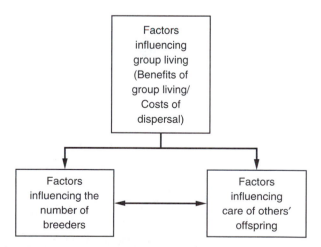

FIGURE 17.29 Interacting factors in the evolution of eusociality. First a stable group with overlapping generations must form. This is influenced by the benefits of group living and the costs of dispersal. Other factors affect whether workers will care for the offspring of another member of the group. Finally, another set of factors will determine the number of breeders in the colony.

his attention to another trait of the hymenopterans that is rare outside that order, their means of sex determination, called haplodiploidy. Only a few other arthropod groups, including the mites, thrips, and whiteflies, display haplodiploidy.

In haplodiploidy, unfertilized eggs usually develop into males and fertilized eggs typically develop into females. The single queen in a hymenopteran nest has a mating flight and stores the sperm obtained at this time for the rest of her ten or more years of life. The queen carefully doles out the sperm. Some, but not all, of the eggs are fertilized. An unfertilized egg develops into a male that has only the single set of chromosomes donated by the mother. A male is, therefore, haploid. A fertilized egg, on the other hand, develops into a female. This means that a female hymenopteran has two sets of chromosomes, one from her mother and the other from her father; she is diploid.

Hamilton (1964) noticed that haplodiploidy can change the rules for determining the degree of relationship among the kin within a colony and results in peculiar asymmetries in the closeness of relationships. For example, a surprising and important outcome of haplodiploidy is that the sister-workers within a colony could be, on the average, more closely related to one another and to the siblings they help to raise than they would be to their own offspring. How can this be? Since the queen is diploid, her eggs are formed by meiosis and are not identical. Each egg contains replicates of 50% of the mother's chromosomes, one of her two copies of each chromosome. In other words, the coefficient of relationship of the mother to her offspring is 0.5. The mother's chromosomes are only half of the genetic legacy of each daughter, and as you see, the sisters share an average of half of the maternal alleles. This means that they have an average of 25% of their alleles in common through their mother.

If only one male fertilizes the queen, then the father's contribution to each of his daughters is identical and accounts for half of her genetic inheritance. Since the father is haploid, identical copies of each of his chromosomes are in each sperm cell. Therefore, the sisters would have 50% of their alleles in common by virtue of their father's input and an average of 25% in common because of their mother's donation (Figure 17.30). As a result, sisters could share an aver-

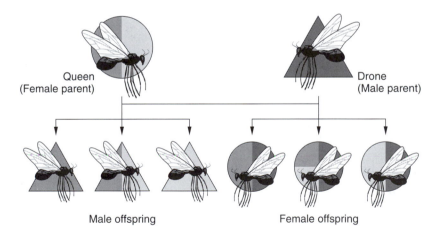

Degrees of relatedness in haplodiploid species

	Daughter	Son	Mother	Father	Sister	Brother
Female	0.5	0.5	0.5	0.5	0.75	0.25
Male	1	0	1	0	0.5	0.5

FIGURE 17.30   The genetic contributions of a male and female of a haplodiploid species to their offspring. The possible combinations are indicated by shading. Female offspring have a 50% chance of sharing a gene from their diploid mother. The maternal contribution is only half of the female offspring's genes. As a result, sisters have 25% of their genes in common through their mother's gametes. The father is haploid, so his sperm are all identical. In other words, sisters of haplodiploid species share 25% of their genes because of their mother's contribution and 50% of their genes because of their father's contribution, or a total of 75% of their genes by common descent. Male offspring are haploid. Each male's entire set of genes comes from his mother. Thus, the average degree of relationship among brothers is 50%.

age of 75% of their alleles by descent, a larger fraction than the 50% that a mother shares with her offspring. Thus, the sister-workers increase their inclusive fitness more by rearing reproductively capable siblings than they would if they produced their own offspring. The greater genetic profit would favor daughters that remain at home to augment their mother's reproductive efforts. In this way, then, the high level of relatedness resulting from haplodiploidy facilitates the evolution of eusociality.

Sex Ratio Data and the Kin Selection Hypothesis Robert Trivers and Hope Hare (1976) noted that another asymmetry in the closeness of relationships provided a way to test the kin selection hypothesis. Because of haplodiploidy, males are more closely related to their mother than they are to their sisters. A result of this asymmetry is that the interests of the royal figure are at odds with those of her daughters. The queen shares an equal number of her alleles with her sons and daughters, so from her perspective the optimal way to perpetuate her alleles is with an equal investment in sons and daughters. In contrast, a female worker is more closely related to her sisters than to her brothers, assuming that her mother (the queen) mated with just one male. Sisters, we know, may share 75% of their alleles by descent. However, they share only 25% of their alleles with their brothers. (Each egg formed by the mother contains copies of half of her alleles. These are the only alleles a son inherits, but a daughter receives an additional set of chromosomes from the father. Therefore, half of the maternal half, or 25%, of a female worker's alleles are likely to be present in her brother.) This asymmetry in relationship led Trivers and Hare to expect a sex ratio skewed toward reproductive females and away from males; a female offspring would enjoy a threefold increase in inclusive fitness if she raised sisters rather than brothers. They calculated that from the female worker's point of view, the optimal sex ratio is 3:1 in favor of females, but from the queen's point of view it is 1:1. Thus there is a conflict of interest between the queen and the workers.

Although it might seem, at first, that the queen must be the winner in this conflict because she could determine the sex of her offspring by choosing whether or not to fertilize the egg, it turns out that the workers have many ways in which to alter the sex ratio. If eusociality among the Hymenoptera truly evolved through kin selection and the sterile workers are actually behaving in ways that increase the frequency of their own alleles, then the combined weight of reproductive females within a hive should be three times greater than the combined weight of reproductive males. This prediction was confirmed. Trivers and Hare (1976) accumulated data on the sex ratios (investment ratios) of 21 ant species. Although there is a good deal of scatter in the data, the investment in reproductively capable females was found to exceed that in reproductive males by an amount close to 3.1. A similar sex ratio was observed in a colony of a eusocial sweat bee, whose genetic relatedness was measured at 0.75 (Packer and Owen 1994).

It is interesting to note that worker hymenopterans may even adjust their investment in brothers or sisters according to their closeness of relationship to each. For example, in primitively eusocial bees, a daughter-worker may ascend to the queen's throne when her mother dies or is removed from the colony. This changes the genetic relationships among the colony's members. The workers would then be sisters of the egg-laying queen. The offspring they raise would be nieces and nephews, with whom they have equal genetic closeness. We would expect, then, that sister-workers would invest equally in males and females. Ulrich Mueller (1991) tested this prediction by comparing the sex ratios in colonies in which the foundress queen was in place with those from colonies from which the foundress queen had been removed. The workers' investment in males reflected the closeness of their genetic relationship with males. In colonies where the workers were raising nieces and nephews, 63% of the weight of all reproductives were males. When the workers were raising brothers and sisters, only 43% of the combined weight of reproductives were males.

A remarkable exception to the usual sex ratio among ants also supports the kin selection hypothesis. It has been argued that sterile castes evolved because the workers increased their inclusive fitness more by raising siblings than by producing their own offspring. As evidence, it was noted that sex ratios are often biased toward females, the members of the colony that share a greater proportion of alleles with the workers that care for them. However, in some ant species the workers that attend to the daily routine of running the nest and caring for the brood are slaves that are genetically unrelated to the colony's members. The soldiers of slave-making species wage war against colonies of other ant species and drag back the pupae or larvae that later hatch to serve their colony as slaves. The slaves work diligently, performing all the duties they are genetically programmed to perform in their home colony. Because the nursemaids that care for the brood are unrelated to the colony, their fitness is not increased by altering the 1:1 sex ratio that is optimal for the queen. Although sex ratio data were available for only two slavemaking species (*Leptothorax duloticus* and *Harpagoxenus sublaevis*), Trivers and Hare (1976) noted that the dry weight of female reproductives approximately equaled the dry weight of male reproductives. The slaves cannot revolt because if a mutation that led to rebellious behavior did occur, it would not be passed on. The future reproductives being nurtured by the slaves bear the queen's alleles, not the slaves'. The 1:1 investment ratio in the two sexes of slave-making ants is expected because it is optimal for the queen mother, and it is her offspring's alleles, not those of the slaves, that are passed into future generations.

Multiple Mating and the Closeness of Relationships Hamilton's (1964) kin selection hypothesis relates the high frequency and the extreme degree of altruism among the Hymenoptera to the unusually close genetic relationship among the sister-workers in the nest. However, since he formulated his ideas, there have been repeated reports of multiple matings and insemination in many hymenopteran species. For example, a queen honeybee (*Apis mellifera*) makes the best of her mating opportunities by mating with an average of 17.25 males (Adams et al. 1977) and by storing most of the 6 million sperm she receives from each (Kerr et al. 1962; Woyke 1964). If the sperm from each male remains clumped, workers of the same age within a colony are likely to have the same father. The rules for calculating relatedness would then remain virtually untarnished (Trivers and Hare 1976). However, biochemical studies show that the sperm of different males mix in honeybee queens (Laidlaw and Page 1984; Page and Metcalf 1982). Likewise, there is evidence in two species of social wasps that the queens mate more than once and use the sperm from different males in relatively constant proportions through time (Ross 1986).

If the sperm of multiple males are used simultaneously, the female workers within a colony may have different fathers. Workers fathered by different males share only 25%, rather than 75%, of their alleles by descent. In this extreme case, the workers would be no more closely related than are half siblings in diploid species such as the ones discussed in previous sections. More realistically, if the queen mates twice and the sperm mix randomly, the workers will share 50% of their alleles by common descent. They will have an average of 25% of their alleles in common from their mother, and they will have a 50% chance of having the paternal half of their alleles donated by the same father. Remember that because males are haploid, all the sperm from a single father are identical (0.25 + 0.25 = 0.5). Obviously, the exact genetic relationship among the sister-workers of a colony will depend on the number of males that mated with their mother, the extent of the mixing of sperm, the utilization of the sperm by the queen, and the outcome of any competition between sperm of different males.

The extent to which multiple matings affect genetic relatedness among the colony's members must be determined by direct measurements. Studies on eusocial wasps and paper wasps showed that the coefficient of relatedness among members is often less than 0.5 and rarely approaches 0.75 (Ross 1986; Strassmann et al. 1989). On the other hand, multiple matings may not be common in most hymenopterans. A review of the studies in which the genetic relatedness of a colony's members was measured (Boomsma and Ratnieks 1996) indicates that multiple mating occurs in only one phylogenetic group in ants (Atta), eusocial bees (Apis), and wasps (Vespula).

Even if it is not widespread, the reality of multiple insemination among some hymenopteran species obviously has ramifications for the kin selection hypothesis. Hamilton (1964) pointed out that multiple insemination would reduce the tendency to evolve sterile castes because the workers would be expected to increase their fitness more by rearing their own offspring, which would share a greater proportion of their alleles than would their reproductive siblings. Thus, haplodiploidy cannot entirely explain the evolution of eusociality in hymenopterans.

OTHER PATHS TO GENETIC CLOSENESS   Although a close genetic relationship is not necessary for the evolution of eusociality, it does seem to be a factor that paves the way. Eusocial species other than the hymenoptera have a close genetic relationship among the colony's members. Two groups of insects, termites and certain aphids, like the hymenoptera, have sterile castes. A high degree of relatedness, perhaps resulting from extreme inbreeding, may predispose termites to eusociality (Bartz 1979; Reilly 1987). Aphids, small insects that obtain nourishment by sucking the juices from plants, may be almost genetically identical to other members of the colony. There are two types of one developmental stage of certain aphids of the family Pemphigidae. One of these, the secondary type of the first instar larva, serves as a soldier acting in the colony's defense (W. A. Foster 1990). When a larva of a syrphid fly, a predator of aphids, enters the colony, the aphid soldiers insert their piercing mouthparts into the predator, killing it. This action could not have evolved by enhancing the soldier's reproductive success since the soldier is pre-programmed to die. Instead, it is thought to be a result of an unusually high genetic relatedness among the members of an aphid colony. Indeed, because the individuals develop from unfertilized eggs of a single female, they are nearly identical genetically (reviewed by Ito 1989).

There is also an unusually high degree of genetic relatedness among the members of any single colony of mole-rats. Free-living colonies of naked mole-rats were sampled and the genetic similarities of the members determined by using the technique of DNA fingerprinting. Even between colonies, there is more genetic similarity than is found among nonkin of other free-living vertebrates. This similarity is thought to be a consequence of the extreme inbreeding within a colony (Faulkes, Abbott, and Mellnor 1990; Reeve et al. 1990).

The inbreeding is promoted by their population structure. Naked mole-rats live for a long time and are prolific breeders. Two females caught in the wild have survived for 16 years in captivity, and they still breed. In nature litters, which may include as many as 12 pups, are born year-round every 70 to 80 days. When the queen dies, one of the larger females assumes her reign, after violent in-fighting among the other eligible females of the colony (Brett 1991). Furthermore, there is little mixing of genes between colonies. Members of different colonies are quite aggressive toward one another (Lacey and Sherman 1991). Intruders may even be killed. New colonies are thought to be formed by "budding": A group may leave the parental colony and seal off the intervening tunnels (Brett 1991). High genetic similarity among mole-rats is not surprising, then, considering the restricted breeding and the lack of intercolony mixing of genes.

Individual Selection Throughout this book, we have seen that a particular behavior evolves when the benefits of performing it outweigh the costs. Indirect fitness gains through kin selection are only one potential benefit of eusociality. For a fuller understanding of the evolution of eusociality, we must consider other possible benefits, as well as costs. So far in this chapter we have seen that there may be several factors, acting in unison or independently, that promote altruism in various species. Therefore, it is wise for us to examine some of the other factors that might favor eusociality. As we saw in the evolution of helping, many of these factors affect the costs and benefits of dispersing or remaining with the natal group.

Special Needs Social grouping may be necessary for the survival of some species. Consider, for instance, the termite. In termites, both sexes are diploid, so the assumed advantages of haplodiploidy do not apply to these insects. Nonetheless, as we have seen, individuals within a termite colony are highly related to one another (Reilly 1987). The high degree of genetic relatedness presumably sets the stage for the evolution of eusociality, but it is not a complete explanation. We may wonder, then, what other factors may have played a role in the evolution of eusociality among termites.

You may also be aware that termites are unable to digest wood, the primary component of their diet. The termite's gut contains symbiotic protozoans that break down the cellulose from the wood in the termite's diet to a soluble carbohydrate. A termite must be supplied with these protozoans after hatching and again after each molt. The protozoans are obtained by licking the anal secretions of other termites in the colony. Since the cellulose-digesting intestinal flagellates are essential and must be obtained from other termites, it is adaptive for these insects to live in groups. Given this restriction

and the fact that a nonreproductive is at least as closely related to a sibling as it would be to its own offspring, the termite's path was paved for social living (Lin and Michener 1972).

It has also been suggested that the transfer of symbionts has played a role in the establishment and maintenance of eusociality in naked mole-rats. In this case, the symbionts are obtained by eating feces. The symbionts themselves may be an important source of protein and other nutrients in a diet that contains very little other protein (Dyer 1998).

The transfer of symbionts could be accomplished simply by extended parental care, and so it cannot completely explain the evolution of eusociality in termites (Thorne 1997) or in any other species. Earlier in this chapter, in our discussions of helping and assistance in mate acquisition, it was suggested that in any situation in which there is a low probability of successfully raising offspring alone it is advantageous for individuals to assist the reproductive efforts of others, especially if those that benefit are relatives. The same contention may be valid when applied to the evolution of eusociality, so let's review some ecological factors that could apply.

Ecological Factors Favoring Eusociality Eusociality in some insects may have developed as a result of some of the same ecological pressures that contribute to social behavior in other species.

Need to Defend the Nest An example of the importance of assistance in protecting the nest against predators is provided by the social wasp (*Mischocyttarus mexicanus*). The nests of this species may be founded by one to several females. This wasp is particularly interesting because the female workers are able to reproduce themselves. However, when there are several foundresses, only one female actually lays eggs. We may wonder, then, why a female worker would give up reproducing herself and raise the queen's offspring. The variation in the number of females that are establishing a nest makes it possible to determine whether lone- or group-nesting strategies are more successful and to isolate some of the factors responsible for success.

Marcia Litte's (1977) comparison of one-foundress nests and multiple-foundress nests demonstrated the importance of a larger group for nest defense against predators. The two important predators of these wasps are birds and ants. Because the sting of this wasp is mild, multifoundress nests were no more effective in driving away birds than were smaller nests. However, the larger nests were more effective in defending the nest against ants. A lone foraging ant was rarely permitted to approach within 4 centimeters of the nest, and those that succeeded in mounting the nest were thrown off. When they are home, all adult female wasps actively defend the nest against predators. However, these females must also forage. While a single foundress is foraging, her nest is left unattended, but in a multifoundress nest, there is always someone available to guard the home front. Undefended nests fail more often than those protected by continuous vigilance. The females in the colony are related, although perhaps not as closely related as Litte believed (Strassmann, Queller, and Solís 1995). Since a female would have a greater genetic gain by successfully rearing relatives than by failing in her attempt to raise her own off-spring, eusociality is favored.

Sometimes the nest must be protected from a takeover by other females of the same species. Conspecific pressure, not predation or parasitism, has been shown to give an advantage to cooperation for the paper wasp (*Polistes metricus*). George Gamboa (1978), for example, did not detect any predation when he studied *Polistes*. Furthermore, he observed that the presence of additional foundresses had no effect on the prevalence or severity of parasitism. However, he did notice that single-foundress nests were usurped by a challenging foundress significantly more frequently than were multifoundress nests: When a single foundress leaves her nest unattended during her foraging trips, there is a greater opportunity for a coup. In addition, a lone foundress may be unable to defend her nest without assistance even if she is at home at the time of assault. A coup did not have the same devastating effect on multifoundress nests. When a takeover was successful in a nest with several foundresses, the queen was not dethroned; rather, a subordinate foundress was replaced.

Need for Assistance in Nest Building During colonial times in the United States, it was hard work to build a house. As a result, children often remained in the home of their parents even when they were married and about to begin their own families. When the dwelling became crowded, an addition was built.

For a eusocial insect, nest building is no less of a chore. The nests are often intricate and take a great deal of time and energy to construct (Figure 17.31). Therefore, in the early stages of sociality, a newly matured adult might have been better off using the mother's nest as a safe haven for her eggs, even if this required enlarging it. Eventually this cooperation may have evolved toward rearing younger siblings rather than one's own offspring (Andersson 1984).

The problem of defending and building a nest makes it difficult for a single parent to raise offspring successfully. In addition, a female that emerges from hibernation late in the season may find it difficult to find a suitable spot for a nest (West-Eberhard 1975). She may be better off staying with a group and assisting someone else's reproductive efforts so that she has a chance to inherit an established productive nest later in the season. When it is difficult to found a new nest, a wasp may be

FIGURE 17.31 **A paper wasp nest illustrates the elaborate nests of social insects. The difficulty of defending against takeover by conspecifics may be a factor that favors eusociality in some species.**

better off accepting small but guaranteed fitness returns rather than gamble on her ability to raise a brood to independence. Consider the fitness consequences to a female if she died before the brood was independent. If she is a worker, she would still derive some fitness because other workers would continue to care for the brood. However, if she is a solitary foundress, she loses everything. Thus, the assured fitness returns may favor eusociality (Gadagar 1990a, b).

Cost of Dispersal Factors that favor eusociality in naked mole-rats are thought to involve the hazards of dispersal from their relatively safe system of underground tunnels (Lovegrove 1991). One factor that makes dispersal a risky business is that the tubers and bulbs mole-rats prefer to eat are distributed unevenly throughout the habitat. The natal colony may have access to a patch of food, but a group that sets off on its own may have to burrow extensively before encountering another rich area. Burrowing uses quite a bit of energy. Consequently, members of a small group might die of starvation before locating a new food resource (Lovegrove 1991; Lovegrove and Wissel 1988).

INCREASED PROBABILITY OF SUCCESSFUL REPRODUC-TION IN THE FUTURE Earlier in this chapter it was argued that helping by some vertebrates could be considered a form of payment for permission to remain on the breeding territory while waiting to take over. A similar situation may hold for some eusocial invertebrates.

Eusociality among some wasps may have been favored by a low probability of reproducing alone coupled with the possibility of taking over a productive nest in the future. Females of various wasp species become workers at established colonies. Sometimes, but not always, these so-called joiners are helping to produce relatives. Marcia Litte (1977) reports that there is a dominance hierarchy among the foundress and the joiners in the wasp species *Mischocyttarus mexicanus*. This hierarchy determines who is next in line to ascend the throne. Litte examined the process of queen replacement by experimentally removing queens from established colonies. In every case, the right of egg laying was assumed by the female just under the former queen in the dominance hierarchy.

Parental Manipulation   For some biologists the explanation for the evolution of eusociality among insects lies in Alexander's (1974) hypothesis of parental manipulation. A parent's concern must be to maximize its overall fitness through all of its reproductive offspring, whereas the interest of any particular offspring is to maximize its own fitness. The reasoning behind Alexander's conclusion that the parent will always win in conflicts over fitness was given earlier in this chapter. The proponents of the parental manipulation hypothesis believe that eusociality is evidence that the queen has won this parent-offspring conflict. The queen's fitness is elevated more if her daughters remain at the nest and help raise additional siblings than if they leave and rear their own offspring. This is true because the queen shares an average of 50% of her alleles with her offspring but only 25% of her alleles with grandchildren. Therefore, selection should favor her turning some of her offspring into workers (Charnov 1978).

How might the queen manipulate her offspring into increasing her fitness at the expense of their own? One way would be by behaviorally dominating her offspring. Indeed, this is exactly what occurs among naked mole-rats. Female workers are coerced into giving up their own reproduction by the aggression of the queen. The queen threatens her workers with throaty, raspy calls. She and her workers engage in nose-to-nose combat. By shoving with her head, the queen can displace a worker by up to a meter. This aggression leads to endocrine changes in workers that suppress reproduction. Because of this bullying, some of the female subordinates even fail to ovulate (Faulkes and Abbott 1997). But, nonbreeding naked mole-rats are not sterile. If the queen dies or is experimentally removed from the colony, the higher-ranking females in the dominance hierarchy often fight to the death to replace her as the breeder. This is the only way in which a female mole-rat can gain direct fitness. The nonbreeding workers gain indirect fitness by raising the queen's offspring, who are their close relatives (Lacey and Sherman 1997; Margulis, Saltzman, and Abbott 1995).

Parental manipulation is also one interpretation of the actions of the queen in the primitively eusocial bee *Lasioglossum zephyrum*. The workers do the labor, but the queen must act as the foreman of the job and direct their activities. When a worker returns from a pollen-collecting trip, the queen usually meets her subject in a tunnel and then backs away. This stimulates the worker to follow her. The queen stops at the entrance of the cell that should be provisioned with the pollen. Without her direction, the workers would be unable to locate an appropriate cell for their pollen and would deposit it in the tunnel. In addition, the presence of the queen stimulates activity in the hive, much as an office is busier when the boss is around. In an experiment, colony activity was measured before and after removing the queen. Sometimes one of the workers immediately ascended the vacant throne. However, when no worker took that role, the activity level in the hive began to decline within 30 minutes after the queen was removed, and it remained significantly lower than it was while the queen was in place (Breed and Gamboa 1976).

Alternatively, the queen could "force" her offspring to assist her reproductive activities by restricting the options available to them through their diet. After all, it is a daughter's diet, not her genes, that determines whether she will be reproductively capable. The queen's domination might have originated with unequal food distribution among her daughters. Because smaller females are less likely to reproduce successfully, selection would favor their staying at home to rear siblings. With the possibility of reproducing on her own restricted or eliminated, it would be to a daughter's advantage to enjoy whatever fitness she gained by helping to rear her reproductively capable siblings. At the same time, the mother would be maximizing her personal fitness.

CONCLUSIONS

There is no explanation for the evolution of altruism that applies to every example. Different life histories and ecological conditions may alter the relative importance of a particular evolutionary mechanism. Furthermore, the mechanisms suggested are not mutually exclusive and may be working simultaneously. To complicate matters, the same data may be consistent with more than one hypothesis. It is difficult to demonstrate conclusively that one particular hypothesis is correct because evolution is the central concern and behaviors are not fossilized. Therefore, there is no behavioral record of the evolutionary steps toward altruism. We must reconstruct what may have occurred by studying living species that display different degrees of cooperation.

SUMMARY

Altruism is the performance of a service that benefits a conspecific at a cost to the one that does the deed.

Strictly speaking, the benefits and costs are measured in units of fitness (the reproductive success of a gene, organism, or behavior). Since changes in fitness are nearly impossible to ascertain, however, the gains and losses are usually arbitrarily defined as certain goods or services that seem to influence the participants' chances of survival.

The prevalence of altruism among animals has puzzled evolutionary biologists. Evolution involves changes in the frequency of certain alleles in the gene pool of a population. If aiding a conspecific costs the altruist, it should be less successful in leaving offspring that bear copies of its alleles than are the recipients of its services. As a result, the alleles for altruism would be expected to decrease in the population. Therefore, the existence of altruism seems at first to contradict evolutionary theory.

The hypotheses for the evolution of altruism can be arbitrarily classified into four overlapping classes.

Individual Selection   The general thrust of these hypotheses is that when the interaction is examined closely enough, the altruist will be found to be gaining, rather than losing, by its actions. The benefit may not be immediate; sometimes the gain is in the individual's future reproductive potential.

Kin Selection   When the beneficiaries of the good deeds are genetically related to the altruist, the enhanced reproductive success they enjoy will perpetuate the alleles that the altruist shares with them by virtue of their common descent. The altruist's relatives are more likely than nonrelatives to carry the alleles that lead to altruism. Kin selection includes (1) direct fitness, which is accrued through the production of one's own offspring, and (2) indirect fitness, which is gained by helping relatives raise more offspring than they could without help. In the latter case, fitness benefits are devalued according the distance of the genetic relationship among the relatives.

How can relatives be identified? There are several possibilities. One way might be to use location as a cue: The individuals that share one's home are likely to be kin. Also, individuals might be identified as kin because they are recognized from prior social contact or as a result of their association with a known relative. Another possible way to recognize a relative is by certain traits that characterize family members. In other words, an image of a family member may be matched or compared to the appearance of a stranger to determine whether it is related. Finally, recognition may be genetically programmed. Perhaps there are alleles that in addition to labeling relatives with a noticeable characteristic cause the altruist to assist others that bear the label.

Reciprocal Altruism   Altruism might also evolve, in spite of the initial cost to the altruist, if the service is repaid with interest. In other words, altruism will be favored if the final gain to the altruist exceeds its initial cost. However, for reciprocal altruism to work, individuals that fail to make restitution must be discriminated against. Because of this requirement, certain factors make reciprocal altruism more likely. First, there should be a good chance that an opportunity for future repayment will arise. Second, the individuals must be able to recognize one another.

Parental Manipulation   According to this hypothesis, there is selection for altruism toward siblings when it maximizes the parents' fitness. Any allele that causes an offspring to behave in a manner that reduces the parents' fitness would lower the selfish sibling's fitness when it becomes a parent and passes the trait on to its own offspring.

No single hypothesis applies to every example of altruistic behavior. In addition, these evolutionary mechanisms are not mutually exclusive; more than one may be responsible for a single example of altruism. To confuse matters even further, similar behaviors may evolve by different mechanisms in different species.

Members of some social species emit alarm calls to warn their neighbors of a predator's approach. For example, many small passerine birds do this. The suggested explanations for this behavior that invoke individual selection are numerous. The caller may benefit if the call discourages the predator's attack. The predator might look for easier victims because the call is notification that it has been detected and the prey are wary. It might also be discouraged because an alarm call causes the prey to scatter, making them more difficult to catch. Another suggestion is that the caller benefits by causing conspecifics to flee to a safer place, where in the anonymity of the flock it can avoid being spotted by the predator. Kin selection and reciprocal altruism have also been invoked to account for avian alarm calling. However, given the natural history of the species that call, these hypotheses seem unlikely.

Ground squirrels emit two types of alarm calls, one in response to terrestrial predators and one in response to aerial predators. The calls seem to have been selected in different ways. Individual selection seems to be the best explanation for the evolution of alarm calls in response to aerial predators. However, kin selection seems to be the most likely mechanism for the evolution of ground squirrels' alarm calls in response to terrestrial predators.

Another form of altruism is helping. A helper is an individual that assists in the rearing of offspring that are not its own, usually by providing food or by protecting the young. In most species helpers are previous offspring that are helping their parents raise their siblings, so kin selection seems to be a reasonable explanation in these cases. The helpers are not always relatives, however. In these cases, helping may be a means of maximizing individual fitness in the future. Helping commonly accompanies ecological conditions that make reproduction difficult or costly. Helping may be a means of obtaining permission to remain on a high-quality territory, of maintaining group or territory cohesiveness, of earning the future assistance of those helped, of obtaining a mate, or of protecting the young from predators.

Another apparently altruistic behavior is helping another individual to acquire a mate. Kin selection is thought to have been important in the evolution of this behavior in some populations of wild turkeys and in lions because the cooperating males are related. However, some lion coalitions include unrelated males, and in other species, such as the long-tailed manakin, the males that display together are unrelated. In such cases, it seems that cooperation increases the nonmating male's future chances of reproduction.

Finally, reciprocal altruism seems to explain the alliances formed by male olive baboons. One male may assist another to win a consort away from a third male. However, at some time in the future, the male that was assisted will have to repay the favor in kind.

Eusocial species are those that have sterile workers, cooperative care of the young, and an overlap of generations so that the colony's labor is a family affair. The eusocial insects behave altruistically in several ways: Food is shared, the members of the colony specialized for defense often die while performing their duty, and some members are sterile but care for the young of the colony's royalty.

Eusociality is common among hymenopterans (e.g., ants, bees, and wasps) but almost nonexistent outside that order. Hamilton (1964) noted that haplodiploidy, which is a means of sex determination in which fertilized eggs develop into females and nonfertilized eggs develop into males, is also common in the order Hymenoptera but rare outside it. He suggested that haplodiploidy may have predisposed the hymenopterans to eusociality because it results in a closer relationship between sister-workers and their siblings than the relationship that would exist between the workers and their own offspring. The female workers are likely to have 75% of their alleles in common with the reproductively capable siblings they help to raise, but they would have only 50% of their alleles in common with offspring they produced. As a result, a female worker makes greater gains in inclusive fitness by raising siblings than she would by producing offspring.

Trivers and Hare (1976) have supported Hamilton's hypothesis with an analysis of the ratio of investment in males and females among various ant species. An optimal sex ratio is one that maximizes inclusive fitness. Since a queen has 50% of her alleles in common with her sons and daughters, the optimal ratio of investment in her reproductive offspring is one male to one female. On the other hand, since the female workers share three times the numbers of alleles with their sisters as with their brothers, the optimal sex ratio from their point of view is three reproductively capable females to one reproductively capable male. In the 21 species of ants analyzed, the ratio of investment in the sexes was biased toward females by an amount close to 3:1.

However, it is now known that a queen may mate with more than one male. If the sperm of more than one male are used simultaneously, the genetic relationships among members of a colony are changed. Half sisters share fewer than 75% of their alleles by common descent. This observation weakens the argument that haplodiploidy strongly favored the evolution of eusociality.

As a result, the role of individual selection has been given more attention in recent years. Termites are eusocial, but both males and females are diploid. The individual termite presumably gains by its close association with others because it must obtain intestinal symbionts from its colony members immediately after hatching and again after each molt. These symbionts are necessary for the insect to be able to break down the cellulose in the wood that forms the main portion of its diet. In addition, certain ecological factors may make eusociality a means of increasing fitness. Cooperative action may be needed to defend a nest against predators, parasites, or even conspecifics. The combined efforts of several individuals may make the chore of nest building easier. In these cases, an individual may maximize its lifetime fitness by postponing immediate reproduction in favor of assisting another individual in order to increase the likelihood of successful reproduction in the future.

Parental manipulation is a third hypothesis for the evolution of eusociality. It has been suggested that the queen forces her offspring to assist in her reproductive efforts because the production of offspring raises her fitness more than the production of grandchildren. She may accomplish this by behaviorally dominating her workers or by restricting their reproductive options.

Naked mole-rats are eusocial mammals. The members of a colony have a high degree of genetic relatedness as a result of inbreeding, and this predisposes them to eusociality. This subterranean lifestyle and the hazards of dispersal are other factors that may have favored the evolution of eusociality in the naked mole-rat.

18

Maintaining Group Cohesion: Description and Functions of Communication and Contact

Whether we walk across the Arctic tundra, the African plains, or the Amazon basin, we see animals in groups. This question arises, then: What holds such groups together? As we will see, the answers are often deceptively simple and interrelated. We will find that social bonds are cemented by both communication and physical contact.

Why is communication so important? Animals, including humans, must make frequent decisions about how to behave, and communication helps them choose the most appropriate action (Figure 18.1). The most beneficial course of action often depends on the circumstances. Should an intruder challenge a territory holder? One factor to consider in the decision is the worthiness of the territory holder—its size, strength, and motivation to defend the territory. Which mate should a female choose? Most importantly, it should be one of her own species, but it would also be in her best interest to pick the highest-quality male—the one with genes that will enhance her offspring's chances of survival or the best provider. Which offspring should a female feed when faced with a hungry brood? The best allocation of resources might be to feed to the one in most need of nourishment. Clearly, the more the individual knows about the details of the situation, the better able it is to make the best choice. Communication is one source of information.

DEFINING COMMUNICATION

The broadest definition of communication is the transfer of information (Batteau 1968). More specifically, communication can be considered to occur when one animal (the sender), provides information to another

a   *b*

FIGURE 18.1   Signals that invite play. The play bow (*a*) used by dogs and other
canines and the play face (*b*) used by many primates indicate that the actions that
follow are playful. During play, the behaviors may contain components of aggressive
or reproductive behaviors. These play signals provide information that helps the
receiver determine that it is playtime so it can respond appropriately to the actions
that follow.

individual (the receiver) that can be used by the receiver to make a decision about the most appropriate action, given the existing circumstances. In many cases, an individual will behave differently, depending on the circumstances; the information provided informs the receiver which of the alternative conditions currently prevails. The sender gains from providing the information if it increases the chances that the receiver will respond in a way that is beneficial to the sender. The receiver responds to the information because it increases the likelihood of choosing the action that will most benefit itself (Bradbury and Vehrencamp 1998).

Animals communicate by sending a signal—a particular trait, posture, movement, sound, or chemical—that has a specific meaning. It provides information to the receiver that will help it choose a response that will benefit both parties. The familiar golden arches of McDonald's are a good example of the use of an arbitrary symbol that is given an invariable and distinct meaning among humans. For someone familiar with the symbol, it does more than just signify the location of a restaurant. The type of restaurant and the menu are standard across the country, so a traveler knows exactly what to expect.

Stereotypy is an important aspect of signals, and we find that throughout an animal population, every member sends the same signal in a remarkably similar way

(E. O. Wilson 1980). This contrasts with human language, in which the same thought can be expressed in remarkably different words by different individuals. For example, if you think back to recent elections, you may remember how the diverse speeches of the candidates made almost identical promises.

Not all of the information provided by a sender is transferred through signals. To qualify as a signal, the sender must benefit *because* the action or trait provides information. The receiver may gather information from many sources, including actions or traits of the sender that serve purposes other than providing information. For example, when a fish cleans its nest, it benefits from the cleaning activities. In the cleaning process, it may create water turbulence that could incidentally provide information to a receiver about the size of the sender. The receiver might benefit by using that information to decide whether to challenge the sender for its territory. However, the sender benefits from good housekeeping, even if no intruder is present. In this case, the information is called a cue, but it isn't truly communication. If the fish were to undulate its body in a manner that created water vibrations, this might provide information about size. Since the only benefit to the sender would be in providing information, the undulations might be considered a communicative signal (Bradbury and Vehrencamp 1998).

Not all situations in which the behavior of one animal alters that of another involve communication. For instance, many people believe that communication is not involved when brute force is used to achieve one's ends (J. M. Cullen 1972; M. S. Dawkins 1986; Wiley 1994).

CHANNELS FOR COMMUNICATION

Communication can involve any of a variety of sensory channels—vision, audition, chemical, touch, and electrical fields. Each channel has both advantages and limitations (Table 18.1). As we will see, the channel used for a particular signal will depend on the biology and habitat of the species, as well as the function of the signal.

VISION

There are two obvious properties of visual signals. The first is ease of localization. If the signal can be seen, the location of the sender is known. For example, when a male is displaying to attract a mate, there is never any doubt about his precise location. The receiver can see him and, therefore, respond to him in terms of his exact location, as well as his general presence. The second property is rapid transmission and fade-out time. The message is sent literally at the speed of light, and as soon as the sender stops displaying the signal is gone. For instance, if a displaying bird suddenly spots a hawk and flees, its position will not be revealed by any lingering images.

In addition, visual systems can provide a rich variety of signals. This diversity is possible because of the number of stimulus variables that can be perceived by most animals. These include brightness and color, as well as spatial and temporal pattern (altered by the animal's movements and posturing).

Visual signals have their disadvantages, of course. The most obvious is, quite simply, that if the sender cannot be seen, its signals are useless; and vision is easily blocked by all sorts of obstructions, from mountains to fog. Also, visual signals are useless at night or in dark places (including the depth of the sea), except for light-producing species. Furthermore, visual powers weaken with distance, so visual signals are not usually used for long-distance communication. As distance increases, visual signals generally become simpler and bolder and necessarily carry less information. A male iguanid lizard (*Anolis auratus*) uses head bobs, up and down movements of the head, in two displays. One is a "challenge" display used in territorial defense. The defender gets within a few centimeters of the intruder, turns sideways, and bobs its head. At such close range, the display is easy to see. The "assertion" display, given from conspicuous places within the territory, is thought to attract females. The intended receiver may be several meters away and inattentive. Although this display also consists of head bobs, unlike the challenge display its beginning movements are at high acceleration and velocity, serving to get the attention of the intended viewer. The movements of the assertion display are also more exaggerated than those of the challenge display, which increases the distance over which the display can be detected (Fleishman 1988).

Some visual signals are badges, aspects of an animal's appearance that have been modified to convey information. Some badges, such as the white dot on the tip of the wings of a male black-winged damselfly (Coenagrionidae; Figure 18.2), are permanent. Others are inconspicuous when the animal is at rest but can be flashed to convey information. An example of this type of badge is the dewlap (a loose fold of skin hanging from the throat) of a male anole, a type of lizard (Figure 18.3). Most of the time the dewlap is out of sight, but it is extended like a broad shield when the male is attracting a female or defending a territory. Other badges, such as the color of an animal, may change with the animal's physiological state or as a form of display behav-

TABLE 18.1 Characteristics of Different Sensory Channels for Communication

Feature	Type of Signal				
	Visual	Auditory	Chemical	Tactile	Electrical
Effective distance	Medium	Long	Long	Short	Short
Localization	High	Medium	Variable	High	High
Abiltiy to go around obstacles	Poor	Good	Good	Good	Good
Rapid exchange	Fast	Fast	Slow	Fast	Fast
Complexity	High	High	Low	Medium	Low
Durability	Variable	Low	High	Low	Low

FIGURE 18.2   The white dot on the tip of a male black-winged damselfly's wings is a permanent visual badge, an aspect of anatomy that serves as a communicative signal. In this case, the badge identifies sex. The female lacks this mark. (Photo by Stephen Goodenough.)

ior. A male scarlet tanager, for example, is only scarlet during the breeding season. Squid and octopi, as well as many species of fish, change color as a display.

AUDITION

Sound signals have a number of advantages. They can be transmitted over long distances, especially in water. Although sound is transmitted at a slower speed than light, it is still a rapid means of sending a message. The transitory nature of sound makes possible a rapid exchange and immediate modification, but it does not permit the signal to linger. After the message has been sent, the signal disappears without a trace. Sound signals have an additional advantage of being able to convey a message when there is limited visibility, such as at night, in water, or in dense vegetation.

Every music lover is aware of the tremendous variety of sound that can be produced by the temporal variation of just two of its parameters: frequency (pitch) and amplitude (loudness). Depending on the species, animals may vary either or both of these aspects of sound, or they may, like a drummer, simply alter the pattern of presentation. Thus, some species, such as birds, sing the melody, and others, such as crickets, provide the beat.

The type of sounds used by a particular species in communication will be determined by how the animal produces them. Sounds may be generated by respiratory structures, beating on objects in the environment, or rubbing appendages together. Many structures specialized for sound production have evolved in association with respiratory structures. Mammals have a larynx and birds a syrinx. You are probably aware that frogs and toads use air sacs or chambers as resonators, but do

FIGURE 18.3   An anole with his dewlap extended. The dewlap is an example of a visual badge, an aspect of anatomy that serves as a communicative signal. This badge can be hidden most of the time and extended to court a female or defend a territory.

you know that similar structures are used by some birds such as frigate birds (*Fregata*), swans (*Cygnus*), and various grouse and even by primates such as howler monkeys (*Alouatta*) and orangutans (*Pongo pygmaeus*)? The environment may also be used to produce auditory signals. Humans often tap their toes when listening to music, but for some animals foot stamping itself is the signal. Rabbits, birds, and some hoofed mammals stamp their feet on the ground as a means of communication. This musical theme has several variations—beavers (*Castor*) slap the water with their tails and woodpeckers drum on trees. Arthropods most commonly produce their auditory signals by rubbing together various parts of their exoskeleton. Crickets, for example, sing by opening and closing their wings. Each wing has a thickened edge, called a scraper, that rubs against a row of ridges, the file, on the underside of the other wing cover. Whenever the scraper is moved across the file, a pulse of sound is produced. This method of generating sounds is called stridulation.

CHEMICAL SENSES

The chemical senses, smell and taste, are a third channel for communication. Information may be carried by chemicals over long distances, especially when assisted by currents of air or water. The rate of transmission and fade-out time are slower than for visual or auditory signals. Depending on the function of the signal, this may be an advantage rather than a drawback. Sometimes, for example in the delineation of territorial boundaries, a durable signal is more efficient because it remains after the signaler has gone. Furthermore, chemical signals can be used in situations in which visibility is limited. The ease with which the sender of a chemical signal can be located varies with the chemical emitted, but it is usually more difficult to locate a signaler that is using chemicals than one using visual or auditory signals.

Chemical signals can be varied to serve different functions by the evolutionary choice of the chemical used. We might expect, for instance, that chemical signals used to attract mates would indicate the sender's species so that only suitable mates would respond. Larger molecules offer greater possibilities for diversity. The more atoms in a molecule, the more ways in which those constituents can be arranged. By combining atoms in different ways, distinctive molecules can serve as sex attractants, identifying different species. A sex attractant is a chemical that attracts suitable mates and guides their approach. We do indeed find that sex attractants usually have 10 to 17 carbon molecules and molecular weights ranging from 180 to 300 (Figure 18.4), a size range that permits species specificity. An added benefit of increased molecular size in sex attractants is that molecules with higher molecular weight are more effective in attracting mates, at least for insects.

However, limits are placed on the size of the chemicals used as signals. Larger molecules are energetically expensive to make and store, and they are only effective over short distances. Territorial markers also tend to be compounds with high molecular weight. These less volatile chemicals persist longer, a characteristic desirable in this situation. On the other hand, the opposite properties would be favored for a chemical signaling alarm. The high volatility conferred by low molecular weight would create a steeper concentration gradient around the caller, thereby making it easier for recruits to locate the signaler. Furthermore, a fast fade-out time is advantageous for an alarm call, so the signal disappears soon after the danger is gone (E. O. Wilson 1965; Wilson and Bossert 1963).

The chemicals used in communication sometimes occur as a blend of several compounds (Rasmussen 1998; Robinson 1996). Changes in the relative concentrations of the ingredients in the mixture can change the message. We see this, for example, in the chemical blend that male cockroaches (*Nauphoeta cinerea*) produce to attract females. A female will mate only with a male that produces all three of the chemical components of the sex attractant. It turns out that this sex attractant also carries a message to other males about

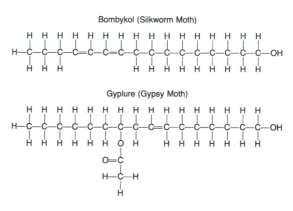

FIGURE 18.4 **Sex attractants of three different insects. Notice that each molecule has between 10 and 17 carbons. The size of the molecule has been adjusted through evolution so that it is large enough to allow for species specificity but not so large that it is difficult to synthesize or that its volatility is severely reduced. (Modified from E. O. Wilson 1963.)**

dominance status. Two of the components signal dominance, and the third signals submissiveness. The exact blend will determine the male's dominance status, and the blend changes with time. Dominant males pump out the two chemicals that signal dominance at a faster rate than the third chemical. Thus, the levels of the two dominance chemicals gradually drop, and with them the male's status (Moore et al. 1997).

Furthermore, the meaning of a particular signal may vary with the context in which it is given. This is the case with a chemical signal sent by the queen honeybee (*Apis* spp.). The chemical trans-9-keto-2-decenoic acid is picked up from the queen as the workers groom her and is distributed throughout the hive, along with the food that is shared by the workers. When attained in this manner, the chemical prevents the rearing of any additional queens. However, the queen also exudes the chemical as she soars skyward on her nuptial flight. In this context, the same chemical causes males to gather around her. In other words, it serves as a queen inhibitor or as a sex attractant, depending on the context (Robinson 1996).

Some species, including certain amphibians, reptiles, and mammals, have a chemosensory structure, called the vomeronasal organ, that is important in chemical communication, especially mate attraction, courtship, parental care, and aggression. It is separate from other chemosensory structures and its neural wiring goes to brain regions other than the main olfactory system (Belluscio et al. 1999). Its nerve fibers project to a region of the brain where there are connections to the hypothalamic-pituitary axis. You may recall from Chapter 7 that this is a region of the brain important in hormonal regulation. It should not be surprising, then, that the vomeronasal system affects and is affected by certain hormones, especially those involved in reproduction (Halpern 1987).

The vomeronasal organ is located in the roof of the mouth or between the nasal cavity and the mouth, and so communicative chemicals must reach it through the nose, mouth, or both. And, since the chemicals that the vomeronasal organ is specialized to detect are nonvolatile, they must be brought to the organ. In a snake, the vomeronasal organ is on the roof of the mouth and the chemicals are delivered to it by the tongue. A mammal, however, must lick or touch its nose to chemicals, which are usually in urine or special body secretions. Following this contact, many mammals make a characteristic facial grimace, known as flehmen, which helps transfer the chemicals to the vomeronasal organ. In flehmen, the head is raised and the lips are curled back (Figure 18.5).

Chemicals produced to convey information to other members of the same species are called pheromones. Some of these, releaser pheromones, have an immediate effect on the recipient's behavior. A good example of a releaser pheromone is a sex attractant, a chemical that attracts suitable mates and guides their approach. The most famous sex attractant is probably that of the female silk moth, *Bombyx mori*. She emits a minuscule amount, only about 0.01 micrograms, of her powerful sex attractant, bombykol, from a small sac at the tip of her abdomen (Figure 18.6). This pheromone, which is carried by the wind to any males in the vicinity, binds to the receptor hairs on the male's antennae. As few as 200 molecules of bombykol have an immediate effect on the male's behavior—he turns and flies upwind in search of the emitting female (Schneider 1974). Other examples of releaser pheromones are trail pheromones, which direct the foraging efforts of others, and alarm substances, which warn others of danger.

Primer pheromones exert their effect more slowly, by altering the physiology and subsequent behavior of the recipient. In insect societies, queens control the

FIGURE 18.5 Flehmen is a characteristic posture in which the head is raised and the upper lip curled back. It serves to deliver nonvolatile communicatory chemicals, such as those found in urine or secretions, to the vomeronasal organ. In many mammals, flehmen helps regulate reproductive activities.

a

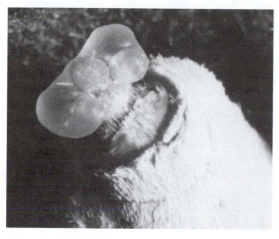

b

FIGURE 18.6 The antennae of a male silk moth (*a*) and the tip of the abdomen of a female silk moth (*b*). Each antennae is about a quarter of an inch long. The feathery appearance is due to numerous branches, each containing many odor-receptor hairs. Half of the odor receptors on the male's antennae are tuned to the female sex attractant, bombykol, which is why the male is sensitive to such minute quantities of it. The tip of the female's abdomen has a pair of glands, shown in an expanded active state here. Within these glands is a small quantity of the sex attractant. This is an example of a releaser pheromone because it causes an immediate effect on the behavior of the male. When he detects the pheromone, he turns and flies upwind in search of the female.

reproductive activities of nestmates largely through primer pheromones. For example, a queen honeybee (*A. mellifera* L.) produces several compounds from her mandibular gland that ensure that she will remain the only reproductive individual in the colony (Robinson 1996; Winston and Slessor 1992). This pheromone coats the queen's entire body surface but is most concentrated on her head and feet. Most of the pheromone

spreads through the colony by the activities of the workers that are attending the queen, but some is spread through the wax of the comb (Naumann et al. 1991). The pheromone prevents the workers from feeding larvae the special diet that would cause them to develop into rival queens. When the queen dies, the inhibiting substance is no longer produced, and new queens can be reared (E. O. Wilson 1968).

In ants, as in honeybees, the queens inhibit the reproductive activities of the workers. It has long been suspected that the queen's control is exerted through primer pheromones, and evidence is now emerging to support this suspicion (Vargo 1997; Vargo and Laurel 1994; Vargo and Passera 1991).

Vertebrates also produce primer pheromones that influence reproductive activity in various ways. They may help regulate reproductive activities so that the activities occur in the appropriate social or physical setting. In species in which there is cooperative care of the young, it may be advantageous for females to give birth synchronously. In others, in which dominant individuals control access to the resources needed to raise offspring, a subordinate female may raise more young if she gives birth at a time when she may have the least competition for the needed resources (McClintock 1983).

TOUCH

Many kinds of animals communicate by physically touching. These tactile signals are obviously only effective over short distances and are not effective around barriers. Tactile messages can be sent quickly, and it is easy to locate the sender, even in the dark. The message may be varied by how and where the recipient is touched. Sometimes the message is simple, perhaps an expression of dominance. For example, a dominant dog may place its leg on the back of a subordinate. However, complex messages may also be communicated by touch. Honeybee scouts inform nestmates of the location of a food source by dancing. The recruits cannot see the choreography because the hive is so dark, but they follow the dancers' movements with their antennae (discussed shortly).

Grooming among certain primates, such as rhesus monkeys (*Macaca mulatta*), is a special form of tactile communication. Assuming that self-grooming has skin care as its primary function, Maria Boccia compared several aspects of social grooming and self-grooming: body site preference, duration (both overall and to specific areas of the body), and method (stroking and/or picking). She reasoned that if the primary function of both social grooming and self-grooming were hygiene, these physical aspects of grooming would be the same in both. However, social grooming was found to be different from self-grooming in each of these respects.

Therefore, she concluded that skin care is not the most important factor in molding the form of social grooming (Boccia 1983). Furthermore, she showed that the message of the tactile signal varies according to the body site being groomed (Boccia 1986). In other words, the monkey's response depended on which part of its body was groomed. The animal being groomed was likely to move away from the groomer when the posterior part of the body was groomed. The responses to the five body regions typically groomed may reflect a continuum, from a tendency to maintain an affiliation at one extreme to a tendency to terminate the interaction at the other.

SUBSTRATE VIBRATIONS

Some signals are encoded in the pattern of vibrations of the environmental substrate, such as the ground or the water surface. These are called seismic signals. For example, a female white-lipped frog (*Leptodactylus albilabris*) of the dense Puerto Rican rainforest literally "feels the earth move" when she is courted since the male sits on the ground or some other solid substrate and thumps his courtship message, causing the substrate to reverberate beneath his foot (Lewis and Narins 1985).

Seismic signals are also used by animals that usually live alone in underground burrows. A male mole-rat attracts a mate from neighboring burrows by drumming with his hind legs, generating a specific pattern of vibration (Narins et al. 1992; Rado, Terkel, and Wollberg 1998). Foot drumming by kangaroo rats (*Dipodomys spectabilis*) declares territory ownership (Randall and Lewis 1997).

Seismic signals travel farther than auditory ones (Randall and Lewis 1997). They are, therefore, well suited for long-distance communication, particularly in areas in which vision is obstructed. (In watching old Westerns, you may have seen the hero place his ear to the ground to determine whether help was on its way by feeling the vibrations caused by distant running horses.)

Tapping on the surface of water creates seismic signals of a slightly different sort—ripples, or surface waves (Figure 18.7). For example, a female water strider (*Rhagadotarsus*) must receive "good vibes" from a courting male before she agrees to mate. Using his legs, the male taps out a patterned sequence of ripples on the water's surface. If a female approaches a calling male, he switches to a courtship vibratory serenade that stimulates her to copulate and release her eggs. A male may defend his signaling site by producing other patterns of surface vibrations that serve as aggressive signals. If an intruder does not retreat when suitably warned, the defender will attack him (Wilcox 1972). Ripple signals have, in fact, been described in a variety of species of water striders. They serve several functions, including

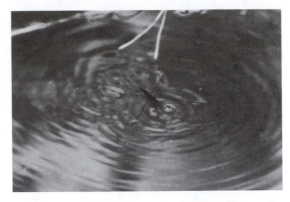

FIGURE 18.7   Water striders communicate by tapping out messages on the surface of the water, thereby creating surface waves. The pattern of the wave determines the message. Here, a male taps out a repel signal.

sex identification, mate attraction, courtship, and territorial defense (Jablonski and Wilcox 1996).

ELECTRICAL FIELDS

Two distantly related groups of tropical freshwater fish produce electrical signals used in both orientation (electrolocation, discussed in Chapters 10 and 12) and communication. These groups are the knife (gymnotid) fish of South America and the elephant-nose (mormyrid) fish of Africa.

The electrical signals are generated by electric organs that are derived from muscle in most species, but in one family of gymnotids they are derived from nerve (Hopkins 1977, 1988). When a normal muscle cell contracts or when a nerve cell generates an impulse, a weak electrical current is generated. The modified muscle (or nerve) cells in an electric organ also generate a weak electrical current, and because they are arranged in stacks their currents are added, resulting in a stronger current. When an electric organ discharges, the tail end of the fish, where the electric organ is located, becomes momentarily negative with respect to the head. Thus, an electrical field is created around the fish (Figure 18.8a). This electrical field is the basis of the signal. A diversity of signals is created by varying the shape of the electrical field, the discharge frequency, and the timing patterns between signals from the sender and receiver, as well as by stopping the electrical discharge.

Two general categories of electrical signals are produced by the weakly electric fish. Mormyrids and some gymnotids produce brief pulses at low frequencies. Other electric fish produce signals continuously at high frequencies (Figure 18.8b). Electrical signals are perceived by specialized electroreceptor organs embedded in the fish's skin (Hopkins 1974).

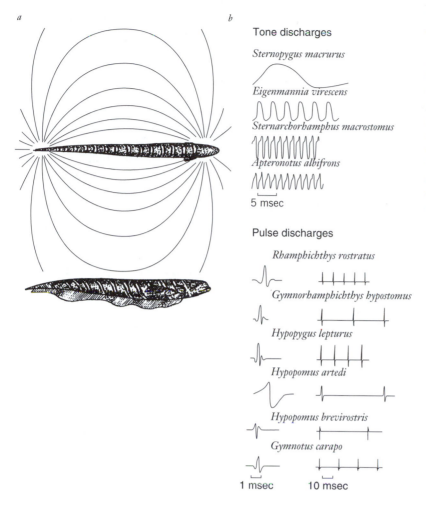

a

b

Tone discharges

Sternopygus macrurus

Eigenmannia virescens

Sternarchorhamphus macrostomus

Apteronotus albifrons

5 msec

Pulse discharges

Rhamphichthys rostratus

Gymnorhamphichthys hypostomus

Hypopygus lepturus

Hypopomus artedi

Hypopomus brevirostris

Gymnotus carapo

1 msec   10 msec

FIGURE 18.8 (*a*) Discharges from an electric organ create an electrical field around a weakly electric fish that is used as a communication signal. The signal can be varied by altering the shape of the electrical field, the waveform of the electrical discharge, the discharge frequency, and the timing patterns between signals from the sender and receiver, as well as by stopping the electrical discharge. (*b*) Some species of weakly electric fish produce electrical signals with a delay between each pulse, and others produce a continuous "buzz" of signals. (From Hopkins 1974.)

What are the characteristics of electrical signals? When the electric organ discharges, an electrical field is created instantaneously. It also disappears at the instant the discharge stops. As a result, electrical signals are ideally suited for transmitting information that fluctuates quickly, such as aggressive tendencies (Hagedorn 1995). An electrical signal does not propagate away from the sender but instead exists as an electrical field around the sender. Because an electrical signal is not propagated, its waveform is not distorted during transmission. As a result, the waveform of the electrical signal may be a reliable indicator of the sender's identity (Hopkins 1986a). Indeed, we find that although the waveform is generally constant for a particular individual, it is different in males and females and among different species (Dunlap, Thomas, and Zakon 1998; Hagedorn 1986; Kramer and Kuhn 1994; Zakon and Dunlap 1999).

Electrical signals are well suited for communication in the environment in which these fish live. Both groups of electric fish are active at night and generally live in muddy tropical rivers and streams, where visibility is poor. Electrical signals can move around obstacles and are undisturbed by the suspended matter that cre-

ates murky water. However, they are effective only over short distances, about 1 to 2 meters, depending on the depth of the water and the relative positions of the sender and receiver (Hopkins 1986b, 1999). The shortness of their effective distance may actually be an advantage. These fish often live in groups that are made up of several species. Since many individuals will be signaling at once, the short effective distance of the signal may reduce electrical "noise."

The weakly electric fish use electrical signals to convey the same messages that other organisms send by alternative channels. Using electrical discharges, these fish can threaten one another or proclaim submission. Males of some species not only advertise their sex and species by electrical signals but also court females by "singing" an electrical courtship song (Hagedorn 1986; Hopkins 1972, 1974, 1986b; Hopkins and Bass 1981). In one species, *Eigenmannia virescens*, interruptions in the electrical discharge, called chirps, are used by males both to defend their territory and to court females (Kramer 1990). In addition, electrical signals are part of the mechanism that promotes schooling in at least one mormyriform fish, *Marcusenius cyprinoides* (Moller 1976).

FUNCTIONS OF COMMUNICATION

Communication is important to both the tugs and the ties of social life. Signals settle conflict and they allow cooperation. Certain signals bring individuals together and others help keep them apart. Some alarm and others calm. Successful sexual reproduction and parenting could not occur without communication. It should not be surprising, then, that we have seen many examples of communication throughout this book. Here we will highlight just a few of the messages that signals convey.

RECOGNITION OF SPECIES

"To each his own" is more than a cliché in the animal world. Some behaviors, particularly aggressive and reproductive acts, are adaptive only when they are directed toward members of the same species. It is disadvantageous to waste time and energy in mating if the production of viable fertile offspring is unlikely. Likewise, time and energy should not be squandered defending a territory from an individual that is not competing for resources or mates. Such competition is usually most intense among members of the same species. Only these individuals would have identical requirements. Therefore, it is often important for an animal to be able to recognize individuals of its own species.

Species recognition is particularly important when there are many closely related species living in the same area. For instance, on one small beach, measuring about 56 square meters, an area slightly smaller than a quarter of a tennis court, 12 species of fiddler crabs (*Uca*) may be courting (Crane 1941). In spite of this density of species, a female fiddler crab still chooses the correct one for a mate. Her discrimination is probably based largely on the distinctive courtship display of the males of each species in which the large claw, or cheliped, is waved about in a specific pattern. One species, *U. rhizophorae*, courts by moving his cheliped up and down. Another fiddler crab, *U. annulipes*, waves his claw in large circles. Another, *U. pugilator*, holds his cheliped out to the side of his body and waves it in tight circles. These differences are easily spotted, even by humans who are not as attuned to subtle differences in cheliped gyrations as female fiddler crabs.

The relationship between sexual behavior and the maintenance of species identity is also obvious among frogs and toads (Figure 18.9). Male frogs usually attract their mates by producing species-specific calls that advertise species identity, sex, reproductive state, and location. Judging by the response of nearby females, these calls seem to be the amphibian equivalent of "I'm over here. Come and get me." Often the males of several species will serenade together in a chorus, and a female must choose one of her "own kind" from the variety of callers at the local pond. To make the proper choice even more difficult, it must often be made in the dark. As a result, selection has favored clear species differences in croaks. The power of selection is most obvious in areas where the ranges of closely related species overlap. For example, in southeast Australia, two such species of treefrogs, *Hyla ewingi* and *H. verreaux*, inhabit some of the same areas. In both species, the male's call is a trill, a rapidly repeated series of sound

FIGURE 18.9   A male American toad (*Bufo americanus*) is calling. Male frogs and toads attract females by croaking a species-specific call. In response, interested females approach the suitor. The male amplifies the sounds produced by his vocal cords by inflating the large vocal sac beneath his chin.

impulses. The calls are so similar in the areas where only one of these species is found that females are not very good at distinguishing between them. Where the species coexist, however, the calls have become different: The *H. ewingi* call is slower and the *H. verreaux* call is faster than the respective calls in areas occupied by only one of the species. Females from the overlap zone easily detect the differences in the calling rate and preferentially approach a male whose call has the right tempo (Littlejohn and Loftus-Hills 1968).

MATE ATTRACTION

Besides being species-specific, the signals that attract a mate must be easy to locate and effective over long distances so that males and females can find each other even if the species members are widely distributed. For this reason, chemical and auditory signals are used commonly, but not exclusively, for attraction. The sex attractant pheromone of the female silk moth (*B. mori*) was described previously in this chapter. If the wind favors the distribution of the pheromone, males from perhaps 100 meters away may be attracted. The sex attractant pheromone of another species, *Actias selene*, is also potent. In one experiment, males that had been experimentally displaced 46 kilometers away were still able to relocate newly hatched females at the original site (Immelmann 1980).

Auditory signals also carry well, especially when amplified by communal displaying or by special anatomical or environmental structures. The courtship songs of crickets (Hoy 1978; Moiseff, Pollack, and Hoy 1978; Popov and Shuvalov 1977) and the calls of frogs and toads (Littlejohn 1977; Rheinlaender, Gerhardt, and Yager 1979; Wells 1977) have been shown to attract females. As you can see in Figure 18.10, female crickets will approach and congregate on a loudspeaker that broadcasts the courtship song of a male of its species.

COURTSHIP

Once the individuals are close enough to interact, they court before committing themselves to mating. There are several functions of courtship: identification of species and sex, reduction of aggressive tendencies, coordination of the mates' behavior and physiology (Etkin, 1964), and assessment of male qualities.

Identification

In his courtship display, a male signals his species, sex, and reproductive condition. It is necessary for a male to identify himself as belonging to a certain species so that any errors made during the initial attraction can be corrected before reproductive effort is wasted on a member of the wrong species. The advertising male must

FIGURE 18.10 Female crickets are aggregating on a loudspeaker that is broadcasting the courtship song of the male of that species. One of the functions of the courtship song is to attract appropriate mates.

also make clear his sex because it is not always easy to tell. Although it is easy to distinguish the sexes of some species, such as dogs, in other species, sparrows for instance, the sexes are remarkably similar.

Reduction of Aggression

Since members of the opposite sex are not always friendly when they first meet, a second function of courtship is to reduce aggressive tendencies so that the couple can approach each other safely. Among certain vertebrates, the female must enter the male's domain. To do so, she must overcome his aggressive tendencies. When a female stickleback (*Gasterosteus aculeatus*) enters a male's territory, for instance, she reduces the probability of attack by assuming a head-up position that displays her egg-swollen abdomen and distinguishes her from an intruding male. The subsequent zigzag courtship dance of the male consists of swimming toward the female, a movement of incipient attack, followed by swimming away as he leads the female toward the nest (N. Tinbergen 1952).

Coordination of Behavior and Physiology

The third function of courtship is to coordinate the couple's behavior and physiology. Among certain species of birds, the male's song may stimulate a female into reproductive condition (see Kroodsma and Byers 1991), and among salamanders, female receptivity may be increased by male courtship pheromones (see Houck 1998; Verrell 1988).

Consider, for example, the mountain dusky salamander (*Desmognathus ochrophaeus*). During courtship, a male applies a courtship pheromone to the female by pulling his lower jaw across the female's back and angling his snout so that his premaxillary teeth scrape the female's skin, thereby "injecting" the pheromone into the female's circulatory system. The female then indicates her receptivity by assuming a tail-straddling position, and the male deposits a packet of sperm called a spermatophore (Figure 18.11*a*).

Lynne Houck and Nancy Reagan (1990) demonstrated that the male's courtship pheromone makes the female more receptive. They staged a total of 200 courtship encounters between 50 pairs of salamanders on each of four nights. By removing the mental gland, which produces the pheromone, from each male, they prevented the males from delivering any pheromone during courtship. These glands were then used to create an elixir containing the courtship pheromone. Thirty minutes before some encounters, each female was treated with this pheromone-containing elixir. Before other encounters, the same female was treated with a saline solution. After a female received a pheromone treatment, she assumed a tail-straddle position, indicating receptivity, 43 minutes (26%) sooner and mated 59 minutes (28%) sooner than she did after receiving a saline injection (Figure 18.11*b*).

Assessment

Finally, courtship may allow a female to judge the qualities of her suitor so that she can choose the one most likely to enhance her own reproductive success. Most commonly, a female is required to invest more time and energy than a male in raising offspring successfully. It is, therefore, important that she choose a mate with the best genes possible. This means that males must advertise and females must discriminate among them. Some aspects of courtship suggest that it provides a means for evaluating the suitor's qualities, including his physical prowess, ability to provide food for the offspring, or even the extent of his commitment. These suggestions stem from several studies, including those on the courtship displays of falcons and eagles. Their aerial acrobatics may allow a female to judge her suitor's speed and coordination, skills important in hunting and, therefore, in his ability to provide food for his offspring. In addition, the courtship of the common tern (*Sterna hirundo*) requires males to catch fish and offer them to the female. She compares the quantity of fish provided by her various suitors and usually chooses the best fisherman. The number of fish a male provides during courtship, the quantity provided to his chicks, and the fledgling success of the clutch are significantly correlated. The quality of the courtship offering, then, is a reliable indicator of the male's ability to provide for the pair's offspring (Wiggins and Morris 1986). Additional discussions on the use of courtship signals in assessing the characteristics of the mate are found in Chapter 14.

a

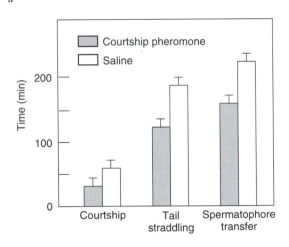

b

FIGURE 18.11 During courtship a male mountain dusky salamander injects a female with his courtship pheromone, which serves to make her more receptive. (*a*) The male (on the left) alternately scrapes the female's back with his teeth and swabs the area with courtship pheromone, which is produced by a gland beneath his chin. The female (on the right) indicates her readiness to mate by assuming a tail-straddling position. (*b*) Courtship behaviors and mating occur more quickly when a female has been treated with courtship pheromone than when she has been treated with saline. Pheromone treatment did not significantly shorten the average time from the beginning of courtship to active courtship. However, after pheromone treatment, the average time until the female is in a receptive, tail-straddling position was 43 minutes (26% shorter), and the average time until mating (spermatophore transfer) was 59 minutes (28% shorter). (From Houck and Reagan 1990.)

ALARM

One of the most fascinating facets of animal communication involves alarms—when one animal warns another of danger. The dangers in the wild are of many stripes. Among the most common are predators. Individuals may even have to be on guard against other members of their species bent on either infanticide or the takeover of their territory.

There are several types of alarm signals. Sometimes, as in ants, the signal serves to enlist support in confronting the danger. In certain passerine birds, alarm calls cause others to flee to a protective area or, if they are already in cover, to freeze. In other species, such as certain deer, an alarm call causes the animals to become quietly alert. The design of the alarm call will be affected by its specific function.

Specificity Among Alarms

Most species use the same signal to indicate any source of danger, although some use specific calls to designate the type of threat. For example, vervet monkeys (*Cercopithecus aethiops*) classify their most common predators into one of at least three groups—snakes (e.g., pythons), mammals (e.g., leopards), or birds (e.g., eagles). Considering the hunting strategy of each type of predator, the characteristics of the alarm call and the response of conspecifics within hearing distance seem to be adaptive. The low-amplitude alarm call emitted when a snake is encountered captures the attention of individuals near the caller that might be in danger from the slow-moving reptile without attracting other predators in the area. Other monkeys respond by looking at the ground, the most likely place to find a snake. However, when a major mammalian predator such as a leopard is seen, very loud, low-pitched, and abrupt chirps are emitted. These properties make the call audible from a great distance and make the caller easy to locate by its fellows. The most common response to the chirp is to scatter and run for cover in the trees, a relatively safe haven from the ambush style of attack characteristic of a leopard. The staccato grunts of the threat-alarm bark that are sounded when an avian predator is spotted are also loud and low-pitched. These features allow the grunts to be easily located and to be transmitted over long distances, thereby broadcasting the position of the predator. When the threat-alarm bark is heard, others will look up and/or hastily retreat to thickets. The dense brush makes it difficult for a swooping eagle to catch them (Struhsaker 1967). The responses just described are also typical of those to playback tapes of these three types of call. The responses, then, are specific to the nature of the alarm call and not to the appearance of the particular type of predator (Seyfarth, Cheney, and Marler 1980).

Similarity Among Species

Alarm signals, particularly of species living in the same area, are often remarkably similar. For example, note the striking resemblance of the alarm calls of several species of passerine birds shown in Figure 18.12. One possible explanation for this similarity is that when species have common predators, it is advantageous to be alerted to danger regardless of who sounds the alarm. It should not matter to a small bird feeding on

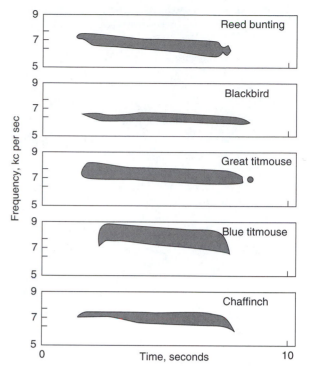

FIGURE 18.12   Alarm calls of several species of small songbirds. Notice the structural similarity of the calls. As a result, an alarm call will alert birds of many species. (Modified from Marler 1959.)

the ground whether or not it is a conspecific that alerts it to an eagle overhead. If the divergence among alarm calls of species living in the same area is slight, the chance of a predator striking unnoticed is decreased. Thus there is selection for convergence of alarm signals (Marler 1973).

A second possible reason for the similarity of alarm calls is that they have the same functions. The source of an alarm pheromone used to recruit assistance should be easily located over a reasonable distance; workers who are too far away to detect the danger themselves should be able to travel to the site in time to be of service. Molecules with a low molecular weight generally meet these requirements well. They are more volatile and, therefore, disperse rapidly over a reasonable area. This creates a steep concentration gradient that facilitates locating the caller. Natural selection has, therefore, chosen chemicals with low molecular weights to serve as alarm substances (E. O. Wilson 1971).

It has also been suggested that, ideally, an alarm that triggers dispersal to safer areas should be difficult to locate (Marler 1957). In that way, the caller would be in less personal danger. Armed with a knowledge of the signal properties known to assist the localization of a sound source, Peter Marler was able to predict the type of bird calls that would be best suited for alarm. The alarm calls typical of small passerine birds shown in Figure 18.12 have the characteristics that would be expected to make them difficult to locate—they begin

FIGURE 18.13 **Threat signal in a baboon. Notice that the animal's weapons, its teeth, are displayed.**

and end gradually and employ only a few wavelengths centered on 8 kHz.

It is interesting that there are some birds that take advantage of the similarity among alarm calls to foil competitors. For instance, in one study (Møller 1988a), 63% of the alarm calls issued by great tits (*Parus major*) were false alarms. These were given when other individuals, particularly sparrows, were in the caller's feeding area but when no predator was in sight. The other individuals generally heed the warning and flee. The result of the false alarm call is that the feeding area is cleared of competitors and the caller can feed undisturbed.

AGONISTIC ENCOUNTERS

In many species, social interaction involves agonistic behavior, that group of actions involved in conflicts that includes aggressive behaviors, such as threats, and submissive behaviors, such as appeasement or avoidance. Since such behavior is usually the result of competition, it is most common between members of the same species and members of the same sex. Once dominance is attained by a particular individual, it is often flaunted with a badge or characteristic stature so that in an established social group, each individual knows its place without continuous testing.

In Chapter 16, we discussed the nature of signals used to settle direct conflicts over a resource. You may recall that in displays used to settle conflicts, the rivals generally show off their resource holding potential (RHP), those characteristics such as strength, size, weapons, or ability to fight that will bear on the ability to win the competition (Figure 18.13). Each rival assesses the RHP of the other. Commonly, one contestant will judge itself to be weaker than the other and back off without a fight. In this way, the loser avoids possible injury in a battle that it could not win.

The displays are related to fighting ability and are difficult to fake (Dawkins and Krebs 1984). This prevents the neighborhood weakling from bluffing its way to triumph. (Factors influencing the honesty of signals are discussed in Chapter 19.)

Some displays emphasize size. Since the bigger animal often wins, size is a reliable indicator to the probability of success and is difficult to fake. Birds emphasize size by fluffing their feathers. Fish often erect their fins and flare their gill covers outward (Figure 18.14), and some even puff themselves up.

FIGURE 18.14 **A threat display by a male Siamese fighting fish (*Betta splendens*). Individuals of many species make themselves appear larger when threatening another. Here the fish flares its opercula (gill covers) outward and spreads its fins, actions that help to intimidate its rival.**

Sometimes an indirect cue provides an indication of size. For example, the pitch of a call is often related to body size. The lower the pitch, the larger one would predict the caller to be. This relationship has been clearly demonstrated for the toad *Bufo bufo*. Furthermore, since physical strength is important in contests over possession of females, larger males are rarely replaced by smaller ones. In these disputes, males clasp onto a female's back, sometimes for several days, until she lays her eggs, which he then fertilizes. Since males outnumber females in the ponds where they mate, there is intense competition for mates. The males will try to dislodge rivals from a female's back by pushing and wrestling. When a male is attacked, he calls. In laboratory studies in which taped croaks were played to male toads, the pitch of the defender's call affected the probability of attack. Challengers were less likely to attack following a low-pitched call (Figure 18.15) (Davies and Halliday 1978).

Other displays are indirect trials of strength or endurance. For example, red deer stags, *Cervus elaphus*, roar at one another while defending their harem

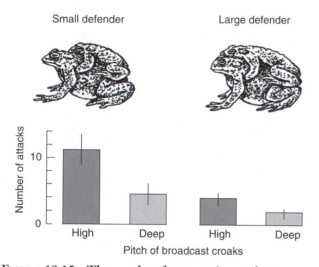

FIGURE 18.15 **The results of an experiment demonstrating that male toads (*Bufo bufo*) use the pitch of their rival's croak to assess his size and fighting ability. When a male toad finds a suitable female, he will ride on her back until she lays eggs that he can fertilize. Rival males may try to dislodge him. When attacked, the defending male calls. In this experiment, medium-sized males attacked either small or large males that were clasping females. The defending male was prevented from croaking by a rubber band through his mouth. During each attack, tape-recorded croaks were played to simulate the call of a defending male. Regardless of the actual size of the defender, he was less likely to be attacked when the pitch of the broadcast croak was deep. The pitch of the call is only one factor used to assess size. Notice that large defenders are attacked less frequently than small males. The strength of the defender's kick may be another indicator. (Modified from Davies and Halliday 1978.)**

(Figure 18.16). The roaring contest seems to follow rules. One stag roars several times and then listens while another answers with his best bellows. Frequently, the contest involves several stags. As would be expected if they were using roaring for assessment, rivals' roars rarely overlap. As the contest progresses, the rate of roaring increases until one competitor gives up (Figure 18.17). Because roaring is so exhausting, the stag's ability to continue the vocal duel is an indication of strength. At the peak of the rut, harem holders often roar day and night, causing them to lose as much as 20% of their body weight. Those that become so exhausted that they can no longer roar lose their females to the challenger.

Several aspects of the roaring duels are consistent with the idea that they are used for assessment. First, roaring ability requires both strength and endurance and is, therefore, a cue that is difficult to fake. Second, the maximum and mean roaring rates of an individual stag are significantly correlated with his fighting ability. Third, on the rare occasions when fights do occur, the combatants have equal roaring skills more often than is expected by chance (Clutton-Brock and Albon 1979; Clutton-Brock et al. 1979).

RECRUITMENT

Reasons

Recruitment is a specific type of assembly in which the members of the group are directed to a specific place to assist in a particular job such as food retrieval, nest construction or defense, or migration. Recruitment is common among social insects (E. O. Wilson 1980). Some species have several recruitment systems, each for a different purpose. The African weaver ant (*Oecophylla longinoda*), for instance, has different signals for attracting others to food or to new terrain, for emigration, and for attacking intruders that are nearby or distant (Hölldobler and Wilson 1978).

Methods

As we will see, social insects may employ a variety of methods to recruit others.

Tandem Running and Odor Trails Tandem running, in which an individual maintains physical contact with the leader, and odor trails in ants will be described in the next chapter. Odor trails are also left by termites, even the primitive species that never leave the nest to forage, such as *Zootermopsis nevadensis*. Alistair Stuart (1963a, b; 1967) has shown that when a *Zootermopsis* nymph detects a change in light intensity or air currents that might indicate a break in the nest wall, it writes a trail in a substance produced by the sternal gland on its fifth abdominal segment as it runs from the site of damage. The nymph gets its nestmates' attention by bump-

FIGURE 18.16   A red deer stag (*Cervus elaphus*) is roaring at a rival. Red deer stags challenge one another by bellowing. In this vocal duel, each male takes a turn at roaring at the other. The pace of the bellowing increases until one contestant gives up. Because roaring is strenuous, it is a reliable indicator for the assessment of a rival's fighting ability.

ing them while jittering from side to side. This bumping alarms the others, which then follow the trail back to the breach in the wall to repair it.

Bee Dancing   The most remarkable example of recruitment communication is in honeybees (*Apis* spp.). Although it may take several days before a new food source is noticed, soon after one honeybee discovers the prize, many recruits join the harvest. The dramatic increase in the number of bees utilizing a new food source is evident in Figure 18.18. We know that the initial discoverer does not lead the others to the food because recruitment is observed even if she is captured as she leaves the hive on a return trip (Maeterlinck 1901).

How does the pioneer scout enlist the assistance of her sisters? The first step in scientifically addressing a

question such as this is to observe the bees' behavior. Observations have revealed that as a scout bee sips the sugary solution provided by either a newly blossoming patch of flowers or an experimental feeding dish, she marks the spot with a chemical from the Nasonoff's gland at the tip of her abdomen. Traces of the flower's odor or a dusting of pollen adhere to her body. The scout then returns to the hive and mounts the vertical comb. When other workers gather around her, she unloads some of the nectar to these nestmates. If the food source was relatively close to the hive, generally less

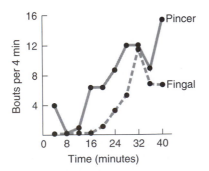

FIGURE 18.17   A roaring contest between two red deer stags. The males defend their harems by roaring at rivals. As the males take turns roaring, the tempo of the contest accelerates. Finally one male gives up and leaves. (Modified from Clutton-Brock and Albon 1979.)

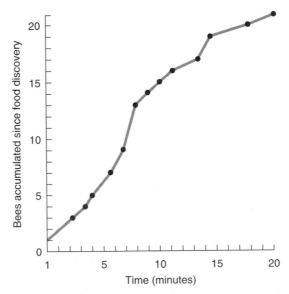

FIGURE 18.18 The number of bees arriving at a newly discovered food site over time. Notice how quickly workers are recruited. (Modified from Ribbands 1955.)

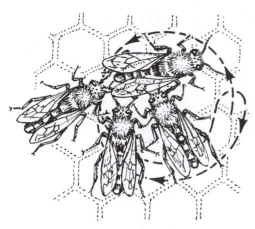

FIGURE 18.19 **The round dance of a honeybee scout. After finding food close to the hive, a scout returns and does a round dance on the vertical surface of the comb. This dance consists of circling alternately on the right and left. It informs the recruits that food can be found within a certain radius of the hive.**

than 50 meters, she begins to dance in circles, once to the left and once to the right. This circling, or round dance (Figure 18.19), is repeated many times. However, if the bonanza was far from the hive, she performs a waggle dance, during which she moves in the pattern of a figure 8. It is a simple dance; straight ahead, circle to the right, then up the center again and loop to the left (Figure 18.20). During the central straight run at about 15 Hz she vibrates or waggles her abdomen and produces a low-frequency buzz with her wings at about 250 Hz. Her sisters, unable to see in the darkness of the hive, crowd around and follow her movements with their

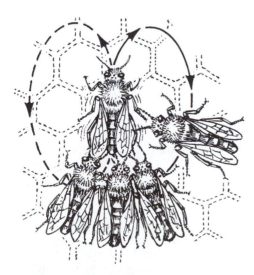

FIGURE 18.20 **The waggle dance of a honeybee scout. When a scout finds food at some distance from the hive, she returns and does a waggle dance. Aspects of the dance correlate with the distance and direction to the food source.**

antennae. Eventually there is a rapid exodus of recruits that later show up at the food site.

Several aspects of the waggle dance are correlated with the location of the food. Direction to the find is indicated by the angle of the waggle run of the dance. There is close agreement between the angle of this run with respect to gravity and the angle that the recruits must assume with the sun's azimuth (direction on the horizon) as they fly to the food source. On the vertical comb within the dark hive, bees substitute straight up for the position of the sun (gravity tells them which end is up), and then the scout waggles at the angle of the flight path to the food relative to the sun. So when the food is to be found in the direction of the sun, the waggle run of the dance is oriented straight up. If the recruits should fly off with the sun directly behind them, the waggle run would point straight down on the comb. Likewise, a food source 20° to the right of the sun would be indicated by a straight run directed 20° to the right of vertical (Figure 18.21). The angle of the dance is adjusted by approximately 15° an hour to compensate for the apparent movement of the sun across the sky.

Distance to the food source is correlated with two other features of the straight run. The longer the duration of the straight run, which is emphasized by the dancer's waggling, the greater the distance to the food (von Frisch 1971). The distance to the food is also correlated with the duration of buzzing during the straight run. As the distance to the food increases, the sound trains of buzzes are longer (Wenner 1964).

Most of what we know about the bee dance comes from experiments performed by Karl von Frisch (summarized in von Frisch 1967, 1971). Most biologists today believe that honeybees communicate the distance and direction to the food source in the waggle dance. Stated more precisely, von Frisch's dance-language hypothesis is that the information in the dance guides recruits to the general area of the food, and then olfactory cues, odor molecules adhering to the body of the scout, and the bee odor left to mark the find help them locate the exact site. Von Frisch provided the following evidence for his hypothesis.

1. The behavior of the recruits depends on the scout bee's dance. When food is located near the hive, the returning scout does a round dance, and the recruited foragers randomly search for the source in the vicinity of the hive. However, when the location of the site exceeds a certain distance, the scout performs a waggle dance and the recruits begin to search preferentially in the direction of the food.

These statements are based on the results of several of Karl von Frisch's experiments. His procedure, which is still used by some investigators, was to establish a feeding station where bees can imbibe a weak, unscented sugar solution. Because the solution is weak,

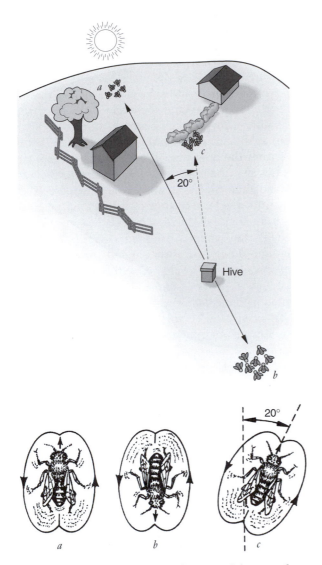

FIGURE 18.21 **Direction correlations of the waggle dance of honeybees. The dancer indicates the direction to the food by the orientation of the waggle run of her dance relative to gravity. When the bee is dancing on the vertical surface of the comb in the dark hive, the direction of the sun is straight up. The angle that recruits should assume with the sun as they fly to the food is equal to the angle of the waggle run relative to vertical. (*a*) The food is in the direction of the sun, so the dancer waggles straight up. (*b*) When the recruits should fly directly away from the sun to reach the food, the waggle run is oriented straight down on the comb. (*c*) A food source located 20° to the right of the sun would be indicated by a dance oriented so that the waggle run is 20° to the right of vertical.**

the returning foragers do not dance. On the test day, a stronger sucrose solution is set out at the feeding station and a scent, perhaps lavender oil, is added to it. The forager that discovers the bounty returns to the hive and dances. During this performance, the hive is closed to prevent bees from exiting. Meanwhile, the

food is removed from the feeding station, and plates emitting lavender fragrance, but containing no reward, are strategically placed. When the hive is reopened, recruits leave and search for the food. When they detect the food odor emanating from a scent plate, they are attracted to it. The direction in which the bees are searching is determined by counting the number of bees arriving at each scent plate.

In one experiment, bees were trained to a feeding station 10 meters to the east of the hive. After sipping the strong scented sucrose on the test day, the scout returned to the hive and did a round dance. While the hive was closed, the feeding station was emptied and scent plates were positioned to the north, south, east, and west, each at a distance of 25 meters. As you can see in Figure 18.22*a*, the recruits arrived at each scent plate in almost equal numbers. Apparently they did not know the direction to the nectar. However, when the same feeding station was moved 100 meters north of the hive, the returning discoverer did a waggle dance. This time almost all the bees arrived at the northern scent plate, indicating that they had direction information (Figure 18.22*b*).

The accuracy of the direction information was further demonstrated in fan experiments. Following his usual procedures, von Frisch trained bees to a feeding station and then placed scent plates so that they were

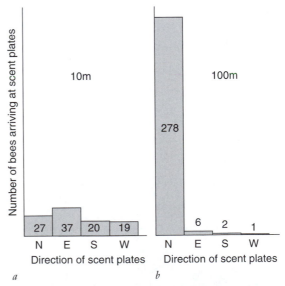

FIGURE 18.22 **Number of bees arriving at scent plates positioned to the north, south, east, and west of the hive when the training station was 10 meters to the east of the hive (*a*) and 100 meters to the north of the hive (*b*). When the food is close to the hive, the recruits search randomly within a certain radius. However, when the food source is distant, they look preferentially in the direction of the training station. (Modified from von Frisch 1971.)**

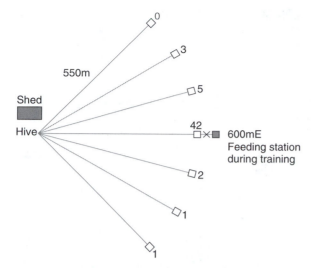

FIGURE 18.23 **Results of a fan experiment showing that after following a waggle dance, the recruits search preferentially in the direction of the training station. Bees were trained to a feeding station 600 meters east of the hive. The greatest number of recruits arrived at the food-free scent plate in the direction of the training station. An "X" denotes the position of the feeding station during training, and squares show the positions of the scent plates. (Modified from von Frisch 1967.)**

arrayed like a fan. Each plate was 550 meters from the hive and separated from its neighbors by 150 meters. The overwhelming majority of recruits appeared at the scent plate nearest to the original feeding station (Figure 18.23). Then von Frisch performed a step experiment to examine the accuracy of the distance information. After training bees to a feeding station, scent plates were placed in line with the empty feeding station at intervals closer or farther away. Again, the

recruits appeared at the scent plates closest to the original feeding station (Figure 18.24).

2. The recruits search for the food in all directions from the hive unless they have attended a waggle dance. Normal waggle dancing can be disoriented by placing the comb in a horizontal position and depriving the bees of a view of the sun or a patch of blue sky. (When dancing on a horizontal surface, the bees orient the dance relative to the sun, and if the sun is not directly visible they determine its position from the pattern of polarization in the sky.) After attending a disoriented dance, recruits search randomly. However, if the hive is restored to its vertical position, both the dance and the search become oriented.

3. When faced with detours, recruits act as if they have direction information. In his detour experiment, von Frisch placed the feeding station so that the scout had to fly around a building to reach it. When she danced, she indicated the direction to the food as a straight line, the bee line. Some of the recruits were observed to fly up and over the building, suggesting that they were using direction information from the dance.

It has even been possible to "talk" to bees in their own language by using a mechanical model of a dancing bee (Figure 18.25). A computer controlled the model's movements and sound production. The dancing model bee fed recruits scented sugar water through a syringe, mimicking the manner in which a real bee would feed the recruits upon returning to the hive. The model danced, indicating a food source 250 meters to the south of the hive. Scent plates without food were placed 250 meters to the north, south, southeast, and southwest of the hive. Although the model's dance was not as effective as a live bee's dance in recruiting others to look for food or in accurately directing them to the correct location, most of the recruits showed up at the

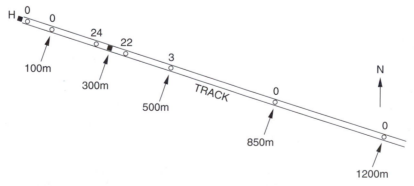

FIGURE 18.24 **The results of an experiment showing that after following a waggle dance, the recruits preferentially search for food at the distance of the training station from the hive. Most of the recruits arrived at the scent plates closest to the original food station. The solid square at 300 meters shows the position of the feeding station during training, and circles indicate the positions of the food-free scent plates. (Modified from von Frisch 1967.)**

FIGURE 18.25 A mechanical bee that can "talk" to live bees is constructed of brass covered in beeswax. On its back is a single wing made from part of a razor blade. An electromagnet causes the wing to vibrate so that it mimics the sound patterns produced by real dancing bees. A tiny plastic tube connected to a syringe releases droplets of scented sugar water to recruit bees, thus simulating regurgitation by live scouts. Computer software choreographed the robot's dance so that it directed the bees to a food source in a direction chosen by the experimenter.

feeding station indicated by the dance (Figure 18.26). Two critical components of the dance are the wagging movements and the buzzing. If either was missing from the model's dance, the recruits showed no preference for the direction indicated (Michelson et al. 1989).

The story, as told so far, seems quite straightforward, but the dance-language hypothesis has, in fact, been controversial. Adrian Wenner and his associates hypothesize that honeybee recruitment is based on odor cues. Olfactory information available to bees includes the scent of both the food and the locality, as well as the bee odor left at the site. Odors brought back on the body of the scout are detected by the recruits' antennae as they follow the dancer. If the recruits have had previous experience with those odors, they can find the food by monitoring all the sites with that scent at which they had been previously successful. It would be possible to find the source of an unfamiliar odor by dropping downwind for about 200 to 300 meters and progressing upwind in search of that odor (Wenner 1971; Wenner and Wells 1990).

Wenner has experimentally demonstrated that bees that are familiar with a food source are able to locate food by using odor cues. They know how far to fly because distance information is available in the buzzing that accompanies waggling. Furthermore, bees can be stimulated to revisit a food dish in response to an injection of the food scent into the hive. Thus, they can learn to associate a particular odor with food in a certain location (Johnson and Wenner 1966; Wenner and Johnson 1966). In addition, one can manipulate the number of new recruits by varying the odor of either the food or its location (Wenner 1971).

It should also be mentioned that Wenner (1971) believes that the results of von Frisch's fan experiments are consistent with the odor hypothesis for recruitment. He believes that the array of scent plates might produce a continuous gradient of odor. If one *assumes* that each bee is searching for the center of the odor field, then the probability that a bee will arrive at a particular plate depends on the distance of that plate from the center of the odor field. With this reasoning, the expected distribution of bees would not be significantly different from that which von Frisch found.

So, with both dance and olfactory information available simultaneously, how does one determine which sources of information are actually being used by the bees? If the odor information and the dance directions send the bees to different sites, one could tell which cue the bees were using by observing where they arrived.

One way to create conflicting information is to get the dancers to "lie." Bees do not normally deceive one another, but they can be made to prevaricate by certain experimental manipulations. If the sun, or a suitable bright light, is seen by a dancing bee, she will orient her waggle run to the light rather than to gravity. Normally, nestmates that are attending this dance also see the light and interpret the movements without confusion. However, because the bees can dance relative to either light or gravity, the experimenter can manipulate the information in the dance at will, simply by changing the bees' sensitivity to light. A bee's ability to detect light is affected by its three ocelli, non-image-forming light detectors located atop her head. When the ocelli are covered with black paint, the bee is six times less sensitive to light. In spite of this, she can still forage and dance.

In a clever experiment (J. L. Gould 1975), the bees were presented with a light that was bright enough to be seen by a normal bee but not by an ocelli-painted one. The ocelli of scouts were painted, and their dance indicated the flight direction relative to gravity. However, the normal recruits interpreted the dance relative to the artificial sun. Since the frame of reference of the dancer was different from that of the recruits

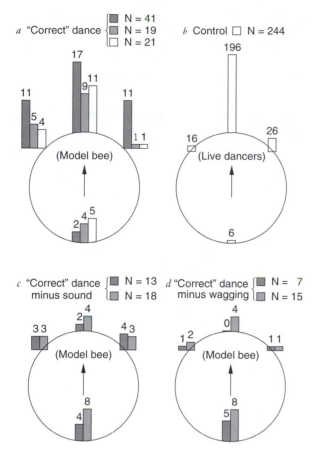

FIGURE 18.26 The dance of a mechanical bee can direct recruits to a distant food source. The figure shows the results of experiments in which the model bee danced by both waggling and buzzing in a manner that mimicked a dance of a live bee directing recruits to a food station 250 meters to the south of the hive. Scent plates were placed 250 meters to the north, south, southeast, and southwest of the hive. Most of the recruits showed up at the scent plate signaled by the model bee's dance (*a*). The number of bees recruited and the ability of the recruits to locate the correct scent plate are somewhat less than that of bees recruited by a live dancer (*b*). When model bees' dances lack accompanying sound patterns (*c*) or waggling (*d*), the recruits show no preference for the direction indicated by the dance. (From Michelsen et al. 1989.)

attending the dance, the message sent was different from the one received. The dance-language hypothesis predicts that, depending on the information in the dance, the misinformation provided by the dance should cause recruits to arrive at a scent plate as far off from the foraging station as the light deviated from the vertical. However, the odor search hypothesis predicts that bees would not be misdirected.

When this experiment was performed using von Frisch's training techniques, the bees were misdirected as predicted. However, when Wenner's methods were used, the bees were not deceived (Figure 18.27). The

difference in the procedures is that Wenner trains his bees on concentrated sucrose that contains the scent that will be used in later tests. Unlike von Frisch's methods, this causes foragers to dance during training and it allows food odor to accumulate in the hive (J. L. Gould 1975, 1976).

Can these conflicting results be reconciled to answer our original question about the information sources actually used by the recruited bees? Perhaps they can if the problem is considered from the bee's point of view. For a bee, collecting nectar is a matter of life and death. She may die if she runs out of energy while searching for food. Therefore, one would expect that a forager would use all information available—both dance and odor (Kirchener and Towne 1994). Certain circumstances might make one of these sources of information easier to use. For example, James Gould (1975) suggests that the different training techniques of von Frisch and Wenner may be sampling different stages in the use of a food source. In early spring, when the frigid siege of winter is ending and bees begin to venture out of the hive in search of nourishment, flower patches are scarce. When a patch first comes into bloom, scouts may dance to guide recruits to the site. However, after the blossoms have been sipped for a few days, their fragrance begins to accumulate in the hive. At that time, odor alone may be sufficient to recruit bees to the site.

Although Gould's (1975) suggestion offered a compromise between opposing viewpoints and many consider the issue settled (Robinson 1986), others recognize that the controversy is still alive (Esch and Burns 1996). The voice of skepticism about the dance-language hypothesis, although small, has yet to be completely silenced (Shannon 1998; Wenner 1998). Gould's misdirection experiments have been criticized as lacking adequate controls (Rosin 1978, 1980a, b, 1984). Furthermore, the results of some experiments of proponents of the dance-language hypothesis do not seem to fit neatly into place. For instance, Adrian Wenner and Patrick Wells (1987) point out that in some experiments cited as indicating that the dance contains information about the direction to the food source (J. L. Gould, Henerey, and MacLeod 1970), 87% of the marked bees that were seen to follow the dance and leave the hive failed to appear at any feeding dish. Those that did arrive at the predicted site took 30 times longer than would have been needed for a direct flight. If nothing else, the odor hypothesis remains a thorn in the foot of the dance-language hypothesis, causing an occasional limp in its otherwise self-confident gait.

In any event, the dance-language controversy should serve as a useful reminder that an answer to a question often depends on how it is asked. We generally find that it is best to design experiments with the natural history of the organism in mind.

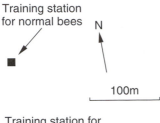

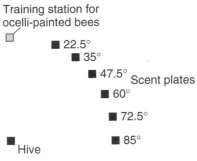

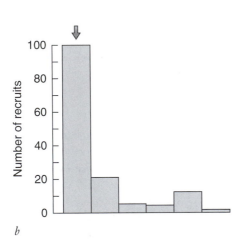

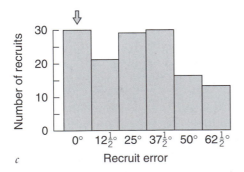

FIGURE 18.27   The results of misdirection experiments using ocelli-painted foragers. The experiment was designed to determine whether bees utilized olfactory or dance information by creating a situation in which these two sources of information would send the recruits in different directions. When a bright light is presented to bees, they interpret it as the sun and orient or interpret dances relative to it. The light sensitivity of some bees was reduced by covering their ocelli with black paint. The ocelli-treated foragers did not see the artificial sun, and their dance indicated the direction to the food source relative to gravity. The untreated recruits interpreted the dance relative to the "sun." (*a*) The experimental setup. By shining a light on the hive from different directions, the experimenters could reorient the dancers. However, the ocelli-painted recruits attending the dances would not see the light and would, therefore, misinterpret the dance information. (*b*) The results of the experiment when the bees were trained on diluted unscented sucrose (von Frisch's techniques). The direction indicated by the dance is shown with an arrow. Notice that the greatest number of bees arrived here. (*c*) The results of the experiment when the bees were trained on concentrated scented sucrose (Wenner's techniques). The direction indicated in the dance is shown with an arrow. Notice that the bees did not show a preference for this station. (J. L. Gould 1975.)

MAINTENANCE OF SOCIAL BONDS BY PHYSICAL CONTACT

It is easy for humans to understand the significance of physical contact as a signal of social bonding, given our penchant for hugging or otherwise holding one another. Social bonds between certain other animals are also cemented by physical contact. These species are called contact animals (Figure 18.28), as opposed to distance animals, which live in groups but prefer to remain at a specified distance from one another. Among contact animals, cuddling, nuzzling, and touching in general tend to firm social bonds (Eibl-Eibesfeldt 1975).

Many of the greeting signals exchanged by some animals as they encounter one another serve as an assurance of nonaggression. Chimpanzees often greet by touching hands or sometimes by placing a hand on the companion's thigh (Goodall 1965). Sea lions rub noses, and lions rub cheeks. African wild dogs greet one another by pushing their muzzles into the corners of each other's mouths (Schaller 1972).

A special type of contact behavior, grooming, is exhibited by both invertebrates and vertebrates. The workers of most social insects groom their nestmates, usually by licking, although they occasionally pick up extraneous material with their mandibles. Grooming probably plays a role in cleanliness, as well as providing a means for disseminating the colony's odor and

FIGURE 18.28   Sea lions are contact animals. Some animals maintain social bonds through bodily contact.

pheromones. The substance produced by the queen honeybee that inhibits the production of other queens is distributed around the colony in this manner.

Grooming is also found in a variety of vertebrates. Birds who prefer bodily contact, such as waxbills (*Estrildidae*), babblers (*Timalidae*), white eyes (*Zosteropidae*), and parrots (*Psittacidae*), groom or preen one another. Rodents groom each other by gently nibbling on each other's fur. But allogrooming (grooming another individual) is most prominent among primates. They usually spend large parts of their day in this activity (Figure 18.29).

Figure 18.29   Tactile communication—social grooming among crab-eating macaques. Although the grooming originally functioned only for skin care, its social functions have now become more important. Primates spend a major portion of their day grooming, which helps form and maintain social bonds.

Grooming probably originated as a means of hygiene, at least among vertebrates. However, allogrooming has become important in signaling conciliation and bonding. Although grooming may still be essential for the care of the skin, these social functions are often more important than hygiene (Boccia 1983).

One social function of grooming is to suppress aggression. When a bird is threatened, for instance, it can often halt the attack if it acts as if it is about to be preened. We also find that allogrooming among rodents is most common in conflict situations. Likewise, mountain sheep in the area south of the Mediterranean, the mouflon (*Ovis ammon*), employ licking, a form of grooming, in a ritualized appeasement ceremony (Pfeffer 1967). For primates, grooming appears to be a very relaxing experience, one that reduces tension and restores relationships.

The formation and expression of social bonds is a second function of primate grooming behavior. Most grooming occurs within close relationships, and in species in which grooming is rare, it occurs most commonly between mothers and their offspring. Furthermore, there is an increase in grooming between males and females when the females are in estrus. This increase reflects consort bonds (Carpenter 1942).

A possible third function of social grooming among primates, enlisting support for agonistic encounters, stems from the two functions just described, the reduction of aggression and the maintenance of social bonds. Robert Seyfarth (1976, 1977) has constructed a model that is useful in predicting which individuals are most likely to groom one another. One of the most important factors is the ability of a monkey to support the groomer in future aggressive encounters. Of course, every individual would want to groom the most dominant animals, so there would be competition for access to these animals. Low-ranking animals would lose the competition for access to the high-ranking animals. Therefore, the model predicts that most grooming will occur between individuals of similar rank. Indeed, a relationship between grooming and agonistic support has sometimes been observed, especially in chimpanzees (de Waal 1978).

SUMMARY

Communication occurs when information is transferred from sender to receiver to their mutual benefit. Animal communication systems differ from human language in many ways. Animal signals are more specific, the meaning of a display is usually constant, and animal displays are stereotyped. In other words, each performance of a display is the same, even if different individuals are displaying.

Any sensory channel may be used for communication. Signals within each channel have characteristic properties that make them more or less useful, depending on the species, the environment, and the function of the signal.

Visual signals are easy to locate, are transmitted quickly, and disappear just as fast. However, visual signals must be seen and are, therefore, only useful when there is enough light and where there are few obstacles to obscure the signal. Visual signals include badges, anatomical specializations that convey a message, as well as movements and postures.

Auditory signals can be transmitted long distances. The rate of transmission and fade-out is rapid. They do not require light and, in fact, work well under water. The sounds may be generated by respiratory structures, the rubbing of appendages, or beating on parts of the environment.

Chemicals can convey messages over great distances, particularly when assisted by currents of air or water. They are transmitted more slowly, are more durable, and are usually more difficult to locate than visual or auditory signals. Furthermore, they are effective in environments with limited visibility. Chemicals used to convey information to conspecifics are called pheromones. When the chemical has an immediate effect on the behavior of the recipient, as occurs with sex attractants and trail and alarm substances, it is called a releaser pheromone. Primer pheromones act slowly, exerting their effect by altering the physiology of the recipient.

Tactile signals are effective only over short distances, and it is easy to locate the sender. These signals are transmitted and disappear rapidly and are effective in areas with limited visibility. Grooming, a special form of tactile communication, is practiced by many species but is especially prominent among primates.

Seismic signals are those caused by vibration of the environment. They are well suited to communication over long distances, particularly when vision is limited.

Electrical fields are used for communication among the mormyriform fish of Africa and the gymnotid fish of South America. Transmission and fade-out are almost instantaneous, but the signals do not travel far. They are effective when visibility is limited.

There are numerous functions for communication. Species recognition is important so that reproductive efforts are not wasted on members of the wrong species and so that aggression is directed toward those individuals that are competing for the same resources.

Sexual reproduction is often dependent on communication. A mate must be attracted and then courted. During courtship the male advertises himself and the female assesses his qualities, aggression is reduced, and the behavior and physiology of the mates are coordinated.

Since animals in nature are surrounded by potential dangers, an individual may alert others when it is alarmed. Usually all sources of danger are signified by the same signal, but sometimes, as in vervet monkeys, a different call is used for each class of predator. The alarm signals of species living in the same environment are usually very similar. One reason might be that the more eyes looking for danger, the more likely it is that the threat will be detected. Or, the similarity may result from the shaping of the signal to suit its purpose.

Communication is also important in aggressive interactions. Each rival signals its resource holding potential, characteristics such as size, strength, and weapons that will influence its ability to win the contest. If one rival assesses the other as a stronger competitor, it usually gives up the contest without a fight.

When individuals are brought together to perform a specific duty, as occurs in social insects, it is called recruitment. Tandem running and odor trails are two means of recruitment, but the most elaborate recruitment signals are those of the honeybees. When a scout finds food, she recruits nestmates to help in the harvest by dancing. A round dance is performed when the food is close to the hive. It informs recruits that food is within a certain distance of the hive but does not specify the location. The waggle dance is performed when food is farther from the hive. Aspects of this dance correlate with the direction and distance to the food source. The angle that the recruits should assume with the sun as they fly to the food is the same as the angle of the waggle run of the dance with respect to gravity. The recruits know how far to fly by the duration of both the waggling and the buzzing that accompany it. Because it is so important for bees to find food, they also use olfactory information to locate it. The spot is marked with a chemical from the Nasonoff's gland, and odors of both the locale and the food are brought back on the scout's body. The relative importance of dance language and odor in honeybee recruitment has been controversial.

Physical contact is another means of maintaining and reaffirming social bonds. In fact, many greeting signals used to reassure individuals when they reunite involve touching. Individuals of many species groom one another, but grooming is probably most obvious among primates. Grooming plays an important role in cementing social bonds.

19

Maintaining Group Cohesion: The Evolution of Communication

Communication, as we've seen, involves animals that share information to the benefit of both the sender and the receiver. The information that is shared can convey many types of messages and can serve many functions. We have also seen that the signals that convey similar messages have different forms in different species. We can often answer the who, what, where, and when of the communication process, but *how* and *why* have the signals taken the various forms that they have? The simple answer is that they have been designed to effec-

tively convey information. Although that answer is true, its simplicity glosses over the many interactions among factors involved in signal design. It ignores the anatomical and physiological constraints on the system and the demands of the job.

So, in this chapter, we will consider the evolution of communication. We will begin by examining some of the hypotheses for the reasons for communication. Surely the reason will affect the costs and benefits of the information exchange to both the sender and the receiver. These will, in turn, affect the design and honesty of the signal. Then we will consider ritualization, the process by which an inadvertent cue to the sender's condition or subsequent action becomes formalized into a communication signal. Next we will look at some of the selective forces that influence signal design. Finally, we will ponder language—how it differs from animal communication systems, whether it is uniquely human, and what the study of communication can tell us about animal cognition.

REASONS FOR COMMUNICATION

SHARING INFORMATION

The traditional ethological view is that communication results in shared information and that this sharing is adaptive. Communication makes private information, such as an animal's internal state or what it is likely to do next, available to others, who may then respond appropriately (J. M. Cullen 1966; W. J. Smith 1977).

Consider how important it is for a courting male orb-web spider (Argiopidae) to properly identify himself and his intentions to a female by plucking on her web with a species-specific rhythm. The females of these species are usually larger than the males and earn their living by eating small arthropods similar to the would-be suitor (Figure 19.1). If the male fails to provide the signals by which the predatory female can distinguish his approach from that of a prey species, the male may quickly turn from suitor to snack (Bristowe 1958). So the male signals the female and communicates both his identity and his intentions.

The sharing of information may also enhance survival by coordinating group activities. For example, when a honeybee informs her fellow workers of the location of a rich nectar source, the food stores of the entire hive, on which *she* also depends, are increased. As long as the interests of the sender and receiver are aligned we expect honest information sharing between them.

It is not difficult to think of situations in which the interests of the sender and the receiver are at odds, however. There are some situations in which the sender might gain by lying. Thus, this question immediately arises: What guarantees are there that animals will tell the truth? In a territorial dispute, for example, the sender might bluff by sending signals that exaggerate its willingness to escalate the contest. Consider, for instance, a female choosing from among several courting males. She would gain the most by choosing the one of highest quality. Therefore, each male would gain by advertising that he would be the best father of her young, and each might try to exaggerate these traits. Also, consider two opponents at a territorial border. The one that displayed greater strength or willingness to fight might have an advantage if the opponent used the display to decide how to react. We might expect, then, that a cheater, one that made false claims about its qualities, resources, or future behavior, would have an advantage. In a population of honest signalers, a mutant cheater would, therefore, be more successful, and the cheating mutation would become more common. In other words, honesty might not be an evolutionarily stable strategy (ESS). (You may recall from Chapter 4 that an ESS is a strategy that cannot be replaced by an alternative one if it is adopted by most members of the population.) It has been argued, therefore, that accurately displaying easily bluffed information, such as motivation or future behavior, is unlikely to be an ESS (Maynard Smith 1972, 1974, 1982). If this argument is pushed to its conclusion, the function of communication must not be to share information.

MANIPULATING OTHERS

What, then, is an alternative explanation for communication? Richard Dawkins and John Krebs (1978) suggest that animals communicate not to convey information but to manipulate the behavior of others to their own advantage. This idea stems from Dawkins' (1976) presentation of an animal as a machine designed to ensure the preservation and propagation of its genes. Simply put, if a particular action by another animal would enhance the spread of one's kinds of genes, then it is beneficial to cause the companion to behave in that way. Although this could be accomplished either by brute force or by persuasion, the latter is often energetically more efficient. As we have seen, animals may be genetically programmed to respond in a standard way to a specific stimulus. The response has been selected through evolution because it enhances survival in most situations. The specific stimulus has been selected from all those available because it is usually a reliable indication of the appropriate situation for that response. In communication, the individual sending a message might be viewed as providing the stimulus to cause the receiver to perform the behavior when it is beneficial to the sender, regardless of its consequences for the receiver.

However, the receiver would be under a different set of selective pressures: to be a mind reader and predict the true future action of the signaler (Krebs and Davies 1984). In this view, the selective pressures on receivers are similar to those suggested by the hypothesis that communication functions in sharing information.

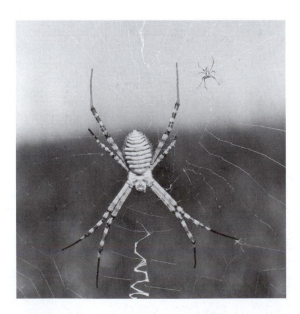

FIGURE 19.1 **The female *Argiope* spider is much larger than the male (upper right). If a courting male fails to identify himself, he could easily be mistaken for prey.**

SIGNALS AND HONESTY

The view that senders are likely to be deceitful manipulators has been questioned by Amotz Zahavi (1975, 1977; Zahavi and Zahavi 1997). He argues that honesty will be favored in a population when an honest signal is costly to the sender. A signal, such as a loud call or a vigorous display, costs energy and may also incur risk. The louder the call and the more vigorous the display, the greater the risk. A signal that says, "I am very strong" might cost more than a signal that says, "I am strong." There are also ostentatious traits, including the huge antlers of red deer and the magnificently patterned tail of a peacock, that are costly in terms of the time and energy required to produce them, even though the cost of sending the signal is cheap. However, the high cost can be borne by a truly strong individual more easily than by a weaker one. Thus, the cost keeps the signal honest because only high-quality signalers have the resources to send a costly signal.

Initially, Zahavi's idea, known as the handicap principle, was not well received because he merely presented examples of signals that could be interpreted as handicaps. However, his verbal arguments paved the way to mathematical models. Now that a mathematical model has been developed that shows that if honesty is costly it can be an ESS (Grafen 1990), the handicap principle has gained some respectability.

Support for the handicap principle requires a signal to meet two criteria. First, it must be costly to send. Second, either the cost or the benefit must depend on the quality or condition of the sender (Figure 19.2) (Johnstone 1997).

Signal cost can take different forms. An obvious one is energy expenditure. Although it is not universally true that displays use energy, there are many examples of displays that do. A way to monitor energy use is by measuring the amount of oxygen consumed during the activity. In insects and frogs, oxygen consumption increases 5 to 30 times during displays (Ryan 1988). Male red deer (*Cervus elaphus*) actually lose weight during the rut because they constantly roar at one another as they compete for mates (Clutton-Brock, Guiness, and Albon 1982). Another type of cost is a lessened ability to move about or forage. For instance, in several bird species, the long tail feathers that males sport as sexual ornaments impair their ability to fly (Jennions 1993).

We see, then, that many signals are costly. But, cost alone doesn't guarantee honesty. The cost could be an inevitable result of the difficulties of producing a clear signal that can be perceived by the receiver in an often noisy environment (Ryan 1998).

However, the chances of honesty go up when the cost of the signal is associated with the quality or con-

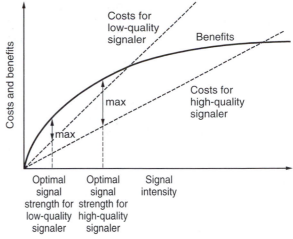

a Honesty due to condition-dependent costs

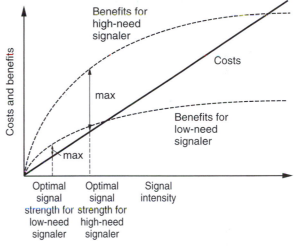

b Honesty due to condition-dependent benefits

Figure 19.2   A signal will remain honest when it is costly to send, as long as either the cost or the benefit to the signaler depends of the condition of the signaler. The optimal signal strength (max) for a given signaler maximizes the net difference between benefits and costs. (*a*) Honesty due to costs is maintained because the costs for a given signal strength is greater for a low-quality signaler than for a high-quality signaler. The benefits are the same for all signalers. (*b*) Honesty due to benefits is maintained because the benefits for a high-need signaler exceed those for a low-need signaler. The costs of the signal are the same for all signalers. (Modified from Johnstone 1997.)

dition of the sender. That's what keeps a poor-quality individual from bluffing: It simply doesn't have what it takes to produce the signal. Among red deer, for example, only the males in top physical condition can continue roaring long enough to win the vocal duel. Roaring uses many of the same muscles and behavior patterns involved in fighting, and so it serves as an honest signal of a male's fighting ability (Bradbury and Vehrencamp 1998).

There are many examples of communication signals that do provide honest, reliable information about the signaler. Consider, for example, the begging of nestling birds—a situation in which the benefits to the signaler are condition-dependent. Nestling birds beg food from their parents by assuming a posture with a widely gaping mouth while emitting calls. The vigor of begging increases with the degree of hunger. Thus, it is a reliable signal of need (Redondo and Castro 1992). Parents provide more food to the chicks that beg more vigorously (Kilner 1995). In this case, the cost is not energy expenditure (McCarty 1996) but the increased risk of predation, which does not depend on the nestling's condition. When tape recordings of begging sounds were broadcast from ground nests (but not tree nests), predation rates increased (Haskell 1994). The hungry young beg more vigorously not because they can afford to pay a higher cost but because they gain a greater benefit from receiving the food.

The aggressive displays of the little blue penguin (*Eudyptula minor*) show a clear relationship between cost and honesty. Little blue penguins have a repertoire of aggressive displays that differ in cost (risk of injury), effectiveness in deterring an opponent, and ability to predict an attack. Although two displays may involve the same ritualized posture, one that involves moving within the rival's striking distance is riskier to perform than one that is performed while stationary and out of the opponent's reach. By its choice of display, a penguin conveys information about both its willingness to sustain injury while performing the display and its willingness to fight. It chooses a display with costs that represent the value it places on the resource. Roughly 10% of penguin interactions are not settled by displays and end in fighting. During battles, losing penguins commonly suffer flesh wounds and sometimes eye loss (Figure 19.3). These injuries could make it more difficult to obtain sufficient food or to breed successfully.

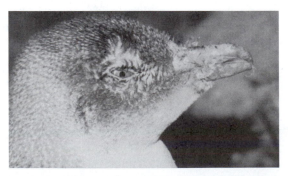

FIGURE 19.3   These eye and head injuries in a little blue penguin resulted from fights. Little blue penguins have an extensive repertoire of aggressive displays, which vary in cost (the risk of injury). By its choice of display, a penguin indicates the extent of its willingness to fight. Signals remain honest because they are costly.

Thus, attempting to intimidate an opponent into retreat by bluffing a strong motivation to attack could be quite costly to the bluffer if the rival called the bluff and a fight ensued (Waas 1991).

The encounters invariably begin with low-risk displays and escalate until one opponent retreats or a fight occurs. The process is somewhat analogous to human actions at an auction—the bids begin low and gradually increase until bidders unwilling to pay the price drop out of the process. The price that little blue penguins must pay is the risk of injury. As a territorial contest escalates and the price of the property increases, one "bidder" usually decides that the territory is not worth that great a risk. For these penguins, the signals remain honest because they are costly, and the cost for a given level of signal is greater for weaker than for stronger individuals.

On the other hand, male pied flycatchers (*Ficedula hypoleuca*) may be deceitful when attracting females. In this species, males may mate with more than one female. When they do, the females have a lower reproductive success than do monogamously mated females because they receive less help in providing food for the young. We would expect, then, that if a female could distinguish between a mated and an unmated male, she would avoid one that was already mated. Males, in contrast, would increase their reproductive success by attracting a second female. It has been suggested that a female that chooses to mate with a previously mated male may do so because she is unaware that he has mated.

One way in which a male may mislead a female about his mating status is to hold two widely separated territories. The second female may then be unaware that he is already mated until she has laid her eggs and it is too late in the season to start over with a new male (Alatalo et al. 1981; Alatalo and Lundberg 1984; Alatalo, Lundberg, and Stahlbrandt 1982). Females may also judge a male's mating status incorrectly because, although unmated males spend more time on their territory and sing more frequently than mated males when no female is present, their behavior is similar when a female is present, making it difficult for a female to distinguish between them (Searcy, Eriksson, and Lundberg 1991).

FACTORS FAVORING SENDER HONESTY

Since we can find examples of both honesty and deceit in animal communication, it may be more fruitful to consider factors that might favor one or the other.

Tight Association with a Physical Attribute

Although a signaler might gain if it could lie, signals closely tied to a physical or physiological condition

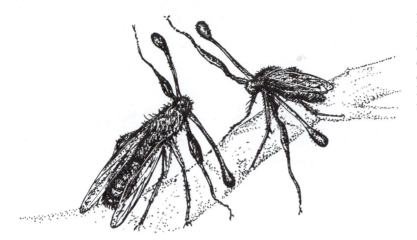

FIGURE 19.4   An aggressive display of male stalk-eyed flies. In this pose, each male can determine the distance between the eyes of its opponent and thereby the rival's size. Size is correlated with fighting success. Since there is little a combatant can do to alter its body size, this is an example of honest signaling.

simply cannot be faked. Since size is usually a good predictor of fighting success and an attribute that is difficult to fake, many displays allow opponents to judge one another's size. For instance, in the threat display, male stalk-eyed flies (Diopsidae) face each other head-to-head, with their forelegs spread outward and parallel to the eyestalks (Figure 19.4). This pose allows each competitor to compare the length of its eyestalks to that of its rival. Eyestalk length increases with body size, and males with shorter eyestalks usually retreat without a fight (Burkhardt and de la Motte 1983; de la Motte and Burkhardt 1983).

Stable Social Unit

A stable social unit also favors honest communication. When some degree of cooperation is essential for an individual's survival, honesty should be favored over dishonesty (Markl 1985; van Rhinjn and Vodegel 1980). One reason to expect honesty is that individuals will both send and receive signals at different times. Thus, the advantages of sending dishonest signals would be reversed when the animal is the receiver. Therefore, the advantages of receiving honest signals might outweigh the advantages of sending dishonest ones, and honesty might come to predominate in the population.

Another reason that honesty should be expected in stable social units is that the members often recognize one another and remember previous interactions. Working with vervet monkeys (*Cercopithecus aethiops*) in Kenya's Amboseli National Park, Dorothy Cheney and Robert Seyfarth (1988) tested the idea that members of a social group would cease to believe a known liar. Vervet monkeys utter different vocal signals, or calls, during different situations. They have two acoustically different calls that warn of the encroachment of another group of monkeys on their territory. One call, the "wrr," is given when another group is spotted in the distance, perhaps as far away as 200 meters. The other, a "chutter," is given if the other group comes so close that there are threats, chases, or actual contact (Figure 19.5). Vervet monkeys also have a stable social group in which the animals recognize each individual and its calls. The question was, what would happen if one individual lied and falsely signaled the approach of another group? Would other monkeys believe the liar again?

To answer this question, Cheney and Seyfarth broadcast tape recordings of the calls of a member of the group to create the illusion that it was lying. The calls were broadcast from a speaker that was concealed where the group might expect that individual to be. First, an individual's "chutter" call was played to see

FIGURE 19.5   Vervet monkey alarm calls that alert group members of the approach of a neighboring troop are of two types. The "chutter" warns that another group is nearby. The "wrr" is given when another group is spotted in the distance. The horizontal lines show frequency in intervals of 1 kHz. (From Cheney and Seyfarth 1991.)

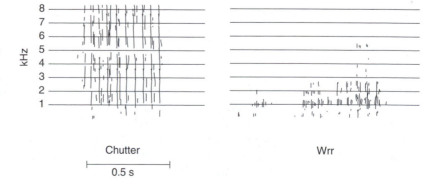

Chutter   Wrr

0.5 s

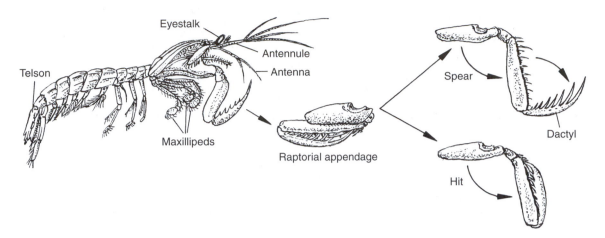

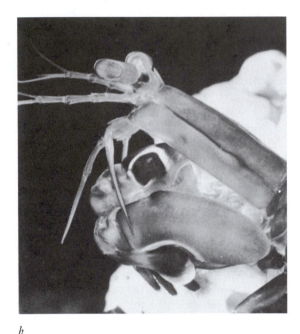

a

b

FIGURE 19.6 (*a*) The mantid shrimp, a stomatopod, has two raptorial appendages that can be used in prey capture or combat. (*b*) The threat display of a stomatopod, called a meral spread, is a good predictor of attack. Here the display is seen from the side.

what the baseline response of the others would be. The next day, when no other group was in sight, they played that individual's "wrr" call eight times, at approximately 20-minute intervals. They found that the other monkeys gradually stopped responding to that individual's warning that another group was in sight. The monkeys no longer believed the liar. However, if another monkey uttered the "wrr" call, the group still believed the warning and responded appropriately. So we see that in a stable social group in which individuals are recognized, others may soon learn not to believe a dishonest signaler.

RECEIVER COSTS AND DISHONESTY

It has also been argued, however, that receivers of signals pay costs as well and that this can corrupt honest signals (Dawkins and Guilford 1991). In some signals, such as the aggressive displays of little blue penguins, the receiver pays for an honest signal by replying with an honest signal, which can be costly to the receiver. In other cases, the cost of assessing the signal may be time. In some species, females use the duration of a male's display to assess his quality. Female field crickets (*Gryllus integer*), for instance, judge a suitor's quality by the length of his song (Hedrick 1986). The time she spends listening to a male's songs could be spent on other activities. Furthermore, calling often attracts parasites. By remaining close to a calling male to assess his quality, the female places herself at risk.

A receiver may pay dearly for challenging the honesty of the signal. If receivers are unwilling to accept the cost of a test, honest signals may be corrupted by occasional dishonest bluffs. For example, stomatopods (*Gonodactylus bredini*), commonly known as mantis shrimps, are marine crustaceans that ferociously defend the burrows and cavities in which they live. Stomatopods have two large forelimbs, called raptorial appendages, that can unfold and shoot forward in a manner similar to that of the praying mantis (Figure 19.6*a*). The raptorial appendages, which are adapted either for spearing or smashing, are used both in prey capture and territorial defense. Therefore, combatants may be seriously injured or even killed during the battles over burrows. Readiness to attack is signaled by a threat display, called a meral spread, in which the raptorial appendages are spread out, thereby exposing a depression called a meral spot, while less dangerous appendages, the maxillipeds, are extended in a circular pattern (Figure 19.6*b*). The meral spread is significantly correlated with an ensuing attack (Caldwell and Dingle 1976; Dingle and Caldwell 1969).

However, a newly molted stomatopod is virtually defenseless with its soft body exposed. In this condition, it may dishonestly use the meral spread and deceive an opponent, causing it to retreat. It can get away with the bluff because the receiver might pay dearly if it chose to test the honesty of the signal and, in fact, the signaler was *not* newly molted (Adams and Caldwell 1990). The same characteristics of a signal that make it costly to send often make it costly to receive as well. As a result, the receiver may settle for a signal that does not guarantee honesty but is less costly to receive.

We see, then, that animals may communicate to share information or to manipulate the behavior of others. Signals may be honest or dishonest. This should be expected. The informational content and relative honesty of a particular signal will be influenced by the social structure and the ecology of the species and the context of the interaction (Hauser 1996). Communication is important both in cooperative interactions and in conflicts. Because the cost of a signal is a result of selection for reliable signals, we might expect it to be higher in situations in which the sender and receiver have conflicting interests and lower when they share common interests (Bradbury and Vehrencamp 1998; Johnstone 1998). Thus we would expect the context of the interaction to affect the design of the signal. We also see that the selective forces that shape a signal will be different for senders and receivers. Keeping these ideas in mind, let's now turn our attention to the evolution of communication signals.

RITUALIZATION

How and from what did the signals used in communication evolve? It is thought that the process began with an association of an incidental cue and a condition, motivational state, or subsequent action. That cue might be an unavoidable part of a behavior that occurs under certain conditions, such as opening the mouth before biting an opponent. If the receiver can perceive the cue and associate it with the condition, it is better able to make decisions about the most appropriate response. If the receiver benefits from the information, evolution will favor an enhanced ability to detect and understand the cue. And, if the sender benefits at the same time, evolution will favor modification of the cue to increase the amount of information transferred (Bradbury and Vehrencamp 1998). During the course of the evolution of communicative signals, behaviors serving as cues lose their flexibility and original function and assume a stereotyped pattern, thus becoming signals. Julian Huxley (1923) called this process ritualization. We will consider the process first from the sender's perspective and then from the receiver's.

THE RAW MATERIAL FOR RITUALIZATION

Even Michelangelo's magnificent *David* began as a block of marble, and the sculptor had to start some-

where. How did the sometimes elaborate communicative signals sculptured by natural selection begin? We will consider three sources of raw material (inadvertent cues): intention movements, displacement activities, and autonomic responses.

Intention Movements

Animals may begin functional behavior patterns with some characteristic action, which may be the incomplete initial motions of an entire pattern or the preparatory phase of the action. By themselves, these actions have little adaptive value. Because it is often possible to judge from these activities just what the animal intends to do, these have been named intention movements (Heinroth 1910). Through evolution, intention movements may be formalized into a display. For example, in species that bite during an attack, such as the ring-billed gull (*Larus delawarensis*), the threat display often emphasizes a widely gaping mouth. However, although jabbing is a movement associated with this gull's attack, in the threat display the bird does not attempt to make contact with its opponent. The gulls' threat display may have originated with the intention movement of opening the mouth, an obvious prerequisite for biting. Initially, an opponent witnessing this intention movement may have anticipated the attack in time to flee. Because injury was avoided, natural selection may have favored this reaction to the open mouth. Then selection could act on the aggressor's signaling, and the gaping mouth became a stereotyped display.

The threat display of the ring-billed gull also illustrates another common process in the ritualization of an action: the development of anatomical features that emphasize the signal. In this case, the widely open mouth reveals a vermilion interior (Moynihan 1958).

Numerous avian displays originated with intention movements for flight or walking (Daanje 1950). A bird about to take flight goes through a sequence of preparatory motions. It will usually begin by crouching, pointing its beak upward, raising its tail, and spreading its wings just slightly. One or more components of the takeoff leap have been ritualized into communicative signals in different species. As shown in Figure 19.7, the part of the courtship display of the blue-footed booby (*Sula nebouxii*) that is aptly named "sky pointing" is a ritualized version of flight intention movements. Notice that the wings are spread and the tip of the beak and the tail point upward. Although this display had its origins in flight intention movements, it is difficult to see how a bird could ever take off from this ritualized pose. Obviously, the movement has changed during the ritualization process.

Displacement Activities

Displacement activities are actions performed by animals in conflict situations that are irrelevant to either of

FIGURE 19.7   Sky pointing by a blue-footed booby. This display is part of courtship and probably evolved from the intention movements of flight. (Photo by John D. Palmer.)

the conflicting tendencies. When an animal cannot "decide" which of two opposing actions to perform, for instance, to fight or to flee, it may perform displacement activities such as preening, nesting, or eating. Similar to intention movements, displacement activities are often incomplete actions.

Courtship is a time of conflicting tendencies. Sexual partners must come together to mate in spite of the aggressive tendencies that tend to keep them apart. Since there are conflicting tendencies to approach and fight during courtship, displacement activities would be expected. It has been suggested that the mock preening of the courtship displays of males of many duck species (Figure 19.8), including the familiar mallard (*Anas platyrhynchos*), originated in displacement preening (Lorenz 1972). Similarly, displacement nesting behavior may have evolved into the portion of the courtship display of the great crested grebe (*Podiceps cristatus*) that was named the "penguin dance" by Julian Huxley

(1914). During this display the partners first shake their heads in a stereotyped motion and then dive for weeds as they would when building a nest. But the weeds are not used to set up housekeeping. Instead, the mates present them to each other (Figure 19.9) in a ritualized behavior.

Autonomic Responses

The autonomic nervous system regulates many of the automatic body functions—digestion, circulatory activities such as heart rate and diameter of the blood vessels, certain hormone levels, and thermoregulation, to name a few. It is thought that many displays originated with autonomic functions. These may be divided into four categories (Morris 1956).

Urination and Defecation   If you are a dog owner, you know about the elaborate territorial urination of dogs. A walk with a male dog is always slow-paced; there is not a tree on the block that he fails to mark. What is the evolutionary origin of this ceremony? Desmond Morris (1956) maintains that it has its roots in the tendency of animals to lose control of their bladder and bowel functions at times of stress. If you have raised a puppy, you probably had to clean up puddles whenever it was excited. A similar response, elicited by the threat of a territorial intruder, may have been the origin of the ritualized marking of territories with urine.

Vasodilation   The dilation of specific blood vessels may cause certain structures to become enlarged, or if the vessels are close to the surface, dilation may increase the coloration of particular areas. A change in the distribution of blood is a common action of the autonomic nervous system at times of stress or conflict.

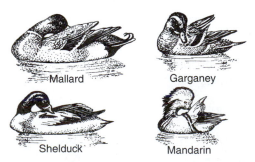

Mallard

Garganey

Shelduck

Mandarin

FIGURE 19.8   Mock preening by courting male ducks, a display thought to have evolved from displacement preening. The movements emphasize the bright markings on the wings. (Modified from N. Tinbergen 1951.)

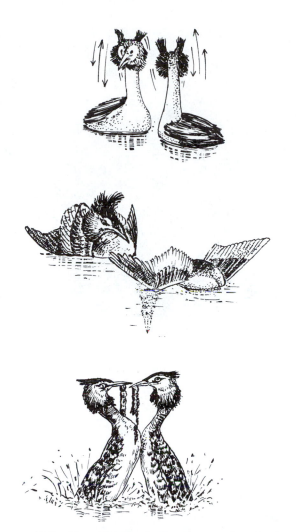

FIGURE 19.9   Courtship in the great crested grebe. This part of the courtship ceremony, the penguin dance, evolved from displacement nest building. The mates dive for weeds to present to one another.

The blushing of embarrassed humans provides a familiar example. Similar circulatory changes, called a sex flush, occur during sexual excitement. On the other hand, we may turn "white as a ghost" when frightened. It has been suggested that similar circulatory responses have become ritualized into communicative signals in some species. In birds such as the turkey, jungle fowl, and bateleur eagle, the naked head or neck skin may flush and fleshy appendages may swell because of vasodilation. These changes now have signal value during courtship and aggressive encounters.

Respiratory Changes   Perhaps the best examples of animal signals that developed from respiratory changes are the inflation displays of birds, during which the males fill pouches on their body with air to attract mates. For instance, the male frigate bird (*Fregata minor*) has a pouch on its throat that is inconspicuous

when deflated, but when inflated its enormous size and brilliant scarlet color attract passing females (Figure 19.10).

Thermoregulatory Responses   In birds and mammals, the erection of feathers or hair helps to adjust body temperature. When a bird fluffs its feathers, it raises each feather just to the point where the tip falls back on the next most posterior feather. This increases the thickness of the insulating layer. When feathers are raised even further, they are described as ruffled. These feathers no longer overlap and entrap air, allowing the bird to cool itself rapidly. The bird can also hold its feathers flat against the body surface, giving it a sleek appearance. This is the typical position of the feathers in an active bird since it reduces resistance to air during flight while still allowing heat to dissipate because the insulating layer of feathers is small. It has been suggested that the thermoregulatory changes in feather position that accompanied social interactions first served as unintended indications of the individual's internal state and then gradually became ritualized into elaborate signals (Morris 1956).

Many displays in numerous bird species are based on the position of feathers. Sleeking, fluffing, and ruffling are common in aggressive and appeasement displays (W. J. Smith 1977). For example, a zebra finch (*Poephila guttata*) that is prevented from escaping from a dominant individual fluffs its feathers as an appeasement signal (Morris 1954). Feather position is also important in courtship displays such as that of the male bicheno finch (*P. bichenovi*). The fluffing of his feathers during courtship accentuates his species-specific black and white markings (Morris 1956).

Sometimes only a specialized region of feathers is erected. Perhaps the most familiar example is the courtship display of the peacock or the turkey, in which the exquisite tail feathers are erected in a beautiful fan shape. Other species have specialized feathers on other parts of the body that serve as a courtship signal when erected. These feathers may form a crest, as in the sulfur-crested cockatoo (*Kakatoe galerita*); an ear tuft, as in several species of owls; a chin growth, as in the capercaillie (*Tetrao urogallus*); a throat plume as in some herons; or eye tufts, such as the shiny green patch between the eyes of the bird of paradise (*Lophorina superba*).

In mammals, the erection of hair, which creates an insulating layer, just as the erection of feathers does in birds, may also be ritualized into communicative displays. Who can mistake the meaning of the hair-on-end posture of a frightened cat? In some species, such as the rufous-naped tamarin (*Saguinus geoffroyi*), a squirrel-sized South American monkey, the meaning of the message varies with the part of the body on which hair is erected. When all the hair is erect, the individual is likely to attack or behave indecisively. However, when

FIGURE 19.10   A courting male frigate bird inflates his huge, brilliant red throat pouch to capture the interest of females. This display originated from respiratory changes controlled by the autonomic nervous system.

only the tail hair is erected, it will probably flee (Moynihan 1970).

Sweating is another means by which the autonomic nervous system of many mammals regulates body temperature. Deodorant ads remind us that perspiration is increased during stress. This response is so common that we often describe an apprehensive condition, such as waiting for exam scores to be posted, as "sweating it out." This autonomic response is thought to have provided the raw material for many animal signals. It has been suggested that the scent glands of mammals may have evolved from sweat glands and that the evolutionary origin of territorial marking may have been the sweating that occurs in frustrating situations (Morris 1956).

Although these three classes of behaviors have traditionally been considered sources of displays, E. O. Wilson (1980) points out that "ritualization is a pervasive, highly opportunistic evolutionary process that can be launched from almost any convenient behavior pattern, anatomical structure or physiological change." For example, the patterns involved in prey catching have been ritualized in the male gray heron (*Ardea cinerea*). During courtship, he erects his crest and certain other body feathers and points his head downward, as if to strike at an object, and snaps his mandibles closed, movements similar to those used during fishing (Verwey 1930). Food exchange has also been ritualized. Billing, the touching of bills, is derived from the parental feeding of young and has taken on a variety of symbolic meanings in different species of birds. It is common in courtship and appeasement displays in which it functions to establish or maintain bonds. Mated pairs of masked lovebirds (*Agapornis personata*) bill to reassure each other during greetings and after a spat (Figure 19.11). The appeasement display of the Canada jay (*Perisoreus canadensis*) also includes billing.

The subordinate bills while squatting and quivering its wings, the posture of a young bird begging for food (Wickler 1972).

Flight is another common behavior that has been ritualized for communication. The yellow-headed blackbird (*Xanthocephalus xanthocephalus*) and the red-winged blackbird (*Agelaius phoeniceus*) perform an aerial

FIGURE 19.11   Lovebirds are billing. This display is derived from parental feeding behavior and now signals nonaggression during greetings and after conflicts.

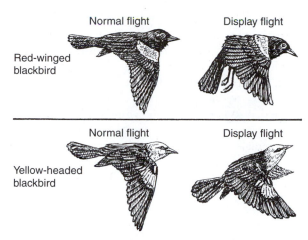

Normal flight   Display flight

Red-winged
blackbird

Normal flight   Display flight

Yellow-headed
blackbird

FIGURE 19.12   Ritualized flight in blackbirds. Almost any behavior can be a starting point for ritualization. Notice the exaggeration of motion that developed during ritualization. (Modified from Orians and Christman 1968.)

display involving ritualized flight patterns to entice females. As you can see in Figure 19.12, ritualized flight is more conspicuous because certain movements are accentuated and because the movements reveal special plumage patterns (Orians and Christman 1968). Portions of certain fiddler crab courtship patterns may have developed from the movement of the male while entering his burrow. It is speculated that this may be the origin of the courtship display of a male *Uca beebei*. When a female approaches him and his burrow, he raises his body and flexed cheliped to an almost vertical position, an exaggeration of his movements into his burrow during the final stages of courtship. This exposes his dark underside, which probably functions to guide a female to his burrow (Christy and Salmon 1991).

THE RITUALIZATION PROCESS

In ritualized patterns, the communicative signal can be distinguished from the behaviors from which it evolved. If one is reasonably certain of the action from which a social signal evolved, a comparison of the initial and present versions of the behavior elucidates the changes that the ancestral action has undergone. The specific changes involved in this behavioral metamorphosis have been categorized in a variety of ways. The scheme presented here is basically that of Niko Tinbergen (1952), elaborated with details and examples from a variety of other studies (Blest 1961; Daanje 1950; Eibl-Eibesfeldt 1975). Keep in mind that the categories may overlap.

Formalization of the Movements

Formalization often results in the exaggeration of all or portions of the original pattern in one or more of five basic ways.

Change in Intensity   The intensity of a behavior refers to features such as the duration of the action or the extent of movement during the performance of a component of the display. As a result, one part of the action is emphasized.

Change in the Rate of Performance   Sometimes a change in the rate at which an action is performed makes it more noticeable. You have probably seen how this works in the theater. Pantomimists often slow down an action so that its meaning is more easily perceived by the audience. On the other hand, a film can be shown at a faster than normal speed to convey increased excitement or confusion. The same basic techniques are apparently also effective in the animal world since changes in the rate of performance are common as a behavior becomes ritualized to function in communication. For example, the rhythmic protective displays of some saturnid moths involve the same wing movements used in flight, but the wings beat at a slower frequency (Blest 1957a).

An extreme degree of a change in the rate of performance is the freezing of activity into a sustained posture. For example, the protective posture of the bittern originated with a series of movements that preceded walking or hopping, but the movements have been ritualized into a display in which the bird sits motionless, with its bill directed upward (Daanje 1950). Saturnid moths also have displays in which they remain stationary and expose their eyespots to frighten predators. It has been suggested that all the wing displays of the saturnid moths evolved from flight movements; these stationary postures are furthest from the original pattern (Blest 1957b).

Development of Rhythmic Repetition   Some displays may have originated as a discrete act but evolved into a repeated performance. The rhythmic waving of the large claw of a male fiddler crab, used to entice females into his burrow, probably began as a single thrusting of the claw in aggressive encounters. Likewise, the rhythmic bioluminescent flashing of a male firefly that now communicates his sex, reproductive intention, and species may have originated as a single flash to illuminate an area in the search for a suitable spot to land. Males may have occasionally flashed as they alighted near females, and this evolved into a display (Lloyd 1968). When woodpeckers are searching for food, the sounds produced by their beaks while pecking a tree vary greatly; however, during their courtship drumming, they have a unique, species-specific pattern.

Change in Components of Original Behavior Pattern   Portions of the original behavior pattern may be deleted, combined with new actions, or performed in a new sequence. For instance, ritualized beak wiping in some bird species consists of little more than the bow that normally precedes the actual cleaning of the beak (Morris 1957). The full forward display of the green heron, during which the bird crouches forward and flips its tail back and forth, is derived from flight intention movements. During actual preflight motions, the tail is never lowered while the bird is crouched (Meyerriecks 1960).

Change in Orientation   Sometimes the display action is not directed toward the stimulus that initiates it, as it was before the behavior became ritualized. For example, consider the inciting behavior of certain species of ducks in which a female urges her mate to act aggressively toward an intruder. In the behavior's original form, a female that felt threatened would run to her drake for protection. She would stop in front of him and look back over her shoulder at the approaching threat. The male would protect the territory and the female by attacking the intruder. Now, in ritualized inciting she still runs to her drake and stares over her shoulder, but the angle at which she holds her neck is always the same. Thus, sometimes she is looking at the interloper, but at other times she is looking at something totally irrelevant to the situation.

Emancipation

During ritualization, behaviors become freed from the internal and external factors that originally caused them. As a result, the ritualized action may appear in a different context than the ancestral form. In other words, it may no longer be elicited by the stimuli that were previously effective. Fluctuations in the tendency to execute the ancestral form do not influence the likelihood of performing the ritualized version. For example, a drake need not be thirsty to engage in display drinking during courtship.

Development of Conspicuous Structures

Sometimes the display behavior is accentuated by anatomical specializations such as brightly colored areas on the body, enlarged claws, antlers, manes, sailfins, and tumescent bodies.

THE RECEIVER'S PRECURSORS

The receiver also plays an important role in the evolution of communicative signals. In some cases, it seems that the receiver has a preexisting sensory bias for the signal. That is, features of the receiver's nervous system may make it more sensitive to the particular type of stimulus used in the signal. This sensory bias may depend on feature detectors similar to the prey detection mechanism of the common toad, described in Chapter 6. According to one model, called sensory

exploitation, the sender would take advantage of the receiver's preexisting sensory biases when new signals are evolving (Ryan 1990; Ryan and Rand 1993b).

The receiver's sensory bias is often used to explain female preferences for certain male signals. For example, the low frequencies in the "chuck" notes that have been added to the end of the whine in the call of male túngara frogs (*Physalaemus pustulosus*) that is used to attract females is thought to have been selected because the female's sensory system is most sensitive to low-frequency sounds (Figure 19.13). The tuning of the female's ears is thought to have evolved before the advent of the chuck call since female sensory systems in two related frog species (*P. pustulosus* and *P. coloradorum*) have similar biases toward lower than average frequencies in the chuck call. However, the chuck notes have evolved only in male *P. pustulosus* (Ryan et al. 1990). Moreover, when female *P. coloradorum* are given a choice between a natural male call of their species, which lacks the chuck notes, and a call to which the chuck notes were synthetically added, they prefer the call with chuck notes (Ryan and Rand 1993a). Thus, the chuck call could not have been responsible for the evolution of the female's bias.

There are also examples of male courtship signals that appear to have exploited the females' sensory adaptations for prey detection. Consider, for example, the water mite (*Neumania papillator*). The eyes of water mites don't form images, and so they must find food and mates by other means. This species of water mite hunts by ambush. A mite waits on aquatic vegetation with its first four legs raised. This position allows it to detect and orient toward passing copepods by the characteristic water vibrations caused by the prey's swimming movements. When a courting male detects a female in the hunting posture, he performs a trembling display by moving his legs. The leg movements cause vibrations that mimic those of the prey, leading the female to grab him as she would a prey item. Food-deprived females are more likely than satiated ones to approach and clutch males. When the female detects that she has grabbed a male and not a meal, she releases the male. He then deposits his spermatophores (packets of sperm) in front of her and fans a pheromone contained in the spermatophore toward the female. The pheromone causes the female to pick up the packets and place them in her genital opening. It is to the male's advantage to elicit the predatory clutch because it allows him to orient to the female so that the spermatophores can be placed directly in front of her (Proctor 1991). Analysis of the evolutionary relationships among water mites suggests that males evolved courtship trembling *after* the sensitivity to water vibrations for prey detection. Thus, it seems that males do, indeed, exploit the prey-detecting mechanisms of the female to their own advantage (Proctor 1992).

FIGURE 19.13   A male túngara frog is calling to attract a mate. It is thought that the chuck notes that are added to the end of the whine part of the courtship call evolved to take advantage of the female's greater sensitivity to low-frequency sounds. Thus it is an example of the sender's exploitation of a sensory bias in the receiver.

SELECTIVE FORCES ACTING ON THE FORM OF THE SIGNAL

Why do birds sing and fireflies flash? Why isn't it the other way around? These questions sound absurdly simple, but in fact they have no simple answers. Trying to give a simple answer to why a signal has a certain form is somewhat like attempting to name the single ingredient responsible for the flavor of a stew. It cannot be done because, in both cases, the final product results from an intermingling of many elements. In communication there are many factors, including the anatomy, physiology, and behavior of the species, as well as features of the habitat, that are important in determining whether or not any given signal is possible and, even more important, which signals can be successful in conveying a message. Nonetheless, it is possible to name the ingredients that blend in a stew, and so it should be possible to label the selective forces that may have had a hand in molding a signal. Throughout the discussion of selective forces that follows, try to remember that the pigeonholing is artificial because the factors that have been separated are, in reality, constantly interacting in the shaping of the signal.

SPECIES CHARACTERISTICS

Anatomy and Physiology

The anatomy and physiology of the sender and receiver will dictate which sensory channels can be most easily exploited for communication. An animal must have the anatomical structures necessary to produce the signal. For example, although it is not the sole factor, anatomy is undoubtedly one reason why ants do not rely on sound for communication. Ants lack an anatomical structure that could produce sounds loud enough to be heard at the distances over which they forage. Anatomy is also the reason why whales have no displays involving fluffed feathers.

Consider how an attribute as simple as body size might influence the form of a display. First, a small body may limit the usefulness of visual signals and favor the use of another sensory channel because small individuals have difficulty seeing and being seen over distances. Is it any wonder, then, that semaphoring plays no role in an ant's long-distance communication? Since the anatomy of ants makes it unlikely that auditory and visual signals can be effective, natural selection has favored other channels of communication, for example, chemicals.

Second, if visual signals are employed by a species, body size may influence the form of the signal. For example, in addition to whole body movements and posturing, two forms of visual display are common among the primates of Central and South America: facial expression and the erection of hair on parts of the body. It has been suggested that these two forms of display are functionally equivalent. The evolutionary choice of form may depend on body size since the raising of long hair on a small body is more visible at greater distances than is a change in expression on a diminutive face (Moynihan 1967). The small New World primates have displays involving the erection of hair but have little variation in facial expression, whereas just the opposite is true among the larger primates. Tamarins and marmosets are squirrel-sized, so subtleties of facial expression would be difficult to see at a distance. However, the erection of tufts and ruffs of hair would increase visibility, and both tamarins and marmosets have long, silky fur that makes their displays more effective (Figure 19.14*a*). Although the New World monkeys tend to be poker-faced compared to their Old World cousins, larger species such as the capuchins (*Cebus*)—best known for soliciting coins for organ-grinders—the common spider monkey (*Ateles*), and the woolly monkey (*Lagothrix*) have a richer variety of facial expressions (Figure 19.14*b*) than do the smaller New World primates such as the tamarins and marmosets. The faces of the larger monkeys are big enough

a

b

FIGURE 19.14 A marmoset (*a*) and a woolly monkey (*b*). Among the New World monkeys, small species, such as the squirrel-sized marmoset, communicate more frequently with displays that involve the erection of hair, whereas larger species, such as the woolly monkey, rely more heavily on facial expressions. This is probably because differences in facial expression would be difficult to see in small species.

for a visual signal as subtle as a facial expression to be seen and deciphered.

Third, body size may also affect the form of visual displays that involve movements of the entire body. For example, it may affect the kinds of movements used in the display. Smaller individuals are often more agile. You may have noticed a relationship between body size and agility among human athletes: Gymnasts are usually more petite than weight lifters. Likewise, in the animal kingdom, smaller species are often better acrobats. Among herons (Meyerriecks 1960) and gulls (Moynihan 1956), for example, the larger species have fewer and less elaborate aerial displays than do the smaller species.

The anatomy and physiology of the receiver are also important forces acting on the design of a signal (Guilford and Dawkins 1991, 1992). We have previously discussed some examples in which senders have taken advantage of the sensory biases of receivers. Even when there is no sensory exploitation, natural selection should favor signals that are easy for a receiver to detect and discriminate, and this will depend on its sensory abilities. For example, it is not surprising that bird songs are more melodious than the courtship songs of insects when the responsiveness of their auditory receptors is considered. The tympanal membranes of insects, unlike the ears of birds and mammals, cannot detect differences in pitch (Marler and Hamilton 1966). An insect suitor must, therefore, identify himself not by his tune but by his rhythm. In the symphony of animal sound, the insect is a percussionist.

Some signals are easier for a receiver to discriminate, that is, to recognize that they belong to a particular category. Warning colors, for instance, help a predator recognize prey as unpalatable (discussed in Chapter 13).

Finally, signals should be easy for the receiver to remember, especially if they are used to assess the qualities of the sender. Some quality of the signal from several individuals must be remembered and compared if the "best" is to be chosen.

Behavior

The behavioral repertoire of a species may also influence the form of communication signals. Some behaviors are too important to stop during communication. It is obvious that a brachiator, such as a gibbon that swings from branch to branch, could not evolve a display such as the chest pounding of the gorilla or it would be constantly plummeting to the ground. Although there may be occasions when communication is so important that it warrants the cessation of other activities, there are times when it is advantageous to continue another behavior during signal emission. Some sensory channels, vision, for example, minimize an animal's freedom for other activities. The sender must employ a major part of its motor equipment, and the receiver must turn its eyes toward the signal and thereby impair the performance of other activities (Marler 1967). Producing and receiving sound or chemical signals interfere much less with other activities.

The relationship between the type of foraging of ants and the nature of their recruitment signal provides a specific example of the way in which the behavioral characteristics of a species may affect the form of a signal. The recruitment signal is nicely suited for the food source exploited by each species. Some species of *Leptothorax* feed on large, stationary items such as dead beetles. For this type of prey, the assistance of a single companion is all that is necessary. A volunteer is solicited by "tandem calling." The successful scout returns to the nest and regurgitates a sample of food to several nestmates. Then it turns around and exudes a droplet of a calling pheromone from the poison gland in its elevated abdomen. The nestmate that touches the caller's hind legs or abdomen with its antennae is led to the food source. The recruit keeps in contact by continually patting the leader on its posterior parts (Figure 19.15, Möglich, Maschwitz, and Hölldobler 1974).

Other species, including the fire ant (*Solenopsis*), relish live insects much larger than themselves. Because of the size of the prey, many workers must be recruited; and because the prey moves about, each new position of the quarry must be indicated. A fire ant scout that has been successful in locating food lays a chemical trail during its return to the nest. Other workers are attracted by this odor and follow the trail to the food. If they are rewarded, they, too, leave a chemical trail on their return trip. Several features of this recruitment policy are well suited to the type of food exploited. First, because workers leave a trail when, and only when, food is found, the intensity of the chemical provides an index of the richness of the food source. The pheromone is volatile—within two minutes after the food is gone, the trail disappears. Because the trail self-destructs, the confusing remains of old trails are eliminated. Furthermore, the trail can change position to track the moving prey (E. O. Wilson 1971).

The leaf cutters (*Atta*) and the seed eaters (*Pogonomyrmex*) exploit more permanent or renewing food sources. Their highways require durable road signs. Again, the recruitment signal is well suited for its purpose. These species lay odor trails with long-lasting chemicals. In addition, they mark their paths with persistent visual cues by cutting the vegetation (Figure 19.16) (Hölldobler 1976).

ENVIRONMENTAL CHARACTERISTICS

Determination of the Sensory Channel Employed for a Signal

It is not surprising that the characteristics of a species' habitat may also mold the form of its signals. An obvious variable of great importance is the amount of light available. Darkness severely limits visual displaying unless the animal sports a biotic lantern. Among bioluminescent organisms, the color of emitted light may be an adaptation to increase the efficiency of the signal in environments with different qualities of prevailing light. For example, among North American fireflies,

FIGURE 19.16 Trails cut by ants through the vegetation leading from the nest to food-collecting areas are marked with durable chemicals. This type of signal is useful when the food source is permanent or renewable.

FIGURE 19.15 Tandem running in ants. This means of recruitment is efficient when a single companion is all that is necessary to exploit a food source. The recruited ant maintains contact with the caller by continuously patting the leader's posterior parts.

most (21 of 23) species active during twilight emit yellow light, whereas the majority (23 of 32) of dark-active species produce green light. The emission of yellow light at dusk is thought to make the signal easier to distinguish from the green light that is still reflected from foliage (Lall et al. 1980).

Night is not the only cause of darkness; vision is also restricted in many aquatic habitats, which thus favor communication through the other sensory channels. Low-frequency sounds are transmitted farther than light in deep sea (R. S. Payne 1972). Therefore, the eerie songs of the finback whale are well designed for long-distance communication. They are extremely low-pitched, loud, and pure in tone. In more precise terms, the frequencies in their calls are restricted to a band only about 3–4 Hz wide, centered at 20 Hz. It has been estimated that before competition from the roar of ships' propellers, these tones could be heard over distances of 1000 to 6500 kilometers. At one time, therefore, the whales' wails may have enabled them to keep in touch with companions over an entire ocean basin (R. S. Payne and Webb 1971).

Both chemical and electrical signals have also been selected for communication in dark aquatic environments. The bullhead catfish, for example, has a sophisticated social system in the murky waters it calls home. Each individual's identity and status are identified by chemical cues (Todd 1971). Electrical signals are another solution to the problem of communicating in perpetually murky waters, but this capability is limited to a few organisms, the gymnotid fish of South America and the mormyrid fish of Africa (Hopkins 1974, 1980).

Influence on the Specific Form of a Signal

Dialects in the Language of Bees   The physical characteristics of the environment may do more than simply favor a particular sensory channel for communication; they may also mold the form of the signal within any given channel. The dialects in the language of bees have been interpreted in this light (J. L. Gould 1982b). When a honeybee scout (*Apis* spp.) discovers a rich food source, she returns to the hive and informs her fellow workers. If the food is more than a certain distance from the hive, she performs a waggle dance (see the section on recruitment in Chapter 18 for a detailed discussion). Her gyrations inform the recruits of the distance and direction to the food source. Distance information is encoded in the duration of waggling during the straight run of the dance. However, among different races of bees, the flight distance represented by each second of waggling varies (Figure 19.17). Why should these differences, or dialects, in the bee language exist?

One reason might be that the climate affects foraging distances, and this in turn was a selective pressure for the divergence of the bee language into dialects

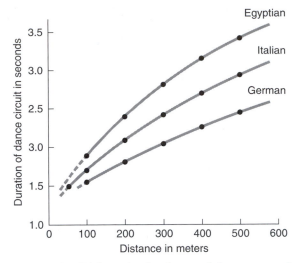

FIGURE 19.17   Dialects in the dance of three races of honeybees. A scout informs other workers of the location of a food source by a waggle dance. The duration of waggling during the straight run of the dance signifies the distance to the food. However, the number of meters designated by each second of waggling varies among the races. These dialects may be related to differences in the environment. Bees that inhabit colder habitats have larger foraging ranges. In their dance, each waggle denotes a greater travel distance than does a waggle in the dances of races with small foraging ranges. (Modified from J. L. Gould 1982b.)

(Gould 1982b; Towne and Gould 1988). Frigidity is a key factor in determining winter survival of a honeybee colony. The nestmates survive the freeze by metabolizing stores of honey to produce heat and by sacrificing outer layers of bees in the colony to provide insulation. Larger colonies would be expected in icier climates because additional members would be required to gather larger larders and to increase the thickness of the living comforter that insulates the colony. The greater demand for food would increase the average foraging distances for those races living in frigid environs. To reduce the time and energy a worker would invest in indicating the location of a food source to recruits, selection might sacrifice some precision in the language of these races. The dance was made more efficient by having a given duration of waggling signify a greater distance. The suggestion that the distance signified by a waggle is influenced by the average distance traveled during foraging is supported by the observation that artificial swarms of German bees (*A. m. carnica*), a race that generally experiences severe winters, causing them to forage over longer distances and dwell in larger colonies, choose larger and more distant nest boxes than their cousins from sunny Italy (*A. m. ligustica*).

It is puzzling to note, however, that the behavior of honeybee species in Thailand does not seem to support the suggestion that differences in climates and foraging ranges lead to dance dialects. Although the three

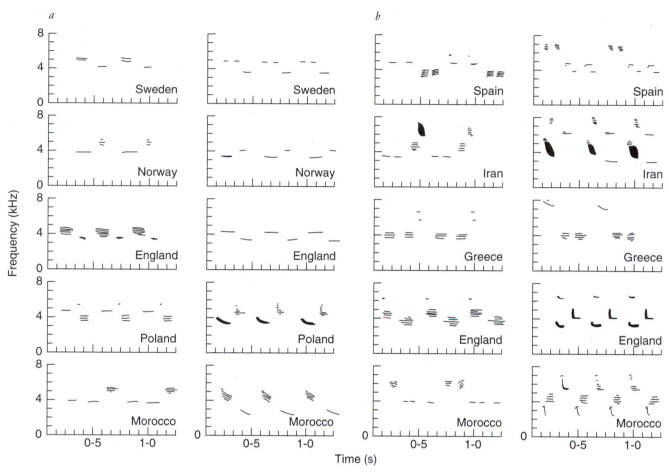

FIGURE 19.18 Song structure of populations of great tits in many countries: (*a*) forest dwellers, (*b*) woodland inhabitants. The left column in each section shows songs that are typical of each area, and the column on the right shows songs that emphasize the characteristic features of the songs of that area. Notice that the songs of the species that live in forests have a lower pitch, fewer frequencies, and fewer notes per phrase than the woodland species. (From Hunter and Krebs 1979.)

species studied (*A. florea*, *A. cerana*, and *A. dorsata*) forage over strikingly different distances, their dance languages indicate distances in much the same way. So, perhaps there is no universal relationship among climates, foraging distances, and dance dialects. Or, it may be that other factors, such as nesting behavior and the cues used to follow the dances, are also important. Although two of the honeybee species in Thailand (*A. florea* and *A. dorsata*) live in open nests and visual signals play a role in their dance communication, those in Europe nest in cavities and use only sound and touch to follow the dances. It may be easier to follow long waggling runs by sight than by sound and touch. Thus, open-nesting Thailand species may be able to use the same dialect no matter how far they usually travel for nectar; greater distance is simply signified by longer waggle runs (Dyer and Seeley 1991).

The Structure of Bird Song Researchers studying bird song have also attempted to correlate the structure of the signal with characteristics of the environment. Comparative studies have demonstrated that differ-

ences in characteristics such as the range of frequencies employed in the song and the occurrence of pure tones or trills (rapid sequence notes) can be correlated with aspects of the species' habitat. For example, an interspecific study in Panama revealed that species singing below the canopy in tropical forests declared their presence with low-pitched whistles, but those inhabiting grasslands used rapid trills centered on higher frequencies (Morton 1975). Even within a species, the type of song has been shown to vary with the habitat of the singer. The song of the great tit (*Parus major*) is ideal for this type of analysis because the species has such a large geographic distribution. Forest dwellers from England, Poland, Sweden, Norway, and Morocco were found to have songs with a lower pitch, a narrower range of frequencies, and fewer notes per phrase than open woodland birds from England, Iran, Greece, Spain, and Morocco (Figure 19.18). In fact, birds from similar habitats in Oxfordshire, England, and Iran, separated by 5000 kilometers, sing more similar melodies than two English populations occupying different habitats (Hunter and Krebs 1979).

Why should these correlations between song structure and type of environment exist? Differences in the transmission of sounds in various habitats may provide a clue. Two processes counteract the transmission of sound. One, attenuation (weakening), affects how far the sound will carry. The second, degradation, determines how distorted the signal becomes during transmission. Exactly how these factors interact to modify the structure of bird song is not certain, but several researchers have presented reasonable hypotheses.

Some investigators emphasize the role of attenuation (Hunter and Krebs 1979; Morton 1975). Eugene Morton measured the attenuation of various frequencies in the tropical rainforest and in the grasslands of Panama by broadcasting sounds on a loudspeaker and recording them on tape recorders positioned at increasing distances from the speaker. High-pitched tones attenuated more rapidly than low-pitched tones in both habitats. A major difference between the habitats in sound attenuation was that in the forest, tones of about 2 kHz carried better than expected. Forest species seem to take advantage of this. Unlike the birds of the grasslands, those in the forest sing in frequencies of 1.5 to 2.5 kHz and rarely employ frequencies above 3.5 kHz. Thus, as seen in Figure 19.19, the distribution of frequencies in the songs of forest birds seems to be ideally suited for maximum projection with minimum effort.

However, other investigators (Marten and Marler 1977; Marten, Quine, and Marler 1977; Wiley and Richards 1978) do not find consistent differences in the attenuation of sounds in grasslands and forests. How, then, can song differences be explained? Mac Hunter and John Krebs (1979) still invoke the frequency-dependent pattern of attenuation (i.e., low frequencies carry farther than high-pitched tones) to explain the song structure of great tits, but they suggest that other factors, such as predation, may explain why low notes are not used in all habitats. The forest populations employ low frequencies, and thus their songs seem designed for long-range communication. Other factors, such as visibility to predators in the open environment, may have selected for higher-frequency songs that carry shorter distances in grassland species.

On the other hand, R. Haven Wiley and Douglas Richards (1978) argue that since attenuation is similar in both environments, one should consider how the other factor that determines how far a sound can be transmitted, degradation, might mold bird songs. A major source of distortion that is present in forests but not in grasslands is reverberation from foliage. This is greater for high-frequency sounds because they have more cycles per second and, therefore, a greater chance of striking small objects such as leaves and branches. Reverberation of high frequencies may deflect them away from the receiver, making them more difficult to detect, or may cause echoes that would scramble complicated song patterns. Therefore, a song designed for maximum detection in a forest should emphasize low frequencies and have few trills or trills with widely spaced notes. The major source of degradation in grasslands would be unpredictable gusts of wind. Since part or all of a song might be gone with the wind at any moment, redundancy would be favored in the songs of grassland species; the repetition of notes such as is found in trills would be expected. The prediction that trills would be more common in grassland species is partially borne out by the observations of Fernando Nottebohm (1975) on the South American rufous-collared sparrows (*Zonotrichia capensis*). Grassland populations do employ faster trills. Another study compared the male territorial songs of 120 species of North American oscine birds. The habitat of the birds could be classified into one of six types: three types of forests and three types of open habitats. The temporal proper-

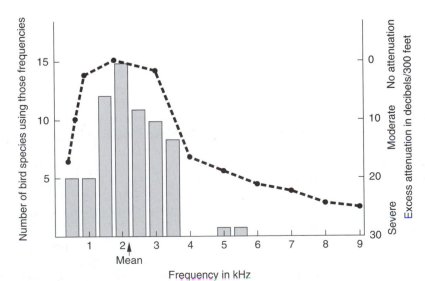

FIGURE 19.19 The relationship between frequencies used in the songs of birds from lowland forest areas of Central America and the attenuation of various frequencies in that environment. The histograms indicate the number of bird species using various frequencies in their songs. The curve indicates the degree of attenuation of sounds of various frequencies in the same habitat. The most common frequencies in bird songs in the area are around 2 kHz, the frequencies that attenuate the least. (Modified from Morton 1975.)

ties of the songs were strongly associated with habitat. Songs of forest-dwelling species had fewer trills, buzzes, and side-bands than those in open habitats. Thus, the temporal properties of oscine bird songs may have evolved to reduce the effects of reverberation in forest habitats (Wiley 1991).

LANGUAGE AND APES

Although animals do not normally use language in the sense that we do, several investigators have studied the ability of apes—chimpanzees (*Pan troglodytes*), gorillas (*Gorilla gorilla*), orangutans (*Pongo*), and especially bonobos (*P. paniscus*)—to learn language. Perhaps these studies have been prompted by a human desire to communicate with apes or to better understand what it means to be human. Regardless of the motivation for the studies, the results have philosophical and practical implications for many areas of study, including linguistics, anthropology, sociology, and neurobiology (Savage-Rumbaugh, Rumbaugh, and Boysen 1980).

Early studies were designed to teach chimpanzees to talk. In the longest and most thorough of these attempts, Keith and Cathy Hayes (1951) managed to teach a chimp named Viki to say three words—"mama," "papa," and "cup"—in a voiceless aspiration. It turns out that these attempts to teach chimps to speak were doomed to failure because chimpanzees lack the necessary vocal apparatus to make the range of sounds of human speech (Lenneberg 1967).

A more fruitful approach to demonstrating the ability of apes to acquire language skills has been to use nonverbal languages. A well-known nonverbal language, and the first to be taught to a chimpanzee, is American Sign Language for the Deaf, ASL. In 1966, Allen and Beatrice Gardner trained Washoe, a young chimpanzee, to communicate using ASL (Figure 19.20). The Gardners believed that an interesting and intellectually stimulating environment would assist the development of language skills. For this reason, Washoe and the chimpanzees trained since then were reared as much as possible like human children. However, spoken English was not permitted around Washoe because it was feared that it might encourage her to ignore signs. After four years of training, Washoe had a reported vocabulary of 132 signs. Her signs were not restricted to requests. She used the signs to refer to more than just the original referent; she applied them correctly to a wide variety of referents. For example, Washoe extended the use of the sign for dog from the particular picture of a dog from which she learned it to all pictures of dogs, living dogs, and even the barking of an unseen dog. She also invented combinations of signs to denote objects for which she had no name. Classic examples are her signing "water bird" for a swan on a lake and "rock berry" for a Brazil nut (Fouts 1974). By the time she knew 8 to 10 signs, Washoe had begun to string them together. Examples of typical early combinations are "please tickle," "gimme food," and "go in" (Gardner and Gardner 1969).

Once the Gardners had demonstrated that chimpanzees could be taught to use a gestural language, other ape-signing projects were begun. Roger Fouts

FIGURE 19.20 Chatting with a chimp in American Sign Language.

(1973) and Herbert Terrace and his coworkers (1979) continued working with chimpanzees, Francine Patterson (1978, 1990) worked with a gorilla named Koko, and Lynn Miles (1990) extended the studies to an orangutan named Chantek. The techniques employed in these signing studies were similar to those originally used by the Gardners, except that spoken English was permitted in the presence of the apes. The emphasis of these studies was on the production of language (the use of signs) and not on the comprehension of language (understanding the meaning of signs) (Rumbaugh and Savage-Rumbaugh 1994).

Although Herbert Terrace and his coworkers (1979) began optimistically believing they could use operant techniques to teach a chimpanzee language, they eventually concluded that an ape cannot create a sentence. Most of their analysis was done on the utterances of Neam Chimpsky, a young male chimp usually called Nim. Videotapes and films of the signing of other chimpanzees were also studied. When all of Nim's multisign combinations were recorded and analyzed, Terrace and his coworkers concluded that Nim did not spontaneously produce sentences characteristic of human language. Their most important criticism of the work with "talking" chimps was that the animals were simply imitating their trainers. Terrace argued that the apes' signs were cued by the trainer and that the trainers were too liberal in their interpretation of the signs.

Needless to say, the conclusion that an ape cannot create a sentence was challenged by others doing ape language research (Marx 1980). Fouts, Patterson, and the Gardners argue that Nim's language abilities were stunted by the operant-conditioning procedures used in his training. Allen Gardner backs his claim "that you can turn it [imitation] on and off, depending on the type of training you give" with a videotape of a chimp who shows little or no imitation of his trainer's signs until the last third of the tape, when operant-conditioning techniques were begun. During this last section of the tape, 70% of the chimp's signs were imitative. The Rumbaughs argue that because of the way in which Nim was trained, he never understood the meaning of words and that is why he was unable to create a sentence. In addition, Nim's trainers changed so often that he may not have had the opportunity to form the relationships claimed to be essential for language development (Marx 1980).

Nonetheless, it became widely accepted that chimpanzees could not learn language, and the later successes of Nim and other chimps received little attention. Project Nim was discontinued, but other trainers began to work with him. His language skills improved impressively and no longer depended on imitation (O'Sullivan and Yeager 1989). The signs of the orangutan, Chantek, were more spontaneous than those of

Nim and could not be attributed to imitation (Miles 1990). Washoe and the other language-trained chimps sign to other animals and objects (Gardner and Gardner 1989) and frequently to themselves (Bodamer et al. 1994). After Washoe's biological infant died, she adopted a ten-month-old infant named Loulis. For the next five years, humans avoided using any sign language in the presence of Loulis. Nonetheless, Loulis learned his first 55 signs during this time by observing other chimps (Gardner and Gardner 1989). Today, Washoe and her family sign to one another during all their daily activities, including playing and eating and even family fights (Fouts and Mills 1997).

At the same time that the Gardners were working with Washoe, David Premack (1976) was training another chimpanzee, Sarah, to use plastic chips of various shapes and colors as words (Figure 19.21). Most of her use of language consisted of using one word from a choice of several to complete a preformed statement or arranged four to five words into a sentence of a specific word order. Premack established certain criteria for accepting that Sarah was using a particular chip as a word. Sarah had to be able to use the plastic chip to request the object it stood for, to select the proper chip when asked the name of the referent, and to "describe" the referent of a particular chip by using other chips. Premack's strategy was to break down linguistic rules into simple units and to teach them to the chimp one at a time. In this way, Sarah was taught not only to name many objects but also to use more complicated relationships such as if-then and same-different.

The more successful aspects of the pioneering studies with Washoe and Sarah were combined in the LANA Project (Rumbaugh 1977; Rumbaugh, Gill, and

FIGURE 19.21 Symbols used as "words" by Sarah, a chimpanzee. Sarah learned to communicate using these plastic shapes.

von Glaserfeld 1973). The chimpanzee Lana was trained to use a computer to communicate. This computerized language system eliminated social cueing and the difficulty of interpreting the symbols, problems that plague the sign language studies.

Lana communicated in a symbolic language, Yerkish, which was invented for the purpose. Yerkish words, called lexigrams, are geometric figures built from combinations of nine simple design elements such as lines, circles, and dots (Figure 19.22). Lana chose words by pressing a computer key labeled with the lexigram. When a key was depressed, it lit up and the lexigram simultaneously appeared on a projection screen. It is significant to note that Lana was required to use lexigrams in an appropriate order. In other words, she had to learn syntax, the rules governing word order in a sentence. For example, she learned that pressing lexigram keys to say, "Please machine give . . ." might be rewarded but that pressing "Please give machine . . ." was not an acceptable way to make a request. Unlike Sarah, who was given a limited choice of words to use at one time, Lana always had her complete vocabulary available to her. Lana developed a large vocabulary and mastered Yerkish grammar. She also coined new words, just as Washoe did. For example, she called an orange soda a "Coke-which-is-orange" (Rumbaugh 1977).

In summary, the ape language projects using Washoe, Sarah, and Lana demonstrated that apes (1) are relatively adept at learning words, (2) readily string words together in short sequences so that the strings adhere to rules of grammar if that is required, and (3) have the ability to coin new words.

These skills are obviously necessary for language, but do they demonstrate the use of true language? Before that question can be answered, a set of criteria to define true language must be established. Although definitions vary, most linguists would agree on two essential elements. First, words or signs must be used as true symbols that can stand for, or take the place of, a real object, event, person, action, or relationship. Symbols should permit reference to objects or events that are not present. Second, words or signs should be combined to form novel phrases or sentences that are understandable to others. This necessitates a knowledge of syntax because a change in the order of symbols can alter the meaning of the message.

The ability of apes to master these two language skills has been questioned. First let's examine the evidence that apes use words as symbols. When a chimp uses a "word" (sign, plastic chip, or lexigram) to name an object, is it used as a symbol that stands for the object, one that can be used to refer to the object even when it is not present, or is the animal producing the "word" because it has been associated with the object through a reward system? When a pigeon pecks a red key for food and a green key for water, no one assumes that the animal is using the key to represent the item. However, when an animal as intelligent as an ape presses a key that is labeled, not with a color, but with a "word," the assumption is often made that it is using language and not simply performing a complicated behavior for a reward. This conclusion is especially tempting when the animal strings words together in a sequence for a reward. How could it be determined whether the animals were using the words as symbols or just mastering a complex conditioned response? It has been argued (Savage-Rumbaugh, Rumbaugh, and Boysen 1980) that in the studies just described, the apes were not required to do anything that eliminates the possibility that they were simply using words as labels rather than as symbols.

However, other experiments might be interpreted as a demonstration of a chimpanzee's ability to use "words" as true symbols. These tests were done with two chimpanzees, Sherman and Austin (Figure 19.23). They communicated information to each other through the use of symbols, information that could not have been communicated without the symbols. They were trained, as Lana was, to communicate by pressing computer keys embossed with lexigrams. The emphasis in this program was on interanimal communication, so the animals were not taught to produce strings of lexigrams

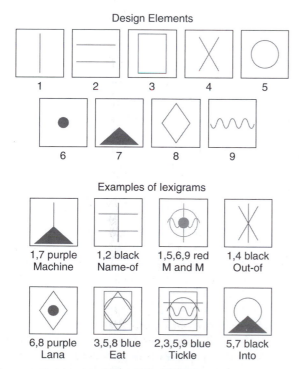

Design Elements

1 2 3 4 5

6 7 8 9

Examples of lexigrams

| 1,7 purple Machine | 1,2 black Name-of | 1,5,6,9 red M and M | 1,4 black Out-of |

| 6,8 purple Lana | 3,5,8 blue Eat | 2,3,5,9 blue Tickle | 5,7 black Into |

FIGURE 19.22 **Symbols of the Yerkish language. Each "word" is a combination of a few geometric shapes embossed on keys of a computer keyboard. Chimpanzees have learned to communicate by pressing the appropriate keys.**

FIGURE 19.23 Sherman and Austin, two chimps, have shown that apes can cooperate with each other to solve problems by using the symbolic language Yerkish. They can communicate only when they have access to the computer keyboard.

or to adhere to grammatical rules. In addition, they were raised in a social, preschool setting. The animals were first taught to name foods by pressing lexigram keys. It is important to note that they were taught to distinguish between the use of a food name as its name and the use of the food name as a request because they were never allowed to eat the same item that they named.

After this training, the animals' ability to communicate with each other symbolically was investigated. In one test, Sherman and Austin were able to specify foods to one another using lexigrams. One of the chimpanzees was taken to a different room, where he watched the experimenter bait a container with one of 11 different foods or drinks. That animal was then led back to the keyboard and asked to name the food in the container. Using the information gained by observing the response of the first animal, the second chimp was permitted to use the keyboard to request the food. If both animals were correct, they were given the food or drink. Sherman and Austin were able to communicate with one another in this manner whether or not they used the same keyboard and even if the experimenter was ignorant of the identity of the food in the container. Also, the animals communicated regardless of which of them was the observer. The animal that did the requesting based on the information provided by his knowledgeable pal could demonstrate that he knew which item he was asking for by selecting its picture from a group. However, when the chimp who knew the identity of the food was prevented from using the keyboard to describe the contents of the container, he could not transmit the information to his buddy (Savage-Rumbaugh, Rumbaugh, and Boysen 1978b).

The chimps also passed the next test—using symbols to inform each other of the appropriate tool to use to solve a problem. The animals were kept in separate rooms. One chimp had to decide which one of six tools he needed to obtain hidden food and then ask the other one for that tool via the keyboard. They could successfully cooperate in this manner only when the keyboard was turned on (Savage-Rumbaugh, Rumbaugh, and Boysen 1978a). Clearly Sherman and Austin were using words as symbols and not simply labeling objects.

The work with Sherman and Austin was important for reasons other than demonstrating that chimpanzees can use symbols in communication. It marked the beginning of a shift in emphasis from demonstrating that apes can *produce* language to showing that they can *understand* the symbols or words of language. In addition, it showed how important the learning environment is in the development of language comprehension.

The change in emphasis and learning environment led to great progress in the study of apes' language abilities. Consider the remarkable abilities of Kanzi and his half-sister, Panbanisha, bonobo or pygmy chimpanzees (*Pan paniscus*). Kanzi, born in 1980, was adopted and raised by Matata, who was part of a language study by Sue Savage-Rumbaugh. For two years, trainers futilely tried to teach Matata to use lexigrams to communicate. Kanzi, who was six months old at the beginning of the study, was always present during Matata's training sessions. But, other than occasionally chasing the symbols projected over the keyboard, he showed little interest. When Kanzi was about 2.5 years old, he was separated from Matata so that she could be bred at another site. Much to the amazement of the experimenters, Kanzi began to use the symbols of the keyboard that they had tried to teach to Matata. Not only did he know the lexigrams, he also knew the English words that the lexigrams represented.

Kanzi had begun to learn communication skills simply by observing his mother's training. He was never trained to use lexigrams. Instead, he picked up the use of language in much the same way as a child would. Once the researchers recognized this, reward-based language training was stopped and replaced with conversation. Kanzi's constant human companions used lexigrams, gestures, and speech to communicate with one another and with him. In this way they served as communicative models. Once he learned to use lexigrams, he began to use them to refer to items like food or objects or to locations that were not in sight (Savage-Rumbaugh 1986; Savage-Rumbaugh and Lewin 1994). Kanzi is now a language star and communicates on a board with 256 lexigrams (Figure 19.24). He has the grammatical skills of a 2.5-year-old human child.

Today, several language-trained chimps and the orangutan Chantek communicate using a voice synthesizer attached to a lexigram keyboard. Now the communicative interactions with humans even *sound* like conversation.

Besides being able to produce language, Kanzi and several other apes have demonstrated that they understand spoken English. In these tests, the words were presented through headphones or from behind a one-way mirror to avoid inadvertently cueing by gestures or facial expressions. The sentences were usually commands to perform some action with one or more objects or people. The person evaluating the response did not know what had been requested. Many of the

requests were so unusual that it would be impossible to have carried them out without actually understanding the language. Consider for example, the directive, "Put the raisins in the shoe." Kanzi responded correctly to 72% of over 600 requests (Savage-Rumbaugh, Shanker, and Taylor 1998). Panbanisha, Kanzi's half-sister, was reared in the same type of learning environment as Kanzi had been. She, too, shows remarkable comprehension of spoken English, re-sponding correctly to 77% of 145 sentences (Williams, Brakke, and Savage-Rumbaugh 1997).

There are still those who doubt that apes can acquire language skills (Kako 1999; Wallman 1992). The idea has, in fact, stirred up a heated debate. Nonetheless, the abilities of Kanzi, Panbanisha, and several other apes have certainly convinced critics that the issue deserves serious consideration (Greenfield and Savage-Rumbaugh 1990; Linden 1992). It has been suggested that instead of focusing on the elements of language that apes "may or may not have mastered, it would be more interesting, and more productive, to focus on the question of what Kanzi can do and how this came about" (Shanker, Savage-Rumbaugh, and Taylor 1999).

Why should there be such a brouhaha about whether apes can acquire language? The answer is that it raises questions about human uniqueness. Language is commonly identified as the characteristic that separates us from other animals, and some of us get nervous as that gap narrows. Researchers still don't agree on how wide the gap is (Pennisi 1999).

COMMUNICATION AND ANIMAL COGNITION

Many people have wondered what it is like to be an animal—whether nonhuman animals have thoughts or subjective feelings. Such musings have led some investigators to seriously consider whether nonhuman animals are cognitive, conscious, aware beings.

Donald Griffin (1978, 1981, 1984) has suggested that tapping animals' communication lines is a way to find out whether animals have conscious thoughts or feelings. After all, the only way we know about the thoughts or feelings of other people is when they *tell* us, through either verbal or nonverbal communication. So, if nonhuman animals also have thoughts and feelings, they probably communicate them to others through their communication signals. If we could learn to speak their language, we could eavesdrop and thereby get a glimpse into the animal mind. We might also learn about the animal mind through interspecies communication—teach the animal a language that we understand and then ask it how and what it thinks (Pepperberg 1993).

Most people agree that one sign of cognition is the ability to form mental representations of objects or

FIGURE 19.24 Kanzi, a pygmy chimpanzee, has demonstrated the most advanced language skills so far. He communicates with a computer keyboard that has over 250 lexigrams. He was not trained by operant-conditioning techniques. Instead, he observed and interacted with humans who used gestures and lexigrams to communicate.

events that are out of sight. So, one way to look for cognition is to ask whether animal signals are symbolic, that is, whether they refer to things that are not present (W. J. Smith 1991). We have seen that certain apes can learn a language that uses symbols, and they can use it to "talk" about things they don't currently see and things that occurred in the past (Savage-Rumbaugh 1986; Savage-Rumbaugh, Shanker, and Taylor 1998).

In Chapter 5, we discussed Alex, an African grey parrot who is able to vocally request more than 80 different items, even if they are out of sight. In addition, he can quantify and categorize these objects. He has shown an understanding of the concepts of color, shape, and same versus different on both familiar and novel objects (Pepperberg 1991). Louis Herman and his colleagues have shown that bottlenosed dolphins (*Tursiops truncatus*) can also learn to understand symbolic languages. In one of these languages, the "words" are gestures, as in sign language. In the other, the words are sounds generated by a computer. The words can refer to objects, actions, and relationships among actions, among other things. In tests of language comprehension, the dolphins show that they understand the experimenter's references to objects that are not present (Herman and Forestell 1985; Herman, Pack, and Morrel-Samuels 1993).

We can also learn about the way animals think from their natural communication systems. The use of language involves the ability to relate words to meanings. In some species, individuals classify signals by their meaning and not by some obvious physical property, such as the way the signal sounds. Consider, for example, the way rhesus monkeys (*Macaca mulatta*) respond to food calls. When a rhesus monkey finds food, it announces this to others with one or more of five different food calls. If the item is really good—of high quality or rare—the monkey will issue a warble, harmonic arch, or chirp. However, a low-quality, common food item is announced with a coo or a grunt. Thus, there are two categories of food calls, and the calls sound distinctively different.

You may recall from Chapter 5 that animals will habituate (gradually stop responding) to a stimulus that is repeated many times without consequence. Marc Hauser (1998) used habituation to determine how the monkeys classify food calls. The reasoning is that if the monkeys classify the call by its sound, after habituating to one type of call they will still respond to any other type of call because it sounds different. However, if they classify the call by its meaning, they will be unresponsive to a call in the same category but remain responsive to a call in a different category. The calls were broadcast through a speaker, and a monkey was said to have responded if it turned its head toward the speaker and stared at it. As you can see in Figure 19.25, rhesus monkeys classify by their meaning. Habituation transferred from a harmonic arch to a warble or vice versa, two calls that advertise high-quality food items. But it did not transfer between calls that have different meanings, from a grunt to a warble or harmonic arch.

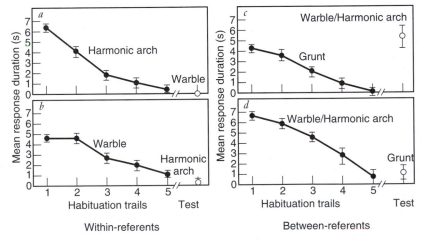

FIGURE 19.25   **Rhesus monkeys relate specific food calls to their meaning. A rhesus monkey that finds food announces this to troop members with a food call. Whereas warbles and harmonic arches announce high-quality, rare food items, grunts are used for low-quality, common items. When a call is broadcast repeatedly through a loudspeaker, monkeys gradually stop responding; that is, they habituate. Although the calls sound different, habituation transfers from a warble to a harmonic arch or vice versa because the calls have the same meaning. However, habituation does not transfer from a grunt to a warble or harmonic arch because the calls have different meanings (Data from Hauser 1998.)**

Another sign of cognition might be whether animals adjust signals according to conditions at the time. We might see such an adjustment if an individual determined whether or not to signal by the composition of its audience. In other words, if an individual sees a predator, does it always sound an alarm or does this depend on its present company? The company does seem to affect the likelihood of alarm calling among domestic chickens (*Gallus gallus*). A cock is more likely to call in the presence of an unalerted companion (Karakashian, Gyger, and Marler 1988). This has been interpreted as an indication that the cock chooses whether or not to call (Marler, Karakashian, and Gyger 1991). The choice of calling or withholding the call would, then, be taken as evidence of cognition.

However, an alternative interpretation of this observation is that the call is not one that warns a companion of a predator but rather one that indicates conflicting tendencies to flee or freeze (W. J. Smith 1991). To avoid detection by a predator, a lone cock might decide to flee for cover if the predator were still at a distance but to freeze if the predator were close at hand. When a cock is alone, then, the only factor it needs to consider when deciding how to avoid detection is the distance of the predator. Since there are no conflicting tendencies, no call is given. However, if a companion is nearby, there is another factor to consider when deciding how to avoid detection—the behavior of the companion. Consider a situation in which a predator has approached to a distance at which the best way for a lone cock to avoid detection is to remain motionless. However, if it is with a companion and the companion moves, this might draw the attention of the predator. In that case, the safest response might be to flee for cover. As a result, there may be a moment of indecision, and the call might be a response to the vacillating tendencies to flee or freeze. The indecision would only occur in the presence of another chicken, so the call would only occur when others were present (W. J. Smith 1991).

Clearly, the question of animal awareness is difficult to answer scientifically. The answer, however, has ethical ramifications, especially for the researchers who study animal behavior. If the line between animals and humans is erased or even smudged a bit, should we rethink the way we treat animals? Should we keep them in zoos? Should apes be tamed for language studies? What about dolphins?

So what's your opinion? Are animals aware, cognitive beings? All of them? Where do we draw the line?

SUMMARY

The traditional ethological view of the selective advantage of communication is that it is beneficial for group members to share information. Interactions can be more appropriate when individuals inform one another of their inner state or their future course of action. Another hypothesis views communication as selfish. A display can be seen as a way to manipulate others to do one's bidding without resort to brute force.

Whereas the information-sharing hypothesis predicts that signals should be honest, the manipulation hypothesis predicts dishonesty. Honest signals can also evolve if the signal is costly. Examples of honesty and deceit in animal signaling are common. We might expect honesty when the signal is tied to a physical attribute and cannot be faked, in stable social units in which members are individually recognized, and when the costs of cheating outweigh the benefits.

The evolution of a signal begins when a receiver makes an association between an unintentional cue and the sender's condition, motivation, or future action. If the receiver benefits, evolution will favor the ability to detect the cue. If the sender benefits, evolution will favor modification of the cue to increase the amount of information transferred. Three evolutionary sources for displays have been recognized. The first category is intention movements, those behavior fragments that may precede a functional action. Displacement activities are a second category. The last group, autonomic responses, includes urination and defecation, vasodilation, respiratory changes, and thermoregulatory responses. Although these are the three most common sources for signals, almost any behavior can be ritualized.

Ritualization, the evolution of a display, changes the form of the flexible ancestral action into a stereotyped signal. If one is reasonably sure of the precursor behavior, it can be compared to the current display to determine the changes in the original form. Typically, all or a part of the ancestral behavior pattern is exaggerated by changes in the duration or extent of movement, by alterations in the rate of performance, or by repetition. In addition, the original actions may be combined in a new order, or some parts may be deleted. The display might also be directed toward a new stimulus. As the signal becomes "emancipated" from the factors that originally caused it, the behavior may be shown in a new context and/or be motivated by different factors. Frequently, these changes in behavior are accentuated by anatomical modifications such as bright colors, antlers, manes, and the like.

The design of the signal may be influenced by a variety of factors. Obviously, the anatomy and physiology of the species is important in directing the evolution of their signals. The sender must be capable of generating the signal, and the recipient must be able to detect it. The species' behavior is also a factor. Some signals can be sent while continuing other behaviors, a benefit to active species. Often, as occurs in ants, a specific behavior, such as the type of prey preferred, will determine the desirable characteristics of a signal.

The physical habitat may also be important in the evolutionary choice of a signal. Visibility is of prime importance. If it is limited by darkness or by obstacles, channels other than vision are favored. The structure of a signal in any sensory channel may be affected by the environment. For example, differences in the average winter temperature may have resulted in the dialects in the language of honeybees. Those races that inhabit colder areas must forage over longer distances. In this situation, some precision in the distance information provided by the waggle dance may have been sacrificed for the sake of efficiency. A single waggle would signify a greater flight distance.

Among other species of honeybees, the nesting habits may be more important than the foraging range in shaping the dance language. Although three species in Thailand have very different foraging distances, their dances indicate distance in similar ways. Perhaps long waggling runs can be followed more easily by open nesting species of Thailand, which can use sight, than by European species that nest in cavities and follow dances by using touch and sound.

Likewise, the lower pitch and scarcity of trills in the songs of birds living in the forest instead of grasslands may be attributed to characteristics of the environment. The two factors that work against the detection of an auditory signal are attenuation, influencing how far the sound will carry, and degradation, determining how distorted the signal becomes during transmission. Both of these factors have been invoked by various researchers to explain the differences in the songs of birds in different habitats.

Animal communication signals are not true language because animals do not use signals as symbols that can take the place of their referent and because they do not string signals together to form novel sentences. However, many workers have been interested in elucidating the language abilities of apes and have, therefore, attempted to teach them language. The early studies designed to train chimpanzees to speak failed because an ape's anatomy does not allow it to make the sounds of words. However, later studies used languages that do not require voice. Chimpanzees and gorillas have learned to communicate with their trainers in American Sign Language. Other chimps communicate by using plastic tokens of different shapes or a computer keyboard that displays geometric configurations that stand for words. These studies have shown that apes can learn "words" and may even coin new ones. They can also learn to string words together into short phrases. If it is required, they may even follow some grammatical rules in constructing "sentences."

In spite of these abilities, the language capacity of apes has been challenged. It has been argued that the chimps have not demonstrated an ability to use the "words" as symbols. However, even if this is a true criticism of the earlier studies with Washoe, Sarah, Lana, and Nim, more recent work with the chimps Sherman and Austin has shown that the apes can communicate with each other to solve problems if, and only if, they have access to their computer keyboards. Since they are able to conduct cognitive operations with only symbolic information available, true symbolization seems to be possible for apes. However, whether apes can achieve the second hallmark of language, the ability to construct novel sentences that follow grammatical rules, is still being debated.

Some people have suggested that knowledge about the communication systems of animals may provide an insight into the question of animal cognition. Some species, primarily apes; Alex, the African grey parrot; and bottlenosed dolphins have been taught to understand signals that represent items that are out of sight. The natural communication systems of some species, such as rhesus monkeys, reveal that they can relate meaning to signals.

Another sign of cognition might be the adjustment of signals according to conditions at the time. Deciding whether animal communication meets this criterion has also been controversial.

The question of animal cognition or awareness is difficult to address scientifically, but the answer has ramifications for how humans view their place in the animal kingdom and the way in which they treat nonhuman animals.

APPENDIX

Magnetoreception

Within the last few decades, magnetic sensitivity has moved out of the realm of mysticism into reality. Effects of magnetism on behavior have been demonstrated in many organisms—from mud-dwelling bacteria to birds and bees and beyond. Many scientists have now, in fact, turned their attention to the nature of the detector, and several pieces of the puzzle are beginning to fit together. Some scientists even speculate that more than one type of receptor may be found in a single organism.

We considered evidence for the effects of magnetism on the behavior of organisms and the way that magnetic cues might be used for orientation and navigation in Chapter 10. Here, we are concerned with the hypothesized mechanisms for magnetoreception, including electroreceptors, magnetite deposits, photoreceptors, and the pineal gland.

ELECTRORECEPTORS

One mechanism by which a magnetic field might be detected is induction. When an electrical conductor moves through a magnetic field, an electrical current is created. Some animals are able to sense electrical fields and could use this sense to detect magnetic fields. Although this mechanism does not seem to be common, presumably because the electrical potentials generated on land or in fresh water would be too small, induction is the mechanism used by the electroreceptive cartilaginous fish of the sea. The motion of a shark, ray, or skate in swimming through a magnetic field induces an electrical field that is well within the range of sensitivity of its electroreceptors. What makes these electrical fields of possible use in orientation is that the potentials generated at the surface of the skin depend on the direction in which the fish is swimming relative to the earth's magnetic field. Notice in Figure A.1 that when a fish is swimming eastward, the potentials on its belly surface would be positive relative to those on its back surface. When the fish is swimming westward, the polarity of the induced field would be reversed. Thus, the potentials on its back surface would be more positive than those on its belly. However, no voltage difference between the back and the belly would be created by swimming north or south. So, just by swimming, the fish gains information that could reveal its compass bearing. But the system may be even more sophisticated than this. It might also indicate the animal's latitude, or north-south position on the earth. This information would be encoded in the intensity of the

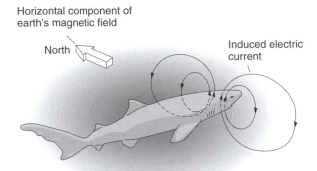

FIGURE A.1   The orientation of the electrical current induced as a shark swims through a magnetic field.

electrical field generated by the swimming movements. Intensity changes with latitude because of the regular variation in the vertical component of the earth's magnetic field (Kalmijn 1978, 1984).

MAGNETITE DEPOSITS

Most organisms with a magnetic sense lack electroreceptors. Many of these have deposits of magnetic material, magnetite, which often forms chains or clumps and in vertebrates are commonly found in the head.

Two types of magnetic behavior of magnetite could make it useful in a magnetoreceptor. First, the magnetite could be a permanent magnet if the crystal is large enough and of the right shape. It could then act as a compass needle, aligning itself with magnetic lines of force. The smallest magnetically stable unit, called a single domain, would rotate as it tracked the external field (Figure A.2). This turning might be detected by the nervous system. One suggested mechanism that might inform the nervous system of the movement of magnetite involves mechanoreception. When the miniature magnets align with a magnetic field, they might stimulate hair cells or pressure receptors. The

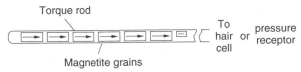

FIGURE A.2   One hypothesis for a mechanism that might link with the nervous system the response of single-domain magnetite particles to changes in the magnetic field. The magnetite grains, which might form a chain, are permanent magnets that would turn like a compass needle in response to alterations in the direction of the magnetic field. This turning might stimulate a mechanoreceptor in a hair cell. (Kirschvink and Gould 1981.)

pattern of stimulation might inform the animal about the strength and direction of the magnetic field. The effect of magnetite rotating would be amplified if the single domains were connected in a chain. Indeed, chains of permanently magnetic grains of magnetite have been discovered in several organisms (Figure A.3).

Alternatively, a magnetite grain could be a superparamagnet. When a grain of magnetite is too small to be magnetically stable, it is still magnetic but the direction of its magnetic field changes with that of an external field. The direction of the magnetic field of individual molecules within the grain are unstable and align themselves with the external field. In this case, the direction of the magnetic field of the magnetite grain could change without rotation of the grain itself.

Perhaps the following analogy will help clarify the difference in the behavior of a permanent magnet and a superparamagnet. Imagine the possible responses of a line of people instructed to always face a beacon of light when the light moved 90° to the right. If the people form a line so that each person stands behind another, with his or her hands on the shoulders of the person in front and never releasing this grip, the entire column would have to rotate as a unit 90° to the right. This would be analogous to the movement of a permanent magnet. However, if each person in line releases his or her hold on the person ahead, each would be free to turn 90° to the right. The individuals would then be standing shoulder to shoulder. The column would not move, but the people in it would rotate. This would be analogous to the behavior of a superparamagnet.

There are several suggestions for mechanisms that might link the nervous system to the behavior of superparamagnetic grains. One proposal is that a series of such grains might be embedded in a rodlike stretch receptor (Figure A.4). When the grains aligned their fields with an external magnetic field perpendicular to the rod, the grains would repel one another and the rod would lengthen and stimulate a stretch receptor. A field parallel to the rod would align the magnetic fields of the grains so that they attracted one another, thereby shortening the rod (Kirschvink and Gould 1981; Towne and Gould 1985).

Some interesting research has been done on magnetoreception in certain mud-dwelling bacteria. The bacterium *Aquaspirillum magnetotactum* apparently has a chain of magnetite particles in its body that acts as a permanent magnet and aligns itself with the external field like a compass needle. The bacterium then swims in that direction. In fact, the bacteria in the Northern Hemisphere move toward magnetic north, and those in the Southern Hemisphere seek magnetic south. The reason for this difference is that the magnets in these bacteria point in opposite directions. In the Northern Hemisphere, the south pole of the bacterial compass

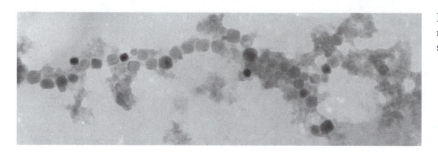

FIGURE A.3   A chain of single-domain magnetite particles from the skull of a sockeye salmon (Mann et al. 1988.)

needle is at the organism's anterior end. However, the north pole of the compass needle of bacteria of the Southern Hemisphere is at the organism's anterior end. The internal magnet of these bacteria is so strong that even when they are dead, they passively align with a magnetic field. This response is appropriate for bottom-seeking organisms. Since the earth's magnetic field dips downward at the poles, becoming almost vertical, the magnetotactic bacteria of both hemispheres always swim downward, toward the muddy bottom they call home (Blakemore 1975; Blakemore and Frankel 1981; Frankel and Blakemore 1980; Frankel, Blakemore, and Wolf 1979).

Magnetite deposits have also been found in many other organisms including insects, birds, fish, and even humans (Gould, Kirschvink, and Deffeyes 1978; Hanson et al. 1984; Kirschvink 1997; Kirschvink et al. 1985; Mann et al. 1988; Walcott, Gould, and Kirschvink 1979; Walker et al. 1997, 1984; Zoeger, Dunn, and Fuller 1981).

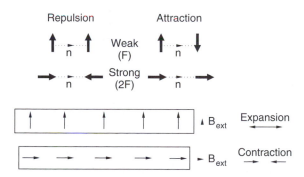

FIGURE A.4   **One hypothesis for a mechanism that might link with the nervous system the response of superparamagnetic magnetite particles to changes in the magnetic field. The directions of the magnetic moments of individual molecules of magnetite within the deposit would track that of the imposed field. As a result, some directions of the magnetic field might cause the magnetite molecules to repel one another. This might cause the expansion of a receptor structure that would in turn stimulate a stretch receptor. When exposed to a different orientation of the magnetic field, the molecules would be attracted to one another and the receptor structure would contract. (Kirschvink and Gould 1981.)**

If the magnetite deposits function as magnetoreceptors in larger organisms, the information they provide would have to be transmitted to the nervous system. Therefore, associations between magnetite and the nervous system are of particular interest. Magnetite deposits in the abdomen of honeybees are in cells closely associated with the nervous system (Kutterbach et al. 1982). In addition, magnetite has been located in the particles that stimulate hair cells within the gravity proprioceptive structures of the guitarfish (O'Leary et al. 1981). Although it has not been found in every bird examined, magnetite has been reported near nerves in the heads of some pigeons (Walcott, Gould, and Kirschvink 1979).

In the bobolink (*Dolichonyx oryzivorus*), a bird that migrates across the equator, magnetite is consistently found in the sheaths around the olfactory nerve and bulb. It is also found in bristles that project into the nasal cavity (Beason and Nichols 1984). Might these bristles function as a hair-cell mechanoreceptor? Branches of the bobolink's trigeminal nerve appear to innervate the region in which magnetite deposits are found. These branches respond to earth-strength changes in the direction of the magnetic field (Beason and Semm 1987; Semm and Beason 1990). Furthermore, if a bobolink is exposed to a strong magnetic pulse, a treatment that would alter the magnetization of magnetite, the bird's orientation is shifted. However, if the anesthetic lidocaine is then applied to the ophthalmic branch of the trigeminal nerve to block impulses, the bobolink orients correctly. These observations are consistent with the idea that magnetite is part of a magnetoreception system and that information from the magnetoreceptors is carried to the brain through the ophthalmic nerve (Beason and Semm 1996).

So far, the closest we have come to identifying the actual magnetoreceptor cells is in the rainbow trout (*Oncorhynchus mykiss*). Michael Walker and his colleagues (1997) first confirmed that the ophthalmic nerve contains fibers that respond to magnetic fields. Then they used a special dye to trace these fibers both to the brain and to cells in the olfactory epithelium in the nose of the trout. These cells, the candidate magnetoreceptor cells, contain small amounts of a material thought to be magnetite.

MAGNETORECEPTION AND LIGHT

It is interesting that in some organisms the reception of light and of magnetic fields seem to be related. Such interactions have been shown in various responses of planaria (F. A. Brown 1971), honeybees (Leucht 1984), crayfish (Sadauskas and Shuranova 1984), turtles (Raybourn 1983), pigeons (Semm and Demaine 1986), quail and chickens (Krause, Cremer-Bartels, and Mitoskas 1985), and rats (Reuss and Olcese 1986).

Some of the pieces of the puzzle concerning light and magnetoreception in vertebrates are falling into place (Figure A.5). The first step in the general scheme occurs in the retina, where some of the photoreceptors also serve as magnetoreceptors. Although several molecular mechanisms by which the photoreceptor molecules might detect magnetic fields have been suggested, the correct one is still not known (Leask 1977a, 1977b; Schulten 1982; Schulten and Windmuth 1986). The basic idea, however, is that photoreceptor molecules absorb light better under certain magnetic conditions. Thus, the amount of light absorption also provides information about the local magnetic field.

The suggestion that animals might somehow "see" the earth's magnetic field seems bizarre, so we might ask whether or not there is support for the idea. Indeed, several observations have suggested to researchers that the visual system is involved in magnetodetection. For example, the electrical activity of certain photoreceptors in the eye of a blowfly (*Calliphora vicina*) is changed by alterations in the magnetic field (Phillips 1987b). Also, the retinas of certain vertebrates respond to light differently when the magnetic field is altered, even when they are removed from the animal (Krause, Cremer-Bartels, and Mitoskas 1985; Raybourn 1983). Furthermore, certain cells in the brain of a pigeon respond both to light and to the orientation of a magnetic field. However, both responses are substantially reduced when the bird is blinded (Demaine and Semm 1985; Semm et al. 1984). Similarly, the effect of magnetic fields on the synthesis of the pineal hormone melatonin is eliminated in blind rats (Oclese, Reuss, and Vollrath 1985).

Information about magnetic fields from the photoreceptors would then be sent to various regions of the brain, where certain centers integrate information on light, magnetic fields, and gravity. This integration might tell the organism the degree of inclination in the lines of force of the earth's magnetic field (discussed in Chapter 10). Since the dip in the geomagnetic field varies regularly with latitude, this information might allow the animal to determine its north-south location. Thus, these integrative brain centers could be the neural basis of the magnetic compass.

The integration of information from the retina about light and magnetic fields with information from gravity receptors takes place in the brain's vestibular system. If the vestibular system is stimulated by tilting the animal, certain neurons in this brain region will respond to changes in the magnetic field. Visual information and magnetic information are also integrated. Certain cells in the accessory optic system (the nucleus of the basal optic root) respond both to light and to directional changes in the natural magnetic field. However, these cells respond to magnetic fields only if the visual system is stimulated and intact (Semm et al. 1984).

The pineal gland in the brain may also play a role in magnetoreception. Not only does the pineal get information about magnetic conditions from the retina, it may also be able to pick up its own information (Demaine and Semm 1985). The pineal then causes changes in the animal's behavior through its neural and endocrine activities. Alterations in the magnetic field cause changes in the electrical activity of certain cells in the pineal (Reuss, Semm, and Vollrath 1983; Semm, Schneider, and Vollrath 1980); changes in the synthesis of the pineal hormone melatonin (Cremer-Bartels et al. 1984; Stehle et al. 1988; Welker et al. 1983); and changes in the amount of cAMP (cyclic adenosine monophosphate), a substance that regulates many cellular activities (Rudolph et al. 1988).

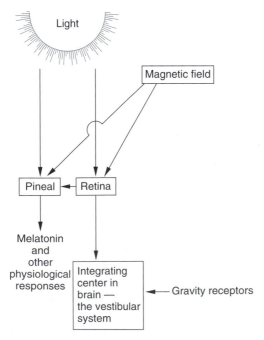

FIGURE A.5 **An overview of the interaction between photoreception and magnetoreception.**

TWO MAGNETORECEPTOR SYSTEMS

Recent studies aimed at exploring the physiological basis for magnetoreception have lent support to the idea that animals might have more than one type of magnetic sensitivity. As we have seen, there are two proposed mechanisms for magnetoreception. One mechanism, which involves light-dependent processes, can be altered by exposure to different wavelengths of light. It is thought that long wavelengths, red, for example, may not contain enough energy to start the processes needed for light-dependent magnetoreception. Thus, light-dependent magnetoreception is suspected when an animal's orientation to magnetic fields is altered by exposure to different wavelengths of light. The other proposed magnetoreception mechanism involves small particles of magnetite. This system can be affected if the animal is exposed to a strong magnetic pulse.

Certain species seem to have both types of magnetoreception systems, each serving a different purpose. For example, the eastern red-spotted newt (*Notophthalmus viridescens*) uses a magnetic compass based on the inclination of the magnetic lines of force when orienting toward the shore. We know this because their orientation was shifted by about 180° when the vertical component of the magnetic field was inverted. Eastern newts are also able to home, that is, to return to the point of origin after being moved to an unfamiliar location. During homing, the newt's orientation is unaffected by an inversion of the vertical component of the magnetic field (Phillips 1986) but is shifted by a change in polarity (Phillips 1987a). Thus, these initial observations suggest that, in the newt at least, the mechanism(s) for magnetoreception involved in homing differs from the one involved in shoreward compass orientation.

The magnetic compass used by the eastern newt when orienting toward the shore is light-dependent (Phillips and Borland 1992). The orientation of newts during homing is also affected by exposure to different wavelengths of light. However, the effects of long wavelengths on homing are different from those on shoreward orientation. Furthermore, light-dependent processes are not expected to respond to polarity of a magnetic field, and we know that a newt's homing ability is sensitive to polarity changes. This again suggests two magnetoreception mechanisms in newts (Phillips and Borland 1994).

Migratory birds may also have two mechanisms of magnetoreception. The orientation of Australian silvereyes (*Zosterops lateralis*) and European robins is temporarily shifted by a magnetic pulse that would alter the magnetization of magnetite. Both species were disoriented in red light but oriented correctly in white, blue, or green light (W. Wiltschko, Munro, Beason, and Wiltschko 1994; W. Wiltschko, Munro, Ford, and Wiltschko 1993; W. Wiltschko and Wiltschko 1995).

It has been suggested that these two mechanisms for magnetic sensitivity serve different functions. The light-dependent mechanism is thought to serve as a magnetic compass. Because a magnetite-based mechanism is theoretically capable of detecting minute variations in the earth's magnetic field, it may be part of the magnetic "map" receptor (Beason, Dussourd, and Deutschlander 1995; R. Wiltschko and Wiltschko 1995). To use the geomagnetic field as a map, an animal might merely compare the local intensity of the field with that at the goal. A receptor system used in a map sense, then, would not have to respond to the direction of the field, but it would be expected to respond to slight variations, less than 0.1%, in the intensity of the magnetic field experienced (Beason and Semm 1991). It is thought that the amount of magnetic material typically found in pigeons' skulls could make up a receptor that would provide enough sensitivity to small differences in the magnetic field to fit the bill (Yorke 1981).

A comparison of the effects of a strong magnetic pulse on the orientation of juvenile and adult Australian silvereyes supports the idea that a magnetite-based receptor system provides at least one coordinate of a "map." It is commonly believed that whereas adult migrants have established a navigational map, juveniles have not. As we have seen, the orientation of adult silvereyes is shifted by a magnetic pulse, presumably because their navigational map is affected. In contrast, the juvenile silvereyes remain oriented in the appropriate migratory direction after a magnetic pulse. It is thought that the magnetic pulse does not affect the orientation of juveniles because they have not yet formed a magnetic map. Instead, their orientation is based on an innate migratory program. They use their magnetic compass, which is based on a light-dependent magnetoreception process, to head in the appropriate direction according to their inherited migratory program (Munro et al. 1997).

REFERENCES

Able, K. P. 1980. Mechanisms of orientation, navigation, and homing. In *Animal Migration Orientation and Navigation*, edited by S. A. Gauthreaux, Jr. New York: Academic Press, pp. 283–373.

Able, K. P. 1982. Field studies of avian nocturnal migratory orientation I. Interaction of sun, wind, and stars as directional cues. *Anim. Behav.* 30:761–767.

Able, K. P. 1991. Common themes and variations in animal orientation systems. *Am. Zool.* 31:157–167.

Able, K. P. 1993. Orientation cues used by migratory birds: A review of cue-conflict experiments. *Trends Ecol. Evol.* 8 (10):367–371.

Able, K. P. 1996. The debate over olfactory navigation by homing pigeons. *J. Exp. Biol.* 199 (1):121–124.

Able, K. P., and M. A. Able. 1993. Daytime calibration of magnetic orientation in a migratory bird requires a view of skylight polarization. *Nature* 364:523–525.

Able, K. P., and M. A. Able. 1996. The flexible migratory orientation system of the Savannah sparrow (*Passerculus sandwichensis*). *J. Exp. Biol.* 199 (1):3–8.

Able, K. P., and V. P. Bingman. 1987. The development of orientation and navigation behavior in birds. *Quart. Rev. Biol.* 62:1–29.

Abu-Gideiri, Y. B. 1966. The behaviour and neuro-anatomy of some developing teleost fishes. *J. Zool. Lond.* 149:215–241.

Aceves-Pina, E. O., and W. G. Quinn. 1979. Learning in normal and mutant *Drosophila*. *Science* 206:93–96.

Adams, J., and R. L. Caldwell. 1990. Deceptive communication in asymmetric fights of the stomatopod crustacean *Gonodactylus bredini*. *Anim. Behav.* 39:706–717.

Adams, J., E. D. Rothman, W. E. Kerr, and Z. L. Paulino. 1977. Estimation of the number of sex alleles and queen mating from diploid male frequencies in a population of *Apis mellifera*. *Genetics* 86:583–596.

Adkins-Regan, E. 1998. Hormonal mechanisms of mate choice. *Am. Zool.* 38 (1):166–178.

Adkins-Regan, E., J.-P. Signoret, and P. Orgeur. 1989. Sexual differentiation of reproductive behavior in pigs: Defeminizing effects of prepubertal estradiol. *Horm. Behav.* 23:290–303.

Adler, K. 1976. Extraocular photoreception in amphibians. *Photochem. Photobiol.* 23:275–298.

Aidley, D. J. 1981. Questions about migration. In *Animal Migration*, edited by D. J. Aidley. New York: Cambridge University Press, pp. 1–8.

Akins, C. K., and T. R. Zentall. 1996. Imitative learning in male Japanese quail (*Coturnix japonica*) using a two-action procedure. *J. Comp. Psychol.* 110:316–320.

Akiyama, M., Y. Kouza, S. Takahashi, H. Wakamatsu, T. Moriya, M. Maetani, S. Watanabe, H. Tei, Y. Sakaki, and S. Shibata. 1999. Inhibition of light- or glutamate-induced mPer1 expression represses the phase shifts in mouse circadian locomotor activity and suprachiasmatic firing rhythms. *J. Neurosci.* 19 (3):1115–1121.

Alatalo, R. V., A. Carlson, A. Lundberg, and S. Ulfstrand. 1981. The conflict between male polygamy and female monogamy: The case of the pied flycatcher *Ficedula hypoleuca*. *Am. Nat.* 117:738–743.

Alatalo, R. V., and P. Helle. 1990. Alarm calling by individual willow tits, *Parus montanus*. *Anim. Behav.* 40 (3):437–442.

Alatalo, R. V., and A. Lundberg. 1984. Polyterritorial polygyny in the pied flycatcher *Ficedula hypoleuca*— Evidence for the deception hypothesis. *Ann. Zool. Fenn.* 21:217–228.

Alatalo, R. V., A. Lundberg, and K. Stahlbrandt. 1982. Why do pied flycatcher females mate with already-mated males? *Anim. Behav.* 30:585–593.

Alatalo, R. V., and O. Rätti. 1995. Sexy son hypothesis— Controversial once more. *Trends Ecol. Evol.* 10 (2):52–53.

Alcock, J., and A. Forsyth. 1988. Post-copulatory aggression toward their mates by males of the rove beetle *Leistotrophus versicolor* (Coleoptera: Staphylinidae). *Behav. Ecol. Sociobiol.* 22:203–308.

Alexander, R. D. 1961. Aggressiveness, territoriality, and sexual behavior in field crickets. *Behaviour* 17:130–223.

Alexander, R. D. 1974. The evolution of social behaviour. *Ann. Rev. Ecol. Syst.* 5:325–383.

Alexander, R. D. 1975. Natural selection and specialized chorusing behavior in acoustical insects. In *Insects, Science, and Society*, edited by D. Pimental. New York: Academic Press, pp. 35–77.

Alexander, R. D. 1979. *Darwinism and Human Affairs*. Seattle: University of Washington Press.

Allada, R., N. E. White, W. V. So, J. C. Hall, and M. Rosbash. 1998. A mutant *Drosophila* homolog of mammalian *CLOCK* disrupts circadian rhythms and transcription of *period* and *timeless*. *Cell* 93:805–814.

Allen, J. A. 1988. Frequency-dependent selection by predators. *Phil. Trans. Roy. Soc. Lond.* B 319:485–503.

Ambrose, H. W. 1972. Effect of habitat familiarity and toe-clipping on rate of owl predation in *Microtus pennsylvanicus*. *J. Mamm.* 53:909–912.

Anderson, J. B., and L. P. Brower. 1996. Freeze-protection of overwintering monarch butterflies in Mexico: Critical role of the forest as a blanket and an umbrella. *Ecol. Entomol.* 21 (2):107–116.

Andersson, M. 1984. The evolution of eusociality. *Ann. Rev. Ecol. Syst.* 15:165–189.

Andersson, M. 1994. *Sexual Selection.* Princeton, NJ: Princeton University Press.

Andrade, M. C. B. 1996. Sexual selection for male sacrifice in the Australian redback spider. *Science* 271 (5245):70–72.

Andrew, R. J. 1986. Evolution of intelligence and vocal mimicking. *Science* 137:585–589.

Aoki, S. 1972. *Colophina clematis* (Homoptera, Pemphigidae), an aphid species with "soldiers." *Kontyu* 45:276–282.

Aoki, S. 1979. Further observations on *Astegopteryx styracicola* (Homoptera, Pemphigidae), an aphid species with soldiers biting man. *Kontyu* 47:99–104.

Aoki, S. 1982. Soldiers and altruistic dispersal in aphids. In *The Biology of Social Insects*, edited by M. D. Breed, C. D. Michener, and H. E. Evans. Boulder, CO: Westview, pp. 154–158.

Arak, A. 1988a. Callers and satellites in the natterjack toad: Evolutionarily stable decision rules. *Anim. Behav.* 36:416–432.

Arak, A. 1988b. Sexual dimorphism in body size: A model and a test. *Evolution* 42:820–825.

Archer, J. 1988. *The Behavioural Biology of Aggression.* Cambridge: Cambridge University Press.

Armitage, K. B. 1986. Marmot polygyny revisited: Determinants of male and female reproductive strategies. In *Ecological Aspects of Social Evolution*, edited by D. I. Rubenstein and R. W. Wrangham. Princeton, NJ: Princeton University Press, pp. 303–331.

Armstrong, D. P. 1991. Levels of cause and effect as organizing principles for research in animal behaviour. *Can. J. Zool.* 69:823–829.

Arnold, A. P., and S. M. Breedlove. 1985. Organizational and activational effects of sex steroids on brain and behavior: A reanalysis. *Horm. Behav.* 19:469–498.

Arnold, E. N. 1988. Caudal autotomy as a defense. In *Biology of the Reptilia*, edited by C. Gans and R. B. Huey. New York: Alan Liss, pp. 235–273.

Arnold, S. J. 1976. Sexual behavior, sexual interference and sexual defense in the salamanders *Ambystoma maculatum*, *Ambystoma tigrinum* and *Plethodon jordani*. *Z. Tierpsychol.* 42:247–300.

Aschoff, J. 1965. Circadian rhythms in man. *Science* 148:1427–1432.

Aschoff, J. 1967. Circadian rhythms. In *Life Sciences and Space Research*, edited by H. Brown and F. Favorite. Amsterdam: North Holland, pp. 159–173.

Axelrod, R. 1984. *The Evolution of Cooperation.* New York: Basic Books.

Axelrod, R., and W. D. Hamilton. 1981. The evolution of cooperation. *Science* 211:1390–1395.

Ayala, F. J., and C. A. Campbell. 1974. Frequency-dependent selection. *Ann. Rev. Ecol. Syst.* 5:115–138.

Bäckman, J., J. Pettersson, and R. Sandberg. 1997. The influence of fat stores on magnetic orientation in day-migrating chaffinch, *Fringilla coelebs*. *Ethology* 103:247–256.

Backwell, P. R. Y., M. D. Jennions, J. H. Christy, and U. Schober. 1995. Pillar building in the crab *Uca beebei*: Evidence for a condition-dependent ornament. *Behav. Ecol. Sociobiol.* 36:185–192.

Bacon, J., and B. Mohl. 1983. The tritocerebral commissure giant (TCG) wind-sensitive interneurone in locust. I. Its activity in straight flight. *J. Comp. Physiol.* 150:439–452.

Baerends, G. P. 1985. Do the dummy experiments with sticklebacks support the IRM concept? *Behaviour* 93:258–277.

Baggerman, B. 1962. Some endocrine aspects of fish migration. *Gen. Comp. Endocrinol. Suppl.* 1:188–205.

Baker, M. C., and M. A. Cunnigham. 1985. The biology of bird-song dialects. *Behav. Brain Sci.* 8:85–133.

Baker, P. J., C. P. J. Robertson, S. M. Funk, and S. Harris. 1998. Potential fitness benefits of group living in the red fox, *Vulpes vulpes*. *Anim. Behav.* 56 (6):1411–1424.

Baker, R. R. 1978. *The Evolutionary Ecology of Animal Migration.* New York: Holmes & Meier.

Baker, R. R. 1980. *The Mystery of Migration.* London: Macdonald.

Baker, R. R., and M. A. Bellis. 1988. "Kamikaze" sperm in mammals? *Anim. Behav.* 36:936–939.

Baker, R. R., and M. A. Bellis. 1989. Elaboration of the kamikaze sperm hypothesis: A reply to Harcourt. *Anim. Behav.* 37:865–867.

Balda, R. P. 1980. Recovery of cached seeds by a captive *Nucifraga caryocatactes*. *Z. Tierpsychol.* 52:331–346.

Balda, R. P., and A. C. Kamil. 1989. A comparative study of cache recovery by three corvid species. *Anim. Behav.* 38:486–495.

Baldaccini, E. N., S. Benvenuti, V. Fiaschi, and F. Papi. 1975. Pigeon navigation: Effects of wind deflection at the home cage on homing behavior. *J. Comp. Physiol.* 99:177–196.

Baldi, R., C. Campagna, S. Pedraza, and B. J. le Boeuf. 1996. Social effects of space availability on the breeding of elephant seals in Patagonia. *Anim. Behav.* 51 (4):717–724.

Balshine-Earn, S., F. C. Neat, H. Reid, and M. Taborsky. 1998. Paying to stay or paying to breed? Field evidence for direct benefits of helping in a cooperatively breeding fish. *Behav. Ecol.* 9 (5):432–438.

Bandura, A. 1962. Social learning through imitation. In *Nebraska Symposium on Motivation*, edited by M. R. Jones. Lincoln: University of Nebraska Press.

Baptista, L. F., and L. Petrinovich. 1984. Social interaction, sensitive phases and the song template hypothesis in the white-crowned sparrow. *Anim. Behav.* 32:172–181.

Baptista, L. F., and L. Petrinovich. 1986. Song development in the white-crowned sparrow: Social factors and sex differences. *Anim. Behav.* 34:1359–1371.

Barash, D. P. 1982. *Sociobiology and Behavior.* 2nd ed. New York: Elsevier.

Bargiello, T. A., F. R. Jackson, and M. W. Young. 1984. Restoration of circadian behavioral rhythms by gene transfer in *Drosophila. Nature* 328:752–754.

Barinaga, M. 1994. From fruit flies, rats, mice: Evidence of genetic influence. *Science* 264:1690–1693.

Barlow, G. W. 1968. Ethological units of behavior. In *The Central Nervous System and Fish Behavior*, edited by D. Ingle. Chicago: University of Chicago Press, pp. 217–232.

Barlow, G. W. 1989. Has sociobiology killed ethology or revitalized it? In *Perspectives in Ethology*, edited by P. P. G. Bateson and P. H. Klopfer. New York: Plenum, pp. 1–45.

Barlow, G. W. 1991. Nature-nurture and the debates surrounding ethology and sociobiology. *Am. Zool.* 31:286–296.

Barnard, C. J., and C. A. J. Brown. 1985. Risk-sensitive foraging in common shrews (*Sorex araneus* L.). *Behav. Ecol. Sociobiol.* 16:161–164.

Bartos, L., and V. Perner. 1998. Distribution of mating across season and reproductive success according to dominance in male red deer. *Folia Zoologia* 47 (1):7–12.

Bartz, S. H. 1979. Evolution of eusociality in termites. *Proc. Nat. Acad. Sci. USA* 76 (11):5764–5768.

Basil, J. A., A. C. Kamil, R. P. Balda, and K. V. Fite. 1996. Differences in hippocampal volume among food storing corvids. *Brain Behav. Evol.* 47:156–164.

Basolo, A. L. 1990. Female preference predates the evolution of the sword in swordtail fish. *Science* 250:808–810.

Basolo, A. L. 1995a. A future examination of a pre-existing bias favouring a sword in the genus *Xiphophorus. Anim. Behav.* 50 (2):365–375.

Basolo, A. L. 1995b. Phylogenetic evidence for the role of a pre-existing bias in sexual selection. *Proc. Roy. Soc. Lond. B* 259 (1356):307–311.

Basolo, A. L. 1998. Evolutionary change in receiver bias: A comparison of female preference functions. *Proc. Roy. Soc. Lond. B* 265 (1411):2223–2228.

Bass, A. H. 1996. Shaping brain sexuality. *Am. Sci.* 84 (4):352–363.

Bass, A. H., D. A. Bodnar, and J. R. McKibben. 1997. From neurons to behavior: Vocal-acoustic communication in teleost fish. *Biol. Bull.* 192:158–160.

Bateman, A. J. 1948. Intra-sexual selection in *Drosophila. Heredity* 2:349–368.

Bates, D. L., and B. M. Fenton. 1990. Aposematism or startle? Predators learn their responses to the defenses of prey. *Can. J. Zool.* 68:49–52.

Bateson, P. 1976. Specificity and the origins of behavior. In *Advances in the Study of Behavior*, edited by J. Rosenblatt, R. A. Hinde, and C. Beer. New York: Academic Press, pp. 1–20.

Bateson, P. 1979. How do sensitive periods arise and what are they for? *Anim. Behav.* 27:470–486.

Bateson, P. 1982. Preferences for cousins in Japanese quail. *Nature* 295:236–237.

Bateson, P. 1983. Optimal outbreeding. In *Mate Choice*, edited by P. Bateson. Cambridge: Cambridge University Press, pp. 257–277.

Bateson, P. 1990. Is imprinting such a special case? *Phil. Trans. Roy. Soc. Lond. B* 329:125–131.

Bateson, P. P. G., and P. H. Klopfer. 1989. Preface. In *Perspectives in Ethology*, vol 8, (*Whither Ethology?*) edited by P. P. G. Bateson and P. H. Klopfer. New York: Plenum, v–viii.

Bateson, P. P. G., and P. H. Klopfer, eds. 1991. *Perspectives in Ethology. Vol. 9 (Human Understanding and Animal Awareness.)* New York: Plenum.

Batteau, D. W. 1968. The world as a source; the world as a sink. In *The Neuropsychology of Spatially Oriented Behavior*, edited by S. J. Freedman. Homewood, IL: Dorsey, pp. 197–203.

Beach, F. A. 1976. Sexual attractivity, proceptivity, and receptivity in female mammals. *Horm. Behav.* 7:105–138.

Beason, R. C., N. Dussourd, and M. E. Deutschlander. 1995. Behavioural evidence for the use of magnetic material in magnetoreception by a migratory bird. *J. Exp. Biol.* 198:141–146.

Beason, R. C., and J. E. Nichols. 1984. Magnetic orientation and magnetically sensitive material in a transequatorial migratory bird. *Nature* 309:151–153.

Beason, R. C., and P. Semm. 1987. Magnetic responses of the trigeminal nerve system of the bobolink (*Dolichonyx oryzivorus*). *Neurosci. Lett.* 80:229–234.

Beason, R. C., and P. Semm. 1991. Neuroethological aspects of avian orientation. In *Orientation in Birds*, edited by P. Berthold. Basel: Birkhäuser Verlag, pp. 106–127.

Beason, R. C., and P. Semm. 1996. Does the ophthalmic nerve carry magnetic navigational information? *J. Exp. Biol.* 199:1241–1244.

Beck, C. D. O., and C. H. Rankin. 1997. Long-term habituation is produced by distributed training at long ISIs and not massed training or short ISIs in *Caenorhabditis elegans. Anim. Learn. Behav.* 25 (4):446–457.

Beck, W., and W. Wiltschko. 1982. The magnetic field as a reference system for the genetically encoded migratory direction in pied flycatcher (*Ficedula hypoleuca* Pallas). *Z. Tierpsychol.* 60:41–46.

Bednarz, J. C. 1988. Cooperative hunting in Harris' hawks (*Parabuteo unicinctus*). *Science* 239:1525–1527.

Bednekoff, P. A. 1997. Mutualism among safe, selfish sentinels: A dynamic game. *Am. Nat.* 150 (3):373–392.

Bednekoff, P. A., A. C. Kamil, and R. P. Balda. 1997. Clark's nutcracker (Aves: Corvidae) spatial memory: Interference effects on cache recovery performance? *Ethology* 103:554–565.

Beecher, M. D. 1990. The evolution of parent-offspring recognition in swallows. In *Contemporary Issues in Comparative Psychology*, edited by D. A. Dewsbury. Sunderland, MA: Sinauer, pp. 360–380.

Beecher, M. D., I. M. Beecher, and S. Hahn. 1981. Parent-offspring recognition in bank swallows (*Riparia riparia*):

Development and acoustic basis. *Anim. Behav.* 29:95–101.

Beehler, M. N., and M. S. Foster. 1988. Hotshots, hotspots, and female preference in the organization of lek mating systems. *Am. Nat.* 131:203–219.

Bekoff, A., and J. A. Kauer. 1984. Neural control of hatching: Gate of the pattern generator for leg movements of hatching in post-hatching chicks. *J. Neurosci.* 4:2659–2666.

Beletsky, L. D., and G. H. Orians. 1987. Territoriality among male red-winged blackbirds. II. Removal experiments and site dominance. *Behav. Ecol. Sociobiol.* 20:339–349.

Beletsky, L. D., and G. H. Orians. 1989. Territoriality among male red-winged blackbirds. III. Testing hypotheses of territorial dominance. *Behav. Ecol. Sociobiol.* 24:333–339.

Beling, I. 1929. Über das Zeitgedächtnis der Bienen. *Z. Vergl. Physiol.* 9:259–338.

Belluscio, L., G. Koentges, R. Axel, and C. Dulac. 1999. A map of pheromone receptor activation in the mammalian brain. *Cell* 97 (2):209.

Belovsky, G. E. 1978. Diet optimization in a generalist herbivore: The moose. *Theoret. Pop. Biol.* 14:105–134.

Benhamou, S. 1989. An olfactory orientation model for mammal's movements in their home ranges. *J. Theor. Biol.* 139:379–388.

Benhamou, S. 1996. No evidence for cognitive mapping in rats. *Anim. Behav.* 52 (1):201–212.

Benhamou, S., and P. Bovet. 1989. How animals use their environment: A new look at kinesis. *Anim. Behav.* 38:375–383.

Benhamou, S., and P. Bovet. 1992. Distinguishing between elementary orientation mechanisms by means of path analysis. *Anim. Behav.* 43:371–377.

Bennett, A. T. D. 1996. Do animals have cognitive maps? *J. Exp. Biol.* 199 (1):219–224.

Bennett, N. C., and J. U. M. Jarvis. 1988. The social structure and reproductive biology of colonies of the mole-rat, *Cryptomys damarensis* (Rodentia, Bathyergidae). *J. Mamm.* 69 (2):293–302.

Bensch, S. 1996. Female mating status and reproductive success in the great reed warbler: Is there a potential cost of polygyny that requires compensation? *J. Anim. Ecol.* 65 (3):283–296.

Bentley, P. J. 1982. *Comparative Vertebrate Endocrinology.* Cambridge: Cambridge University Press.

Bergmann, G. 1965. Der sexuelle Grössendimorphismus der Anatiden als Anpassung an das Höhlenbrüten. *Commentat. Biol.* 28:1–28.

Bermant, G. 1976. Sexual behavior: Hard times with the Coolidge Effect. In *Psychological Research: The Inside Story*, edited by M. H. Siegel and H. P. Zeigler. New York: Harper & Row, pp. 76–103.

Bernstein, I. S. 1981. Dominance: The baby and the bathwater. *Behav. Brain Sci.* 4:419–457.

Berthold, P. 1991. Spatiotemporal programmes and genetics of orientation. In *Orientation in Birds*, edited by P. Berthold. Basel: Birkhäuser Verlag, pp. 86–105.

Berthold, P. 1993. *Bird Migration*, translated by H.-G. Bauer and T. Tomlinson. Oxford: Oxford University Press.

Bertram, B. C. R. 1975. Social factors influencing reproduction in wild lions. *J. Zool. Lond.* 177:463–482.

Bertram, B. C. R. 1976. Kin selection in lions and evolution. In *Growing Points in Ethology*, edited by P. P. G. Bateson and R. A. Hinde. New York: Cambridge University Press, pp. 281–301.

Biben, M. 1998. Squirrel monkey playfighting: Making the case for a cognitive training function for play. In *Animal Play: Evolutionary, Comparative, and Ecological Perspectives*, edited by M. Bekoff and J. A. Byers. Cambridge: Cambridge University Press, pp. 161–182.

Bicker, G., and I. Hähnlein. 1994. Long-term habituation of an appetitive reflex in the honeybee. *Neuroreport* 6:54–56.

Bingman, V. P. 1984. Night-sky orientation of migratory Pied Flycatchers raised in different magnetic fields. *Behav. Ecol. Sociobiol.* 15:77–80.

Bingman, V. P., and K. P. Able. 1979. The sun as a cue in the orientation of the white-throated sparrow, a nocturnal migrant bird. *Anim. Behav.* 27:621–622.

Binkley, S. 1988. *The Pineal: Endocrine and Neuroendocrine Function.* Upper Saddle River, NJ: Prentice Hall.

Binkley, S. 1993. Structures and molecules involved in generation and regulation of biological rhythms in vertebrates and invertebrates. *Experientia* 49:648–653.

Birkhead, T. R. 1996. Mechanisms of sperm competition in birds. *Am. Sci.* 84 (3):254–262.

Birkhead, T. R. 1998. Cryptic female choice: Criteria for establishing female sperm choice. *Evolution* 52 (4):1212–1218.

Birkhead, T. R., and G. A. Parker. 1997. Sperm competition and mating systems. In *Behavioural Ecology: An Evolutionary Approach*, edited by J. R. Krebs and N. B. Davies. Oxford: Blackwell Science, pp. 121–145.

Bischof, H.-J. 1994. Sexual imprinting as a two-stage process. In *Causal Mechanisms of Behavioural Development*, edited by J. A. Hogan and J. J. Bolhuis. Cambridge: Cambridge University Press, pp. 82–97.

Bitran, D., E. M. Hull, G. M. Holmes, and K. J. Lookingland. 1988. Regulation of male rat copulatory behavior by preoptic incertohypothalamic dopamine neurons. *Brain Res. Bull.* 20:323–331.

Black, G. A., and J. B. Dempsey. 1986. A test for the hypothesis of pheromone attraction in salmonid migration. *Environ. Biol. Fishes* 15:229–235.

Blakemore, R. P. 1975. Magnetotactic bacteria. *Science* 190:377–379.

Blakemore, R. P., and R. B. Frankel. 1981. Magnetic navigation in bacteria. *Sci. Am.* 245 (Dec.):58–65.

Blaustein, A. R., M. Bekoff, and T. J. Daniels. 1987. Kin recognition in vertebrates (excluding primates). In *Kin*

Recognition in Animals, edited by D. J. C. Fletcher and C. D. Michener. New York: Wiley, pp. 287–357.

Blaustein, A. R., and R. K. O'Hara. 1982. Kin recognition in *Rana cascadae* tadpoles: Maternal and paternal effects. *Anim. Behav.* 30:1151–1157.

Blaustein, A. R., and R. K. O'Hara. 1986. Kin recognition in tadpoles. *Sci. Am.* 254 (Jan.):108–116.

Blaustein, A. R., and D. S. Waldman. 1992. Kin recognition in anuran amphibians. *Anim. Behav.* 44:207–221.

Blest, A. D. 1957a. The evolution of protective displays in the Saturnidae and Sphingidae (Lepidoptera). *Behaviour* 11:257–309.

Blest, A. D. 1957b. The function of eyespot patterns in the Lepidoptera. *Behaviour* 11:209–257.

Blest, A. D. 1961. The concept of ritualization. In *Current Problems in Animal Behavior*, edited by W. H. Thorpe and O. L. Zangwill. London: Cambridge University Press, pp. 102–124.

Bletz, H., P. Weindler, R. Wiltschko, W. Wiltschko, and P. Berthold. 1996. The magnetic field as reference for the innate migratory direction in blackcaps, *Sylvia atricapilla*. *Naturwissenschaften* 83:430–432.

Blough, P. M. 1989. Attentional priming and visual search in pigeons. *J. Exp. Psych.: Anim. Behav. Proc.* 15:358–365.

Blough, P. M. 1991. Selective attention and search images in pigeons. *J. Exp. Psych.: Anim. Behav. Proc.* 17:292–298.

Blumstein, D. T., J. Steinmetz, K. B. Armitage, and J. C. Daniel. 1997. Alarm calling in yellow-bellied marmots: II. The importance of direct fitness. *Anim. Behav.* 53:173–184.

Boccia, M. L. 1983. A functional analysis of social grooming patterns through direct comparison with self grooming in rhesus monkeys. *Int. J. Primatol.* 4:399–418.

Boccia, M. L. 1986. Grooming site preferences as a form of tactile communication and their role in the social relations of rhesus monkeys. In *Current Perspectives in Primate Social Dynamics*, edited by D. M. Taub and F. A. King. New York: Van Nostrand Reinhold, pp. 505–518.

Bodamer, M. D., D. H. Fouts, R. S. Fouts, and M. L. A. Jensvold. 1994. Functional analysis of chimpanzee (*Pan troglodytes*) private signing. *Human Evol.* 9 (4):281–296.

Böhner, J. 1983. Song learning in the zebra finch (*Taeniopygia guttata*): Selectivity in the choice of tutor and accuracy of song copies. *Anim. Behav.* 31:231–237.

Böhner, J. 1990. Early acquisition of song in the zebra finch, *Taeniopygia guttata*. *Anim. Behav.* 39:369–374.

Bollenbacher, W. E., S. L. Smith, W. Goodman, and L. I. Gilbert. 1981. Ecdysone titer during larval-pupal-adult development in the tobacco hornworm, *Manduca sexta*. *Gen. Comp. Endocrinol.* 44:302–306.

Bolles, R. C. 1969. Avoidance and escape learning: Simultaneous acquisition of different responses. *J. Comp. Physiol. Psychol.* 68:355–358.

Bonardi, C., D. Guthrie, and G. Hall. 1991. The effect of a retention interval on habituation of the neophobic response. *Anim. Learn. Behav.* 19 (1):11–17.

Bond, A. B., and A. C. Kamil. 1998. Apostatic selection by blue jays produces balanced polymorphisms in virtual prey. *Nature* 395:594–596.

Bond, A. B., and D. A. Riley. 1991. Searching image in the pigeon: A test of three hypothetical mechanisms. *Ethology* 87:203–224.

Boomsma, J. J., and F. L. W. Ratnieks. 1996. Paternity in eusocial Hymenoptera. *Phil. Trans. Royal Soc. Lond. B* 351 (1342):947–975.

Borgia, G. 1986. Satin bowerbird parasites: A test of the bright male hypothesis. *Behav. Ecol. Sociobiol.* 19:355–358.

Borgia, G., I. M. Kaatz, and R. Condit. 1987. Flower choice and bower decoration in the satin bowerbird *Ptilonorhyncus violaceus*: A test of the hypothesis for the evolution of male display. *Anim. Behav.* 35:1129–1139.

Borries, C., K. Launhardt, C. Epplen, J. T. Epplen, and P. Winkler. 1999. DNA analyses support the hypothesis that infanticide is adaptive in langur monkeys. *Proc. Roy. Soc. Lond. B* 266 (1422):901–904.

Bourtchuladze, R., B. Frenguelli, J. Blendy, D. Cioffi, G. Schutz, and A. Silva. 1994. Deficient long-term memory in mice with a targeted mutation of the cAMP-responsive element-binding site. *Cell* 79:59–68.

Bowmaker, J. K., V. I. Govardovskii, S. A. Shukolykov, L. V. Zueva, D. M. Hunt, V. G. Sideleva, and O. G. Smirnova. 1994. Visual pigments and the photic environment: The cottoid fish of Lake Baikal. *Vision Res.* 34:591–605.

Boyan, G. S., and J. H. Fullard. 1986. Interneurons responding to sound in the tobacco budworm moth *Heliothis virescens* (Noctuidae): Morphological and physiological characteristics. *J. Comp. Physiol. A* 158:391–404.

Boyan, G. S., and J. H. Fullard. 1988. Information processing at a central synapse suggests a noise filter in the auditory pathway of the noctuid moth. *J. Comp. Physiol. A* 164 (2):251–258.

Boyan, G. S., and L. A. Miller. 1991. Parallel processing of afferent input by identified interneurones in the auditory pathway of the noctuid moth *Noctua pronumba*. *J. Comp. Physiol. A* 168:727–738.

Bradbury, J. W. 1977. Lek mating in the hammer-headed bat. *Z. Tierpsychol.* 45:225–255.

Bradbury, J. W. 1981. The evolution of leks. In *Natural Selection and Social Behavior*, edited by R. D. Alexander and D. W. Tinkle. New York: Chiron, pp. 138–172.

Bradbury, J. W., R. M. Gibson, and M. B. Andersson. 1986. Hotspots and the evolution of leks. *Anim. Behav.* 34:1694–1709.

Bradbury, J. W., and S. L. Vehrencamp. 1998. *Principles of Animal Communication*. Sunderland, MA: Sinauer.

Brannon, E. L., and T. P. Quinn. 1990. A field test for the pheromone hypothesis for homing by Pacific salmon. *J. Chem. Ecol.* 16:603–609.

Breed, M. D. 1998. Recognition pheromones of the honeybee. *BioScience* 48 (6):463–470.

Breed, M. D., and G. J. Gamboa. 1976. Behavioral control of workers by queens in primitively eusocial bees. *Science* 195:694–696.

Breedlove, S. M., and A. P. Arnold. 1983. Hormonal control of a developing neuromuscular system. II. Sensitive periods for the androgen-induced masculinization of the rat spinal nucleus of the bulbocavernosus. *J. Neurosci.* 3:424–432.

Breland, K., and M. Breland. 1961. The misbehavior of organisms. *Am. Psychol.* 16:681–684.

Brett, R. A. 1991. The population structure of naked mole-rat colonies. In *The Biology of the Naked Mole-Rat*, edited by P. W. Sherman, J. U. M. Jarvis, and R. D. Alexander. Princeton, NJ: Princeton University Press, pp. 97–136.

Brines, M. L. 1980. Dynamic patterns of skylight polarization as clock and compass. *J. Theor. Biol.* 86:507–512.

Bristowe, W. S. 1958. *The World of Spiders*. London: Collins.

Brockmann, J., A. Grafen, and R. Dawkins. 1979. Evolutionarily stable nesting strategy in a digger wasp. *J. Theoret. Biol.* 77:473–496.

Brodbeck, D. R. 1994. Memory for spatial and local cues: A comparison of storing and nonstoring species. *Anim. Learn. Behav.* 22:119–133.

Brosset, A. 1997. Aggressive mimicry by the characid fish *Erythrinus erythrinus*. *Ethology* 103 (11):926–934.

Brotherton, P. N. M., and M. B. Manser. 1997. Female dispersion and the evolution of monogamy in the dik-dik. *Anim. Behav.* 54 (6):1413–1424.

Brotherton, P. N. M., J. M. Pemberton, P. E. Komers, and G. Malarky. 1997. Genetic and behavioural evidence of monogamy in a mammal, Kirk's dik-dik (*Madoqua kirkii*). *Proc. Roy. Soc. Lond. B* 264 (1382):675–681.

Brotherton, P. N. M., and A. Rhodes. 1996. Monogamy without biparental care in a dwarf antelope. *Proc. Roy. Soc. Lond. B* 263 (1366):23–29.

Brower, L. P., and W. H. Calvert. 1985. Foraging dynamics of bird predators on overwintering monarch butterflies. *Evolution* 39:852–868.

Brower, L. P., W. N. Ryerson, L. L. Coppinger, and S. C. Glazier. 1968. Ecological chemistry and palatability spectrum. *Science* 161:1349–1351.

Brown, F. A., Jr., 1971. Some orientational influences of non-visual terrestrial electromagnetic fields. *Ann. NY Acad. Sci.* 188:224–241.

Brown, F. A., Jr., J. W. Hastings, and J. D. Palmer. 1970. *The Biological Clock: Two Views*. New York: Academic Press.

Brown, J. L. 1964. The evolution of diversity in avian territorial systems. *Wilson Bull.* 76:160–169.

Brown, J. L. 1975. Helpers among Arabian babblers *Turdoides squamiceps*. *Ibis* 117:243–244.

Brown, J. L. 1982. The adaptationist program. *Science* 217:884–886.

Brown, J. L. 1987. *Helping and Communal Breeding in Birds*. Princeton, NJ: Princeton University Press.

Brown, J. L., and E. R. Brown. 1998. Are inbred offspring less fit? Survival in a natural population of Mexican jays. *Behav. Ecol.* 9 (1):60–63.

Brown, J. L., E. R. Brown, S. D. Brown, and D. D. Dow. 1982. Helpers: Effects of experimental removal on reproductive success. *Science* 215:421–422.

Brown, J. L., and A. Eklund. 1994. Kin recognition and the major histocompatibility complex: An integrative review. *Am. Nat.* 143:435–461.

Brown, J. R., H. Ye, R. T. Bronson, P. Dikkes, and M. E. Greenberg. 1996. A defect in nurturing in mice lacking the immediate early gene *fosB*. *Cell* 86:296–309.

Brown, K. S. 1996. Infanticide as a way to get ahead. *BioScience* 46 (3):174–176.

Brownell, P. H. 1984. Prey detection by the scorpion. *Sci. Am.* (Dec.):86–98.

Bruce, H. M. 1959. An exteroceptive block to pregnancy in the mouse. *Nature* 184:105.

Bruce, H. M. 1961. Time relations in the pregnancy block induced in mice by strange males. *J. Reprod. Fert.* 2:138–142.

Bruce, H. M. 1963. Olfactory block to pregnancy among grouped mice. *J. Reprod. Fert.* 6:451–460.

Brunton, D. H. 1990. The effects of nesting stage, sex, and type of predator on parental defense by killdeer (*Charadrius vociferus*): Testing models of parental defense. *Behav. Ecol. Sociobiol.* 26:181–190.

Budzynski, C. A., R. Strasser, and V. P. Bingman. 1998. The effects of zinc sulfate anosmia on homing pigeons, *Columba livia*, in a homing and a non-homing experiment. *Ethology* 104:111–118.

Bugnyar, T., and L. Huber. 1997. Push or pull: An experimental study on imitation in marmosets. *Anim. Behav.* 54:817–831.

Bullock, T. H. 1973. Seeing the world through a new sense. *Am. Sci.* 61:315–325.

Bullock, T. H., and T. Szabo. 1986. Introduction. In *Electroreception*, edited by T. H. Bullock and W. Heiligenberg. New York: Wiley, pp. 1–12.

Bult, A., and C. B. Lynch. 1996. Multiple selection responses in house mice bidirectionally selected for thermoregulatory nest-building behavior: Crosses of replicate lines. *Behav. Genet.* 26 (4):439–446.

Bult, A., and C. B. Lynch. 1997. Nesting and fitness: Lifetime reproductive success in house mice bidirectionally selected for thermoregulatory nest-building behavior. *Behav. Genet.* 27 (3):231–240.

Burghardt, G. M. 1998. The evolutionary origins of play revisited. In *Animal Play: Evolutionary, Comparative, and Ecological Perspectives*, edited by M. Bekoff and J. A. Byers. Cambridge: Cambridge University Press, pp. 1–26.

Burghardt, G. M., and H. W. Greene. 1988. Predator simulation and duration of death feigning in neonate hognose snakes. *Anim. Behav.* 36:1842–1844.

Burkhardt, D., and I. de la Motte. 1983. How stalk-eyed flies view stalk-eyed flies: Observations and measurements of the eyes of *Cyrtodiopsis whitei* (Diopsidea, Diptera). *J. Comp. Physiol. A* 151:407–421.

Burrows, M. T. 1945. Periodic spawning of the Palolo worms in Pacific waters. *Nature* 289:79–81.

Byers, D., R. L. Davis, and J. A. Kiger, Jr. 1981. Defect in cyclic AMP phosphodiesterase due to the *dunce* mutation of learning in *Drosophila melanogaster*. *Nature* 289:79–81.

Byers, J. A., and C. B. Walker. 1995. Refining the motor training hypothesis for the evolution of play. *Am. Nat.* 146:25–40.

Byrne, J. H. 1987. Cellular analysis of associative learning. *Physiol. Rev.* 67:329–439.

Cahalane, V. H. 1961. *Mammals of North America*. New York: Macmillan.

Caldwell, R. L., and H. Dingle. 1976. Stomatopods. *Sci. Am.* 234 (Jan.):80–89.

Calvert, W. H., and L. P. Brower. 1986. The location of the monarch butterfly (*Danaus plexippus* L.). *J. Lepid. Soc.* 40:164–187.

Calvert, W. H., L. E. Hedrick, and L. P. Brower, 1979. Mortality of the monarch butterfly (*Danaus plexippus* L.): Avian predation at five overwintering sites in Mexico. *Science* 204:847–851.

Camhi, J. M. 1980. The escape system of the cockroach. *Sci. Am.* 248 (Dec.):158–172.

Camhi, J. M. 1984. *Neuroethology*. Sunderland, MA.: Sinauer.

Camhi, J. M. 1988. Escape behavior in the cockroach: Distributed neural processing. *Experientia* 44:401–408.

Camhi, J. M., W. Tom, and S. Volman. 1978. The escape behavior of the cockroach *Periplaneta americana*. II. Detection of natural predators by air displacement. *J. Comp. Physiol.* 128:203–212.

Campbell, D. B., and E. J. Hess. 1996. Chromosomal localization of the neurological mouse mutations tottering (*tg*), Purkinje cell degeneration (*pcd*), and nervous (*nr*). *Mol. Brain Res.* 37:79–84.

Cant, M. A. 1998. A model for the evolution of reproductive skew without reproductive suppression. *Anim. Behav.* 55 (1):163–169.

Caplan, A. L. 1978. *The Sociobiology Debate*. New York: Harper & Row.

Caputi, A. A., R. Budelli, K. Grant, and C. C. Bell. 1998. The electrical image in weakly electric fish: Physical images of resistive objects in *Gnathonemus petersii*. *J. Exp. Biol.* 201:2115–2128.

Caraco, T. 1981. Energy budgets, risk and foraging preferences in dark-eyed juncos (*Junco hyemalis*). *Behav. Ecol. Sociobiol.* 8:213–217.

Caraco, T. 1983. White-crowned sparrows (*Zonotrichia leucophrys*): Foraging preferences in a risky environment. *Behav. Ecol. Sociobiol.* 12:63–69.

Carew, T. J., and E. R. Kandel. 1973. Acquisition and retention of long-term habituation in *Aplysia*: Correlation of behavioral and cellular processes. *Science* 182:1158–1160.

Carew, T. J., H. M. Pinsker, and E. R. Kandel. 1972. Long-term habituation of a defensive withdrawal reflex in *Aplysia*. *Science* 175:451–454.

Carlier, P., and L. Lefebvre. 1997. Ecological differences in social learning between adjacent, mixing populations of Zenaida doves. *Ethology* 103 (9):772–784.

Caro, T. M. 1986a. The functions of stotting: A review of hypotheses. *Anim. Behav.* 34:649–662.

Caro, T. M. 1986b. The functions of stotting in Thomson's gazelles: Some tests of the predictions. *Anim. Behav.* 34:663–684.

Caro, T. M. 1988. Adaptive significance of play: Are we getting closer? *Trends Ecol. Evol.* 88:50–54.

Caro, T. M. 1995. Short-term costs and correlates of play in cheetahs. *Anim. Behav.* 49 (2):333–345.

Carpenter, C. R. 1942. Sexual behavior of free-ranging rhesus monkeys (*Macaca mulatta*) I. Specimens, procedures and behavioral characteristics of estrus. *J. Comp. Physiol. Psychol.* 33:113–142.

Carpenter, F. L., D. C. Paton, and M. A. Hixon. 1983. Weight gain and adjustment of feeding territory size in migrant hummingbirds. *Proc. Nat. Acad. Sci. USA* 80:7259–7263.

Cartar, R. V., and L. M. Dill, 1990. Why are bumblebees risk-sensitive foragers? *Behav. Ecol. Sociobiol.* 26:121–127.

Carter, C. S., and L. L. Getz. 1993. Monogamy and the prairie vole. *Sci. Am.* 268 (June):100–106.

Cassone, V. M. 1998. Melatonin's role in vertebrate circadian rhythms. *Chronobiol. Int.* 15 (5):457–473.

Castellucci, V., and E. R. Kandel. 1974. A quantal analysis of the synaptic depression underlying the gill-withdrawal reflex in *Aplysia*. *Proc. Nat. Acad. Sci. USA* 71:5004–5008.

Catania, K. C., and J. H. Kaas. 1996. The unusual nose and brain of the star-nosed mole. *BioScience* 46 (8):578–586.

Ceriani, M. F., T. K. Darlington, D. Staknis, P. Más, A. A. Petti, C. J. Weitz, and S. A. Kay. 1999. Light-dependent sequestration of TIMELESS by CRYPTOCHROME. *Science* 285:553–556.

Champalbert, A., and P.-P. Lachaud. 1990. Existence of a sensitive period during ontogenesis of social behavior in a primitive ant. *Anim. Behav.* 39:850–859.

Charnov, E. L. 1976. Optimal foraging, the marginal value theorem. *Theoret. Pop. Biol.* 9:129–136.

Charnov, E. L. 1978. Evolution of eusocial behavior: Offspring choice or parental parasitism? *J. Theoret. Biol.* 75:451–465.

Charnov, E. L., and J. R. Krebs. 1975. The evolution of alarm calls: Altruism or manipulation? *Am. Nat.* 109:107–112.

Chase, I. D., C. Bartolomeo, and L. A. Dugatkin. 1994. Aggressive interactions and inter-contest interval: How long do winners keep winning? *Anim. Behav.* 48 (2):393–400.

Chase, J. M. 1998. Central-place forager effects on food web dynamics and spatial pattern in Northern California meadows. *Ecology* 79 (4):1236–1245.

Cheney, D. L., and R. M. Seyfarth. 1988. Assessment of meaning and detection of unreliable signals by vervet monkeys. *Anim. Behav.* 36:477–486.

Cheney, D. L., and R. M. Seyfarth. 1991. Truth and deception in animal communication. In *Cognitive Ethology: The Minds of Other Animals*, edited by C. A. Ristau. Hillsdale, NJ: Erlbaum, pp. 127–151.

Cherry, J. D. 1984. Migratory orientation of tree sparrows. M. S. thesis, State University of New York, Albany.

Chipman, R. K., and K. A. Fox. 1966. Factors in pregnancy blocking: Age and reproductive background of females: Numbers of strange males. *J. Reprod. Fert.* 12:399–403.

Christy, J. H., and M. Salmon. 1991. Comparative studies of reproductive behavior in mantis shrimps and fiddler crabs. *Am. Zool.* 31:329–337.

Clark, D. R. 1971. The strategy of tail-autotomy in the ground skink, *Lygosoma laterale*. *J. Exp. Zool.* 176:295–302.

Clark, R. B. 1960. Habituation of the polychaete *Nereis* to sudden stimuli. 1. General properties of the habituation process. *Anim. Behav.* 8:82–91.

Clarke, A. L., B.-R. Sæther, and E. Røskaft. 1997. Sex biases in avian dispersal: A reappraisal. *Oikos* 79:429–438.

Clarke, A. M., and A. D. B. Clarke. 1976. *Early Experience: Myth and Evidence*. London: Open Books.

Clarke, B. 1969. The evidence for apostatic selection. *Heredity* 24:347–352.

Clayton, N. S. 1987. Song tutor choice in zebra finches. *Anim. Behav.* 35:714–721.

Clayton, N. S., and A. Dickinson. 1998. Episodic-like memory during cache recovery by scrub jays. *Nature* 395:272–274.

Clayton, N. S., and J. R. Krebs. 1994a. Memory for spatial and object-specific cues in food-storing and nonstoring birds. *J. Comp. Physiol. A* 174:371–379.

Clayton, N. S., and J. R. Krebs. 1994b. One-trial associative memory: Comparisons of food-storing and non-storing species of birds. *Anim. Learn. Behav.* 22:366–372.

Clutton-Brock, T. H. 1989a. Female transfer and inbreeding avoidance in social mammals. *Nature* 337:70–72.

Clutton-Brock, T. H. 1989b. Mammalian mating systems. *Proc. Roy. Soc. Lond. B* 236:339–372.

Clutton-Brock, T. H. 1991. *The Evolution of Parental Care*. Princeton, NJ: Princeton University Press.

Clutton-Brock, T. H., and S. D. Albon. 1979. The roaring of red deer and the evolution of honest advertisement. *Behaviour* 69:145–170.

Clutton-Brock, T. H., S. D. Albon, R. M. Gibson, and F. E. Guinness. 1979. The logical stag: Adaptive aspects of fighting in red deer (*Cervus elaphus* L.). *Anim. Behav.* 27:211–225.

Clutton-Brock, T., and C. Godfray. 1991. Parental investment. In *Behavioural Ecology: An Evolutionary Approach*, 3rd ed., edited by J. R. Krebs and N. B. Davies. Oxford: Blackwell Scientific, pp. 7–29.

Clutton-Brock, T. H., F. E. Guiness, and S. D. Albon. 1982. *Red Deer: The Behaviour and Ecology of Two Sexes*. Chicago: University of Chicago Press.

Clutton-Brock, T. H., and P. H. Harvey. 1979. Comparison and adaptation. *Proc. Roy. Soc. Lond. B* 205:547–565.

Clutton-Brock, T. H., and P. H. Harvey. 1984. Comparative approaches to studying adaptation. In *Behavioural Ecology: An Evolutionary Approach*, 2nd ed., edited by J. R. Krebs and N. B. Davies. Sunderland, MA: Sinauer, pp. 7–29.

Clutton-Brock, T. H., M. J. O'Riain, P. N. M. Brotherton, D. Gaynor, R. Kansky, A. S. Griffin, and M. Manser. 1999. Selfish sentinels in cooperative mammals. *Science* 284:1640–1644.

Cockburn, A. 1988. *Social Behaviour in Fluctuating Populations*. New York: Croom Helm.

Cockburn, A. 1998. Evolution of helping behavior in cooperatively breeding birds. *Ann. Rev. Ecol. Syst.* 29:141–177.

Collett, M., T. S. Collett, S. Bisch, and R. Wehner. 1998. Local and global vectors in desert ant navigation. *Nature* 394:269–272.

Consoulas, C., and R. B. Levine. 1998. Presynaptic function during muscle remodeling in insect metamorphosis. *J. Neurosci.* 18 (15):5817–5831.

Cooper, J. B. 1985. Comparative psychology and ethology. In *Topics in the History of Psychology*, edited by G. A. Kimble and K. Schlesinger. Hillsdale, NJ: Erlbaum, pp. 135–164.

Cooper, J., A. T. Scholz, R. M. Horral, A. D. Hasler, and D. M. Madison. 1976. Experimental confirmation of the olfactory hypothesis with homing, artificially imprinted coho salmon (*Oncorhynchus kisutch*). *J. Fish. Res. Board Can.* 28:703–710.

Cordoba-Aguilar, A. 1999. Male copulatory sensory stimulation induces female ejection of rival sperm in a damselfly. *Proc. Roy. Soc. Lond. B* 266 (1421):779–784.

Cott, H. B. 1940. *Adaptive Colouration in Animals*. London: Methuen.

Coulson, J. C. 1966. The influence of the pair bond and age on the breeding biology of the kittiwake gull (*Risa tridactyla*). *J. Anim. Biol.* 35:269–279.

Courtenay, S. C., T. P. Quinn, H. M. C. Dupuis, C. Groot, and P. A. Larkin. 1997. Factors affecting the recognition of population-specific odours by juvenile coho salmon. *J. Fish Biol.* 50:1042–1060.

Cowlishaw, G. 1997. Trade-offs between foraging and predation risk determine habitat use in a desert baboon population. *Anim. Behav.* 53 (4):667–686.

Cox, C. R., and B. J. LeBoeuf. 1977. Female incitation of male competition: A mechanism of mate selection. *Am. Nat.* 111:317–335.

Craig, C. L. 1994a. Limits to learning: effects of predator pattern and colour on perception and avoidance-learning by prey. *Anim. Behav.* 47:1087–1099.

Craig, C. L. 1994b. Predator foraging behavior in response to perception and learning by its prey: Interactions between orb-spinning spiders and stingless bees. *Behav. Ecol. Sociobiol.* 35 (1):45–52.

Craig, C. L., and G. D. Bernard. 1990. Insect attraction to ultra-violet reflecting spider webs and web decorations. *Ecology* 71:616–623.

Craig, C. L., R. S. Weber, and G. D. Bernard. 1996. Evolution of predator-prey systems: Spider foraging plasticity in response to the visual ecology of prey. *Am. Nat.* 147 (2):205–229.

Craig, W. 1918. Appetites and aversions as constituents of insects. *Biol. Bull.* 34:91–107.

Crane, J. 1941. Eastern Pacific expeditions of the New York Zoological Society XXVI: Crabs of the genus *Uca* from the West Coast of Central America. *Zoologica* 26:145–203.

Crawley, J. N. 1996. Unusual behavioral phenotypes of inbred mouse strains. *Trends Neurosci.* 19 (5):181–182.

Creel, S., N. M. Creel, M. G. L. Mills, and S. L. Monfort. 1997. Rank and reproduction in cooperatively breeding African wild dogs—Behavioral and endocrine correlates. *Behav. Ecol.* 8 (3):298–306.

Creel, S. R., and P. M. Waser. 1997. Variation in reproductive suppression among dwarf mongooses: Interplay between mechanisms and evolution. In *Cooperative Breeding in Mammals*, edited by N. G. Solomon and J. A. French. Cambridge: Cambridge University Press, pp. 150–169.

Cremer-Bartels, G., K. Krause, G. Mitoskas, and D. Broderson. 1984. Magnetic field of the earth as additional zeitgeber for endogenous rhythms? *Naturwissenschaften* 71:567–574.

Crespi, B. J. 1992. Eusociality in Australian gall thrips. *Nature* 359:724–726.

Crespi, B. J., and D. Yanega. 1995. The definition of eusociality. *Behav. Ecol.* 6 (1):109–115.

Crews, D. 1974. Effects of castration and subsequent androgen replacement therapy on male courtship behavior and environmentally induced ovarian recrudescence in the lizard, *Anolis carolinensis*. *J. Comp. Physiol.* 87:963–969.

Crews, D. 1979. Neuroendocrinology of lizard reproduction. *Biol. Reprod.* 20:51–73.

Crews, D. 1983. Control of male sexual behavior in the Canadian red-sided garter snake. In *Hormones and Behavior in Higher Animals*, edited by J. Balthazart, E. Prove, and R. Gilles. New York: Springer-Verlag, pp. 398–406.

Crews, D. 1984. Gamete production, sex hormone secretion, and mating behavior uncoupled. *Horm. Behav.* 18:22–28.

Crews, D. 1987. Diversity and evolution of behavioral controlling mechanisms. In *Psychobiology of Reproductive Behavior*, edited by D. Crews. Upper Saddle River, NJ: Prentice Hall, pp. 88–119.

Crews, D., V. Hingorani, and R. J. Nelson. 1988. Role of the pineal gland in the control of annual reproductive behavioral and physiological cycles in the red-sided garter snake (*Thamnophis sirtalis parietalis*). *J. Biol. Rhythms.* 3:293–302.

Crews, D., V. Tranina, F. T. Wetzel, and C. Muller. 1978. Hormone control of male reproductive behavior in the lizard *Anolis carolinensis*: Role of testosterone, dihydrotestosterone, and estradiol. *Endocrinology* 103:1814–1821.

Cristol, D. A. 1995. The coat-tail effect in merged flocks of dark-eyed juncos: Social status depends on familiarity. *Anim. Behav.* 50:151–159.

Cristol, D. A, and P. V. Switzer. 1999. Avian prey dropping behavior. II. American crows and walnuts. *Behav. Ecol.* 10 (8):220–226.

Crocker, G., and T. Day. 1987. An advantage to mate choice in the seaweed fly, *Coleopa frigida*. *Behav. Ecol. Sociobiol.* 20:295–301.

Crook, J. H. 1964. The evolution of social organisation and visual communication in the weaver bird (Ploceinae). *Behaviour Suppl.* 10:1–178.

Crook, J. H. 1970. The socio-ecology of primates. In *Social Behavior in Birds and Mammals*, edited by J. H. Crook. London: Academic Press, pp. 103–106.

Crowley, P. H., D. R. DeVries, and A. Sih. 1990. Inadvertent errors and error-constrained optimization: Fallible foraging by the blue-gill sunfish. *Behav. Ecol. Sociobiol.* 27:135–144.

Croze, H. 1979. Searching image in carrion crows. *Z. Tierpsychol. Suppl.* 5:1–86.

Cullen, E. 1957a. Adaptations in the kittiwake in the British Isles. *Bird Study* 10:147–179.

Cullen, E. 1957b. Adaptations in the kittiwake to cliff nesting. *Ibis* 99:275–302.

Cullen, J. M. 1966. The ritualization of animal activities in relation to phylogeny, speciation and ecology: Reduction of ambiguity through ritualization. *Phil. Trans. Roy. Soc. Lond.* B 241:363–374.

Cullen, J. M. 1972. Some principles of animal communication. In *Non-verbal Communication*, edited by R. A. Hinde. Cambridge: Cambridge University Press, pp. 101–122.

Curio, E. 1978. The adaptive significance of avian mobbing. I. Teleonomic hypotheses and predictions. *Ethology* 48:175–183.

Curio, E., and K. Regelmann. 1986. Predator harassment implies a deadly risk: A reply to Hennessy. *Ethology* 72:75–78.

Daan, S., and J. Aschoff. 1982. Circadian contributions to survival. In *Vertebrate Circadian Systems, Structure and Function*, edited by J. Aschoff, S. Daan, and G. A. Groos. Berlin: Springer-Verlag, pp. 305–321.

Daanje, A. 1950. On locomotor movements in birds and the intention movement derived from them. *Behaviour* 3:48–98.

Daly, M. 1979. Why don't male mammals lactate? *J. Theoret. Biol.* 78:325–345.

Daly, M., and M. Wilson. 1983. *Sex, Evolution and Behavior.* 2nd ed. Boston: PWS-Kent.

D'Arcangelo, G., G. G. Miao, S.-C. Chen, H. D. Soares, J. I. Morgan, and T. Curran. 1995. A protein related to extracellular matrix proteins deleted in mouse mutant *reeler*. *Nature* 374:719–723.

Darlington, T. K., K. Wager-Smith, M. F. Ceriani, D. Staknis, N. Gekakis, T. D. L. Steeves, C. J. Weitz, J. S.

Takahashi, and S. A. Kay. 1998. Closing the circadian loop: CLOCK-induced transcription of its own inhibitors per and tim. *Science* 280 (5369):1599–1603.

Darwin, C. 1859. *On the Origin of Species by Natural Selection, or The Preservation of Favored Races in the Struggle for Life*. London: Murray.

Darwin, C. 1871. *The Descent of Man and Selection in Relation to Sex*. London: Murray.

Darwin, C. 1873. *Expression of the Emotions in Man and Animals*. London: Murray.

Dash, P. K., B. Hochner, and E. R. Kandel. 1990. Injection of the cAMP-responsive element into the nucleus of *Aplysia* sensory neurons blocks long-term facilitation. *Nature* 345:718–721.

Davidson, J. M., C. A. Camargo, and E. R. Smith. 1979. Effects of androgen on sexual behavior in hypogonadal men. *J. Clin. Endocrinol. Metab.* 48:955–958.

Davies, N. B. 1983. Polyandry, cloaca-pecking and sperm competition in dunnocks. *Nature* 302:334–336.

Davies, N. B. 1991. Mating systems. In *Behavioural Ecology: An Evolutionary Approach*, edited by J. R. Krebs and N. B. Davies. Oxford: Blackwell Scientific, pp. 263–294.

Davies, N. B., and M. de L. Brooke. 1988. Cuckoo versus reed warblers: Adaptations and counter adaptations. *Anim. Behav.* 36:262–284.

Davies, N. B., and T. R. Halliday. 1978. Deep croaks and fighting assessment in toads (*Bufo bufo*). *Nature* 274:683–685.

Davies, N. B., and A. I. Houston. 1984. Territory economics. In *Behavioural Ecology: An Evolutionary Approach*, edited by J. R. Krebs and N. B. Davies. Sunderland, MA: Sinauer, pp. 148–169.

Davis, R. L., J. Cherry, B. Dauwalder, P.-L. Han, and E. Skoulakis. 1995. The cyclic AMP system and *Drosophila* learning. *Mol. Cell. Biochem.* 149:271–278.

Dawkins, M. 1971. Perceptual changes in chicks: Another look at the "search image" concept. *Anim. Behav.* 19:566–574.

Dawkins, M. S. 1986. *Unravelling Animal Behavior*. Essex: Longman.

Dawkins, M. S. 1989. The future of ethology: How many legs are we standing on? In *Perspectives in Ethology*, edited by P. P. G. Bateson and P. H. Klopfer. New York: Plenum, pp. 47–54.

Dawkins, M. S., and T. Guilford. 1991. The corruption of honest signaling. *Anim. Behav.* 41:865–873.

Dawkins, R. 1976. *The Selfish Gene*. New York: Oxford University Press.

Dawkins, R. 1980. Good strategy or evolutionarily stable strategy? In *Sociobiology: Beyond Nature/Nurture*, edited by G. Barlow and R. Silverberg. Boulder, CO: Westview, pp. 331–367.

Dawkins, R. 1982. *The Extended Phenotype*. Oxford: Freeman.

Dawkins, R. 1989. *The Selfish Gene*. 2nd ed. New York: Oxford University Press.

Dawkins, R., and T. R. Carlisle. 1976. Parental investment, mate desertion and a fallacy. *Nature* 262:131–133.

Dawkins, R., and J. R. Krebs. 1978. Animal signals: Information or manipulation? In *Behavioural Ecology: An Evolutionary Approach*, edited by J. R. Krebs and N. B. Davies. Sunderland, MA: Sinauer, pp. 282–309.

Dawkins, R., and J. R. Krebs. 1984. Animal signals: Mind-reading and manipulation? In *Behavioural Ecology: An Evolutionary Approach*, edited by J. R. Krebs and N. B. Davies. Sunderland, MA: Sinauer, pp. 380–402.

Dawson, B. V., and B. M. Foss. 1965. Observational learning in budgerigars. *Anim. Behav.* 13:470–474.

Dean, J., D. J. Aneshansley, H. E. Edgerton, and T. Eisner. 1990. Defensive spray of the bombardier beetle: A biological pulse jet. *Science* 248:1219–1221.

de Belle, J. S., A. J. Hilliker, and M. B. Sokolowski. 1989. Genetic localization of *foraging* (*for*): A major gene for larval behavior in *Drosophila melanogaster*. *Genetics* 123:157–163.

de Belle, J. S., and M. B. Sokolowski. 1987. Heredity of *rover/sitter*: Alternative foraging strategies of *Drosophila melanogaster* larvae. *Heredity* 59:73–83.

DeCoursey, P. J., and J. R. Krulas. 1998. Behavior of SCN-lesioned chipmunks in natural habitat: A pilot study. *J. Biol. Rhythms.* 13 (3):229–244.

DeCoursey, P. J., J. R. Krulas, G. Mele, and D. C. Holley. 1997. Circadian performance of suprachiasmatic nuclei (SCN)-lesioned antelope ground squirrels in a desert enclosure. *Physiol. Behav.* 625:1099–1108.

DeFries, J. C., and G. E. McClearn. 1970. Social dominance and Darwinian fitness in the laboratory mouse. *Am. Nat.* 104 (938):408–411.

Deguchi, T. 1979. Circadian rhythm of serotonin N-acetyltransferase activity in organ culture. *Science* 203:1245–1247.

Dehn, M. M. 1990. Vigilance for predators: Detection and dilution effects. *Behav. Ecol. Sociobiol.* 26:337–342.

de la Motte, I., and D. Burkhardt. 1983. Portrait of an Asian stalk-eyed fly. *Naturwissenschaften* 70:451–461.

Delcomyn, F. 1980. Neural basis of rhythmic behavior in animals. *Science* 210:492–498.

Demaine, C., and P. Semm. 1985. The avian pineal gland as an independent magnetic sensor. *Neurosci. Lett.* 62:119–122.

DeNault, L. K., and D. A. McFarlane. 1995. Reciprocal altruism between male vampire bats, Desmodus rotundus. *Anim. Behav.* 49 (3):855–856.

Devenport, L., J. Devenport, and C. Kokesh. 1999. The role of urine marking in the foraging behavior of least chipmunks. *Anim. Behav.* 57 (3):557–563.

Devine, M. C. 1977. Copulatory plugs, restricted mating opportunities and reproductive competition among male garter snakes. *Nature* 267:345–346.

DeVoogd, T. J. 1990. Recent findings on the development of dimorphic anatomy in the avian song system. *J. Exp. Zool. Suppl.* 4:183–186.

DeVries, D. R., R. A. Stein, and P. L. Chesson. 1989. Sunfish foraging among patches: The patch departure decision. *Anim. Behav.* 37:455–467.

de Waal, F. B. M. 1978. Exploitative and familiarity-dependent support strategies in a colony of semi-free living chimpanzees. *Behaviour* 66:268–312.

de Waal, F. B. M. 1991. Rank distance as a central feature of rhesus monkey social organization: A sociometric analysis. *Anim. Behav.* 41 (3):383–395.

Dewsbury, D. A. 1978. What is (was?) the "Fixed Action Pattern"? *Anim. Behav.* 26:310–311.

Dewsbury, D. A. 1982a. Dominance rank, copulatory behavior, and differential reproduction. *Quart. Rev. Biol.* 57:135–158.

Dewsbury, D. A. 1982b. Ejaculate cost and male choice. *Am. Nat.* 119:601–610.

Dewsbury, D. A. 1984a. *Comparative Psychology in the Twentieth Century.* Stroudsburg, PA: Hutchison Ross.

Dewsbury, D. A. 1984b. Sperm competition in muroid rodents. In *Sperm Competition and the Evolution of Animal Mating Systems*, edited by R. L. Smith. Orlando, FL: Academic Press, pp. 547–571.

Dewsbury, D. A. 1988. A test of the role of copulatory plugs in sperm competition in deer mice (*Peromyscus maniculatus*). *J. Mamm.* 69:854–857.

Dewsbury, D. A. 1989. A brief history in the study of animal behavior in North America. In *Perspectives in Ethology* Vol. 8, (*Whither ethology?*) edited by P. P. G. Klopfer and P. H. Klopfer. New York: Plenum, pp. 85–122.

Dial, B. E., and L. C. Fitzpatrick. 1983. Lizard tail autotomy: Function and energetics of postautotomy tail movement in *Scinella lateralis*. *Science* 219:391–393.

Diesel, R. 1990. Sperm competition and reproductive success in the decapod *Inachus phalangium* (Majiidae): A male ghost spider crab that seals off rival's sperm. *J. Zool. Lond.* 220:213–223.

Dill, P. A. 1971. Perception of polarized light by yearling salmon (*Oncorhynchus kisutch*). *Behav. Ecol. Sociobiol.* 16:65–71.

Dill, P. A. 1977. Development of behavior in alevins of Atlantic salmon, *Salmo salar*, and rainbow trout, *S. gairdneri. Anim. Behav.* 25:116–121.

Dill, P. A., and A. H. G. Fraser. 1984. Risk of predation and the feeding behavior of juvenile coho salmon (*Oncorhynchus kisutch*). *Behav. Ecol. Sociobiol.* 16:65–71.

Dingle, H. 1980. Ecology and evolution of migration. In *Animal Migration, Orientation, and Navigation*, edited by S. A. Gauthreaux. New York: Academic Press, pp. 1–101.

Dingle, H., and R. Caldwell. 1969. The aggressive and territorial behavior of the mantis shrimp *Gonodactylus bredini* Manning (Crustacea: Stomatopoda). *Behaviour* 33:115–136.

Dittman, A. H., and T. P. Quinn. 1996. Homing in Pacific salmon: Mechanisms and ecological basis. *J. Exp. Biol.* 199 (1):83–91.

Dittman, A. H., T. P. Quinn, and G. A. Nevitt. 1996. Timing of imprinting to natural and artificial odors by coho salmon (*Oncorhynchus kisutch*). *Can. J. Fish. Aquat. Sci.* 53:434–442.

Dittrich, G., F. Gilbert, P. Green, P. McGregor, and D. Grewcock. 1993. Imperfect mimicry—a pigeon's perspective. *Proc. Roy. Soc. Lond. B* 251 (1332):195.

Dixson, A. F. 1998. *Primate Sexuality: Comparative Studies of Prosimians, Monkeys, Apes, and Human Beings.* Oxford: Oxford University Press.

Dobson, F. S. 1982. Competition for mates and predominant male dispersal in mammals. *Anim. Behav.* 30:1183–1192.

Dobson, F. S., and W. T. Jones. 1985. Multiple causes of dispersal. *Am. Nat.* 126:855–858.

Dole, J. W., and P. Durant. 1974. Movements and seasonal activity of *Atelopus oxyrhynchus* (Anura: Atelopodidae) in a Venezuelan cloud forest. *Copeia* 1974:230–235.

Dorries, K. M., E. Adkins-Regan, and B. P. Halpern. 1995. Olfactory sensitivity to the pheromone androsterone is sexually dimorphic in the pig. *Physiol. Behav.* 57:255–259.

Downhower, J. F., and K. B. Armitage. 1971. The yellow-bellied marmot and the evolution of polygyny. *Am. Nat.* 105:355–370.

Drummond, H., and C. Canales. 1998. Dominance between booby nestlings involves winner and loser effects. *Anim. Behav.* 55 (6):1669–1676.

Dudai, Y. 1979. Behavioral plasticity in a *Drosophila* mutant, dunce DB276. *J. Comp. Physiol.* 130:271–275.

Dudai, Y., Y.-N. Jan, D. Byers, W. G. Quinn, and S. Benzer. 1976. *Dunce,* a mutant of *Drosophila* deficient in learning. *Proc. Nat. Acad. Sci. USA* 73:1684–1688.

Dufty, A. M. 1989. Testosterone and survival: A cost of aggressiveness? *Horm. Behav.* 23:185–193.

Dugatkin, L. A. 1992. Sexual selection and imitation: Females copy the mate choice of others. *Am. Nat.* 139 (6):1384–1389.

Dugatkin, L. A. 1996. Interface between culturally based preferences and genetic preferences: Female mate choice in *Poecilia reticulata. Proc. Nat. Acad. Sci. USA* 93 (7):2770–2773.

Dugatkin, L. A. 1997. *Cooperation Among Animals: An Evolutionary Perspective.* New York: Oxford University Press.

Dugatkin, L. A. 1998. Genes, copying, and female mate choice: Shifting threshold. *Behav. Ecol.* 9 (4):323–327.

Dugatkin, L. A., and J.-G. J. Godin. 1992. Reversal of female mate choice by copying in the guppy (*Poecilia reticulata*). *Proc. Roy. Soc. Lond. B* 249:179–184.

Dugatkin, L. A., and J.-G. J. Godin. 1993. Female mate copying in the guppy (*Poecilia reticulata*)—Age dependent effects. *Behav. Ecol.* 4 (4):289–292.

Dugatkin, L. A., and J.-G. J. Godin. 1998. How females choose their mates. *Sci. Am.* 278 (Apr.):56–61.

Dunbrack, R. I., and L. M. Dill. 1983. A model of size dependent surface feeding in a stream dwelling salmonid. *Environ. Biol. Fishes* 8:203–216.

Dunford, C. 1977. Kin selection for ground squirrel alarm calls. *Am. Nat.* 111:782–785.

Dunham, P. J. 1986. Mate guarding in amphipods: A role for brood patch stimuli. *Biol. Bull.* 170:526–531.

Dunham, P. J., and A. Hurshman. 1990. Precopulatory mate guarding in the amphipod, *Gammarus lawrencianus*: Effects of social stimulation during the post-copulation interval. *Anim. Behav.* 39:976–979.

Dunlap, J. C. 1999. Molecular bases for circadian clocks. *Cell* 96:271–290.

Dunlap, K. D., P. Thomas, and H. H. Zakon. 1998. Diversity of sexual dimorphism in electrocommunication signals and its androgen regulation in a genus of electric fish, *Apteronotus*. *J. Comp. Physiol. A* 183 (1):77–86.

Dyer, A. B., R. Lickliter, and G. Gottlieb. 1989. Maternal peer imprinting in mallard ducklings under experimentally simulated natural social conditions. *Dev. Psychobiol.* 22:463–475.

Dyer, B. D. 1998. A hypothesis about the significance of symbionts as a source of protein in the evolution of eusociality in naked mole rats. *Symbiosis* 24 (3):369–383.

Dyer, F. C. 1991. Bees acquire route-based memories but not cognitive maps in a familiar landscape. *Anim. Behav.* 41:239–246.

Dyer, F. C., and T. D. Seeley. 1991. Dance dialects and foraging range in three Asian honey bee species. *Behav. Ecol. Sociobiol.* 28 (4):227–234.

Eadie, J. M., F. P. Kehoe, and T. D. Nudds. 1988. Pre-hatch and post-hatch brood amalgamation in North American Anatidae: A review of hypotheses. *Can. J. Zool.* 66:1709–1721.

Eales, L. A. 1985. Song learning in zebra finches: Some effects of song model availability on what is learnt and when. *Anim. Behav.* 33:1293–1300.

Earle, S. A. 1979. The gentle whales. *Nat. Geogr. Mag.* 155 (1):2–17.

Eberhard, W. G. 1980. Horned beetles. *Sci. Am.* 242 (Mar.):166–182.

Eberhard, W. G. 1996. *Female Control: Sexual Selection by Cryptic Female Choice.* Princeton, NJ: Princeton University Press.

Ebihara, S., and H. Kawamura. 1981. The role of the pineal organ and the suprachiasmatic nucleus in the control of the circadian locomotor rhythms in the Java sparrow, *Paddua oryzivora*. *J. Comp. Physiol.* 141:207–214.

Ebihara, S., K. Uchiyama, and I. Oshima. 1984. Circadian organization in the pigeon, *Columba livia*, the role of the pineal and the eye. *J. Comp. Physiol. A* 154:59–69.

Eckert, C. G., and P. J. Weatherhead. 1987. Male characteristics, parental care and the study of mate choice in the red-winged blackbird (*Agelaius phoeniceus*). *Behav. Ecol. Sociobiol.* 20:35–42.

Edmunds, M. 1974. *Defence in Animals.* New York: Longman.

Edney, E. B. 1954. Woodlice and the land habitat. *Biol. Rev.* 29:185–219.

Eggert, A.-K., and S. K. Sakaluk. 1994. Sexual cannibalism and its relation to male mating success in sagebrush crickets, *Cyphoderris strepitans* (Haglidae: Orthoptera). *Anim. Behav.* 47 (5):1171–1177.

Eibl-Eibesfeldt, I. 1966. Das Vertridigen der Eiablageplätze bei der Hood-Meerechse (*Amblyrhynchus cristatus venustissimus*). *Z. Tierpsychol.* 23:627–631.

Eibl-Eibesfeldt, I. 1975. *Ethology: The Biology of Behavior.* 2nd ed. New York: Holt, Rinehart & Winston.

Eisenberg, J. F., N. A. Muckenhirn, and R. Rudran. 1972. The relation between ecology and social structure in primates. *Science* 176:863–874.

Eisner, T. 1958. The protective role of the spray mechanism of the bombardier beetle, *Brachynus ballistarius* Lec. *J. Insect Physiol.* 2:215–220.

Eisner, T., W. E. Conner, K. Hicks, K. R. Dodge, H. I. Rosenberg, T. H. Jones, M. Cohen, and J. Meinwald. 1977. Stink of the stinkpot turtle identified: W-phenylalkanoic acids. *Science* 196:1347–1349.

Eisner, T., and J. Meinwald. 1995. The chemistry of sexual selection. *Proc. Nat. Acad. Sci. USA* 92:50–55.

Elliot, P. F. 1975. Longevity and the evolution of polygamy. *Am. Nat.* 109:281–287.

Elwood, R. W., and H. F. Kennedy. 1990. The relationship between infanticide and pregnancy block in mice. *Behav. Neur. Biol.* 53:277–283.

Emlen, J. M. 1973. *Ecology: An Evolutionary Approach.* Reading, MA: Addison-Wesley.

Emlen, S. T. 1967a. Migratory orientation in the indigo bunting (*Passerina cyanea*). Part I. Evidence for the use of celestial cues. *Auk* 84:309–342.

Emlen, S. T. 1967b. Migratory orientation in the indigo bunting (*Passerina cyanea*). Part II. Mechanisms of celestial orientation. *Auk* 84:463–489.

Emlen, S. T. 1969. The development of migratory orientation in young indigo buntings. *Living Bird* 8:113–126.

Emlen, S. T. 1970. Celestial rotation: Its importance in the development of migratory orientation. *Science* 170:1198–1201.

Emlen, S. T. 1972. The ontogenetic development of orientation capabilities. *NASA Spec. Publ.* 262:191–210.

Emlen, S. T. 1975. Migration, orientation, navigation. In *Avian Biology*, edited by D. Farner. New York: Academic Press, pp. 129–219.

Emlen, S. T. 1982a. The evolution of helping. I. An ecological constraints model. *Am. Nat.* 119:29–39.

Emlen, S. T. 1982b. The evolution of helping. II. The role of behavioral conflict. *Am. Nat.* 119:40–53.

Emlen, S. T. 1991. Evolution of cooperative breeding in birds and mammals. In *Behavioural Ecology: An Evolutionary Approach*, edited by J. R. Krebs and N. B. Davies. Oxford: Blackwell Scientific. Publications, pp. 301–337.

Emlen, S. T. 1997. Predicting family dynamics in social vertebrates. In *Behavioural Ecology: An Evolutionary Approach*, 4th ed., edited by J. R. Krebs and N. B. Davies. Oxford: Blackwell Science, pp. 228–253.

Emlen, S. T., N. J. Demong, and D. J. Emlen. 1989. Experimental induction of infanticide in female wattled jacanas. *Auk* 106:1–7.

Emlen, S. T., and L. W. Oring. 1977. Ecology, sexual selection, and the evolution of mating systems. *Science* 197:215–223.

Emlen, S. T., H. K. Reeve, P. W. Sherman, P. H. Wrege, F. L. W. Ratnieks, and J. Shellman-Reeve. 1991. Adaptive versus nonadaptive explanations of behavior: The case of alloparental helping. *Am. Nat.* 138 (1):259–261, 270.

Emlen, S. T., and S. L. Vehrencamp. 1983. Cooperative breeding strategies among birds. In *Perspectives in Ornithology*, edited by A. H. Brush and G. A. Clark, Jr. Cambridge: Cambridge University Press, pp. 93–133.

Emlen, S. T., and P. H. Wrege. 1989. A test of alternative hypotheses for helping behavior in white-fronted bee-eaters of Kenya. *Behav. Ecol. Sociobiol.* 25:303–319.

Endler, J. A. 1978. A predator's view of animal color patterns. *Evol. Biol.* 11:319–364.

Endler, J. A., and A. E. Houde. 1995. Geographic variation in female preferences for male traits in *Poecilia reticulata*. *Evolution* 49:456–468.

Enquist, M., and O. Leimar. 1990. The evolution of fatal fighting. *Anim. Behav.* 39:1–9.

Enright, J. T. 1970. Ecological aspects of endogenous rhythmicity. *Ann. Rev. Ecol. Syst.* 1:221–238.

Enright, J. T. 1972. A virtuoso isopod. Circa-lunar rhythms and their tidal fine structure. *J. Comp. Physiol.* 77:141–162.

Epstein, R., C. E. Kirshnit, R. P. Lanza, and L. C. Rubin. 1984. "Insight" in the pigeon: Antecedents and determinants of intelligent performance. *Nature* 308:61–62.

Esch, H. E., and J. E. Burns. 1996. Distance estimation by foraging honeybees. *J. Exp. Biol.* 199 (1):155–162.

Estep, D. Q. 1988. Copulations by other males shorten the post-ejaculatory intervals of pairs of roof rats, *Rattus rattus. Anim. Behav.* 36:299–300.

Estep, D. Q., K. Nieuwenhuijsen, K. E. M. Bruce, K. J. De Neef, P. A. Walters, S. C. Baker, and A. K. Slob. 1988. Inhibition of sexual behavior among subordinate stump-tail macaques, *Macaca arctoides. Anim. Behav.* 36:854–864.

Etkin, W. 1964. *Social Behavior and Organization Among Vertebrates*. Chicago: University of Chicago Press.

Ewer, J., S. C. Gammie, and J. W. Truman. 1997. Control of insect ecdysis by a positive feedback endocrine system: Roles of eclosion hormone and ecdysis triggering hormone. *J. Exp. Biol.* 200:869–881.

Ewer, J., M. Rosbash, and J. C. Hall. 1988. An inducible promoter to the period gene in *Drosophila* conditionally rescues adult *per*-mutant arrhythmicity. *Nature* 333 (6168):82–84.

Ewert, J.-P. 1967. Aktivierung der verhaltensfolge beim beutefang der erdkröte (*Bufo bufo* L.) durch elektrische mittelhirm-reizung. *Z. vergl. Physiol.* 54:455–481.

Ewert, J.-P. 1969. Quantitative analyze von Reiz-Reaktionsbeziehungen bei visuellen Auslosen der Beutefang-Wendereaktion der Erdkröte (*Bufo bufo* L.). *Pflugers Arch.* 308:225–243.

Ewert, J.-P. 1983. Neuroethological analysis of the innate relasing mechanism for prey-catching behavior in toads. In *Advances in Vertebrate Neuroethology*, edited by J.-P. Ewert, R. R. Capranica, and D. J. Ingle. New York: Plenum, pp. 701–730.

Ewert, J.-P. 1987. Neuroethology of releasing mechanisms: Prey catching in toads. *Behav. Brain Sci.* 10:337–405.

Ewert, J.-P., E. M. Framing, E. Schurg-Pfeifer, and A. Weerasuriya. 1990. Responses of medullary neurons to moving visual stimuli in the common toad. I. Characterization of medial reticular neurons by extracellular recording. *J. Comp. Physiol. A* 167:495–508.

Fagan, R. 1981. *Animal Play Behavior*. New York: Oxford University Press.

Fahrbach, S. E. 1997. Regulation of age polyethism in bees and wasps by juvenile hormones. In *Advances in the Study of Behavior*. New York: Academic Press, pp. 285–316.

Fairbanks, L. A. 1990. Reciprocal benefits of allomothering for female vervet monkeys. *Anim. Behav.* 40:553–562.

Faulkes, C. G., and D. H. Abbott. 1997. The physiology of a reproductive dictatorship: Regulation of male and female reproduction by a single breeding female in colonies of naked mole-rats. In *Cooperative Breeding in Mammals*, edited by N. G. Solomon and J. A. French. Cambridge: Cambridge University Press, pp. 302–334.

Faulkes, C. G., D. H. Abbott, and A. L. Mellnor. 1990. Investigation of genetic diversity in wild colonies of naked mole-rats (*Heterocephalus glaber*) by DNA fingerprinting. *J. Zool. Lond.* 221:87–97.

Feltmate, B. W., and D. Dudley Williams. 1989. A test of crypsis and predator avoidance in the stonefly *Paragnetina media* (Plecooptera: Perlidae). *Anim. Behav.* 37:992–999.

Fenton, M. B. 1983. *Just Bats*. Toronto: University of Toronto Press.

Fenton, M. B. 1992. *Bats*. New York: Facts on File.

Ferkin, M. H., E. S. Sorokin, M. W. Renfroe, and R. E. Johnston. 1994. Attractiveness of male odors to females varies directly with plasma testosterone concentration in meadow voles. *Physiol. Behav.* 55 (2):347–353.

Ferkin, M. H., and I. Zucker. 1991. Seasonal control of odor preferences of meadow voles (*Microtus pennsylvanicus*) by photoperiod and ovarian hormones. *J. Reprod. Fert.* 92:433–441.

Ferster, C. B., and B. F. Skinner. 1957. *Schedules of Reinforcement*. New York: Appleton-Century-Crofts.

Ferveur, J.-F., K. F. Störtkuhl, R. F. Stocker, and R. J. Greenspan. 1995. Genetic feminization of brain structures and changed sexual orientation in male *Drosophila. Science* 267:902–904.

Fields, R. D., and K. Itoh. 1996. Neural cell adhesion molecules in activity-dependent development and synaptic plasticity. *Trends Neurosci.* 19 (11):473–480.

Fink, L. S., and L. P. Brower. 1981. Birds can overcome the cardenolide defence of monarch butterflies in Mexico. *Nature* 291:67–70.

Finley, J., D. Ireton, W. M. Schleidt, and T. A. Thompson. 1983. A new look at the features of mallard courtship displays. *Anim. Behav.* 31:348–354.

Fisher, A. C., Jr. 1979. The mysteries of animal migration. *Nat. Geogr. Mag.* 156 (2):154–193.

Fisher, A. E. 1962. Effects of stimulus variation on sexual satiation in the male rat. *J. Comp. Physiol. Psychol.* 55:614–620.

Fisher, J., and R. A. Hinde. 1949. The opening of milk bottles by birds. *Brit. Birds* 42:347–357.

Fisher, R. A. 1930. *The Genetical Theory of Natural Selection.* Oxford: Oxford University Press.

FitzGerald, R. W., and D. M. Madison. 1983. Social organization of a free-ranging population of pine voles, *Microtus pinetorium. Behav. Ecol. Sociobiol.* 13:183–187.

FitzGibbon, C. D. 1990. Mixed-species grouping in Thomson's and Grant's gazelles: The antipredator benefits. *Anim. Behav.* 39:1116–1126.

FitzGibbon, C. D., and J. H. Fanshawe. 1988. Stotting in Thomson's gazelles: An honest signal of condition. *Behav. Ecol. Sociobiol.* 23:69–74.

Fleishman, L. J. 1988. Sensory influences on physical design of a visual display. *Anim. Behav.* 36:1420–1424.

Flinn, M. V., and B. S. Low. 1986. Resource distribution, social competition, and mating patterns in human societies. In *Ecological Aspects of Social Evolution*, edited by D. I. Rubenstein and R. W. Wrangham. Princeton, NJ: Princeton University Press, pp. 217–243.

Forrester, G. E. 1991. Social rank, individual size, and group composition as determinants of food composition by humbug damselfish, *Dascyllus aruanus. Anim. Behav.* 42:701–711.

Foster, M. S. 1977. Odd couples in manakins: A study of social organization and cooperative breeding in *Chiroxiphia linearis. Am. Nat.* 111:845–853.

Foster, W. A. 1990. Experimental evidence for effective and altruistic colony defense against natural predators by soldiers of the gall-forming aphid *Pemphigus spyrothecae* (Hemiptera: Pemphigidae). *Behav. Ecol. Sociobiol.* 27 (6):421–430.

Foulkes, N. S., J. Borjigin, S. H. Snyder, and P. Sassone-Corsi. 1997. Rhythmic transcription: The molecular basis of circadian melatonin synthesis. *Trends Neurosci.* 20 (10):487–492.

Fournier, F., and M. Festa-Bianchet. 1995. Social dominance in adult female mountain goats. *Anim. Behav.* 49:1449–1459.

Fouts, R. S. 1973. Acquisition and testing of gestural signs in four young chimpanzees. *Science* 180:978–980.

Fouts, R. S. 1974. Language: Origins, definition and chimpanzees. *J. Human Evol.* 3:475–482.

Fouts, R., and S. T. Mills. 1997. *Next of Kin: What Chimpanzees Have Taught Me About Who We Are.* New York: Morrow.

Fox, S. F., and M. A. Rostker. 1982. Social cost of tail loss in *Uta stansburiana. Science* 218:692–693.

Fraenkel, G. S., and D. L. Gunn. 1961. *The Orientation of Animals.* New York: Dover.

Frank, L. G., S. E. Glickman, and P. Lichte. 1991. Fatal sibling aggression, precocial development, and androgens in neonatal spotted hyaenas. *Science* 252:702–704.

Frank, L. G., H. E. Holekamp, and L. Smale. 1995. Dominance, demographics and reproductive success in female spotted hyaenas: A long-term study. In *Serengeti II: Dynamics, Management, and Conservation of an Ecosystem*, edited by A. R. E. Sinclair and P. Arcese. Chicago: University of Chicago Press, pp. 364–384.

Frankel, R. B., and R. P. Blakemore. 1980. Navigational compass in magnetotactic bacteria. *J. Magn. Mater.* 18:1562–1564.

Frankel, R. B., R. P. Blakemore, and R. S. Wolf. 1979. Magnetite in magnetotactic bacteria. *Science* 203: 1355.

Freedman, M. S., R. J. Lucas, B. Soni, M. von Schantz, M. Munoz, Z. David-Gray, and R. Foster. 1999. Regulation of circadian behavior by non-rod, non-cone, ocular photoreceptors. *Science* 284 (5413):502–504.

Frei, U. 1982. Homing pigeons behavior in the irregular magnetic field of western Switzerland. In *Avian Navigation*, edited by F. Papi and H. G. Wallraff. Berlin: Springer-Verlag, pp. 129–139.

Frei, U., and G. Wagner. 1976. Die Anfangsorientierung von Brieftauben im erdmagnetisch gestörten Gebiet des Mont Jorat. *Rev. Suisse Zool.* 83:891–897.

Frey-Roos, F., P. A. Brodmann, and H. U. Reyer. 1995. Relationships between food resources, foraging patterns, and reproductive success in the water pipit (*Anthus sp. spinoletta*). *Behav. Ecol.* 6 (3):287–295.

Fryxell, J. M., J. Greever, and A. R. E. Sinclair. 1988. Why are migratory ungulates so abundant? *Am. Nat.* 131:781–798.

Fullard, J. H., E. Forrest, and A. Surlykke. 1998. Intensity responses of the single auditory receptor of notodontid moths: A test of the peripheral interaction hypothesis in moth ears. *J. Exp. Biol.* 201 (24):3419–3424.

Gadagar, R. 1990a. Evolution of eusociality: The advantage of assured fitness returns. *Phil. Trans. Roy. Soc. Lond. B.* 329:17–25.

Gadagar, R. 1990b. Origin and evolution of eusociality: A perspective from studying eusocial wasps. *J. Genet.* 69:113–125.

Galef, B. G., Jr. 1976. Social transmission of acquired behavior: A discussion of tradition and social learning in vertebrates. In *Advances in the Study of Behavior*, edited by J. S. Rosenblatt, R. A. Hinde, E. Shaw, and C. Beer. New York: Academic Press, pp. 77–100.

Galef, B. G., Jr. 1988. Imitation in animals: History, definition, and interpretation of data from the psychological laboratory. In *Social Learning: Psychological and Biological Perspectives*, edited by T. R. Zentall and B. G. J. Galef. Hillsdale, NJ: Erlbaum, pp. 3–28.

Galef, B. G., Jr. 1990. An adaptationist perspective on social learning, social feeding, and social foraging in Norway rats. In *Contemporary Issues in Comparative Psychology,* edited by D. A. Dewsbury. Sunderland, MA: Sinauer, pp. 55–79.

Galef, B. G., Jr. 1993. Functions of social learning about food: A causal analysis of effects of diet novelty on preference transmission. *Anim. Behav.* 46:257–265.

Galef, B. G., Jr. 1995. Why behaviour patterns that animals learn socially are locally adaptive. *Anim. Behav.* 49:1325–1334.

Galef, B. G., Jr. 1996. The adaptive value of social learning; a reply to Laland. *Anim. Behav.* 52:641–644.

Galef, B. G., Jr., L. A. Manzig, and R. M. Field. 1986. Imitation learning in budgerigars: Dawson and Foss (1965) revisited. *Behav. Process* 13:191–202.

Galef, B. G., Jr., and T. J. Wright. 1995. Groups of naive rats learn to select nutritionally adequate foods faster than do isolated naive rats. *Anim. Behav.* 49 (2):403–409.

Gamboa, G. J. 1978. Intraspecific defense: Advantage of social cooperation among paper wasp foundresses. *Science* 199:1463–1465.

Gammie, S. C., and J. W. Truman. 1997. Neuropeptide hierarchies and the activation of sequential motor behaviors in the hawkmoth, *Manduca sexta*. *J. Neurosci.* 17 (11):4389–4397.

Gamow, R. I., and J. F. Harris. 1973. The infrared receptors of snakes. *Sci. Am.* 228 (May):94–100.

Garcia, J., and R. A. Koelling. 1966. Relation of cue to consequence in avoidance learning. *Psychol. Sci.* 4:123–124.

Gardner, R. A., and B. T. Gardner. 1969. Teaching sign language to a chimpanzee. *Science* 165:664–672.

Gardner, R. A., and B. T. Gardner. 1989. A cross-fostering laboratory. In *Teaching Sign Language to Chimpanzees*, edited by R. A. Gardner, B. T. Gardner, and T. E. Van Cantfort. Albany: State University of New York Press, pp. 1–28.

Gaston, S., and M. Menaker. 1968. Pineal function. The biological clock in the sparrow. *Science* 160:1125–1127.

Geist, V. 1971. *Mountain Sheep*. Chicago: University of Chicago Press.

Gekakis, N., and C. J. Witz. 1998. Role of the CLOCK protein in the mammalian circadian mechanism. *Science* 280 (5369):1564–1569.

Gendron, R. P. 1986. Searching for cryptic prey: Evidence for optimal search rates and the formation of search images in quail. *Anim. Behav.* 34:898–912.

Gendron, R. P., and J. E. R. Staddon. 1983. Searching for cryptic prey: The effect of search rate. *Am. Nat.* 121:172–186.

Gendron, R. P., and J. E. R. Staddon. 1984. A laboratory simulation of foraging behavior: The effect of search rate on the probability of detecting prey. *Am. Nat.* 124:407–415.

Gerlai, R. 1996. Gene-targeting studies of mammalian behavior: Is it the mutation or the background genotype? *Trends Neurosci.* 19 (5):177–181.

Getz, L. L., B. McGuire, J. Hofmann, T Pizzuto, and B. Frase. 1993. Social organization in the prairie vole (*Microtus ochrogaster*). *J. Mamm.* 74:44–58.

Gilbert, S. F. 1988. *Developmental Biology*. Sunderland, MA: Sinauer.

Gill, F. B., and L. L. Wolf. 1975. Economics of feeding territoriality in the golden-winged sunbird. *Ecology* 56:33–45.

Gillette, R., M. U. Gillette, D. J. Green, and R. Huang. 1989. The neuromodulatory response: Integrating second messenger pathways. *Am. Zool.* 29:1275–1286.

Giraldeau, L.-A. 1997. The ecology of information use. In *Behavioural Ecology: An Evolutionary Approach*, edited by J. R. Krebs and N. B. Davies. Oxford: Blackwell Science, pp. 42–68.

Giraldeau, L.-A., and D. L. Kramer. 1982. The marginal value theorem: A quantitative test using load size variation in a central place forager, the eastern chipmunk, *Tamias striatus*. *Anim. Behav.* 30:1036–1042.

Gittleman, J. L. 1989. The comparative approach in ethology: Aims and limitations. In *Perspectives in Ethology*, edited by P. P. G. Bateson and P. H. Klopfer. New York: Plenum, pp. 55–83.

Goddard, S. M., and R. B. Forward, Jr. 1991. The role of the underwater polarized light pattern, in sun compass navigation of the grass shrimp, *Palaemonetes vulgaris*. *J. Comp. Physiol. A* 169:479–491.

Godin, J.-G. J., and L. A. Dugatkin. 1996. Female mating preference for bold males in the guppy, *Poecilia reticulata*. *Proc. Nat. Acad. Sci. USA* 93 (19):10262–10267.

Goff, M., M. Salmon, and K. J. Lohmann. 1998. Hatchling sea turtles use surface waves to establish a magnetic compass direction. *Anim. Behav.* 55 (1):69–77.

Golden, S., C. H. Johnson, and T. Kondo. 1998. The cyanobacterial circadian system: A clock apart. *Curr. Opin. Microbiol.* 1:693–697.

Gonzalez, A., C. Rossini, M. Eisner, and T Eisner. 1999. Sexually transmitted chemical defense in a moth. *Proc. Nat. Acad. Sci. USA* 96 (10):5570–5574.

Goodall, J. 1965. Chimpanzees of the Gombe Stream Reserve. In *Primate Behavior*, edited by I. DeVore. New York: Holt, Rinehart & Winston, pp. 425–473.

Goossens, B., L. Graziani, L. P. Waits, E. Farand, S. Magnolon, J. Coulon, M.-C. Bel, P. Taberlet, and D. Allainé. 1998. Extra-pair paternity in the monogamous Alpine marmot revealed by nuclear DNA microsatellite analysis. *Behav. Ecol. Sociobiol.* 43 (4–5):281–288.

Gorzula, S. J. 1978. An ecological study of *Caiman crocodilus* inhabiting savanna lagoons in Venezuelan Guayana. *Oecologia* 35:21–34.

Goss-Custard, J. D. 1977. Optimal foraging and the size selection of worms by redshank, *Tringa totanus*, in the field. *Anim. Behav.* 25:10–29.

Goss-Custard, J. D., J. T. Cayford, and S. E. G. Lea. 1998. The changing trade-off between food finding and food stealing in juvenile oystercatchers. *Anim. Behav.* 55 (3):745–760.

Gottlieb, G. 1965. Imprinting in relation to parental and species identification by avian neonates. *J. Comp. Physiol. Psychol.* 59:345–356.

Gottlieb, G. 1968. Prenatal behavior of birds. *Quart. Rev. Biol.* 43:148–174.

Gottlieb, G. 1978. Development of species identification in ducklings. IV. Change in species-specific perception caused by auditory deprivation. *J. Comp. Physiol. Psychol.* 92:375–387.

Gottlieb, G. 1985. Development of species identification in ducklings. XI. Embryonic critical period for species-typical perception in the hatchling. *Anim. Behav.* 33:225–233.

Gould, J. L. 1975. Honey bee recruitment: The dance-language controversy. *Science* 189:685–692.

Gould, J. L. 1976. The dance-language controversy. *Quart. Rev. Biol.* 51 (2):211–244.

Gould, J. L. 1980. The case for magnetic sensitivity in birds and bees (such as it is). *Am. Sci.* 68:256–267.

Gould, J. L. 1982a. The map sense of pigeons. *Nature* 296:205–211.

Gould, J. L. 1982b. Why do honey bees have dialects? *Behav. Ecol. Sociobiol.* 10:53–56.

Gould, J. L. 1986. The locale map of honey bees. Do insects have cognitive maps? *Science* 23:861–863.

Gould, J. L., M. Henerey, and M. C. MacLeod. 1970. Communication of direction by the honey bee. *Science* 169:544–554.

Gould, J. L., J. L. Kirschvink, and K. S. Deffeyes. 1978. Bees have magnetic resonance. *Science* 201:1026–1028.

Gould, S. J., and R. C. Lewontin. 1979. The spandrels of San Marco and the Panglossian paradigm: A critique of the adaptationist programme. *Proc. Roy. Soc. Lond. B* 205:581–598.

Gould-Beierle, K. L., and A. C. Kamil. 1998. Use of landmarks in three species of food-storing corvids. *Ethology* 104:361–378.

Gowaty, P. A. 1983. Male parental care and apparent monogamy in eastern bluebirds (*Sialia sialis*). *Am. Nat.* 121:149–157.

Gowaty, P. A. 1996. Multiple mating by females selects for males that stay: Another hypothesis for social monogamy in passerine birds. *Anim. Behav.* 51:482–484.

Graf, S., and M. B. Sokolowski. 1989. The effect of development, food patch quality and starvation on *Drosophila melanogaster* foraging behavior. *J. Insect Behav.* 2:301–313.

Grafen, A. 1982. How not to measure inclusive fitness. *Nature* 298:425–426.

Grafen, A. 1984. Natural selection, kin selection, and group selection. In *Behavioural Ecology: An Evolutionary Approach*, edited by J. R. Krebs and N. B. Davies. Sunderland, MA: Sinauer, pp. 62–84.

Grafen, A. 1990. Biological signals as handicaps. *J. Theoret. Biol.* 144:544–546.

Grant, B. S., A. D. Cook, C. A. Clarke, and D. F. Owen. 1998. Geographic and temporal variation in the incidence of melanism in peppered moth populations in America and Britain. *J. Hered.* 89:465–471.

Grant, B. S., D. F. Owen, and C. A. Clarke. 1996. Parallel rise and fall of melanic peppered moths in America and Britain. *J. Hered.* 87:351–357.

Graul, W. D., S. R. Derrickson, and D. W. Mock. 1977. The evolution of avian polyandry. *Am. Nat.* 111:812–816.

Gray, E. M. 1997. Do female red-winged blackbirds benefit genetically from seeking extra-pair copulations? *Anim. Behav.* 54 (3):605–623.

Grayson, J., and M. Edmunds. 1989. The causes of colour and colour change in caterpillars of the poplar and eyed hawkmoths (*Laothoe populi* and *Smerinthus ocellatus*). *Biol. J. Linn. Soc.* 37:263–279.

Green, C. B. 1998. How cells tell time. *Trends Cell Biol.* 8:224–230.

Green, S. 1975. Dialects in Japanese monkeys: Vocal learning and cultural transmission of locale-specific vocal behavior? *Z. Tierpsychol.* 38:305–314.

Greenfield, P. M., and E. S. Savage-Rumbaugh. 1990. Grammatical combination in *Pan panicus*: Processes of learning and invention in the evolution and development of language. In *"Language" and Intelligence in Monkeys and Apes: Comparative Developmental Perspectives*, edited by S. T. Parker and K. R. Gibson. Cambridge: Cambridge University Press, pp. 540–578.

Greenwood, J. J. D. 1984. The functional basis of frequency-dependent food selection. *Biol. J. Linn. Soc.* 23:177–199.

Greenwood, J. J. D., P. A. Cotton, and D. M. Wilson. 1989. Frequency-dependent selection on aposematic prey: Some experiments. *Biol. J. Linn. Soc.* 36:213–226.

Greenwood, P. J. 1980. Mating systems, philopatry and dispersal in birds and mammals. *Anim. Behav.* 28:1140–1162.

Griffin, D. R. 1955. Bird navigation. In *Recent Studies in Avian Biology*, edited by A. Wolfson. Urbana: University of Illinois Press, pp. 154–197.

Griffin, D. R. 1978. Prospects for a cognitive ethology. *Behav. Brain Sci.* 1:527–538.

Griffin, D. R. 1981. *The Question of Animal Awareness*. 2nd ed. New York: Rockefeller University Press.

Griffin, D. R. 1982. *Animal Mind—Human Mind*. Berlin: Springer-Verlag.

Griffin, D. R. 1984. *Animal Thinking*. Cambridge, MA: Harvard University Press.

Griffin, D. R. 1991. Progress toward a cognitive ethology. In *Cognitive Ethology, the Minds of Other Animals*, edited by C. A. Ristau. Hillsdale, NJ: Erlbaum, pp. 3–17.

Griffin, D. R., F. A. Webster, and C. R. Michael. 1960. The echolocation of flying insects by bats. *Anim. Behav.* 8:141–154.

Griffin, E. A., D. Staknis, and C. J. Weitz. 1999. Light-independent role of CRY1 and CRY2 in the mammalian circadian clock. *Science* 286 (5440):768–771.

Griffiths, M. 1988. The platypus. *Sci. Am.* 258 (May):84–91.

Grosberg, R. K., and J. F. Quinn. 1986. The genetic control and consequences of kin recognition by the larvae of a colonial marine invertebrate. *Nature* 322:456–458.

Gross, M. R. 1982. Sneakers, satellites and parentals: Polymorphic mating strategies in North American sunfishes. *Z. Tierpsychol.* 60:1–26.

Gross, M. S., and R. Shine. 1981. Parental care and mode of fertilization in ectothermic vertebrates. *Evolution* 35:775–793.

Gubernick, D. J. 1980. Maternal "imprinting" or maternal "labelling" in goats? *Anim. Behav.* 28:124–129.

Gubernick, D. J. 1981. Parent and infant attachment in mammals. In *Parental Care in Mammals*, edited by D. J. Gubernick and P. H. Klopfer. New York: Plenum, pp. 243–306.

Gubernick, D. J., K. C. Jones, and P. H. Klopfer. 1979. Maternal "imprinting" in goats? *Anim. Behav.* 27:314–315.

Gubernick, D. J., S. L. Wright, and R. E. Brown. 1993. The significance of father's presence for offspring survival in the monogamous California mouse, *Peromyscus californicus. Anim. Behav.* 46:539–546.

Guilford, T., and M. S. Dawkins. 1987. Search images not proven: A reappraisal of recent evidence. *Anim. Behav.* 35:1838–1845.

Guilford, T., and M. S. Dawkins. 1991. Receiver psychology and the evolution of animal signals. *Anim. Behav.* 42 (1):1–14.

Guilford, T., and M. S. Dawkins. 1992. Understanding signal design: A reply to Blumberg & Alberts. *Anim. Behav.* 44:384–385.

Gunn, D. L. 1937. The humidity reactions of the woodlouse, *Porcellio scaber. J. Exp. Biol.* 14:178–186.

Gunn, D. L., and J. S. Kennedy. 1936. Apparatus for investigating the reactions of land arthropods to humidity. *J. Exp. Biol.* 13:450–459.

Gurney, M. E. 1981. Hormonal control of cell form and number in zebra finch song system. *J. Neurosci.* 1:658–673.

Gurney, M. E., and M. Konishi. 1980. Hormone induced sexual differentiation in brain and behavior in zebra finches. *Science* 208:1380–1382.

Gwinner, E. 1978. Effects of pinealectomy on circadian locomotor activity rhythms in European starlings, *Sturnus vulgaris. J. Comp. Physiol.* 126:123–129.

Gwinner, E. 1996. Circadian and circannual programmes in avian migration. *J. Exp. Biol.* 199:39–48.

Gwinner, E., and I. Benzinger. 1978. Synchronization of a circadian rhythm in European starlings by daily injections of melatonin. *J. Comp. Physiol.* 127:209–213.

Gwinner, E., and W. Wiltschko. 1978. Endogenously controlled changes in migratory direction of the garden warbler, *Sylvia borin. J. Comp. Physiol.* 125:267–273.

Gwynne, D. T., and W. D. Brown. 1994. Mate feeding, offspring investment, and sexual differences in katydids (Orthoptera: Tettigoniidae). *Behav. Ecol.* 5:267–272.

Gwynne, D. T., and L. W. Simmons. 1990. Experimental reversal of courtship roles in an insect. *Nature* 346:172–174.

Hagedorn, M. 1986. The ecology, courtship and mating of gymnotiform electric fish. In *Electroreception*, edited by T. H. Bullock and W. Heiligenberg. New York: Wiley, pp. 497–525.

Hagedorn, M. 1995. The electric fish *Hypopomus occidentalis* can rapidly modulate the amplitude and duration of its electric organ discharges. *Anim. Behav.* 49:1409–1413.

Hagen, J. B. 1999. Retelling experiments: H. B. D. Kettlewell's studies of industrial melanism in peppered moths. *Biol. Philos.* 14 (1):39–54.

Hailman, J. P. 1967a. Cliff-nesting adaptations of the Galapagos swallow-tailed gull. *Wilson Bull.* 77:346–362.

Hailman, J. P. 1967b. The ontogeny of an instinct: The pecking response in chicks of the laughing gull (*Larus atricilla* L.) and related species. *Behaviour Suppl.* 15:1–159.

Hailman, J. P. 1969. How an instinct is learned. *Sci. Am.* 221 (Dec.):98–106.

Hall, J. C. 1977. Portions of the central nervous system controlling reproductive behavior in *Drosophila melanogaster. Behav. Genet.* 7:291–312.

Hall, J. C. 1979. Control of male reproductive behavior by the central nervous system of *Drosophila*: Dissection of a courtship pathway by genetic mosaics. *Genetics* 92:437–457.

Hall, J. C. 1994. The mating of a fly. *Science* 264:1703–1714.

Halliday, T. R. 1983. The study of mate choice. In *Mate Choice*, edited by P. Bateson. Cambridge: Cambridge University Press, pp. 3–32.

Halpern, N. 1987. The organization and function of the vomeronasal system. *Ann. Rev. Neurosci.* 10:325–362.

Halpin, Z. T. 1991. Introduction to the symposium: Animal behavior: Past, present, and future. *Am. Zool.* 31:283–285.

Hamilton, W. D. 1964. The genetical evolution of social behaviour. *J. Theoret. Biol.* 7:1–52.

Hamilton, W. D. 1971. Geometry for the selfish herd. *J. Theoret. Biol.* 31:295–311.

Hamilton, W. D., and M. Zuk. 1982. Heritable true fitness and bright birds: A role for parasites? *Science* 218:384–387.

Hammerstein, P., and G. A. Parker. 1987. Sexual selection: Games between the sexes. In *Sexual Selection: Testing the Alternatives*, edited by J. W. Bradbury and M. B. Andersson. New York: Wiley, pp. 119–142.

Hanson, M., G. Wirmark, M. Oblad, and L. Strid. 1984. Iron-rich particles in European eel (*Anguilla anguilla* L.). *Comp. Biochem. Physiol. A* 79:221–224.

Hardcastle, A. 1925. Young cuckoo fed by several birds. *Brit. Birds* 19:100.

Harden-Jones, F. R. 1968. *Fish Migration*. New York: St. Martin's Press.

Harder, L. D., and L. A. Real. 1987. Why are bumblebees risk averse? *Ecology* 68:1104–1108.

Hardin, P. E., and N. R. J. Glossop. 1999. The CRYs of flies and mice. *Science* 286:2460–2461.

Harlow, H. F. 1949. The formation of learning sets. *Psychol. Rev.* 56:51–65.

Harlow, H. F., and M. K. Harlow. 1962. Social deprivation in monkeys. *Sci. Am.* 207 (Nov.):136–146.

Harlow, H. F., M. K. Harlow, and S. J. Suomi. 1971. From thought to therapy: Lessons from a primate laboratory. *Am. Sci.* 59:539–549.

Harris, W. A. 1980. The effects of eliminating impulse activity on the development of the retinotectal projection in salamanders. *J. Comp. Neurol.* 194:303–317.

Hart, B. 1980. Neonatal spinal transection in male rats: Differential effects on penile reflexes and other reflexes. *Brain. Res.* 185:423–428.

Haskell, D. 1994. Experimental evidence that nestling begging behavior incurs cost due to nest predation. *Proc. Roy. Soc. Lond. B* 257:161–164.

Hasler, A. D., and A. T. Scholz. 1983. *Olfactory Imprinting in Homing Salmon*. Berlin: Springer-Verlag.

Hasler, A. D., A. T. Scholz, and R. M. Horrall. 1978. Olfactory imprinting and homing in salmon. *Am. Sci.* 66:347–355.

Hasler, A. D., and W. J. Wisby. 1951. Discrimination of stream odors by fishes in relation to parent stream behavior. *Am. Nat.* 85:223–238.

Hasselquist, D. 1998. Polygyny in great reed warblers: A long term study of factor contributing to male fitness. *Ecology* 79 (7):2376–2390.

Hasselquist, D., S. Bensch, and T. vonSchantz. 1996. Correlation between male song repertoire, extra-pair paternity and offspring survival in the great reed warbler. *Nature* 381 (6579):229–232.

Hastings, J. W., B. Rusak, and Z. Boulos. 1991. Cellular-biochemical clock hypotheses. In *Neural and Integrative Animal Physiology*, edited by C. L. Prosser. New York: Wiley-Liss, pp. 435–546.

Hastings, M. H. 1997. Central clocking. *Trends Neurosci.* 20 (10):459–464.

Hauser, M. D. 1996. *The Evolution of Communication*. Cambridge, MA: MIT Press.

Hauser, M. D. 1998. Functional referents and acoustic similarity: Field playback experiments with rhesus monkeys. *Anim. Behav.* 55 (6):1647–1658.

Hausfater, G., and S. B. Hrdy. 1984. *Infanticide: Comparative and Evolutionary Perspectives*. New York: Aldine.

Hausfater, G., H. C. Gerhardt, and G. M. Klump. 1990. Parasites and mate choice in gray treefrogs, *Hyla versicolor. Am. Zool.* 30:299–312.

Hausheer-Zarmakupi, Z., D. P. Wolfer, M.-C. Leisinger-Trigona, and H.-P. Lipp. 1996. Selective breeding for extremes in open-field activity of mice entails a differentiation of hippocampal mossy fibers. *Behav. Genet.* 26 (2):167–176.

Haverkamp, L. J. 1986. Anatomical and physiological development of the *Xenopus* embryonic motor system in the absence of neural activity. *J. Neurosci.* 6:1338–1348.

Haverkamp, L. J., and R. W. Oppenheim. 1986. Behavioral development in the absence of neural activity: Effects of chronic immobilization on amphibian embryos. *J. Neurosci.* 6:1332–1337.

Hawryshyn, C. W. 1992. Polarized light vision in fish. *Am. Sci.* 80:164–175.

Hayes, K. J., and C. Hayes. 1951. The intellectual development of a home-raised chimpanzee. *Proc. Am. Phil. Soc.* 95:105.

Hedrick, A. V. 1986. Female preference for male calling bout duration in a field cricket. *Behav. Ecol. Sociobiol.* 19:73–77.

Hedrick, A. V., and E. J. Temeles. 1989. The evolution of sexual dimorphism in animals: Hypotheses and tests. *Trends Ecol. Evol.* 4:136–138.

Hedwig, B., and R. Heinrich. 1997. Identified descending brain neurons control different stridulatory motor patterns in an acridid grasshopper. *J. Comp Physiol. A* 180 (3):285–294.

Hegstrom, C. D., L. M. Riddiford, and J. W. Truman. 1998. Steroid and neuronal regulation of ecdysone receptor expression during metamorphosis of muscle in the moth, *Manduca sexta. J. Neurosci.* 18 (5):1786–1794.

Heinrich, B. 1979. Foraging strategies of caterpillars. Leaf damage and possible predator avoidance strategies. *Oecologia* 42:325–337.

Heinrich, B. 1995. An experimental investigation of insight in common ravens (*Corvus corax*). *Auk* 112 (4):994–1003.

Heinrich, B., and S. L. Collins. 1983. Caterpillar leaf damage and possible predator avoidance strategies. *Ecology* 64:592–602.

Heinroth, O. 1910. Beiträge zur Bilogie, insbiesonder Psychologie und Ethologie der Anatiden, *In Proceedings of the 5th International Ornithological Congress, Berlin*, pp. 589–702.

Heinsohn, R. G. 1991. Kidnapping and reciprocity in cooperatively breeding white-winged choughs. *Anim. Behav.* 41:1097–1100.

Heinsohn, R. G., A. Cockburn, and R. A. Mulder. 1990. Avian cooperative breeding: Old hypotheses and new directions. *Trends Ecol. Evol.* 5:403–407.

Heinsohn, R., and S. Legge. 1999. The cost of helping. *Trends Ecol. Evol.* 14 (2):53–57.

Helbig, A. J. 1991. Inheritance of migratory direction in a bird species: A cross-breeding experiment with SE- and SW- migrating blackcaps (*Sylvia atricapilla*). *Behav. Ecol. Sociobiol.* 28:9–12.

Helbig, A. J., P. Berthold, and W. Wiltschko. 1989. Migratory orientation of blackcaps (*Sylvia atricapilla*): Population specific shifts of direction during the autumn. *Ethology* 82:307–315.

Hennessy, D. F. 1986. On the deadly risk of predator harassment. *Ethology* 72:72–74.

Herberstein, M. E., C. L. Craig, and M. A. Elgar. 2000. Foraging strategies and feeding regimes: Web and decoration investment in *Argiope keyserlingi* Karsch (Araneae: Araneidae). *Evol. Ecol. Res.* 2 (1):69–80.

Herman, L. M., and P. H. Forestell. 1985. Reporting presence or absence of named objects by a language-trained dolphin. *Neurosci. Biobehav. Rev.* 9:667–681.

Herman, L. M., A. A. Pack, and P. Morrel-Samuels. 1993. Representational and conceptual skills of dolphins. In *Language and Communication: Comparative Perspectives*, edited by H. L. Roitblat, L. M. Herman, and P. E. Nachtigall. Hillsdale, NJ: Erlbaum, pp. 403–442.

Herrnstein, R. J., D. H. Loveland, and C. Cable. 1976. Natural concepts in pigeons. *J. Exp. Psych.: Anim. Behav. Proc.* 2:285–302.

Heth, G., J. Todrank, and R. E. Johnston. 1998. Kin recognition in golden hamsters: Evidence for phenotype matching. *Anim. Behav.* 56 (2):409–417.

Hews, D. K. 1988. Alarm response in larval western toads, *Bufo boreas*: Release of larval chemical by a natural predator and its effect on predator capture efficiency. *Anim. Behav.* 36:125–133.

Heyes, C. M. 1993. Imitation, culture and cognition. *Anim. Behav.* 46:999–1010.

Heyes, C. M. 1994. Social learning in animals: Categories and mechanisms. *Biol. Rev.* 69:207–231.

Heyes, C. M., and G. R. Dawson. 1990. A demonstration of observational learning using a bidirectional control. *Quart. J. Exp. Psychol.* 42B:59–71.

Heyes, C. M., G. R. Dawson, and T. Nokes. 1992. Imitation in rats: Initial responding and transfer evidence. *Quart. J. Exp. Psychol.* 45B:81–92.

Hilgard, E. R. 1956. *Theories of Learning*. New York: Appleton-Century-Crofts.

Hill, G. E. 1990. Female house finches prefer colourful males: Sexual selection for a condition-dependent trait. *Anim. Behav.* 40 (3):563–572.

Hill, G. E. 1991. Plumage coloration as a sexually selected indicator of male quality. *Nature* 350:337–339.

Hill, G. E. 1995. Ornamental traits as indicators of environmental health. *BioScience* 45 (1):25–31.

Hinde, R. A. 1970. *Animal Behavior: A Synthesis of Ethology and Comparative Psychology*. 2nd ed. New York: McGraw-Hill.

Hinde, R. A. 1982. *Ethology*. Oxford: Oxford University Press.

Hingston, R. W. G. 1927a. Field observations on spider mimics. *Proc. Zool. Soc. Lond.* 1927:841–859.

Hingston, R. W. G. 1927b. Protective devices in spiders' snares, with a description of seven new species of orb-weaving spiders. *Proc. Zool. Soc. Lond.* 1927:259–293.

Hoage, R. J., and L. Goldman, eds. 1986. *Animal Intelligence, Insights into the Animal Mind*. Washington DC: Smithsonian Institution Press.

Hochner, B., M. Klein, S. Schacher, and E. R. Kandel. 1986. Additional component in the cellular mechanism of presynaptic facilitation contributes to behavioral dishabituation in *Aplysia*. *Proc. Nat. Acad. Sci. USA* 83:8794–8798.

Hocking, B. 1964. Fire melanism in some African grasshoppers. *Evolution* 18:332–335.

Hodos, W. C., and C. B. G. Campbell. 1969. Scala naturae: Why there is no theory in comparative psychology. *Psychol. Rev.* 76:337–350.

Hoffmann, A. A., and Z. Cacoyianni. 1990. Territoriality in *Drosophila melanogaster* as a conditional strategy. *Anim. Behav.* 40:526–537.

Hoffmann, K. 1954. Versuche zu der im Richtungsfinden der Vögel enthaltenen Zeitschätzung. *Z. Tierpsychol.* 11:453–475.

Hogg, J. T. 1988. Copulatory tactics in relation to sperm competition in Rocky Mountain bighorn sheep. *Behav. Ecol. Sociobiol.* 22:49–59.

Hogg, J. T., and S. H. Forbes. 1997. Mating in bighorn sheep: Frequent male reproduction via a high-risk "unconventional" tactic. *Behav. Ecol. Sociobiol.* 41 (1):33–48.

Hoglund, L. B., and M. Astrand. 1973. Preferences among juvenile char (*Salvelinus alpinus* L.) to intraspecific odours and water current studied with the fluiarium technique. In *Institute of Freshwater Research Drottningholm Report*, pp. 21–30.

Hogstad, O. 1995. Alarm calling by willow tits, *Parus montanus*, as mate investment. *Anim. Behav.* 49 (1):209–219.

Holder, C. F. 1901. A curious means of defense. *Sci. Am.* 2 (Sept.):186–187.

Hölldobler, B. 1976. Recruitment behavior. Home range orientation and territoriality in harvester ants, *Pogonomyrmes*. *Behav. Ecol. Sociobiol.* 1:3–44.

Hölldobler, B., and E. O. Wilson. 1978. The multiple recruitment systems of the African weaver ant *Oecophylla longinoda* (Latreille) (Hymenoptera: Formicidae). *Behav. Ecol. Sociobiol.* 3:19–60.

Holley, J. F. 1984. Adoption, parent-chick recognition and maladaptation in the herring gull *Larus argentatus*. *Z. Tierpsychol.* 64:9–14.

Hollis, K. L. 1984. The biological function of Pavlovian conditioning: The best defense is a good offense. *J. Exp. Psych.: Anim. Behav. Process.* 10:413–425.

Hollis, K. L. 1990. The role of Pavlovian conditioning in territorial aggression and reproduction. In *Contemporary Issues in Comparative Psychology*, edited by D. A. Dewsbury. Sunderland, MA: Sinauer, pp. 197–219.

Hollis, K. L., M. J. Dumas, P. Singh, and P. Fackelman. 1995. Pavlovian conditioning of aggressive behavior in blue gourami fish (*Trichogaster trichopterus*)—Winners become winners and losers stay losers. *J. Comp. Psychol.* 109 (2):123–133.

Hollis, K. L., V. L. Pharr, M. J. Dumas, G. B. Britton, and J. Field. 1997. Classical conditioning provides paternity advantage for territorial male blue gouramis (*Trichogaster trichopterus*). *J. Comp. Psychol.* 111 (3):219–225.

Holmes, W. 1940. The colour changes and colour patterns of *Sepia officinalis* L. *Proc. Zool. Soc. Lond.* 110:17–35.

Holmes, W. G., and P. W. Sherman. 1982. The ontogeny of kin recognition in two species of ground squirrels. *Am. Zool.* 22:491–517.

Holmes, W. G., and P. W. Sherman. 1983. Kin recognition in animals. *Am. Sci.* 71:46–55.

Holmgren, N. A., and M. Enquist. 1999. Dynamics of mimicry evolution. *Biol. J. Linn. Soc.* 66:145–158.

Holtkamp-Rötzler, E., G. Fleissner, M. Hanzlik, W. Wiltschko, and N. Petersen. 1997. Mechanoreceptors in the upper beak of homing pigeons (*Columba livia*) as putative structural candidates for magnetoreception. *Verh. Dtsch. Zool. Ges.* 90:290.

Honeycutt, R. L. 1992. Naked mole-rats. *Am. Sci.* 80:43–53.

Hoogland, J. L., and P. W. Sherman. 1976. Advantages and disadvantages of bank swallow (*Riparia riparia*) coloniality. *Ecol. Monogr.* 46:33–58.

Hopkins, C. D. 1972. Sex differences in electric signaling in an electric fish. *Science* 176:1035–1037.

Hopkins, C. D. 1974. Electric communication in fish. *Am. Sci.* 62:426–437.

Hopkins, C. D. 1977. Electric communication. In *How Animals Communicate*, edited by T. A. Sebeok. Bloomington: Indiana University Press, pp. 263–289.

Hopkins, C. D. 1980. Evolution of electric communication channels of mormyrids. *Behav. Ecol. Sociobiol.* 7:1–13.

Hopkins, C. D. 1986a. Behavior of Mormyridae. In *Electroreception*, edited by T. H. Bullock and W. Heiligenberg. New York: Wiley, pp. 527–576.

Hopkins, C. D. 1986b. Temporal structure of nonpropagated electric communication signals. *Brain Behav. Evol.* 28:43–59.

Hopkins, C. D. 1988. Neuroethology of electric communication. *Ann. Rev. Neurosci.* 11:497–535.

Hopkins, C. D. 1999. Design features for electric communication. *J. Exp. Biol.* 202:1217–1228.

Hopkins, C. D., and A. H. Bass. 1981. Temporal coding of species recognition signals in an electric fish. *Science* 212:855–857.

Horsmann, U., H. G. Heinzel, and G. Wendler. 1983. The phasic influence of self-generated air current modulations on the locust flight motor. *J. Comp. Physiol.* 150:427–438.

Hotta, Y., and S. Benzer. 1972. Mapping of behavior in *Drosophila* mosaics. *Nature* 240:527–535.

Houck, L. D. 1988. Effect of body size on male courtship success in a plethodontid salamander. *Anim. Behav.* 39:837–842.

Houck, L. D. 1998. Integrative studies of amphibians: From molecules to mating. *Am. Zool.* 38 (1):108–117.

Houck, L. D., and N. L. Reagan. 1990. Male courtship pheromones increase female receptivity in a plethodontid salamander. *Anim. Behav.* 39:729–734.

Houston, A. I., and J. M. McNamara. 1988. Fighting for food: A dynamic version of the hawk-dove game. *Evol. Ecol.* 2:51–64.

Houtman, A. M., and J. B. Falls. 1994. Negative assortative mating in the white-throated sparrow, *Zonotrichia albicollis*: The role of mate choice and intra-sexual competition. *Anim. Behav.* 48 (2):377–383.

Howard, W. E. 1960. Innate and environmental dispersal of individual vertebrates. *Am. Mid. Nat.* 63:152–161.

Howell, B. W., R. Hawkes, P. Soriano, and J. A. Cooper. 1997. Neuronal position in the developing brain is regulated by *mouse disabled-1*. *Nature* 389:733–737.

Howlett, R. J., and M. E. J. Majerus. 1987. The understanding of industrial melanism in the peppered moth (*Biston betularia*) (Lepidoptera: Geometridae). *Biol. J. Linn. Soc.* 30:31–44.

Hoy, R. R. 1978. Acoustic communication in crickets. A model system for the study of feature detection. *Fed. Proc.* 37:2316–2323.

Hsu, U., and L. L. Wolf. 1999. The winner and loser effect: Integrating multiple experiences. *Anim. Behav.* 57 (4):903–910.

Huang, Z.-Y., E. Plettner, and G. E. Robinson. 1998. Effects of social environment and worker mandibular glands on endocrine-mediated behavioral development in honey bees. *J. Comp. Physiol. A* 183 (2):143–152.

Huang, Z.-Y., and G. E. Robinson. 1992. Honeybee colony integration: worker-worker interactions mediate hormonally regulated plasticity in division of labor. *Proc. Nat. Acad. Sci. USA* 89:11726–11729.

Huang, Z.-Y., and G. E. Robinson. 1996. Regulation of honey bee division of labor by colony age demography. *Behav. Ecol. Sociobiol.* 39 (3):147–158.

Hughes, K. A., L. Du, F. H. Rodd, and D. N. Reznick. 1999. Familiarity leads to female mate preference for novel males in the guppy, *Poecilia reticulata*. *Anim. Behav.* 58 (4):907–916.

Hunter, M. L., and J. R. Krebs. 1979. Geographical variation in the song of the great tit *Parus major* in relation to ecological factors. *J. Anim. Ecol.* 48:759–785.

Huntingford, F. A. 1986. Development of behaviour in fish. In *The Behavior of Teleost Fishes*, edited by T. J. Pitcher. Baltimore, MD: Johns Hopkins Press, pp. 47–68.

Huxley, J. 1914. The courtship habits of the great crested grebe *Podiceps cristatus*. *Proc. Zool. Soc. Lond.* 2:491–562.

Huxley, J. 1923. Courtship activities in the red-throated diver *Colymbus stellatus pontopp*; together with a discussion on the evolution of courtship in birds. *J. Linn. Soc. Lond.* 2:491–562.

Iguchi, K., and T. Hino. 1996. Effect of competitor abundance on feeding territoriality in a grazing fish, the ayu *Plecoglossus altivelis*. *Ecol. Res.* 11 (2):165–173.

Immelmann, K. 1963. Drought adaptations in Australian desert birds. In *Proceedings of the 13th International Ornithological Congress, 1962*, pp. 649–657.

Immelmann, K. 1969. Über den Einfluss frühkinlicher Erfahrungen auf die geschlechtliche Objektfixierung bei Estrildiden. *Z. Tierpsychol.* 26:677–691.

Immelmann, K. 1972. The influence of early experience upon the development of social behavior in estrildine finches. In *Proceedings of the 15th International Ornithological Congress, The Hague, 1970*, pp. 316–338.

Immelmann, K. 1980. *Introduction to Ethology*. New York: Plenum.

Immelmann, K., and S. J. Suomi. 1981. Sensitive phases in development. In *Behavioral Development*, edited by K. Immelmann, W. Barlow, L. Petrinovich, and M. Main. Cambridge: Cambridge University Press, pp. 395–431.

Ims, R. A. 1987a. Male spacing systems in microtine rodents. *Am. Nat.* 130:475–485.

Ims, R. A. 1987b. Responses in spatial organization and behaviour to manipulations of the food resource in the vole *Clethrionomys rufocanus*. *J. Anim. Ecol.* 56:585–596.

Ims, R. A. 1988. Spatial clumping of sexually receptive females induces space sharing among male voles. *Nature* 335:541–543.

Inouye, S.-I. T., and H. Kawamura. 1979. Persistence of circadian rhythmicity in mammalian hypothalamic "island" containing the suprachiasmatic nucleus. *Proc. Nat. Acad. Sci. USA* 76:5962–5966.

Inouye, S.-I. T., and S. Shibata. 1994. Neurochemical organization of circadian rhythm in the suprachiasmatic nucleus. *Neurosci. Res.* 20:109–130.

Insel, T. R., and T. J. Hulihan. 1995. A gender-specific mechanism for pair-bonding—Oxytocin and partner preference formation in monogamous voles. *Behav. Neurosci.* 109 (4):782–789.

Ioalè, P., M. Nozzolini, and F. Papi. 1990. Homing pigeons do extract directional information from olfactory stimuli. *Behav. Ecol. Sociobiol.* 26:301–305.

Ito, Yosiaki. 1989. The evolutionary biology of sterile soldiers in aphids. *Trends Ecol. Evol.* 4 (3):69–73.

Jablonski, P. G., and R. S. Wilcox. 1996. Signalling asymmetry in the communication of the water strider *Aquarius remigis* in the context of dominance and spacing in the non-mating season. *Ethology* 102 (5):353–359.

Jackson, R. R., and R. S. Wilcox. 1990. Aggressive mimicry, prey-specific predatory behaviour and predator-recognition in the predator-prey interactions of *Portia fimbriata* and *Euryattus* sp., jumping spiders from Queensland. *Behav. Ecol. Sociobiol.* 26:111–119.

Jakob, E. M. 1991. Costs and benefits of group living for pholcid spiderlings: Losing food, saving silk. *Anim. Behav.* 41:711–722.

Jamieson, I. G. 1986. The functional approach to behavior: Is it useful? *Am. Nat.* 127:195–208.

Jamieson, I. G. 1989. Behavioral heterochrony and the evolution of birds' helping at the nest: An unselected consequence of communal breeding? *Am. Nat.* 133:394–406.

Jamieson, I. G. 1991. The unselected hypothesis for the evolution of helping behavior: Too much or too little emphasis on natural selection? *Am. Nat.* 138 (1):271–282.

Jamieson, I. G., and J. L. Craig. 1987. Critique of helping in birds: A departure from functional explanations. In *Perspectives in Ethology*, edited by P. P. G. Bateson and P. Klopfer. New York: Plenum, pp.79–98.

Jan, Y., L. Jan, and M. Dennis. 1977. Two mutations of synaptic transmission in *Drosophila*. *Proc. Roy. Soc. Lond.* 198:87–108.

Janetos, A. C. 1980. Strategies of female mate choice: A theoretical analysis. *Behav. Ecol. Sociobiol.* 7:107–112.

Janus, C. 1988. The development of responses to naturally occurring odors in spiny mice *Acomys cahirinus*. *Anim. Behav.* 36:1400–1406.

Jarosz, S. J., T. J. Kuehl, and W. R. Dukelow. 1977. Vaginal cytology, induced ovulation, and gestation in the squirrel monkey (*Saimiri sciureus*). *Biol. Reprod.* 16:97–103.

Jarvis, E. D., H. Schwabl, S. Ribeiro, and C. V. Mello. 1997. Brain gene regulation by territorial singing behavior in freely ranging songbirds. *Neuroreport* 8 (8):2073–2077.

Jarvis, J. U. M. 1981. Eusociality in a mammal: Cooperative breeding in naked mole-rat colonies. *Science* 212:571–573.

Jaynes, J. 1969. The historical origins of "ethology" and comparative psychology. *Anim. Behav.* 17:601–606.

Jenni, D. A., and B. J. Betts. 1978. Sex differences in nest construction, incubation, and parental behavior in the polyandrous American jacana (*Jacana spinosa*). *Anim. Behav.* 26:207–218.

Jenni, D. S., and G. Collier. 1972. Polyandry in the American jacana (*Jacana spinosa*). *Auk* 89:743–765.

Jennings, H. S. 1906. *Behavior of the Lower Organisms*. New York: Columbia University Press.

Jennions, M. D. 1993. Female choice in birds and the cost of long tails. *Trends Ecol. Evol.* 8:230–232.

Jivoff, P. 1997. The relative roles of predation and sperm competition on the duration of the post-copulatory association between the sexes in the blue crab, *Callinectes sapidus*. *Behav. Ecol. Sociobiol.* 40 (3):175–186.

Joels, M., and E. R. De Kloet. 1989. Effects of glucocorticoids and norepinephrine on the excitability in the hippocampus. *Science* 245:1502–1505.

Johnsgard, P. A. 1967. *Animal Behavior*. Dubuque, IA: Brown.

Johnson, C. H., S. S. Golden, and T. Kondo. 1998. Adaptive significance of circadian programs in cyanobacteria. *Trends Microbiol.* 6 (10):407–410.

Johnson, D. F., and G. Collier. 1989. Patch choice and meal size in foraging rats as a function of the profitability of food. *Anim. Behav.* 38:285–297.

Johnson, D. L., and A. M. Wenner. 1966. A relationship between conditioning and communication in honey bees. *Anim. Behav.* 14:261–265.

Johnson, P. B., and A. D. Hasler. 1980. The use of chemical cues in the upstream migration of coho salmon, *Oncorhynchus kisutch*. Wildbaum. *J. Fish. Biol.* 17:67–73.

Johnston, T. D. 1988. Developmental explanation and the ontogeny of birdsong: Nature/nurture redux. *Behav. Brain Sci.* 11:617–663.

Johnstone, R. A. 1997. The evolution of animal signals. In *Behavioural Ecology*, edited by J. R. Krebs and N. B. Davies. Oxford: Blackwell Science, pp. 155–178.

Johnstone, R. A. 1998. Game theory and communication. In *Game Theory and Animal Behavior*, edited by L. A. Dugatkin and H. K. Reeve. New York: Oxford University Press, pp. 94–117.

Jones, F. R. H. 1955. Photokinesis in the ammocoete larva of the brook lamprey. *J. Exp. Biol.* 34:492–503.

Joron, M., and J. L. B. Mallet. 1998. Diversity in mimicry: Paradox or paradigm. *Trends Ecol. Evol.* 13 (11):461–466.

Joule, J., and G. N. Cameron. 1975. Species removal studies. I. Dispersal strategies of sympatric *Sigmodon hispidus* and *Reithrodontomys fulvescens*. *J. Mamm.* 56:378–396.

Judd, W. W. 1955. Observations on the blue-tailed skink, *Eumeces fasiatus*, captured in Randeau Park, Ontario and kept in captivity over winter. *Copeia* 1955:135–136.

Kako, E. 1999. Elements of syntax in the systems of three language-trained animals. *Anim. Learn. Behav.* 27 (1):1–14.

Kalmijn, A. J. 1966. Electro-perception in sharks and rays. *Nature* 212:1232–1233.

Kalmijn, A. J. 1971. The electric sense of sharks and rays. *J. Exp. Biol.* 55:371–383.

Kalmijn, A. J. 1978. Experimental evidence of geomagnetic orientation in elasmobranch fishes. In *Animal Migration, Navigation and Homing*, edited by K. Schmidt-Koenig and W. T. Keeton. New York: Springer-Verlag, pp. 347–353.

Kalmijn, A. J. 1984. Theory of electromagnetic orientation: A further analysis. In *Comparative Physiology of Sensory Systems*, edited by L. Bolis, D. Keynes and S. H. P. Maddrell. Cambridge: Cambridge University Press, pp. 525–560.

Kamil, A. C., and J. E. Jones. 1997. The seed-storing corvid Clark's nutcracker learns geometric relationships among landmarks. *Nature* 390:276–279.

Kamil, A. C., and J. E. Mauldin. 1988. A comparative-ecological approach to the study of learning. In *Evolution and Learning*, edited by R. C. Bolles and M. D. Beecher. Hillsdale, NJ: Erlbaum, pp. 117–133.

Kamil, A. C., and S. I. Yoerg. 1982. Learning and foraging behavior. In *Perspectives on Ethology*, edited by P. P. G. Bateson. New York: Plenum, pp. 325–364.

Kandel, E. 1976. *The Cellular Basis of Behavior: An Introduction to Behavioral Neurobiology*. San Francisco: Freeman.

Kandel, E. R. 1979a. *The Behavioral Biology of Aplysia*. San Francisco: Freeman.

Kandel, E. R. 1979b. Small systems of neurons. *Sci. Am.* 241 (Mar.):66–76.

Kandel, E. R., and J. H. Schwartz. 1982. Molecular biology of learning: Modulation of transmitter release. *Science* 218:433–443.

Kaneko, M. 1998. Neural substrates of *Drosophila* rhythms revealed by mutants and molecular manipulations. *Curr. Opin. Neurobiol.* 8:652–658.

Kankel, D. R., and J. C. Hall. 1976. Fate mapping of the nervous system and other internal tissues in genetic mosaics of *Drosophila melanogaster*. *Dev. Biol.* 48:1–24.

Kaplan, A., and W. E. Trout III. 1969. The behavior of four neurological mutants of *Drosophila*. *Genetics* 61:399–409.

Karakashian, S. J., M. Gyger, and P. Marler. 1988. Audience effects on alarm calling in chickens (*Gallus gallus*). *J. Comp. Psychol.* 102:129–135.

Kasal, C., M. Menaker, and R. Perez-Polo. 1979. Circadian clock in culture: N-acetyltransferase activity of chick pineal glands oscillates *in vitro*. *Science* 203:656.

Katz, Y. B. 1985. Sunset and the orientation of European robins (*Erithacus rubecula*). *Anim. Behav.* 33:825–828.

Kaufmann, J. H. 1983. On definitions and functions of dominance and territoriality. *Biol. Rev.* 58:1–20.

Kawai, M. 1965. Newly acquired and pre-cultural behavior of the natural troop of Japanese monkeys on Koshima Islet. *Primates* 6:111–148.

Kawamura, S. 1959. Sub-culture propagation among Japanese macaques. *Primates* 2:43–60.

Kawata, M. 1985. Mating system and reproductive success in a spring population of the red-backed vole, *Clethrionomys rufocanus* bedfordiae. *Oikos* 45:181–190.

Keenleyside, M. H. A. 1979. *Diversity and Adaptation in Fish Behavior*. New York: Springer-Verlag.

Keeton, W. T. 1971. Magnets interfere with pigeon homing. *Proc. Nat. Acad. Sci. USA.* 68:102–106.

Keeton, W. T., T. S. Larkin, and D. M. Windsor. 1974. Normal fluctuations in the earth's magnetic field influence pigeon orientation. *J. Comp. Physiol.* 95:95–103.

Keller, F. S. 1941. Light aversion in the white rat. *Psychol. Rec.* 4:235–250.

Keller, L., and K. G. Ross. 1998. Selfish genes: A green beard in the red fire ant. *Nature* 394:573–575.

Kelley, D. B. 1988. Sexually dimorphic behaviors. *Ann. Rev. Neurosci.* 11:225–251.

Kelley, D. B., and D. I. Gorlick. 1990. Sexual selection and the nervous system. *BioScience* 40 (Apr.):275–283.

Kemp, A. C. 1971. Some observations on the sealed-in nesting method of hornbills (Family: Bucerotidae). *Ostrich* Suppl. 8:149–155.

Kemp, A. C. 1976. A study of the ecology, behaviour, and systematics of *Tockus* hornbills (Aves: Bucerotidae). *Transvaal Mus. Monogr.* 20, 125 pp.

Kent, D. S., and J. A. Simpson. 1992. Eusociality in the beetle *Austroplatypus incompertus* (Coleoptera: Curculionidae). *Naturwissenschaften* 79:86–87.

Kerr, W. E., R. Zucchi, J. T. Nakakaira, and J. E. Butolo. 1962. Reproduction in the social bees. *J. NY Entomol. Soc.* 70:265–270.

Kessel, E. L. 1955. Mating activities of balloon flies. *Syst. Zool.* 4:97–104.

Ketterson, E. D., and V. Nolan. 1992. Hormones and life histories: An integrative approach. *Am. Nat.* Suppl. 140:33–62.

Kettlewell, H. B. D. 1955. Selection experiments on industrial melanism in the Lepidoptera. *Heredity* 9:323–343.

Kettlewell, H. B. D. 1956. Further selection experiments on industrial melanism in the Lepidoptera. *Heredity* 10:287–301.

Kettlewell, H. B. D. 1973. *The Evolution of Melanism*. Oxford: Clarendon Press.

Kiepenheuer, J. 1978. A repetition of the deflector loft experiment. *Behav. Ecol. Sociobiol.* 3:393–395.

Kiepenheuer, J. 1979. Pigeon homing: Deprivation of olfactory information does not affect the deflector effect. *Behav. Ecol. Sociobiol.* 6:11–22.

Kiepenheuer, J. 1982. The effect of magnetic anomalies on the homing behavior of pigeons. In *Avian Navigation*, edited by F. Papi and H. G. Wallraff. Berlin: Springer-Verlag, pp. 120–128.

Kilner, R. 1995. When do canary parents respond to nestling signals of need? *Proc. Roy. Soc. Lond. B* 260:343–348.

Kiltie, R. A. 1988. Countershading: Universally deceptive or deceptively universal? *Trends Ecol. Evol.* 3:21–23.

Kiltie, R. A. 1989. Wildfire and the evolution of dorsal melanism in fox squirrels, *Sciurus niger. J. Mamm.* 70:726–739.

Kimble, G. A. 1961. *Hilgard and Marquis' Conditioning and Learning.* New York: Appleton-Century-Crofts.

King, A. P., and M. J. West. 1983. Epigenesis of cowbird song—A joint endeavor of males and females. *Nature* 305:704–706.

Kirchner, W. H., and W. F. Towne. 1994. The sensory basis of the honeybee's dance language. *Sci. Am.* 270 (June):74–80.

Kirchner, W. H., and U. Braun. 1994. Dancing honey bees indicate the location of food sources using path integration rather than cognitive maps. *Anim. Behav.* 48 (6):1437–1441.

Kirkpatrick, M. 1982. Sexual selection and the evolution of female choice. *Evolution* 36:1–12.

Kirschvink, J. L. 1997. Homing in on vertebrates. *Nature* 390:339–341.

Kirschvink, J. L., and J. L. Gould. 1981. Biogenic magnetite as a basis for magnetic field detection in animals. *Biosystems* 13:181–201.

Kirschvink, J. L., S. Padmanabha, C. K. Boyce, and J. Oglesby. 1997. Measurement of the threshold of honeybees to weak, extremely low-frequency magnetic fields. *J. Exp. Biol.* 200:1363–1368.

Kirschvink, J. L., M. M. Walker, S. B. Chang, A. E. Dizon, and K. A. Peterson. 1985. Chains of single-domain magnetite particles in chinook salmon, *Oncorhynchus tshawytscha. J. Comp. Physiol. A* 157:375–381.

Klaassen, M. 1996. Metabolic constraints on long-distance migration in birds. *J. Exp. Biol.* 199 (1):57–64.

Klapow, L. A. 1972. Natural and artificial rephasing of a tidal rhythm. *J. Comp. Physiol.* 79:233–258.

Kleiman, D. G. 1977. Monogamy in mammals. *Quart. Rev. Biol.* 52:39–69.

Klein, M., and E. R. Kandel. 1978. Presynaptic modulation of voltage-dependent Ca^{2+} current: Mechanism for behavioral sensitization in *Aplysia californica. Proc. Nat. Acad. Sci. USA* 75:3512–3516.

Klopfer, P. 1988. Metaphors for development: How important are experiences early in life? *Dev. Psychobiol.* 21:671–678.

Klopfer, P. H. 1971. Mother love: What turns it on? *Am. Sci.* 59:404–407.

Klopfer, P. H. 1996. "Mother love" revisited: On the use of animal models. *Am. Sci.* 84 (4):319–321.

Knowlton, N. 1979. Reproductive synchrony, parental investment, and the evolutionary dynamics of sexual selection. *Anim. Behav.* 27:1022–1033.

Knudsen, E. I. 1981. The hearing of the barn owl. *Sci. Am.* 245 (June):113–125.

Knudsen, E. I. 1982. Auditory and visual maps of space in the optic tectum of the owl. *J. Neurosci.* 2:1177–1194.

Knudsen, E. I., and M. Konishi. 1978. A neural map of auditory space in the owl. *Science* 200:795–797.

Koenig, W. D., and P. B. Stacey. 1990. Acorn woodpeckers: Group-living and food storage under contrasting ecological conditions. In *Cooperative Breeding in Birds*, edited by P. B. Stacey and W. D. Koenig. Cambridge: Cambridge University Press, pp. 415–453.

Köhler, W. 1927. *The Mentality of Apes.* New York: Harcourt Brace.

Kolata, G. 1984. Studying learning in the womb. *Science* 225:302–303.

Koltermann, R. 1971. 24-std-Periodik in der Langzeiterinnerung an Duft-und Farbsignale bei der Honigbiene. *Z. Vergl. Physiol.* 75:49–68.

Komers, P. E. 1997. Property rites. *Nat. Hist.* 106 (2):28–31.

Konishi, M. 1965. The role of auditory feedback in the control of vocalization in the white-crowned sparrow. *Z. Tierpsychol.* 22:770–783.

Konishi, M. 1973. Locatable and nonlocatable acoustic signals for barn owls. *Am. Nat.* 107:775–785.

Konishi, M. 1993a. Listening with two ears. *Sci. Am.* 286 (Apr.):66–73.

Konishi, M. 1993b. The neuroethology of sound localization in the owl. *J. Comp. Physiol. A* 173:3–7.

Konopka, R. J., and S. Benzer. 1971. Clock mutants of *Drosophila melanogaster. Proc. Nat. Acad. Sci. USA* 68:2112–2116.

Kowalski, U., R. Wiltschko, and E. Fuller. 1988. Normal fluctuations of the geomagnetic field may affect initial orientation in pigeons. *J. Comp. Physiol. A* 163:593–600.

Kramer, B. 1990. Sexual signals in electric fishes. *Trends Ecol. Evol.* 5:247–250.

Kramer, B., and B. Kuhn. 1994. Species recognition by the sequence of discharge intervals in weakly electric fishes of the genus *Campylomormyrus* (Mormyridae, Teleostei). *Anim. Behav.* 48 (2):435–445.

Kramer, D. L., and D. M. Weary. 1991. Exploration versus exploitation: A field study of time allocation to environmental tracking by foraging chipmunks. *Anim. Behav.* 41:443–449.

Kramer, G. 1949. Über Richtungstendenzen bei der nächtlichen Zugunruhe gekäfigter Vögel. In *Ornithologie als biologische Wissenschaft*, edited by E. Mayr and E. Schüz. Heidelberg: Carl Winter, pp. 269–283.

Kramer, G. 1950. Weitere Analyse der Faktoren, welche die Zugaktivität des gekäfigten Vogels orientieren. *Naturwissenschaften* 37:377–378.

Kramer, G. 1951. Eine neue Methode zur Erforschung der Zugorientierung und die bisher damit erzielten Ergebnisse. In *Proceedings of the 10th International Ornithological Congress, Uppsala*, 1950. pp. 269–280.

Krause, J. 1993. The effect of "Schreckstoff" on the shoaling behavior of the minnow: A test of Hamilton's selfish herd theory. *Anim. Behav.* 45 (5):1019–1024.

Krause, K., G. Cremer-Bartels, and G. Mitoskas. 1985. Effects of low magnetic field on human and avian retina. In *The Pineal Gland: Endocrine Aspects. Advances in the Biosciences*, edited by G. M. Brown and S. D. Wainwright. New York: Pergamon Press, pp. 209–215.

Krebs, E. A. 1999. Last but not least: Nestling growth and survival in asynchronously hatching crimson rosellas. *J. Anim. Ecol.* 68 (2):266–281.

Krebs, E. A., R. B. Cunnigham, and C. F. Donnely. 1999. Complex food allocation in asynchronously hatching broods of crimson rosellas. *Anim. Behav.* 57 (4):753–763.

Krebs, J. R. 1985. Sociobiology ten years on. *New Sci.* 1476:40–43.

Krebs, J. R., and M. I. Avery. 1984. Chick growth and prey quality in the European bee-eater (*Merops apiaster*). *Oecologia* 64:363–368.

Krebs, J. R., N. S. Clayton, S. D. Healy, D. A. Cristol, S. N. Patel, and A. R. Jolliffe. 1996. The ecology of the avian brain: Food-storing memory and the hippocampus. *Ibis* 138:34–46.

Krebs, J. R., and N. B. Davies. 1997. *Behavioural Ecology: An Evolutionary Approach*. 4th ed. Oxford: Blackwell Science.

Krebs, J. R., and R. Davies. 1984. Animal signals: Mind-reading and manipulation. In *Behavioural Ecology: An Evolutionary Approach*, edited by J. R. Krebs and N. B. Davies. Sunderland MA: Sinauer, pp. 380–402.

Krebs, J. R., J. T. Erichsen, M. J. Webber, and E. L. Charnov. 1977. Optimal prey selection in the great tit (*Parus major*). *Anim. Behav.* 25:30–38.

Krebs, J. R., D. F. Sherry, S. D. Healy, V. H. Perry, and A. L. Vaccarino. 1989. Hippocampal specialization of food storing birds. *Proc. Nat. Acad. Sci. USA* 86:1388–1392.

Kreithen, M. L., and D. B. Quine. 1979. Infrasound detection by the homing pigeon: A behavioral audiogram. *J. Comp. Physiol.* 129:1–4.

Kreulen, D. 1975. Wildebeest habitat selection on the Serengeti plains, Tanzania, in relation to calcium and lactation: A preliminary report. *E. Afr. Wildl. J.* 13:297–304.

Krohmer, R. W., and D. Crews. 1987. Temperature activation of courtship behavior in the male red-sided garter snake (*Thamnophis sirtalis parietalis*): Role of the anterior hypothalamus-preoptic area. *Behav. Neurosci.* 101:228–236.

Kroodsma, D. E. 1974. Song learning, dialects, and dispersal in Bewick's wren. *Z. Tierpsychol.* 66:189–226.

Kroodsma, D. E., and B. Byers. 1991. The functions of bird song. *Am. Zool.* 31:318–328.

Kroodsma, D. E., and M. Konishi. 1991. A suboseine bird (eastern phoebé, *Sayornis phoebe*) develops normal song without auditory feedback. *Anim. Behav.* 42:477–487.

Kroodsma, D. E., and R. Pickert. 1980. Environmentally dependent sensitive periods for avian vocal learning. *Nature* 288:477–479.

Kume, K., M. J. Zylka, S. Sriram, L. P. Shearman, D. R. Weaver, X. Jin, E. S. Maywood, M. H. Hastings, and S. M. Reppert. 1999. mCRY1 and mCRY2 are essential components of the negative limb of the circadian clock feedback loop. *Cell* 98:193–205.

Kummer, H., W. Gotz, and W. Angst. 1974. Triadic differentiation: An inhibitory process protecting pair bonds in baboons. *Behaviour* 49:62–87.

Kupfermann, I., V. Castellucci, H. Pinsker, and E. Kandel. 1970. Neuronal correlates of habituation and dishabituation of the gill withdrawal reflex in *Aplysia*. *Science* 167:1743–1745.

Kutterbach, D. A., C. Walcott, R. J. Reeder, and R. B. Frankel. 1982. Iron-containing cells in the honey bee (*Apis mellifera*). *Science* 218:695–697.

Kyriacou, C. P., and J. C. Hall. 1980. Circadian rhythm mutations in *Drosophila melanogaster* affect short term fluctuations in the male's courtship song. *Proc. Nat. Acad. Sci. USA* 77:6729–6733.

Labov, J. B. 1981. Pregnancy blocking in rodents: Adaptive advantages for females. *Am. Nat.* 118:361–371.

Lacey, E. A., and P. W. Sherman. 1991. Social organization of naked mole-rat colonies: Evidence of division of labor. In *The Biology of the Naked Mole-Rat*, edited by P. W. Sherman, J. U. M. Jarvis, and R. D. Alexander. Princeton, NJ: Princeton University Press, pp. 275–336.

Lacey, E. A., and P. W. Sherman. 1997. Cooperative breeding in naked mole-rats: Implications for vertebrate and invertebrate sociality. In *Cooperative Breeding in Mammals*, edited by N. G. Solomon and J. A. French. Cambridge: Cambridge University Press, pp. 267–301.

Lack, D. 1939. The display of the black cock. *Brit. Birds* 32:290–303.

Lack, D. 1943. *The Life of the Robin*. London: Penguin.

Lack, D. 1968a. Bird migration and natural selection. *Oikos* 19:1–9.

Lack, D. 1968b. *Ecological Adaptations for Breeding Birds*. London: Chapman & Hall.

Lacy, R. C., and P. W. Sherman. 1983. Kin recognition by phenotype matching. *Am. Nat.* 121:489–512.

Laidlaw, H. H., Jr., and R. E. Page, Jr. 1984. Polyandry in honey bees (*Apis mellifera* L.). Sperm utilization and intracolony genetic relationships. *Genetics* 108:985–997.

Laland, K. N., and K. Williams. 1997. Shoaling generates social learning of foraging information in guppies. *Anim. Behav.* 53:1149–1159.

Lall, A. B., H. H. Seliger, W. H. Biggley, and J. E. Lloyd. 1980. The ecology of colors of firefly bioluminescence. *Science* 210:560–562.

Lambert, S., and G. M. Ferguson. 1985. Blood ejection frequency by *Phrynosoma cornutum* (Iguanidae). *SW Nat.* 30:616–617.

Lande, R. 1981. Models of speciation by sexual selection on polygenic traits. *Proc. Nat. Acad. Sci. USA* 78:3721–3725.

Lang, A. 1996. Silk investment in gifts by males of the nuptial feeding spider *Pisaura mirabilis* (Araneae, Pisauridae). *Behaviour* 133 (9–10):697–716.

Lashley, K. 1950. In search of the engram. *Symp. Soc. Exp. Biol.* 4:454–482.

Lawes, M. J., and M. R. Perin. 1995. Risk sensitive foraging behavior of the round-eared elephant shrew (*Macroscelides proboscideus*). *Behav. Ecol. Sociobiol.* 37 (1):31–37.

Lawrence, P. A. 1981. A general cell marker for the clonal analysis of *Drosophila* development. *J. Embryo. Exper. Morphol.* 64:321–332.

Leake, L. D. 1986. Leech Retzius cells and 5–hydroxytryptamine. *Comp. Biochem. Physiol.* 83C:229–239.

Leask, M. J. M. 1977a. A physico-chemical mechanism for magnetic field detection by migratory birds and homing pigeons. *Nature* 267:144–145.

Leask, M. J. M. 1977b. Primitive models of magnetoreception. In *Animal Migration, Navigation and Homing*, edited by K. Schmidt-Koenig and W. T. Keeton. New York: Springer-Verlag, pp. 318–322.

LeBoeuf, B. J. 1974. Male-male competition and reproductive success in elephant seals. *Am. Zool.* 14:163–176.

Lednor, A. J. 1982. Magnetic navigation in pigeons: Possibilities and problems. In *Avian Navigation*, edited by F. Papi and H. G. Wallraff. Berlin: Springer-Verlag, pp. 109–119.

Lee, C. G., V. Parikh, T. Itusukaichi, K. Bae, and I. Edery. 1996. Resetting the *Drosophila* clock by photic regulation of PER and a PER-TIM complex. *Science* 271 (5256):1740–1744.

Lefebvre, L. 1995. Culturally-transmitted feeding behaviour in primates: Evidence for accelerating learning rates. *Primates* 36 (2):227–239.

Lefebvre, L., J. Templeton, K. Brown, and M. Koelle. 1997. Carib grackles imitate conspecific and Zenaida dove tutors. *Behaviour* 134 (13–14):1003–1017.

Lehman, M. N., R. Silver, W. R. Gladstone, R. M. Kahn, M. L. Gibson, and E. L. Bittman. 1987. Circadian rhythmicity restored by neural transplant. Immunocytochemical characterization of the graft and its integration with the host. *J. Neurosci.* 7:1626–1639.

Lehrman, D. S. 1953. A critique of Konrad Lorenz's theory of instinctive behavior. *Quart. Rev. Biol.* 28:337–363.

Lehrman, D. S. 1970. Semantic and conceptual issues in the nature-nurture problem. In *Development and Evolution of Behavior; Essays in Honor of T. C. Schneirla*, edited by L. R. Aronson, E. Tobach, D. S. Lehrman and J. S. Rosenblatt. San Francisco: Freeman, pp. 17–52.

Leibrecht, B. C., and H. R. Askew. 1980. Habituation from a comparative perspective. In *Comparative Psychology: An Evolutionary Analysis of Animal Behavior*, edited by M. R. Denny. New York: Wiley, pp. 208–229.

Leighton, M. 1986. Hornbill social dispersion: Variations on a monogamous theme. In *Ecological Aspects of Social Evolution*, edited by D. I. Rubenstein and R. W. Wrangham. Princeton, NJ: Princeton University Press, pp. 108–130.

Lenneberg, E. H. 1967. *Biological Foundations of Language*. New York: Wiley.

Lent, C. M., M. H. Dickinson, and C. G. Marshall. 1989. Serotonin and leech feeding behavior: Obligatory neuromodulation. *Am. Zool.* 29:1241–1254.

Lent, C. M., and W. H. Watson III. 1989. Introduction to the symposium: Behavioral neuromodulators: Cellular, comparative and evolutionary patterns. *Am. Zool.* 29:1211–1212.

LeSauter, J., and R. Silver. 1998. Output signals of the SCN. *Chronobiol. Int.* 15 (5):535–550.

Leucht, T. 1984. Responses to light under varying magnetic conditions in the honeybee, *Apis mellifera. J. Comp. Physiol. A* 154:865–870.

Levin, L. R., P.-L. Han, P. M. Hwaung, P. G. Feinstein, R. L. Davis, and R. R. Reed. 1992. The *Drosophila* learning and memory gene *rutabaga* encodes a Ca^{+2}/ calmodulin-adenylyl cyclase. *Cell* 68:479–489.

Lewis, E. R., and P. M. Narins. 1985. Do frogs communicate with seismic signals? *Science* 227:187–189.

Liberg, O., and T. von Schantz. 1985. Sex-biased philopatry and dispersal in birds and mammals: The Oedipus hypothesis. *Am. Nat.* 126:129–135.

Lickliter, R., and G. Gottlieb. 1988. Social specificity: Interaction with own species is necessary to foster species-specific maternal preferences in ducklings. *Dev. Psychobiol.* 21:311–321.

Liebenthal, E., O. Uhlmann, and J. M. Camhi. 1994. Critical parameters of the spike trains in a cell assembly: Coding of turn direction by the giant interneurons of the cockroach. *J. Comp. Physiol. A* 174:281–296.

Light, P., M. Salmon, and K. J. Lohmann. 1993. Geomagnetic orientation of loggerhead sea turtles: Evidence for an inclination compass. *J. Exp. Biol.* 182:1–10.

Ligon, J. D., and P. B. Stacey. 1991. The origin and maintenance of helping behavior in birds. *Am. Nat.* 138:254–258.

Lin, N., and C. D. Michener. 1972. Evolution of sociality in insects. *Quart. Rev. Biol.* 47:131–159.

Linden, E. 1992. A curious kinship: Apes and humans. *Nat. Geogr. Mag.* 181 (3):2–45.

Linden, M., and A. P. Møller. 1989. Cost of reproduction and covariation of life history traits in birds. *Trends Ecol. Evol.* 4:367–371.

Lindström, Å., and T. Allerstam. 1986. The adaptive significance of reoriented migration of chaffinches *Fringilla coelebs* and bramblings *F. montifringilla* during autumn migration in southern Sweden. *Behav. Ecol. Sociobiol.* 19:417–424.

Lindström, Å., D. Hasselquist, S. Bensch, and M. Grahn. 1990. Asymmetric contests over resources for survival and migration: A field experiment with bluethroats. *Anim. Behav.* 40:453–461.

Lindstrom, L., R. V. Alatalo, and J. Mappes. 1997. Imperfect Batesian mimicry—The effects of the frequency and the distastefulness of the model. *Proc. Roy. Soc. Lond. B* 264 (1379):149–153.

Litte, M. 1977. Behavioral ecology of the social wasp, *Mischocyttarus mexicanus. Behav. Ecol. Sociobiol.* 2:229–246.

Littlejohn, M. J. 1977. Long-range acoustic communication in anurans: Integrated evolutionary approach. In *The*

Reproductive Biology of Amphibians, edited by D. H. Taylor and S. I. Guttman. New York: Plenum, pp. 263–294.

Littlejohn, M. J., and J. J. Loftus-Hills. 1968. An experimental evaluation of premating isolation in the *Hyla ewingi* complex (Anura: Hylidae). *Evolution* 22:659–663.

Liu, X., L. Lorenz, Q. Yu, J. C. Hall, and M. Rosbash. 1988. Spatial and temporal expression of the *period* gene in *Drosophila melanogaster. Genes Dev.* 2:228–238.

Lloyd, J. E. 1968. Illumination, another function of firefly flashes? *Entomol. News* 79:265–268.

Loeb, J. 1918. *Forced Movements, Tropisms and Animal Conduct.* Philadelphia: Lippincott.

Lohmann, K. J. 1991. Magnetic orientation by hatchling loggerhead sea turtles (*Caretta caretta*). *J. Exp. Biol.* 155:37–49.

Lohmann, K. J. 1992. How sea turtles navigate. *Sci. Am.* 266 (Jan.):100–106.

Lohmann, K. J., and C. M. F. Lohmann. 1996a. Detection of magnetic field intensity by sea turtles. *Nature* 380:59–61.

Lohmann, K. J., and C. M. F. Lohmann. 1996b. Orientation and open-sea navigation in sea turtles. *J. Exp. Biol.* 199 (1):73–81.

Lohmann, K. J., N. D. Pentcheff, G. A. Nevitt, G. D. Stetten, R. K. Zimmer-Faust, H. E. Jarrard, and L. C. Boles. 1995. Magnetic orientation of spiny lobsters in the ocean: Experiments with undersea coil systems. *J. Exp. Biol.* 198:2041–2048.

Lorenz, K. 1935. Der Kumpan in der Umwelt des Vogels. *J. Ornithol.* 83:137–213.

Lorenz, K. 1937. Über die Bildung des Instinkbegriffes. *Naturwissenschaften* 25:289–331.

Lorenz, K. 1950. The comparative method in studying innate behavior patterns. *Symp. Soc. Exp. Biol.* 4:221–268.

Lorenz, K. 1952. *King Solomon's Ring.* New York: Crowell.

Lorenz, K. 1958. The evolution of behavior. *Sci. Am.* 199 (Dec.):67–78.

Lorenz, K. 1972. Comparative studies on the behavior of Anatinae. In *Function and Evolution of Behavior: An Historical Sample from the Pens of Ethologists*, edited by P. H. Klopfer and J. P. Hailman. Reading, MA: Addison-Wesley, pp. 231–258.

Lorenz, K. 1981. *The Foundations of Ethology.* New York: Springer-Verlag.

Lorenz, K., and N. Tinbergen. 1938. Taxis und Instinkhandlung in der Eirollbewegung der Graugans. *Z. Tierpsychol.* 2:1–29.

Lovegrove, B. G. 1991. The evolution of eusociality in molerats (Bathyergidae): A question of risks, numbers, and costs. *Behav. Ecol. Sociobiol.* 28 (1):37–46.

Lovegrove, B. G., and C. Wissel. 1988. Sociality in molerats: Metabolic scaling and the role of risk sensitivity. *Oecologia* 74:600–606.

Low, B. S. 1990. Marriage systems and pathogen stress in human societies. *Am. Zool.* 30:325–339.

Lucia, C. M., and D. R. Osborne. 1983. Sunset as an orientation cue in the white-throated sparrow. *Ohio J. Sci.* 83:185–188.

Lupi, D., H. M. Cooper, A. Froehlich, L. Standford, M. A. McCall, and R. G. Foster. 1999. Transgenic ablation of rod photoreceptors alters the circadian phenotype of mice. *Neuroscience* 89 (2):363–374.

Lynch, C. B. 1980. Response to divergent selection for nesting behavior in *Mus musculus. Genetics* 96:757–765.

Lynch, C. B. 1992. Clinal variation in cold adaptation in *Mus domesticus:* Verification of predictions from laboratory populations. *Am. Nat.* 139:1219–1236.

Lythgoe, J. N., W. R. A. Muntz, J. C. Partridge, J. Shand, and D. McB Williams. 1994. The ecology of visual pigments of snappers (Lutjanidae) on the Great Barrier Reef. *J. Comp. Physiol. A* 174:255–260.

MacRoberts, M. H., and B. R. MacRoberts. 1976. Social organization and behavior of the acorn woodpecker in central coastal California. *Ornithol. Monogr.* 21:1–115.

Maddock, L. 1979. The "migration" and grazing succession. In *Serengeti: Dynamics of an Ecosystem*, edited by A. R. E. Sinclair and M. Norton-Griffiths. Chicago: University of Chicago Press, pp. 104–129.

Madison, D. M., A. T. Scholz, J. C. Cooper, R. M. Horral, and A. D. Hasler. 1973. Olfactory hypotheses and salmon migration: A synopsis of recent findings. *Fish. Res. Bd. Can. Tech. Report* 414.

Madsen, T., and R. Shine. 1996. Seasonal migrations of predators and prey—A study of pythons and rats in tropical Australia. *Ecology* 77 (1):149–156.

Maeterlinck, M. 1901. *The Life of the Bee.* New York: Dodd, Mead.

Maher, C. R., and D. F. Lott. 1995. Definitions of territoriality used in the study of variation in vertebrate spacing systems. *Anim. Behav.* 49 (6):1581–1597.

Maldonado, H., A. Romano, and D. Tomsie. 1997. Long-term habituation (LTH) in the crab *Chasmagnathus*: A model for behavioral and mechanistic studies of memory. *Braz. J. Med. Res.* 30 (7):813–826.

Mallot, R. W., and J. W. Siddall. 1972. Acquisition of the people concept in pigeons. *Psychol. Rep.* 31:3–13.

Manger, P. R., and J. D. Pettigrew. 1995. Electroreception and the feeding behavior of the platypus (*Ornithorhynchus anatinus*, Monotremata, Mammalia). *Phil. Trans. Roy. Soc. Lond. B* 347 (1322):359–381.

Mann, N. I., and P. J. B. Slater. 1994. What causes young male zebra finches, *Taeniopygia guttata*, to choose their father as song tutor? *Anim. Behav.* 47:671–677.

Mann, N. I., and P. J. B. Slater. 1995. Song tutor choice by zebra finches in aviaries. *Anim. Behav.* 49 (3):811–820.

Mann, S., N. H. C. Sparks, M. M. Walker, and J. L. Kirschvink. 1988. Ultrastructure, morphology and organization of biogenic magnetite from sockeye salmon, *Oncorhynchus nerka*: Implications for magnetoreception. *J. Exp. Biol.* 140:35–49.

Manzure, M., and E. Fuentes. 1979. Polygyny and agnostic behaviour in the tree-dwelling lizard, *Liolaemus tenuis. Behav. Ecol. Sociobiol.* 6:23–28.

Mappes, J., and R. V. Alatalo. 1997. Batesian mimicry and signal accuracy. *Evolution* 51 (6):2050–2053.

Margulis, S. W., W. Saltzman, and D. H. Abbott. 1995. Behavioral and hormonal changes in female naked mole-rats (*Heterocephalus glaber*) following removal of the breeding female from the colony. *Horm. Behav.* 29:227–247.

Marhold, S., W. Wiltschko, and H. Burda. 1997. A magnetic polarity compass for direction finding in a subterranean mammal. *Naturwissenschaften* 84 (9):421–423.

Markel, R. W. 1994. An adaptive value of spatial learning and memory in the blackeye goby, *Coryphopterus nicholsi*. *Anim. Behav.* 47 (6):1462–1464.

Markl, H. 1985. Manipulation, modulation, information, cognition: Some of the riddles of communication. In *Experimental Behavioral Ecology and Sociobiology*, edited by B. Hölldobler and M. Lindauer. Sunderland, MA: Sinauer, pp. 163–194.

Markow, T. A., M. Quaid, and S. Kerr. 1978. Male mating experience and competitive courtship success in *Drosophila melanogaster*. *Nature* 276:821–822.

Marler, P. 1955. Characteristics of some animal calls. *Nature* 176:6–7.

Marler, P. 1957. Specific distinctiveness in the communication signals of birds. *Behaviour* 11:13–39.

Marler, P. 1967. Animal communication signals. *Science* 157:769–774.

Marler, P. 1970. A comparative approach to vocal learning: Song development in white-crowned sparrows. *J. Comp. Physiol. Psychol. Monogr.* 71:1–25.

Marler, P. 1973. A comparison of vocalizations of red-tailed monkeys and blue monkeys *Cercopithecus ascanius* and *C. mitis* in Uganda. *Z. Tierpsychol.* 33:223–247.

Marler, P. 1996. Social cognition: Are primates smarter than birds? In *Current Ornithology*, edited by V. J. Nolan and E. D. Ketterson. New York: Plenum, pp. 1–33.

Marler, P., and W. J. Hamilton III. 1966. *Mechanisms of Animal Behavior*. New York: Wiley.

Marler, P., S. Karakashian, and M. Gyger. 1991. Do animals have the option of withholding signals when communication is inappropriate? The audience effect. In *Cognitive Ethology, the Minds of Other Animals*, edited by C. A. Ristau. Hillsdale, NJ: Erlbaum, pp. 187–208.

Marler, P., and M. Tamura. 1962. Song "dialects" in three populations of white-crowned sparrows. *Condor* 64:368–377.

Marler, P., and M. Tamura. 1964. Culturally transmitted patterns of vocal behavior in sparrows. *Science* 146:1483–1486.

Martan, J., and B. A. Shepherd. 1976. The role of the copulatory plug in reproduction of the guinea pig. *J. Exp. Zool.* 196:79–84.

Marten, K., and P. Marler. 1977. Sound transmission and its significance for animal vocalization. I. Temperate habitats. *Behav. Ecol. Sociobiol.* 2:271–290.

Marten, K., D. Quine, and P. Marler. 1977. Sound transmission and its significance for animal vocalization. II. Tropical forest habitats. *Behav. Ecol. Sociobiol.* 2:291–302.

Marx, J. L. 1980. Ape-language controversy flares up. *Science* 207:1330–1333.

Massey, A. 1988. Sexual interactions in red-spotted newt populations. *Anim. Behav.* 36:205–210.

Mast, S. O. 1938. Factors involved in the process of orientation of lower organisms in light. *Biol. Rev.* 13:186–224.

Mathis, A., D. P. Chivers, and R. J. F. Smith. 1996. Cultural transmission of predator recognition in fishes: Intraspecific and interspecific learning. *Anim. Behav.* 51:185–201.

Mauck, R. A., and T. C. Grubb, Jr. 1995. Petrel parents shunt all experimentally increased reproductive costs to their offspring. *Anim. Behav.* 49 (4):999–1008.

Maxson, S. C. 1996. Searching for candidate genes with effects on an agonistic behavior, offense, in mice. *Behav. Genet.* 26 (5):471–476.

Maxson, S. J., and L. W. Oring. 1980. Breeding season time and energy budgets of the polyandrous spotted sandpiper. *Behaviour* 74:200–263.

Maynard Smith, J. 1965. The evolution of alarm calls. *Am. Nat.* 99:59–63.

Maynard Smith, J. 1972. *On Evolution*. Edinburgh: Edinburgh University Press.

Maynard Smith, J. 1974. The theory of games and the evolution of animal conflicts. *J. Theoret. Biol.* 47:209–221.

Maynard Smith, J. 1976. Evolution and the theory of games. *Am. Sci.* 64:41–45.

Maynard Smith, J. 1977. Parental investment: A prospective analysis. *Anim. Behav.* 25:1–9.

Maynard Smith, J. 1982. *Evolution and the Theory of Games*. Cambridge: Cambridge University Press.

Maynard Smith, J., and G. A. Parker. 1976. The logic of asymmetric contests. *Anim. Behav.* 24:159–175.

Maynard Smith, J., and G. R. Price. 1973. The logic of animal conflict. *Nature* 246:15–18.

Mayr, E. 1983. How to carry out the adaptationist program? *Am. Nat.* 121:295–310.

McCann, T. S. 1981. Aggression and sexual activity of male southern elephant seals. *J. Zool. Lond.* 195:295–310.

McCarty, J. P. 1996. The energetic costs of begging in nestling passerines. *Auk* 113:178–188.

McClintock, M. K. 1983. Pheromonal regulation of the ovarian cycle: Enhancement, suppression, and synchrony. In *Pheromones and Reproduction in Mammals*, edited by J. G. Vandenbergh. New York: Academic Press, pp. 113–149.

McCracken, G. F. 1984. Communal nursing in Mexican free-tailed bat maternity colonies. *Science* 223:1090–1091.

McDonald, D. B., and W. K. Potts. 1994. Cooperative display and relatedness among males in a lek-mating bird. *Science* 266:1030–1032.

McEwen, B. S. 1976. Interactions between hormones and nerve tissue. *Sci. Am.* 235 (July):48–58.

McFall-Ngal, M. J. 1990. Crypsis in the pelagic environment. *Am. Zool.* 30:175–188.

McGowan, K. J., and G. E. Woolfenden. 1989. A sentinel system in the Florida scrub jay. *Anim. Behav.* 37:1000–1006.

McGuire, B. 1988. The effects of cross-fostering on parental behavior of meadow voles (*Microtus pennsylvanicus*). *J. Mamm.* 69:332–341.

McKaye, K. R. 1977. Competition for breeding sites between the cichlid fishes of Lake Jiloa, Nicaragua. *Ecology* 58:291–302.

McKaye, K. R. 1981. Natural selection and the evolution of interspecific brood care in fishes. In *Natural Selection and Social Behavior*, edited by R. D. Alexander and D. W. Tinkle. New York: Chiron, pp. 173–183.

McKaye, K. R., and N. M. McKaye. 1977. Communal care and kidnapping of young by parental cichlids. *Evolution* 31:674–681.

McLaren, A., and D. Michie. 1960. Control of prenatal growth in mammals. *Nature* 187:363–365.

McMillan, J. P., H. C. Keatts, and M. Menaker. 1975. On the role of the eyes and brain photoreceptors in the sparrow: Entrainment to light cycles. *J. Comp. Physiol.* 102:251–256.

McNamara, J. M., S. Mermad, and A. Houston. 1991. A model for risk-sensitive foraging for a reproducing animal. *Anim. Behav.* 41:787–792.

McShea, W. J. 1989. Reproductive synchrony and home range size in a territorial microtine. *Oikos* 56:182–186.

Meisel, R., and D. W. Pfaff. 1984. RNA and protein synthesis inhibitors: Effects on sexual behaviors in female rats. *Brain Res. Bull.* 12:187–193.

Mellgren, R. L., ed. 1983. *Animal Cognition and Behavior*. Amsterdam: North Holland.

Mello, C., F. Nottebohm, and D. Clayton. 1995. Repeated exposure to one song leads to rapid and persistent decline in an immediate early gene's response to that song in zebra finch telencephalon. *J. Neurosci.* 15 (10):6919–6925.

Mello, C., D. S. Vicario, and D. F. Clayton. 1992. Song presentation induces gene expression in the songbird forebrain. *Proc. Nat. Acad. Sci. USA* 89:6818–6822.

Meltzoff, A. N. 1988. The human infant as *Homo imitans*. In *Social Learning: Psychological and Biological Perspectives*, edited by T. R. Zentall and B. G. Galef, Jr. Hillsdale, NJ: Erlbaum, pp. 319–341.

Menaker, M. 1968. Light reception by the extra-retinal receptors in the brain of the sparrow. In *Proceedings of the 76th Annual Convention of the American Psychological Association*, pp. 299–300.

Menaker, M. 1997. Commentary: What does melatonin do and how does it do it? *J. Biol. Rhythms.* 12 (6):532–534.

Mendoza, S. P., C. I. Coe, E. L. Lowe, and S. Levine. 1979. The physiological response to group formation in adult squirrel monkeys. *Psychoneuroendocrinology* 3:221–229.

Mendoza, S. P., and W. A. Mason. 1989. Behavioral and endocrine consequences of heterosexual pair formation in squirrel monkeys. *Physiol. Behav.* 46:597–603.

Menzel, R. 1989. Bee-havior and the neural systems and behavior course. In *Perspectives in Neural Systems and Behavior, MBL Lectures in Biology*, edited by T. J. Carew and D. B. Kelley. New York: Alan Liss, pp. 249–266.

Menzel, R., L. Chittka, S. Eichmüller, S. Geiger, K. Pietsch, and P. Knoll. 1990. Dominance of celestial cues over landmarks disproves map-like orientation in honey bees. *Z. Naturforsch.* 45c:723–726.

Metzgar, L. H. 1967. An experimental comparison of screech owl predation on resident and transient white-footed mice (*Peromyscus leucopus*). *J. Mamm.* 48:387–391.

Meyerriecks, A. J. 1960. Comparative breeding behavior of four species of N. American herons. *Nuttall Ornithol. Club Pub.* 2:1–158.

Michelson, A., B. B. Andersen, W. H. Kirchner, and M. Lindauer. 1989. Honeybees can be recruited by a mechanical model of a dancing bee. *Naturwissenschaften* 76:277–280.

Mikkola, K. 1979. Resting site selection by *Oligia* and *Biston* moths (Lepidoptera: Noctuidae and Geometridae). *Ann. Entomol. Fenn.* 45:81–87.

Mikkola, K. 1984. On the selective forces acting in industrial melanism of *Biston* and *Oligia* moths (Lepidoptera: Noctuidae and Geometridae). *Biol. J. Linn. Soc.* 21:409–421.

Miklósi, Á., V. Csányi, and R. Gerlai. 1997. Antipredator behavior in Paradise fish (*Macropodus opercularis*) larvae: The role of genetic factors and paternal influence. *Behav. Genet.* 27 (3):191–200.

Miles, H. L. W. 1990. The cognitive foundations for reference in a signing orangutan. In *"Language" and Intelligence in Monkeys and Apes: Comparative Developmental Perspectives*, edited by S. T. Parker and K. R. Gibson. New York: Cambridge University Press, pp. 511–539.

Milinski, M., and T. C. M. Bakker. 1990. Female sticklebacks use male coloration in mate choice and hence avoid parasitized males. *Nature* 344:330–333.

Miller, D. B. 1980. Maternal vocal control of behavioral inhibition in mallard ducklings (*Anas platyrhynchos*). *J. Comp. Physiol. Psychol.* 94:606–623.

Miller, D. B., and G. Gottlieb. 1978. Maternal vocalizations of mallard ducks (*Anas platyrhynchos*). *Anim. Behav.* 26:1178–1194.

Miller, J. D. 1998. The SCN is comprised of a population of coupled oscillators. *Chronobiol. Int.* 15 (5):489–511.

Miller, R. C. 1922. The significance of the gregarious habit. *Ecology* 3:122–126.

Mineka, S., and M. Cook. 1988. Social learning and the acquisition of snake fear in monkeys. In *Social Learning: Psychological and Biological Perspectives*, edited by T. R. Zentall and B. G. J. Galef. Hillsdale, NJ: Erlbaum, pp. 51–73.

Mock, D. W., H. Drummond, and C. H. Stinson. 1990. Avian siblicide. *Am. Sci.* 78:438–449.

Mock, D. W., and G. A. Parker. 1998. Siblicide, family conflict and the evolutionary limits of selfishness. *Anim. Behav.* 56 (1):1–10.

Moehlman, P. D. 1979. Jackal helpers and pup survival. *Nature* 277:382–383.

Moehlman, P. D. 1986. Ecology and cooperation in canids. In *Ecological Aspects of Social Evolution*, edited by D. I. Rubenstein and R. W. Wrangham. Princeton, NJ: Princeton University Press, pp. 64–86.

Möglich, M., V. Maschwitz, and B. Hölldobler. 1974. Tandem calling: A new kind of signal in ant communication. *Science* 186:1046–1047.

Mohl, B. 1985. The role of proprioception in locust flight control. II. Information signalled by forewing stretch receptors during flight. *J. Comp. Physiol. A* 156:103–166.

Moiseff, A., G. S. Pollack, and R. R. Hoy. 1978. Steering responses of flying crickets to sound and ultrasound: Mate attraction and predator avoidance. *Proc. Nat. Acad. Sci. USA* 74:4025–4056.

Molina, A., A. G. Castellano, and J. López-Barneo. 1997. Pore mutations in *Shaker* K⁺ channels distinguish between the sites of tetraethylammonium blockade and C-type inactivation. *J. Physiol.* 499 (2):361–367.

Møller, A. P. 1988a. False alarms as a means of resource usurpation in the great tit *Parus major*. *Ethology* 79:25–30.

Møller, A. P. 1988b. Female choice selects for male sexual tail ornaments in the monogamous swallow. *Nature* 322:640–642.

Møller, A. P. 1990. Effects of a haemotaphagus mite on the barn swallow (*Hirundo rustica*): A test of the Hamilton and Zuk hypothesis. *Evolution* 44:771–784.

Møller, A. P. 1991. Parasite load reduces song output in a passerine bird. *Anim. Behav.* 41:723–730.

Møller, A. P. 1992. Parasites differentially increase the degree of fluctuating asymmetry in secondary sexual characters. *J. Evol. Biol.* 5:691–699.

Möller, A. P., and T. R. Birkhead. 1989. Copulation behavior in mammals: Evidence that sperm competition is widespread. *Biol. J. Linn. Soc.* 38:119–131.

Møller, A. P., P. Christie, and E. Lux. 1999. Parasitism, host immune function, and sexual selection. *Quart. Rev. Biol.* 74 (1):3–20.

Møller, A. P., and A. Pomiankowski. 1993. Fluctuating asymmetry and sexual selection. *Genetica* 89:267–279.

Møller, A. P., and J. P. Swaddle. 1997. *Fluctuating Asymmetry, Developmental Stability and Evolution*. Oxford: Oxford University Press.

Moller, P. 1976. Electric signals and schooling behavior in weakly electric fish, *Mardusentus cyprinoides* L. (Mormyriformes). *Science* 193:697–699.

Moment, G. B. 1962. Reflexive selection: A possible answer to an old question. *Science* 136:262–263.

Montgomery, J. C. 1984. Noise cancellation in the electrosensory system of the thornback ray; common mode rejection of input produced by the animal's own ventilatory movement. *J. Comp. Physiol. A* 155:103–111.

Moore, A. J., and P. J. Moore. 1988. Female strategy during mate choice: Threshold assessment. *Evolution* 42:387–391.

Moore, B. R. 1980. Is the homing pigeon's map geomagnetic? *Nature* 285:69–70.

Moore, F. E., and P. A. Simm. 1986. Risk-sensitive foraging by a migratory bird (*Dendroica coronata*). *Experientia* 42:1054–1056.

Moore, F. R. 1978. Sunset and the orientation of a nocturnally migrant bird. *Nature* 274:154–156.

Moore, F. R. 1980. Solar cues in the orientation of the Savannah sparrow (*Passerculus sandwichensis*). *Anim. Behav.* 33:657–663.

Moore, F. R. 1986. Sunrise, skylight polarization, and the early morning orientation of night-migrating warblers. *Condor* 88:493–498.

Moore, F. R. 1987. Sunset and the orientation behaviour of migrating birds. *Biol. Rev.* 62:65–86.

Moore, F. R. 1994. Resumption of feeding under risk of predation: Effect of migratory condition. *Anim. Behav.* 48 (4):975–977.

Moore, M. C., D. K. Hews, and R. Knapp. 1998. Hormonal control and evolution of alternative male phenotypes: Generalizations of models for sexual differentiation. *Am. Zool.* 38 (1):133–151.

Moore, P. J., N. L. Regan-Wallin, K. F. Haynes, and A. J. Moore. 1997. Odour conveys status on cockroaches. *Nature* 389 (6646):25.

Moore, R. Y. 1979. The retinohypothalamic tract, suprachiasmatic hypothalamic nucleus and central neural mechanisms of circadian rhythm regulation. In *Biological Rhythms and Their Central Mechanisms*, edited by M. Suda, O. Hayaishi, and H. Nakagawa. Amsterdam: Elsevier/North Holland Biomedical Press, pp. 343–354.

Moore, R. Y., and V. B. Eichler. 1972. Loss of a circadian adrenal corticosterone rhythm following suprachiasmatic lesion in the rat. *Brain Res.* 42:201–206.

Moore, R. Y., and R. Silver. 1998. Suprachiasmatic nucleus organization. *Chronobiol. Int.* 15 (5):475–487.

Moran, G. 1984. Vigilance behavior and alarm calls in a captive group of meerkats, *Suricata suricatta*. *Z. Tierpsychol.* 65:228–240.

Morell, V. 1998. A new look at monogamy. *Science* 281:1982–1983.

Morgan, C. L. 1894. *An Introduction to Comparative Psychology*. New York: Scribner.

Morris, D. 1954. The reproductive behavior of the zebra fish *Peophilia guttata* with special reference to pseudofemale behavior and displacement activities. *Behaviour* 6:271–322.

Morris, D. 1956. The feather postures of birds and the problem of the origin of social signals. *Behaviour* 9:75–113.

Morris, D. 1957. Typical intensity and its relation to the problem of ritualization. *Behaviour* 11:1–12.

Morris, D. 1958. The reproductive behavior of the tenspined stickleback (*Pygosteus pungitius* L.). *Behaviour* Suppl. 6:1–154.

Morton, E. S. 1975. Ecological sources of selection in avian sounds. *Am. Nat.* 109:17–34.

Moum, S. E., and R. L. Baker. 1990. Colour change and substrate selection in larval *Ischnura verticalis*. *Can. J. Zool.* 68:221–224.

Mowrer, O. H. 1940. An experimental analogue of "regression" with incidental observations on "reaction-formation." *J. Abn. Soc. Psychol.* 35:56–87.

Moynihan, M. 1956. Notes on the behavior of some North American gulls. I. Aerial hostile behavior. *Behaviour* 10:126–178.

Moynihan, M. 1958. Notes on the behavior of some North American gulls. II. Nonaerial hostile behavior of adults. *Behaviour* 12:95–182.

Moynihan, M. 1967. Comparative aspects of communication in New World primates. In *Primate Ethology*, edited by D. Morris. London: Weidenfeld & Nicholson, pp. 236–266.

Moynihan, M. 1970. Some behavior patterns of platyrrhine monkeys. II. *Saguinus geoffroyi* and some other tamarins. *Smithsonian Contrib. Zool.* 28:1–77.

Mueller, U. G. 1991. Haplodiploidy and the evolution of facultative sex ratios in a primitive eusocial bee. *Science* 254:442–444.

Mullen, R. 1977a. Genetic dissection of the CNS with mutant-normal mouse and rat chimeras *Vol. 2*. In *Society for Neuroscience Symposia*, edited by W. M. Cowan and J. A. Ferrendelli. Bethesda, MD: Society of Neuroscience, pp. 47–65.

Mullen, R. 1977b. Site of *pcd* gene action and Purkinje cell mosaicism in cerebella of chimeric mice. *Nature* 270:245–247.

Munro, U., J. A. Munro, J. B. Phillips, R. Wiltschko, and W. Wiltschko. 1997. Evidence for a magnetite-based navigational "map" in birds. *Naturwissenschaften* 84 (1):26–28.

Murton, R. K., A. J. Isaacson, and N. J. Westwood. 1966. The relationships between wood-pigeons and their clover supply and the mechanism of population control. *J. App. Ecol.* 3 (1):55–96.

Myers, C. W., and J. W. Daly. 1983. Dart-poison frogs. *Sci. Am.* 248 (Feb.):120–133.

Narins, P. M., O. J. Reichman, U. M. Jarvis, and E. R. Lewis. 1992. Seismic signal transmission between burrows of the Cape mole-rat, *Georychus capensis*. *J. Comp. Physiol. A* 170:13–21.

Naumann, K., M. L. Winston, K. N. Slessor, G. D. Prestwich, and F. X. Webster. 1991. Production and transmission of honey bee queen (*Apis mellifera* L.) mandibular gland pheromone. *Behav. Ecol. Sociobiol.* 29 (5):321–332.

Negro, J. J., J. Bustmante, J. Milward, and D. M. Bird. 1996. Captive fledgling American kestrels prefer to play with objects resembling natural prey. *Anim. Behav.* 52:707–714.

Neill, S. R. St. J., and J. M. Cullen. 1974. Experiments on whether schooling by their prey affects the hunting behavior of cephalopods and fish predators. *J. Zool. Lond.* 172:549–569.

Nelson, D. A. 1994. Selection-based learning in bird song development. *Proc. Nat. Acad. Sci. USA* 91:10498–10501.

Nelson, D. A. 1998. External validity and experimental design: The sensitive phase for song learning. *Anim. Behav.* 56 (2):487–491.

Nelson, D. A., P. Marler, and M. L. Morton. 1996. Overproduction in song development—An evolutionary correlate with migration. *Anim. Behav.* 51 (5):1127–1140.

Nelson, J. B. 1967. Coloniality and cliff-nesting in the gannet. *Ardea* 55:60–90.

Nelson, M. E., and M. A. Maciver. 1999. Prey capture in the weakly electric fish *Apteronotus albifrons*: Sensory acquisition strategies and electrosensory consequences. *J. Exp. Biol.* 202:1195–1203.

Nelson, R. J. 1995. *An Introduction to Behavioral Endocrinology*. Sunderland, MA: Sinauer.

Nelson, R. J. 1997. The use of genetic "knockout" mice in behavioral endocrinology. *Horm. Behav.* 31:188–196.

Nelson, R. J., and K. A. Young. 1998. Behavior in mice with targeted disruption of single genes. *Neurosci. Biobehav. Rev.* 22 (3):453–462.

Neumann, D. 1976. Entrainment of a semi-lunar rhythm. In *Biological Rhythms in the Marine Environment*, edited by P. DeCoursey. Columbia: University of South Carolina Press, pp. 115–127.

Neuweiler, G. 1984. Foraging, echolocation and audition in bats. *Naturwissenschaften* 71:446–455.

Neuweiler, G. 1990. Auditory adaptations for prey capture in echolocating bats. *Physiol. Rev.* 70:615–641.

Nevitt, G. A. 1999a. Foraging by seabirds on an olfactory landscape. *Am. Sci.* 87 (1):46–53.

Nevitt, G. A. 1999b. Olfactory foraging in Antarctic seabirds: A species-specific attraction to krill odors. *Marine Ecol. Prog. Ser.* 177:235–241.

Nevitt, G. A., R. R. Veit, and P. Kareiva. 1995. Dimethyl sulfide as a foraging cue for Antarctic procellariiform seabirds. *Nature* 376:680–682.

Nevitt, G. A., A. H. Dittman, T. P. Quinn, and W. J. Moody, Jr. 1994. Evidence for a peripheral olfactory memory in imprinted salmon. *Proc. Nat. Acad. Sci. USA* 91:4288–4292.

New, J. G., and D. Bodznick. 1990. Medullary electrosensory processing in the little skate. II. Suppression of self-generated electrosensory interference during respiration. *J. Comp. Physiol. A* 167:295–307.

Newcombe, C., and G. Hartman. 1973. Some chemical signals in the spawning behavior of rainbow trout (*S. gairdneri*). *J. Fish. Res. Board Can.* 30:995–997.

Newman, E. A., and P. H. Hartline. 1982. The infrared "vision" of snakes. *Sci. Am.* 246 (Mar.):116–124.

Newton, P. N. 1986. Infanticide in an undisturbed population of Hanuman langurs, *Presbytis entellus*. *Anim. Behav.* 34:785–789.

Nicolai, J. 1964. Der Brutparasitismus der Viduinae als ethologisches problem. *Z. Tierpsychol* 21:9–204.

Noble, G. K. 1936. Courtship and sexual selection in the flicker (*Colaptes auratus luteus*). *Auk* 53:269–282.

Noble, G. K. 1939. The role of dominance in the social life of birds. *Auk* 56:263–273.

Noble, G. K., and E. J. Farris. 1929. The method of sex recognition in the wood frog, *Rana sylvatica* Le Conte. *Am. Mus. Novit.* 363:1–17.

Noë, R., and A. A. Sluijter. 1990. Reproductive tactics of male savanna baboons. *Behaviour* 113:117–170.

Nordeen, E. J., and K. W. Nordeen. 1990. Neurogenesis and sensitive periods in avian song learning. *Trends Neurosci.* 13:31–36.

Nordeen, K. W., E. J. Nordeen, and A. P. Arnold. 1986. Estrogen establishes sex differences in androgen accumulation in zebra finch brain. *J. Neurosci.* 6:734–738.

Nordell, S. E., and T. J. Valone. 1998. Mate choice copying as public information. *Ecol. Lett.* 1 (2):74–76.

Nordeng, H. 1971. Is the local orientation of anadromous fishes determined by pheromones? *Nature* 233:411–413.

Nordeng, H. 1977. A pheromone hypothesis for homeward migration in anadromous salmonids. *Oikos* 28:155–159.

Norris, K. S., and C. H. Lowe. 1964. An analysis of background color-matching in amphibians and reptiles. *Ecology* 45:565–580.

Nottebohm, F. 1975. Continental patterns of song variability in *Zonotrichia capensis*: Some possible ecological correlates. *Am. Nat.* 109:605–624.

Nottebohm, F., and J. P. Arnold. 1976. Sexual dimorphism in vocal control areas of the songbird brain. *Science* 194:211–213.

Oclese, J., S. Reuss, and L. Vollrath. 1985. Evidence for the involvement of the visual system in mediating magnetic field effects on pineal melatonin synthesis in the rat. *Brain Res.* 333:382–384.

Oetting, S., E. Pröve, and H.-J. Bischof. 1995. Sexual imprinting as a two-stage process: Mechanisms of information storage and stabilization. *Anim. Behav.* 50:393–403.

O'Hara, R. K., and A. R. Blaustein. 1981. An investigation of sibling recognition in *Rana cascadae* tadpoles. *Anim. Behav.* 29:1121–1126.

Okamura, H., S. Miyake, Y. Sumi, S. Yamaguchi, A. Yasui, M. Muijtjens, J. H. J. Hoeilmakers, and G. T. J. van der Horst. 1999. Photic induction of mPer 1 and mPer 2 in *Cry*-deficient mice lacking a biological clock. *Science* 286:2531–2534.

O'Leary, D. P., J. Vilches-Troya, D. F. Dunn, and A. Campoz-Munoz. 1981. Magnets in guitarfish vestibular receptors. *Experientia* 37:86–88.

Olson, D. J. 1991. Species differences in spatial memory among Clark's nutcrackers, scrub jays, and pigeons. *J. Exp. Psych.: Anim. Behav. Proc.* 17:363–376.

Olson, D. J., A. C. Kamil, R. P. Balda, and P. J. Nims. 1995. Performance of four seed-caching corvid species in operant tests of nonspatial and spatial memory. *J. Comp. Psychol.* 109 (2):173–181.

Orchinik, M., P. Licht, and D. Crews. 1988. Plasma steroid concentrations change in response to sexual behavior in *Bufo marinus*. *Horm. Behav.* 22:338–350.

Orians, G. H. 1969. On the evolution of mating systems of birds and mammals. *Am. Nat.* 103:589–603.

Orians, G. H., and G. M. Christman. 1968. A comparative study of red-winged, tri-colored and yellow-headed blackbirds. *University of California Publications in Zoology* No. 84.

Oring, L. W., and D. B. Lank. 1986. Polyandry in spotted sandpipers: The impact of environment and experience. In *Ecological Aspects of Social Evolution*, edited by D. I. Rubenstein and R. W. Wrangham. Princeton, NJ: Princeton University Press, pp. 21–42.

Ostfeld, R. S. 1985. Limiting resources and territoriality in microtine rodents. *Am. Nat.* 126:1–15.

Ostfeld, R. S. 1990. The ecology of territoriality in small mammals. *Trends Ecol. Evol.* 5:411–415.

O'Sullivan, C., and C. Yeager. 1989. Communicative context and linguistic competence. In *Teaching Sign Language in Chimpanzees*, edited by R. G. Gardner, B. T. Gardner, and T. E. Van Cantfort. Albany: State University of New York Press, pp. 269–279.

Owen, D. 1980. *Camouflage and Mimicry*. Chicago: University of Chicago Press.

Owen, P. C., and S. A. Perrill. 1998. Habituation in the green frog, *Rana clamitans*. *Behav. Ecol. Sociobiol.* 44 (3):209–213.

Owens, D., and M. Owens. 1996. Social dominance and reproductive patterns in brown hyaenas, *Hyaena brunnea*, of the central Kalahari desert. *Anim. Behav.* 51 (3):535–551.

Packer, C. 1977. Reciprocal altruism in *Papio anubis*. *Nature* 265:441–443.

Packer, C. 1986. The ecology of felid sociality. In *Ecological Aspects of Social Evolution*, edited by D. J. Rubenstein and R. W. Wrangham. Princeton NJ: Princeton University Press, pp. 429–451.

Packer, C., D. A. Gilbert, A. E. Pusey, and S. J. O'Brien. 1991. A molecular genetic analysis of kinship and cooperation in African lions. *Nature*, 351:562–565.

Packer, C., L. Herbst, A. E. Pusey, J. D. Bygott, J. P. Hanby, S. J. Cairns, and M. B. Mulder. 1988. Reproductive success in lions. In *Reproductive Success, Studies of Individual Variation in Contrasting Breeding Systems*, edited by T. H. Clutton-Brock. Chicago: University of Chicago Press, pp. 363–383.

Packer, C., and A. E. Pusey. 1982. Cooperation and competition within coalitions of male lions: Kin selection or game theory? *Nature* 296:740–742.

Packer, C., and A. E. Pusey. 1987. Intrasexual cooperation and the sex ratio in African lions. *Am. Nat.* 130:636–642.

Packer, C., and L. Ruttan. 1988. The evolution of cooperative hunting. *Am. Nat.* 132:159–198.

Packer, C., D. Scheel, and A. E. Pusey. 1990. Why lions form groups: Food is not enough. *Am. Nat.* 119:263–281.

Packer, L., and R. Owen. 1994. Relatedness and sex ratio in a primitively eusocial halictine bee. *Behav. Ecol. Sociobiol.* 34:1–10.

Page, R. E., Jr., and R. A. Metcalf. 1982. Multiple mating, sperm utilization, and social evolution. *Am. Nat.* 119:263–281.

Page, T. L. 1983. Regeneration of rhythms in lobectomized roaches. *J. Comp. Physiol.* 152:231–240.

Page, T. L. 1985. Circadian organization in cockroaches: Effects of temperature cycles on locomotor activity. *J. Insect Physiol.* 31:235–242.

Palmer, J. D. 1966. How a bird can tell the time of day. *Nat. Hist.* 75:48–53.

Palmer, J. D. 1990. The rhythmic lives of crabs. *BioScience* 40:352–357.

Palmer, J. D. 1995. *The Biological Rhythms and Clocks of Intertidal Animals*. New York: Oxford University Press.

Pankiw, T., Z. Y. Huang, M. L. Winston, and G. E. Robinson. 1998. Queen mandibular gland pheromone influences worker honey bee (*Apis mellifera* L.) foraging ontogeny and juvenile hormone titers. *J. Insect Physiol.* 44 (7–8):685–692.

Papi, F. 1986. Pigeon navigation: Solved problems and open questions. *Monit. Zool. Ital. (N.S.)* 20:471–517.

Papi, F. 1995. Recent experiments on pigeon navigation. In *Behavioural Brain Research in Naturalistic and Semi-naturalistic Settings*, edited by E. Alleva, A. Fasolo, H. P. Lipp, L. Nadel, and L. Ricceri. Dordrecht: Kluwer, pp. 225–238.

Papi, F., L. Fiore, V. Fiaschi, and S. Benvenuti. 1972. Olfaction and homing in pigeons. *Monit. Zool. Ital. (N.S.)* 6:85–95.

Parish, D. M. B., and J. C. Coulson. 1998. Parental investment, reproductive success and polygyny in the lapwing, *Vanellus vanellus*. *Anim. Behav.* 56 (5):1161–1167.

Parker, G. A. 1970. Sperm competition and its evolutionary consequences in the insects. *Biol. Rev.* 45:525–567.

Parker, G. A. 1974. Assessment strategy and the evolution of fighting behavior. *J. Theoret. Biol.* 47:223–243.

Parker, G. A. 1984. Sperm competition and the evolution of mating strategies. In *Sperm Competition and the Evolution of Animal Mating Systems*, edited by R. L. Smith. Orlando, FL: Academic Press, pp. 2–60.

Parkes, A. S., and H. M. Bruce. 1961. Olfactory stimuli in mammalian reproduction. *Science* 134:1049–1054.

Parrish, J. K. 1989. Re-examining the selfish herd: Are central fish safer? *Anim. Behav.* 38:1048–1053.

Partridge, L. 1980. Mate choice influences a component of offspring fitness in fruit flies. *Nature* 283:290–291.

Partridge, L. 1988. The rare-male effect: What is its evolutionary significance. *Phil. Trans. Roy. Soc. Lond. B* 319:525–539.

Partridge, L., and W. G. Hill. 1984. Mechanisms for frequency-dependent mating success. *Biol. J. Linn. Soc.* 23:113–132.

Patterson, F. 1978. The gestures of a gorilla: Sign language acquisition in another pongid species. *Brain Lang.* 5:72–97.

Patterson, F. L. 1990. Language acquisition by a lowland gorilla; Koko's first ten years. *Word* 41 (2):97–143.

Pavlov, I. 1927. *The Conditioned Reflex*. Translated by G. V. Anrep. London: Oxford University Press.

Payne, R. B. 1962. How the barn owl locates its prey by hearing. *Living Bird* 1:151–159.

Payne, R. B. 1977. The ecology of brood parasitism in birds. *Ann. Rev. Ecol. Syst.* 8:1–28.

Payne, R. B., and K. Payne. 1977. Social organization and mating success in local song populations of village indigobirds (*Vidua chalybeata*). *Z. Tierpsychol.* 45:113–173.

Payne, R. S. 1972. The song of the whale. In *The Marvels of Animal Behavior*. Washington DC: National Geographic Society, pp. 144–167.

Payne, R. S., and D. Webb. 1971. Orientation by means of long range acoustic signaling in baleen whales. *Ann. NY Acad. Sci.* 188:110–142.

Pearson, K. G., D. N. Reye, and R. M. Robertson. 1983. Phase-dependent influences of wing stretch receptors on flight rhythm in the locust. *J. Neurophysiol.* 49:1168–1181.

Peeke, H. V. S. 1995. Habituation of a predatory response in the stickleback (*Gasterosteus aculeatus*). *Behaviour* 132 (15–16):1255–1266.

Peeke, H. V. S, and M. H. Figler. 1997. Form and function of habituation and sensitization of male courtship in the three-spined stickleback (*Gasterosteus aculeatus* L.). *Behaviour* 134 (15–16):1273–1287.

Pelkwijk, J. J. ter, and N. Tinbergen. 1937. Eine reizbiologische Analyse einiger Verhaltensweisen von *Gasterosteus aculeatus* L. *Z. Tierpsychol* 1:193–200.

Pengelley, E. T. 1975. *Circannual Clocks: Annual Biological Rhythms*. New York: Academic Press.

Penn, D., and W. K. Potts. 1998. How do major histocompatibility genes influence odor and mating preferences? *Adv. Immunol.* 69:411–435.

Penn, D. J., and W. K. Potts. 1999. The evolution of mating preferences and the major histocompatibility complex genes. *Am. Nat.* 153 (2):145–164.

Pennisi, E. 1999. Are our primate cousins "conscious"? *Science* 284:2073–2076.

Pepperberg, I. M. 1987a. Acquisition of same/different concept by an African grey parrot (*Psittacus erithacus*): Learning with respect to categories of color, shape and material. *Anim. Learn. Behav.* 15:423–432.

Pepperberg, I. M. 1987b. Evidence for conceptual quantitative abilities in the African grey parrot: Labeling of cardinal sets. *Ethology* 75:37–61.

Pepperberg, I. M. 1991. A communicative approach to animal cognition: A study of conceptual abilities of an African grey parrot. In *Cognitive Ethology, the Minds of Other Animals*, edited by C. A. Ristau. Hillsdale, NJ: Erlbaum, pp. 153–186.

Pepperberg, I. M. 1993. Cognition and communication in an African grey parrot (*Psittacus erithacus*). In *Language and Communication, Comparative Perspectives*, edited by H. L. Roitblat, L. M. Herman, and P. E. Nachtigall. Hillsdale, NJ: Lawrence Erlbaum, pp. 221–248.

Pepperberg, I. M. 1994. Numerical competence in an African grey parrot (*Psittacus erithacus*). *J. Comp. Psychol.* 108:36–44.

Pepperberg, I. M. 2000. *The Alex Studies: Cognitive and Communicative Abilities of Grey Parrots*. Cambridge, MA: Harvard University Press.

Perdeck, A. C. 1958. Two types of orientation in migrating starlings, *Sturnus vulgaris* L., and chaffinches, *Fringilla coelebs* L., as revealed by displacement experiments. *Ardea* 46:1–37.

Perdeck, A. C. 1967. Orientation of starlings after displacement to Spain. *Ardea* 51:91–104.

Pereira, H. S., D. E. MacDonald, A. J. Hilliker, and M. B. Sokolowski. 1995. *Chaser* (*Csr*), a new gene affecting larval foraging behavior in *Drosophila melanogaster*. *Genetics* 141:263–270.

Pereira, M. E., and M. K. Izard. 1989. Lactation and care for unrelated infants in forest-living ringtailed lemurs. *Am. J. Primatol.* 18:101–108.

Pérez, M., and F. Coro. 1984. Physiological characteristics of the tympanic organ in noctuid moths. I. Responses to brief acoustic pulses. *J. Comp. Physiol. A* 154:441–447.

Petrie, M. 1992. Peacocks with low mating success are more likely to suffer predation. *Anim. Behav.* 44 (3):585–586.

Petrie, M. 1994. Improved growth and survival of offspring of peacocks with more elaborate trains. *Nature* 371:598–599.

Petrie, M., and T. Halliday. 1994. Experimental and natural change in the peacock's (*Pavo cristatus*) train can affect mating success. *Behav. Ecol. Sociobiol.* 35 (3):213–217.

Petrie, M., T. Halliday, and C. Sanders. 1991. Peahens prefer peacocks with elaborate trains. *Anim. Behav.* 41 (2):323–331.

Petrinovich, L. 1988. Individual stability, local variability and the cultural transmission of song in white-crowned sparrows. *Behaviour* 107:208–240.

Petrinovich, L., and L. F. Baptista. 1987. Song development in the white-crowned sparrow: Modification of learned song. *Anim. Behav.* 35 (4):961–974.

Pettigrew, J. D., P. R. Manger, and S. L. B. Fine. 1998. The sensory world of the platypus. *Phil. Trans. Roy. Soc. Lond. B* 353 (1372):1199–1210.

Pfeffer, P. 1967. Le moufflon de Corse (*Ovis ammon musimom* Schreber 1782) position systématique, écologie et éthology canparées. *Mammalia* Suppl. 31. 262 pp.

Pfennig, D. W., G. J. Gamboa, H. K. Reeve, J. S. Reeve, and I. D. Ferguson. 1983. The mechanism of nestmate discrimination in social wasps (*Polistes*, Hymenoptera: Vespidae). *Behav. Ecol. Sociobiol.* 13:299–305.

Pfennig, D. W., H. K. Reeve, and P. W. Sherman. 1993. Kin recognition and cannibalism in spadefoot tadpoles. *Anim. Behav.* 46 (1):87–94.

Phillips, J. B. 1986. Two magnetoreception pathways in a migratory salamander. *Science* 233:765–766.

Phillips, J. B. 1987a. Laboratory studies of homing orientation in the eastern red-spotted newt, *Notophthalmus viridescens*. *J. Exp. Biol.* 131:215–229.

Phillips, J. B. 1987b. Specialized visual receptors respond to magnetic field alignment in the blowfly (*Calliphora vicina*). *Soc. Neurosci. Abstr.* 13:397.

Phillips, J. B., and S. C. Borland. 1992. Behavioral evidence for the use of a light-dependent magnetoreception mechanism by a vertebrate. *Nature* 359:142–144.

Phillips, J. B., and S. C. Borland. 1994. Use of a specialized magnetoreception system for homing by the Eastern red-spotted newt *Notophthalmus viridescens*. *J. Exp. Biol.* 188:275–291.

Phillips, J. B., and J. A. Waldvogel. 1982. Reflected light cues generate the short-term deflector loft effect. In *Avian Navigation*, edited by F. Papi and H. G. Wallraff. Berlin: Springer-Verlag, pp. 190–202.

Phillips, J. B., and J. A. Waldvogel. 1988. Celestial polarized light as a calibration reference for sun compass of homing pigeons. *J. Theoret. Biol.* 131:55–67.

Phoenix, C., R. Goy, A. Gerall, and W. Young. 1959. Organizing action of prenatally administered testosterone propionate on the tissues mediating mating behavior in the female guinea pig. *Endocrinology* 65:369–382.

Pickard, G. E., and F. W. Turek. 1982. Splitting of the circadian rhythm is abolished by unilateral lesions of the suprachiasmatic nuclei. *Science* 215:119–121.

Pietrewicz, A. T., and A. C. Kamil. 1979. Search image formation in the blue jay (*Cyanocitta cristata*). *Science* 204:1332–1333.

Pietrewicz, A. T., and A. C. Kamil. 1981. Search images and the detection of cryptic prey: An operant approach. In *Foraging Behavior: Ecological, Ethological and Psychological Approaches*, edited by A. C. Kamil and T. D. Sargent. New York: Garland STPM Press, pp. 311–331.

Pinsker, H., I. Kupfermann, V. Castellucci, and E. R. Kandel. 1970. Habituation and dishabituation of the gill-withdrawal reflex in *Aplysia*. *Science* 167:1740–1742.

Plaisted, K. C., and N. J. Mackintosh. 1995. Visual search for cryptic stimuli in pigeons: Implications for the search image and search rate hypotheses. *Anim. Behav.* 50 (5):1219–1232.

Platt, M. L., E. M. Brannon, T. L. Briese, and J. A. French. 1996. Differences in feeding ecology predict differences in performance between golden lion tamarins (*Leontopithecus rosalia*) and Wied's marmosets (*Callithrix kuhli*) on spatial and visual memory tasks. *Anim. Learn. Behav.* 24 (4):384–393.

Plautz, J. D., M. Kaneko, J. C. Hall, and S. A. Kay. 1997. Independent photoreceptive circadian clocks throughout *Drosophila*. *Science* 278:1632–1635.

Plautz, J. D., M. Straume, R. Stanewsky, C. F. Jamieson, C. Brandes, H. B. Dowse, J. C. Hall, and S. A. Kay. 1997. Quantitative analysis of *Drosophila period* gene transcription in living animals. *J. Biol. Rhythms* 12:204–217.

Plomin, R., J. C. DeFries, and G. E. McClearn. 1980. *Behavioral Genetics—A Primer*. San Francisco: Freeman.

Plummer, M. R., and J. M. Camhi. 1981. Discrimination of sensory signal from noise in the escape system of the cockroach: The role of wind acceleration. *J. Comp. Physiol.* 142:347–357.

Poole, T. B. 1966. Aggressive play of polecats. *Symp. Zool. Soc. Lond.* 18:23–24.

Popov, A. V., and V. F. Shuvalov. 1977. Phonotactic behavior of crickets. *J. Comp. Physiol.* 119:111–126.

Popp, J. W. 1987. Resource value and dominance among American goldfinches. *Bird Behav.* 7:73–77.

Porter, R. H., V. J. Tepper, and D. M. White. 1981. Experimental influences on the development of huddling preferences and "sibling" recognition in spiny mice. *Dev. Psychobiol.* 14:375–382.

Potts, W. K., C. J. Manning, and E. K. Wakeland. 1991. Mating patterns in seminatural populations of mice influenced by MHC genotype. *Nature* 352:619–621.

Pough, F. H. 1976. Multiple cryptic effects of cross-banded and ringed patterns of snakes. *Copeia* 1976:834–836.

Premack, D. 1976. *Intelligence in Ape and Man*. Hillsdale, NJ: Erlbaum.

Premack, D. 1978. On the abstractness of human concepts: Why it would be difficult to talk to a pigeon. In *Cognitive Processes in Animal Behavior*, edited by S. H. Hulse, H. Fowler, and W. K. Honig. Hillsdale, NJ: Erlbaum, p. 423–451.

Price, J. L., J. Blau, A. Rothenfluh, M. Abodeely, and M. W. Young. 1998. *Double-time* is a novel *Drosophila* clock gene that regulates PERIOD protein accumulation. *Cell* 94 (1):83–95.

Proctor, H. C. 1991. Courtship in the water mite *Neumania papillator*: Males capitalize on female adaptations for predation. *Anim. Behav.* 42 (4):589–598.

Proctor, H. C. 1992. Sensory exploitation and the evolution of male mating behavior: A cladistic test. *Anim. Behav.* 44 (4):745–752.

Pruett-Jones, S. 1992. Independent versus non-independent mate choice: Do females copy each other? *Am. Nat.* 140:1000–1009.

Pusey, A. E., and C. Packer. 1997. The ecology of relationships. In *Behavioural Ecology: An Evolutionary Approach*, edited by J. R. Krebs and N. B. Davies. Oxford: Blackwell Science, pp. 254–283.

Pyke, G. H., H. R. Pulliam, and E. L. Charnov. 1977. Optimal foraging: A selective review of theory and tests. *Quart. Rev. Biol.* 52:137–154.

Pyle, D. W., and M. H. Gromko. 1978. Repeated mating by female *Drosophila melanogaster*: The adaptive importance. *Experientia* 34:449–450.

Quinlan, R. J., and J. M. Cherrett. 1977. The role of substrate preparation in the symbiosis between the leaf-cutting ant *Acromyrmex octospinosus* (Reich) and its food fungus. *Ecol. Entomol.* 2:161–170.

Quinn, T. P. 1980. Evidence for celestial and magnetic compass orientation in lake migration sockeye salmon fry. *J. Comp. Physiol.* 137:243–248.

Quinn, T. P., and C. A. Busak. 1985. Chemosensory recognition of siblings in juvenile coho salmon (*Oncorhynchus kisutch*). *Anim. Behav.* 33:51–56.

Quinn, T. P., and A. H. Dittman. 1990. Pacific salmon migrations and homing: Mechanisms and adaptive significance. *Trends Ecol. Evol.* 5:174–177.

Quinn, T. P., and T. J. Hara. 1986. Sibling recognition and olfactory sensitivity in juvenile coho salmon (*Oncorhynchus kisutch*). *Can. J. Zool.* 64:921–925.

Quinn, T. P., R. T. Merrill, and E. L. Brannon. 1981. Magnetic field detection in sockeye salmon. *J. Exp. Zool.* 217:137–142.

Quinn, T. P., and G. M. Tolson. 1986. Evidence of chemically mediated population recognition in coho salmon (*Oncorhynchus kisutch*). *Can. J. Zool.* 64:84–87.

Quinn, W. G., W. A. Harris, and S. Benzer. 1974. Conditioned behavior in *Drosophila melanogaster*. *Proc. Nat. Acad. Sci. USA* 71:708–712.

Rabenoid, K. N. 1990. *Campylorhynchus* wrens: The ecology of delayed dispersal and cooperation in the Venezuela savanna. In *Cooperative Breeding in Birds*, edited by P. B. Stacey and W. D. Koenig. Cambridge: Cambridge University Press, pp. 159–196.

Rado, R., J. Terkel, and Z. Wollberg. 1998. Seismic communication signals in the blind mole-rat (*Spalax ehrenbergi*): Electrophysiological and behavioral evidence for their processing by the auditory system. *J. Comp. Physiol. A* 183 (4):503–511.

Raisman, G., and K. Brown-Grant. 1977. The "suprachiasmatic syndrome": Endocrine and behavioral abnormalities following lesions in the suprachiasmatic nuclei in the female rat. *Proc. Roy. Soc. Lond. B* 198:297–314.

Ralph, M. R., R. G. Foster, F. C. Davis, and M. Menaker. 1990. Transplanted suprachiasmatic nucleus determines circadian rhythm. *Science* 247:975–978.

Ramirez, J. M., and K. G. Pearson. 1990. Chemical deafferentation of the locust flight system by phentolamine. *J. Comp. Physiol. A* 167:485–494.

Randall, J. A., and E. R. Lewis. 1997. Seismic communication between the burrows of kangaroo rats, *Dipodomys spectabilis*. *J. Comp. Physiol. A* 181 (5):525–531.

Rasa, O. A. E. 1986. Coordinated vigilance in dwarf mongoose family groups: The "Watchman's Song" hypothesis and the costs of guarding. *Ethology* 71:340–344.

Rasa, O. A. E., S. Bisch, and T. Teichner. 1998. Female mate choice in a subsocial beetle: Male phenotype correlates with helping potential and offspring survival. *Anim. Behav.* 56 (5):1213–1220.

Rasmussen, L. E. L. 1998. Chemical communication: An integral part of functional Asian elephant (*Elephas maximus*) society. *Ecoscience* 5 (3):410–426.

Raybourn, M. S. 1983. The effects of direct-current magnetic fields on turtle retinas in vitro. *Science* 220:715–717.

Raynor, L. S., and G. W. Uetz. 1990. Trade-offs in foraging success and predation risk with spatial position in colonial spiders. *Behav. Ecol. Sociobiol.* 27:77–85.

Real, L. A. 1981. Uncertainty and pollinator-plant interactions: The foraging behavior of bees and wasps on artificial flowers. *Ecology* 62:20–26.

Real, L. A., J. Ott, and E. Silverfine. 1982. Optimal foraging: Some simple stochastic models. *Behav. Ecol. Sociobiol.* 10:251–263.

Reavis, R. H. 1997. The natural history of a monogamous coral-reef fish, *Valenciennea strigata* (Gobiidae). 2. Behavior, mate fidelity, and reproductive success. *Environ. Biol. Fishes* 49:247–257.

Reavis, R. H., and G. W. Barlow. 1998. Why is the coral-reef fish *Valenciennea strigata* (Gobiidae) monogamous? *Behav. Ecol. Sociobiol.* 43 (4–5):229–237.

Redondo, T., and F. Castro. 1992. Signalling of nutritional need by magpie nestlings. *Ethology* 92:193–204.

Rees, H. D., and H. E. Gray. 1984. Glucocorticoids and mineralocorticoids. In *Peptides, Hormones, and Behavior*, edited by C. B. Nemeroff and A. J. Dunn. New York: Spectrum, pp. 579–644.

Reeve, H. K. 1989. The evolution of conspecific acceptance thresholds. *Am. Nat.* 133:407–435.

Reeve, H. K., and L. Keller. 1995. Partitioning reproduction in mother-daughter versus sibling associations: A test of optimal skew theory. *Am. Nat.* 145:119–132.

Reeve, H. K., D. F. Westneat, W. A. Noon, and P. W. Sherman. 1990. DNA "fingerprinting" reveals high levels of inbreeding in colonies of the eusocial naked mole-rat. *Proc. Nat. Acad. Sci. USA* 87:2496–2500.

Reichert, H., C. H. F. Rowell, and C. Griss. 1985. Course correction circuitry translates feature detection into behavioral action in locusts. *Nature* 315:142–144.

Reilly, L. M. 1987. Measurements of inbreeding and average relatedness in a termite population. *Am. Nat.* 130:339–349.

Reimchen, T. E. 1989. Shell color ontogeny and tubeworm mimicry in a marine gastropod *Litorina mariae*. *Biol. J. Linn. Soc.* 36:97–109.

Reiter, R. J. 1993. The melatonin rhythm: Both clock and calendar. *Experientia* 49:654–663.

Rensch, B., and G. Ducker. 1959. Die Spiele von Mungo und Ichneumon. *Behaviour* 14:185–213.

Rescorla, R. A. 1988a. Behavioral studies of Pavlovian conditioning. *Ann. Rev. Neurosci.* 11:329–352.

Rescorla, R. A. 1988b. Pavlovian conditioning: It's not what you think it is. *Am. Psychol.* 43:151–160.

Reuss, S., and J. Olcese. 1986. Magnetic field effects on the rat pineal gland: Role of retinal activation by light. *Neurosci. Lett.* 64:97–101.

Reuss, S., P. Semm, and L. Vollrath. 1983. Different types of magnetically sensitive cells in the rat pineal gland. *Neurosci. Lett.* 40:23–26.

Reye, D. N., and K. G. Pearson. 1988. Entrainment of the locust central flight oscillator by wing stretch receptor stimulation. *J. Comp. Physiol. A* 162:77–89.

Reyer, H.-U. 1980. Flexible helper structure as an ecological adaptation in the red Kingfisher (*Ceryle rudis rudis* L.). *Behav. Ecol. Sociobiol.* 6:219–227.

Reyer, H.-U. 1984. Investment and relatedness: A cost/benefit analysis of breeding and helping in the pied king-fisher (*Ceryle rudis*). *Anim. Behav.* 32:1163–1178.

Reyer, H.-U. 1986. Breeder-helper interactions in the pied kingfisher reflect costs and benefits of cooperative breeding. *Behaviour* 96:277–303.

Reynolds, J. D., and M. R. Gross. 1992. Female mate preference enhances offspring growth and reproduction in a fish, *Poecilia reticulata*. *Proc. Roy. Soc. Lond. B* 250:57–62.

Rheinlaender, J. H., C. Gerhardt, and D. D. Yager. 1979. Accuracy of phonotaxis by the green treefrog (*Hyla cinerea*). *J. Comp. Physiol.* 133:247–255.

Ribbink, A. J. 1977. Cuckoo among Lake Malawi cichlid fish. *Nature* 267:243–244.

Ribble, D. O. 1991. The monogamous mating system of *Peromyscus californicus* as revealed by DNA fingerprinting. *Behav. Ecol. Sociobiol.* 29:161–166.

Ricciutti, E. 1978. Night of the grunting fish. *Audubon* 80 (4):92–97.

Riddiford, L. M., and J. W. Truman. 1993. Hormone receptors and the regulation of insect metamorphosis. *Am. Zool.* 33 (3):340–347.

Ridley, M. 1978. Paternal care. *Anim. Behav.* 26:904–932.

Riechert, S. E. 1979. Games spiders play. II. Resource assessment strategies. *Behav. Ecol. Sociobiol.* 6:121–128.

Riechert, S. E. 1981. The consequences of being territorial: Spiders, a case study. *Am. Nat.* 117 (6):871–892.

Riechert, S. E. 1982. Spider interaction strategies: Communication vs. coercion. In *Spider Communication: Mechanisms and Ecological Significance*, edited by P. N. Witt and J. Rovner. Princeton, NJ: Princeton University Press, pp. 281–315.

Riechert, S. E. 1984. Games spiders play III. Cues underlying context-associated changes in agnostic behavior. *Anim. Behav.* 32:1–15.

Riechert, S. E. 1986a. Between population variation in spider territorial behavior: Hybrid-pure population line comparisons. In *Evolutionary Genetics of Invertebrate Behaviour: Progress and Prospects*, edited by M. D. Huettel. New York: Plenum, pp. 33–42.

Riechert, S. E. 1986b. Spider fights as a test of evolutionary game theory. *Am. Sci.* 74:604–610.

Riechert, S. E. 1993a. Investigation of potential gene flow limitation of behavioral adaptation in an aridlands spider. *Behav. Ecol. Sociobiol.* 32 (5):355–364.

Riechert, S. E. 1993b. Evolution of behavioral phenotypes. In *Advances in the Study of Behavior*. New York: Academic Press, pp. 103–134.

Riechert, S. E., and T. R. Tracy. 1975. Thermal balance and prey availability: Bases for a model relating web-site characteristics to spider reproductive success. *Ecology* 56:265–285.

Ristau, C. A., ed. 1991. *Cognitive Ethology, the Minds of Other Animals*. Hillsdale, NJ: Erlbaum.

Ritzmann, R. E., and A. J. Pollack. 1986. Identification of thoracic interneurons that mediate giant interneuron-to-motor pathways. *J. Comp. Physiol. A* 159:639–654.

Robbins, R. K. 1981. The "false head" hypothesis: Predation and wing pattern variation of Lycaenid butterflies. *Am. Nat.* 118:770–775.

Roberts, G., and T. N. Sherratt. 1998. Development of cooperative relationships through increasing investment. *Nature* 394 (6689):175–179.

Roberts, J. L. 1984. The biosynthesis of peptide hormones. In *Peptides, Hormones, and Behavior*, edited by C. B. Nemeroff and A. J. Dunn. New York: Spectrum, pp. 99–118.

Roberts, S. 1965. Photoreception and entrainment of cockroach activity rhythms. *Science* 148:958–959.

Roberts, S. 1974. Circadian rhythms in cockroaches. Effects of optic lobe lesions. *J. Comp. Physiol.* 88:21–30.

Robertson, L., and J. Takahashi. 1988a. Circadian clock in cell culture. I. Oscillation of melatonin release from dissociated chick pineal cells in flow-through microcarrier culture. *J. Neurosci.* 8:12–21.

Robertson, L., and J. Takahashi. 1988b. Circadian clock in cell culture. II. In vitro photic entrainment of melatonin oscillation from dissociated chick pineal cells. *J. Neurosci.* 8:22–30.

Robertson, R. J., and B. J. Stutchbury. 1988. Experimental evidence for sexually selected infanticide in tree swallows. *Anim. Behav.* 36:749–753.

Robertson, R. M., and K. G. Pearson. 1983. Interneurons in the flight system of the locust: Distributions, connections, and resetting properties. *J. Comp. Neurol.* 215:33–50.

Robertson, R. M., and K. G. Pearson. 1984. Interneuronal organization in the flight system of the locust. *J. Neurophysiol.* 30:95–101.

Robertson, R. M., and K. G. Pearson. 1985. Neural circuits in the flight system of the locust. *J. Insect Physiol.* 53:110–128.

Robinow, S., W. S. Talbot, D. S. Hogness, and J. W. Truman. 1993. Programmed cell death in the Drosophila CNS is ecdysone regulated and coupled with a specific ecdysone receptor isoform. *Development* 119 (4):1251–1259.

Robinson, G. E. 1986. The dance language of the honey bee: The controversy and its resolution. *Am. Bee J.* 126:184–189.

Robinson, G. E. 1987. Regulation of honey bee age polyethism by juvenile hormone. *Behav. Ecol. Sociobiol.* 20:329–338.

Robinson, G. E. 1996. Chemical communication in honeybees. *Science* 271:1824–1825.

Robinson, G. E. 1998. From society to genes with the honey bee. *Am. Sci.* 86 (5):456–462.

Robinson, G. E., A. Strambi, C. Strambi, Z. L. Paulino-Simoes, S. O. Tozeto, and J. M. Negraes Barbosa. 1987. Juvenile hormone titers in European and Africanized honey bees in Brazil. *Gen. Comp. Endocrinol.* 66:457–459.

Rodda, G. H. 1984. The orientation and navigation of juvenile alligators: Evidence of magnetic sensitivity. *J. Comp. Physiol. A* 154:649–658.

Roeder, K. D. 1967. *Nerve Cells and Insect Behavior.* Cambridge, MA.: Harvard University Press.

Roeder, K. D., and R. S. Payne. 1966. Acoustic orientation of a moth in flight by means of two sense cells. *Symp. Soc. Exp. Biol.* 20:251–272.

Roeder, K. D., and A. E. Treat. 1957. Ultrasonic reception by the tympanic organ of noctuid moths. *J. Exp. Biol.* 134:127–157.

Roeder, K. D., and A. E. Treat. 1961. The detection and evasion of bats by moths. *Am. Sci.* 49:135–148.

Romanes, G. J. 1882. *Animal Intelligence.* New York: Appleton.

Romanes, G. J. 1884. *Mental Evolution in Animals.* London: Keegan, Paul, Trench.

Romanes, G. J. 1889. *Mental Evolution in Man.* New York: Appleton.

Romey, W. L. 1995. Position preferences within groups: do whirligigs select positions which balance feeding opportunities with predator avoidance? *Behav. Ecol. Sociobiol.* 37 (3):195–200.

Rood, J. P. 1990. Group size, survival, reproduction, and routes to breeding in dwarf mongooses. *Anim. Behav.* 39 (3):566–572.

Rosen, R. A., and D. C. Hales. 1981. Feeding of paddlefish, *Polyodon spathula. Copeia* 1981: 441–455.

Rosenthal, G. G., and C. S. Evans. 1998. Female preference for swords in *Xiphophorus helleri* reflects a bias for large apparent size. *Proc. Nat. Acad. Sci. USA* 95 (8):4431–4436.

Rosin, R. 1978. The honey bee "language" controversy. *J. Theoret. Biol.* 72:589–602.

Rosin, R. 1980a. The honey-bee "dance language" hypothesis and the foundations of biology and behavior. *J. Theoret. Biol.* 87:457–481.

Rosin, R. 1980b. Paradoxes of the honey-bee "dance language" hypothesis. *J. Theoret. Biol.* 84:775–800.

Rosin, R. 1984. Further analysis of the honey bee "dance language" controversy. I. Presumed proofs for the "dance language" hypothesis, by Soviet scientists. *J. Theoret. Biol.* 107:417–442.

Ross, K. G. 1986. Kin selection and the problem of sperm utilization in social insects. *Nature* 323:799–800.

Rowell, C. H. F. 1989. Descending interneurones of the locust reporting deviation from flight course: What is their role in steering? *J. Exp. Biol.* 146:177–194.

Rowell, T. E. 1974. The concept of social dominance. *Behav. Biol.* 11:131–154.

Rowland, W. J., and P. Sevenster. 1985. Sign stimuli in the threespine stickleback (*Gasterosteus aculeatus*): A re-examination and extension of some classic experiments. *Behaviour* 93:241–257.

Rowley, I. C. R. 1981. The communal way of life in the splendid wren, *Malurus splendens. Z. Tierpsychol.* 55:228–267.

Rowley, I. C. R., and E. M. Russell. 1990. Splendid fairy-wrens: Demonstrating the importance of longevity. In *Cooperative Breeding in Birds*, edited by P. B. Stacey and W. D. Koenig. Cambridge: Cambridge University Press, pp. 3–30.

Rubenstein, D. I. 1981. Population density, resource patterning and territoriality in the Everglades pygmy sunfish. *Anim. Behav.* 29:155–172.

Rudge, D. W. 1999. Taking the peppered moth with a grain of salt. *Biol. Philos.* 14 (1):9–37.

Rudolph, K., A. Wirz-Justice, K. Krauchi, and H. Feer. 1988. Static magnetic fields decrease nocturnal pineal cAMP in the rat. *Brain Res.* 446:159–160.

Rumbaugh, D. M. 1977. The emergence and the state of ape language research. In *Progress in Ape Research*, edited by G. H. Bourne. New York: Academic Press, pp. 75–83.

Rumbaugh, D. M., T. V. Gill, and E. C. von Glaserfeld. 1973. Reading and sentence completion by a chimpanzee. *Science* 182:731–733.

Rumbaugh, D. M., and E. S. Savage-Rumbaugh. 1994. Language in comparative perspective. In *Animal Learning and Cognition*, edited by N. J. Macintosh. San Diego, CA: Academic Press, pp. 307–333.

Rushforth, N. B. 1965. Behavioral studies of the coelenterate *Hydra pirardi* Brien. *Anim. Behav. Suppl.* 1:30–42.

Russell, E. M., and I. C. R. Rowley. 1988. Helper contributions to reproductive success in the splendid fairy-wren *Malurus splendens*. *Behav. Ecol. Sociobiol.* 22:131–140.

Ryan, M. J. 1988. Energy, calling, and selection. *Am. Zool.* 28:885–898.

Ryan, M. J. 1990. Sexual selection, sensory systems, and sensory exploitation. *Oxford Surv. Evol. Biol.* 7:157–195.

Ryan, M. J. 1997. Sexual selection and mate choice. In *Behavioural Ecology: An Evolutionary Approach*, edited by J. R. Krebs and N. B. Davies. Oxford: Blackwell Science, pp. 179–202.

Ryan, M. J. 1998a. *The Handicap Principle: A Missing Piece of Darwin's Puzzle* (book review). *Quart. Rev. Biol.* 73 (4):477.

Ryan, M. J. 1998b. Sexual selection, receiver biases, and the evolution of sex differences. *Science* 281 (5385):1999–2003.

Ryan, M. J., J. H. Fox, W. Wilczynski, and A. S. Rand. 1990. Sexual selection for sensory exploitation in the frog *Physalaemus pustulosus*. *Nature* 343:66–67.

Ryan, M. J., and A. S. Rand. 1993a. Phylogenetic patterns of behavioral mate recognition systems in the *Physalaemus pustulosus* species group (Anura: Leptodactylidae): The role of ancestral and derived characters and sensory exploitation. In *Evolutionary Patterns and Processes*, edited by D. R. Lees and D. Edwards. London: Academic Press, pp. 251–267.

Ryan, M. J., and A. S. Rand. 1993b. Sexual selection and signal evolution: The ghost of biases past. *Phil. Trans. Roy. Soc. Lond. B* 340:187–195.

Sadauskas, K. K., and Z. H. Shuranova. 1984. Effects of pulsed magnetic field on electrical activity of crayfish neurons. *Biofizika* 29:681–683.

Sagrillo, C. A., D. R. Grattan, M. M. McCarthy, and M. Selmanoff. 1996. Hormonal and neurotransmitter relation of GnRH gene expression and related reproductive behaviors. *Behav. Genet.* 26 (3):241–277.

Sandberg, R. 1994. Interaction of body condition and magnetic orientation in autumn migrating robins, *Erithacus rubecula*. *Anim. Behav.* 47:679–686.

Sandberg, R., J. Bäckman, and U. Ottosson. 1998. Orientation of snow buntings (*Plectrophenax nivalis*) close to the magnetic north pole. *J. Exp. Biol.* 201:1859–1870.

Sandberg, R., and F. R. Moore. 1996. Migratory orientation of red-eyed vireos, *Vireo olivaceus*, in relation to energetic condition and ecological context. *Behav. Ecol. Sociobiol.* 39 (1):1–10.

Santschi, F. 1911. Le mechanisme d'orientation chez les fourmis. *Rev. Suisse Zool.* 19:117–134.

Sargent, T. D. 1968. Cryptic moths: Effects on background selection of painting the circumocular scales. *Science* 159:100–101.

Sargent, T. D. 1969. Behavioural adaptations of cryptic moths. III. Resting attitudes of two bark-like species, *Melanolophia canadaria* and *Catocala ultronia*. *Anim. Behav.* 17:670–672.

Sargent, T. D. 1976. *Legion of the Night*. Amherst: University of Massachusetts Press.

Sargent, T. D., C. D. Millar, and D. M. Lambert. 1998. The "classical" explanation of industrial melanism: Assessing the evidence. *Evol. Biol.* 30:229–322.

Sassoon, D., and D. Kelley. 1986. The sexually dimorphic larynx of *Xenopus laevis*: Development and androgen regulation. *Am. J. Anat.* 177:457–472.

Sato, T. 1994. Active accumulation of spawning substrate: A determinate of extreme polygyny in a shell-brooding cichlid fish. *Anim. Behav.* 48:669–678.

Sauer, E. G. F. 1957. Die Sternorientierung nächtlich ziehender Grasmücken (*Sylvia atriciapilla, borin and curruca*). *Z. Tierpsychol.* 14:29–70.

Sauer, E. G. F. 1961. Further studies on the stellar orientation of nocturnally migrating birds. *Psychol. Forschung* 26:224–244.

Sauer, E. G. F., and E. M. Sauer. 1960. Star navigation in nocturnal migrating birds. The 1958 planetarium experiments. *Cold Spring Harbor Symp. Quant. Biol.* 25:463–473.

Saunders, D. S. 1977. *An Introduction to Biological Rhythms*. New York: Wiley.

Savage-Rumbaugh, E. S. 1986. *Ape Language from Conditioned Response to Symbol*. New York: Columbia University Press.

Savage-Rumbaugh, E. S., and R. Lewin. 1994. *Kanzi, the Ape at the Brink of the Human Mind*. New York: Wiley.

Savage-Rumbaugh, E. S., D. M. Rumbaugh, and S. Boysen. 1978a. Linguistically mediated tool use and exchange by chimpanzees *Pan troglodytes*. *Behav. Brain Sci.* 1:539–554.

Savage-Rumbaugh, E. S., D. M. Rumbaugh, and S. Boysen. 1978b. Symbolic communication between two chimpanzees (*Pan troglodytes*). *Science* 201:641–644.

Savage-Rumbaugh, E. S., D. M. Rumbaugh, and S. Boysen. 1980. Do apes use language? *Am. Sci.* 68:49–61.

Savage-Rumbaugh, S., S. G. Shanker, and T. J. Taylor. 1998. *Apes, Language, and the Human Mind*. New York: Oxford University Press.

Sawaki, Y., I. Nihonmatsu, and H. Kawamura. 1984. Transplantation of the neonatal suprachiasmatic nuclei into rats with complete bilateral suprachiasmatic lesions. *Neurosci. Res.* 1:67–72.

Schaefer, P., G. V. Kondagunta, and R. E. Ritzman. 1994. Motion analysis of escape movements evoked by tactile stimulation in the cockroach *Periplaneta americana*. *J. Exp. Biol.* 190:287–294.

Schaller, G. B. 1972. *The Serengeti Lion*. Chicago: University of Chicago Press.

Scheich, H., G. Langner, C. Tidemann, R. B. Coles, and A. Guppy. 1986. Electroreception and electrolocation in platypus. *Nature* 319 (30):401–402.

Scheller, R. H., and R. Axel. 1984. How genes control an innate behavior. *Sci. Am.* 250 (Mar.):54–62.

Schermuly, L., and R. Klinke. 1990. Infrasound sensitive neurones in the pigeon cochlear ganglion. *J. Comp. Physiol. A* 166:355–363.

Schjelderup-Ebbe, T. 1922. Beiträge zur Sozialpsychologie des Haushuhns. *Z. Psychol.* 88 (3–5):225–252.

Schleidt, W. 1961a. Reaktionen von Truthühnern auf fliegende Raubvögel und Veruche zur Analyse ihrer AAM's. *Z. Tierpsychol* 18:534–560.

Schleidt, W. 1961b. Über die Auslösung des Kollern beim Truthahan (*Meleagris galopavo*). *Z. Tierpsychol.* 11:417–435.

Schlenoff, D. H. 1985. The startle responses of blue jays to *Catocala* (Lepidoptera: Noctuidae) prey models. *Anim. Behav.* 33:1057–1067.

Schluter, A., J. Parzefall, and I. Schlupp. 1998. Female preference for symmetrical vertical bars in male sailfin mollies. *Anim. Behav.* 56 (1):147–153.

Schmidt-Koenig, K. 1987. Bird navigation: Has olfactory orientation solved the problem? *Quart. Rev. Biol.* 62:31–47.

Schmidt-Koenig, K., and H. J. Schlichte. 1972. Homing in pigeons with impaired vision. *Proc. Nat. Acad. Sci. USA* 69 (9):2446–2447.

Schneider, D. 1974. The sex-attractant receptor of moths. *Sci. Am.* 231 (July):28–35.

Schoech, S. J. 1998. Physiology of helping in Florida scrub jays. *Am. Sci.* 86 (1):70–77.

Schoech, S. J., R. L. Mumme, and J. C. Wingfield. 1996. Prolactin and helping behaviour in the cooperatively breeding Florida scrub-jay, *Aphelocoma coerulescens*. *Anim. Behav.* 52:445–456.

Schoener, T. W. 1968. Sizes of feeding territories among birds. *Ecology* 49:123–141.

Schoener, T. W. 1983. Simple models of optimal feeding-territory size: A reconciliation. *Am. Nat.* 121:608–629.

Schoener, T. W. 1987. A brief history of optimal foraging ecology. In *Foraging Behavior*, edited by A. C. Kamil, J. R. Krebs and H. R. Pulliam. New York: Plenum, pp. 5–67.

Scholz, A. T., R. M. Horrall, J. C. Cooper, and A. D. Hasler. 1976. Imprinting to chemical cues: The basis for home stream selection in salmon. *Science* 192:1247–1249.

Scholz, A. T., R. M. Horrall, J. C. Cooper, A. D. Hasler, D. M. Madison, J. J. Popp, and R. Daly. 1975. *Artificial Imprinting of Salmon and Trout in Lake Michigan*, Report 80. Madison, WI: Department of Natural Research and Fishery Management. 45 pp.

Schöne, H. 1984. *Spatial Orientation*. Princeton, NJ: Princeton University Press.

Schuett, G. W. 1997. Body size and agnostic experience affect dominance and mating success in male copperheads. *Anim. Behav.* 54 (1):213–224.

Schuller, G., K. Beuter, and R. Rübsamen. 1975. Dynamic properties of the compensation system for Doppler shifts in the bat, *Rhinolophus ferrumequinum*. *J. Comp. Physiol.* 97:113–125.

Schulten, K. 1982. Magnetic field effects in chemistry and aging. *Adv. Solid-State Phys.* 22:61–83.

Schulten, K., and A. Windmuth. 1986. Model for a physiological magnetic compass. In *Biophysical Effects of Steady Magnetic Fields*, edited by M. G. Kiepenheuer and J. Boccasra. New York: Springer-Verlag, pp. 96–106.

Schulz, D. J., Z. Y. Huang, and G. E. Robinson. 1998. Effects of colony food shortage on behavioral development in honey bees. *Behav. Ecol. Sociobiol.* 42 (5):295–303.

Schutz, von F. 1965. Sexuelle Prägung bei Anatiden. *Z. Tierpsychol.* 22:50–103.

Schüz, E. 1949. Die Spät-Auflassung ostpreussischer Jungstörche in West-Deutschland durch die Vogelwarte Rossiten 1933. *Vogelwarte* 15:63–78.

Schwagmeyer, P. L. 1979. The Bruce effect: An evaluation of male/female advantages. *Am. Nat.* 114:932–938.

Schwagmeyer, P. L., and G. A. Parker. 1990. Male mate choice as predicted by sperm competition in thirteen-lined ground squirrels. *Nature* 348:62–64.

Schwind, R. 1983. Zonation of the optical environment and zonation in the rhabdom structure within the eye of the backswimmer, *Notonecta glauca*. *Cell Tissue Res.* 232:53–62.

Schwind, R. 1991. Polarization vision in water insects and insects living on a moist substrate. *J. Comp. Physiol. A* 169:531–540.

Schwippert, W. W., T. W. Beneke, and J.-P. Ewert. 1990. Response of medullary neurons to visual stimuli in the common toad. II. An intracellular recording and cobalt-lysine labeling study. *J. Comp. Physiol. A* 167:509–520.

Scott, J. P. 1962. Critical periods in behavioral development. *Science* 138:949–958.

Scott, J. P., and J. L. Fuller. 1965. *Genetics and the Social Behavior of the Dog*. Chicago: University of Chicago Press.

Searcy, W. A. 1979. Female choice of mates: A general model for birds and its application to red-winged blackbirds (*Agelaius phoeniceus*). *Am. Nat.* 114:77–100.

Searcy, W. A., D. Eriksson, and A. Lundberg. 1991. Deceptive behavior in pied flycatchers. *Behav. Ecol. Sociobiol.* 29 (3):167–176.

Searcy, W. A., and K. Yasukawa. 1996. Song and female choice. In *Ecology and Evolution of Acoustic Communication in Birds*, edited by D. E. Kroodsma and E. H. Miller. Ithaca, NY: Cornell University Press, pp. 454–473.

Seibt, U., and W. Wickler. 1987. Gerontophagy versus cannibalism in the social spiders *Stegodyphus mimosarium* and *Stegodyphus dumicola* Popcock. *Anim. Behav.* 35:1903–1905.

Seitz, A. 1940. Die Paarbildung bei einigen Cichliden. *Z. Tierpsychol.* 4:40–84.

Selander, R. K. 1966. Sexual dimorphism and differential niche utilization in birds. *Condor* 68:113–151.

Seligman, M. E. P. 1970. On the generality of the laws of learning. *Psychol. Rev.* 77:406–418.

Selset, R., and K. B. Doving. 1980. Behavior of mature anadromous char (*Salmo alpinus* L.) toward odorants produced by smolts of their own population. *Acta Physiol. Scand.* 108:113–122.

Selverston, A. 1995. Modulation of circuits underlying rhythmic behaviors. *J. Comp. Physiol. A* 176:139–144.

Semm, P., and R. C. Beason. 1990. Responses to small magnetic variations by the trigeminal system of the bobolink. *Brain Res. Bull.* 25:735–740.

Semm, P., and C. Demaine. 1986. Neurophysiological properties of magnetic cells in the pigeon's visual system. *J. Comp. Physiol. A* 159:619–625.

Semm, P., D. Nohr, C. Demaine, and W. Wiltschko. 1984. Neural basis of the magnetic compass: Interactions of visual, magnetic and vestibular inputs in the pigeon's brain. *J. Comp. Physiol. A* 155:283–288.

Semm, P., T. Schneider, and L. Vollrath. 1980. Effects of an earth-strength magnetic field on electrical activity of pineal cells. *Nature* 288:607–608.

Serventy, D. L. 1971. Biology of desert birds. In *Avian Biology*, edited by D. S. Farner, J. R. King, and K. C. Parkes. New York: Academic Press, pp. 287–339.

Seyfarth, R. M. 1976. Social relationships among adult female baboons. *Anim. Behav.* 24:917–938.

Seyfarth, R. M. 1977. A model of social grooming among adult female monkeys. *J. Theoret. Biol.* 65:671–698.

Seyfarth, R. M., D. L. Cheney, and P. Marler. 1980. Monkey responses to three different alarm calls: Evidence of predator classification and semantic communication. *Science* 210:801–803.

Shalter, M. D. 1978. Localization of passerine seet and mobbing calls by goshawks and pygmy owls. *Z. Tierpsychol.* 46:260–267.

Shalter, M. D., and W. M. Schleidt. 1977. The ability of barn owls, *Tyto alba*, to discriminate and localize avian alarm calls. *Ibis* 119:22–27.

Shanker, S. G., E. S. Savage-Rumbaugh, and T. J. Taylor. 1999. Kanzi: A new beginning. *Anim. Learn. Behav.* 27 (1):24–25.

Shannon, D. C. 1998. Bee dance language questioned. *Am. Bee J.* 138 (10):710.

Shearman, L. P., M. J. Zylka, D. R. Weaver, L. F. Kolakowski, Jr., and S. M. Reppert. 1997. Two period homologs: Circadian expression and photic regulation in the suprachiasmatic nuclei. *Neuron* 19:1261–1269.

Sheldon, M., D. S. Rice, G. D'Arcangelo, H. Yoneshima, K. Nakajima, K. Mikoshiba, B. W. Howell, J. A. Cooper, D. Goldwitz, and T. Curran. 1997. *Scrambler* and *yotari* disrupt the *disabled* gene and produce a *reeler*-like phenotype in mice. *Nature* 389:730–733.

Sherman, P. M. 1994. The orb-web: An energetic and behavioural estimator of a spider's dynamic foraging and reproductive strategies. *Anim. Behav.* 48 (1):19–34.

Sherman, P. W. 1977. Nepotism and the evolution of alarm calls. *Science* 197:1246–1253.

Sherman, P. W. 1980a. The limits of ground squirrel nepotism. In *Sociobiology: Beyond Nature/Nurture?* edited by G. W. Barlow and J. Silverberg. Bolder CO: Westview, pp. 505–544.

Sherman, P. W. 1980b. The meaning of nepotism. *Am. Nat.* 116:604–606.

Sherman, P. W. 1985. Alarm calls of Belding's ground squirrels to aerial predators: Nepotism or self preservation? *Behav. Ecol. Sociobiol.* 17:313–323.

Sherman, P. W. 1988. The levels of analysis. *Anim. Behav.* 36:616–619.

Sherman, P. W., J. U. M. Jarvis, and S. H. Braude. 1992. Naked mole-rats. *Sci. Am.* 267 (Aug.):72–78.

Sherman, P. W., E. A. Lacey, H. K. Reeve, and L. Keller. 1995. The eusociality continuum. *Behav. Ecol.* 6:102–108.

Sherman, P. W., H. K. Reeve, and D. W. Pfennig. 1997. Recognition systems. In *Behavioural Ecology*, edited by J. R. Krebs and N. B. Davies. Oxford: Blackwell Science, pp. 69–96.

Sherratt, T. N., and G. Roberts. 1998. The evolution of generosity and choosiness in cooperative exchanges. *J. Theoret. Biol.* 193 (1):167–177.

Sherry, D. F., A. L. Vaccarino, K. Buckenham, and R. Herz. 1989. The hippocampal complex of food storing birds. *Brain Behav. Evol.* 34:308–317.

Shettleworth, S. J. 1995. Comparative studies of memory in food storing birds: From the field to the Skinner box. In *Behavioral Brain Research in Naturalistic and Semi-naturalistic Settings*, NATO ASI Series, edited by E. Alleva, A. Fasolo, H. P. Lipp, L. Nadel, and L. Ricceri. Dordrecht: Kluwer, pp. 159–192.

Shibata, S., and R. Y. Moore. 1988. Electrical and metabolic activity of suprachiasmatic nucleus neurons in hamster hypothalamic slices. *Brain Res.* 438:374–378.

Shields, W. M. 1982. *Philopatry, Inbreeding, and the Evolution of Sex.* Albany: State University of New York Press.

Shields, W. M. 1984. Barn swallow mobbing: Self-defence, collateral kin defence, group defence, or parental care? *Anim. Behav.* 32:132–148.

Shields, W. M. 1987. Dispersal and mating systems: Investigating their causal connections. In *Mammalian Dispersal Patterns*, edited by B. D. Chepko-Sade and Z. T. Halpin. Chicago: University of Chicago Press, pp. 3–24.

Shigeyoshi, Y., K. Taguchi, S. Yamamoto, S. Takekida, L. Yan, H. Tei, T. Moriya, S. Shibata, J. J. Loros, and J. C. Dunlap. 1997. Light-induced resetting of a mammalian circadian clock is associated with rapid induction of the mPer1 transcript. *Cell* 91 (7):1043–1053.

Shiotsuka, R., J. Jovonovich, and J. A. Jovonovich. 1974. In vitro data on drug sensitivity: Circadian and ultradian corticosterone rhythms in adrenal organ cultures. In *Chronobiological Aspects of Endocrinology*, edited by J. Aschoff, F. Ceresa, and F. Halberg. Stuttgard: Schattauer-verlag, pp. 255–267.

Sidman, R. L., M. C. Green, and S. H. Appel. 1965. *Catalog of the Neurological Mutants of the Mouse*. Cambridge, MA: Harvard University Press.

Siegel, R. G., and W. K. Honig. 1970. Pigeon concept formation: Successive and simultaneous acquisition. *J. Exp. Anal. Behav.* 13:385–390.

Siegel, R. W., J. C. Hall, D. A. Gailey, and C. P. Kyriacou. 1984. Genetic elements of courtship in *Drosophila*: Mosaics and learning mutants. *Behav. Genet.* 14:383–410.

Sigg, H., and J. Falett. 1985. Experiments on respect of possession and property in hamadryas baboons (*Papio hamadryas*). *Anim. Behav.* 33:978–984.

Silberglied, R. E., J. G. Shepherd, and J. L. Dickinson. 1984. Eunuchs: The role of apyrene sperm in Lepidoptera. *Am. Nat.* 123:255–265.

Silva, A. J., R. Paylor, J. M. Wehner, and S. Tonegawa. 1992. Impaired spatial learning in alpha-calcium-calmodulin kinase II mutant mice. *Science* 257:206–211.

Silver, R., J. LeSauter, P. A. Tresco, and M. N. Lehman. 1996. A diffusible coupling signal from the transplanted suprachiasmatic nucleus controlling circadian locomotor rhythms. *Nature* 382:810–813.

Silver, R., P. Witkovsky, P. Horvath, V. Alones, C. J. Barnstable, and M. N. Lehman. 1988. Coexpression of opsin- and VIP-like immunoreactivity in CSF-containing neurons of the avian brain. *Cell Tissue Res.* 253:189–198.

Silverin, B. 1980. Effects of long-acting testosterone treatment on free-living pied flycatchers, *Fidedula hypoleuca*, during the breeding period. *Anim. Behav.* 28:906–912.

Simmons, J. A. 1973. The resolution of target range by echolocating bats. *J. Acoust. Soc. Am.* 54:157–173.

Simmons, J. A. 1974. The response of the Doppler shift echolocation system in the bat *Rhinolophus ferrumequinum*. *J. Acoust. Soc. Am.* 56:672–682.

Simmons, J. A., B. M. Fenton, and M. J. O'Farrell. 1979. Echolocation and pursuit of prey by bats. *Science* 203:16–21.

Simmons, J. A., D. J. Howell, and N. Suga. 1975. Information content of bat sonar echoes. *Am. Sci.* 63:204–215.

Simmons, L. W. 1988. Male size, mating potential and lifetime reproductive success in the field cricket, *Gryllus bimaculatus* (De Geer). *Anim. Behav.* 36:372–379.

Simpson, S. M., and B. K. Follett. 1981. Pineal and hypothalamic pacemakers: Their role in regulating circadian rhythmicity in Japanese quail. *J. Comp. Physiol.* 144:381–389.

Sinclair, A. R. E. 1983. The function of distance movements in vertebrates. In *The Ecology of Animal Movement*, edited by I. R. Swingland and P. J. Greenwood. Oxford: Clarendon, pp. 240–258.

Sinervo, B., and C. M. Lively. 1996. The rock-paper-scissors game and the evolution of alternative male strategies. *Nature* 380 (6571):240–243.

Siviy, S. M. 1998. Neurobiological substrates of play behavior: Glimpses into the structure and function of mammalian playfulness. In *Animal Play: Evolutionary, Comparative and Ecological Perspectives*, edited by M. Bekoff and J. A. Byers. Cambridge: Cambridge University Press, pp. 221–242.

Siwicki, K. K., C. Eastman, G. Petersen, M. Rosbash, and J. C. Hall, 1988. Antibodies to the period gene product of *Drosophila* reveal diverse tissue distribution and rhythmic changes in the visual system. *Neuron* 1:141–150.

Skinner, B. F. 1953. *Science and Human Behavior*. New York: Free Press, Macmillan.

Skipper, R., and G. Skipper. 1957. Those British-bred pompadours—The story completed. *Water Life Aquar. World* 12 (2):63–64.

Skutch, A. F. 1935. Helpers at the nest. *Auk* 52:257–273.

Skutch, A. F. 1961. Helpers among birds. *Condor* 63:198–226.

Skutch, A. F. 1987. *Helpers at Birds Nests*. Iowa City: University of Iowa Press.

Slagsvold, T., and J. T. Lifjeld. 1988. Ultimate adjustment of clutch size to parent feeding capacity in a passerine bird. *Ecology* 69:1918–1922.

Slagsvold, T., and J. T. Lifjeld. 1990. Influence of male and female quality on clutch size in tits (*Parus* spp.). *Ecology* 71:1258–1266.

Slater, P. J. B., L. A. Eales, and N. S. Clayton. 1988. Song learning in zebra finches (*Taeniopygia guttata*): Progress and prospects. In *Advances in the Study of Behavior*, edited by J. S. Rosenblatt, C. Beer, M. Busnel, and P. J. B. Slater. New York: Academic Press, pp. 1–34.

Smale, L., K. E. Holekamp, and P. A. White. 1999. Siblicide revisited in the spotted hyaena: Does it conform to obligate or facultative models? *Anim. Behav.* 58 (3):545–551.

Smedley, S. R., and T. Eisner. 1996. Sodium: A male moth's gift to its offspring. *Proc. Nat. Acad. Sci. USA* 93 (2):809–813.

Smiseth, P. T., T. Amundsen, and L. T. T. Hansen. 1998. Do males and females differ in the feeding of large and small siblings? An experiment with the bluethroat. *Behav. Ecol. Sociobiol.* 42 (5):321–328.

Smith, J. N. M. 1974. The food searching behavior of two European thrushes. II. The adaptiveness of the search patterns. *Behaviour* 49:1–61.

Smith, J. N. M., and H. P. A. Sweatman. 1974. Food searching behavior of titmice in patchy environments. *Ecology* 55:1216–1232.

Smith, R. J. F. 1982. The adaptive significance of the alarm substance—fright reaction system. In *Chemoreception in Fishes*, edited by T. J. Hara. New York: Elsevier, pp. 327–342.

Smith, R. J. F. 1989. The response of *Asterropteryx semipunctatus* to chemical stimuli from injured conspecifics, an alarm response in gobies. *Ethology* 81:279–290.

Smith, R. L. 1979. Repeated copulation and sperm precedence: Paternity assurance for a male brooding water bug. *Science* 205:1029–1031.

Smith, W. J. 1977. *The Behavior of Communicating, An Ethological Approach*. Cambridge, MA: Harvard University Press.

Smith, W. J. 1991. Animal communication and the study of cognition. In *Cognitive Ethology, the Minds of Other Animals*, edited by C. A. Ristau. Hillsdale, NJ: Erlbaum, pp. 209–230.

Smotherman, W. P. 1982. Odor aversion learning by the rat fetus. *Physiol. Behav.* 29:769–771.

Sneddon, L. U., F. A. Huntingford, and A. C. Taylor. 1997. Weapon size versus body size as a predictor of winning in fights between shore crabs, *Carcinus maenus* (L.). *Behav. Ecol. Sociobiol.* 41 (4):237–242.

Snow, B. K. 1974. Lek behavior and breeding of Guy's hermit hummingbird *Phaethornis guy*. *Ibis* 116:278–297.

Snow, D. W. 1963. The evolution of manakin displays. In *Proceedings of the 13th International Ornithological Congress*. 1:553–561.

Sokolowski, M. B., and K. P. Hansell. 1992. The *foraging* locus: Behavioral tests for normal muscle movement in rover and sitter *Drosophila melanogaster* larvae. *Genetica* 85:205–209.

Solomon, D. J. 1973. Evidence for pheromone-influenced homing by migrating Atlantic salmon, *Salmo salar* (L.). *Nature* 244:231–232.

Sommer, V. 1993. Infanticide among the langurs of Jodhpur: Testing the sexual selection hypothesis. In *Infanticide and Parental Care*, edited by S. Parmigiani and F. vom Saal. London: Harwood Academic Press, pp. 155–198.

Song, P.-S., and K. L. Poff. 1989. Photomovement. In *The Science of Photobiology*, edited by K. C. Smith. New York: Plenum, pp. 305–346.

Southern, H. N. 1954. Mimicry in cuckoos' eggs. In *Evolution as a Process*, edited by J. Hurley, A. C. Hardy, and E. B. Ford. London: Allen & Unwin, pp. 219–232.

Speed, M. P., and J. R. G. Turner. 1999. Learning and memory in mimicry: Do we understand the mimicry spectrum? *Biol. J. Linn. Soc.* 67 (3):281–312.

Spencer, H. 1855. *Principles of Psychology*. New York: D. Appleton.

Srinivasan, M. V. 1998. Ants match as they march. *Nature* 392:660–661.

Stacey, P. B. 1979. Habitat saturation and communal breeding in the acorn woodpecker. *Anim. Behav.* 27:1153–1166.

Stacey, P. B., and C. E. Bock. 1978. Social plasticity in the acorn woodpecker. *Science* 202:1298–1300.

Stacey, P. B., and J. D. Ligon. 1987. Territory quality and dispersal options in the acorn woodpecker, and a challenge to the habitat saturation model of cooperative breeding. *Am. Nat.* 120:654–676.

Stallcup, J. A., and G. E. Woolfenden. 1978. Family status and contributions to breeding by Florida scrub jays. *Anim. Behav.* 26:1144–1156.

Stamps, J. A. 1991. Why evolutionary issues are reviving interest in proximate behavioral mechanisms. *Am. Zool.* 31:338–348.

Stanewsky, R., M. Kaneko, P. Emery, B. Beretta, K. Wager-Smith, S. A. Kay, M. Rosbash, and J. C. Hall. 1998. The *cry(b)* mutation identifies cryptochrome as a circadian photoreceptor in *Drosophila*. *Cell* 95 (5):681–692.

Starks, P. T., D. J. Fischer, R. E. Watson, G. L. Melikian, and S. D. Nath. 1998. Context-dependent nestmate-discrimination in the paper wasp, *Polistes dominulus*: A critical test of the optimal acceptance threshold model. *Anim. Behav.* 56 (2):449–458.

Stehle, J., S. Reuss, H. Schröeder, M. Henschel, and L. Vollrath. 1988. Magnetic field effects on pineal N-acetyltransferase activity and melatonin content in the gerbil—Role of pigmentation and sex. *Physiol. Behav.* 44:91–94.

Stehn, R. A., and M. E. Richmond. 1975. Male induced pregnancy termination in the prairie vole, *Microtus ochrogaster*. *Science* 187:1210–1211.

Stent, G. S. 1981. Strength and weakness of the genetic approach to the development of the nervous system. *Ann. Rev. Neurosci.* 4:163–194.

Stephens, D. W. 1987. On economically tracking a variable environment. *Theoret. Pop. Biol.* 32:15–25.

Stephens, D. W., and E. L. Charnov. 1982. Optimal foraging: Some simple stochastic models. *Behav. Ecol. Sociobiol.* 10:251–263.

Stephens, D. W., and J. R. Krebs. 1986. *Foraging Theory*. Princeton, NJ: Princeton University Press.

Stephens, D. W., and S. R. Paton. 1986. How constant is the risk-aversion? *Anim. Behav.* 34:1659–1667.

Stetson, M., and M. Watson-Whitmyre. 1976. Nucleus suprachiasmaticus. The biological clock in the hamster? *Science* 191:197–199.

Stevenson, P. A., and W. Kutsch. 1987. A reconsideration of the central pattern generator concept for locust flight. *J. Comp. Physiol. A* 161:115–129.

Stöhr, S. 1998. Evolution of mate-choice copying: A dynamic model. *Anim. Behav.* 55 (4):893–903.

Strassmann, J. E., C. R. Hughes, D. C. Queller, S. Turlillazzi, R. Cervo, S. K. Davies, and K. F. Goodnight. 1989. Genetic relatedness in primitively eusocial wasps. *Nature* 342:268–269.

Strassmann, J. E., D. C. Queller, and C. R. Solís. 1995. Genetic relatedness and population structure in the social wasp *Mischocyttarus mexicanus* (Hymenoptera: Vespidae). *Insectes Sociaux* 42:379–383.

Struhsaker, T. T. 1967. Auditory communication among vervet monkeys (*Cercopithecus aethiops*). In *Social Communication Among Primates*, edited by S. A. Altmann. Chicago: University of Chicago Press, pp. 281–324.

Stuart, A. M. 1963a. Origin of the trail in the termites *Nasutitermes corniger* (Motschulsky) and *Zootermopsis nevadensis* (Hagen), Isoptera. *Physiol. Zool.* 36:69–84.

Stuart, A. M. 1963b. Studies on the communication of alarm in the termite *Zootermopsis nevadensis* (Hagen), Isoptera. *Physiol. Zool.* 36:85–95.

Stuart, A. M. 1967. Alarm, defence, and construction behavior relationships in termites (Isoptera). *Science* 156:1123–1125.

Stutt, A. D., and P. Willmer. 1998. Territorial defence in speckled wood butterflies: Do the hottest males always win? *Anim. Behav.* 55 (5):1341–1347.

Suga, N. 1990. Biosonar and neural computation in bats. *Sci. Am.* 262 (June):60–68.

Sugiyama, Y. 1965. Behavioral development and social structure in two troops of hanuman langurs (*Presbytis entellus*). *Primates* 6:213–247.

Sugiyama, Y. 1984. Proximate factors of infanticide among langurs at Dharwar: A reply to Bogess. In *Infanticide: Comparative and Evolutionary Perspectives*, edited by G. Hausfater and S. B. Hrdy. New York: Aldine, pp. 311–314.

Sullivan, J. P., O. Jassim, S. E. Fahrbach, and G. E. Robinson. 2000. Juvenile hormone paces behavioral development in the adult worker honey bee. *Horm. Behav.* 37 (1):1–14.

Sullivan, J. P., O. Jassim, G. E. Robinson, and S. E. Fahrbach. 1996. Foraging behavior and mushroom bodies in allaectomized honey bees. *Soc. Neurosci. Abst.* 22:1144.

Suri, V. P., Z. W. Qian, J. C. Hall, and M. Rosbash. 1998. Evidence that the TIM light response is relevant to light-induced phase shifts in *Drosophila melanogaster*. *Neuron* 21 (1):225–234.

Sutherland, W. 1998. The importance of behavioural studies in conservation biology. *Anim. Behav.* 56 (4):801–809.

Svare, B. 1988. Some trends in the responses studied and the species employed by behavioral endocrinologists. *Horm. Behav.* 22:139–142.

Taborsky, M. 1984. Broodcare helpers in the cichlid fish *Lamprologus brichardi*: Their costs and benefits. *Anim. Behav.* 32:1236–1252.

Taborsky, M. 1985. Breeder-helper conflict in a cichlid fish with broodcare helpers: An experimental analysis. *Behaviour* 95:45–75.

Taborsky, M., and D. Limberger. 1981. Helpers in fish. *Behav. Ecol. Sociobiol.* 8:143–145.

Takahashi, J. S., H. Hamm, and M. Menaker. 1980. Circadian rhythms of melatonin release from individual superfused chicken pineal glands *in vitro*. *Proc. Nat. Acad. Sci. USA* 77:2319–2322.

Takahashi, J. S., and M. Menaker. 1979. Brain mechanisms in avian circadian systems. In *Biological Rhythms and Their Central Mechanisms*, edited by O. Suda, O. Hayaishi and H. Nakagawa. Amsterdam: Elsevier/North Holland, pp. 95–109.

Tamura, N. 1989. Snake-directed mobbing by the Formosan squirrel *Callosciurus erythraeus thaiwanensis*. *Behav. Ecol. Sociobiol.* 24:175–180.

Tang, Y. P., E. Shimizu, G. R. Dube, C. Rampon, G. A. Kerchner, M. Zhuo, G. S. Liu, and J. Z. Tsien. 1999. Genetic enhancement of learning and memory in mice. *Nature* 401 (6748):63–69.

Tanouye, M. A., C. A. Ferrus, and S. C. Fujita. 1981. Abnormal action potentials associated with *Shaker* complex locus in *Drosophila*. *Proc. Nat. Acad. Sci. USA* 78:6548–6552.

Tautz, J. 1979. Reception of particle oscillation in medium—An unorthodox sensory capacity. *Naturwissenschaften* 66:452–461.

Tautz, J., and H. Markl. 1978. Caterpillars detect flying wasps by hairs sensitive to airborne vibration. *Behav. Ecol. Sociobiol.* 4:101–110.

Temeles, E. J. 1989. Effect of prey consumption on foraging activity of northern barriers. *Auk* 106:353–357.

Templeton, J. J., and L.-A. Giraldeau. 1995. Patch assessment in foraging flocks of European starlings: Evidence for public information use. *Behav. Ecol.* 6:65–72.

Templeton, J. J., and L.-A. Giraldeau. 1996. Vicarious sampling: The use of personal and public information by starlings foraging in a simple patchy environment. *Behav. Ecol. Sociobiol.* 38 (2):105–114.

Tennesen, M. 1999. Testing the depths of life. *Nat. Wildlf.* (Feb.):1.

Tepperman, J. 1980. *Metabolic and Endocrine Physiology*. 4th ed. Chicago: Year Book Medical.

Terborgh, J., and A. W. Goldizen. 1985. On the mating system of the cooperatively breeding saddle-backed tamarin (*Saquinus fuscicollis*). *Behav. Ecol. Sociobiol.* 16:293–299.

Terkel, J. 1972. A chronic cross-transfusion technique in freely behaving rats. *J. Comp. Physiol.* 80:365–371.

Terkel, J., and J. S. Rosenblatt. 1972. Humoral factors underlying maternal behavior at parturition: Cross-transfusion between freely moving rats. *J. Comp. Physiol. Psych.* 80:365–371.

Terman, M., and J. S. Terman. 1970. Circadian rhythm of brain self-stimulation behavior. *Science* 168:1242–1244.

Ternes, J., D. Farner, and M. Deavors. 1967. A free-running circadian operant behavior rhythm in the chimpanzee. U.S. Air Force Tech. Doc. ARL-TDR 67-13, pp. 1–18.

Terrace, H. S., L. A. Petitto, R. J. Sanders, and T. G. Bever. 1979. Can an ape create a sentence? *Science* 206:891–900.

Terrick, T. D., R. L. Mumme, and G. M. Burghardt. 1995. Aposematic coloration enhances chemosensory recognition of noxious prey in the garter snake *Thamnophis radix*. *Anim. Behav.* 49 (4):857–866.

Terrill, S. B. 1991. Evolutionary aspects of orientation and migration in birds. In *Orientation in Birds*, edited by P. Berthold. Basel: Birkhäuser Verlag, pp. 180–201.

Theodorakis, C. W. 1989. Size segregation and effects of oddity on predation risk in minnow schools. *Anim. Behav.* 38:496–502.

Thompson, C. W., N. Hillgarth, M. Leu, and H. E. McClure. 1997. High parasite load in house finches (*Carpodacus mexicanus*) is correlated with reduced expression of a sexually selected trait. *Am. Nat.* 149 (2):270–294.

Thompson, K. V. 1996. Play-partner preferences and the function of social play in infant sable antelope, *Hippotragus niger. Anim. Behav.* 52:1143–1155.

Thorndike, E. L. 1898. Animal intelligence: An experimental study of the associative process in animals. *Psychol. Monogr.* 2(8):1–109.

Thorndike, E. L. 1911. *Animal Intelligence: Experimental Studies.* New York: Macmillan.

Thorne, B. L. 1997. Evolution of eusociality in termites. *Ann. Rev. Ecol. Syst.* 28:27–54.

Thornhill, R. 1976. Sexual selection and nuptial feeding behavior in *Bittacus apicalis* (Insecta: Mecoptera). *Am. Nat.* 110:529–548.

Thornhill, R. 1980. Mate choice in *Hylobittacus apicalis* (Insecta: Macoptera) and its relation to some models of female choice. *Evolution* 34:519–538.

Thornhill, R. 1981. *Panorpa* (Mecoptera: Panorpidae) scorpionflies: Systems for understanding resource-defense polygyny. *Ann. Rev. Ecol. Syst.* 12:355–386.

Thornhill, R., and J. Alcock. 1983. *The Evolution of Insect Mating Systems.* Cambridge, MA: Harvard University Press.

Thorpe, W. H. 1963. *Learning and Instinct in Animals.* 2nd ed. London: Methuen.

Thorpe, W. H., and F. G. W. Jones. 1937. Olfactory conditioning in a parasitic insect and its relation to the problem of host selection. *Proc. Roy. Soc. Lond. B* 124:56–81.

Tinbergen, L. 1960. The natural control of insects in pine woods I. Factors influencing the intensity of predation by song birds. *Arch. Neérl. Zool.* 13:265–343.

Tinbergen, N. 1942. An objectivistic study of the innate behaviour of animals. *Biblioth. Biotheor.* 1:37–98.

Tinbergen, N. 1948. Social releasers and the experimental method required for their study. *Wilson Bull.* 60:5–52.

Tinbergen, N. 1951. *The Study of Instinct.* London: Oxford Clarendon Press.

Tinbergen, N. 1952. The curious behavior of the stickleback. *Sci. Am.* 187:2–6.

Tinbergen, N. 1963. On aims and methods of ethology. *Z. Tierpsychol.* 20:410–433.

Tinbergen, N. 1965. Behavior and natural selection. In *Ideas in Evolution and Behavior*, edited by A. J. Moore. New York: Natural History Press, pp. 519–542.

Tinbergen, N. 1989. The Study of Instinct. New York: Oxford University Press.

Tinbergen, N., G. J. Broekhuysen, F. Feekes, J. C. W. Houghton, H. Kruuk, and E. Szulc. 1962. Egg-shell removal by the black-headed gull, *Larus ridibundus* L., a behaviour component of camouflage. *Behaviour* 19:74–118.

Tinbergen, N., and W. Kruyt. 1938. Über die Oreintierrung des Bienenwolfes (*Philanthus triangulum* Fabr.) III. Die Bevorzugung bestimmter Wegmarken. *Z. vergl. Physiol.* 25:292–334.

Tinbergen, N., and A. C. Perdeck. 1950. On the stimulus situation releasing the begging response in the newly hatched herring gull chick (*Larus argentatus argentatus* Pont.). *Behaviour* 3:1–38.

Todd, J. H. 1971. The chemical language of fish. *Sci. Am.* 224 (May):98–108.

Todrank, J., G. Heth, and R. E. Johnston. 1998. Kin recognition in golden hamsters: Evidence for kinship odours. *Anim. Behav.* 55 (2):377–386.

Tonegawa, S. 1994. Gene targeting: A new approach for the analysis of mammalian memory and learning. *Progr. Clin. Biol. Res.* 390:5–18.

Tonegawa, S. 1995. Mammlian learning and memory studied by gene targeting. *Ann. NY Acad. Sci.* 758:213–217.

Tosini, G., and M. Menaker. 1996. Circadian rhythms in cultured mammalian retina. *Science* 272:419–421.

Tougaard, J. 1996. Energy detection and temporal integration in the noctuid A1 auditory receptor. *J. Comp. Physiol. A* 178 (5):669–677.

Towne, W. F., and J. L. Gould. 1985. Magnetic field sensitivity in honeybees. In *Magnetite, Biomineralization and Magnetoreception in Organisms—A New Biomagnetism*, edited by J. H. Kirschvink, D. S. Jones, and B. J. MacFadden. New York: Plenum, pp. 385–406.

Towne, W. F., and J. L. Gould. 1988. The spatial precision of honeybees' dance communication. *J. Insect Behav.* 1:129–155.

Traniello, J. F. A., and S. N. Beshers. 1991. Maximization of foraging efficiency and resource defense by group retrieval in the ant *Formica schaufussi. Behav. Ecol. Sociobiol.* 29 (4):283–290.

Trivers, R. L. 1971. The evolution of reciprocal altruism. *Quart. Rev. Biol.* 46:35–57.

Trivers, R. L. 1972. Parental investment and sexual selection. In *Sexual Selection and the Descent of Man, 1871–1971*, edited by B. Campbell. Chicago: Aldine, pp. 136–179.

Trivers, R. L. 1974. The parent-offspring conflict. *Am. Zool.* 14:249–264.

Trivers, R. L., and H. Hare. 1976. Haplodiploidy and the evolution of the social insects. *Science* 191:249–263.

Truman, J. W. 1972. Physiology of insect rhythms. II. The silk moth brain as the location of the biological clock controlling eclosion. *J. Comp. Physiol.* 81:99–114.

Truman, J. W. 1996. Steroid receptors and nervous system metamorphosis in insects. *Develop. Neurosci.* 18 (1–2):87–101.

Truman, J. W., and L. M. Riddiford. 1970. Neuroendocrine control of ecdysis in silk moths. *Science* 167:1624–1626.

Turchin, P., and P. Kareiva. 1989. Aggregation in *Aphis varians*: An effective strategy for reducing predation risk. *Ecology* 70:1008–1016.

Turek, F. W., J. P. McMillan, and M. Menaker. 1976. Melatonin: Effects on the circadian locomotor rhythm of sparrows. *Science* 194:1441–1443.

Turner, J. R. G. 1977. Butterfly mimicry: The genetical evolution of an adaptation. *Evol. Biol.* 10:163–206.

Tychsen, P. H., and B. S. Fletcher. 1971. Studies on the rhythm of mating in the Queensland fruitfly, *Dacus tryoni*. *J. Insect Physiol.* 17:2139–2156.

Uesugi, K. 1996. The adaptive significance of Batesian mimicry in the swallowtail butterfly, *Papio polytes* (Insecta, Papilionidae): Associative learning in a predator. *Ethology* 102 (9):762–775.

Underwood, H., and T. Siopes. 1984. Circadian organization in quail. *J. Exp. Zool.* 232:557–566.

Underwood, H., R. K. Barrett, and T. Siopes. 1990a. Melatonin does not link the eyes to the rest of the circadian system in quail: A neural pathway is involved. *J. Biol. Rhythms.* 5:349–361.

Underwood, H., R. K. Barrett, and T. Siopes. 1990b. The quail's eye: A biological clock. *J. Biol. Rhythms.* 5:257–265.

Urquhart, F. A. 1987. *The Monarch Butterfly: International Traveler*. Chicago: Nelson-Hall.

Vahed, K. 1998. The function of nuptial feeding in insects: A review of empirical studies. *Biol. Rev.* 73:43–78.

Van den Assem, J., and J. van der Molen. 1969. The waning of the aggressive response of the three-spined stickleback upon constant exposure to a conspecific. 1. A preliminary analysis of the phenomenon. *Behaviour* 34:286–324.

van der Horst, G., M. Muijtiens, K. Kobayashi, R. Takano, A. Kanno, M. Takao, J. de Wit, A. Verkerk, A. P. M. Eker, D. van Leenan, R. Buijus, D. Bootsma, J. H. J. Hoeijmakers, and A. Yasui. 1999. Mammalian Cry1 and Cry2 are essential for the maintenance of circadian rhythms. *Nature* 398:627–630.

Vander Wall, S. B. 1982. An experimental analysis of cache recovery in Clark's nutcracker. *Anim. Behav.* 30:84–94.

Vander Wall, S. B., and R. P. Balda. 1977. Coadaptations of the Clark's nutcracker and the pinyon pine for efficient seed harvest and dispersal. *Ecol. Monogr.* 47:89–111.

Vander Wall, S. B., and H. E. Hutchins. 1983. Dependence of Clark's nutcracker, *Nucifraga columbiana*, on conifer seeds during the postfledgling period. *Can. Field Nat.* 97:208–214.

Vaněček, J., A. Pavlik, and H. Illnerová. 1987. Hypothalamic melatonin receptor sites revealed by autoradiography. *Brain Res.* 435:359–362.

van Lawick-Goodall, J. 1968. Behavior of free-living chimpanzees of the Gombe Stream area. *Anim. Behav. Monogr.* 1:165–311.

van Rhinjn, J., and R. Vodegel. 1980. Being honest about one's intentions: An evolutionarily stable strategy for animal conflicts. *J. Theoret. Biol.* 85:623–641.

van Staaden, M. J. 1998. Ethology: At 50 and beyond. *Trends Ecol. Evol.* 13 (1):6–8.

Van Valen, L. 1973. A new evolutionary law. *Evol. Theory* 1:1–30.

Vargo, E. L. 1997. Poison gland of queen fire ants (*Solenopsis invicta*) is the source of a primer pheromone. *Naturwissenschaften* 84 (11):507–510.

Vargo, E. L., and M. Laurel. 1994. Studies on the mode of action of a queen primer pheromone of the fire ant *Solenopsis invicta*. *J. Insect Physiol.* 40 (7):601–610.

Vargo, E. L., and L. Passera. 1991. Pheromonal and behavioral queen control over the production of gynes in the Argentine ant *Iridomyrmex humilis* (Mayr). *Behav. Ecol. Sociobiol.* 28:161–170.

Vaughan, T. A. 1986. *Mammalogy*. Philadelphia: Saunders College.

Verner, J., and M. F. Willson. 1966. The influence of habitats on mating systems of North American passerine birds. *Ecology* 47:143–147.

Verrell, P. A. 1988. The chemistry of sexual persuasion. *New Sci.* 118:40–43.

Verwey, J. 1930. Die Paarungsbiologie des Fischreihers. *Abteilungen Physiol.* 48:1–120.

Villee, C. A., E. P. Solomon, and P. W. Davis. 1985. *Biology*. Philadelphia: Saunders College.

Visalberghi, E., and E. Alleva. 1979. Magnetic influences on pigeon homing. *Biol. Bull.* 125:246–256.

Visalberghi, E., and C. DeLillo. 1995. Understanding primate behavior: A cooperative effort of field and laboratory research. In *Behavioral Brain Research in Naturalistic and Semi-naturalistic Settings*, edited by E. Alleva, H. Lipp, L. Nadel, and L. Riceri. Dordrecht: Kluwer, pp. 413–424.

vom Saal, F. S. 1981. Variation in phenotype due to random intrauterine positioning of male and female fetuses in rodents. *J. Reprod. Fert.* 62:633–650.

vom Saal, F. S. 1983. Variation in infanticide and parental behavior in male mice due to prior intrauterine proximity to female fetuses: Elimination by prenatal stress. *Physiol. Behav.* 30:675–681.

vom Saal, F. S., and F. H. Bronson. 1978. In utero proximity of female mouse fetuses to males: Effect on reproductive performance during later life. *Biol. Reprod.* 19:842–853.

vom Saal, F. S., and F. H. Bronson. 1980a. Sexual characteristics of adult female mice are correlated with their blood testosterone levels during prenatal development. *Science* 208:597–599.

vom Saal, F. S., and F. H. Bronson. 1980b. Variation in length of estrous cycle in mice due to former intrauterine proximity to male fetuses. *Biol. Reprod.* 22:77–78.

von der Emde, G. 1999. Active electrolocation of objects in weakly electric fish. *J. Exp. Biol.* 202:1205–1215.

von Frisch, K. 1950. Die Sonne als Kompass im Leben der Bienen. *Experientia* 6:210–221.

von Frisch, K. 1967. *The Dance Language and Orientation of Bees*. Cambridge, MA: Harvard University Press.

von Frisch, K. 1971. *Bees: Their Vision, Chemical Senses and Language*. 2nd ed. Ithaca, NY: Cornell University Press.

Vos, G. E. de. 1979. Adaptedness of arena behavior in black grouse (*Tetrao tetrix*) and other grouse species (Tetraonidae). *Behaviour* 68:277–314.

Waage, J. K. 1979. Dual function of the damselfly penis: Sperm removal and transfer. *Science* 203:916–918.

Waas, J. R. 1991. The risks and benefits of signalling aggressive motivation: A study of cave-dwelling little blue penguins. *Behav. Ecol. Sociobiol.* 29:139–146.

Wachowitz, S., and J.-P. Ewert. 1996. A key by which the toad's visual system gets access to the domain of prey. *Physiol. Behav.* 60 (3):877–887.

Wagner, G. 1976. Das Orientierungverhalten von Brieftauben im erdmagnetisch gestörten Gebiete des Chasseral. *Rev. Suisse Zool.* 83:883–890.

Wagner, H., and H. Luksch. 1998. Effect of ecological pressures on brains: Examples from avian neuroethology and general meanings. *Z. Naturforsch.* 53 (5–8):560–581.

Walcott, C. 1977. Magnetic fields and the orientation of homing pigeons under sun. *J. Exp. Biol.* 70:105–123.

Walcott, C. 1978. Anomalies in the earth's magnetic field increase scatter of pigeon vanishing bearings. In *Animal Migration, Navigation and Homing*, edited by K. Schmidt-Koenig and W. T. Keeton. Berlin: Springer-Verlag, pp. 143–151.

Walcott, C. 1980. Homing pigeons vanishing directions at magnetic anomalies are not altered by magnets. *J. Exp. Biol.* 86:349–352.

Walcott, C. 1982. Is there evidence for a magnetic map in homing pigeons? In *Avian Navigation*, edited by F. Papi and H. G. Wallraff. Berlin: Springer-Verlag, pp. 99–108.

Walcott, C., J. L. Gould, and J. L. Kirschvink. 1979. Pigeons have magnets. *Science* 205:1027–1029.

Walcott, C., and R. P. Green. 1974. Orientation of homing pigeons altered by a change in the direction of an applied magnetic field. *Science* 184:180–182.

Waldvogel, J. A., S. Benvenuti, W. T. Keeton, and F. Papi. 1978. Homing pigeon orientation influenced by deflected winds at home loft? *J. Comp. Physiol.* 128:297–301.

Waldvogel, J. A., and J. B. Phillips. 1982. New experiments involving permanent-resident deflector loft birds. In *Avian Navigation*, edited by F. Papi and H. G. Wallraff. Berlin: Springer-Verlag, pp. 179–189.

Waldvogel, J. A., J. B. Phillips, and A. I. Brown. 1988. Changes in short-term deflector loft effect are linked to the sun compass of homing pigeons. *Anim. Behav.* 36:150–158.

Walker, M. M., and M. E. Bitterman. 1989. Honeybees can be trained to respond to very small changes in geomagnetic intensity. *J. Exp. Biol.* 145:489–494.

Walker, M. M., C. E. Diebel, C. V. Haugh, P. M. Pankhurst, and J. C. Montgomery. 1997. Structure and function of the vertebrate magnetic sense. *Nature* 390:371–376.

Walker, M. M., J. L. Kirschvink, R. Chang Shih-Bin, and A. E. Dizon. 1984. A candidate magnetic sense organ in yellowfin tuna, *Thunnus albacares*. *Science* 224:751–753.

Wallman, J. 1992. *Aping Language*. Cambridge: Cambridge University Press.

Wallraff, H. G. 1980. Olfaction and homing in pigeons: Nerve section experiments, critique, hypotheses. *J. Comp. Physiol.* 139:209–224.

Wallraff, H. G. 1981. The olfactory component of pigeon navigation: Steps of analysis. *J. Comp. Physiol.* 143:411–422.

Wallraff, H. G. 1996. Seven theses on pigeon homing deduced from empirical findings. *J. Exp. Biol.* 199 (1):105–111.

Walter, H. 1979. *Eleanora's Falcon: Adaptations to Prey and Habitat in a Social Raptor*. Chicago: University of Chicago Press.

Wang, Z. X., C. F. Ferris, and G. J. DeVries. 1994. Role of septal vasopressin innervation in paternal behavior in prairie voles (*Microtus ochrogaster*). *Proc. Nat. Acad. Sci. USA* 91 (1):400–404.

Wang, Z. X., L. J. Young, Y. Liu, and T. R. Insel. 1997. Species differences in vasopressin receptor binding are evident early in development: Comparative anatomic studies in prairie and montane voles. *J. Comp. Neurol.* 378 (4):535–546.

Waser, P. M., and W. T. Jones. 1983. Natal philopatry among solitary mammals. *Quart. Rev. Biol.* 58:355–390.

Wasserman, E. A. 1995. The conceptual abilities of pigeons. *Am. Sci.* 83 (3):246–255.

Wasserman, E. A., J. A. Hugart, and K. Kirkpatrick-Steger. 1995. Pigeons show same-different conceptualization after training with complex visual stimuli. *J. Exp. Psych.: Anim. Behav. Proc.* 21:248–252.

Watanabe, S. 1997. Visual discrimination of real objects and pictures in pigeons. *Anim. Learn. Behav.* 25(2):185–192.

Waterman, T. H. 1989. *Animal Navigation*. New York: Freeman.

Waters, D. A. 1996. The peripheral auditory characteristics of noctuid moths: Information coding and endogenous noise. *J. Exp. Biol.* 199 (4):857–868.

Watson, J. B. 1930. *Behaviorism*. New York: Norton.

Watt, M., C. S. Evans, and J. M. P. Joss. 1999. Use of electroreception during foraging by the Australian lungfish. *Anim. Behav.* 58 (5):1039–1045.

Watt, P. J., S. F. Nottingham, and S. Young. 1997. Toad tadpole aggregation behaviour: Evidence for a predator avoidance function. *Anim. Behav.* 54 (4):865–872.

Watts, C. R., and A. W. Stokes. 1971. The social order of turkeys. *Sci. Am.* 224:112–118.

Weatherhead, P. J., and R. J. Robertson. 1979. Offspring quality and the polygyny threshold: The "sexy son" hypothesis. *Am. Nat.* 113:201–208.

Weatherhead, P. J., and R. J. Roberston. 1981. In defence of the "sexy son" hypothesis. *Am. Nat.* 117:349–356.

Weaver, D. R. 1998. The suprachiasmatic nucleus: A 25–year retrospective. *J. Biol. Rhythms.* 13 (2):100–112.

Weber, N. A. 1972. The attines: The fungus-culturing ants. *Am. Sci.* 60:448–456.

Weeks, J. C., G. A. Jacobs, and C. I. Miles. 1989. Hormonally mediated modifications of neuronal structure, synaptic connectivity, and behavior during metamorphosis of the tobacco hornwom, *Manduca sexta*. *Am. Zool.* 29:1331–1344.

Weeks, J. C., and R. B. Levine. 1990. Postembryonic neuronal plasticity and its hormonal control during insect metamorphosis. *Ann. Rev. Neurosci.* 13:183–194.

Weeks, J. C., and J. W. Truman. 1984. Neural organization of peptide-activated ecdysis behaviors during metamorphosis of *Manduca sexta*. II. Retention of the proleg motor pattern despite the loss of prolegs at pupation. *J. Comp. Physiol. A* 155:423–433.

Wehner, R. 1981. Spatial vision in arthropods. In *Handbook of Sensory Psychology*, edited by H. Autrum. Berlin: Springer-Verlag, pp. 287–417.

Wehner, R. 1983. Celestial and terrestrial navigation: Human strategies–insect strategies. In *Neuroethologic and Behavioral Physiology*, edited by F. Huber and H. Markel. Berlin: Springer-Verlag, pp. 336–381.

Wehner, R. 1987. "Matched filters"—Neural models of the external world. *J. Comp. Physiol. A* 87:511–531.

Wehner, R., and I. Flatt. 1972. The visual orientation of desert ants, *Cataglyphis bicolor* (Hymenoptera: Formicidae). In *Information Processing in Visual Systems of Arthropods*, edited by R. Wehner. Berlin: Springer-Verlag, pp. 295–302.

Wehner, R., B. Michel, and P. Antonsen. 1996. Visual navigation in insects: Coupling of egocentric and geocentric information. *J. Exp. Biol.* 199 (1):129–140.

Wehner, R., and F. Raber. 1979. Visual spatial memory in desert ants (*Cataglyphis bicolor*) (Hymenoptera, Formicidae). *Experientia* 35:1569–1571.

Wehner, R., and M. V. Srinivasan. 1981. Searching behavior of desert ants, genus *Cataglyphis* (Formicidae: Hymenoptera). *J. Comp. Physiol.* 142:315–338.

Weigmann, C., and J. Lambrect. 1991. Intraspecific nest parasitism in bar-headed geese, *Anser indicus*. *Anim. Behav.* 41:677–688.

Weindler, P., F. Böhme, V. Liepa, and W. Wiltschko. 1998. The role of daytime cues in the development of magnetic orientation in a night-migrating bird. *Behav. Ecol. Sociobiol.* 42 (4):289–294.

Weindler, P., R. Wiltschko, and W. Wiltschko. 1996. Magnetic information affects stellar orientation in young bird migrants. *Nature* 383:158–160.

Weintraub, A., R. Bockman, and D. Kelley. 1985. Prostaglandin E2 induces sexual receptivity in female *Xenopus laevis*. *Horm. Behav.* 19:386–399.

Weisner, P. 1992. A strong case for role reversal. *BioScience* 42 (4):327.

Welker, H. A., P. Semm, R. P. Willig, J. C. Commentz, W. Wiltschko, and L. Vollrath. 1983. Effects of an artificial magnetic field on serotonin N-acetyltransferase activity and melatonin content of the rat pineal gland. *Exp. Brain Res.* 50:426–432.

Wells, K. D. 1977. The courtship of frogs. In *The Reproductive Biology of Amphibians*, edited by D. H. Taylor and S. I. Guttman. New York: Plenum, pp. 233–262.

Welsh, D. K., D. E. Logothetis, M. Meister, and S. M. Reppert. 1995. Individual neurons dissociated from rat suprachiasmatic nucleus express independently phased circadian firing patterns. *Neuron* 14:697–706.

Welty, J. C. 1962. *The Life of Birds*. 2nd ed. Philadelphia: Saunders.

Wenner, A. M. 1964. Sound communication in honeybees. *Sci. Am.* 210 (Apr.):116–125.

Wenner, A. M. 1971. *The Bee Language Controversy—An Experience in Science*. Boulder, CO: Educational Programs Improvement.

Wenner, A. M. 1998. Odors, wind and colony foraging. II. Insights from beehunting. *Am. Bee J.* 138 (12):897–899.

Wenner, A. M., and D. L. Johnson. 1966. Simple conditioning in honeybees. *Anim. Behav.* 14:149–155.

Wenner, A. M., and P. H. Wells. 1987. The honey bee dance language controversy: The search for "truth" vs. the search for useful information. *Am. Bee J.* 127:130–131.

Wenner, A. M., and P. H. Wells. 1990. *Anatomy of a Controversy—The Question of a "Language" Among Bees*. New York: Columbia University Press.

Werren, J. H., M. R. Gross, and R. Shine. 1980. Paternity and the evolution of male parental care. *J. Theoret. Biol.* 82:619–631.

West, M. J., and M. J. King. 1988. Female visual displays affect the development of male song in the cowbird. *Nature* 334:244–246.

West, M. J., A. P. King, and D. H. Eastzer. 1981. The cowbird: Reflections on development from an unlikely source. *Am. Sci.* 69:56–66.

West-Eberhard, M. J. 1975. The evolution of social behavior by kin selection. *Quart. Rev. Biol.* 50:1–33.

Wetzel, D,, and D. Kelley. 1983. Androgen and gonadotropin control of the mate calls of male South African clawed frogs, *Xenopus laevis*. *Horm. Behav.* 17:388–404.

White, C. S., D, M. Lambert, C. D. Millar, and P. M. Stevens. 1991. Is helping behavior a consequence of natural selection? *Am. Nat.* 133:246–253.

Whitehouse, M. E. A. 1997. Experience influences male-male contests in the spider *Argyrodes antipodiana*. *Anim. Behav.* 53 (5):913–923.

Whiten, A., D. M. Custance, J.-C. Gomez, P. Teixidor, and K. A. Bard. 1996. Imitative learning in artificial fruit processing in children (*Homo sapiens*) and chimpanzees (*Pan troglodytes*). *J. Comp. Psychol.* 110:3–14.

Whitten, P. L., D. K. Brockman, and R. C. Stavisky. 1998. Recent advances in noninvasive techniques to monitor hormone-behavior interactions. *Yrbk. Phys. Anthropol.* 41:1–23.

Wickler, W. 1968. *Mimicry*. New York: McGraw-Hill.

Wickler, W. 1972. *Mimicry*. 2nd ed. New York: World University Library, McGraw-Hill.

Wicksten, M. K. 1980. Decorator crabs. *Sci. Am.* 242 (Feb.):146–154.

Wiedenfeld, D. A., and M. G. Wiedenfeld. 1995. Large kill of neotropical migrants by tornado and storm in Louisiana. *J. Field Ornithol.* 66 (1):70–80.

Wiggins, D. A., and R. D. Morris. 1986. Criteria for female choice of mates: Courtship feeding and paternal care in the common tern. *Am. Nat.* 128:126–129.

Wilcox, R. S. 1972. Communication by surface waves: Mating behavior of a water strider (Gerridae). *J. Comp. Physiol.* 80:255–266.

Wilcox, R. S., R. R. Jackson, and K. Gentile. 1996. Spiderweb smokescreens—Spider trickster uses background noise to mask stalking movements. *Anim. Behav.* 51 (2):313–326.

Wilcox, R. S., and T. Ruckdeschel. 1982. Food threshold territoriality in a water strider (*Gerris remigis*). *Behav. Ecol. Sociobiol.* 11:85–90.

Wiley, R. H. 1974. Evolution of social organization and life history patterns among grouse (Aves: Tetraonidae). *Quart. Rev. Biol.* 49:209–227.

Wiley, R. H. 1991. Associations of song properties with habitats for territorial oscine birds of eastern North America. *Am. Nat.* 38:973–993.

Wiley, R. H. 1994. Errors, exaggeration, and deception in animal communication. In *Behavioral Mechanism in Evolutionary Ecology*, edited by L. Real. Chicago: University of Chicago Press, pp. 157–189.

Wiley, R. H., and D. G. Richards. 1978. Physical constraints on acoustic communication in the atmosphere: Implications for the evolution of animal vocalizations. *Behav. Ecol. Sociobiol.* 3:69–94.

Wilkinson, G. S. 1984. Reciprocal food sharing in the vampire bat. *Nature* 308:181–184.

Wilkinson, G. S. 1990. Food sharing in vampire bats. *Sci. Am.* 262 (Feb.):76–82.

Wilkinson, G. S., and P. R. Reillo. 1994. Female choice to artificial selection on an exaggerated male trait in a stalk-eyed fly. *Proc. Roy. Soc. Lond. B* 255 (1342):1–6.

Williams, C. B. 1965. *Insect Migration*. London: Collins.

Williams, G. C. 1966. *Adaptation and Natural Selection*. Princeton, NJ: Princeton University Press.

Williams, G. C. 1975. *Sex and Evolution*. Princeton, NJ: Princeton University Press.

Williams, H. 1990. Models for song learning in the zebra finch: Fathers or others? *Anim. Behav.* 39:745–757.

Williams, J. R., T. R. Insel, C. R. Harbargh, and C. S. Carter. 1994. Oxytocin administered centrally facilitates formation of a partner preference in female prairie voles (*Microtus ochrogaster*). *J. Neuroendocrinol.* 6 (3):247–250.

Williams, S. L., K. E. Brakke, and E. S. Savage-Rumbaugh. 1997. Comprehension skills of language-competent and nonlanguage-competent apes. *Lang. Commun.* 17 (4):301–317.

Williams, T. C., and J. M. Williams. 1978. An oceanic mass migration of land birds. *Sci. Am.* 239 (Oct.):166–176.

Wilson, D. M. 1961. The central nervous control of flight in a locust. *J. Exp. Biol.* 38:471–490.

Wilson, D. M., and T. Weis-Fogh. 1962. Patterned activity of co-ordinated motor units studied in flying locusts. *J. Exp. Biol.* 40:643–647.

Wilson, E. O. 1965. Chemical communication in the social insects. *Science* 149:1064–1071.

Wilson, E. O. 1968. Chemical systems. In *Animal Communication: Techniques of Study and Results of Research*, edited by T. A. Sebeok. Bloomington: Indiana University Press, pp. 75–102.

Wilson, E. O. 1971. *The Insect Societies*. Cambridge, MA: Belknap/Harvard University Press.

Wilson, E. O. 1975. *Sociobiology*. Cambridge, MA: Belknap/Harvard University Press.

Wilson, E. O. 1980. *Sociobiology, the Abridged Version*. Cambridge, MA: Belknap/Harvard University Press.

Wilson, E. O., and W. H. Bossert. 1963. Chemical communication among animals. *Recent Prog. Horm. Res.* 19:673–716.

Wiltschko, R. 1983. The ontogeny of orientation in young pigeons. *Comp. Biochem. Physiol. A* 76:701–708.

Wiltschko, R. 1996. The function of olfactory input in pigeon orientation: Does it provide navigational information or play another role? *J. Exp. Biol.* 199 (1):113–199.

Wiltschko, R., R. Kumpfmüller, R. Muth, and W. Wiltschko. 1994. Pigeon homing: The effect of a clock-shift is often smaller than predicted. *Behav. Ecol. Sociobiol.* 35:63–73.

Wiltschko, R., and W. Wiltschko. 1981. The development of sun compass orientation in young homing pigeons. *Behav. Ecol. Sociobiol.* 9:135–141.

Wiltschko, R., and W. Wiltschko. 1999a. The orientation system of birds. I. Compass mechanisms. *J. Ornithol.* 140 (1):1–40.

Wiltschko, R., and W. Wiltschko. 1999b. The orientation system of birds. III. Migratory orientation. *J. Ornithol.* 140 (3):273–308.

Wiltschko, W. 1978. Further analysis of the magnetic compass of migratory birds. In *Animal Migration, Navigation and Homing*, edited by K. Schmidt-Koenig and W. T. Keeton. Berlin: Springer-Verlag, pp. 302–310.

Wiltschko, W. 1982. The migratory orientation of garden warblers *Sylvia borin*. In *Avian Navigation*, edited by F. Papi and H. G. Wallraff. Berlin: Springer-Verlag, pp. 50–58.

Wiltschko, W., P. Daum, A. Fergenbauer-Kimmel, and R. Wiltschko. 1987. The development of the star compass in garden warblers. *Ethology* 74:285–292.

Wiltschko, W., and E. Gwinner. 1974. Evidence for an innate magnetic compass in garden warblers. *Naturwissenschaften* 61:406.

Wiltschko, W., U. Munro, R. C. Beason, and R. Wiltschko. 1994. A magnetic pulse leads to a temporary deflection in the orientation of migratory birds. *Experientia* 50:697–700.

Wiltschko, W., U. Munro, H. Ford, and R. Wiltschko. 1993. Red light disrupts magnetic orientation of migratory birds. *Nature* 364:525–527.

Wiltschko, W., U. Munro, H. Ford, and R. Wiltschko. 1998. Effect of a magnetic pulse on the orientation of silvereyes, *Zosterops l. lateralis*, during spring migration. *J. Exp. Biol.* 201:3257–3261.

Wiltschko, W., and R. Wiltschko. 1972. Magnetic compass of European robins. *Science* 176:62–64.

Wiltschko, W., and R. Wiltschko. 1991. Magnetic orientation and celestial cues in migratory orientation. In *Orientation in Birds*, edited by P. Berthold. Basil: Birkhäuser Verlag, pp. 16–37.

Wiltschko, W., and R. Wiltschko. 1995. Migratory orientation in European robins is affected by wavelength of light as well as by magnetic pulse. *J. Comp. Physiol. A* 177:363–369.

Wiltschko, W., and R. Wiltschko. 1996. Magnetic orientation in birds. *J. Exp. Biol.* 199 (1):29–38.

Wiltschko, W., R. Wiltschko, W. T. Keeton, and R. Madden. 1983. Growing up in an altered magnetic field affects the initial orientation of young homing pigeons. *Behav. Ecol. Sociobiol.* 12:135–142.

Wiltschko, W., R. Wiltschko, U. Munro, and H. Ford. 1998. Magnetic versus celestial cues—Cue-conflict experiments with migrating silvereyes at dusk. *J. Comp. Physiol. A* 182 (4):521–529.

Wingfield, J. C. 1984a. Environmental and endocrine control of reproduction in the song sparrow, *Melospiza melodia*. I. Temporal organization of the breeding cycle. *Gen. Comp. Endocrinol.* 56:406–416.

Wingfield, J. C. 1984b. Environmental and endocrine control of reproduction in the song sparrow, *Melospiza melodia*. II. Agonistic interactions as environmental information stimulating secretion of testosterone. *Gen. Comp. Endocrinol.* 56:417–424.

Wingfield, J. C., G. F. Ball, A. M. Dufty, Jr., R. E. Hegner, and M. Ramenofsky. 1987. Testosterone and aggression in birds. *Am. Sci.* 75:602–608.

Wingfield, J. C., and M. C. Moore. 1987. Hormonal, social, and environmental factors in the reproductive biology of free-living male birds. In *Psychobiology of Reproductive Behavior*, edited by D. Crews. Upper Saddle River, NJ: Prentice Hall, pp. 148–175.

Winslow, J. T., N. Hastings, C. S. Carter, C. R. Harbaugh, and T. R. Insel. 1993. A role for central vasopressin in pair bonding in monogamous prairie voles. *Nature* 365 (6446):545–548.

Winston, M. L. 1987. *The Biology of the Honey Bee*. Cambridge, MA: Harvard Univeristy Press.

Winston, M. L., and K. N. Slessor. 1992. The essence of royalty: Honey bee queen pheromone. *Am. Sci.* 80:374–385.

Wisby, W. J., and A. D. Hasler. 1954. The effect of olfactory occlusion on migrating silver salmon (*O. kisutch*). *J. Fish. Res. Board Can.* 11:472–478.

Withers, G. S., S. E. Fahrbach, and G. E. Robinson. 1993. Selective neuroanatomical plasticity and division of labor in the honeybee. *Nature* 364:238–240.

Withers, G. S., S. E. Fahrbach, and G. E. Robinson. 1995. Effects of experience and juvenile hormone on the organization of the mushroom bodies of honey bees. *J. Neurobiol.* 26:130–144.

Wittenberger, J. F. 1978. The evolution of mating systems in grouse. *Condor* 80:126–137.

Wittenberger, J. F. 1979. The evolution of mating systems in birds and mammals. In *Handbook of Behavioral Neurobiology*, edited by P. Marler and J. G. Vandenbergh. New York: Plenum, pp. 271–349.

Wittenberger, J. F. 1981. *Animal Social Behavior*. Boston: Duxbury.

Wolf, L. L. 1975. "Prostitution" behavior in a tropical hummingbird. *Condor* 77:140–144.

Wolff, J. O. 1994. More on juvenile dispersal in mammals. *Oikos* 71 (2):349–352.

Wolff, J. O. 1997. Population regulation in mammals: An evolutionary perspective. *J. Anim. Ecol.* 66 (1):1–13.

Wolff, J. O., and J. H. Plissner. 1998. Sex biases in avian natal dispersal: An extension of the mammalian model. *Oikos* 83 (2):327–330.

Wood, D. E. 1995. Neuromodulation of rhythmic patterns in the blue crab, *Callinectes sapidus* by amines and the peptide proctolin. *J. Comp. Physiol. A* 177:335–349.

Wood, D. E., and C. D. Derby. 1995. Coordination and neuromuscular control of rhythmic behaviors in the blue crab, *Callinectes sapidus*. *J. Comp. Physiol. A* 177:307–319.

Wood, D. E., R. A. Gleeson, and C. D. Derby. 1995. Modulation of behavior by biogenic amines and peptides in the blue crab, *Callinectes sapidus*. *J. Comp. Physiol. A* 177:321–333.

Wood, D. E., M. Nishikawa, and C. D. Derby. 1996. Proctolinlike immunoreactivity and identified neurosecretory cells as putative substrates for modulation of courtship display behavior in the blue crab, *Callinectes sapidus*. *J. Comp. Physiol. A* 368 (1):153–163.

Woolfenden, G. E. 1975. Florida scrub jay helpers at the nest. *Auk* 92:1–15.

Woolfenden, G. E., and J. W. Fitzpatrick. 1978. The inheritance of territory in group-breeding birds. *BioScience* 28:104–108.

Woolfenden, G. E., and J. W. Fitzpatrick. 1990. Florida scrub jays: A synopsis after 18 years of study. In *Cooperative Breeding in Birds*, edited by P. B. Stacey and W. D. Koenig. Cambridge: Cambridge University Press, pp. 241–266.

Wourms, M. K., and F. E. Wasserman. 1985. Butterfly wing markings are more advantageous during handling than during the initial strike of an avian predator. *Evolution* 39:845–851.

Woyke, J. 1964. Causes of repeated mating flights by queen honeybees. *J. Apic. Res.* 3:17–23.

Wrangham, R. D. 1980. Female choice of least costly males; a possible factor in the evolution of leks. *Z. Tierpsychol.* 54:357–367.

Wright, J. 1997. Helping-at-the-nest in Arabian babblers: Signalling social status or sensible investment in chicks? *Anim. Behav.* 54 (6):1439–1448.

Wright, J., and I. Cuthill. 1990. Manipulation of sex difference in parental care: effect of brood size. *Anim. Behav.* 40:462–471.

Wyers, E. J., V. S. Peeke, and M. J. Herz. 1973. Behavioral habituation in invertebrates. In *Habituation*, edited by V. S. Peeke and M. J. Herz. New York: Academic Press, pp. 1–57.

Wyllie, I. 1988. *The Cuckoo*. London: Batsford.

Yamazaki, K., A. Singer, and G. K. Beauchamp. 1998. Origin, functions, and chemistry of H-2 regulated odorants. *Genetica* 104 (3):235–240.

Yamazaki, K., M. Yamaguchi, E. A. Boyse, and L. Thomas. 1980. The major histocompatibility complex as a source of odors imparting individuality among mice. In *Chemical Signals*, edited by D. Müller-Schwartze and R. M. Silverstein. New York: Plenum, pp. 267–273.

Yan, O. Y., C. R. Andersson, T. Kondo, S. S. Golden, and C. H. Johnson. 1998. Resonating circadian clocks enhance fitness in cyanobacteria. *Proc. Nat. Acad. Sci. USA* 95 (15):8660–8664.

Yasukawa, K. 1981. Male quality and female choice of mate in the red-winged blackbird (*Agelaius phoeniceus*). *Ecology* 62:922–929.

Ydenberg, R. C. 1984. The conflict between feeding and territorial defense in the great tit. *Behav. Ecol. Sociobiol.* 15:103–108.

Ydenberg, R. C., and J. R. Krebs. 1987. The trade-off between territorial defense and foraging in the great tit (*Parus major*). *Am. Zool.* 27:337–346.

Ylonen, H., T. Kojola, and J. Vitala. 1988. Changing female spacing behavior and demography in an enclosed population of *Clethrionomys glareolus*. *Holarct. Ecol.* 11:286–292.

Yorke, E. 1981. Sensitivity of pigeons to small magnetic field variations. *J. Theoret. Biol.* 89:533–537.

Young, D. 1989. *Nerve Cells and Animal Behaviour*. Cambridge: Cambridge University Press.

Young, L. J., Z. Wang, and T. R. Insel. 1998. Neuroendocrine bases of monogamy. *Trends Neurosci.* 21:71–75.

Young, M. E., and E. A. Wasserman. 1997. Entropy detection by pigeons: Responses to mixed visual displays after same-different discrimination training. *J. Exp. Psych.: Anim. Behav. Proc.* 23 (2):157–170.

Young, M. W. 1998. The molecular control of circadian behavioral rhythms and their entrainment in *Drosophila*. *Ann. Rev. Biochem.* 67:135–152.

Young, M. W. 2000. The tick-tock of the biological clock. *Sci. Am.* 282 (Mar.):64–71.

Young, W. S., E. Shepard, A. C. DeVries, A. Zimmer, M. E. LaMarca, E. I. Ginns, J. Amico, R. J. Nelson, L. Henninghausen, and K. U. Wagner. 1998. Targeted reduction of oxytocin expression provides insights into its physiological roles. *Adv. Exp. Med. Biol.* 449:231–240.

Youthed, G. J., and R. C. Moran. 1969. The lunar day activity rhythm of myrmeleontid larvae. *J. Insect Physiol.* 15:1259–1271.

Zach, R. 1978. Selection and dropping of whelks by Northwestern crows. *Behaviour* 67:134–148.

Zach, R. 1979. Shell-dropping: Decision making and optimal foraging in Northwestern crows. *Behaviour* 67:106–117.

Zagotta, W. N., S. Germeraad, S. S. Garber, T. Hoshi, and R. W. Aldrich. 1989. Properties of ShB A-type potassium channels expressed in *Shaker* mutant *Drosophila* by germline transformation. *Neuron* 3:773–782.

Zahavi, A. 1974. Communal nesting by the Arabian babbler: A case of individual selection. *Ibis* 116:84–87.

Zahavi, A. 1975. Mate selection—A selection for a handicap. *J. Theoret. Biol.* 53:205–214.

Zahavi, A. 1977. Reliability in communication systems and the evolution of altruism. In *Evolutionary Ecology*, edited by B. Stonehouse. London: Macmillan, pp. 253–259.

Zahavi, A. 1995. Altruism as a handicap: Limitations of kin selection and reciprocity. *J. Avian Biol.* 26:1–3.

Zahavi, A., and A. Zahavi. 1997. *The Handicap Principle: A Missing Piece of Darwin's Puzzle.* Oxford: Oxford University Press.

Zakon, H. H., and K. D. Dunlap. 1999. Sex steroids and communication signals in electric fish. *Brain Behav. Evol.* 54 (1):61–69.

Zentall, T. R., J. Sutton, and L. M. Shelburne. 1996. True imitative learning in pigeons. *Psychol. Sci.* 7:343–346.

Zhang, F., S. Endo, L. J. Cleary, A. Eskin, and J. H. Byrne. 1997. Role of transforming growth factor-ß in long-term synaptic facilitation in *Aplysia*. *Science* 275:1318–1319.

Zimmerman, N. H., and M. Menaker. 1979. The pineal gland: A pacemaker within the circadian system of the sparrow. *Proc. Nat. Acad. Sci. USA* 76:999–1003.

Zippelius, H. 1972. Die Karawanenbildung bei Feld-und Hausspitzmaus. *Z. Tierpsychol.* 30:305–320.

Zoeger, J., J. R. Dunn, and M. Fuller. 1981. Magnetic material in the head of the common Pacific dolphin. *Science* 213:892–894.

Zuk, M. 1994. Immunology and the evolution of behavior. In *Behavioral Mechanisms in Evolutionary Ecology*, edited by L. A. Real. Chicago: University of Chicago Press, pp. 354–368.

PHOTO CREDITS

Chapter 1
Figure 1.1: Comstock, Inc. Figure 1.2: Milton H. Tierney, Jr./Visuals Unlimited. Figure 1.3: Nigel J. Dennis/Photo Researchers.

Chapter 2
Figure 2.1: Courtesy American Museum of Natural History. Figure 2.3, left & top, right: Culver Pictures. Figure 2.3 top, right: AP/Wide World Photos. Figure 2.8: Thomas D. McAvoy/LIFE Magazine, Time, Inc. Figure 2.15: Corbis-Bettmann. Figure 2.16: Nina Leen/LIFE Magazine, Time, Inc. Figure 2.17: Underwood & Underwood/Corbis.

Chapter 3
Figure 3.1: Hans Pfletschinger/Peter Arnold. Figure 3.6: From *Animal Behavior*, 5th Ed., by Lee Drickamer, 1997, McGraw-Hill Companies. Reproduced with the permission of Lee Drickamer and the McGraw-Hill Companies. Figure 3.14: Courtesy Jeff Hall, Dept. of Biology, Brandeis University. Figure 3.21: Photo courtesy Jennifer Brown. Reproduced with permission of Jennifer Brown, Michael Greenberg and *Cell*.

Chapter 4
Figure 4.2: Len Rue, Jr./Photo Researchers. Figure 4.4: Walter Chandoha. Figure 4.5 & Figure 4.6 bottom: Courtesy Susan Reichert, Dept. of Zoology, University of Tennessee. Figure 4.6 top, right: Glynne Robinson Betts/Photo Researchers. Figure 4.7: Fred Bavendam/Peter Arnold. Figure 4.8: Klaus D. Francke/Peter Arnold. Figure 4.9: David Overcash/Bruce Coleman, Inc. Figure 4.11: Eric Hosking/Photo Researchers. Figure 4.13: Courtesy Jane Brockman, University of Florida, Gainsville.

Chapter 5
Figure 5.1a: Allan D. Cruickshank/Photo Researchers. Figure 5.1b: Helen Williams/Photo Researchers. Figure 5.1c: Allan D. Cruickshank/Photo Researchers. Figure 5.6: Nina Leen/LIFE Magazine, Time, Inc. Figure 5.7: From *The Mentality of Apes* by W. Kohler, 1926 © by Harcourt, Brace & Co., Routeledge. Figure 5.8: Courtesy Irene Pepperberg, Ecology Dept., University of Arizona. Figure 5.9: Frank Lane Picture Agency. Figure 5.10: Courtesy Masao Kawai, Primate Research Istitute, Kyoto University.

Chapter 6
Figure 6.20: G. Ronald Austing/Photo Researchers. Figure 6.21: Hal Harrison/Grant Heilman Photography. Figure 6.23: F.A.O Photo. Figure 6.24: Stephen Dalton/Photo Researchers.

Chapter 7
Figure 7.8a: Scott Camazine/Photo Researchers. Figure 7.8b: Steve Ross/Photo Researchers. Figure 7.8c: Matt Meadows/Peter Arnold, Inc. Figure 7.14: Courtesy Stephen Goodenough. Figure 7.16: Courtesy Ronald Barfield. Figure 7.23: David Crews, Departments of Zoology and Psychology, Universtiy of Texas. Figure 7.25: Dwight Kuhn/Bruce Coleman, Inc.

Chapter 8
Figure 8.3: Courtesy W. E. Bemis. Figure 8.4: Stephen J. Krasemann/DRK Photo. Figure 8.10: Arthur Morris/Visuals Unlimited. Figure 8.13: Mary M. Thacher/Photo Researchers. Figure 8.14: Nina Leen/LIFE Magazine, Time Inc. Figure 8.17: Jen and Des Bartlett/Photo Researchers. Figure 8.18: Courtesy Peter H. Klopfer, Dept. of Zoology, Duke University. Figure 8.20 & Figure 8.22: Courtesy Harlow Primate Lab. Figure 8.21: Science Vu/Visuals Unlimited. Figure 8.23: Courtesy William A. Harris, Department of Biology, University of California, San Diego.

Chapter 9
Figure 9.2a: James L. Amos/Peter Arnold, Inc. Figure 9.4a: Courtesy Elliot S. Valenstew. Figure 9.8: Jeff Foott/DRK Photo. Figure 9.9: Don Smetzer/Tony Stone World Wide. Figure 9.10: Bert L. Dunne/Bruce Coleman, Inc. Figure 9.15: Stephen Dalton/NHPA. Figure 9.17 : Courtesy Steve A. Kay, Jeffrey Plants/Scripps Research Institute. Figure 9.18: Ben Rose. Figure 9.22: Courtesy Rae Silver.

Chapter 10
Figure 10.1: M. W. F. Tweedie/Photo Researchers. Figure 10.7: Courtesy K. Schmidt-Koenig, Dept. of Zoology, Duke University. Figure 10.8: James L. Amos/Peter Arnold, Inc. Figure 10.11: Karl H. Maslonski/Photo Researchers. Figure 10.14: Johnathan Blair/Woodfin Camp & Associates. Figure 10.18b: Courtesy T. H. Waterman, Biology Dept., Yale University. Figure 10.19: David Dennis/Tom Stack & Associates. Figure 10.21: Courtesy Charles Walcott, Laboratory of Ornithology, Cornell University. Figure 10.22: Dean Lee/The Wildlife Collection. Figure 10.29: Toni Angermayer/Photo Researchers.

Chapter 11
Figure 11.1: Gregory K. Scott/Photo Researchers. Figure 11.5: Courtesy Betty McGuire, University of Massachusetts, Amherst. Figure 11.6: Charles Ott/Photo Researchers. Figure 11.7: Gregory Dimijian/Photo Researchers. Figure 11.9: Thomas D. Mangelsen/Images of Nature. Figure 11.10: Russ Kinne/Comstock, Inc.

Chapter 12
Figure 12.1: Ross Hutchins/Photo Researchers. Figure 12.2: Francois Gohier/Ardea London. Figure 12.3: Catherine Craig/Sinauer Associates. Figure 12.4a: Tom McHugh/Photo Researchers. Figure 12.4b: Z. Leszczynski/Animals Animals.

Figure 12.5: Courtesy Robert Jackson, University of Canterbury, New Zealand. Figure 12.6a: Joe McDonald/Animals Animals. Figure 12.6b: From R. Igor Gamow and John F. Harris, *Scientific American*, May 1973. Figure 12.9: Tom McHugh/Photo Researchers. Figure 12.10a: Courtesy Barth Falkenberg, School of Biological Sciences, University of Nebraska, Lincoln. Figure 12.10b: Photo by Barth Faulkenberg, courtesy Al Kamil, School of Biological Sciences, University of Nebraska. Figure 12.10c: Photo by H. J. Vermes, courtesy Ted Sargent. Figure 12.14: J. B. & S. Bottomley/Ardea London. Figure 12.15: Courtesy Paul V. Switzer. Figure 12.16: Breck Kent/Animals Animals. Figure 12.17: Z. Leszczynki/Animals Animals. Figure 12.19: Runk Schoenberger/Grant Heilman Photography. Figure 12.20: David Hosking/Photo Researchers.

Chapter 13

Figure 13.1: L. P. Brower, Sweet Briar College. Figure 13.2: Robert and Linda Mitchell. Figure 13.8: Courtesy H. B. Kettlewell. Figure 13.10: Photo by H. J. Vermes, courtesy Ted Sargent. Figure 13.12: Courtesy F. Harvey Pough, Cornell University. Figure 13.18: Courtesy T. E. Reimchen, Department of Biology, University of Victoria. Figure 13.20: Leonard Lee Rue, National Audubon Society/Photo Researchers. Figure 13.23: Courtesy Edmund D. Brodie, Jr., Dept. of Biology, University of Texas at Arlington. Figure 13.24: Courtesy Thomas Eisner & Daniel Aneshansley, Cornell University. Figure 13.26: Anup & Manoj Sham/Animals Animals.

Chapter 14

Figure 14.2: Courtesy Darryl T. Gwynne, Biology Department, University of Toronto. Figure 14.3: Courtesy H. Carl Gerhardt. Figure 14.4: David Woodfall/NHPA. Figure 14.6a: Hans Pfletschinger/Peter Arnold, Inc. Figure 14.6b: Jonathan Waage, Division of Biology and Medicine, Brown University. Figure 14.7: Irene Vandermolen/Photo Researchers. Figure 14.8: Clark & Handley, National Audubon Society/Photo Researchers. Figure 14.9: Courtesy Mark A. Bellis. Figure 14.10: S. J. Arnold, Oregon State University. Figure 14.11: Richard Matthews/Planet Earth Pictures. Figure 14.12: Tom McHugh/Photo Researchers. Figure 14.14: Courtesy Stephen Goodenough. Figure 14.15: Eric Hosking. Figure 14.17: John Cancalosi/DRK Photo. Figure 14.18: Phil Savoie/BBC Hulton Picture Library. Figure 14.19: Howard Earl Uible/Photo Researchers.

Chapter 15

Figure 15.1: Courtesy Aaron Norman. Figure 15.2: Courtesy Robert L. Smith, Department of Entomology, University of Arizona. Figure 15.3: Courtesy Terrance Mace, University of Puget Sound. Figure 15.4: Eric & David Hosking. Figure 15.5: Stephen Dalton/Photo Researchers. Figure 15.7: P. Ward/Bruce Coleman, Inc. Figure 15.8: Tom McHugh/Photo Researchers. Figure 15.10: Al Lowry/Photo Researchers. Figure 15.11: Verna R. Johnston/Photo Researchers. Figure 15.13: C. Behnke/Animals Animals. Figure 15.14: David Overcash/Bruce Coleman, Inc.

Chapter 16

Figure 16.1: Courtesy Simon Pollard, Dept. of Entomology, University of Alberta, Edmonton. Figure 16.3: Courtesy Gary McCracken, Department of Zoology, University of Tennessee, Knoxville. Figure 16.6: Courtesy W. E. Bemis. Figure 16.7: Walt Anderson/Visuals Unlimited. Figure 16.8: Clem Haagner/Photo Researchers. Figure 16.12: Frank Lane Picture Agency. Figure 16.13: Courtesy Richard Howard/Purdue University. Figure 16.15: Leonard Lee Rue/Photo Researchers.

Chapter 17

Figure 17.1: Leonard Lee Rue/Animals Animals. Figure 17.2: From *Gerhard marcuse cichlid of Lake Malawi* , *12th ed.* by H. Axelrod & Warren Burgess, 1988, TFH Publishing. Figure 17.6: Kenneth W. Fink/Bruce Coleman. Figure 17.10: Richard R. Hansen/Photo Researchers. Figure 17.13: Ian Cleghorn/Photo Researchers. Figure 17.15: Leanne Nash. Figure 17.16: Courtesy Patricia D. Moehlman/Wildlife Conservation International. Figure 17.19: Tom J. Ulrich/Visuals Unlimited. Figure 17.21: M. K. & I. M. Morcombe/NHPA. Figure 17.22: Nigel Dennis/NHPA. Figure 17.26: R. Williamson/Visuals Unlimited. Figure 17.27: Raymond A. Mendez/Animals Animals. Figure 17.31: Courtesy Scott Camazine, Neurobiology and Behavior, Cornell University.

Chapter 18

Figure 18.1a: Townsend P. Dickinson/Comstock, Inc. Figure 18.1b: Courtesy Frans B. M. de Waal, Yerkes Primate Center. Figure 18.2: Courtesy Stephen Goodenough. Figure 18.3: Runk/Schoenberger/Grant Heilman Photography. Figure 18.5: Leonard Lee Rue/Photo Researchers. Figure 18.6: Courtesy Dietrich Schneider, Max-Planck Institute. Figure 18.7: Courtesy Stimson Wilcox, SUNY Binghampton. Figure 18.9: R. M. Meadows/Peter Arnold. Figure 18.10: Pierre Boulat/Paris. Figure 18.11: Courtesy Lynne D. Houck and S.J. Arnold, Department of Ecology and Evolution, University of Chicago. Figure 18.13: Photo Researchers. Figure 18.14: Gene Wolfsheimer/National Audubon Society/Photo Researchers. Figure 18.16: Fritz Siedel/Frank Lane Picture Agency. Figure 18.25: Mark Moffett/Minden Pictures, Inc. Figure 18.28: Art Gingert/Comstock, Inc. Figure 18.29: Roy P. Fontaine/Photo Researchers.

Chapter 19

Figure 19.1: Anthony Bannister/The Anthony Bannister Photo Library. Figure 19.3: Courtesy Joseph Waas. Figure 19.6b: Roy L. Caldwell, Department of Intergative Biology, University of California. Figure 19.7: Courtesy John D. Palmer. Figure 19.10: Jen and Des Bartlett/Photo Researchers. Figure 19.11: Toni Angermayer/Photo Researchers. Figure 19.13: Phil Devries/Oxford Scientific Films . Figure 19.14: Arthur Ambler, National Audubon Society/Photo Researchers. Figure 19.15: Michael Moblick. Figure 19.16: Raymond A. Mendez/Animals Animals. Figure 19.20: Susan Kuklin/Photo Researchers. Figure 19.23 & Figure 19.24: Courtesy Georgia State University & Yerkes, Emory.

Appendix

Figure A.3: Courtesy Stephen Mann, School of Chemistry, University of Bath.

PERMISSION CREDITS

Chapter 1

Quote: Perspectives in Ethology, Dawkins, M.S., Plenum Press, 1989. Quote: Animal Intelligence: Experimental Studies, Thorndike, E.L., Macmillan Publishing Co., 1911. Quote: Quarterly Rev. Biol. 28, Lehrman, D.S., StonyBrook Foundation, 1953. Quote: Behaviorism, Watson, J.B., W.W. Norton & Company, 1930.

Chapter 2

Table 2.1 From Psychonomic Science 4:123-124, Garcia, J. And R.Λ. Koelling, reprinted by permission of Psychonomic Society, Inc., 1966. Fig. 2.2 Mental Evolution in Man: Origin of Human Faculty, Romanes, G.J., Appleton and Lange, 1889. Fig. 2.4 Systematic Zoology 4, Kessel, E.L., Society of Systematic Zoology, 1955. Fig. 2.5 Zeitschrift fur Tierpsychologie 2, Lorenz, K. And N. Tinbergen, Paul Parey Scientific Publications, 1938. Fig. 2.7 Wilson Bulletin 60, Tinbergen, N., Wilson Ornithological Society, 1948. Fig. 2.10 Courtship ritual of stickleback, Oxford University Press. Fig. 2.11 L'Instinct dans le comportement des animaux et de l'homme, Desmond Morris, Masson Editeurs, 1956. Fig. 2.12 Physiological Mechanisms in Animal Behavior. Symposium Society Experimental Biology IV, Lorenz, K., Company of Biologists, Ltd., 1950. Fig. 2.13 Bibliotheca Biotheoretica 1, Tinbergen, N., E.J. Brill, 1942. Fig. 2.14 Animal Intelligence: Experimental Studies, Thorndike, E.L., Macmillan Publishing Company, 1911. Fig. 2.22 Reprinted by permission of the publisher from SOCIOBIOLOGY: THE NEW SYNTHESIS by Edward O. Wilson, Cambridge, Mass.: The Belknap Press of Harvard University Press, Copyright © 1975 by the President and Fellows of Harvard College.

Chapter 3

Fig. 3.3 Proc. Nat. Acad. Sci. USA 89, Mello, C., and D.S. Vicario, Proc. Nat. Acad. Sci., 1992. Fig. 3.4 Journal of Neuroscience 15, Mello, C., and F. Nottebohm, Society for Neuroscience, 1995. Fig. 3.5 Behavior Genetics 27(3), Miklosi, A.V., V. Csanyi, and R. Gerlai, Behavior Genetics, Plenum Publishing, 1997. Fig. 3.7 Genetics, Lynch, C.B., Genetics Society of America, 1980. Fig. 3.8 Heredity 59, de Belle, J.S. and M.B. Sokolowski, Blackwell Science Ltd., 1987. Fig. 3.15 Proc. Nat. Acad. Sci. USA 68, Konopka, R.J., and S. Benzer, Proc. Nat. Acad. Sci., 1971. Fig. 3.17 Reprinted by permission from Nature 401, Tang, Y.P., E. Shimizu, G.R. Dube, C. Rampon, G.A. Kerchner, M. Zhuo, G.S. Liu, and J.Z. Tsien, 1999, Macmillan Magazines Ltd. Fig. 3.18 Scientific American 250(3) , Scheller, R.H. and R. Axel, Scientific American, Inc.

Chapter 4

Fig. 4.14 Journal of Theoretical Biology 77, Brockmann, J., A. Grafen and R. Dawkins, Academic Press Ltd., 1979.

Chapter 5

Fig. 5.2 Animal Behaviour, Balda, R.P. and A.C. Kamil, Academic Press, 1989. Fig. 5.3 Journal of Comparative Psychology, Olson, D.J., A.C. Kamil, R.P. Balda, and P.J. Nims, Copyright © 1995 by the American Psychological Association. Reprinted (or Adapted) with permission. Fig. 5.4 Animal Behaviour, Clark, R.B., Academic Press, 1960. Fig. 5.5 Journal of Comparative Psychology, Hollis, K.L., V.L. Pharr, M.J. Dumas, G.B. Britton, and J. Field. Copyright © 1997 by the American Psychological Association. Reprinted (or Adapted) with permission. Fig. 5.11 Reprinted by permission from Nature, Clayton, N.S., and A. Dickinson, 1998, Macmillan Magazine Ltd.

Chapter 6

Quote: Insect Migration, Williams, C.B., 1965. Fig. 6.2 Journal of Comparative Physiology 159A, Ritzmann, R.E. and A.J. Pollack, Springer-Verlag, 1986. Fig. 6.12 American Zoologist 29, Lent, C.M., M.H. Dickinson, and C.G. Marshall, American Society of Zoologists, 1989. Fig. 6.14 American Zoologist 29, Lent, C.M., M.H. Dickinson, and C.G. Marshall, American Society of Zoologists, 1989. Fig. 6.15(a) Journal of Comparative Physiology A155, Montgomery, J.C., Springer-Verlag, 1984. Fig. 6.16(a) Zeitschrift fur vergleichende Physiologie 54, Ewert, J.-P., Springer-Verlag, 1967. Fig. 6-16(b) Pflugers Archiv 308, Ewert, J.-P., Springer-Verlag, 1969. Fig. 6.19 Animal Behavior, Alcock, J., Sinauer Associates, 1975. Fig. 6.22 Reprinted with permission from Science, Knudsen and Konishi. Copyright 1978, American Association for the Advancement of Science. Fig. 6.25(a) Journal of Comparative Physiology, Horsmann, Springer-Verlag, 1983. Fig. 6.25(a) Philosophical Transactions Royal Society of London (Series B), Weis-Fogh, T., The Royal Society, 1956. Fig 6.25(a) Journal of Comparative Physiology, Horsmann, Springer-Verlag, 1983. Fig. 6.25(b,c) Journal of Experimental Biology 39, Wilson, D.M. and T. Weis-Fogh, Company of Biologists, Ltd., 1962. Fig. 6.26 Journal of Neurophysiology 53, Robertson, R.M. and K.G. Pearson, American Physiological Society, 1985. Fig. 6.27 Journal of Comparative Physiology, Mohl, Springer-Verlag, 1985. Fig. 6.27 Journal of Comparative Physiology, Mohl, Springer-Verlag, 1985.

Chapter 7

Fig. 7.1 Nature 187, McLaren, A. And D. Michie, Macmillan Magazines Ltd., 1960. Fig. 7.5 Bioscience 40, Kelley, D.B. and D.L. Gorlick, American Institute of Biological Sciences, 1990. Fig. 7.7 General and Comparative Endocrinology 66, Robinson, G.E., Academic Press, Inc., 1987., Fig. 7.10 General and Comparative Endocrinology 44, Bollenbacher, W.E., Academic Press, Inc. 1981. Fig. 7.10 Journal of Comparative Physiology 155, Weeks, J.C. and J.W. Truman,

Chapter 8

Chapter 9

Chapter 10

and F. Papi, Springer-Verlag, 1990. Fig. 10.30 Science 203, Simmons, J.A., M.B. Fenton, and M.J. O'Farrell, American Association for the Advancement of Science, 1979. Fig. 10.31 American Scientist 63, Simmons, J.A., D.J. Howell, and N. Suga, Sigma Xi Scientific Research Society, 1975. Fig. 10.32, 10.33 and 10.34 Journal of Experimental Biology, von der Emde, G., The Company of Biologists Ltd., 1999.

Chapter 11
Table 11.1 Reprinted from Animal Behavior 28, Greenwood, P.J., Mating systems, philopatry and dispersal in birds and mammals, 1140-1162, 1980, by permission of the publisher Academic Press London. Table 11.2 Reprinted from Animal Behavior 30, Dobson, F.S., Competition for mates and predominant juvenile male dispersal in mammals, 1183-1192, 1982, by permission of the publisher Academic Press London. Table 11.3 American Naturalist 126, Liberg, O., and T. Von Schantz, The University of Chicago Press, 1985. Table 11.4 American Naturalist 126, Ostfeld, R.S., The University of Chicago Press, 1985. Fig. 11.2 Journal of Animal Ecology 56, Ims, R.A., Blackwell Scientific Publications Ltd. Fig. 11.3 Trends in Ecology and Evolution 5, Ostfeld, R.S., Elsevier Trends Journals, 1990. Fig. 11.4 American Naturalist 130, Ims, R.A., The University of Chicago Press, 1987. Fig. 11.8 Ecological Entomology 21(2), Anderson, J.B., and L.P. Brower, Blackwell Scientific Ltd., 1996.

Chapter 12
Table 12.1 Acknowledgment to the original publication. Zach, R. 1978. Behaviour. 67, E.J. Brill. Fig. 12.8 Reprinted from Animal Behaviour 58, Watt, M., C.S. Evans, and J.M.P. Joss, Use of electroreception during foraging by the Australian lungfish, 1039-1045, 1999, by permission of the publisher Academic Press Ltd., 1999. Fig. 12.11 Foraging Behavior: Ecological, Ethological and Psychological Approaches, Pietrewics, A.T., and A.C. Kamil, Garland STPM Press, 1981. Fig. 12.12 Reprinted from Animal Behaviour 50, Plaisted, K.C., and N.J. Mackintosh, Visual search for cryptic stimuli in pigeons, 1219-1232, 1995, by permission of the publisher Academic Press Ltd. Fig. 12.13 Behaviour 68, Zach, R., E.J. Brill Leiden, 1979. Fig. 12.18 Behavioural Ecology and Sociobiology 38(2), Templeton, J.J., and L.-A. Giraldeau, Springer-Verlag, 1996. Fig. 12.20(b) Reprinted from Animal Behaviour 53(4), Cowlishaw, G., Trade-offs between foraging and predation risk determine habitat use in a desert baboon population, 667-686, 1997, by permission of the publisher Academic Press Ltd. Fig. 12.21 Behavioral Ecology and Sociobiology 16, Dill, L.M., and A.H.G. Fraser, Springer-Verlag, 1984.

Chapter 13
Table 13.1 Heredity, Kettlewell, Blackwell Scientific Publications Ltd., 1955, 1956. Table 13.2 Zeitschrift fur Tierpsychologie Supplement 5, Croze, H., Paul Parey Scientific Publications, 1970. Table 13.3 Reprinted with permission from Science 219, Dial, B.E., and L.C. Fitzpatrick, Copyright 1983, American Association for the Advancement of Science. Table 13.4 Reprinted from Animal Behaviour, Caro, 1986, by permission of the publisher Academic Press Ltd. Table 13.5 Reprinted from Animal Behaviour, Caro,

1986, by permission of the publisher Academic Press London. Fig. 13.5 Proceedings of the Zoological Society of London, Holmes, Cambridge University Press, 1940. Fig. 13.6 Journal of Mammalogy 70, Kiltie, R.A., American Society of Mammalogists, 1989. Fig. 13.7 Journal of Mammalogy 70, Kiltie, R.A., American Society of Mammalogists, 1989. Fig. 13.9 Journal of Heredity, Grant, B.S., D.F. Owen, and C.A. Clarke, American Genetics Association, 1996. Fig. 13.11 Reprinted from Animal Behaviour, Feltmate and Williams, 1989, by permission of the publisher Academic Press London. Fig. 13.14 Evolutionary Biology 11, Endler, J.A., Plenum Publishing Co., 1978. Fig. 13.15 Reprinted with permission from Science, Moment, G., Copyright 1962, American Association for the Advancement of Science. Fig. 13.16 Reprinted from Animal Behaviour 49(4), Terrick, T.D., R.L. Mumme, and G.M. Burghardt, Aposematic coloration enhances chemosensory recognition of noxiour prey in the garter snake, 857-866, 1995, by permission of the publisher Academic Press Ltd. Fig. 13.17 Proceedings of the Zoological Society of London, Hinston, Cambridge University Press, 1927. Fig. 13.21 Reprinted from Animal Behaviour, Burghardt and Greene, 1988, by permission of the publisher Academic Press London. Fig. 13.22 Animal Behavior, Johnsgard, P.A., William C. Brown Publishers, 1967. Fig. 13.25 Reprinted from Animal Behaviour, Schenoff, 1985, by permission of the publisher Academic Press London. Fig. 13.27 Reprinted from Animal Behaviour, Hews, 1988, by permission of the publisher Academic Press London. Fig. 13.28 Ecology 70, Turchin, P., and P. Kareiva, Ecological Society of America, 1989. Fig. 13.29 Journal of Zoology 172, Neill, S.R. St. J., and J.M. Cullen, Reprinted with the permission of Cambridge University Press, 1974.

Chapter 14
Table 14.1 Journal of Comparative and Physiological Psychology 55, Fisher, A.E., American Psychological Association, 1962. Table 14.2 Journal Comp. Physiol. Psychol. 55, Fisher, A.E., Copyright 1962 by the American Psychological Association. Reprinted (or Adapted) by permission. Fig. 14.13 Reprinted from Animal Behaviour 47(5), Eggert, A.-K., and S.K. Sakaluk, Sexual cannibalism and its relation to male mating success in sagebrush crickets, 1171-1177, 1994, by permission of the publisher Academic Press Ltd. Fig. 14.16 Reprinted from Animal Behaviour 56(5), Rasa, O.A.E., S. Bisch, and T. Teichner, Female mate choice in a subsocial beetle, 1213-1220, 1998, by permission of the publisher Academic Press Ltd. Fig. 14.20 Reprinted from Animal Behaviour 50(2), Basolo, A.L., A further examination of a pre-existing bias favouring a sword in the genus Xiphophorus, 365-375, 1995, by permission of the publisher Academic Press Ltd.

Chapter 15
Table 15.1 In Natural Selection on Social Behavior, McKaye, K.R., Chron Press: New York, 1977. Table 15.2 Reprinted by permission of the publisher from THE EVOLUTION OF INSECT MATING SYSTEMS by R. Thornhill and J. Atlock, Cambridge, Mass.: Harvard University Press, Copyright © 1983 by the President and Fellows of Harvard College. Fig. 15.3 Handbook of Behavioral Neurobiology, Wittenberger, J.F., Plenum Publishing Corporation, 1979.

Fig. 15.6 Behavioural Ecology and Sociobiology 43(4-5), Goossens, B., L. Graziani, L.P. Waits, E. Farand, S. Magnolon, J. Coulon, M.-C. Bel, P. Taberlet, and D. Allaine, Springer-Verlag, 1998. Fig. 15.9 Handbook of Behavioral Neurobiology 3: Social Behaviour and Communication, Wittenberger, J.F., Plenum Publishing Corporation, 1979. Fig. 15.9 American Naturalist 103, Orians, The University of Chicago Press, 1969. Fig. 15.12 American Naturalist 105, Downhower, J.F., and K.B. Armitage, The University of Chicago Press, 1971. Fig. 15.16 Ecological Aspects of Social Behavior, Oring, L.W., and D.B. Lank, Princeton University Press, 1986.

Chapter 16
Fig. 16.4 Science 239, Bednarz, J.C., American Association for the Advancement of Science, 1988. Fig. 16.5 Behavioural Ecology and Sociobiology 29, Traniello, J.F.A., and S.N. Beshers, Springer-Verlag, 1991. Fig. 16.9 American Zoologist 27, Ydenberg, R.C., and J.R. Krebs, American Society of Zoologists, 1987. Fig. 16.11 Proc. Nat. Acad. Sci. 80, Carpenter, F.L., D.C. Paton, and M.A. Hixon, Proc. Nat. Acad. Sci., 1983. Fig. 16.14 Red Deer: The Behavior and Ecology of the Two Sexes, Clutton-Brock, T.H., F.E. Guiness, and S.D. Albon, The University of Chicago Press, 1982.

Chapter 17
Table 17.1 Behavioral Ecology and Sociobiology 17, Sherman, P.W., Springer-Verlag, 1985. Table 17.2 Behavioral Ecology and Sociobiology 6, Reyer, H.-U, Springer-Verlag, 1980. Table 17.3 Behavioral Ecology and Sociobiology 25, Emlen, S.T., and P.H. Wrege, Springer-Verlag, 1989. Fig. 17.4 Reprinted from Animal Behaviour 56(2), Heth, G., J. Todrank, and R.E. Johnston, Kin recognition in golden hamsters: evidence for phenotype matching, 409-417, 1998, by permission of the publisher Academic Press Ltd. Fig. 17.8 Scientific American 262(2), Wilkinson, G.S., Scientific American, Inc., 1990. Fig. 17.9 American Naturalist 107, Konishi, The University of Chicago Press, 1973. Fig. 17.11 Behavioral Ecology and Sociobiology 17, Sherman, P.W., Springer-Verlag, 1985. Fig. 17.12 Behavioral Ecology and Sociobiology 17, Sherman, P.W., Springer-Verlag, 1985. Fig. 17.14 Reproductive Success, Packer, C., L. Herbst, A.E. Pussey, J.D. Bygott, J.P. Hanby, S.J. Cairns, and M.B. Mulder, The University of Chicago Press, 1988. Fig. 17.17 The Auk 92, Woolfenden, G.E., The American Ornitholigists Union, 1975. Fig. 17.18 Reprinted by permission from Nature 277, Moehlman, P.D., 1979, Macmillan Magazines Ltd. Fig. 17.20 American Naturalist 119(1), Emlen, S.T., The University of Chicago Press, 1982. Fig. 17.23 Fig. 17.23 American Naturalist 119(1), Emlen, S.T., The University of Chicago Press, 1982. Fig. 17.24 Trends in Ecology and Evolution 5(2), Heinsohn, R.G., A. Cockburn and R.A. Mulder, Elsevier Science Publishers Ltd., 1990. Fig. 17.25 Cooperative Breeding in Mammals, Lacey, E.A., and P.W. Sherman, Reprinted with the permission of Cambridge University Press, 1997. Fig. 17.28 The Naked Mole-Rat, Lacey, E.A., and P.W. Sherman, Princeton University Press, 1991.

Chapter 18
Fig. 18.8(a&b) American Scientist 62, Hopkins, C., Sigma Xi Scientific Research Society, 1974. Fig. 18.11(b) Reprinted from Animal Behaviour, Houck and Reagan, 1990, by permission of the publisher Academic Press Ltd. Fig. 18.12 Darwin's Biological Work, Marler, P., Reprinted with the permission of Cambridge University Press, 1959. Fig. 18.15 Reprinted by permission from Nature 274, Davies, N.B., and T.R. Halliday, 1978, Macmillan Magazines Ltd. Fig. 18.17 Behaviour 69, Clutton-Brock, T.H. and S.D. Albon, E.J. Brill, 1979. Fig. 18.18 Proc. Roy. Soc. Lond. A30, Ribbands, C.R., The Royal Society, 1955. 18.22 Reprinted from Karl von Frisch: Bees: Their Vision, Chemical Senses, & Language, revised edition. Copyright © 1950, 1971, by Cornell University. Used by permission of the publisher, Cornell University Press. Fig. 18.23 & 24 Reprinted by permission of the publisher from THE DANCE LANGUAGE AND ORIENTATION OF BEES by K. Von Frisch, Cambridge, Mass.; The Belknap Press of Harvard University Press, Copyright © 1967 by the President and Fellows of Harvard College. Fig. 18.26 Naturiwissenschaften 76, Michelsen, A.B., B. Anderson, W.H. Kirchner, and M. Lindauer, Springer-Verlag, 1989. Fig. 18.27 Science, 189, Gould, J., American Association for the Advancement of Science, 1975.

Chapter 19
Fig. 19.2 Behavioural Ecology, Johnstone, R.A., Blackwell Science Ltd., 1997. Fig. 19.5 Reprinted from Animal Behaviour 36, Cheny D.L., and R.M. Seyfarth, Assessment of meaning and the detection of unreliable signals by vervet monkeys, 477-486, 1988, by permission of the publisher Academic Press London. Fig. 19.8 The of Instinct, Tinbergen, Oxford Clarendon, 1951. Fig. 19.12 Zoology 84, Orians, G.H., and G.M. Christman, University of California Press, 1968. Fig. 19.17 Behavioral Ecology and Sociobiology 10, Gould, J., Springer-Verlag, 1982. Fig. 19.18 Journal of Animal Ecology 48, Hunter, M.L., and J.R. Krebs, Blackwell Scientific Publications Ltd., 1979. Fig. 19.19 American Naturalist 109, Morton, E.S., American Society of Zoologists, 1975. Fig. 19.25 Reprinted from Animal Behaviour 55(6), Hauser, M.D., Functional referents and acoustic similarity: field playback experiments with rhesus monkeys, 1647-1658, 1998, by permission of the publisher Academic Press London.

Appendix
Figure A2 and A4: BioSystems 18, Kirschvink and Gould, Elsevier Scientific Publishers Ireland Ltd., 1981.

INDEX

Page numbers in *italics* indicate illustrations.